P9-BHY-589

Algorithms and Combinatorics 6

András Recski

Matroid Theory and its Applications in Electric Network Theory and in Statics

Springer-Verlag Berlin Heidelberg New York
London Paris Tokyo

András Recski
Eötvös Loránd University
Budapest, VIII. Múzeum krt. 6/8
Hungary

Joint edition published by
Springer-Verlag Berlin Heidelberg New York London Paris Tokyo
and
Akadémiai Kiadó, Budapest, Hungary

Sole distribution rights for the non-socialist countries:
Springer-Verlag Berlin Heidelberg New York
London Paris Tokyo

Mathematics Subject Classification (1980): 05B35

ISBN 3-540-15285-7 Springer-Verlag Berlin Heidelberg New York
ISBN 0-387-15285-7 Springer-Verlag New York Berlin Heidelberg

Printed in Hungary

Printing: Akadémiai Kiadó és Nyomda Vállalat, Budapest
Binding: J. Schäffer, Grünstadt
2141/3140-543210

To the memory of my parents

Introduction

I. The topics of this book

The concept of a matroid has been known for more than five decades. Whitney (1935) introduced *it as a common generalization of graphs and matrices.* In the last two decades, it has become clear how important the concept is, for the following reasons:

(1) Combinatorics (or discrete mathematics) was considered by many to be a collection of interesting, sometimes deep, but mostly unrelated ideas. However, like other branches of mathematics, combinatorics also encompasses some general tools that can be learned and then applied, to various problems. Matroid theory is one of these tools.

(2) Within combinatorics, the relative importance of algorithms has increased with the spread of computers. Classical analysis did not even consider problems where "only" a finite number of cases were to be studied. Now such problems are not only considered, but their complexity is often analyzed in considerable detail. Some questions of this type (for example, the determination of when the so called "greedy" algorithm is optimal) cannot even be answered without matroidal tools.

(3) In various engineering problems, part of the information is given by graphs, part of it by matrices. For example, the interconnection of the elements of an electric network or a bar-and-joint framework can conveniently be described by a graph, while the electric properties of the individual devices (or the rigidity properties of the individual bars) are naturally given in terms of vectors or matrices. It is thus reasonable to expect that a common generalization of graphs and matrices will be useful in treating certain problems related to such structures. Indeed, this is the case for the qualitative problems of deciding solvability of an electrical network or deciding rigidity of a framework.

Accordingly, *this book gives an introduction to matroid theory (after presenting the necessary background about graphs and algebra); with special emphasis on its algorithmic aspects and on the applications in electrical engineering and in statics.*

II. Who should read this book?

This book is intended for mathematicians and engineers alike. Since their motivations and backgrounds are usually different, some unusual methods are applied in this book.

(1) Mathematical results are presented in the odd-numbered chapters only. Even-numbered chapters describe associated applications. Those mathematicians not interested in these applications can simply read the odd-numbered chapters. On the other hand, in order to read and understand the applications, the reader should have some familiarity with at least the basic ideas from the odd-numbered chapters.

(2) The chapters consist of 2 to 6 sections. Each section contains a basic description, followed by exercises and problems. The basic descriptions are intended for every reader; new definitions and theorems are introduced with examples involving as few technical details as possible. *Exercises* are generally easy, sometimes routine tasks, intended to check the readers' understanding of the basic ideas and techniques. *Problems* are more difficult, usually requiring longer calculations or proofs.

(3) Mathematicians interested in the applications will find all the necessary engineering concepts in the even-numbered chapters, hence *no background knowledge from engineering is assumed.*

(4) Engineers will surely observe that the examples from electricity or from statics in the even-numbered chapters are sometimes unusual, even absurd. This book is certainly not intended to teach the design of a TV set or a house. Engineers have many "case studies" in their texts to choose from, during their everyday design activities. The mathematical tools of those solutions may be very similar or even identical. Our engineering examples are mainly meant to illustrate the mathematical features (including the illustration of the limits of the validity of a mathematical theorem via absurd examples), hence they are not supposed to replace examples in conventional engineering texts.

However, even these absurd or "pathological" examples have some applicability, besides their didactical purposes. CAD (Computer Aided Design) program packages must have some built-in subroutines to check whether the given specifications are correct. Absurd initial specifications can arise from bad modelling assumptions as well as from input errors. Engineers could immediately find these mistakes in small examples by intuition, but in the case of large scale problems this is left to the computer. Therefore if one tests such a CAD package, pathological problems (in addition to numerous real-life problems) are of practical importance.

III. Relations to other books on matroids

(1) The books of Welsh (1976) and Lawler (1976) are excellent texts on matroid theory. The former included almost every branch of matroid theory while the latter concentrated on a few chapters only, but in greater detail, with much emphasis on the algorithmic aspects. The breadth of the mathematical part of this book is between the above two, and it is also of algorithmic character. Of course, this book also covers many results of the last decade as well.

The more recent book edited by White (1986) can serve as a very good reference if one is familiar with matroids already. However, that book does not emphasize the algorithmic aspects and some of its chapters may prove to be too abstract, especially for readers with engineering background.

(2) Apart from the recently published text of Murota (1987) which concentrates on the structural analysis of large-scale systems, the applications of matroid theory have apparently never been covered in any book so far. In this book about 40% of the presented material is on applications.

(3) Unlike graphs, matroids cannot be visualized in a straightforward way. Therefore a great number of examples from graph theory, algebra and affine geometry are presented in this book; each time when a new concept of matroid theory is introduced, we try to explain its meaning (if any) in graph theory, in algebra and in geometry. We hope that during matroidal studies drawing will be as natural for the reader as it clearly is in graph theory. (This way of presentation can be found in some chapters of White (1986) as well.)

(4) Collections of exercises and problems are very important for studying any branch of mathematics. Such collections are available for graph theory, see (Lovász, 1979a), for example, but apparently not for matroid theory. This book contains nearly 800 exercises and problems with full solutions, including more than 500 mathematical ones in the odd-numbered chapters. More than half of these mathematical problems refer to matroids.

(5) Another, probably unique, feature of this book is that applications in electric and in civil engineering are presented in a parallel way (as the necessary mathematical tools become available), hence relations among these two branches are also illustrated at some instances.

IV. Some remarks of technical character

(1) Theorems, statements etc. are numbered continuously within each section of each chapter. (For example, Section 3 of Chapter 12 contains Statement 12.3.1, then Theorems 12.3.2 and 12.3.3, followed by Corollary 12.3.4 etc.) Exercises and problems are also continuously numbered within each section. (For example, 9.1.1–9.1.12 are exercises and 9.1.13–9.1.20 are problems.)

(2) If the number of an exercise or a problem is in boldface then the result or proof will be needed later in the book. Hence the reader should study the solution even if he/she did not solve it.

(3) If the number of an exercise or a problem is followed by an asterisk (*) then the given solution is followed by some additional information (e.g. a stronger theorem or a conjecture is presented). Hence these solutions should also be read.

(4) Although the book contains nearly 600 figures, the reader is urged to draw further ones. Even if the given solution of a problem does not contain any illustration, one should follow it by drawing in many cases.

(5) About 400 works are listed as references. Many of them are explicitly quoted in the text, the others are suggested for further study.

V. Acknowledgements

I received much help from a number of friends, colleagues and students while writing this book.

László Lovász called my attention to matroids as early as in 1969, when we were still students. Since that time he has given me much help, as the head of our department, as one of the referees of this book and as one of the editors of the series "Algorithms and Combinatorics".

I learned much combinatorics from Vera T. Sós and network theory from Árpád Csurgay. I am very thankful to them, to L. Lovász and to the other referee of this book, András Frank, for their remarks, for the uncountably many discussions about the new results, etc.

István Hegedűs, János Ladvánszky, Rabe von Randow, Béla Roller, János Solymosi and Éva Tardos also read great portions of the manuscript, found several mistakes and had valuable remarks on terminology as well.

Thanks to the honouring invitations of W. R. Pulleyblank, R.G. Bland and B. Korte, I had the possibility of giving courses on matroids and their applications at the University of Waterloo (Waterloo, Ontario, Canada), at Cornell University (Ithaca, New York, USA) and at the University of Bonn (FRG), respectively. I received many remarks and suggestions from them and their colleagues and students.

Special thanks are due to Masao Iri from the University of Tokyo where I spent a highly fruitful period in the academic year 1978/79, and to my former working place, the Research Institute for Telecommunication. Support of the Hungarian Academy of Sciences and the Alexander-von-Humboldt Stiftung, as well as useful remarks of the following colleagues, are also gratefully acknowledged.

B. Andrásfai, L. Babai, J. Baracs, A. Baranyi, R. Bixby, R. Brualdi, P. Bryant, H. Carlin, I. Cederbaum, Y. Ceyhun, R. Conelly, H. Crapo, W. Cunningham, A. Dervişoğlu, R. Duke, J. Edmonds, A. Fettweis, S. Fujishige, I. Gobbi, M. Golumbic, M. Grötschel, A. Horváth, H. Imai, A. Ingleton, R. Kalman, G. Katona, K. Kawakita, E. Korach, M. Las Vergnas, J. Mason, W. Mayeda, D. Miklós, M. Milić, M. Nakamura, F. Nielsen, Y. Oono, T. Ozawa, B. Petersen, M. Plummer, A. Radványi, T. Roska, L. Schrijver, A. Sebő, M. Simonovits, J. Takács, T. Tarnai, Y. Tokad, K. Truemper, D. Welsh, W. Whiteley and V. Zoller.

Special thanks are due to I. Eszterág, T. Fadgyas and J. Selwood, who typed the manuscript. Finally I wish to thank for the help of the Publishing House of the Hungarian Academy of Sciences, especially G. Csikós–Kugler for her careful editorial work and M. Dobiecki for her beautiful artwork, as well as for the help of Springer Verlag; and, last but not least, to my wife and children for their love and patience.

Contents

PART TWO

PART ONE

Chapter 1

Basic concepts from graph theory

Section 1.1 Graphs; points, edges, paths, circuits, connectivity

Both in mathematics and in its applications, one frequently considers certain pairs of elements of a set. Then it is quite natural to draw the elements of the set as different points of the plane and indicate the considered pairs by lines between the corresponding points. The position of the points in the plane is of no interest, neither is the shape of the lines; all that one must avoid is that a line between the points x, y passes through a third point, say z, of the set. For example, all the drawings in Fig. 1.1 are considered the same; four points are in the set and all pairs but $\{c, d\}$ occur.

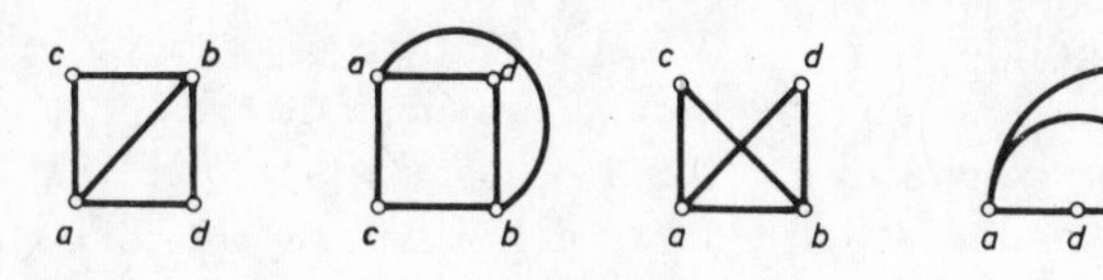

Fig. 1.1

Such systems will be called graphs; the elements of the set are the points and the considered pairs are the edges. Formally, a *graph* is a system $G = (V, E)$ where V is a nonempty set and E is a collection of pairs of (not necessarily distinct) elements of V. The elements of V and E will be called *points* and *edges*, respectively.

When we speak about a graph G, its point set and edge set will be denoted by $V(G)$ and $E(G)$, respectively; their cardinalities (i.e. the number of points and edges of G) will be denoted by $v(G)$ and $e(G)$, respectively.

If an edge $e \in E$ is a pair $\{v_1, v_2\}$ of points then these points are called the *end points* of e. If $v_1 = v_2$ then e is a *loop*. If two nonloop edges have the same set of end points, they are called *parallel* or *multiple* edges. Graphs with no loops or multiple edges are called *simple*.

If $e, f \in E$ are edges with end points $\{v_1, v_2\}$ and $\{w_1, w_2\}$, respectively, and $\{v_1, v_2\} \cap \{w_1, w_2\} \neq \emptyset$ then e, f are *adjacent* edges. Two distinct points v_1, v_2

are *adjacent* if there exists an edge $\{v_1, v_2\}$ in the graph. A point v and an edge e are *incident* if v is an end point of e. A point is *isolated* if it is incident to no edges. The number of incident edges is called the *degree* of a point; possible loops are counted twice.

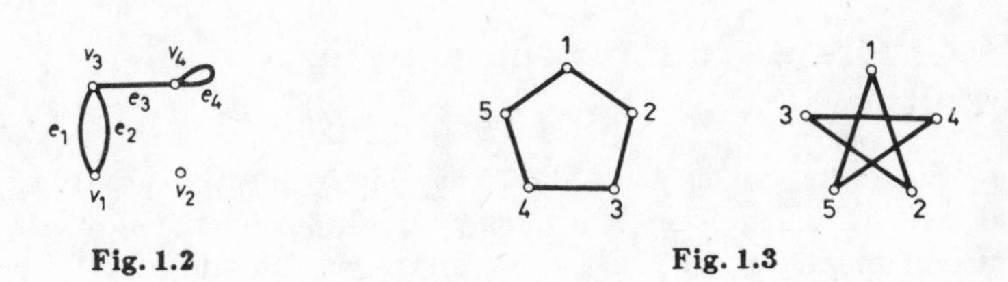

Fig. 1.2 **Fig. 1.3**

For example, Fig. 1.2 shows a graph $G = (V, E)$ with $V = \{v_1, v_2, v_3, v_4\}$; $E = \{e_1, e_2, e_3, e_4\}$.

The end points of e_1, e_2, e_3 and e_4 are $\{v_1, v_3\}, \{v_1, v_3\}, \{v_3, v_4\}$ and $\{v_4\}$, respectively. e_4 is a loop, e_1, e_2 are parallel edges. e_3 is adjacent to every other edge, v_3 is incident to every edge but e_4, and v_3 is adjacent to every point but v_2. The point v_2 is isolated; the degrees of v_1, v_2, v_3, v_4 are $2, 0, 3$ and 3, respectively.

We have seen that different drawings can describe the same graph. The two drawings of Fig. 1.3 obviously correspond to the same graph but whether it is true for some graphs of Fig. 1.4 is not that obvious. It takes some time to recognize that the first and the second drawings describe the same graph (A, B, C, D, E, F correspond to $1, 3, 5, 2, 4, 6$, respectively) and that the graphs described by the second and by the third drawings must be different (since in the second drawing no three points are pairwise adjacent).

Formally, the graphs $G = (V, E)$ and $G' = (V', E')$ are *isomorphic* if there exists a bijection $\pi : V \to V'$ (i.e. a one–one correspondence between V and V') and a bijection $\rho : E \to E'$ so that $v \in V$ and $e \in E$ are incident in G if and only if $\pi(v)$ and $\rho(e)$ are incident in G'. In the case of simple graphs it suffices to say that G, G' are isomorphic if there exists a bijection $\pi : V \to V'$ so that $v, w \in V$ are adjacent in G if and only if $\pi(v)$ and $\pi(w)$ are adjacent in G'.

Although algorithmic aspects of graph theory will be considered in detail later, we observe here that *checking* isomorphism is very easy if someone has already found this bijection, but *finding* this bijection (or *proving* nonisomorphism) is in general a much more complicated task, unless we try the $v!$ possible bijections (a task which might take some thousand years with a computer of 10^9 operations per second for a graph with as few as $v = 20$ points).

The graph $G' = (V', E')$ is a *subgraph* of $G = (V, E)$ if $V' \subseteq V, E' \subseteq E$ and $v \in V'$ and $e \in E'$ are incident in G' if and only if they were incident in G. Observe that we required G' be a graph and did not simply choose subsets of V

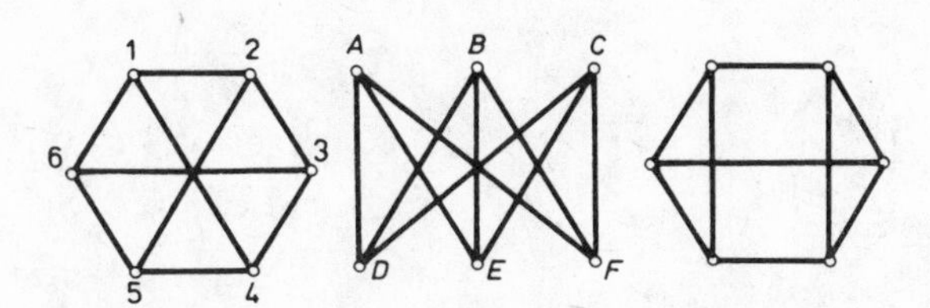

Fig. 1.4

and E; otherwise it could happen that an end point of some edge from E' does not belong to V'.

If E' consists of those edges from E both of whose end points are in V' then G' is called the *induced* subgraph (more exactly, the subgraph of G, induced by V'). For example, if G is the road map of Canada (points and edges correspond to cities and to roads, respectively) then the map showing the freeways only is a subgraph; the road map of Ontario is an induced subgraph.

If $V' = V$ and $E' = E - \{e\}$ for some edge $e \in E$ then G' will be denoted by $G - e$. If $V'' = V - \{v\}$ for some point $v \in V$ and E'' consists of those edges from E which are not incident to v then G'' will be denoted by $G - v$. These processes are called the *deletion* of an edge or a point. If $X \subseteq E$ or $Y \subseteq V$ then $G - X$ or $G - Y$ denotes the successive deletion of every edge of X (every point of Y, respectively) from the graph G. The result can be proved to be the same for any order of the deletions of the elements.

A simple graph is *complete* if any pair of points is adjacent. The complete graph with n points will be denoted by K_n. Every simple graph with v points can be obtained from K_v by deleting some edges. Every induced subgraph of a complete graph is also complete.

If H is a subgraph of G then the graph formed by every point of G and by the edge set $E(G) - E(H)$ is called *the complement of H with respect to G*. On the other hand, if we speak about the *complement of a* (simple) *graph H* (without saying anything else) then we mean the complement with respect to the complete graph with the same point set. For example, consider the graphs in Fig. 1.5. The complement of (b) with respect to (a) is (c); however, if we speak about (b) itself, and not about a subgraph of some other graph, then its complement is (d).

If we consider a graph as a road map, we may wish to know whether any city can be reached from any other one; i.e. whether the graph is connected or rather a collection of smaller graphs, drawn "accidentally" on the same sheet of paper.

Let us call an ordered system $[v_0, e_1, v_1, e_2, \ldots, v_{k-1}, e_k, v_k]$ an *edge sequence* if $v_0, v_1, \ldots, v_k \in V(G), e_1, e_2, \ldots, e_k \in E(G)$ and e_i joins v_{i-1} and v_i for every $i = 1, 2, \ldots, k$. Such a sequence certainly gives a connection between v_0 and v_k,

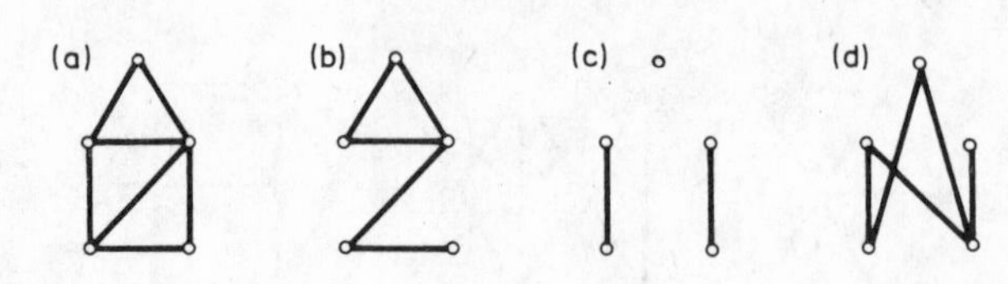

Fig. 1.5

but it may be very complicated (for example if $v_{i-1} = v_i$ and e_i is a loop). An edge sequence is called a *path* if all the points of it are different. The *length* of a path is the number of its edges. For example, $[v_3, e_2, v_1, e_1, v_3, e_3, v_4, e_4, v_4]$ is an edge sequence in the graph of Fig. 1.2, while $[v_1, e_1, v_3, e_3, v_4]$ is a path of length 2.

An edge sequence as above is called *closed* if $v_0 = v_k$. If all the other points are different from each other and from v_0 then this closed edge sequence is a *circuit*. The *length* of a circuit is the number of its edges. For example, the graph of Fig. 1.2 contains two circuits $[v_1, e_1, v_3, e_2, v_1]$ and $[v_4, e_4, v_4]$ with lengths 2 and 1, respectively. A graph is simple if and only if it does not have any circuit of length 1 or 2.

A graph is *connected* if any point of it can be reached along a path from any other point. If there is a path P_1 from v_1 to v_2 and another path P_2 from v_2 to v_3 then v_3 can also be reached along a path from v_1 (though not necessarily by joining P_1 and P_2 since this could lead to nonpath edge sequences, see Exercise 1.1.6). Observe, furthermore, that if v_1 can be reached along a path from v_2 then so can v_2 from v_1. Hence we may say that certain points v_i, v_j of a graph are in the same class if and only if either $v_i = v_j$ or v_i can be reached along a path from v_j. These classes determine the *connected components* of the graph; the number of these components will be denoted by $c(G)$. A graph is connected if and only if $c(G) = 1$.

How can we check if a graph G is connected? The following algorithm seems to be straightforward:

Algorithm 1.1 *Checking the connectivity of a graph.*

Input A graph G with at least one point.

Output The connected component H of G, containing the first point of G. (If $G = H$, the graph is connected. Otherwise, repeating the algorithm with input $G - H$ one can stepwise obtain every component of G.)

Step 1 Mark the first point of G.

Step 2 If none of the unmarked points is adjacent to any of the marked points, go to **End**. If an unmarked point v is adjacent to a marked point, go to **Step 3**.

Step 3 Mark v. If every point of G is marked, go to **End**, otherwise go to **Step 2**.

End The algorithm terminates, H is the subgraph induced by the marked points.

One can soon discover, however, that this is not a very quick algorithm. If a graph with n points is connected, both Steps 2 and 3 must be executed $n-1$ times. With some bad luck point v in Step 2 might even be the last among the unmarked points we check. Then the total number of steps is more than $(n-1)+(n-2)+\ldots+2+1 \sim n^2/2$. (We shall consider questions like how to store a graph, how to check adjacency etc. in Section 1.4. Here we considered adjacency checks and markings as elementary steps.) This can really happen, e.g. if G is a path with its vertices numbered by $1, n, n-1, n-2, \ldots, 3, 2$.

Hence the following algorithm seems to be better:

Algorithm 1.2 *Checking the connectivity of a graph.*

Input and **Output** as in Algorithm 1.1

Step 1 Prepare three files. (File 1 will contain the vertices which have already been studied or scanned, File 2 some vertices which are labelled but unscanned, and File 3 the unlabelled vertices.) Initially, let File 1 be empty, put the first point of G at the beginning of File 2 and all the other points in File 3.

Step 2 Let v be the first point in File 2. Check every point adjacent to v. If some of them are in File 3, put them at the end of File 2 (and delete from File 3). When ready, delete v from File 2 and put in File 1.

Step 3 If File 2 or File 3 or both are empty, go to **End**, otherwise go to **Step 2**.

End The algorithm terminates, H is the subgraph induced by the points of Files 1 and 2.

Suppose that G is stored in such a way that the operation "consider every point adjacent to v" requires only as many steps as the number of these "neighbours" (and not $n-1$ steps for checking adjacency with every other point). Then, although both Steps 2 and 3 might be executed $n-1$ times as above, one execution of Step 2 requires only as many steps as the degree of v. Hence the total number of steps is proportional to the sum of the degrees of the points, i.e. to the number of edges (see Problem 1.1.14)[†]. (More precisely, the total number of steps is proportional to max (n, e), where n and e are the number of points and edges of the graph, respectively, since $e=0$ is possible.)

Hence this algorithm (and also Algorithm 1.3 in Problem 1.1.32) are best possible in a sense since even inputting the graph requires a number of steps, proportional to max (n, e). In particular, this algorithm is much better than Algorithm 1.1 for *sparse* graphs (where the number of edges is significantly less than $n^2/2$), which is the case in many applications. For example, Algorithm 1.1

[†] A function $f(n)$ is proportional to $g(n)$ if there are positive constants c_1, c_2 so that $c_1 g(n) < f(n) < c_2 g(n)$ holds for every sufficiently large n.

might (with usual data structures, see Section 1.4) use fifty times more steps than Algorithm 1.2 for a graph with 1000 points and 5000 edges.

We close this section with two further concepts. Let us suppose that point a can be reached along paths from point b, but every such path passes through point c (with $a \neq c, b \neq c$). Then c is called *separating point.* Similarly, if every $a - b$ path uses the edge e then e is called *separating edge.* In particular, a loop is never separating. Deleting a separating point or edge always increases the number of connected components of the graph. For example, v_3, v_4 and v_5 are separating points, e_6 and e_7 are separating edges in the graph of Fig. 1.6 (choose $a = v_1, b = v_6$ in all the cases).

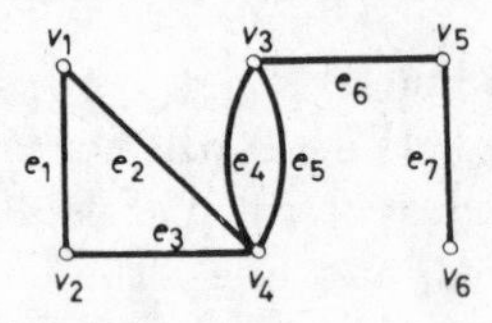

Fig. 1.6

Exercises for Section 1.1

1.1.1 Are the two graphs of Fig. 1.7 isomorphic?

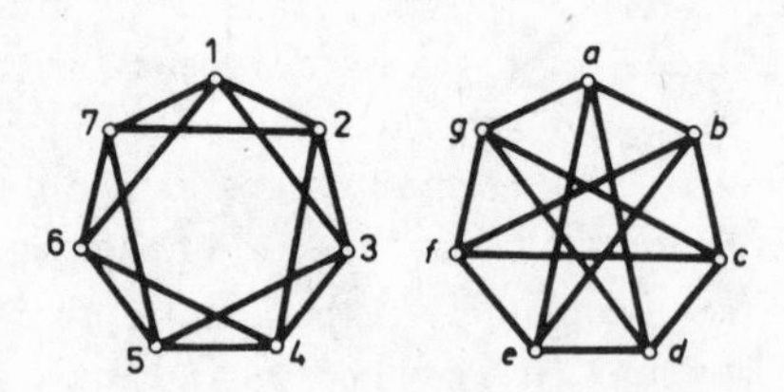

Fig. 1.7

1.1.2 Give two isomorphic graphs G, H where the bijection $\pi : V(G) \to V(H)$ of the definition is unique. Give another pair where it is not.

1.1.3 Let the 2-element subsets of a 5-element set be the points of a graph G and let two points be adjacent in G if the corresponding subsets are disjoint. Is G isomorphic to the graph of Fig. 1.8 or to that of Fig. 1.9?

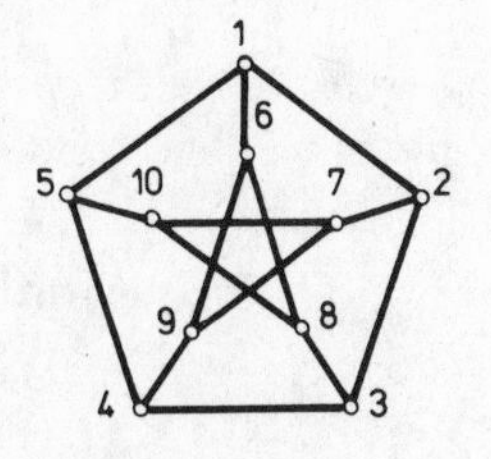

Fig. 1.8

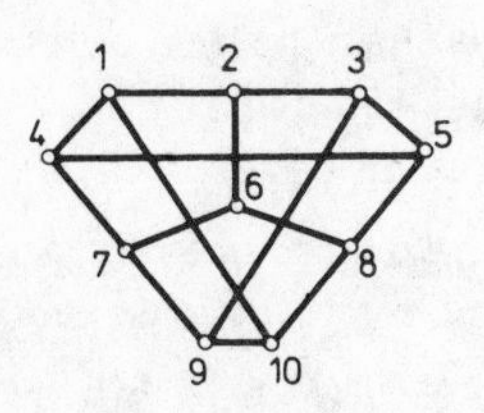

Fig. 1.9

1.1.4 Draw a graph G so that G and its complement H
(a) both are connected
(b) both are disconnected.

1.1.5 Find a graph which is isomorphic to its complement and has
(a) 5 vertices
(b) 6 vertices.

1.1.6 Let $[v_0, e_1, v_1, \ldots, v_{k-1}, e_k, v_k]$ be an edge sequence with $v_0 \neq v_k$. Prove that there exists a path between v_0 and v_k, consisting of certain edges of this sequence. If the sequence passes through a point $v_i (0 < i < < k)$, does there also exist a path between v_0 and v_k, passing through v_i?

1.1.7 Let the graph G contain two distinct paths between the points v_1 and v_2. Prove that G contains a circuit. Does this circuit pass through v_1 and v_2? Prove that the statement is not necessarily true with two distinct edge sequences.

1.1.8 The first paper on graph theory [Euler, 1736] proved that the citizens of Königsberg (today Kaliningrad) cannot walk over all the bridges of the river Pregel so that each bridge is passed exactly once (Fig. 1.10). Prove this.

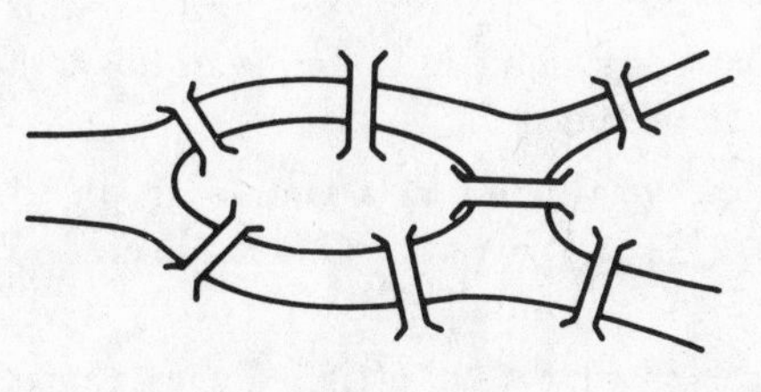
Fig. 1.10

1.1.9 Prove that an edge is separating if and only if it is not contained in any circuit.

1.1.10 Is it true that a point is separating if and only if it is not contained in any circuit?

1.1.11 Prove that the end points of the separating edges are either separating points or of degree one. Is every separating point incident to at least one separating edge?

1.1.12 Let the degree of every point be three. Prove that every separating point is incident to at least one separating edge.

1.1.13 Let $e \in E(G)$ be arbitrary. Prove that $c(G) \leq c(G-e) \leq c(G)+1$. Is a similar statement true for $v \in V(G)$ as well?

Problems for Section 1.1

1.1.14 Prove that the sum of the degrees of the points of G is $2e(G)$.

1.1.15 In a larger party one does not necessarily know everybody else, but any acquaintance is supposed to be mutual. Prove that the number of those people who know an odd number of others is even.

1.1.16 Let the degree of every point of a connected graph G be odd. Prove that deleting any separating edge of G the resulting subgraphs have an odd number of points.

1.1.17 Is there any simple graph where all points have different degrees?

1.1.18 Describe those simple graphs where exactly two points have the same degree.

1.1.19 Let $d_1, d_2, \ldots, d_n$ be positive integers. We are looking for a graph with n points so that these numbers are the degrees of the points.
(a) Prove that this is always possible if $d_1 + d_2 + \ldots + d_n$ is even.
(b) Give a degree sequence which cannot be realized with loop-free graphs.
(c) Give a degree sequence which can be realized with loop-free graphs but cannot with simple graphs.

1.1.20 Let the degree of every point be at least 2. Prove that the graph has at least one circuit.

1.1.21 Let the degree of every point be even. Prove that the edge set of the graph is the union of edge-disjoint circuits.

1.1.22 Prove that G has at least one circuit if $e(G) \geq v(G)$.

1.1.23 Let the degree of every point in a simple graph G be at least k. Prove that G has a path of length k.

1.1.24 Let the degree of every point in a simple graph G with n vertices be greater than $\lfloor n/2 \rfloor - 1$. Prove that G is connected. What happens if we change "greater" to "not smaller"?

1.1.25 Let G be connected and $k < v(G)$. Prove that G has a connected subgraph with k points.

1.1.26 What is the maximal number of separating points in a graph with n points?

1.1.27 Prove that any two longest paths of a connected graph share a point (which may be different for different pairs, see Problem 1.2.20 as well).

1.1.28 We have seen in Exercise 1.1.8 that, in order to have a closed edge sequence which covers each edge exactly once, every point must be of even degree. Prove that this condition is also sufficient for connected graphs.

1.1.29 Let every point of a connected graph have even degree, with exactly two exceptions. Prove that there exists an (obviously open) edge sequence which covers each edge exactly once.

1.1.30 Prove that every connected graph with e edges contains a closed edge sequence of $2e$ edges, which covers each edge at least once.

1.1.31 Change Step 3 in Algorithm 1.2 to the following:
Step 3'. If File 2 is empty, go to **End**, otherwise go to **Step 2**.
(a) Show that the algorithm remains correct and its total number of steps is still proportional to max (n, e).
(b) Give a graph where the modified algorithm terminates much later than the original one.

1.1.32 Prove that the following algorithm also works and estimate the total number of required steps:

Algorithm 1.3 *Checking the connectivity of a graph.*

Input and **Output** as in Algorithm 1.1.

Step 1 Prepare a vector $a = (a_1, a_2, \ldots, a_n)$ of length $n = v(G)$. Initially let $a_i = 0$ for every $i = 1, 2, \ldots, n$; later a_i will denote the name of that point of G (the "father" of point i) from which we first moved to i. Initially let $x = v_1$ (the first point of G); later x will denote the point reached.

Step 2 Mark x. If no unmarked point is adjacent to x, go to **Step 4**.

Step 3 (We move forward). If x is adjacent to some unmarked points, choose one, say y. Put $a_y = x$, then change the value of x to y and go to **Step 2**.

Step 4 (We move back). If $x = v_1$, go to **End**. Otherwise read $a_x = z$, change the value of x to z and go to **Step 2**.

End The algorithm terminates, H is the subgraph induced by the marked points.

Section 1.2 Trees, forests, cut sets

We have seen (Exercise 1.1.9) that the deletion of an edge e disconnects the graph if and only if e was not contained in any circuit. Hence, exactly those connected graphs have a minimal number of edges which do not contain circuits. The connected and circuit-free graphs will be called *trees*.

The word "minimal" above means that deleting any edge from such a graph disconnects the graph. The word "minimum" would mean instead that the graph has the smallest possible number of edges among all connected graphs (with a given number of points). One must, in general, distinguish between these two words (a function has only one minimum value A in an interval but can have several minimal values $A, B, C, \ldots$ see Fig. 1.11). However, in this context, every minimal graph is also minimum.

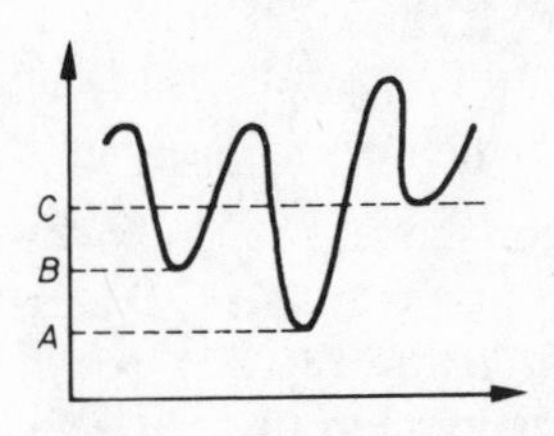

Fig. 1.11

Proposition 1.2.1 *In a connected graph G with n points, every tree has $n-1$ edges.*

Proof. The statement is trivial for $n = 2$. Suppose we have proved it for every $n \leq n_0$ and proceed to $n_0 + 1$. Every tree has points with degree one (since, by Problem 1.1.20, it would contain a circuit otherwise). Deleting such a point leads to a tree with n_0 points and, by the induction hypothesis, with $n_0 - 1$ edges.

□

If G is a connected graph, we may step by step delete those edges which are contained in circuits. The result is certainly a tree, containing every point of G. Such a tree will be called a *spanning tree* of G. If v is a point of degree one in a spanning tree T then $T - v$ is still a tree but not spanning any more.

Proposition 1.2.2 *If G is a connected graph with $v(G)$ points and $e(G)$ edges then it contains at least $e(G) - v(G) + 1$ circuits.*

Proof. Let T be a spanning tree of G. It contains $v(G) - 1$ edges. If e is any edge from the complement of T with respect to G then $T \cup \{e\}$ must contain at least one circuit C_e. Repeat the same construction for every edge $e, f, g, \ldots$ $\ldots \in E(G) - E(T)$. The resulting circuits $C_e, C_f, C_g, \ldots$ will all be different, since only C_e will contain e, only C_f will contain f etc.

□

This simple lower bound might be sharp (see Exercise 1.2.7). In particular, any subgraph of form $T \cup \{e\}$ contains exactly one circuit. Indeed, suppose the contrary. If $e = \{a, b\}$ is an edge and C_1, C_2 are distinct circuits of $T \cup \{e\}$ then both $C_1 - e$ and $C_2 - e$ must contain a path in T between the points a, b. But then T contains a circuit, by Exercise 1.1.7, a contradiction.

Hence the above system $C_e, C_f, C_g, \ldots$ is uniquely defined. In what follows, this system will be called a *fundamental system of circuits* of G with respect to T.

If a (perhaps disconnected) graph is circuit-free, its components are trees. Such a graph is called a *forest.* A subgraph of G, containing a maximal number of edges and every point of G but no circuits, is called a *spanning forest* of G. Every connected component of a spanning forest of G is a spanning tree of the corresponding component of G. Along the same line of reasoning as above, we can prove that a forest F has $v(F) - c(F)$ edges, where $c(F)$ is the number of connected components of F, hence a *fundamental system of circuits* of a graph G with respect to a spanning forest has $e(G) - v(G) + c(G)$ circuits.

The quantities $v(G) - c(G)$ and $e(G) - v(G) + c(G)$ will frequently arise. The former will be called the *rank* of the graph G and will be abbreviated by $r(G)$; the latter is denoted by $n(G)$ and is called the *nullity* of G.

One can easily modify any algorithm of the previous section to find a spanning tree or forest (see Exercise 1.2.9) or even to find a fundamental system of circuits (see Problem 1.2.16).

We have seen that separating edges disconnected a connected graph (or, in general, increased the number of connected components by one). In general, a set of edges of G will be called *separating* if its deletion increases $c(G)$. If a separating set is minimal (i.e. if no proper subset of it is separating any more) then it will be called a *cut set.* If a graph has separating edges, they are of course single element cut sets. For example, both $\{e_2, e_3\}$ and $\{e_7\}$ are cut sets in the graph of Figure 1.6. The deletion of the former leads to two components with point sets $\{v_1, v_2\}$ and $\{v_3, v_4, v_5, v_6\}$; that of the latter to $\{v_1, v_2, \ldots, v_5\}$ and $\{v_6\}$.

Proposition 1.2.3 *Every graph G has at least $v(G) - c(G)$ cut sets.*

Proof. Let F be a spanning forest of G with edges $e, f, g, \ldots$ We shall construct cut sets $Q_e, Q_f, Q_g, \ldots$ so that only Q_e will contain e, only Q_f will contain f etc. Let T be a connected component of F (i.e. a spanning tree of a connected component H of G) and let $e \in E(T)$. The complement of T with respect to H is not yet separating (since T is connected) but adding edge e to this complement leads to a separating set of edges, which then contains at least one cut set Q_e. □

In fact, this Q_e is also unique since the deletion of e from T disconnects the graph into two point sets A, B of $V(H)$; exactly those edges are in Q_e which join a point of A to a point of B. For example, in the graph G of Fig. 1.12 let $\{1, 3, 5\}$ be a spanning tree; then the corresponding cut sets are shown in Fig. 1.13.

Such a system of cut sets will be called a *fundamental system of cut sets* with respect to the given spanning forest.

We close the section with a basic result.

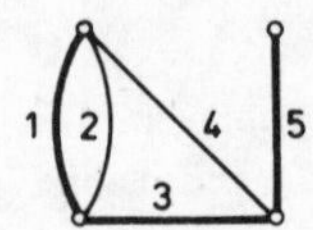

Fig. 1.12

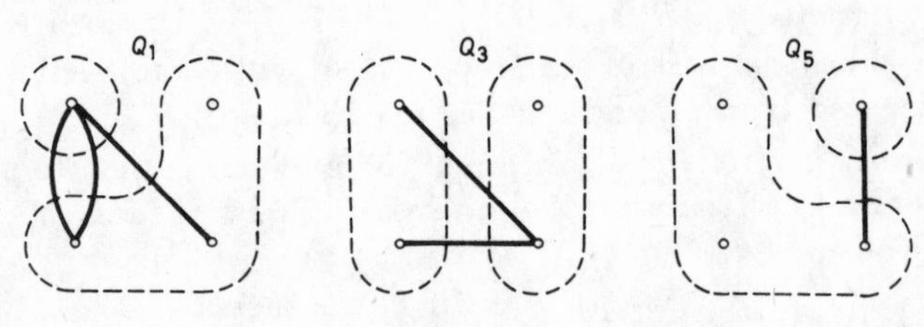

Fig. 1.13

Theorem 1.2.4 *If C is an arbitrary circuit and Q is an arbitrary cut set in a graph then the number of common edges, $|E(C) \cap E(Q)|$, of C and Q is even.*

Proof. If C and Q intersect, they are in the same connected component H of G. The deletion of $E(Q)$ from $E(H)$ disconnects the graph H into two point sets A, B of $V(H)$. If $V(C) \cap A$ or $V(C) \cap B$ is empty, the above quantity is zero, hence even. Otherwise, since C is a closed edge sequence, each time it leaves A for B it must return as well, and since it uses always different edges of Q, the cardinality of the intersection must be even.

□

Exercises for Section 1.2

1.2.1 Prove that G is a tree if and only if any one of the four statements below is true.

(1) G is connected and $e(G) = v(G) - 1$.

(2) G is circuit-free and $e(G) = v(G) - 1$.

(3) There is exactly one path between any pair of points of G.

(4) Adding any edge e to G the resulting graph contains exactly one circuit.

1.2.2 Prove that G is a forest if and only if every edge of G and every point of G with degree at least 2 is separating.

1.2.3 Prove that $c(G) + e(G) \geq v(G)$ for every graph G.

1.2.4 Give lower and upper bounds to the number of points with degree one in a tree T. Give stronger lower bound if T has a point with degree k.

1.2.5 Characterize those graphs G where $r(G)$ or $n(G)$ is zero.

1.2.6 Characterize those edges of a graph which are contained in every spanning forest and those which are not contained in any spanning forest.

1.2.7 Characterize those graphs G which have exactly $n(G)$ circuits.

1.2.8 Let $e, f \in E(G)$. Does there exist a spanning forest T such that e, f are contained in the same circuit of the fundamental set of circuits with respect to T?

1.2.9 Modify the algorithms of Section 1.1 to find a spanning forest of the input graph as well. Estimate the total number of operations required.

1.2.10 Is every edge contained in at least one cut set?

1.2.11 A *star* is a set of edges, incident to a given point. Is every star a cut set?

1.2.12 Prove that every cut set free subgraph can be extended to the complement of a spanning forest.

1.2.13 Give an example where cut sets, i.e. minimal separating sets, are not minimum.

1.2.14 Let C be a circuit of G and e, f two distinct edges of C. Prove that there exists a cut set Q so that $E(C) \cap E(Q) = \{e, f\}$.

1.2.15 Let F be a spanning forest of $G, x \notin E(F)$ and $y \in E(F)$. Let C_x and Q_y denote the circuit contained in $F \cup \{x\}$, and the cut set disjoint from $F - y$, respectively. Prove that the following three statements are equivalent.
(1) $x \in E(Q_y)$.
(2) $y \in E(C_x)$.
(3) $(F - y) \cup \{x\}$ is also a spanning forest of G.

Problems for Section 1.2

1.2.16 Modify Algorithm 1.3 to find a fundamental set of circuits with respect to a spanning forest. Estimate the total number of operations required.

1.2.17 (continued) Check your solution to the previous problem on a complete graph with n points.

1.2.18 Prove that an edge set X in a graph G is a circuit if and only if X is a minimal set which intersects the complement of every spanning forest of G.

1.2.19 Prove that an edge set X in a graph G is the complement of a spanning forest if and only if X is a minimal set which intersects every circuit of G.

1.2.20 Prove that every longest path of a tree shares a common point (cf., Problem 1.1.27 as well).

1.2.21 Let $d_1, d_2, \ldots, d_n$ be a sequence of positive integers with $d_1 + d_2 + \ldots + {} + d_n = 2n - 2$. Prove that it can be realized as the degree sequence of a tree with n points (c.f. Problem 1.1.19 as well).

1.2.22 Assign the numbers $1, 2, \ldots, n$ to the points of a tree T and call the nth point the root. T has at least one point with degree one, different from the root (see Exercise 1.2.4); let v_1 be such a point with minimal number assigned. Delete v_1 and its incident edge $\{v_1, v_2\}$ and record v_2. Find a point in the new tree, which again is of degree one and different from the root; let v_3 be such a point with minimal number assigned ($v_3 = v_2$ is also possible). Delete v_3 and its incident edge $\{v_3, v_4\}$ and record v_4. Continue this process: in the $(n-1)^{th}$ step record the root. This record of length $n-1$ consisting of not necessarily distinct numbers, is called the *Prüfer-code* of T, see Prüfer [1918].
(a) Find the Prüfer-code of a path of length $n-1$.
(b) Find the Prüfer-code of a star (where all the $n-1$ edges are incident to a common point).

1.2.23 (continued) Show that the tree can uniquely be reconstructed from its Prüfer-code. Then prove the theorem of Cayley [1889]: The number of distinct trees with n points is n^{n-2} (if two isomorphic trees with distinct numberings are considered as distinct).

1.2.24 (continued) Give a new solution to Problem 1.2.21.

1.2.25 Prove that an edge set X in a graph G is a cut set if and only if X is a minimal set which intersects every spanning forest of G.

1.2.26 Prove that an edge set X in a graph G is a spanning forest if and only if X is a minimal set which intersects every cut set of G.

1.2.27 Prove that a nonempty edge set X in a graph G is a circuit if and only if X is a minimal set which intersects every cut set of G by an even number of edges.

1.2.28 Prove that a nonempty edge set X in a graph G is a cut set if and only if X is a minimal set which intersects every circuit of G by an even number of edges.

1.2.29 Let X, Y be two disjoint edge sets in G. Let X be circuit free and Y be cut set free. Prove that G has such a spanning forest F that $X \subseteq E(F)$ and $Y \cap E(F) = \emptyset$.

1.2.30 Let $V(G)$ be the disjoint union of two subsets A, B and let X be the set of those edges which join a point of A to a point of B. Prove that X is the disjoint union of cut sets of G.

Section 1.3 Directed graphs

Let us recall that graphs were defined as systems $G = (V, E)$ where E was a collection of pairs of elements of V. If $e \in E$ was an edge, we wrote $e = \{u, v\}$ with $u, v \in V$. The order of u and v was irrelevant.

In many cases one might prefer to define a system where the order of the elements within the pairs does play a role. For example, if our "graph" is the model of a road map (points represent cities and edges represent direct road connections) then our graph model was useful. But if points represent special places in a city and edges represent street connections then, due to the one-way streets, we cannot say that a street is a connection between x and y, rather that it is a way from x to y (or vice versa, as the case may be).

Hence we introduce the concept of *directed graphs* as systems $\vec{G} = (V, \vec{E})$ where V is a nonempty set of *points* and $\vec{E}$ is a collection of *ordered* pairs of points called *directed edges*. An edge $e = (u, v)$ can be drawn as an arrow from u to v, u is the *tail* and v is the *head* of e. Loops are permitted; then $u = v$ and the order is irrelevant. The edges $e = (u, v)$ and $f = (u, v)$ are parallel; however, e and $g = (v, u)$ are not. Hence, on the directed graph of Fig. 1.14, the only pair of parallel edges is c, d.

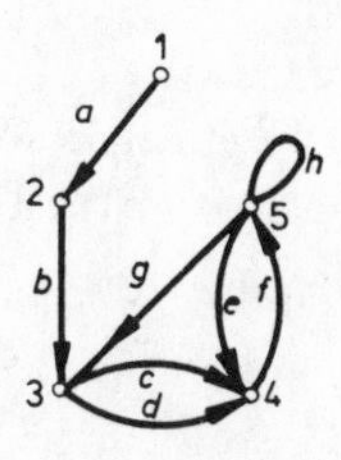

Fig. 1.14

The number of edges leaving a point u is the *outdegree* of u, the number of those entering u is the *indegree* of it. A loop contributes one both to the outdegree and to the indegree. Points with indegree zero are *sources*, with outdegree zero are *sinks*. For example, in the directed graph of Fig. 1.14, point 1 is a source, there is no sink, and the outdegree of the points $1, 2, 3, 4, 5$ are $1, 1, 2, 1, 3$, respectively.

If we become uninterested in the order of the points in the edge (i.e. if we change the arrows of a directed graph to usual lines) then we obtain the *underlying undirected graph* of our directed graph. On the other hand, if we assign an ordering to every nonloop edge of an undirected graph (i.e. if we change lines to arrows) then we *assign an orientation* to the undirected graph to get a directed one.

If a sequence of directed edges corresponds to a path or a circuit in the underlying undirected graph, it does not necessarily mean that we can really "drive" along these directed edges as on one-way streets. If yes, we speak about *directed paths* and *directed circuits*, respectively. For example, the sequences (a, b, d, f) and (a, b, g, e) of edges are paths in the underlying graph of the directed graph of Fig. 1.14 but only the former is a directed path. Similarly, (e, f) or (c, f, g) are directed circuits but (c, d) or (d, e, g) are not.

If the directed graph is *connected*, that is, if the underlying undirected graph is connected, we might still be unable to reach any point from any other along directed paths. If we can, the directed graph will be called *strongly connected.* Our example in Fig. 1.14 is not strongly connected (e.g., 1 cannot be reached from any other point along directed paths) but by deleting the first two points, the rest is strongly connected. Let us say that two points u, v are in the same class if either $u = v$ or if any can be reached from the other along directed paths. These classes determine the *strongly connected components* of the directed graph. Note that these components are disjoint, and together include every point of the directed graph but not necessarily all of its directed edges. For example, the strongly connected components of the directed graph of Fig. 1.14 are determined by the point sets $\{1\}, \{2\}, \{3,4,5\}$ and the edges a, b are not included in any component.

In this book directed graphs play a minor role only. Hence, unless we explicitly state the contrary, graphs will always mean undirected graphs.

Exercises for Section 1.3

1.3.1 Prove that the sum of the indegrees of the points in a directed graph equals to the sum of their outdegrees. What is this common value?

1.3.2 What is the minimum number of edges in a strongly connected directed graph with n points?

1.3.3 Prove that if a directed graph has no directed circuits, it has sinks and sources.

1.3.4 How many directed circuits are in a directed graph if every point has outdegree 1?

1.3.5 Show that the analogy of Problem 1.1.23 is not necessarily true, i.e. there need not exist a point in a strongly connected graph so that after its deletion the result is still strongly connected.

1.3.6 Prove that $\vec{G} = (V, \vec{E})$ is strongly connected if and only if, for every proper subset A of V, there is a directed edge with tail in A and head in $V - A$.

1.3.7 A cut set Q, separating the points of A from the points of $V - A$ is *directed* if either every edge of Q has tail in A and head in $V - A$ or *vice versa.* Prove that a connected directed graph is strongly connected if and only if it has no directed cut sets.

1.3.8 Prove that every directed edge is contained either in a directed circuit or in a directed cut set.

Problems for Section 1.3

1.3.9 Delete the sources and the sinks from the directed graph. Repeat the procedure for the new graph etc. as long as there are sources and/or sinks. Prove that the result is the "empty graph" (with no points) if and only if the original graph had no directed circuits.

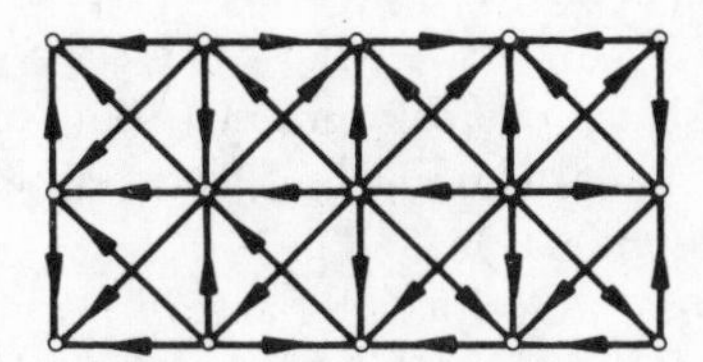

Fig. 1.15

1.3.10 Does the directed graph of Fig. 1.15 contain directed circuits?

1.3.11 Contract the strongly connected components $P_1, P_2, \ldots$ of a directed graph into points $p_1, p_2, \ldots$ In the new directed graph let $e = (p_i, p_j)$ be a directed edge if there was at least one directed edge in the original directed graph with tail in $V(P_i)$ and head in $V(P_j)$. Prove that the new directed graph has no directed circuits.

1.3.12 Assign orientation to every edge of a complete graph. The resulting directed graph is called a *tournament*.

(a) Prove that every tournament has a point u so that every point can be reached from u along a directed path of length 1 or 2.

(b) Prove that every tournament T has a directed path passing through all the points of T [Rédei, 1934].

(c) Prove that if a tournament T is strongly connected, then it has a directed circuit passing through all the points of T [Camion, 1959].

(d) Prove that if a tournament T is not strongly connected then it has a directed edge so that reversing the orientation of this single edge T becomes strongly connected.

1.3.13 Suppose $\overrightarrow{G}$ is a strongly connected directed graph and deleting a suitable set of at most k edges will destroy its strong connectedness. Prove that the same destruction can be performed by reversing the orientation of at most k edges as well.

1.3.14 Let every edge of a directed graph be coloured by either red or blue or green. Let e be a green edge. Prove that e is either contained in a circuit of red and green edges only, where the orientation of every green edge is the same along the circuit, or e is contained in a cut set of blue and green edges only, where the orientation of every green edge is the same in the cut set [Minty, 1966].

1.3.15 Prove that for every undirected graph, an orientation can be assigned to the edges so that the difference between the outdegree and indegree is at most one at every point.

Section 1.4 Graph theory algorithms I: Matrix representations of a graph and related data structures

While we considered various algorithms in the previous sections, we postponed one important question until now, namely how to store a graph. We shall present now seven different representations of a graph. Some of them will be of theoretical importance (for proving theorems etc), some others are everyday tools of programmers for real life applications.

Perhaps the most natural description of a graph $G = (V, E)$ is the *incidence matrix* of G. This matrix $\mathbf{B}(G)$ has $v(G)$ rows and $e(G)$ columns, and entry b_{ij} (i.e. the element in the intersection of row i and column j) equals to 1 if edge j is incident to point i and $b_{ij} = 0$ otherwise. For example, if G is the graph of Fig. 1.16 then $\mathbf{B}(G)$ is

	1	2	3	4	5	6	7	8
1	1	1	1	1	0	0	0	0
2	1	1	0	0	1	0	0	0
3	0	0	1	0	1	1	0	0
4	0	0	0	1	0	1	1	0
5	0	0	0	0	0	0	1	1

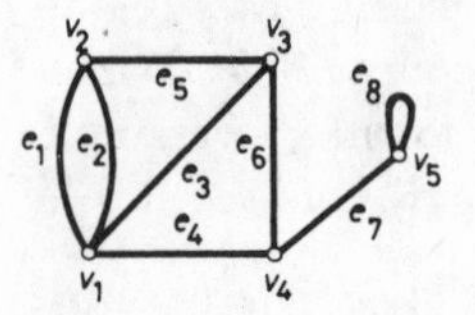

Fig. 1.16

Another natural description is the *adjacency matrix* $\mathbf{A}(G)$ with $v(G)$ rows and $v(G)$ columns. An entry a_{ij} (for $i \neq j$) is the number of edges joining point i to point j and a_{ii} is the number of loops incident to point i. For example, $\mathbf{A}(G)$ for the above graph of Fig. 1.16 is

	1	2	3	4	5
1	0	2	1	1	0
2	2	0	1	0	0
3	1	1	0	1	0
4	1	0	1	0	1
5	0	0	0	1	1

The storage requirement for these matrices is proportional to $e(G) \cdot v(G)$ and to $[v(G)]^2$, respectively. However, we store a large number of "unnecessary" data as well. For example, there are only two ones in any column of $\mathbf{B}$ (only one for loops), the rest carries no information. Should we store the positions of these ones only, our storage requirement would be as little as about $2 \cdot e(G)$, yet our graph would remain reconstructible. For example, the graph of Fig. 1.16 would then be described as

$$(1,2;1,2;1,3;1,4;2,3;3,4;4,5;5,5)$$

In fact we must study such "more compact" storage ideas. Otherwise all our efforts to reduce the number of operations for a connectivity test (from one proportional to $[v(G)]^2$ to another proportional to max $[e(G), v(G)]$, see Section 1.1) would be of no use if inputting the graph itself required at least $[v(G)]^2$ operations. Quite a few other problems in graph theory can also be solved in time proportional to max $[e(G), v(G)]$, see especially Section 3.4 which further emphasizes the importance of these considerations.

Although the above vector of length $2 \cdot e(G)$ is a compact description, yet it is not very practical. Should we have a command like "list all the points, adjacent to a given point u" in an algorithm, we had to check the whole vector again and again for every instance of u; thus the total number of operations in Algorithm 1.2 were proportional to $e(G) \cdot v(G)$, i.e. again, much more than max $[e(G), v(G)]$.

Hence, we should rather store for every point, the list of the points adjacent to it. Although every edge (except the loops) will thus be recorded twice, the above difficulty disappears. Now, our graph of Fig. 1.16 could look like this

1	2	2	3	4
2	1	1	3	
3	1	2	4	
4	1	3	5	
5	4	5		

The only disadvantage yet is that such a matrix-like array would require again much space; about $v(G) \cdot d$ where d is the maximum degree in G. However, we can arrange the rows of this "matrix" continuously like

$$\mathbf{e}(G) = (2,2,3,4,1,1,3,1,2,4,1,3,5,4,5).$$

Then we need an additional vector of "pointers"

$$\mathbf{v}(G) = (1,5,8,11,14,16);$$

this is a vector $\mathbf{v} = (v_1, v_2, \ldots)$ which shows that in order to list the points, adjacent to point i, we should start reading $\mathbf{e}$ at entry v_i and finish at $v_{i+1} - 1$.

This pair $[\mathbf{e}(G), \mathbf{v}(G)]$ is called the *adjacency list* of G, its storage requirement is $2e(G) + v(G) - 1$, hence it is proportional to max $[e(G), v(G)]$. Using this list our previous estimations on the maximal number of operations for Algorithms 1.2–1.3 are realizable.

We briefly mention two other data structures as well. The *ordered adjacency list* is also a pair $[\mathbf{e}_0(G), \mathbf{v}(G)]$ with the only difference that the entries $e_{v_i}, e_{v_i+1}, \ldots, e_{v_{i+1}-1}$ (i.e. the neighbours of point i) are arranged in increasing order. Although this ordering might take some extra time, it might be advantageous in certain cases, see **Box 1.2.**

The *chain adjacency list* appears to be much more complicated at first. It is a triple $[\mathbf{e}(G), \mathbf{e}_c(G), \mathbf{v}(G)]$. Entry v_i shows the location where the first neighbour of point i is listed in $\mathbf{e}(G)$. However, the position of the next neighbour is not necessarily the next entry in $\mathbf{e}(G)$; this position is shown as the value of the v_i^{th} entry of $\mathbf{e}_c(G)$. For example, our graph (Fig. 1.16) looks now like this

$$\mathbf{v} = (3, 9, 4, 13, 15)$$
$$\mathbf{e} = (2, 1, 2, 2, 4, 3, 1, 3, 1, 4, 4, 3, 1, 5, 5)$$
$$\mathbf{e}_c = (5, ^*, 1, 7, 8, 2, 11, ^*, 6, ^*, ^*, ^*, 14, 12, 10)$$

Then the neighbourhood of, say, point 3 can be obtained as follows. Read $v_3 = 4$ at first. Then $e_4 = 2$ is a neighbour and the fourth entry of $\mathbf{e}_c$ shows that we should go to entry 7. Now $e_7 = 1$ is a neighbour again, and since the 7th entry of $\mathbf{e}_c$ is 11, we turn to entry 11. By $e_{11} = 4$ we obtain a new neighbour and the * in the corresponding entry of $\mathbf{e}_c$ shows that no more point is adjacent to point 3.

This list is longer and requires slightly more operations as well. However, it has advantages as well. If we wish to delete, say, the edge $\{1, 3\}$ from our graph, all we have to do is to change the fifth entry of $\mathbf{e}_c$ to * and the fourth entry of $\mathbf{e}_c$ to 11. In comparison, by using usual (or ordered) adjacency lists we should change the position of almost every entry.

Box 1.1 summarizes the five different representations of the graphs we have seen so far, and two others (to be defined later). Any of these 5 representations can be transformed into any other and — except if $\mathbf{B}(G)$ is involved — no such transformation requires more operations than constant times $[v(G)]^2$, see Exercise 1.4.6 and Problems 1.4.18–19. Hence we can conclude:

Statement 1.4.1 *If the number of operations for a graph theory algorithm is proportional to $[v(G)]^2$ or is more, then the choice of the proper data structure is of secondary importance.*

But in the case of "quick" algorithms (like those in Sections 1.1 or 3.4) such a choice might be crucial. Then the proper choice should also be influenced by the actual problem since various operations in graph theory might be especially quick or especially slow in certain data structures. Such information is presented in **Box 1.2** (partly taken from [Golumbic, 1980]), the proofs are Exercise 1.4.4 and Problem 1.4.20 It is important to realize now that

"Statement" 1.4.2 *None of the known data structures is optimal for every possible task.*

Box 1.1. Various representations of an undirected graph

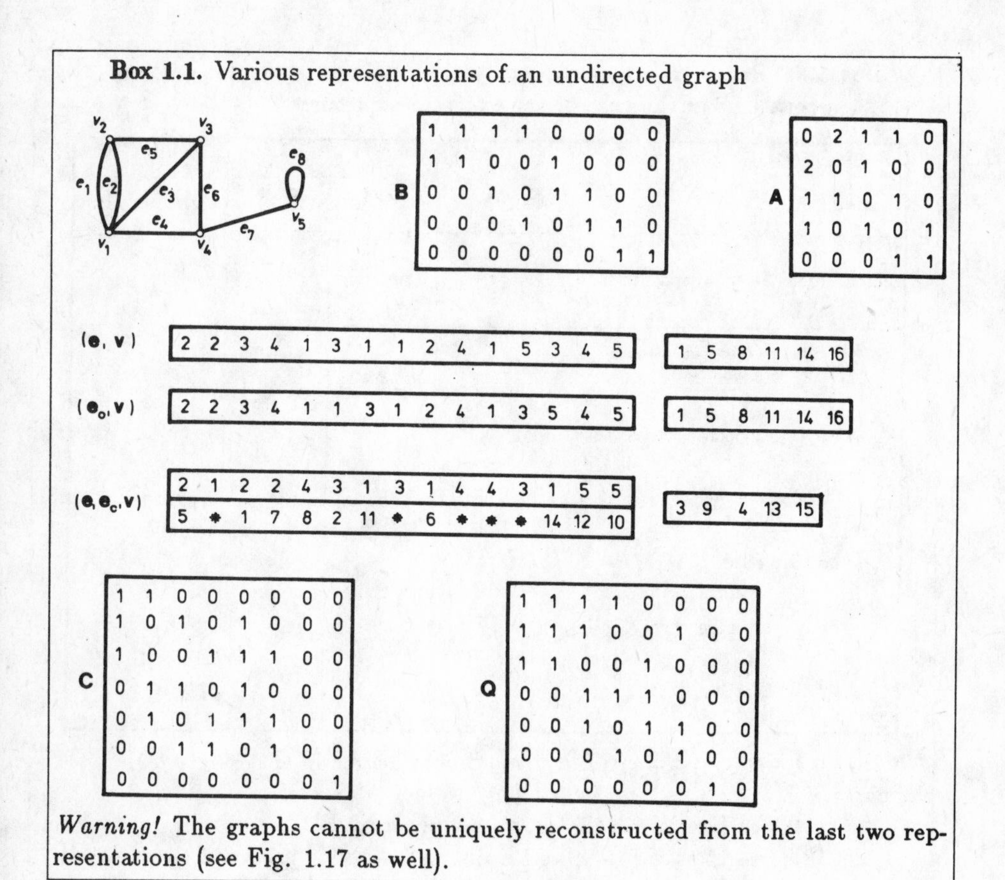

Warning! The graphs cannot be uniquely reconstructed from the last two representations (see Fig. 1.17 as well).

At the end of this section some further properties of the incidence matrix $\mathbf{B}(G)$ will be presented. But now we define two other matrices of the graph at first. The *circuit matrix* $\mathbf{C}(G)$ has $e(G)$ columns, its rows correspond to the circuits of G, and an entry c_{ij} equals 1 if edge j belongs to circuit i; while $c_{ij} = 0$ otherwise. The *cut set matrix* $\mathbf{Q}(G)$ is defined analogously. Examples can be seen in **Box 1.1.**

These matrices are not appropriate for data storage. Observe, for example, that if we change our graph of Fig. 1.16 (and in **Box 1.1**) to any graph of Fig. 1.17 then neither $\mathbf{C}$ nor $\mathbf{Q}$ will change. Hence no unique reconstruction is possible from these matrices. Their importance is that their linear algebraic properties are strongly related to the structure of the graph.

Throughout this book, we shall frequently use concepts like matrix, rank, linear independence, determinants, etc. All these are defined in vector spaces over fields. In most of the cases our fields are either the field $\mathbb{R}$ of the reals, the field $\mathbb{Q}$ of the rationals, or the binary field $\mathbf{B}$ (sometimes denoted by GF(2) as

Box 1.2. Requirement of storage and number of steps for certain graph operations in the case of various data-structures

then \ If Graph is given by		**A**	$(\mathbf{e}, \mathbf{v})$	$(\mathbf{e}_o, \mathbf{v})$	$(\mathbf{e}, \mathbf{e}_c, \mathbf{v})$
Storage requirement proportional to		v^2	$2e + v$	$2e + v$	$4e + v$
Number of required steps is proportional to	Decide whether two given points are adjacent	1	d	$\log d$	d
	Mark neighbours of given point	v	d	d	d
	Mark every edge	v^2	e	e	e
	Add an edge	1	e	e	1
	Delete an edge	1	e	e	d
	Delete a vertex	v	e	e	$\min(e, d^2)$

Notations: $v = v(G)$; $e = e(G)$; $d =$ maximum degree in G

Remarks:

(1) See Exercise 1.4.5 for graphs given by **B**

(2) See Exercise 1.4.4 and Problem 1.4.18 for proofs of these data

(3) See Exercise 1.4.19 for combination of the last two data-structures

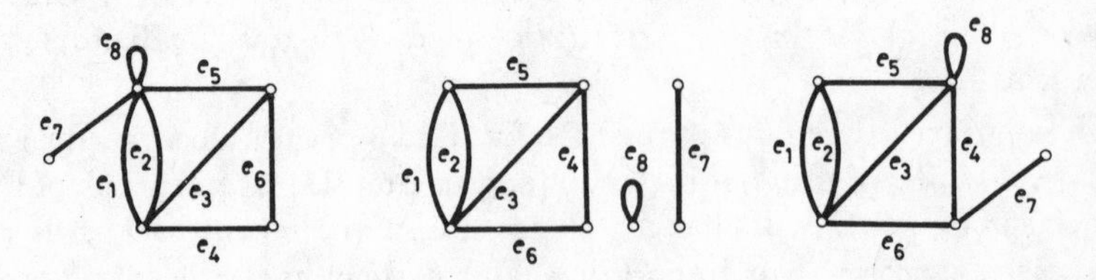

Fig. 1.17

well). This latter has two elements, 0 and 1, and apart from the rule $1 + 1 = 0$, both addition and multiplication are "usual".

It is crucial to realize that concepts of linear algebra are always defined with respect to the underlying field of the vector space. For example, the rank of the matrix $\begin{pmatrix} 1 & 1 & 0 \\ 1 & 0 & 1 \\ 0 & 1 & 1 \end{pmatrix}$ is 2 over $\mathbb{B}$ but 3 over $\mathbb{R}$; or the inverse of $\begin{pmatrix} 1 & 0 & 1 \\ 1 & 1 & 0 \\ 1 & 0 & 0 \end{pmatrix}$ is $\begin{pmatrix} 0 & 0 & 1 \\ 0 & 1 & 1 \\ 1 & 0 & 1 \end{pmatrix}$ over $\mathbb{B}$ but $\begin{pmatrix} 0 & 0 & 1 \\ 0 & 1 & -1 \\ 1 & 0 & -1 \end{pmatrix}$ over $\mathbb{R}$, see Exercise 1.4.7.

Statement 1.4.3 *Let* $\mathbf{C}$ *and* $\mathbf{Q}$ *be the circuit and cut set matrix of a graph, respectively, so that columns are numbered in the same way in both matrices. Then* $\mathbf{C} \cdot \mathbf{Q}^T = \mathbf{O}$ *over* $\mathbb{B}$.

Proof. Let $\mathbf{CQ}^T = \mathbf{R}$ and prove that for any entry $r_{ij} = 0$. Since r_{ij} is the scalar product of the ith row of $\mathbf{C}$ and the jth column of $\mathbf{Q}^T$ (i.e. the jth row of $\mathbf{Q}$), r_{ij} is simply the number of common edges in the ith circuit and j^{th} cut set of the graph, modulo 2 (since we work over $\mathbb{B}$). But the number of these common edges is even (Theorem 1.2.4), so the statement follows.

□

Theorem 1.4.4 *The rank of the circuit and the cut set matrices of a graph* G *over* $\mathbb{B}$ *are* $n(G)$ *and* $r(G)$, *respectively.*

Proof. Let F be a spanning forest of G and suppose that the edges of G are numbered so that edges in $E(F)$ receive numbers $1, 2, \ldots, r(G)$ and the rest receive the numbers $r(G) + 1, r(G) + 2, \ldots, e(G)$. The fundamental system of circuits with respect to T receives numbers $1, 2, \ldots, n(G)$ and the rest receive $n(G)+1, n(G)+2, \ldots$, etc. Similarly, let $1, 2, \ldots, r(G)$ be the numbers for members of the fundamental system of cut sets with respect to T and the next higher numbers for the other cut sets. Then $\mathbf{C}$ and $\mathbf{Q}$ decompose as

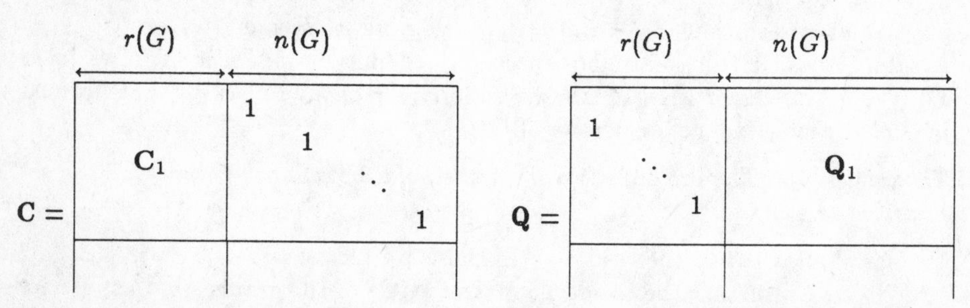

showing that the ranks of $\mathbf{C}$ and $\mathbf{Q}$ are at least $n(G)$ and at least $r(G)$, respectively. $r(\mathbf{C}) \leq n(G)$ directly follows from Exercise 1.4.10(b). $r(\mathbf{Q}) \leq r(G)$ could be proved similarly. Alternatively, the sum of these two $(r(\mathbf{C})+r(\mathbf{Q}) \leq e(G))$ can be obtained from a theorem of Sylvester (Problem 1.4.24) and from Statement 1.4.3.

□

Corollary 1.4.5 *In the above decomposition,* $\mathbf{C}_1 = \mathbf{Q}_1^T$ *over* $\mathbb{B}$.

Proof. If we perform the multiplication $\mathbf{CQ}^T$, we should obtain $\mathbf{O}$ by Statement 1.4.3. The upper-left $n(G)\times r(G)$ block of the result should be, on the other hand,

$(\mathbf{C}_1|\mathbf{E}) \cdot (\mathbf{E}|\mathbf{Q}_1)^T = \mathbf{C}_1 + \mathbf{Q}_1^T$, where $\mathbf{E}$ denoted unity matrices of appropriate size. Hence $\mathbf{C}_1 = -\mathbf{Q}_1^T$, as requested, (since $\mathbf{M} = -\mathbf{M}$ over $\mathbb{B}$). □

Another proof could be obtained, using Exercise 1.2.15.

We have seen that the rank of $\mathbf{C}(G)$ is $n(G)$. Hence, if we choose $n(G)$ columns among the total number of $e(G)$ columns of $\mathbf{C}(G)$, we get a set of linearly independent vectors (over $\mathbb{B}$) for at least one choice. However, a much stronger statement can be proved.

Theorem 1.4.6 *Let $X \subseteq E(G)$, $|X| = n(G)$. Consider the column vectors of $\mathbf{C}(G)$ corresponding to the elements of X. These vectors are linearly independent over $\mathbb{B}$ if and only if $E(G) - X$ is the edge set of a spanning forest of G.*

Proof. Let $E(G) - X$ be the edge set of a spanning forest F at first, and consider the fundamental set of circuits with respect to F. These circuits determine $n(G)$ rows in $\mathbf{C}(G)$ which form a unity matrix with the X-columns, proving that these columns are linearly independent.

Now let $E(G) - X$ be the edge set of a subgraph H which is not a spanning forest. Since $e(H) = r(G)$, H is disconnected, hence X contains the edge set of a cut set Q of G. The columns of $\mathbf{C}(G)$ corresponding to the edges of Q are linearly dependent over $\mathbf{B}$ (see Statement 1.4.3), hence so are the total set of columns corresponding to X. □

In exactly the same way one can prove

Theorem 1.4.7 *Let $X \subseteq E(G)$, $|X| = r(G)$. Consider the column vectors of $\mathbf{Q}(G)$, corresponding to the elements of X. These vectors are linearly independent over $\mathbb{B}$, if and only if X is the edge set of a spanning forest of G.*

We have seen that $\mathbf{C}(G)$ and $\mathbf{Q}(G)$ have rank $n(G)$ and $r(G)$, respectively. A submatrix of $\mathbf{C}(G)$ containing exactly $n(G)$ linearly independent rows (over $\mathbb{B}$) is called a *reduced circuit matrix*. Similarly, a *reduced cut set matrix* is formed by $r(G)$ linearly independent rows of $\mathbf{Q}(G)$.

Theorem 1.4.8 *The previous two theorems hold for reduced circuit and cut set matrices as well.*

Its *proof* is left to the reader as Exercise 1.4.11.

Now we return to the incidence matrix $\mathbf{B}(G)$ of the graphs. Let us restrict ourselves to loopless graphs. Then every column of $\mathbf{B}(G)$ contains exactly two ones and $v(G) - 2$ zeroes.

Theorem 1.4.9 *The rank of $\mathbf{B}(G)$ is $r(G)$; the relation $\mathbf{C}\mathbf{B}^T = \mathbf{O}$ holds and if $X \subseteq E(G)$, $|X| = r(G)$, then the column vectors of $\mathbf{B}(G)$ corresponding to the elements of X are linearly independent if and only if X is the edge set of a spanning forest of G. All these statements are in the vector space over $\mathbb{B}$.*

Its *proof* is also left to the reader as Problem 1.4.28.

We close this section by briefly reviewing the same results for directed graphs. The ways how directed graphs can be stored are summarized in **Box 1.3.**

The modification for the incidence and adjacency matrices are straightforward. The adjacency lists are now shorter since only directed edges with tail at the given point are listed. However, in most cases a second set of adjacency lists (directed edges listed by heads) is added for making some operations easier.

Box 1.3. Various representations of a directed graph

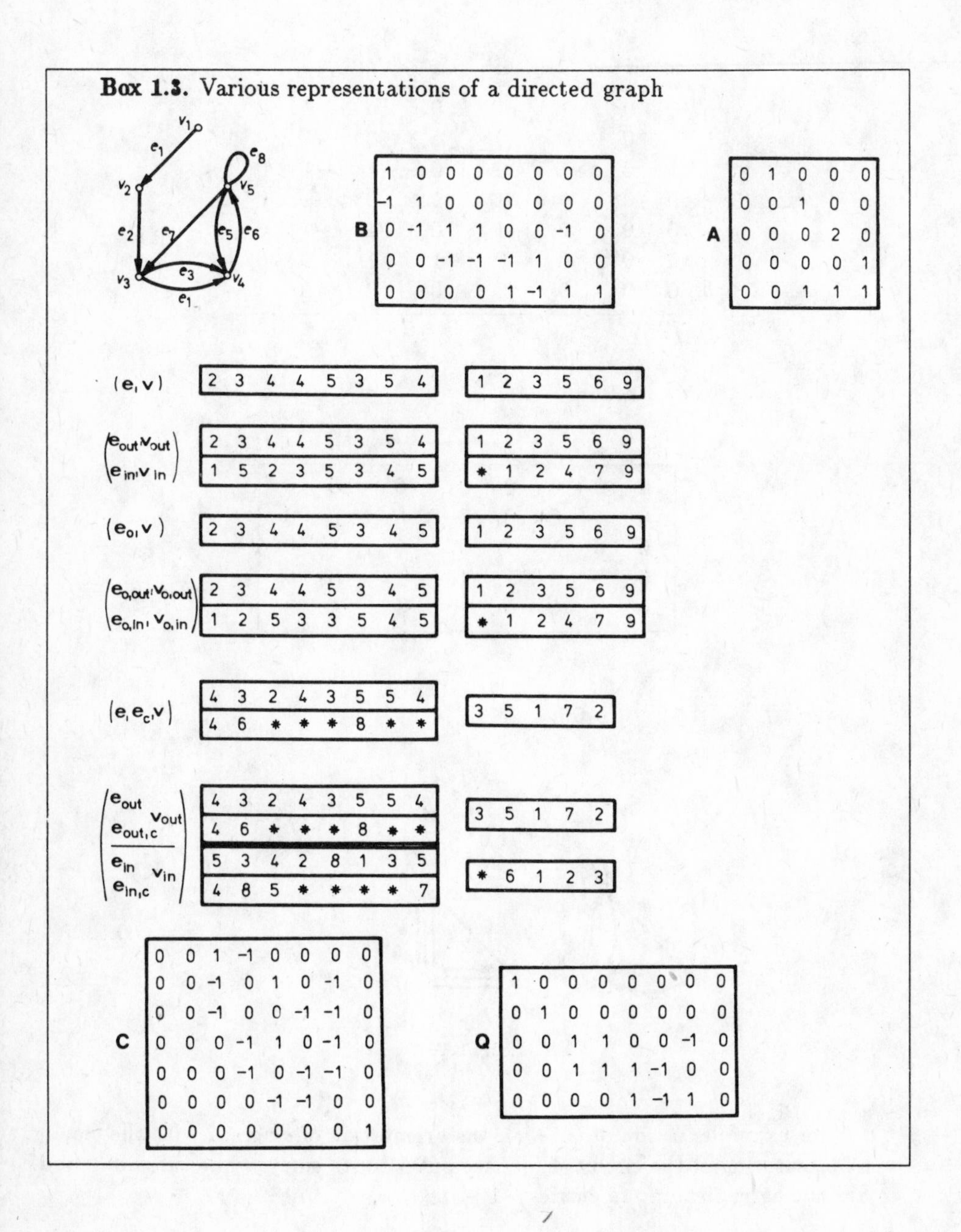

In order to define the circuit and the cut set matrices of a graph, we should assign orientations to the circuits and to the cut sets, respectively. For example, circuits might be considered clockwise, cut sets from top to bottom and from left to right, see Fig. 1.18 for orientations of the circuits and cut sets of the directed graph of Fig. 1.14. Then the circuit matrix and the cut set matrix of this directed graph are as follows.

a	b	c	d	e	f	g	h
0	0	1	−1	0	0	0	0
0	0	−1	0	1	0	−1	0
0	0	−1	0	0	−1	−1	0
0	0	0	−1	1	0	−1	0
0	0	0	−1	0	−1	−1	0
0	0	0	0	−1	−1	0	0
0	0	0	0	0	0	0	1

and

a	b	c	d	e	f	g	h
1	0	0	0	0	0	0	0
0	1	0	0	0	0	0	0
0	0	1	1	0	0	−1	0
0	0	1	1	1	−1	0	0
0	0	0	0	1	−1	1	0

Fig. 1.18

(For example, in circuit $\{c, e, g\}$, the orientation of e agreed with the clockwise orientation of the circuit, hence we put 1 to its entry, while those of c and g did not agree, leading to entries -1.)

Finally, all the statements and theorems 1.4.3 through 1.4.9 are valid for matrices of directed graphs as well, if we change the field $\mathbb{B}$ to field $\mathbb{Q}$ of the rationals or to field $\mathbb{R}$ of the reals (or, as a matter of fact, to any field with more than two elements). The only exception is Corollary 1.4.5, where we have $\mathbf{C}_1 = -\mathbf{Q}_1^T$. The reason why we need a field of cardinality of at least three elements is that 1 and -1 should be distinguished. In fact, the undirected case could be considered as a special case of the directed one where the underlying field is not arbitrary but $\mathbb{B}$. Verification of these theorems for directed graphs is left for the reader (see the solution of Exercise 1.4.16 as the first step).

Exercises for Section 1.4

1.4.1 Let G be a graph, $\mathbf{X}_k = [\mathbf{A}(G)]^k$ and $\mathbf{Y} = \mathbf{B} \cdot \mathbf{B}^T$ (operations performed over $\mathbb{Q}$). What is the graph-theory meaning of the entries of $\mathbf{X}_k$ and $\mathbf{Y}$?

1.4.2 How can we recognize loops and separating edges in a graph given by $\mathbf{A}, \mathbf{B}, \mathbf{C}$ or $\mathbf{Q}$?

1.4.3 A vector of length p contains p integers in increasing order. We want to decide whether a given number x is among them. Give an algorithm with total number of operations proportional to $\log p$ only. What will be the base of the logarithm?

1.4.4 Verify the first three columns in **Box 1.2.**

1.4.5 Among the six operations in **Box 1.2,** is there any which can be performed more quickly if the graph is given by $\mathbf{B}$ rather than by $\mathbf{A}$?

1.4.6 Estimate the total number of steps required to convert data structure $(\mathbf{e}, \mathbf{v})$ into $\mathbf{A}$ or into $(\mathbf{e}, \mathbf{e}_c, \mathbf{v})$.

1.4.7 Verify the numerical results on the 3×3 matrices, given before Statement 1.4.3.

1.4.8 Which, if any, among the following matrices are circuit matrices of suitable graphs?

$$\mathbf{M}_1 = \begin{pmatrix} 1 & 1 & 1 & 0 \\ 1 & 1 & 0 & 1 \\ 1 & 0 & 1 & 1 \end{pmatrix}; \mathbf{M}_2 = \begin{pmatrix} 1 & 1 & 1 \\ 1 & 1 & 0 \\ 1 & 0 & 1 \end{pmatrix}; \mathbf{M}_3 = \begin{pmatrix} 1 & 1 & 0 \\ 1 & 0 & 1 \\ 0 & 1 & 1 \end{pmatrix}$$

1.4.9 Give a graph and a maximal set of linearly independent circuits so that they do not form the fundamental system of circuits with respect to any spanning forest. Give the same with cut sets.

1.4.10 (a) Consider the sum (over $\mathbb{B}$) of two rows of $\mathbf{C}(G)$. The edges corresponding to the ones of the result determine a subgraph of G. Characterize this subgraph.

(b) Using this result prove that the rows of $\mathbf{C}(G)$ corresponding to the fundamental system of circuits with respect to a spanning forest generate all the other rows (over $\mathbf{B}$).

1.4.11 Prove Theorem 1.4.8.

1.4.12 Why did we exclude the loops of G in Theorem 1.4.9?

1.4.13 Let $\mathbf{A}, \mathbf{B}$ be the adjacency and the incidence matrices of a directed graph G, respectively. Determine $\mathbf{A} + \mathbf{B}\mathbf{B}^T$.

1.4.14 Prove that the incidence matrix $\mathbf{B}$ of a directed graph is *totally unimodular* (over $\mathbb{Q}$), i.e. the determinant of every square submatrix of $\mathbf{B}$ is $0, +1$ or -1.

1.4.15 Show that similar statement does not always hold for $\mathbf{C}$ or $\mathbf{Q}$ of a directed graph.

1.4.16 Prove the analogue of Statement 1.4.3 for directed graphs.

Problems for Section 1.4

1.4.17 The *trace* tr $\mathbf{M}$ of a square matrix $\mathbf{M}$ is the sum of its diagonal entries. Let $\mathbf{M} = [\mathbf{A}(G)]^3$ (operations performed over $\mathbb{Q}$). What is the graph theory meaning of tr $\mathbf{M}$ if G is loop-free?

1.4.18 The number of comparisons required to arrange the entries of a vector of length p in increasing order, can easily be estimated by $p \log p$, see Exercise 1.4.3. In fact, the total number of operations for this ordering is also known [Williams, 1964; Aho et al., 1974] to be proportional to $p \log p$. Using this prove that data structure $(\mathbf{e}, \mathbf{v})$ can be converted into $(\mathbf{e}_0, \mathbf{v})$ so that the number of required operations is proportional to $e \log d$.

1.4.19 (continued) Give another algorithm where this number is proportional to max (e, v) only.

1.4.20 Verify the fourth column in **Box 1.2.**

1.4.21 Try to unit the advantages of the data structures $(\mathbf{e}_0, \mathbf{v})$ and $(\mathbf{e}, \mathbf{e}_c, \mathbf{v})$.

1.4.22 Which, if any, among the matrices of Exercise 1.4.8 can be submatrices of circuit matrices of suitable graphs?

1.4.23 How does the deletion of an edge of G modify $\mathbf{C}(G)$ or $\mathbf{Q}(G)$?

1.4.24 Prove the following theorem of *Sylvester*: If $\mathbf{M}$ and $\mathbf{N}$ are $a \times b$ and $b \times c$ matrices and ρ denotes rank, then $\rho(\mathbf{M}) + \rho(\mathbf{N}) \leq b + \rho(\mathbf{MN})$.

1.4.25 Quite a few properties of the vector spaces over $\mathbb{B}$ are unusual. Prove that here one can find
(a) a nonzero vector $\mathbf{x}$ which is perpendicular to itself (i.e., $\mathbf{x}^T\mathbf{x} = 0$);
(b) a proper subspace M which is contained in $M^\perp = \{\mathbf{x} | \mathbf{x}$ is perpendicular to every $\mathbf{y} \in M\}$;
(c) a subspace M so that $M \cup M^\perp$ does not generate the whole vector space.

1.4.26 (continued) On the other hand, prove that
(a) dim $M+$ dim $M^\perp$ = the dimension of the whole vector space [Lovász, 1979a];
(b) the vector $(1, 1, \ldots, 1)^T$ arises as linear combinations of vectors from $M \cup M^\perp$ [Gallai; Chen, 1970b]

1.4.27 Let $\mathbf{X}$ be a square matrix with $0, +1$ and -1 entries so that the number of nonzero entries in every row is even. Prove that det $\mathbf{X}$ is even (over $\mathbb{Q}$).

1.4.28 Prove Theorem 1.4.9.

1.4.29 Call a point in a directed graph a supersource if its indegree is zero and there is a directed edge from it to every other point. Give a quick algorithm to decide whether a directed graph (given by a suitable data structure) has a supersource. What is the remarkable feature of this rather artificial problem?

1.4.30 Prove that even among the directed circuits of a directed graph G, one can choose $n(G)$ linearly independent ones (over $\mathbb{Q}$).

1.4.31 Let $\mathbf{M}, \mathbf{N}$ be two matrices of size $k \times \ell$ and $\ell \times k$, respectively, with $k \le \ell$. A theorem of *Binet and Cauchy* states that det $\mathbf{MN}^T$ is the sum of $\binom{\ell}{k}$ products of form det $\mathbf{M}'$ det $\mathbf{N}'$, where $\mathbf{M}'$ is formed by all rows and k columns of $\mathbf{M}$ (in every possible way) and $\mathbf{N}'$ is formed by all columns and the corresponding rows of $\mathbf{N}$. For example,

$$\left|\begin{pmatrix} a & b & c \\ d & e & f \end{pmatrix}\begin{pmatrix} g & h \\ i & j \\ k & \ell \end{pmatrix}\right| = \begin{vmatrix} a & b \\ d & e \end{vmatrix}\begin{vmatrix} g & h \\ i & j \end{vmatrix} + \begin{vmatrix} a & c \\ d & f \end{vmatrix}\begin{vmatrix} g & h \\ k & \ell \end{vmatrix} + \begin{vmatrix} b & c \\ e & f \end{vmatrix}\begin{vmatrix} i & j \\ k & \ell \end{vmatrix}$$

Using this theorem prove that, if $\mathbf{M}$ is a submatrix of $\mathbf{B}(G)$, formed by $r(G)$ linearly independent rows of the incidence matrix of a directed graph G, then det $\mathbf{MM}^T$ equals the number of spanning forests of G (over $\mathbb{Q}$).

1.4.32 (continued) Give a new proof to Cayley's theorem (cf., Problem 1.2.23).

1.4.33 We found that $\mathbf{C}$ is not totally unimodular, in general, see Exercise 1.4.15. Show that the submatrix of $\mathbf{C}$, consisting of the rows which correspond to members of a fundamental system of circuits with respect to a spanning forest, is totally unimodular [Cederbaum, 1956].

1.4.34 (continued) Let us now consider a submatrix $\mathbf{X}$ of $\mathbf{C}$, consisting of $n(G)$ linearly independent rows (over $\mathbb{Q}$). Prove that the determinant of every maximal nonsingular submatrix of $\mathbf{X}$ is $\pm k$ (where k does depend on $\mathbf{X}$ only, but is the same for every such determinant).

Section 1.5 Greedy algorithms I: Graphs

Let us assign weights, i.e. real numbers, to the edges of a simple graph G. For $e \in E(G)$, let $w(e)$ denote the weight of e. For $X \subseteq E(G)$, define the weight of the subset X as $\sum_{e \in X} w(e)$. Find the spanning forest of G with minimum weight. Throughout this section, all weights are nonnegative.

If, for example, the points a, b, c, d, e, f of Fig. 1.19 are cities and an edge in the graph means that it is possible to build a direct road between the corresponding cities then the weights on the edges may be imagined as the cost of

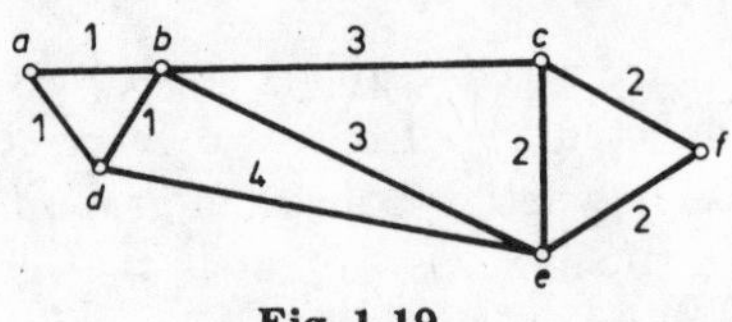

Fig. 1.19

building that road (measured in an appropriate unit of money). Then our problem means that a minimal cost road net should be found which already connects any two cities.

Figure 1.20 shows the five stages of constructing a "cheap" road net. Our idea was simple: build the cheapest road at first, then the cheapest one among the rest of the roads etc. but without forming circuits. For example, edge $\{c, e\}$ was preferred to $\{b, d\}$ in the third step (though its weight is larger) to avoid a circuit. If several possible roads with the same cost were available, we had simply a random choice.

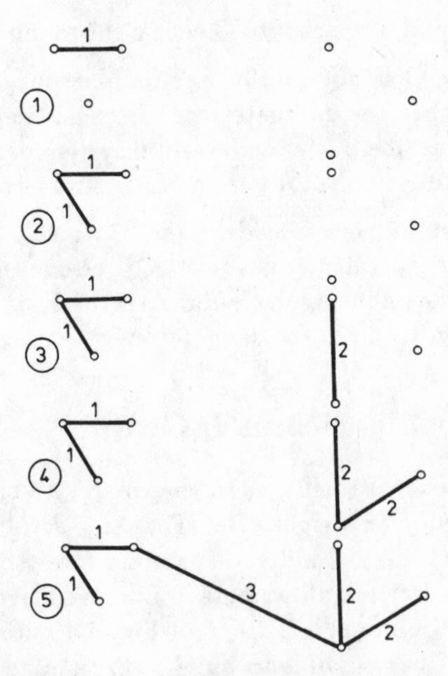

Fig. 1.20

One can easily check that our "idea" worked well on this example; the total cost of 9 units is really best possible. We shall prove that this strategy always leads to a minimum weight spanning forest (for every graph and for every weighting); a result which is of essential importance in matroid theory as well.

Theorem 1.5.1 *For an arbitrary graph G and for an arbitrary weighting of $E(G)$, Algorithm 1.4 gives a minimum weight spanning forest:*

Algorithm 1.4 *Minimum weight spanning forest.*

Input A graph G and nonnegative weights $w(e)$ for every $e \in E(G)$.

Output A spanning forest F of G so that $w(F) \leq w(T)$ for every spanning forest T of G.

Step 1 Prepare File 1 for the edges which will form the required minimum weight spanning forest F. Initially let File 1 be empty.

Step 2 Prepare File 2 for those edges which can be applied to extend the already existing subgraph. Formally, let H be the subgraph, formed by the edges of File 1 and put into File 2 the set $Y = \{e | e \in \in E(G) - E(H); H \cup \{e\}$ has no circuits $\}$.

Step 3 If $Y = \emptyset$, go to **End**. Otherwise choose an element $e \in Y$ with minimum weight (if there are several such elements, choose randomly among them), add e to File 1 and go to **Step 2**.

End The algorithm terminates, F is the subgraph formed by the edges of File 1.

Proof. Output F is obviously a spanning forest. Suppose there is a minimum weight spanning forest F_0 with $w(F_0) < w(F)$. If there are several such minimum weight "counterexamples", choose one with as many edges common with F as possible. Let $e_0 \in E(F_0) - E(F)$.

If we add e_0 to F, a circuit C is obtained, where every edge $e \in E(C) - \{e_0\}$ has $w(e) \leq w(e_0)$ (since if $w(e) > w(e_0)$, we would choose e_0 rather than e in Step 3). There is at least one such edge $e_1 \in E(C) - \{e_0\}$ which connects the two components of $F_0 - e_o$; hence $F_1 = (F_0 - e_0) \cup \{e_1\}$ is also a spanning forest. We have seen that $w(e_1) > w(e_0)$ is impossible. So is $w(e_1) < w(e_0)$, since then $w(F_1) < w(F_0)$ would contradict the minimality of F_0. Finally $w(e_1) = w(e_0)$ is impossible since $|E(F_1) \cap E(F)| > |E(F_0) \cap E(F)|$.

□

If a spanning forest of *maximum* weight should be found, we could multiply every weight by -1 and use Algorithm 1.4. This means, we choose an edge with the largest (optimal) weight, then one of the second largest ones etc. Roughly speaking, this is a "greedy" method: we increase the weight of the forest with the possible maximal amount at each step (all we check is to avoid circuits), without caring of the future.

In general, an algorithm is called *greedy* if, in every step, it tries to increase the value of the objective function with as much as possible (and stops when no such "one-step-increase" is possible any more). Normally such greedy algorithms are not optimal. For example, if one has to choose a maximum weight circuit of length four in the graph of Fig. 1.21 then a greedy algorithm would pick up the edges with weights 100 and 99 first, resulting in a total weight of 201 at the end, while the maximum is obviously 297.

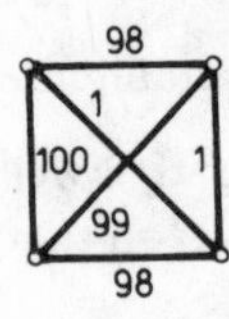

Fig. 1.21

It is a very interesting question of combinatorial optimization, which problems can be solved in such a greedy way. We return to this question in Section 7.4 (see also Exercise 1.5.7 and Problems 1.5.15–16).

Exercises for Section 1.5

1.5.1 Let F be a maximum weight spanning forest, C be an arbitrary circuit. If e denotes an edge of C with minimum weight, can e belong to $E(F)$?

1.5.2 Prove that the maximum weight spanning forest is unique if every edge has different weight.

1.5.3 Show that the following algorithm is also correct.

Algorithm 1.5 *Minimum weight spanning forest.*

Input and **Output** as in Algorithm 1.4.

Step 1 Prepare two files. File 1 will contain the edge set of F; File 2 will contain the edges which might belong to F at the end. Initially let File 1 be empty and File 2 contain all edges of G.

Step 2 Let e be an edge with maximum weight among the elements of File 2. (If File 2 were empty, go to **End**.) If e is a separating edge of G then put e in File 1.

Step 3 Delete e from File 2 and also from the graph G. Go to **Step 2**.

End The algorithm terminates, F is the subgraph, formed by the edges in File 1.

1.5.4 Let G be a graph with nonnegative weights w on its edge set. Let X be a proper subset of $V(G)$ which induces a connected subgraph H of G. Let F be a minimum weight spanning tree of H. Give an example that there might exist a connected subgraph T of G so that $V(H) \subset V(T)$ and $w(T) < w(F)$.

1.5.5* (continued) Let G be the complete graph with n points. Place the points to the plane and define the weights of the edges as distances between the corresponding points. The minimum weight spanning trees correspond to the "cheapest" road nets among the n cities. Show that even cheaper solutions are possible if junctions are allowed outside the cities as well. (Such trees (with extra points) are the *Steiner-trees*).

1.5.6 Let Q be a cut set in the connected graph G. Let F_1 and F_2 be minimum weight spanning trees in the two components of $G - Q$. Let e be the minimum weight edge in $E(Q)$. What can be said about $F_1 \cup F_2 \cup \{e\}$?

1.5.7 Imagine an $n \times n$ "chessboard" with numbers written on each square. We wish to place n "castles" on the board so that no two of them may be in the same row or in the same column; among these "acceptable" configurations we wish to choose the one where the sum of the numbers written on the squares of the castles is maximum. Show that this problem cannot be generally solved by a greedy algorithm.

Problems for Section 1.5

1.5.8 Estimate the total number of operations required for Algorithm 1.4. (Take the remarks at the beginning of Problem 1.4.18 into consideration. Compare the Problems 1.5.10 and 1.5.12 below as well.)

1.5.9 Show that the following algorithm is also correct if the input graph G is connected.

Algorithm 1.6 *Minimum weight spanning forest.*

Input and **Output** as in Algorithm 1.4; G must be connected.

Step 1 Prepare two files. File 1 will contain the edge set of F; File 2 will contain those points which are already interconnected. Initially, let File 1 be empty and File 2 contain the first point of G.

Step 2 Let e have minimum weight among those edges which join a point u from File 2 to a point v of G, not in File 2. (If every point of G is already in File 2, go to **End**.) Add e to File 1 and v to File 2. Go to **Step 2**.

End The algorithm terminates, F is the subgraph formed by the edges of File 1.

1.5.10 (continued) The total number of operations required for Algorithm 1.6 can easily be estimated as proportional to v^3 at most. Prove that the stronger estimation v^2 is also valid (and is, for some classes of graphs, better than that of Problem 1.5.8).

1.5.11 Show that the weight function on the set of edges can be modified in such a way that every edge has different weight and the resulting minimum weight spanning forest (which, by Exercise 1.5.2, is now unique) is also optimal for the original weighting.

1.5.12 Consider the points and the edges of the graph G as cities and possible interconnecting roads (with given costs). If G is connected and every edge has different weight then show that the following algorithm is also correct (see Fig. 1.22 as well).

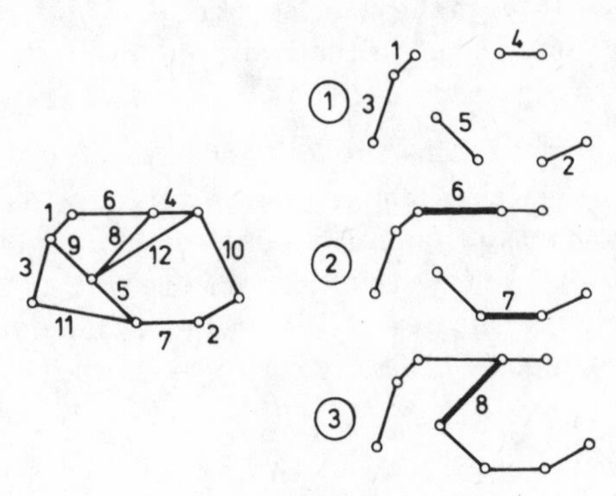

Fig. 1.22

Algorithm 1.7 *Minimum weight spanning forest.*

Input and **Output** as in Algorithm 1.4; G must be connected, edge weights must be different.

Step 1 Every city starts building the cheapest possible road interconnecting it with another city. (Some roads will be built from both ends, e.g. the cheapest surely will.)

Step 2 If the resulting road net is connected, go to **End**. Otherwise call the connected components as cities now and go to **Step 1**.

End The algorithm terminates, F is the subgraph formed by the finished roads.

We mention that, using more sophisticated data structures, this algorithm can be realized so that the total number of operations is proportional to $e \log \log v$ only, which is better than the results in Problem 1.5.8.

1.5.13 Explain the reason why should all weights be different in Algorithm 1.7. Give a weaker condition.

1.5.14 The points of H_G are the spanning forests of another graph G; two points a, b of H_G are adjacent if and only if, for the corresponding spanning forest A, B of $G, |E(A) - E(B)| = |E(B) - E(A)| = 1$. Prove that H_G is connected for any G.

1.5.15 Instead of maximum weight spanning forests, find a maximum weight subgraph containing at most one circuit. Does the greedy algorithm always work?

1.5.16 The same question if the considered subgraphs contain at most two circuits.

Chapter 2

Applications

Section 2.1 Kirchhoff-equations; the basic problem of network analysis

Electric circuits are formed by interconnecting various devices. In electric circuit theory one prepares mathematical models of these devices; describes the interconnection also in a mathematical way, and wishes to determine the mathematical model of the whole circuit from those of the devices and interconnection:

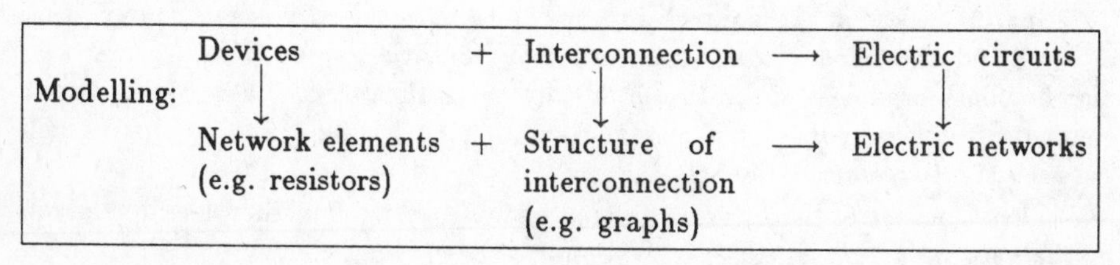

For example, let us operate a usual electric lamp with $u = 220$ Volt, $P = 100$ Watt. This circuit has three devices; the energy source, the lamp and the wire. We model the energy source as a voltage source (with a given voltage $u = 220$ Volt, with no internal resistance, and hence with an arbitrary current), we model the lamp as a resistor (which has arbitrary voltage u and arbitrary current i, subject to the condition $u = Ri$ only, where R is the resistance of the lamp, $u^2/P = 484$ Ohm in our case), and we suppose that the wire has no effect whatsoever. Hence we describe the structure of interconnection by saying that the resistor is parallel to the voltage source. Then we obtain a network and applying some rule (e.g. that parallel devices have the same voltage) we can calculate that the current i of the resistor should be $u/R = 0.45$ Amps.

Our modelling assumptions are valid to a certain extent only (e.g. the energy source must have some internal resistance, the wire might have some effect, the resistance of the lamp is not a constant but depends on outside temperature etc.) but we accept them since the obtained results are in good agreement with possible measurements performed on the real circuit. Even the basic assumptions that modelling the devices is independent of the way of their interconnection and that modelling the interconnection is independent of the actual devices, are also valid to a certain extent only.

In what follows, we first consider networks composed of resistors, voltage and current sources (with respective notations in Fig. 2.1). All of them can be

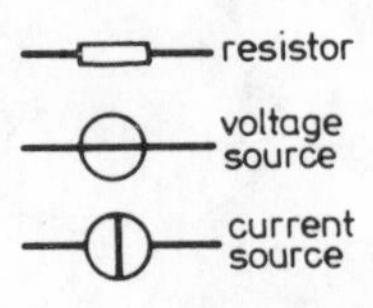

Fig. 2.1

described by their voltages u and currents i; in the case of a *resistor* both are arbitrary, subject to $u = Ri$ with a given constant R; in the case of a *voltage source* i is arbitrary, u is given (as, say, a given function of time); and in the case of a *current source* u is arbitrary and i is given. In the case of a resistor, the relation $u = Ri$ is called *Ohm's Law*, the constant R is called the *resistance* or *impedance* of the resistor.

Special cases of the sources are the *short circuit* (a voltage source with $u = 0$) and the *open circuit* (a current source with $i = 0$). The former can also be considered as a special resistor with resistance zero. These two network elements might help to memorize the symbols of Fig. 2.1 and to distinguish between the two sorts of sources.

The structure of the interconnection of these network elements will be given by the *network graph*; its edges correspond to the network elements and two edges are adjacent if the corresponding network elements have a common node (see Fig. 2.2). Observe that, in our model, the interconnecting wires have no effect, hence they "become" points of the network graph only, no matter how long they are in the circuit schematics.

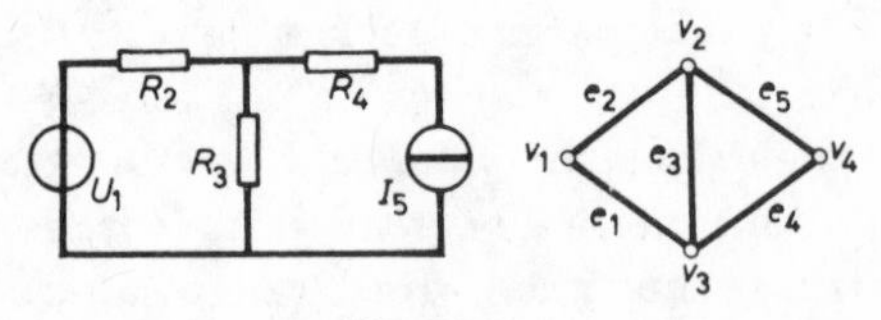

Fig. 2.2

The two laws of Kirchhoff describe the relations among the voltages (currents, respectively) of the interconnected devices. *Kirchhoff's Voltage Law* states that the sum of the voltages along any circuit of the network graph is zero. *Kirchhoff's Current Law* can be formulated so that the sum of the currents along any cut set of the network graph is zero (see Exercise 2.1.4 as well).

For example, here are some voltage and current equations for the network of Fig. 2.3.

$$\begin{aligned} u_1 + u_2 &= u_3 \\ u_1 + u_2 &= u_4 + u_5 \end{aligned} \qquad \begin{aligned} i_1 + i_3 + i_5 &= 0 \\ i_2 + i_3 + i_5 &= 0 \\ i_4 &= i_5 \end{aligned} \tag{2.1}$$

In this example we have five network elements; u_1 and i_5 are given, hence the number of unknown voltages and currents is 8. These five Kirchhoff equations above plus three Ohm-equations ($u_j = R_j i_j$ for $j = 2, 3, 4$) together can be seen to be just enough to uniquely determine all the quantities as functions of u_1, i_5, R_2, R_3 and R_4. For example, $i_3 = (R_2 + R_3)^{-1}(u_1 - R_2 i_5)$.

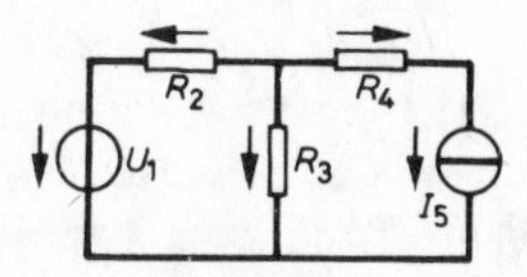

Fig. 2.3

Let us study Eqs (2.1) in more detail. The only difference between Fig. 2.3 and the left hand side of Fig. 2.2 is that small arrows were added for each network element. They might be called reference directions for measuring voltage and current; from the mathematical point of view we simply assigned orientation to every edge of the network graph. Now the voltage and current equations can be rewritten in matrix form.

$$\begin{pmatrix} -1 & -1 & 1 & 0 & 0 \\ -1 & -1 & 0 & 1 & 1 \\ 0 & 0 & -1 & 1 & 1 \end{pmatrix} \begin{pmatrix} u_1 \\ u_2 \\ u_3 \\ u_4 \\ u_5 \end{pmatrix} = \begin{pmatrix} 0 \\ 0 \\ 0 \end{pmatrix} \quad \text{and}$$

$$\begin{pmatrix} 1 & -1 & 0 & 0 & 0 \\ 1 & 0 & 1 & 1 & 0 \\ 1 & 0 & 1 & 0 & 1 \\ 0 & 1 & 1 & 1 & 0 \\ 0 & -1 & -1 & 0 & -1 \\ 0 & 0 & 0 & 1 & -1 \end{pmatrix} \begin{pmatrix} i_1 \\ i_2 \\ i_3 \\ i_4 \\ i_5 \end{pmatrix} = \begin{pmatrix} 0 \\ 0 \\ 0 \\ 0 \\ 0 \\ 0 \end{pmatrix} \tag{2.2}$$

and the coefficient matrices are just **C** and **Q**, i.e. the circuit and cut set matrices of the directed graph. (Eqs (2.2) contain all the possible Kirchhoff's equations, while (2.1) was just a maximal set of linearly independent equations.)

It is very important to see that the "reference directions of the network elements" (i.e. the orientations assigned to the edges of the network graph) are of no interest. Should one reverse the arrow of, say, resistor R_2 of Fig. 2.3, the only difference in Eqs (2.2) were that the second columns of **C** and **Q** should be multiplied by -1; then the solution of the system of network equations would give the same result, only the sign of u_2 and i_2 would be changed.

We conclude that Kirchhoff's Voltage and Current Laws can shortly be written as

$$\mathbf{Cu} = \mathbf{0}; \ \mathbf{Qi} = \mathbf{0} \tag{2.3}$$

where $\mathbf{C}, \mathbf{Q}$ are the circuit and cut set matrices of the directed graph, obtained from the network graph G when orientations are assigned to the edges, and $\mathbf{u}, \mathbf{i}$ are vectors, containing the voltages and currents, respectively, of the network elements. It is certainly enough to use the reduced circuit and cut set matrices since the equations, determined by the other rows, were consequences of the previous ones. Hence the number of linearly independent voltage equations is $n(G)$, that of the current equations is $r(G)$, making a total of $e(G)$ Kirchhoff equations (c.f. Theorem 1.4.4).

If the network contains e_U voltage sources, e_I current sources and e_R resistors (with $e_U + e_I + e_R = e(G)$) then we have furthermore e_R equations from Ohm's laws. The total number of equations is $e(G) + e_R$.

What is the total number of unknown quantities? The voltages of the voltage sources and the currents of the current sources are known, all the other $e(G) - e_U$ voltages and $e(G) - e_I$ currents are unknown, making a total of $2e(G) - e_U - e_I = e(G) + e_R$.

Hence the system of network equations, formed by (2.3) and by the Ohm's laws, has the same number of equations and unknown variables. If the matrix **W** of the coefficients of this system is nonsingular, the network is uniquely solvable.

Checking the nonsingularity of **W** and, in the case of an affirmative answer, finding the unique solution of the network, will be called the *basic problem of network analysis.*

Is it possible that a network has no unique solution? Theoretically, if the existing electric circuit does have a unique solution and all the modelling assumptions are correct, then the network should be uniquely solvable. However, if an electric circuit should be designed and therefore network elements are interconnected to find out whether the resulting network meets the specifications, then all sort of "absurd" situations can arise (especially if the interconnection is made by a computer, with lack of engineering intuition). But even if an existing circuit is analyzed, the modelling assumptions might be wrong (e.g. the resistance of a long wire must also be taken into consideration) or even input errors in a computer aided network analysis program can arise.

For example, the network of Fig. 2.4 is not uniquely solvable. Even without preparing the system of equations it is intuitively clear that two voltage sources cannot form a circuit in the network graph since if u_1 and u_2 are independently given, the Kirchhoff equation $u_1 = u_2$ might lead to a contradiction. (Even if somehow $u_1 = u_2$ were met, their currents i_1 and i_2, respectively, could not be uniquely determined.) This observation can be generalized as well.

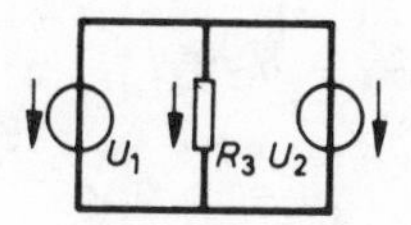

Fig. 2.4

Statement 2.1.1 *A necessary condition for the unique solvability of a network, consisting of resistors, voltage and current sources is that the subset of edges of the network graph, determined by the voltage sources, must not contain any circuit and the subset of edges, corresponding to the current sources, must not contain any cut set.*

Proof. Suppose the network has a unique solution $\mathbf{u}, \mathbf{i}$. If C were a circuit with edges $\{e_1, e_2, \ldots, e_k\}$, each corresponding to a voltage source, then consider the currents $i_1, i_2, \ldots, i_k$ of these edges. If we add the same nonzero quantity i_0 to all these currents (measured in such a way that i_0 is considered clockwise, i.e. if i_j was measured clockwise, put $i_j + i_0$, otherwise $i_j - i_0$) then none of the equations becomes violated. (Kirchhoff's Voltage Laws are of course not effected, neither are Ohm's Laws, since these edges corresponded to voltage sources, not to resistors; while Kirchhoff's Current Laws are related to cut sets, meeting C by an even number of edges and with an appropriate sign combination.) Hence if there were a solution, it could not be unique. Cut sets of current sources would lead to a contradiction in a similar way changing $\mathbf{u}$ on the cut set edges.

□

But even if these two necessary conditions are met, the network may not be uniquely solvable. We shall see many such examples later, only the simplest reason is presented now. For example, the network in Fig. 2.2 satisfies the above conditions, yet if $R_2 + R_3 = 0$ then it has no unique solution. Although we mostly imagine resistors with positive resistances only, the cases $R = 0$ and $R < 0$ should also be taken into consideration. $R = 0$ (i.e. a short circuit) could be modelled as a voltage source, and really, if $R_2 = R_3 = 0$ then $\{1, 2, 3\}$ is a circuit of voltage sources, violating Statement 2.1.1. But $R < 0$ raises new problems. Such a device (a *negative resistor*) cannot be so easily imagined as a positive one, it does not consume, rather produces energy (more about this in Section 8.1). The classical results on unique network solvability simply excluded negative resistors; so we do the same until Section 6.1. Then the above two conditions are sufficient.

Theorem 2.1.2 *If a network consists of positive resistors, voltage and current sources, then the two necessary conditions of Statement 2.1.1 are already sufficient for unique solvability.*

Its *proof* is fairly involved, we postpone it to the next section.

Exercises for Section 2.1

2.1.1 Prove that if two resistors R_1 and R_2 are in *series* or in *parallel* (see Fig. 2.5) then they can be substitued by a single resistor with value $R_s = R_1 + R_2$ or $R_p = (R_1 + R_2)^{-1} R_1 R_2$, respectively (provided, in the second case, that $R_1 + R_2 \neq 0$).

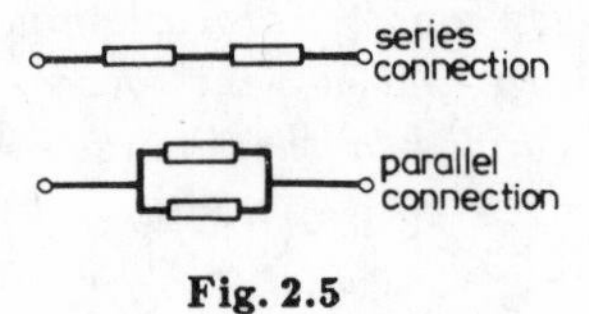

Fig. 2.5

2.1.2 Consider the system of Fig. 2.6 as a single network element. Prove that either the voltage source can be substituted by a short circuit or the current source by an open circuit (and the value of the remaining source modified) so that the resulting system, as a single network element, does not change.

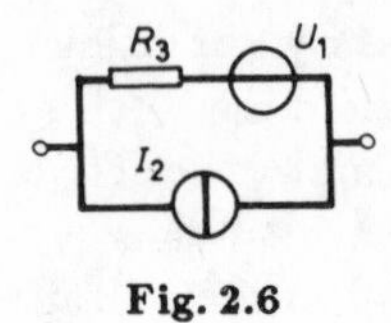

Fig. 2.6

2.1.3 Give a similar statement for the system of Fig. 2.7.

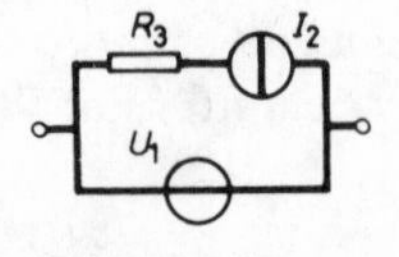

Fig. 2.7

2.1.4 Kirchhoff's Current Law is usually stated in the following form: The sum of the currents on edges, incident to a given point, is zero. (Summation is performed so that edges with reference directions towards the point receive plus signs and those, away from the point, receive minus signs). Show that this Law in its present form and in the other form (formulated with cut sets) are equivalent.

2.1.5 Prove that the two conditions in Statement 2.1.1 together are equivalent to the following one: The network graph must have a spanning forest, containing all the edges, corresponding to voltage sources and none of the edges, corresponding to current sources.

2.1.6 Exercise 2.1.2 reduced the system of Fig. 2.6 to one of those in Fig. 2.8. These are the models of (not idealized) energy sources. Explain why is the resistor parallel in the first case and series in the second and never *vice versa*.

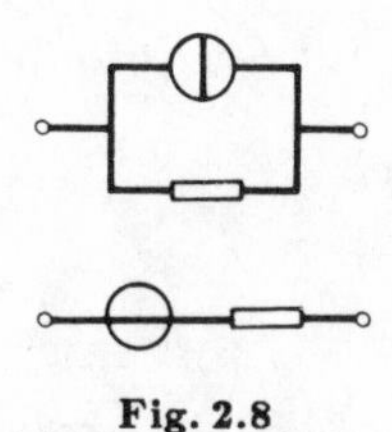

Fig. 2.8

2.1.7 Prove that if every voltage source has a series resistor and every current source has a parallel resistor then
(a) the network can be modelled by using only one type of source (voltage or current);
(b) the conditions of Statement 2.1.1 are always met.

Problems for Section 2.1

2.1.8 Figure 2.9 shows a system of four resistors and a switch. Intuitively, it is clear that closing the switch cannot increase the resistance of the whole system (considered as a single network element). Using this observation show [Lehman, 1979] that

$$\frac{(a+b)(c+d)}{a+b+c+d} \geq \frac{ac}{a+c} + \frac{bd}{b+d} \quad \text{for } a,b,c,d > 0.$$

2.1.9 Consider each system in Fig. 2.10 as a single network element and determine their resistances. Find a general rule [Morgan-Voyce, 1959].

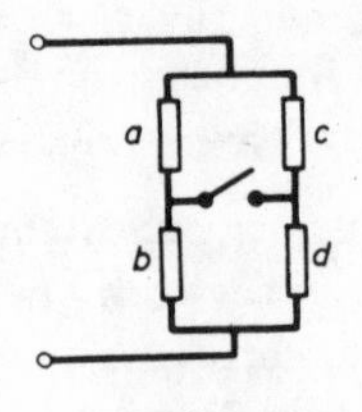

Fig. 2.9

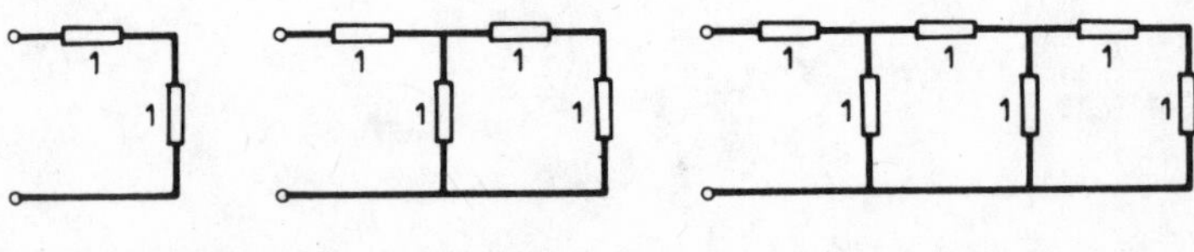

Fig. 2.10

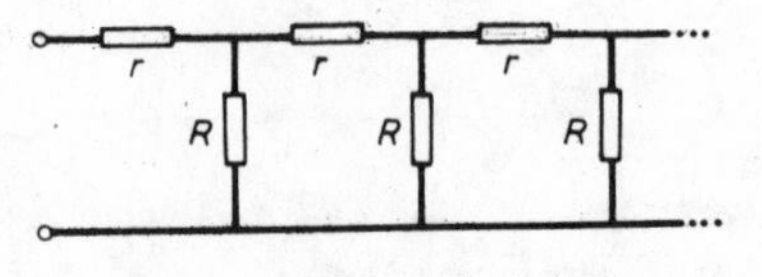

Fig. 2.11

2.1.10 Determine the resistance of the "infinite ladder" of Fig. 2.11. (Such networks can model certain "distributed" networks like transmission lines, see **Box 2.2.**)

2.1.11 Suppose the wires in Fig. 2.12 have resistance R each and their resistance is "uniformly distributed along the length of the wires". Once the right hand ends of these wires are interconnected by a resistor R_0, we should expect resistance $2R + R_0$ on the left hand side. However, a "short circuit-like" connection of resistance R_1 has arisen somewhere along the wires, hence we measure resistance R_2 on the left hand side. Locate the failure.

2.1.12 Put one unit resistor to every edge of a cube (Fig. 2.13). What is the resistance if we consider the entire system as a single resistor

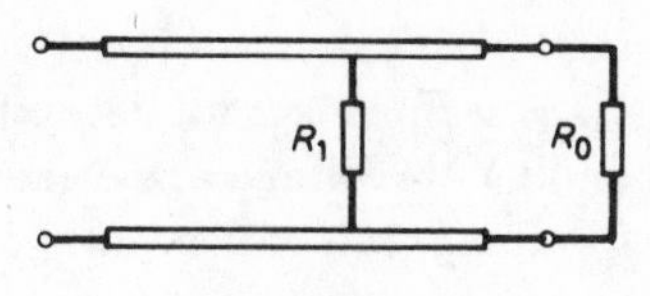

Fig. 2.12

(a) between points A, B;
(b) between points A, C;
(c) between points A, D?

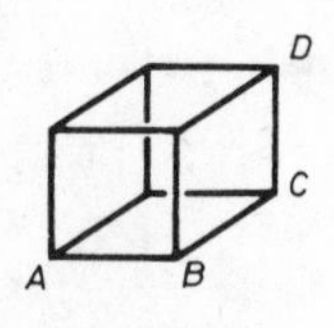

Fig. 2.13

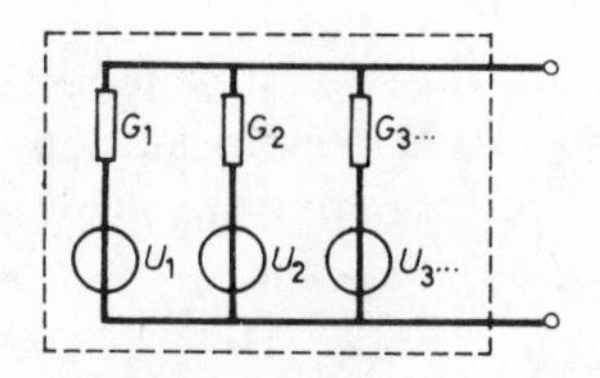

Fig. 2.14

2.1.13 Let us connect voltage sources and their series resistors in parallel (see Fig. 2.14 where not the resistances $R_1, R_2, \dots$ but their reciprocal values, $G_1, G_2, \dots$ (the so called *admittances*) are given). Prove *Millman's* theorem that the voltage of the resulting system is $(\sum G_i U_i)/(\sum G_i)$.

2.1.14 Some old network theory textbooks wrote: When writing up all the equations for a network, be sure that the voltage and current of every network element arise. Is it possible that the union of the edge set of $n(G)$ linearly independent circuits or that of $r(G)$ linearly independent cut sets does not contain every edge of the graph G?

Section 2.2 Unique solvability of resistive networks

In this section, we prove Theorem 2.1.2. Using the result of Exercise 2.1.5, we prefer the following modified form.

Theorem 2.2.1 *Let a network N consist of positive resistors, voltage and current sources. Then the necessary and sufficient condition for its unique solvability is that its network graph G must have at least one spanning forest F, containing* *every edge corresponding to voltage sources, and none of the edges corresponding to current sources.*

Proof. The necessity has already been proved (Statement 2.1.1). For the proof of the sufficiency, a detailed analysis of the system of network equations is required.

Let $E(G) = E_U \cup E_X \cup E_Y \cup E_I$ be such a decomposition that E_U and E_I is the set of edges, corresponding to voltage and current sources, respectively, while the set of resistors is decomposed so that E_X is in $E(F)$ and E_Y is in $E(G) - E(F)$.

Write the Kirchhoff's Laws in a form $\mathbf{Cu} = \mathbf{0}$ and $\mathbf{Qi} = \mathbf{0}$ so that the rows of $\mathbf{C}$ and $\mathbf{Q}$ correspond to the fundamental systems of circuits and cut sets, respectively, with respect to F. If the voltages and currents of the elements in E_U, E_X, E_Y and E_I are collected to vectors $\mathbf{u}_U, \mathbf{u}_X, \mathbf{u}_Y, \mathbf{u}_I$ and $\mathbf{i}_U, \mathbf{i}_X, \mathbf{i}_Y, \mathbf{i}_I$ respectively, then instead of the compact form of Eqs (2.3) we can write

$$\begin{pmatrix} \mathbf{C}_{11} & \mathbf{C}_{12} & \mathbf{E} & \mathbf{O} \\ \mathbf{C}_{21} & \mathbf{C}_{22} & \mathbf{O} & \mathbf{E} \end{pmatrix} \begin{pmatrix} \mathbf{u}_U \\ \mathbf{u}_X \\ \mathbf{u}_Y \\ \mathbf{u}_I \end{pmatrix} = \mathbf{O} \text{ and } \begin{pmatrix} \mathbf{E} & \mathbf{O} & \mathbf{Q}_{11} & \mathbf{Q}_{12} \\ \mathbf{O} & \mathbf{E} & \mathbf{Q}_{21} & \mathbf{Q}_{22} \end{pmatrix} \begin{pmatrix} \mathbf{i}_U \\ \mathbf{i}_X \\ \mathbf{i}_Y \\ \mathbf{i}_I \end{pmatrix} = \mathbf{O}.$$

Here we applied the special structure of $\mathbf{C}$ and $\mathbf{Q}$, as has been found in the proof of Theorem 1.4.4; $\mathbf{E}$ denotes unity matrices of appropriate sizes.

It might be more convenient to write instead

$$\begin{pmatrix} \mathbf{u}_Y \\ \mathbf{u}_I \end{pmatrix} = - \begin{pmatrix} \mathbf{C}_{11} & \mathbf{C}_{12} \\ \mathbf{C}_{21} & \mathbf{C}_{22} \end{pmatrix} \begin{pmatrix} \mathbf{u}_U \\ \mathbf{u}_X \end{pmatrix} \text{ and } \begin{pmatrix} \mathbf{i}_U \\ \mathbf{i}_X \end{pmatrix} = - \begin{pmatrix} \mathbf{Q}_{11} & \mathbf{Q}_{12} \\ \mathbf{Q}_{21} & \mathbf{Q}_{22} \end{pmatrix} \begin{pmatrix} \mathbf{i}_Y \\ \mathbf{i}_I \end{pmatrix} \tag{2.4}$$

Before proceeding further, let us see an example. The network of Fig. 2.2 is redrawn in Fig. 2.15, together with its network graph and with an indication of the fundamental system of circuits and cut sets with respect to the spanning tree $\{e_1, e_2, e_4\}$. Then $E_X = \{e_2, e_4\}$; $E_Y = \{e_3\}$ and the systems of equations, corresponding to Eqs (2.4) are

$$\begin{pmatrix} u_3 \\ u_5 \end{pmatrix} = - \left(\begin{array}{c|cc} -1 & -1 & 0 \\ \hline -1 & -1 & 1 \end{array}\right) \begin{pmatrix} u_1 \\ u_2 \\ u_4 \end{pmatrix} \text{ and } \begin{pmatrix} i_1 \\ i_2 \\ i_4 \end{pmatrix} = - \left(\begin{array}{c|c} 1 & 1 \\ \hline 1 & 1 \\ 0 & -1 \end{array}\right) \begin{pmatrix} i_3 \\ i_5 \end{pmatrix}$$

where straight lines indicate the blocks $\mathbf{C}_{ij}, \mathbf{Q}_{ij}$ of Eqs (2.4).

We have not yet used Ohm's Law. In order to keep the matrix formalism, collect the resistances into two diagonal matrices $\mathbf{D}_X$ and $\mathbf{D}_Y$, one containing the resistances of the resistors with corresponding edges in E_X, the other containing those in E_Y. Then

$$\mathbf{u}_X = \mathbf{D}_X \mathbf{i}_X \text{ and } \mathbf{u}_Y = \mathbf{D}_Y \mathbf{i}_Y \tag{2.5}$$

Substituting these equations to Eqs (2.4) we obtain

$$\mathbf{u}_Y = \mathbf{D}_Y \mathbf{i}_Y = -\mathbf{C}_{11}\mathbf{u}_U - \mathbf{C}_{12}\mathbf{u}_X = -\mathbf{C}_{11}\mathbf{u}_U - \mathbf{C}_{12}\mathbf{D}_X\mathbf{i}_X =$$

$$= -\mathbf{C}_{11}\mathbf{u}_U - \mathbf{C}_{12}\mathbf{D}_X(-\mathbf{Q}_{21}\mathbf{i}_Y - \mathbf{Q}_{22}\mathbf{i}_I)$$

or

$$[\mathbf{D}_Y - \mathbf{C}_{12}\mathbf{D}_X\mathbf{Q}_{21}]\mathbf{i}_Y = -\mathbf{C}_{11}\mathbf{u}_U + \mathbf{C}_{12}\mathbf{D}_X\mathbf{Q}_{22}\mathbf{i}_I. \tag{2.6}$$

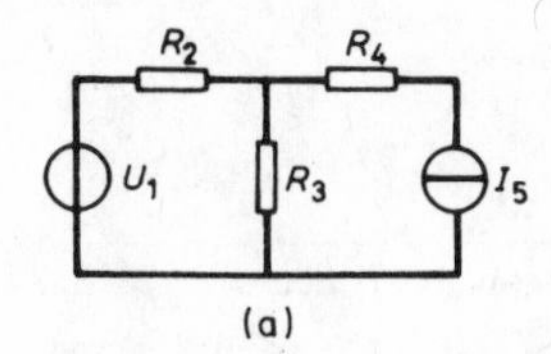

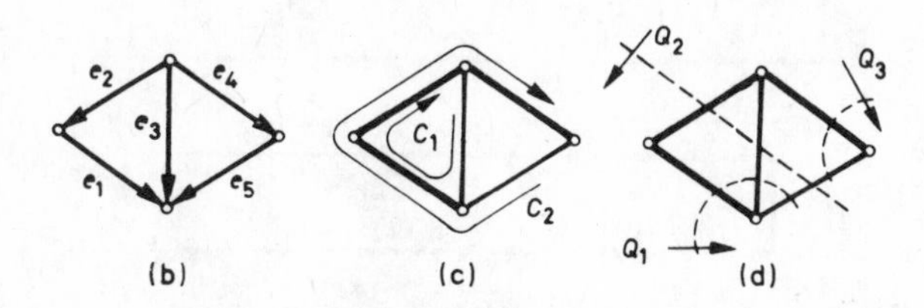

Fig. 2.15

Since $\mathbf{u}_U$ and $\mathbf{i}_I$ are known (they are the voltages of the voltage sources and the currents of the current sources), we conclude that $\mathbf{i}_Y$ can be determined if its coefficient (the matrix in square bracket in Eq. (2.6)) is invertible. Once we have $\mathbf{i}_Y$, we have $\mathbf{i}_U$ and $\mathbf{i}_X$ from Eq. (2.4), then $\mathbf{u}_X$ and $\mathbf{u}_Y$ from (2.5), and finally $\mathbf{u}_I$ from (2.4) again.

Recollect that $\mathbf{C}_{12} = -\mathbf{Q}_{21}^T$ (by the directed graph version of Corollary 1.4.5) hence the coefficient of $\mathbf{i}_Y$ is $\mathbf{D}_Y + \mathbf{C}_{12}\mathbf{D}_X\mathbf{C}_{12}^T$. Since $\mathbf{D}_X$ was a diagonal matrix with positive entries, we might write $\mathbf{D}_X = \mathbf{D}\mathbf{D}^T$ (where $\mathbf{D}$ is a diagonal matrix, whose entries are the square roots of those of $\mathbf{D}_X$). Then $\mathbf{C}_{12}\mathbf{D}_X\mathbf{C}_{12}^T =$ $= (\mathbf{C}_{12}\mathbf{D}) \cdot (\mathbf{C}_{12}\mathbf{D})^T$, i.e. is positive semidefinite. $\mathbf{D}_Y$ is positive definite (since every entry of it is positive), so their sum is positive definite — hence invertible — as well.

□

Using still the same notations, the calculations during the proof are briefly summarized in the left hand side of **Box 2.1.** Reading the right hand side leads to another way of the calculation. Since the matrices to be inverted may be of different sizes in case of these two methods, the application of one or the other might be advantageous in certain cases. This symmetry will further be studied in Section 4.1.

Exercises for Section 2.2

2.2.1 Repeat the proof of the theorem in the special case if $E_X = E_Y = \emptyset$.

2.2.2 Prove that no matrix should be inverted if $E_X = \emptyset$ or if $E_Y = \emptyset$.

2.2.3 Find the system of equations, corresponding to Eq. (2.6), in the case of the network in Fig. 2.2. What happens if $R_2 + R_3 = 0$?

Problems for Section 2.2

2.2.4 Give examples for networks where the "left hand side" or the "right hand side" method (as indicated in **Box 2.1**) is significantly better than the other.

Box 2.1. Analysis of resistive networks

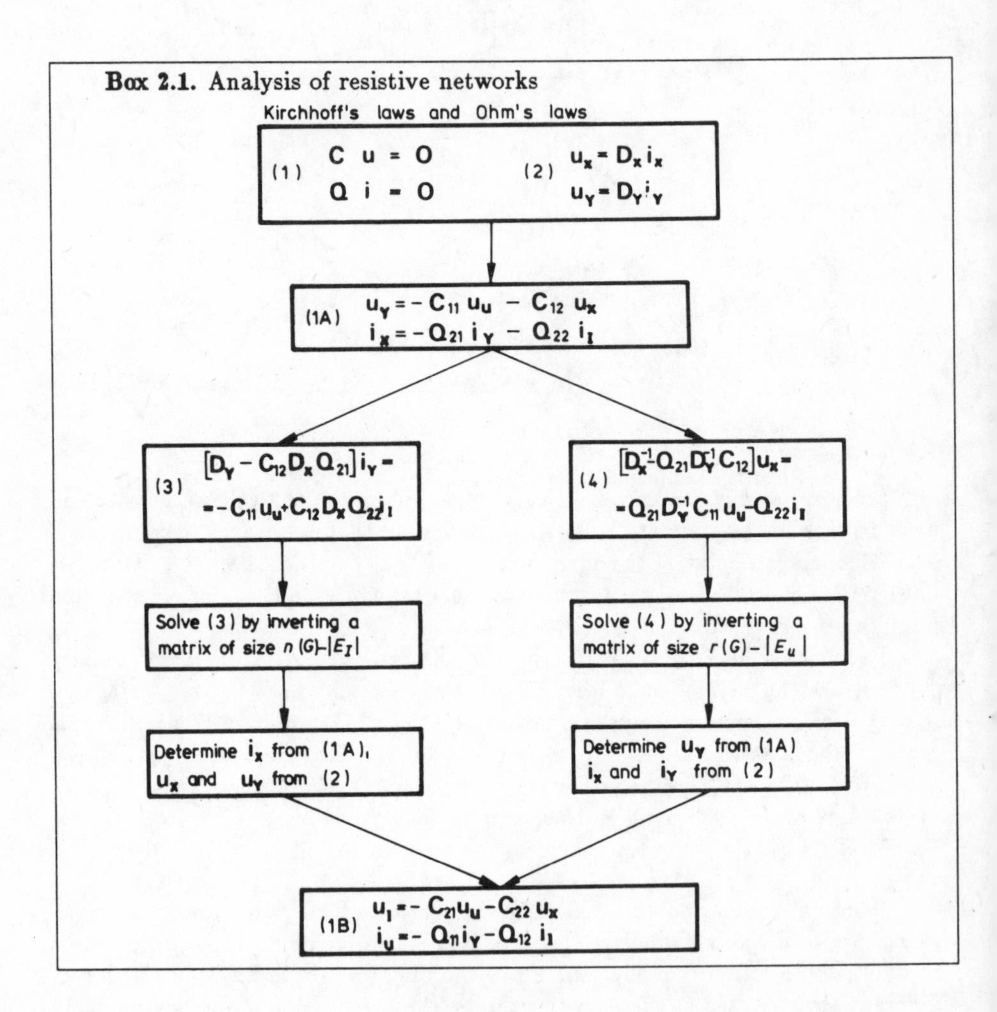

Kirchhoff's laws and Ohm's laws
(1)
C u = O
Q i = O
(2)
$u_x = D_x i_x$
$u_Y = D_Y i_Y$
(1A)
$u_Y = -C_{11} u_u - C_{12} u_x$
$i_x = -Q_{21} i_Y - Q_{22} i_I$
(3)
$[D_Y - C_{12} D_x Q_{21}] i_Y =$
$= -C_{11} u_u + C_{12} D_x Q_{22} i_I$
(4)
$[D_x^{-1} - Q_{21} D_Y^{-1} C_{12}] u_x =$
$= Q_{21} D_Y^{-1} C_{11} u_u - Q_{22} i_I$
Solve (3) by inverting a matrix of size $n(G) - |E_I|$
Solve (4) by inverting a matrix of size $r(G) - |E_u|$
Determine i_x from (1A), u_x and u_Y from (2)
Determine u_Y from (1A) i_x and i_Y from (2)
(1B)
$u_I = -C_{21} u_u - C_{22} u_x$
$i_u = -Q_{11} i_Y - Q_{12} i_I$

2.2.5 Give an example when none of these methods is the "best".

2.2.6 Let e be an edge, corresponding to a voltage source, and Q be a cut set in the network graph containing e. Replace the voltage source by a short circuit and add with the same value a voltage source in series to each of the edges of $Q - \{e\}$. Prove that this does not affect the voltages and currents of the other network elements.

2.2.7 (continued) What is the corresponding statement for current sources?

2.2.8 What is wrong in the following reasoning? Applying Problems 2.2.6–2.2.7 we have resistors series to every voltage source and parallel to every current source, hence the conditions of Statement 2.1.1 are always met by Exercise 2.1.7(b).

Section 2.3 Capacitors, inductors; state equations

The results of the previous two sections (and those in Section 2.5 below) were already known in the last century. Networks containing positive resistors and voltage and current sources only, are very rare in recent applications; we have to broaden the set of network elements to be able to model more complicated circuits as well.

The networks we have studied so far are time-invariant, lumped, linear and memoryless, see **Box 2.2** for the definitions of these concepts. In this section we show that the last restriction is not important; we can generalize our results to networks, containing positive resistors, capacitors, inductors and voltage and current sources.

Box 2.2. Some important concepts of network theory

Time-invariant networks: The way of the interconnection does not change. (For example, the network in Fig. 2.9 is *time-variant*, since it has a switch.)

Lumped networks: A finite number of network elements are interconnected by "ideal" wires; the finiteness of the speed of electromagnetic energy need not be taken into consideration. (For example, a transmission line is usually modelled as a *distributed* network.)

Linear networks: The mathematical models of the devices are linear functions. (For example, a diode with large resistance in one direction and small in the other, is *nonlinear*.)

Memoryless networks: The devices are modelled by algebraic functions among voltages and currents. (For example, a capacitor is described by a differential equation, in accordance with the fact that it can store energy.)

See **Box 8.1** for further important concepts.

In the electric engineering applications presented in later chapters of this book, we shall always require the networks to be time-invariant, lumped and linear (although some of the presented results can be applied for distributed or nonlinear networks as well, see [Csurgay *et al.*, 1974 and Iri *et al.*, 1982], respectively. The set of network elements will be broadened in a different way: We shall not restrict ourselves to "two-terminal devices", modelled by edges of a graph; hence "multiterminal devices" (like transistors, transformers etc.) can also be applied, see Chapters 6 through 18.

Recall that *capacitors* and *inductors* are defined by $i = C\dfrac{du}{dt}$ and by $u = L\dfrac{di}{dt}$ respectively; see their symbols in Fig. 2.16. The coefficients C and L will be called *capacitance* and *inductance*, respectively.

Let us study the network of Fig. 2.17 in some detail. If $u_1(t)$ denotes the voltage of the voltage source U_1 (as function of time) then the voltage of the capacitor satisfies

$$u_1(t) = u_2(t) + C_2 R_3 \frac{d}{dt} u_2(t) \tag{2.7}$$

In order to determine $u_2(t)$, we must know its initial value $u_2(t_0)$ in the moment t_0 of the interconnection of these devices. Then we obtain

$$u_2(t) = u_2(t_0) + r^{-1} \int_{t_0}^{t} e^{t/r} u_1(t)dt \tag{2.8}$$

where $r = C_2 R_3$. If we wish to know the current of the voltage source as well, we obtain

$$i_1(t) = R_3^{-1}[u_2(t) - u_1(t)]. \tag{2.9}$$

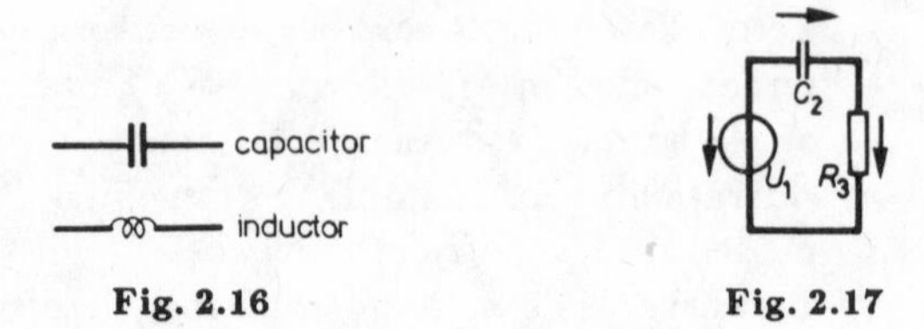

Fig. 2.16 Fig. 2.17

Speaking quite generally for some time, such a network is an example for a general thing called "system" which receives "inputs" from the surrounding world and gives "outputs" to it as well. Some systems (the memoryless ones) have such outputs which are functions of the actual input only, but most systems have "memory", i.e. their output depends both on the actual input and on the past history of the system. This past history gives the "state" of the system. Then it is reasonable to suppose that

(a) the recent state depends on the initial state and on the inputs received since that time; and that
(b) the output depends on the recent state and on the recent input.

In our example $u_1(t)$ is the input, $i_1(t)$ is the output, $u_2(t)$ is the state. The network has memory, hence $i_1(t)$ cannot be obtained from $u_1(t)$ directly, only via $u_2(t)$. Hence Eqs (2.8) and (2.9) express the above statements (a) and (b), respectively.

Following the notations of system theory, let the input, the output and the state be denoted by u, y and x, respectively. Statement (b) above is then $y = g(x, u)$ and in our example statement (a) can be written as $\frac{dx}{dt} = f(x, u)$, see Eq. (2.7). A system is linear if f and g are linear functions; in many cases an even more special form arises

$$\begin{aligned} \frac{dx}{dt} &= Ax + Bu \\ y &= Cx + Du \end{aligned} \tag{2.10}$$

which is called *canonical state equations.* Equations (2.7) and (2.9) are such examples. (Here x, y and u may be vectors and A, B, C and D matrices.)

It is easy to see that networks containing resistors, capacitors, inductors and voltage and current sources only (in what follows, shortly RLC-*networks*) can store energy in the capacitors and inductors only. Hence if we choose the state vector x to contain the voltages of every capacitor and the currents of every inductor, the statements (a) and (b) will certainly hold. However, in this way we usually do not get canonical state equations. In particular, the order of the differential equation to be solved may be smaller than the number of the LC-elements.

For example, if a capacitor C_2 is parallel to a voltage source U_1 then $u_2(t) = = u_1(t)$ and the output of the system is $i_1(t) = -C\frac{d}{dt}u_1(t)$; i.e. we need not solve any differential equation whatsoever. Similarly, if some capacitors form a circuit in the network graph then the voltage of one of them is the sum of the others, having an algebraic (and not differential) equation. Hence we have basically seen the necessity part in the following result:

Theorem 2.3.1 *Consider a network containing positive resistors, capacitors and inductors, and voltage and current sources. We wish to have the network uniquely solvable and wish to prescribe the voltage of each capacitor and the current of each inductor independently, as initial values for the numerical solution of the differential equations. Then its necessary and sufficient condition is that the network graph contains a spanning forest F so that all the edges, corresponding to voltage sources or capacitors, are in $E(F)$ and none of the edges, corresponding to current sources or inductors, is in $E(F)$.*

The *proof* of the sufficiency (actually the explicit construction of the canonical state equation) follows the line of the proof of Theorem 2.2.1 and is therefore left to the reader as Problem 2.3.9.

Suppose now that the conditions of this theorem are not met. If even the set of edges, corresponding to the voltage sources, contains a circuit, or if even the current source edges contain a cut set, then the network has certainly no unique solution (the proof of Statement 2.1.1 still works). However, if voltage sources and capacitors together form a circuit (or current sources and inductors together form a cut set) then the network is still uniquely solvable, only the number of those capacitor voltages (or inductor currents, respectively) decreases which can independently be prescribed. In this case the state equations may or may not be canonical (see Exercise 2.3.3, but we do not go into further details).

However, the main question is not the canonicity of the state equations but the unique solvability itself. This can be answered more easily. Let E_C and E_L denote the sets of edges, corresponding to capacitors and to inductors, respectively.

Theorem 2.3.2 *A network, containing positive resistors, capacitors and inductors, and voltage and current sources, is uniquely solvable if and only if the network graph G contains a spanning forest F so that all the voltage source edges are in $E(F)$ and none of the current source edges is in $E(F)$. If this condition is met and F_0 is a spanning forest with $|E_C \cap E(F)| + |E_L \cap [E(G) - E(F)]|$ maximum among all spanning forests satisfying the above properties, then the voltages of the capacitors in $E(F_0)$ and the currents of the inductors not in $E(F_0)$ can independently be prescribed, as initial values for the numerical solution of the differential equation.*

The *proof* essentially follows the line of that of Theorem 2.1.1 but is much longer. The interesting steps of it can be seen while solving Problems 2.3.10–11. For a complete proof the reader is referred to [Rohrer, 1970].

Exercises for Section 2.3

2.3.1 What is the analogue of Exercise 2.1.1 if two capacitors or two inductors are in series or in parallel?

2.3.2 The networks in Fig. 2.18 satisfy the condition of Theorem 2.3.1. Determine their canonical state equations.

Fig. 2.18

Fig. 2.19

2.3.3 The networks in Fig. 2.19 do not satisfy the condition of Theorem 2.3.1, only the weaker condition of Theorem 2.3.2. Determine their equations and decide whether they are canonical.

2.3.4 Illustrate by an example that in Theorem 2.3.2 not only the resistors but the capacitors and the inductors must also be positive.

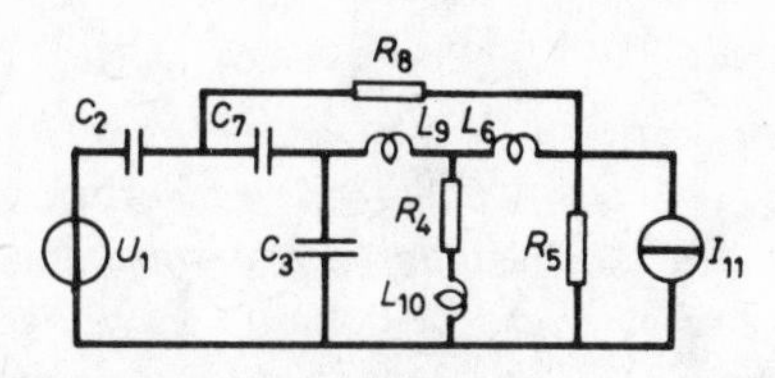

Fig. 2.20

2.3.5 Consider the network of Fig. 2.20 [Balabanian and Bickart, 1969]. Draw its network graph and determine the order of the differential equation which arises for its solution.

Problems for Section 2.3

2.3.6 Show that the system in Fig. 2.21 is equivalent to a simple resistor (i.e. it is memoryless), if the value of all the four network elements is unity.

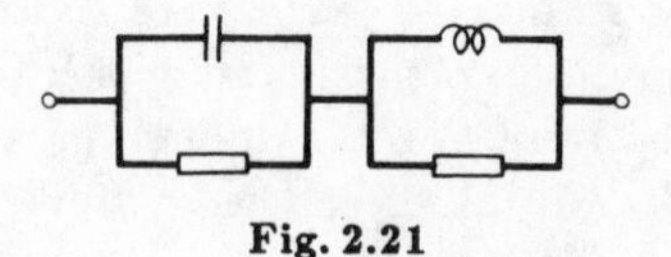

Fig. 2.21

2.3.7 If two capacitors are in parallel or series, they can be substituted by a single one (Exercise 2.3.1). Hence there is essentially no difference between the two networks of Fig. 2.18. How is it possible then that the order of the respective differential equations are different?

2.3.8 (continued) In general, what happens if capacitors form a cut set or inductors form a circuit?

2.3.9 Prove the sufficiency part of Theorem 2.3.1.

2.3.10 Prove Theorem 2.3.2 in the special case if the network contains capacitors and voltage and current sources only.

2.3.11 For the proof of Theorem 2.2.1 the matrices **C** and **Q** were decomposed into eight blocks each. Perform the analogous decompositions for the proof of Theorem 2.3.2 and determine the zero-blocks.

Section 2.4 How to check the conditions

The conditions of Theorems 2.2.1, 2.3.1 and 2.3.2 were the existences of certain spanning forests. In this section algorithms are given to find such forests (or to prove that no such forest exists).

Statement 2.4.1 *Let a network N consist of e_U voltage sources, e_R positive resistors and e_I current sources. Consider the network graph G of N and assign weights to its edges so that the weights of the voltage source edges, the resistor edges and the current source edges be $0, 1$ and 2, respectively. Find a spanning forest F_0 with minimum weight in G by Algorithm 1.4 or 1.5. Then N is uniquely solvable if and only if $w = r(G) - e_U$ and, in case of equality, F_0 can play the role of F in Theorem 2.2.1.*

Proof. The greedy algorithm does not choose any of the resistors or the current sources as long as voltage sources are available, neither does any of the current sources as long as resistors are available. Hence if N is uniquely solvable, the greedy algorithm finds a spanning forest with e_U voltage sources and $r(G) - e_U$ resistors, thus $w = r(G) - e_U$. On the other hand, if N is not uniquely solvable, then either the voltage source edges contain a circuit, thus more than $r(G) - e_U$ edges with positive weight arise in F_0, or the current source edges contain a cut set, thus at least one of the $r(G) - e_U$ nonzero-weight edges of F_0 has weight 2. □

Statement 2.4.2 *Let a network N consist of positive resistors, capacitors, inductors, and voltage and current sources. Consider the network graph G of N and assign weights to its edges so that the weights of the voltage source edges, the capacitor edges, the resistor edges, the inductor edges and the current source edges be $0, 1, 2, 3, 4$, respectively. Find a minimum weight spanning forest F in G. Then N is uniquely solvable if and only if every voltage source edge is in $E(F)$ and none of the current source edges is in $E(F)$. In case of unique solvability, F can play the role of F_0 in Theorem 2.3.2.*

The *proof* is almost the same as above, so it is left to the reader. Observe that here any strictly increasing sequence of weights (instead of $0, 1, 2, 3, 4$) would be appropriate.

Exercises for Section 2.4

2.4.1 Suppose that among the number of capacitors, inductors, resistors, voltage and current sources, none is greater than 9. Instead of $0, 1, 2, 3, 4$, use the weights $1, 10, 100, 1000, 10000$ in Statement 2.4.2. What further information can be obtained from the weight of F?

2.4.2 Someone has found us a maximum number of capacitors in a network so that they and the voltage sources together form a circuit-free subgraph H_1 and therefore she wishes to prescribe their voltages independently of each other, as initial conditions for the numerical solution of the differential equation. Somebody else has found a maximum number of inductors so that they and the current sources together form a cut set free subgraph H_2 and wishes to prescribe their currents independently. Can both wishes be simultaneously satisfied?

Problems for Section 2.4

2.4.3 Suppose the "greedy algorithm subroutine" of our computer cannot handle more than three different weights. Hence we can apply it for memoryless networks (see Statement 2.4.1). Modify Statement 2.4.2 so that its algorithm should give the right answer with using this subroutine only.

2.4.4 In a 2-person game Player 1 knows k nonnegative integers $x_1, x_2, \ldots, x_k$. Player 2 must find them out. The only type of "question" Player 2 may pose is a k-tuple $(a_1, a_2, \ldots, a_k)$ and then the "answer" is $\sum_{i=1}^{k} a_i x_i$. Of course, if Player 2 poses k questions $(1,0,0,\ldots,0), (0,1,0,\ldots,0)$ etc. then the solution is trivial but try to give a strategy with less than k questions.

Section 2.5 The topological formulae of Kirchhoff and Maxwell

In this section we prove Theorem 2.2.1 again, essentially in the same way as Kirchhoff [1847] did. Although this proof does not directly generalize to RLC networks, it has other advantages.

First of all recall that if $\mathbf{Mx} = \mathbf{0}$ and $\mathbf{x} = \begin{pmatrix} \mathbf{x}_1 \\ \mathbf{x}_2 \end{pmatrix}$ where the quantities, contained in $\mathbf{x}_1$ are unknown, and those, contained in $\mathbf{x}_2$ are known, and if the number of rows of $\mathbf{M}$ equals that of the entries of $\mathbf{x}_1$, then $\mathbf{M}$ should be decomposed into $(\mathbf{M}_1|\mathbf{M}_2)$ and $\mathbf{x}_1$ can be uniquely determined if and only if $\mathbf{M}_1$ is nonsingular. Then $\mathbf{x}_1 = -\mathbf{M}_1^{-1}\mathbf{M}_2\mathbf{x}_2$.

Hence, if all the Kirchhoff's and Ohm's laws of a network are considered then, with the same notation as in Section 2.2, all we have to prove is that the square submatrix $\mathbf{W}$, indicated by heavy frame, is nonsingular (see Fig. 2.22).

For example, in the network of Fig. 2.2, the system of equations is

$$\begin{pmatrix} 1 & 1 & 0 & -1 & 0 & 0 & 0 & 0 & 0 & 0 \\ 1 & 1 & -1 & 0 & -1 & 0 & 0 & 0 & 0 & 0 \\ 0 & 0 & 0 & 0 & 0 & -1 & 0 & 0 & -1 & -1 \\ 0 & 0 & 0 & 0 & 0 & 0 & -1 & 0 & -1 & -1 \\ 0 & 0 & 0 & 0 & 0 & 0 & 0 & -1 & 0 & 1 \\ 0 & -1 & 0 & 0 & 0 & 0 & R_2 & 0 & 0 & 0 \\ 0 & 0 & -1 & 0 & 0 & 0 & 0 & R_4 & 0 & 0 \\ 0 & 0 & 0 & -1 & 0 & 0 & 0 & 0 & R_3 & 0 \end{pmatrix} \begin{pmatrix} u_1 \\ u_2 \\ u_4 \\ u_3 \\ u_5 \\ i_1 \\ i_2 \\ i_4 \\ i_3 \\ i_5 \end{pmatrix} = \begin{pmatrix} 0 \\ 0 \\ 0 \\ 0 \\ 0 \\ 0 \\ 0 \\ 0 \end{pmatrix} \quad (2.11)$$

and deleting the first and last columns of the coefficient matrix, we obtain a nonsingular submatrix with determinant $-(R_2 + R_3)$.

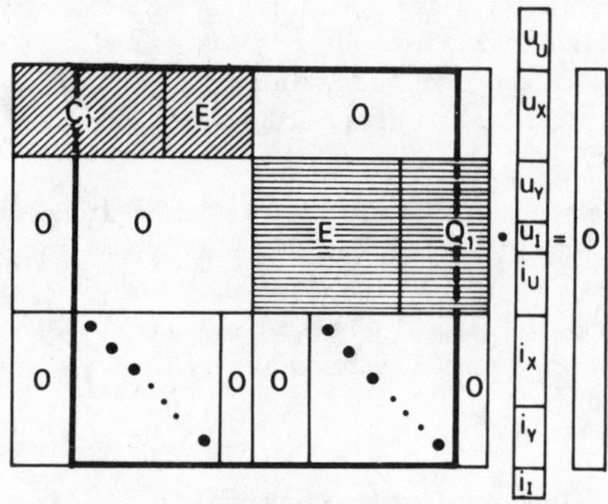

Fig. 2.22

We shall need the *Laplace-expansion* of a determinant. If a determinant has order $(a+b)$ then choose a columns in all the possible $\binom{a+b}{a}$ ways to form a square submatrix with the first a rows (and then another one with the remaining columns and with the last b rows). Multiply the $a \times a$ and the $b \times b$ determinants and assign $+$ or $-$ sign to the product in the following way: Consider the chosen a columns at first and then the rest, as a permutation of the total $(a+b)$- element set and the parity p of this permutation gives the required sign $(-1)^p$. Finally, add these products, the result is just the determinant. For example, if $a = b = 2$ then

$$\begin{array}{c}\begin{matrix}1&2&3&4\end{matrix}\\ \begin{vmatrix}a&b&c&d\\e&f&g&h\\i&j&k&\ell\\m&n&o&p\end{vmatrix}\end{array} = \begin{array}{c}\begin{matrix}1&2\end{matrix}\\ \begin{vmatrix}a&b\\e&f\end{vmatrix}\end{array}\begin{array}{c}\begin{matrix}3&4\end{matrix}\\ \begin{vmatrix}k&\ell\\o&p\end{vmatrix}\end{array} - \begin{array}{c}\begin{matrix}1&3\end{matrix}\\ \begin{vmatrix}a&c\\e&g\end{vmatrix}\end{array}\begin{array}{c}\begin{matrix}2&4\end{matrix}\\ \begin{vmatrix}j&\ell\\n&p\end{vmatrix}\end{array} + \begin{array}{c}\begin{matrix}1&4\end{matrix}\\ \begin{vmatrix}a&d\\e&h\end{vmatrix}\end{array}\begin{array}{c}\begin{matrix}2&3\end{matrix}\\ \begin{vmatrix}j&k\\n&o\end{vmatrix}\end{array} +$$

$$+ \begin{array}{c}\begin{matrix}2&2\end{matrix}\\ \begin{vmatrix}b&c\\f&g\end{vmatrix}\end{array}\begin{array}{c}\begin{matrix}1&4\end{matrix}\\ \begin{vmatrix}i&\ell\\m&p\end{vmatrix}\end{array} - \begin{array}{c}\begin{matrix}2&4\end{matrix}\\ \begin{vmatrix}b&d\\f&h\end{vmatrix}\end{array}\begin{array}{c}\begin{matrix}1&3\end{matrix}\\ \begin{vmatrix}i&k\\m&o\end{vmatrix}\end{array} + \begin{array}{c}\begin{matrix}3&4\end{matrix}\\ \begin{vmatrix}c&d\\g&h\end{vmatrix}\end{array}\begin{array}{c}\begin{matrix}1&2\end{matrix}\\ \begin{vmatrix}i&j\\m&n\end{vmatrix}\end{array}$$

Let us expand the determinant $\mathbf{W}$ of Fig. 2.22 so that the first $e(G)$ rows can be considered as a form of one group. Due to the special structure of the last e_R rows (two diagonal matrices) the only way to obtain nonsingular $e_R \times e_R$ matrices in the "lower" part is if from each resistor-edge, either the voltage or the current copy is chosen (but not both). Hence among the $e(G)$ columns left for the first $e(G)$ rows, we have all the columns corresponding to the voltages of the current sources or to the currents of the voltage sources, plus either the voltage or the current of each resistor. This $e(G) \times e(G)$ matrix is nonsingular if and only if the considered resistors and the voltage sources together form a circuit free subgraph of G and the other resistors and the current sources together form a cut set free subgraph of G (see Theorems 1.4.6–1.4.8).

For example, if we consider the square matrix in Eq. (2.11), we must choose either -1 or $R_j (j = 2, 3, 4)$ from each of the last three rows. Of course, we must choose -1 exactly twice and R_j exactly once (since we should later choose two entries from the first two rows and three entries from the "middle"). This can happen in three different ways, the first two correspond to nonzero expansion members (see Fig. 2.23) in accordance with the fact that $\{e_1, e_2, e_4\}$ and $\{e_1, e_3, e_4\}$ are spanning trees of the network graph while $\{e_1, e_2, e_3\}$ is not.

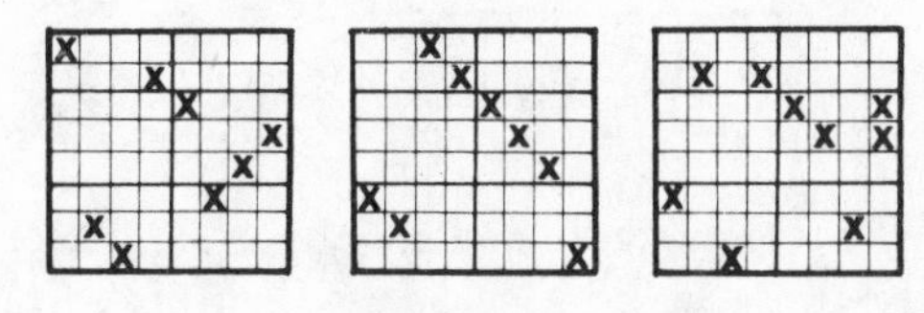

Fig. 2.23

Hence we conclude that the nonzero expansion members of det $\mathbf{W}$ are in one–one correspondence with those spanning forests of the network graph which contain every voltage source edge but none of the current source edges. Each such member is obtained as a product of the determinant of three submatrices. Two of them are submatrices of $\mathbf{C}$ and $\mathbf{Q}$ hence, if they were obtained from fundamental systems of circuits and cut sets, respectively, then their determinant is ± 1 (see Problem 1.4.31). The third is just the product of those resistors which together with the current sources form the complement of a spanning forest.

Theorem 2.5.1 *Prepare the Ohm's Laws and the Kirchhoff's Voltage and Current Laws with respect to fundamental systems of circuits and cut sets for a network N consisting of resistors, voltage and current sources. The determinant of the coefficient matrix $\mathbf{W}$ of the system of equations to be solved is*

$$\pm \sum_F \prod_{e_j \notin F} R_j,$$

where the summation is over every such spanning forest F of the network graph, which contains every voltage source edge but none of the current source edges, and the products are over those resistor edges which are not in F.

Proof. We have already seen that det $\mathbf{W} = \sum_F \left(\pm \prod_{e_j \notin F} R_j \right)$, all what we still have to prove is that each member has the same sign. This can be done by the reader, see Problem 2.5.5.

□

For example, the determinant of the network in Fig. 2.24 is

$$\pm(R_3R_5 + R_3R_4 + R_2R_5 + R_2R_4 + R_2R_3)$$

since the spanning trees, containing edge 1 but not edge 6, in the network graph are

$$\{e_1, e_2, e_4\}, \{e_1, e_2, e_5\}, \{e_1, e_3, e_4\}, \{e_1, e_3, e_5\} \text{ and } \{e_1, e_4, e_5\}.$$

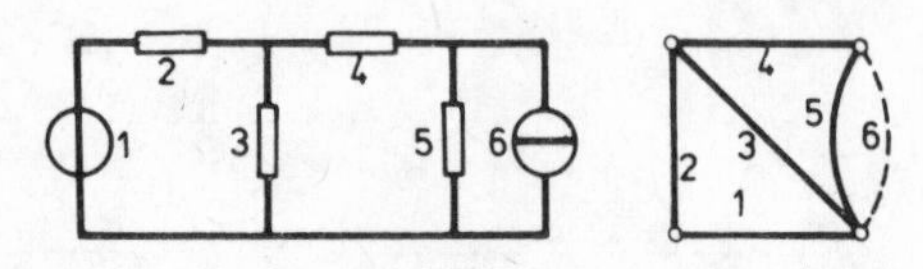

Fig. 2.24

Once we have this theorem, the nontrivial part of Theorem 2.2.1 immediately follows. If the resistors are positive, so are their products as well, and their sum cannot be zero. Hence the network is uniquely solvable. Essentially this was the original idea of Kirchhoff.

Theorem 2.5.1 is sometimes called *topological formula*. This expression means that det $\mathbf{W}$ can be obtained directly from the network graph (i.e. from "topological information") and from the network elements, without really generating the system of equations. Not only det $\mathbf{W}$ but the complete solution of the system of equations can be obtained in similar way, see Problem 2.5.6.

Observe, however (Problem 2.5.12) that solving a network in this way would require a very large number of operations in case of large networks. Hence this method is not suggested for practical numerical calculations.

Exercises for Section 2.5

2.5.1 Suppose that the network contains two resistors in series (which can be replaced by a single one). How is this fact reflected in Theorem 2.5.1?

2.5.2 Let N be a resistive network, R_0 an arbitrary resistor of it. Put open circuit or short circuit instead of R_0 to obtain two new networks N_o and N_s respectively. If the determinant (as in Theorem 2.5.1) of N, N_o and N_s are denoted by D, D_o and D_s, respectively then prove that $D = D_s + R_0 D_o$ [Feussner, 1902, 1904].

2.5.3 (continued) Determine det $\mathbf{W}$ for the network of Fig. 2.25.

2.5.4 Deduce the theorem of Binet and Cauchy (see Problem 1.4.31) from the rule of the Laplace-expansion. (Hint: Expand the determinant of the matrix $\begin{pmatrix} \mathbf{O} & \mathbf{A} \\ \mathbf{B} & \mathbf{E} \end{pmatrix}$.)

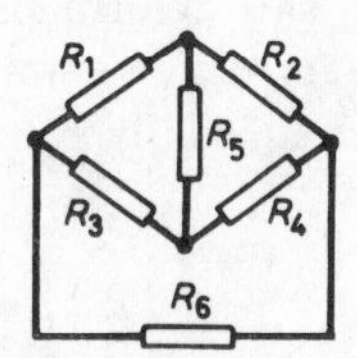

Fig. 2.25

Problems for Section 2.5

2.5.5 Finish the proof of Theorem 2.5.1.

2.5.6 Let N be a connected resistor network, x and y two points of its network graph. Add a short circuit between x and y (if they were adjacent, a loop arises). Let D and D_{xy} denote the determinant of N and that of the new network N_{xy}, respectively. Prove that the resistance we could measure between the points x, y in N is just D_{xy}/D.

2.5.7 (continued) Determine the total resistance between the two end points of the resistor R_6 in Fig. 2.25.

2.5.8 Let $R_1, R_2, \ldots, R_n$ be positive resistors, forming a network N. If we measure total resistance between the end points of these resistors, the results $r_1, r_2, \ldots, r_n$ will obviously be smaller, since r_j is obtained from R_j plus a subnetwork, parallel to it. Prove [Foster, 1949] that $\sum_{j=1}^{n}(r_j/R_j)$ depends on the network graph only, but not on the values of the resistances. Determine this value.

2.5.9 (continued) Imagine an infinite square grid in the 2-dimensional plane and put 1 Ohm resistance into each edge. Take always larger finite pieces and "disregard" the rest, hence give a heuristic proof that the total resistance between any two adjacent points in the grid is 1/2 Ohm.

2.5.10 (continued) Solve the analogous problems for triangular and hexagonal grids as well.

2.5.11 Let G be a connected graph with v points and e edges. Suppose that every edge is contained in the same number of spanning trees. Show that if each edge is replaced with a resistor of 1 Ohm then the total resistance between any two adjacent points will be $(v-1)/e$.

2.5.12 An algorithm generates all the spanning trees of a graph in order to use the topological formulae. Estimate the number of operations required.

Section 2.6 How to brace a square grid using diagonal rods and/or cables

One of the simplest structures in statics are the frameworks consisting of rigid rods and rotatable joints. For example, if we consider the planar structure of

Fig. 2.26 (with four rods joined to form a square and joints A, D fixed to the plane) and a horizontal force F attacks this system at B, then it is intuitively clear that a deformation like in Fig. 2.27 will arise. In case of a fifth rod between the joints A, C were also present, this deformation could perhaps be prevented (if the rods are "strong" and the force F is "not too strong"). Really, if we see frameworks in real life, such squares usually contain some diagonals as well, see Fig. 2.29.

Fig. 2.26 **Fig. 2.27**

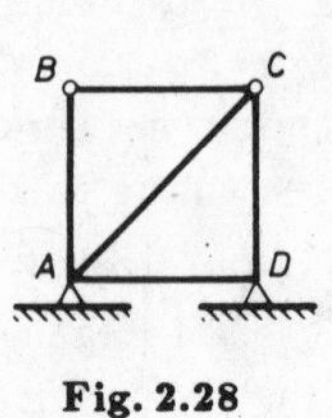

Fig. 2.28

Our modelling assumption is that the rods are relatively strong and rigid while their joints are not so reliable with respect to rotation. Hence we model the *rods* as absolutely rigid and the *joints* as absolutely rotatable. Their interconnection will be called a *framework*.

In the case of such simple frameworks like in Figs 2.26 and 2.28, one can immediately recognize that the latter is "rigid" while the former is not. Until the formal definition of rigidity in Section 6.3 we intuitively say that a planar system is *rigid* if, after fixing two of its joints to the plane, the position of all the other joints are uniquely determined. A planar system is nonrigid if after these fixings some other joints still can move (like in Fig. 2.27); such systems are also called dynamically underdetermined (more about this in Section 4.2).

If the systems are more complex, recognizing rigidity might be more complicated. For example, one of the two systems in Fig. 2.30 is rigid, the other not (see Exercise 2.6.1).

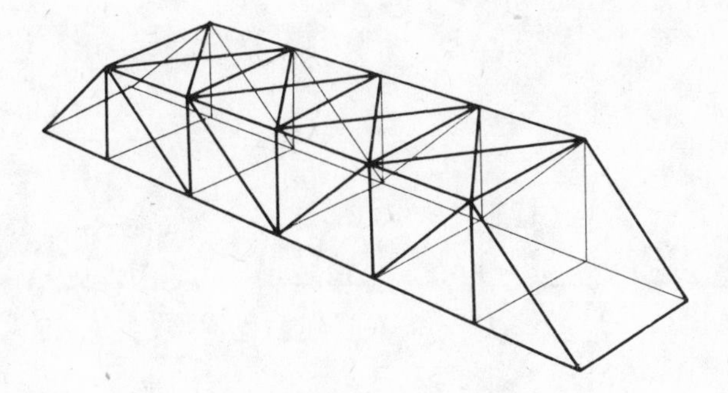

Fig. 2.29

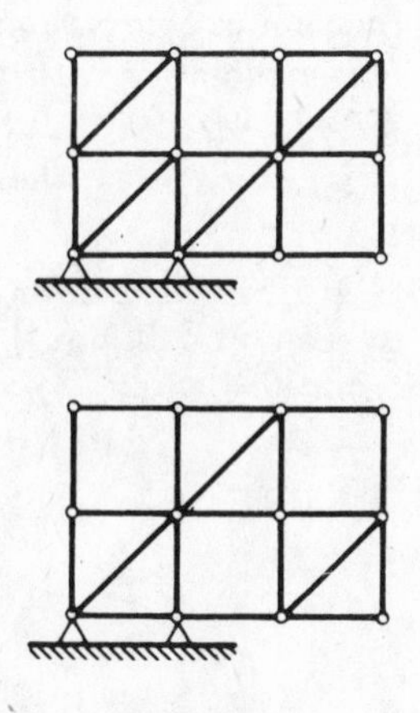

Fig. 2.30

In this section we consider the rigidity of very simple frameworks only; namely square grids with some diagonals added (as in Fig. 2.30). This will help us to consider one-story buildings later on (see Section 10.2).

First of all observe that — under our assumption — the direction of a diagonal does not count. The difference between diagonal AC (in Fig. 2.28) and a possible other diagonal BD is that, in the case of the attacking force F (Fig. 2.26) the reaction of the diagonal brace will be tension or compression. However, since we assumed that the rods are absolutely rigid, we can simply say that any solution prevents the deformation as shown in Fig. 2.27. (More about this in Exercises 2.6.5–9 and in Problems 2.6.11–14, see also Sections 4.2 and 18.3).

Let us study a series of deformations as shown in Fig. 2.31. In the first step the horizontal "row 1 of rooms" was deformed, then the horizontal row 2. In the third step the vertical "column b of rooms" was deformed, followed by the vertical column a.

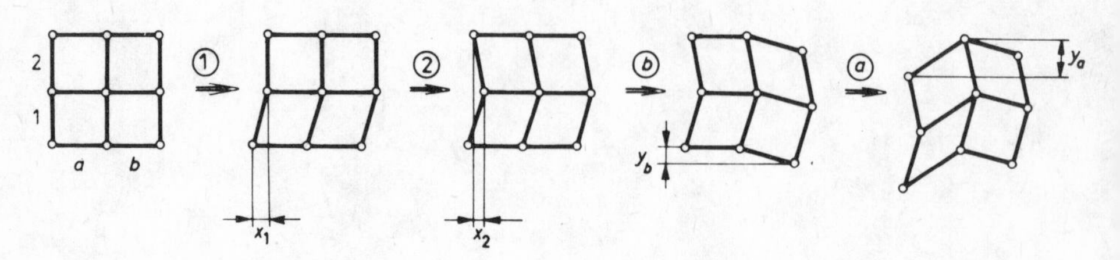

Fig. 2.31

Observe that every deformation of a square grid can be obtained by these steps since each square becomes a rhombus with parallelism preserved. Hence if rows and columns are numbered by $1, 2, \ldots$ and by $a, b, \ldots$, respectively and we know the deformations $x_1, x_2, \ldots, y_a, y_b, \ldots$ then the final shape of the entire system is uniquely determined.

What happens if we have a diagonal rod in, say, the "room $a2$" (that is, in the square, belonging to row 2 and column a)? It does not prevent deformations but ensures that deformations x_a and y_2 will have the same magnitude (Fig. 2.32).

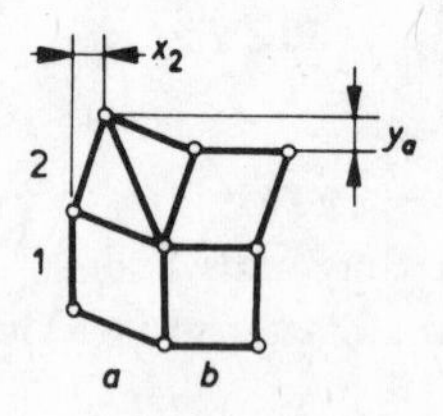

Fig. 2.32

Finally, what happens if we apply sufficiently many diagonal rods to ensure $x_1 = x_2 = \ldots = y_a = y_b = \ldots$? It means that no square will deform relative to the others, hence the entire system is rigid (it only could "turn around" if this common value is nonzero, but this is blocked if two joints are fixed to the plane).

Theorem 2.6.1 [Bolker and Crapo, 1977]. *Let a square grid contain k rows and ℓ columns of rooms. Let us define a graph $G = (V, E)$ with $V = \{v_1, v_2, \ldots, v_k, w_1, w_2, \ldots w_\ell\}$ and join v_i with w_j if there is a diagonal rod in the square where row i and column j intersect. The square grid with this system of diagonals is rigid if and only if G is connected.*

Proof. Since the points of G correspond to the quantities $x_1, x_2, \ldots, y_a, y_b \ldots$ and adjacency corresponds to equality, the statement trivially follows from the above observations.

□

For example, the 3×3 system in Fig. 2.33 is not rigid since the corresponding graph is disconnected. A possible deformation of it is also shown in the figure.

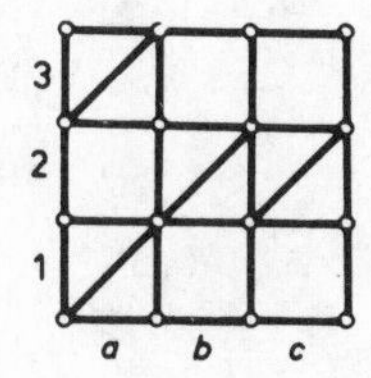

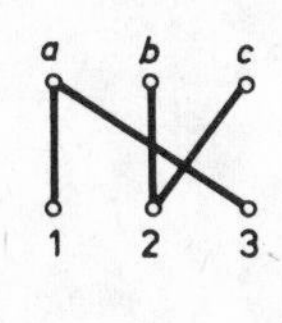

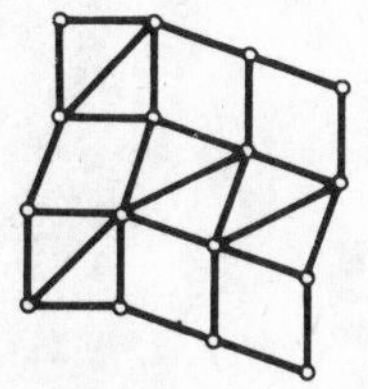

Fig. 2.33

Corollary 2.6.2 [Bolker and Crapo, 1977]. *In order to make the $k \times \ell$ system of squares rigid, we need at least $k + \ell - 1$ diagonal rods. If we have a system of exactly $k + \ell - 1$ diagonal rods then the entire system is rigid if and only if the corresponding graph is a spanning tree.*

Although this trivially follows from Theorem 2.6.1, we also mention a heuristic "proof" of the first part. At the definition of rigidity two joints were fixed to the plane. If there are, say, two lower joints of "room" $a1$ then the last deformation in Fig. 2.31 is not possible. All the horizontal deformations $x_1, x_2, \ldots$ and all but one vertical deformations $y_b, y_c, \ldots$ give $k+\ell-1$ "degrees of freedom" for this system, hence at least as many constraints are required to make the system rigid.

One interesting corollary of these observations is that since rigidity depends on the graph only, the permutation of the names of the points does not change anything. For example, both systems of Fig. 2.34 are rigid.

If one compares the two systems of Fig. 2.35, a significant difference can be found. Both systems are rigid, in fact the number of diagonal braces is more than $3 + 3 - 1$, hence there is some redundancy. However, the first system is more "reliable" since if any one diagonal rod is "broken" the rest is still rigid, while in the second system the rods in rooms $c2$ and $b3$ are "critical". Let us

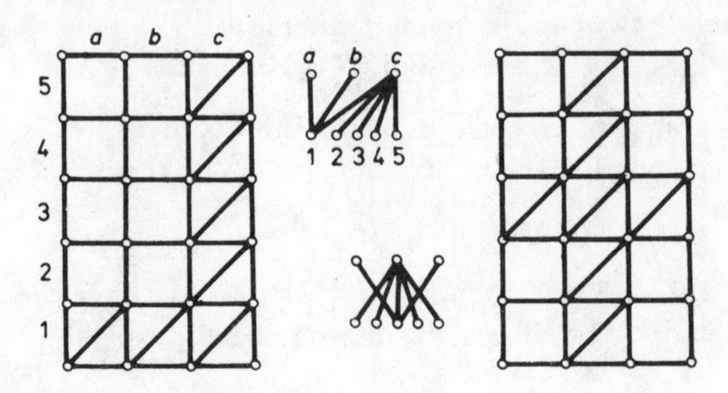

Fig. 2.34

call a diagonal rod of a rigid system *critical* if its removal leads to a nonrigid system. Then the following observation is again straightforward.

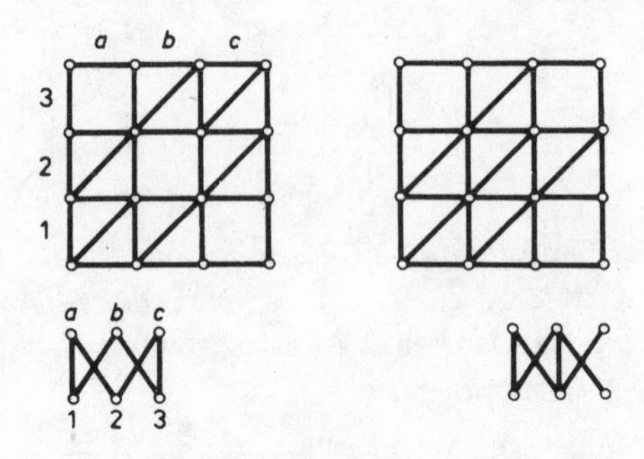

Fig. 2.35

Corollary 2.6.3 *A diagonal rod is critical if and only if the corresponding edge is a separating edge in the graph.*

One of our modelling assumptions was that in the case of a square $ABCD$ (like in Fig. 2.26) if we apply a diagonal rod, its direction AC or BD is of no importance. In the real life some rigid rods are more reliable under tension than under compression. Then this direction of the diagonal may be an important question.

In order to study these problems, let us introduce another concept. A "rod", which preserves its length hence absolutely resists tension but is not "rigid" hence cannot accept any compression, is called a *cable*. For example, if we apply a diagonal cable in the square grid $ABCD$ of Fig. 2.26 between the points A and C then the deformation shown in Fig. 2.27 is impossible; it is not, however, if the cable is between B and D. As another example, the first system in Fig. 2.36 is rigid, the second one is not, see a deformation in Fig. 2.37. (In what follows, rods will be denoted by continuous and cables by broken lines.)

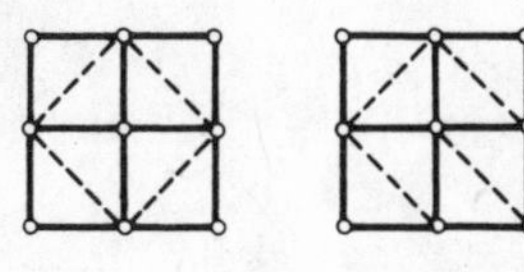

Fig. 2.36

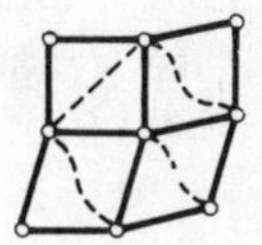

Fig. 2.37

Consider the "mixed" systems of Fig. 2.38, each containing two diagonal rods and two diagonal cables. It is not difficult to see that the first system is rigid, the second is not. Let us study them, however, in more detail.

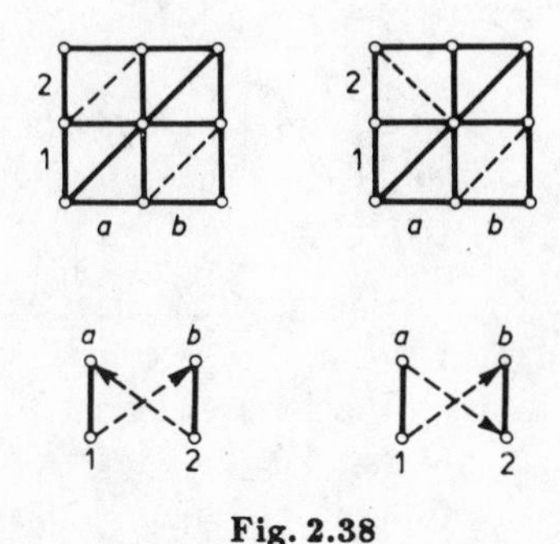

Fig. 2.38

When we introduced the quantities $x_1, x_2, \ldots, y_a, y_b, \ldots$ for denoting the deformations, we did not speak about their signs. Let us decide that x_j is positive if, say, the upper "horizontal" edges of the squares move to the left (relative to the lower edges) and negative otherwise; and that y_k is positive if the right hand side "vertical" edges move up relative to the left hand side ones. Then a "Northwest-Southeast" cable in, say, the square $a2$ ensures $x_2 \leq y_a$ while a "Northeast-Southwest" cable ensures $x_2 \geq y_a$.

Returning to Fig. 2.38, the relations

$$x_1 = y_a \text{ and } x_2 = y_b$$

are obvious, due to the two diagonal rods. However, taking the two cables also into consideration, two further inequalities

$x_1 \geq y_b$ and $x_2 \geq y_a$ are obtained in the first system, leading to $x_1 = y_a \leq x_2 = y_b \leq x_1$. This is possible only if all these quantities are equal hence this system is rigid.

$x_1 \geq y_b$ and $x_2 \leq y_a$ are obtained in the second system, leading to $x_1 = y_a \geq x_2 = y_b \leq x_1$. This can be met by $x_2 = y_b < x_1 = y_a$ as well, hence a deformation like in Fig. 2.39 can arise.

Hence all we have to do is to modify our graph model by assigning orientations to the edges. If a cable gives a constraint $x_i \geq y_j$ then let the edge (i, j) have tail in i and head in j, if it gives $x_i \leq y_j$ then *vice versa*. If a diagonal rod gives a constraint $x_i = y_j$, then we use two directed edges (i, j) and (j, i) with opposite orientations.

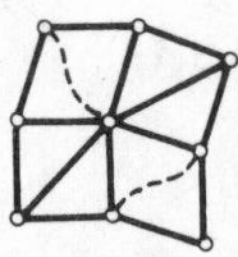

Fig. 2.39

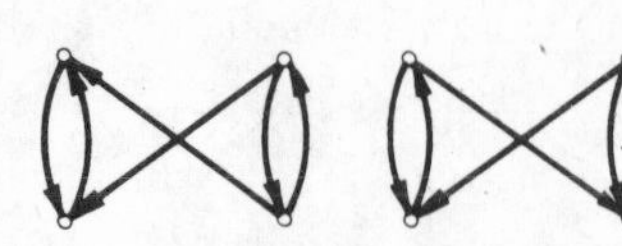

Fig. 2.40

Theorem 2.6.4 [Baglivo and Graver, 1983]. *The system of diagonal rods and cables in a square grid is rigid if and only if the corresponding directed graph is strongly connected.*

Proof. Use the same way of reasoning as in the last example. If the directed graph is strongly connected then every edge is contained in a directed circuit, along which we obtain equalities. If the directed graph is not strongly connected, then we can find a subset of points with no edges directed into this subset. Choosing the x and y values in this subset larger than in the rest, we obtain a deformation.

□

For example, the directed graphs of the systems of Fig. 2.38 are shown in Fig. 2.40 The first one is strongly connected, the second one is not.

Corollary 2.6.5 [Baglivo and Graver, 1983]. *If only diagonal cables are allowed, then at least* $2\max(k, \ell)$ *cables are required to make a* $k \times \ell$ *square grid rigid. This number is sufficient in appropriate arrangements.*

Its *proof* is left to the reader as Problem 2.6.12.

Since every diagonal rod can theoretically be replaced by two perpendicular diagonal cables, Corollary 2.6.2 would give an upper bound $2(k+\ell-1)$. Observe that our last result is much better if both k and ℓ is large.

Finally we mention that cables, too, can be *critical* if their removal lead to a nonrigid system, see Problem 2.6.14.

Exercises for Section 2.6

2.6.1 Determine which system in Fig. 2.30 is rigid. For the nonrigid one, draw a deformation as well.

2.6.2 Suppose that a room has no diagonal rod (i.e., it either has a cable or nothing). Determine whether it can be deformed by suitable deformations (provided the whole system is nonrigid).

2.6.3 We used some diagonal rods in a system and the resulting graph has two connected components. Show that if a room can be deformed at all then it will be deformed under every global deformation.

2.6.4 Suppose rooms $a1$ and $b2$ have diagonal rods. Prove that if rooms $a2$ and $b1$ can be deformed, they always stay congruent.

2.6.5 Consider the system of Fig. 2.41. Show that it is not rigid. Put further cables to make it rigid.

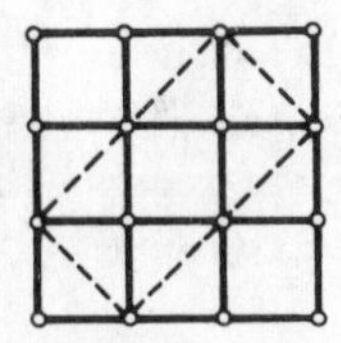

Fig. 2.41

2.6.6 Is the system of Fig. 2.42 rigid? Can you put cable instead of the only diagonal rod? If yes, what should its direction be?

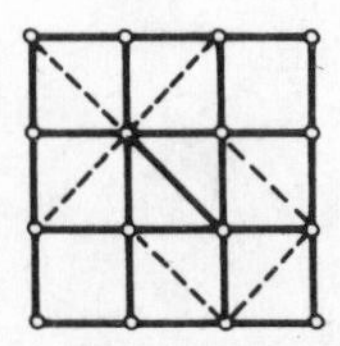

Fig. 2.42

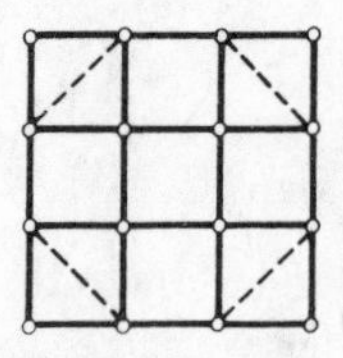

Fig. 2.43

2.6.7 Add a minimal number of cables to the system of Fig. 2.43 to make it rigid.

2.6.8 Deduce Theorem 2.6.1 from Theorem 2.6.4.

2.6.9 Change every diagonal cable to one with perpendicular direction. Show that this does not change rigidity.

Problems for Section 2.6

2.6.10 Generalize Exercise 2.6.4. Show that if two opposite corners of a square grid cannot be deformed at all then the other two always stay congruent.

2.6.11 We used some diagonal rods in a system and the resulting graph has two connected components. Now put diagonal rods to those rooms which can be deformed but remove the old diagonal rods. Determine the number of connected components in the new graph.

2.6.12 Prove Corollary 2.6.5.

2.6.13 A system is rigid and contains some diagonal cables and a single diagonal rod. Is it true that this rod can also be changed to a cable (of suitable direction) with preserving rigidity?

2.6.14 Let $k, \ell \geq 2$. If we use $k + \ell$ diagonal rods, show that at least four of them are noncritical. What is the corresponding statement if we use $2\max(k, \ell) + 1$ diagonal cables?

Chapter 3

Planar graphs and duality

Section 3.1 Duality and 2-isomorphism

We have seen several pairs of different theorems with identical logical structure in Chapter 1. For example, spanning forests (and their complements) were characterized as maximum size subgraphs without containing any circuit (any cut set, respectively). As another example, maximum size collections of linearly independent columns of the circuit matrix (and of the cut set matrix) were shown to be in one–one correspondence with the complements of the spanning forests (with the spanning forests, respectively).

Apparently, truth or falsity of a graph theory statement does not change if circuits and cut sets, furthermore spanning forests and their complements are exchanged. This sentence — although incorrect, see Exercise 3.2.6, for example — expresses a very important idea whose historical roots are worth of a few sentences.

A polygon is regular if every edge of it is equal and every angle of it is equal. A polyhedron is regular if it is bounded by congruent regular polygons. There are infinitely many regular polygons (regular triangle, square, regular pentagon, hexagon, etc.) but only five regular polyhedra, see Fig. 3.1. They were known to the ancient Greeks already who could even prove that no more regular polyhedra exist. If the centre of each bounding polygon of such a regular polyhedron is considered, they determine a new regular polyhedron. For example, an octahedron can be obtained from a cube or another tetrahedron from a tetrahedron, see Fig. 3.2. The arrangement of the five bodies in Fig. 3.1 already followed this pattern.

The class of convex polyhedra satisfies *Euler's formula*: If the number of vertices, edges and faces of the polyhedron are denoted by v, e and f, respectively, then $v+f=e+2$. Related to this 18th century result, a paper of Lhuilier around 1812 applied already the construction, shown in Fig. 3.2, for general (i.e., not necessarily regular) polyhedra.

More exactly, he spoke about pairs of polyhedra where vertices of the first polyhedron correspond to the faces of the second and *vice versa.* Two vertices a, b are adjacent in the first (i.e., there is an edge e_{ab} between them) if and only if the corresponding faces A, B are adjacent in the second one (i.e., there is an edge e_{AB} separating them). The main observation is that a one–one correspondence is thus established between the edge sets.

Let us draw again the "graphs" of the cube and the octahedron (Fig. 3.3) and determine this correspondence between the edge sets. Spanning trees of

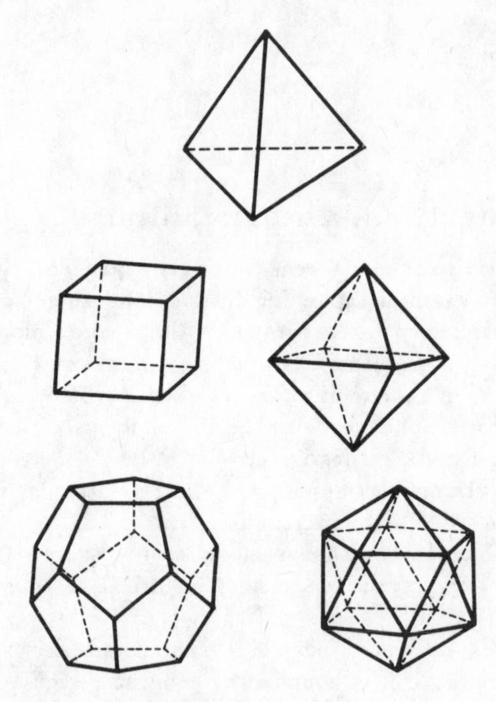

Fig. 3.1

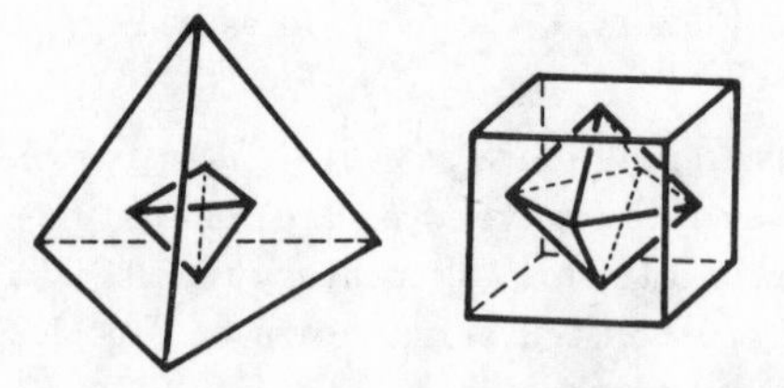

Fig. 3.2

one of these graphs correspond to complements of spanning trees of the other (Fig. 3.4), circuits of one correspond to cut sets of the other (Fig. 3.5).

If our graphs can be drawn to the plane without intersecting edges then this one–one correspondence can be obtained without any reference to polyhedra. If the edges do not intersect in a planar drawing then the plane is split into regions.

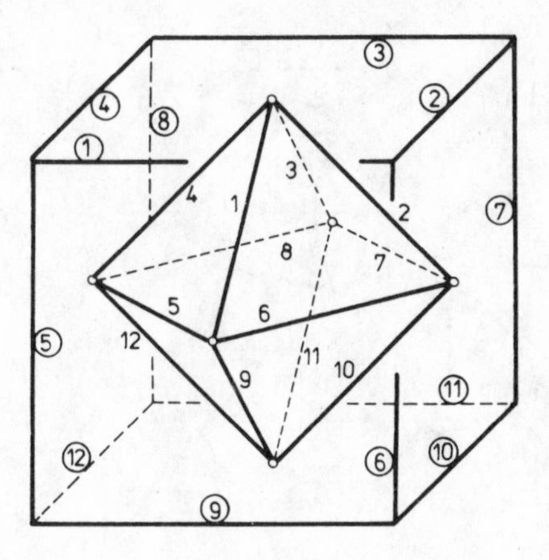

Fig. 3.3

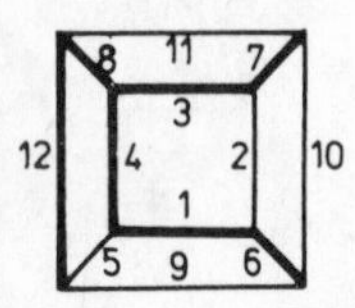

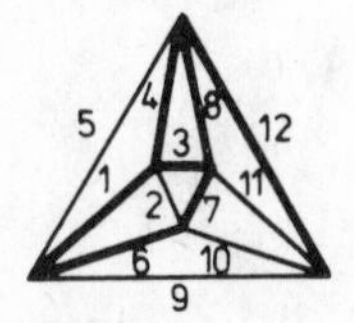

Fig. 3.4

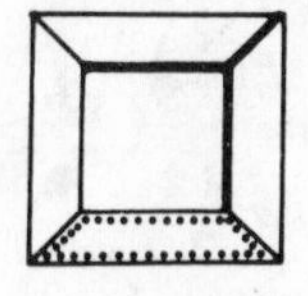

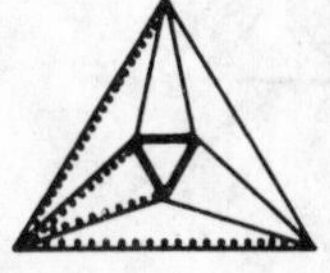

Fig. 3.5

These regions — determined by the drawing of the first graph — will be the points of the second graph, and two points in the second graph will be adjacent if the regions shared a common boundary edge in the first graph.

In this way graphs with loops, multiple edges, with points of degree one or two can also be handled (see Fig. 3.6) although such phenomena are impossible if the graphs correspond to polyhedra.

If a graph G can be drawn in the plane without intersecting edges then this method always leads to a graph G^* which can also be drawn in the plane in the required way (in fact we obtain the planar drawing of G^* rather than the abstract graph). It would be natural to call this graph the dual of G.

Unfortunately, this definition would raise serious problems. What we obtained is determined not by G but by its actual planar drawing. Different drawings of the same graph can lead to different duals. Both in Fig. 3.7 and Fig. 3.8 the graphs in the first lines are isomorphic while their "duals" (in the second lines) are not.

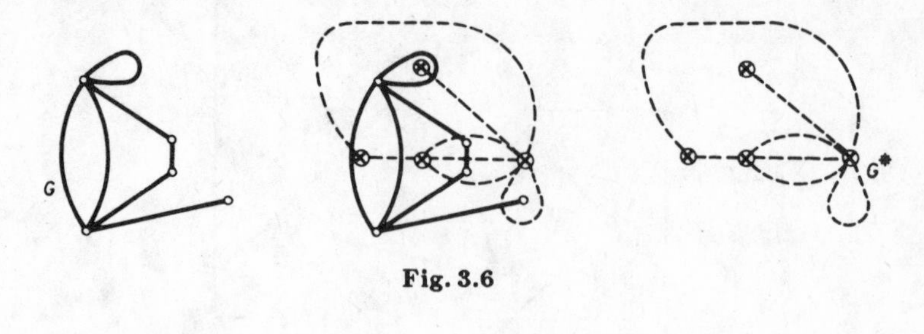

Fig. 3.6

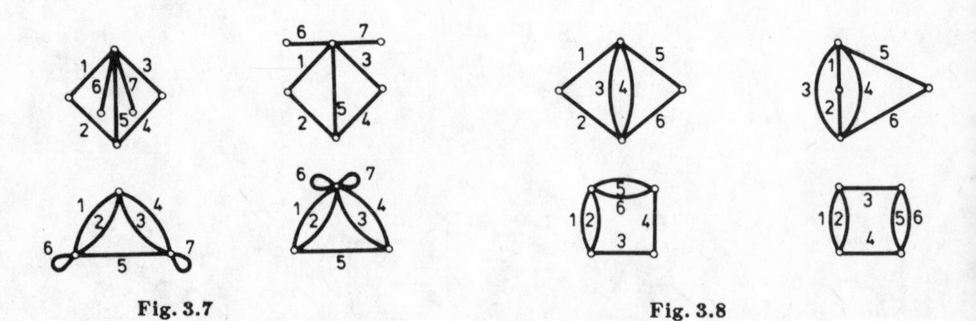

Fig. 3.7

Fig. 3.8

This difficulty can be overcome in two different ways. Either we distinguish, from now on, between the drawings (in the first lines) of the same graph, or we do not distinguish, in a sense, between the nonisomorphic graphs of the second lines. The first solution is unacceptable since considering different drawings of the same abstract graph as equivalent, was a basic assumption in graph theory. For the second solution we must weaken our concept of isomorphism.

Recall that two graphs are isomorphic if one–one correspondence can be given both between their point sets and edge sets, preserving incidence. We call two graphs *2-isomorphic* [Whitney, 1933b] if there is one–one correspondence between their edge sets which preserves circuits. More exactly, a subset of edges form a circuit in one of the graphs if and only if the corresponding edges form a circuit (possibly in a different order) in the other graph.

For example, the graphs of the second line of Fig. 3.7 are 2-isomorphic. The subsets $\{6\}, \{7\}, \{1,2\}, \{3,4\}, \{1,3,5\}, \{1,4,5\}, \{2,3,5\}, \{2,4,5\}$ are circuits in both graphs, other subsets of edges are not circuits in either graph.

Every graph is 2-isomorphic to itself, the definition is symmetric, and if G_1 and G_2 are 2-isomorphic to G_2 and G_3, respectively, then G_1 and G_3 are also 2-isomorphic. Hence we may speak about classes of 2-isomorphic graphs and shall define duality among these classes.

Two such classes $\mathbf{G}_1$ and $\mathbf{G}_2$ of graphs are called *dual* to each other if, for every graph $G_1 \in \mathbf{G}_1$ and $G_2 \in \mathbf{G}_2$ there is a one–one correspondence between the edge sets of G_1 and G_2 so that the circuits of G_1 are cut sets of G_2 and *vice versa*. The dual class $\mathbf{G}_2$ of the class $\mathbf{G}_1$ will be denoted by $\mathbf{G}_1^*$ as well.

Theorem 3.1.1 *(a) For every such class $\mathbf{G}$, the relation $(\mathbf{G}^*)^* = \mathbf{G}$ holds.*
(b) If an arbitrary graph $G \in \mathbf{G}$ is drawn to the plane with no intersecting edges and its "dual" H is prepared as in Fig. 3.6 then $H \in \mathbf{G}^*$.

Proof. We shall need the following two lemmata:

Lemma 3.1.2 *If G and G' are 2-isomorphic and G can be drawn to the plane without intersecting edges then so can G'.*

Lemma 3.1.3 *If G and G' are 2-isomorphic then the circuit-preserving one–one correspondence between their edge sets preserves cut sets as well. On the other hand, if the edge sets of two graphs are in one–one correspondence and cut sets of one graph correspond to cut sets of the other then the two graphs are 2-isomorphic (and this same correspondence will be circuit-preserving as well).*

The proof of Lemma 3.1.2 is postponed to a later section (Problem 3.3.19) and that of Lemma 3.1.3 is left to the reader (Problem 3.1.9).

Let G be an arbitrary graph from the class $\mathbf{G}$ and let us draw it to the plane with no intersecting edges. The resulting regions of the plane will be the points of the graph H. If C is a circuit of G then certain regions $T_1, T_2, \ldots$ are inside C while the others $S_1, S_2, \ldots$ (including the unbounded region) are outside C, see Fig. 3.9. The corresponding point sets $\{t_1, t_2, \ldots\}$ and $\{s_1, s_2, \ldots\}$ determine two connected subgraphs of H (since we can move from one region to another in G without intersecting C). The edges of H, corresponding to the edges of C in G, are just of form $\{t_i, s_j\}$, hence they form a cut set of H.

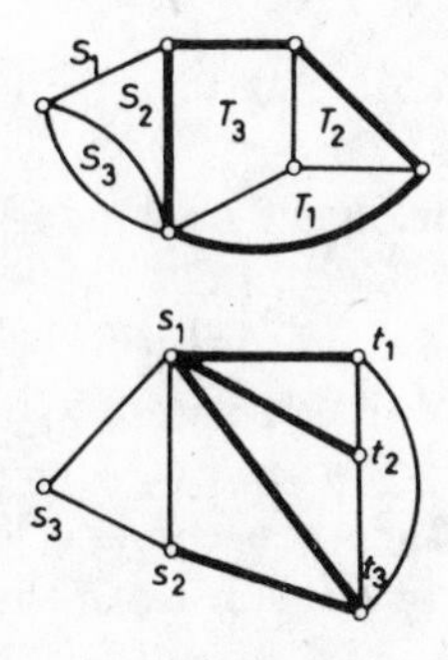

Fig. 3.9

Conversely, if $X \subseteq E(G)$ and the corresponding edges in H form a cut set Q then X must be a circuit in G. Since Q separates two point sets from each other, the corresponding regions must be separated by X. Hence X contains a circuit of G and if it were properly contained then, by the above reasoning, a proper subset of Q would also be a cut set, a contradiction. Hence part (b) of the theorem is proved while part (a) is obvious.

□

Once we have this exact concept of duality (among classes $\mathbf{G}_1$ and $\mathbf{G}_2$ of graphs), we may occasionally say that a graph $G_1 \in \mathbf{G}_1$ and another graph $G_2 \in \mathbf{G}_2$ are "dual" to each other but then we should remember that preparing a graph G_1^* from G_1 in the way shown in Fig. 3.6 the result will, in general, be only 2-isomorphic to G_2.

The concept of duality determines a symmetry among certain concepts of graph theory. This is summarized in **Box 3.1.** Observe that edges correspond to themselves and "nothing corresponds" to points.

Box 3.1. Correspondence among graph theory concepts, related to the duality of planar graphs

G	G^*
Edge	Edge
Circuit	Cut set
—loop	—separating edge
—parallel edges	—series edges
Spanning forest	Complement of a spanning forest
—fundamental system of circuits	—fundamental system of cut sets
—nullity	—rank
—circuit matrix	—cut set matrix
Contraction of an edge	Deletion of an edge

See Problem 5.2.13 as well.

There are two new concepts in **Box 3.1.** Two edges $e, f \in E(G)$ are called *series* if $\{e, f\}$ is a cut set of G. Observe that the definition of series edges in electric network theory is somewhat different (see Exercise 3.1.4).

Let $e \in E(G)$ be an edge $\{x_1, x_2\}$ of G. The *contraction* of e leads to a new graph G/e where the points x_1, x_2 are identified (and denoted by, say, x) and any edge, joining some point to x_1 or x_2 in G will join that point to x in G/e. The edge e itself is then deleted (hence $E(G/e) = E(G) - \{e\}$ holds) but if f was an edge of G parallel to e, then it becomes a loop, incident to x.

If several edges are contracted, the order of these contractions is irrelevant. Hence we shall use the notation G/X as well, for $X \subset E(G)$. Furthermore, one can easily prove that, for $e, f \in E(G), e \neq f$, the relation $(G/e) - f = (G - f)/e$

holds and similarly the deletion and contraction of disjoint subsets of edges also commute.

Exercises for Section 3.1

3.1.1 Find isomorphic or 2-isomorphic pairs of graphs in Fig. 3.10.

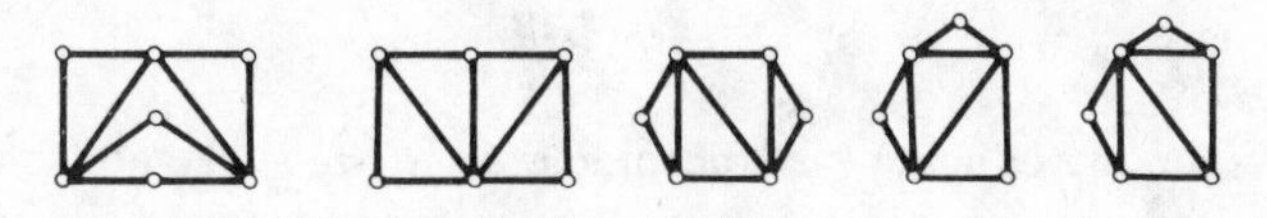

Fig. 3.10

3.1.2 Let $x, y \in V(G)$ be nonseparating points in a connected graph G but suppose $G - \{x, y\}$ contains two components. Join these components again but twist around one component first (see Fig. 3.11) to get a new graph H. Prove that G and H are 2-isomorphic (see Problem 3.3.20 as well).

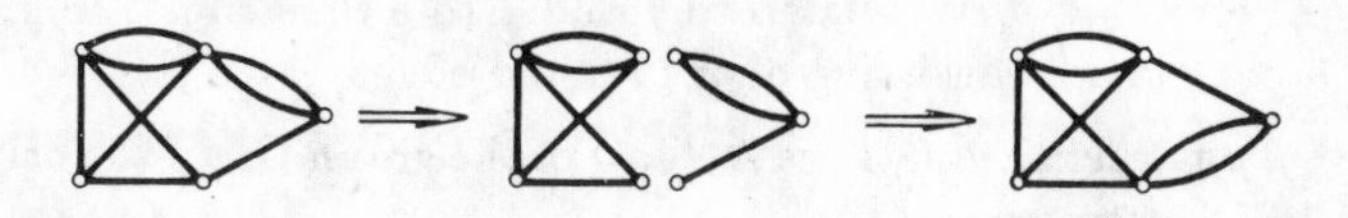

Fig. 3.11

3.1.3 Are the two graphs in Fig. 3.12 2-isomorphic?

3.1.4 Two edges are called "series" in electric network theory if they are incident to a point of degree 2. What is the relation of this to our definition?

3.1.5 Recognize series or parallel edges if the graph is given by the **B**, **C** or **Q** matrix.

3.1.6 Let $t(G)$ denote the number of spanning forests of the graph G. Suppose $e \in E(G)$ is neither a loop, nor a separating edge. Prove that $t(G) = = t(G - e) + t(G/e)$.

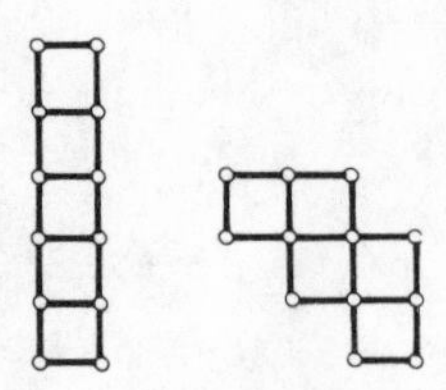

Fig. 3.12

3.1.7 When introducing 2-isomorphic graphs, we spoke about edges and circuits only, but not about points. In a similar way (i.e., without using the concept of points), give new definitions to the following concepts: loop, parallel edges, separating edge, series edges, cut set, spanning forest, rank, nullity.

3.1.8 Let $e \in E(G)$. If a path of length two is put instead of the place of e (thus creating a new point of degree 2), the new graph will be called the *series extension* of G along e. The *parallel extension* of G along e is simply the addition of a new edge parallel to e. Determine how the matrices $\mathbf{A}, \mathbf{B}, \mathbf{C}, \mathbf{Q}$ and the rank and the nullity of the graph change by these extensions.

Problems for Section 3.1

3.1.9 Prove Lemma 3.1.3.

3.1.10* Two edges $e, f \in E(G)$ are called "related" if $e = f$ or if they are series. Show that if e is related to f and f to g then so is e to g. Characterize the classes consisting of the related edges.

3.1.11 Construct the matrices $\mathbf{A}, \mathbf{C}, \mathbf{Q}$ of the graph G/e if the original matrices of G were given.

3.1.12* (continued) Give an algorithm which decides from the matrix $\mathbf{A}$ whether the graph is connected.

3.1.13 Let $A, B \subseteq E(G)$ be two disjoint subsets so that A contains no circuit and B contains no cut set. Prove that $F \subseteq E(G)$ is a spanning forest in G, containing A and being disjoint of B, if and only if $F - A$ is a spanning forest of $(G - B)/A$.

3.1.14 Using **Box 3.1** give the "dual" of Algorithm 1.5.

Section 3.2 Planar graphs

Drawing the graphs to the plane without intersecting edges played already some role in the previous section. This means we associate different points $P_1, P_2, \ldots$ of the plane to the points $v_1, v_2 \ldots$ of the graph and if $e = \{v_i, v_j\}$ was an edge then P_i and P_j are connected by a continuous line which does not contain any other point P_k (not even P_i and P_j as internal points) and which has no common point with any other line (except from possible common end points). If this is possible, the graph is called *planar*.

For example, K_5 is nonplanar, see below. K_4 is planar, see the second drawing of Fig. 3.13. The first drawing is inappropriate since lines $\{1,3\}$ and $\{2,4\}$ intersect. Observe that planarity is a property of the graph and not of its illustration.

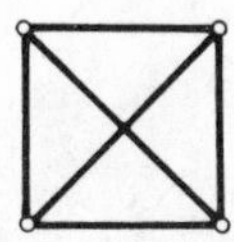

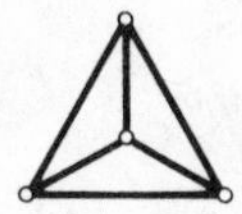

Fig. 3.13

Before discussing planarity in some detail, we briefly mention that the questions which graphs can be drawn to various surfaces (not only to planes) are also interesting, see Exercise 3.2.2 and Problems 3.2.9–10, for example.

Statement 3.2.1 *A graph can be drawn to the surface of a sphere if and only if it is planar.*

Proof. The so called *stereographic projection* is shown in Fig. 3.14. The "Globe" touches a plane P in its "Southern Pole" S. Let N denote the opposite point of the sphere (the "Northern Pole"). If V is an arbitrary point of the sphere, different from N, then let us associate to V the unique intersection V' of P with the line NV. This method, starting from any planar drawing, leads to a drawing on the sphere, and starting from any spherical drawing which does not contain N, leads to a planar drawing.

□

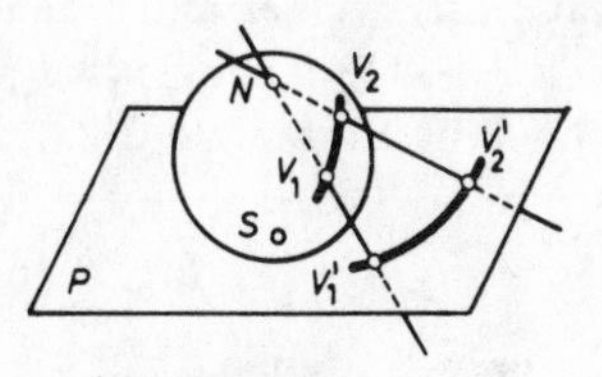

Fig. 3.14

Statement 3.2.2 *Let G be a planar graph and let C be a circuit, formed by the boundary of one of the bounded regions of a particular planar drawing of G. Then there exists another planar drawing of G, having C as the boundary of the unbounded region.*

Proof. Project the original planar drawing to the sphere and consider the region bounded by the image of C. Rotate the sphere to have the Northern Pole N inside this region and project the graph back to the plane. □

For example, three bounded regions in the drawing on part (a) of Fig. 3.15 were denoted by T_1, T_2 and T_3. The other three drawings of Fig. 3.15 show the same planar graph by making these regions unbounded.

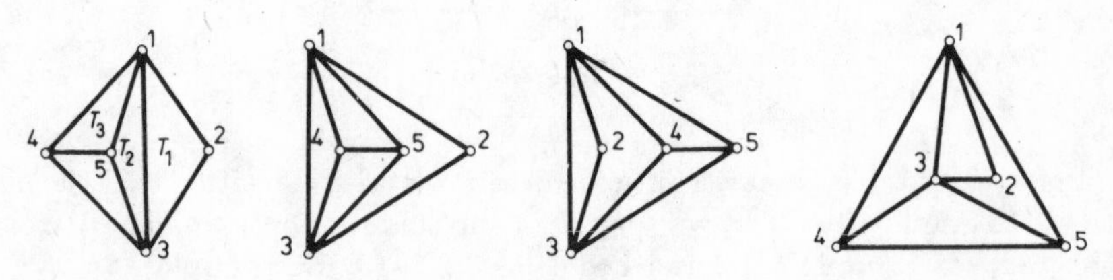

Fig. 3.15

Let us return to the question of planarity.

Statement 3.2.3 *If G is planar and connected, all of its correct drawings determine $e(G) - v(G) + 2$ regions in the plane.*

Proof. Let G be drawn to the plane together with its "dual" G^* as obtained by the process shown in Fig. 3.6. Any spanning tree T of G has $v(G) - 1$ edges. The remaining $e(G) - v(G) + 1$ edges determine a spanning tree of G^*. Hence G^* has $e(G) - v(G) + 2$ points. □

Statement 3.2.4 *Let G be a simple connected planar graph and suppose that every circuit of G has at least k edges. Then*

$$e(G) \leq \frac{k}{k-2}v(G) - \frac{2k}{k-2}.$$

Proof. If every circuit has at least k edges and t denotes the number of regions in a planar drawing of G then $2e(G) \geq k.t$ (every edge is counted twice). On the other hand $t = e(G) - v(G) + 2$ by Statement 3.2.3, hence the result follows. □

Corollary 3.2.5 *None of the two graphs in Fig. 3.16 is planar.*

Proof. In the case of K_5, apply Statement 3.2.4 with $k = 3$ to obtain $10 \leq$ $\leq 3 \times 5 - 6$, a contradiction. The other graph has no circuit of length three, hence the stronger upper bound with $k = 4$ can be applied to yield a contradiction by $9 \leq 2 \times 6 - 4$.

□

The two graphs in Fig. 3.16 will be called *Kuratowski graphs.* Obviously, a graph having them (or their series extensions) as subgraphs, cannot be planar.

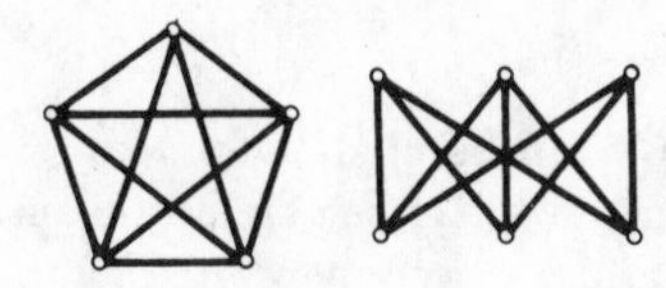

Fig. 3.16

Theorem 3.2.6 [Kuratowski, 1930]. *A graph is planar if and only if it does not contain as subgraphs any copy or any series extension of the two graphs of Fig. 3.16.*

While the *proof* of the necessity of this condition was fairly simple, its sufficiency is more complicated and will follow after the solutions of Exercises 3.2.3, 3.3.4, 3.3.13 and Problems 3.2.13–14, 3.3.14–16 only.

We close with an interesting link between this and the previous sections. We recall that a graph G had a dual H if there is a one–one correspondence between $E(G)$ and $E(H)$ which transforms circuits into cut sets and *vice versa.*

Theorem 3.2.7 [Whitney, 1932] *A graph has a dual if and only if it is planar.*

Proof. The sufficiency was part (b) of Theorem 3.1.1. The proof of the necessity is left to the reader as Problem 3.2.12.

Exercises for Section 3.2

3.2.1 Is the set of edges, bounding a region in a planar drawing of G, necessarily a circuit of G?

3.2.2 Two interesting surfaces which are significantly different from the plane or from the sphere are the torus and the Möbius-surface (Fig. 3.17). The former is like the surface of a doughnut, the latter is obtained if ends of a ribbon are fixed after a twist. Show that the Kuratowski graphs can be drawn to these surfaces without intersecting edges.

3.2.3* Instead of Statement 3.2.2 (any region can be made outside-region) a more realistic requirement in the layout design of printed circuit boards or integrated circuits may be that certain points of the planar graph be outside (i.e., on the unbounded region) for access from other devices. Prove that if G is planar then

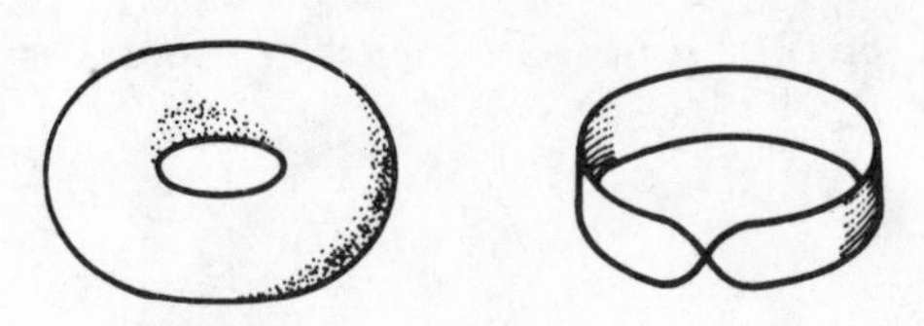

Fig. 3.17

(a) any $v \in V(G)$ can be made outside;
(b) two points $u, v \in V(G)$ can simultaneously be made outside if and only if G, extended by a new edge $\{u, v\}$ is planar;
(c) every point can simultaneously be made outside if and only if G, extended by a new point, adjacent to every old point, is planar.

3.2.4 Prove that the Kuratowski graphs have no duals.

3.2.5 Find graphs which are isomorphic to their own dual.

3.2.6 Give a true graph theory statement which becomes false if "dualized" according to **Box 3.1.**

3.2.7 Draw the graphs of Fig. 3.18 into the plane with edges as straight line segments. (Every simple planar graph can be drawn in this way [Wagner, 1936; Fáry, 1948], see also pp. 266–267 in [Lovász, 1979a].)

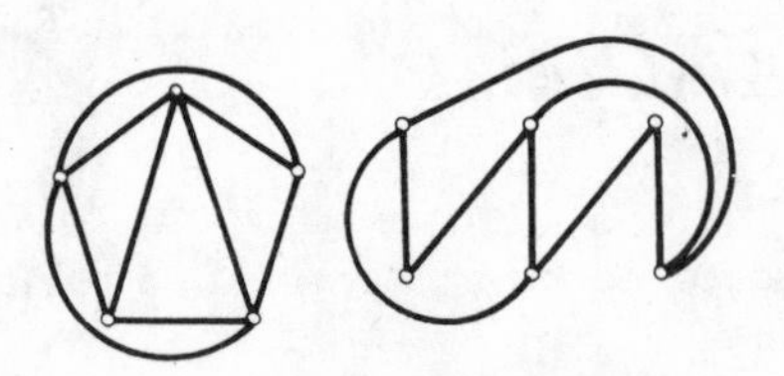

Fig. 3.18

Problems for Section 3.2

3.2.8 By a theorem of [Wagner, 1937] a graph is planar if and only if it cannot lead to one of the Kuratowski graphs by deleting and contracting some edges. (The necessity is obvious, the sufficiency easily follows from Theorem 3.2.6.) In order to understand the difference between Kuratowski's and Wagner's conditions give a nonplanar graph which can be contracted to K_5 but contains neither K_5 nor its series extensions.

3.2.9 Show that K_7 and the graph of Fig. 1.8 (the so called *Petersen graph*) can be drawn to the torus without intersecting edges.

3.2.10 The method of drawing a "dual" as shown in Fig. 3.6 could be applied on the torus as well. Applying Exercise 3.2.2 obtain a "dual-like" graph for K_5. Is there no contradiction with Exercise 3.2.4?

3.2.11 Let k denote the minimum degree in a graph G. Give an upper bound for k if G is planar. If this bound is sharp, give a lower bound for the number of those points which have degree k.

3.2.12 Prove Theorem 3.2.7.

The following two problems are preparations for a proof of Theorem 3.2.6 in the next section.

3.2.13 Let C be a circuit of a simple graph G. Two edges of $E(G) - E(C)$ are called *C-equivalent* if either $e = f$ or G has a path P so that e and f are the first and the last edges of P and no internal point of P lies on C. Determine the equivalence classes on the graph of Fig. 3.19 with respect to the circuit of the heavy edges. Show that this relation always leads to equivalence classes (the so called *C-components* or *C-bridges*).

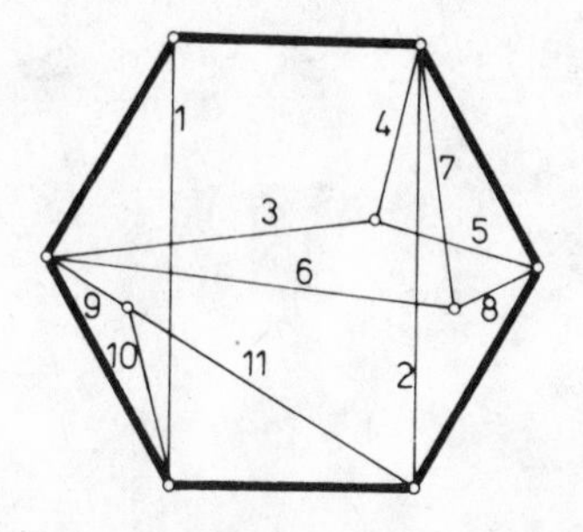

Fig. 3.19

3.2.14 (continued) Let H_1, H_2 be two C-components. Denote $V(H_1) \cap V(C)$ and $V(H_2) \cap V(C)$ by V_1 and V_2, respectively. Prove that the mutual relation of H_1 and H_2 can be exactly one of the following three:

(1) They may be disjoint (like $\{3,4,5\}$ and $\{9,10,11\}$ in Fig. 3.19) if C can be divided into two arcs so that V_1 is on one of the arcs and V_2 on the other.

(2) They may be crossing (like $\{1\}$ and $\{3,4,5\}$) if C has four points a, b, c, d (in this cyclic order) so that $a, c \in V_1$ and $b, d \in V_2$.

(3) They may be equivalent (like $\{3,4,5\}$ and $\{6,7,8\}$) if $V_1 = V_2$ and $|V_1| = |V_2| = 3$.

Section 3.3 2-Connected graphs

The strong connectivity of directed graphs (Section 1.1) was a stronger property than "usual" connectivity, e.g. it excluded separating edges. However, separating points could still be present although one might have the intuitive feeling that those, too, make the connections between various parts of the graph "more difficult".

An undirected graph G (with no loops and with at least 3 points) is called *2-connected* if $G - v$ is connected for every $v \in V(G)$. Among the six graphs in Fig. 3.20, G_1 is not connected at all, G_2 is connected but no orientation can be assigned to it to make it strongly connected. G_3 can be made strongly connected but is not 2-connected while the other three graphs are 2-connected.

More generally, a graph with no loops and with at least $k+1$ points is called *k-connected* if $G - X$ is connected for every $X \subset V(G), |X| < k$. Among the graphs of Fig. 3.20 only the last one is 3-connected. Let G be a connected graph. Let $e, f \in E(G)$ be called equivalent if either $e = f$ or G has a circuit containing e and f.

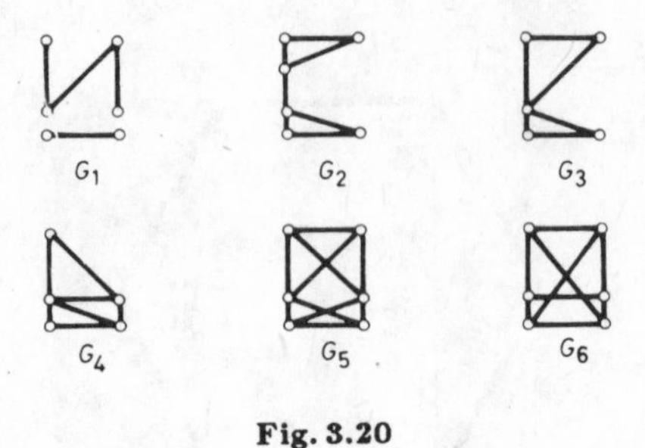

Fig. 3.20

Lemma 3.3.1 *If e is equivalent to f and f is equivalent to g then e is equivalent to g.*

Proof. Since the relation is symmetric, the statement is trivial except if $e \neq f \neq$ $\neq g \neq e$. In this case let C_{ef} be a circuit containing e and f, C_{fg} be a circuit containing f and g. Let $g = \{v_1, v_2\}$. Let us proceed from v_1 to f on C_{fg}. The resulting path P_1 might have several common points with C_{ef}, let x_1 be the first one along our walk. Similarly, let x_2 be the first point along the path P_2 between v_2 and f on C_{fg}, which belongs to $V(C_{ef})$ as well (Fig. 3.21). Since P_1 and P_2 had no common point, $x_1 \neq x_2$. Finally consider the arc x_1, x_2 of C_{ef}, containing e. This together with the x_1v_1 part of P_1, with the x_2v_2 part of P_2, and with the edge g, forms a circuit as requested.

□

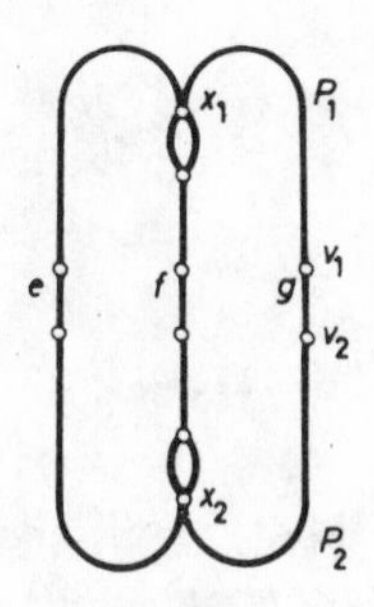

Fig. 3.21

The equivalence classes of this relation are the *blocks* or *2-connected components* of G. Observe that if G has separating edges, they are single element blocks. We shall see later that each subgraph of G, determined by such a block, either consists of two points or is a maximal 2-connected subgraph of G.

Theorem 3.3.2 *Let G have no loops and at least 3 points. G is 2-connected if and only if for every pair $a, b \in V(G)$ there are two paths between a and b with no other common point.*

Two such paths P_1 and P_2 satisfy $V(P_1) \cap V(P_2) = \{a, b\}$. Since the internal points of these paths are different, only the terminal points are not, we shall call such paths *internally disjoint*, for brevity.

Observe that one of these paths may be a single edge only (e.g., let G be a circuit and a, b two adjacent points of it).

Proof. If G is not 2-connected, it has a separating point contained in every path between two suitable points. Hence only the sufficiency of the condition is to be proved. We apply induction on the *distance* d_{ab} of a and b (i.e., the length of the shortest path between them). The statement is trivial if $d_{ab} = 1$ since the edge $\{a, b\}$ could not be separating.

Let P_0 be the shortest path between a and b and let c be the point in $V(P_0)$, adjacent to b. There are two internally disjoint paths P_1 and P_2 between a and c since $d_{ac} < d_{ab}$. By the 2-connectivity of G the subgraph $G - c$ is connected hence there exists a path P_3 between a and b avoiding c (see the dotted line in Fig. 3.22). By walking on P_3 from b towards a let d denote the first point belonging to $V(P_1) \cup V(P_2)$. The case $d = a$ is also possible. Suppose $d \in V(P_1)$ by symmetry. Then take the ad part of P_1 together with the db part of P_3 on the one hand and P_2 and the edge $\{c, b\}$ on the other hand, as the requested paths between a and b.

□

The following alternative of this theorem is frequently useful:

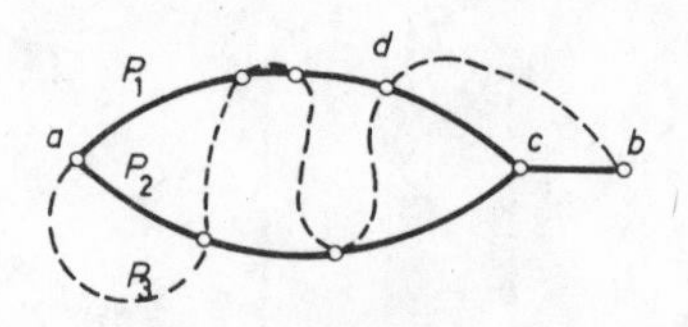

Fig. 3.22

Theorem 3.3.3 *A graph G without loops and with at least 3 points is 2-connected if and only if for every pair $a, b \in V(G)$ there is a circuit of G containing both a and b.*

The following "stronger" statement is not simply a reformulation:

Theorem 3.3.4 *A graph G without isolated points and with at least 3 points is 2-connected if and only if for every pair $e, f \in E(G)$ there is a circuit of G containing both e and f.*

Proof. The sufficiency is obvious: The condition of Theorem 3.3.3 is implied (take two arbitrary edges e, f which are incident to the two given points a, b, respectively).

For the necessity let $e, f \in E(G)$. Apply series extension to both of these edges. The resulting points of degree 2 are denoted by x, y, respectively. These extensions do not change 2-connectedness, hence Theorem 3.3.3 can be applied with the choice $a = x, b = y$. The resulting circuit will obviously be appropriate. □

Several equivalent characterizations of 2-connectivity are summarized in **Box 3.2.** Only a part of these statements can be generalized to k-connected graphs. We shall see in Section 5.2 that the analogues of Theorems 3.3.2 and 3.3.3 are true (Theorem 5.2.6 and Problem 5.2.18, respectively) while that of Theorem 3.3.4 is not (see Exercise 3.3.6).

Exercises for Section 3.3

3.3.1 Verify the statements in **Box 3.2.**

3.3.2 Prove that a graph without loops and with at least 3 points is 2-connected if and only if any two edges of the graph are contained in a cut set.

3.3.3 Let P be an ab path in the 2-connected graph G. Does there exist another ab path, internally disjoint to P?

3.3.4 Let $B_1, B_2, \ldots, B_k$ be the blocks of the connected graph G. Let $P_1, P_2, \ldots, P_\ell$ denote the separating points of G. Prepare a graph $H(G)$ with $k+\ell$ points so that a point b_i, corresponding to B_i is adjacent to a point p_j, corresponding to P_j, if and only if $P_j \in V(B_i)$. (No two

Box 3.2. Equivalent characterizations of 2-connectivity

Let G be loopless and contain at least 3 points. G is 2-connected if and only if

(a) for every $v_1, v_2 \in V(G)$ there is a circuit of G containing v_1 and v_2;

(b) $E(G) \neq \emptyset$ and for every $v \in V(G)$ and $e \in E(G)$ there is a circuit of G containing v and e;

(c) G has no isolated points and for every $e_1, e_2 \in E(G)$ there is a circuit containing e_1 and e_2;

(d) for every $v_1, v_2, v_3 \in V(G)$ there is a path from v_1 to v_2 containing v_3;

(e) for every $v_1, v_2, v_3 \in V(G)$ there is a path from v_1 to v_2 not containing v_3;

(f) $E(G) \neq \emptyset$ and for every $v_1, v_2 \in V(G)$ and $e \in E(G)$ there is a path from v_1 to v_2 containing e.

points of form b_i or no two points of form p_j are adjacent.) Prove that $H(G)$ is a tree.

3.3.5 Let e be an edge in a 2-connected graph G with at least 4 points. Prove that at least one of $G - e$ and G/e is 2-connected.

3.3.6 Give a 3-connected graph with 3 edges, not contained in one and the same circuit.

3.3.7 Show that for every $e \in E(G)$ in a k-connected graph G, the graph $G - e$ is $(k-1)$-connected.

3.3.8 G is called *k-edge-connected* if $G - X$ is connected for every $X \subset E(G)$, $|X| < k$. Show that such a graph need not be k-connected.

3.3.9 Prove that G is 2-edge-connected if and only if G is connected with no separating edges.

3.3.10 Delete an edge and all the others, series to it, from a 2-edge-connected graph. Prove that the components of the resulting graph will be 2-edge-connected.

3.3.11 Prove that an orientation can be assigned to G so that the resulting directed graph be strongly connected if and only if G is 2-edge-connected.

3.3.12 Prove that two graphs are 2-isomorphic if and only if their blocks are pairwise 2-isomorphic.

3.3.13 Show that Kuratowski's theorem should be proved for 2-connected graphs only.

Problems for Section 3.3

3.3.14 Show that Kuratowski's theorem should be proved for 3-connected graphs only.

3.3.15 Let G be a *minimal nonplanar graph* (i.e., G is nonplanar and $G - e$ is planar for every $e \in E(G)$). Suppose that every point of G has degree at least 3. Prove that
(a) G is 3-connected;
(b) G has a circuit with a chord also in G (an edge $e = \{x, y\}$ is a *chord* of the circuit C if $e \notin E(C)$ yet $x, y \in V(C)$ are met);
(c) G is one of the two Kuratowski graphs.

3.3.16 Prove Kuratowski's theorem.

3.3.17 Let $e = \{x, y\}$ and $f = \{x, z\}$ be adjacent edges in a graph G. Suppose G and G' are 2-isomorphic but the corresponding edges e' and f' are nonadjacent in G'. Show that $G - x$ has a separating point which is contained in every yz path of $G - x$.

3.3.18 Let G and G' be 2-isomorphic and G be 3-connected. Show that G and G' are isomorphic.

3.3.19 Let G and G' be 2-isomorphic and G planar. Show that so is G'.

3.3.20 We have seen in Exercise 3.1.2 a "twisting" operation which preserved 2-isomorphism. Using this operation for a 2-connected graph G several times, every graph H can be obtained if G and H were 2-isomorphic, by a deep theorem of [Whitney, 1933b]. Using this result, give a new solution to the previous problem.

Section 3.4 Graph theory algorithms II: DFS–BFS techniques

We have studied several data structures in Section 1.4 to store our graphs efficiently. Now we present some algorithms which can then really quickly be performed. The total number of steps will be proportional to max (e, v) in most of these algorithms.

In order to check the connectivity of a graph, we used Algorithm 1.2 or Algorithm 1.3. Both of them visited every point in the graph (or in a connected component if the graph was disconnected) but in different order. The "strategy" of Algorithm 1.2 was to collect all neighbours of the initial point at first, then all the neighbours of these points etc. On the other hand, Algorithm 1.3 had a different "strategy", it tried to find a long path from the initial point, stepped back only if it had no other choice but even then tried to build a long path from the new point etc. For example, in the case of the graph in Fig. 3.23, by the first strategy we visit the points in the order $1, 2, 3, 4, 5, 6, 7$ while by the second in $1, 2, 6, 3, 7, 4, 5$.

Recall that these algorithms were modified (Exercise 1.2.9) to construct a spanning forest of the input graph. The above difference between Algorithms 1.2 and 1.3 can much better be visualized if we draw the resulting spanning forests. In the case of Algorithm 1.2 the forests are "broad", while in the case of Algorithm 1.3 they are "deep", see Fig. 3.24 for the graph of Fig. 3.23. Accordingly, the strategy of Algorithm 1.2 is called *breadth-first-search* (BFS) and that of Algorithm 1.3 is called *depth-first-search* (DFS).

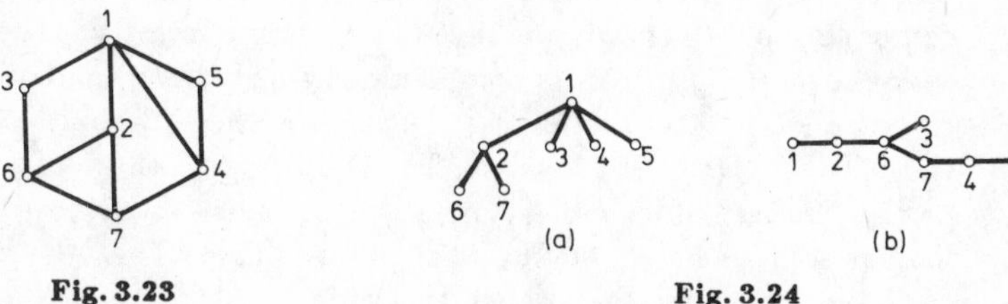

Fig. 3.23 **Fig. 3.24**

Both BFS and DFS visit the points of the graph very quickly; the total number of operations is proportional to max (e, v). In this way, we could give two quick algorithms both for the connectivity check (Algorithms 1.2 and 1.3) and for finding a spanning forest (Algorithm 1.8 and 1.9). In many other problems, only DFS or only BFS leads to quick algorithms.

Consider for example, the solution of Problem 1.2.16 again. Recall that if an edge $\{p, q\}$ was joined to the spanning forest (and we moved from p to q) then p was called "father" of q and a_q was defined to be p. Let us, furthermore, count the points of the graph G, i.e. let us define a function f so that $f(v_1) = 1$ (for the initial point), $f(p) = 2$ for the first "son" of v_1, $f(q) = 3$ if q was marked at the third instance of Step 2 of Algorithm 1.3 etc.

If, for a moment, we imagine directed edges (p, q) if q is the "son" of p then those points r can be reached along directed paths from p which are the "descendants" of p. If r is a descendant of p then obviously $f(p) < f(r)$.

Lemma 3.4.1 *For every edge $\{u, v\}$ of G, whether it is in the DFS spanning forest or not, either u or v is a descendant of the other.*

Proof. Let, say, $f(u) < f(v)$. Then v is a descendant of u since otherwise we should move back from u to its father before reaching v in some other branch. But such a backtrack is impossible if u still has an unmarked neighbour v.

□

Now we are in a position to construct the fundamental system of circuits.

Algorithm 3.1 *Fundamental system of circuits by DFS.*

Input A graph G with at least one point.

Output A spanning forest F and a fundamental system $C_e, C_f, C_g, \ldots$ of circuits, where $\{e, f, g, \ldots\} = E(G) - E(F)$.

Step 1 Same as Step 1 of Algorithm 1.3. In order to prepare the function $f(p)$ as well, introduce a "counter" c. Initially let $c = 0$.

Step 2 Let $c = c + 1$ (i.e., increase the value of the counter by 1). Mark x and put $f(x) = c$.

Step 3 If no unmarked point is adjacent to x, go to **Step 5**.

Step 4 (Forward step) Choose an unmarked neighbour y of x. Put $a_y = x$, record the edge $\{x, y\}$, change the value of x to y and go to **Step 2**.

Step 5 (Backtrack) If $x = v_1$, go to **Step 6**. Otherwise read $a_x = z$, change the value of x to z and go to **Step 3**.

Step 6 If every point is marked, the recorded edges form a spanning tree. Otherwise this part of the algorithm should be repeated for every component of G. Finally we have the spanning forest F. Then for every edge $e, f, g, \ldots$ of its complement we obtain a fundamental circuit $C_e, C_f, C_g, \ldots$ as follows: Let u, v be the end points of e so that $f(u) < f(v)$. Let $w_0 = v$; let w_1 be the "father" of w_0; w_2 be the "father" of w_1 etc. Prepare this sequence $w_0, w_1, w_2, \ldots$ as long as some member, say w_k of it just equals u. Then the points of C_e are $w_0, w_1, w_2, \ldots, w_k$ (in this order).

Using this function $f(p)$ a large number of other problems can effectively be solved by DFS, e.g. assigning orientations to the edges of a graph to make it strongly connected, or deciding whether a given directed graph has directed circuits (Problems 3.4.8 and 3.4.6, respectively). Many other problems (like recognizing separating points, separating edges, blocks, even planarity check and a drawing if the graph is planar) can be performed in time proportional to max (e, v), see [Hopcroft and Tarjan, 1973, 1974], [Booth and Leuker, 1976], [Reingold *et al.*, 1977], [Aho *et al.*, 1974] and [Even, 1979]. Such problems are listed in **Box 3.3.**

Box 3.3. Quick DFS–BFS algorithms

Depth-First-Search	Breadth-First-Search
Checking connectivity (Algorithm 1.3) Spanning forest construction (Algorithm 1.9) Fundamental system of circuits (Algorithm 3.1)	Checking connectivity (Algorithm 1.2) Spanning forest construction (Algorithm 1.8)
Checking 2-connectivity, 2-edge-connectivity Checking planarity Finding strongly connected orientation (Algorithm 3.6) Deciding whether a directed graph —has directed circuits (Algorithm 3.5) —is strongly connected	Determining distance between points and finding shortest paths (Algorithm 3.2) Finding shortest directed paths in directed graphs (Algorithm 3.4) Various problems related to network flows (see Section 5.2)

We close this section with an algorithm where special features of BFS are applied. Let us define the *distance* of two points in a connected graph as the length of the minimum paths between them. If $v_0 \in V(G)$ is a given point then its neighbours are of distance 1 from v_0; a point which is nonadjacent to v_0 but adjacent to a point of distance 1 from v_0 is of distance 2 from v_0 etc. Hence BFS starting from v_0 can determine the distance of every point from v_0.

More generally let us assign nonnegative weights $w(u, v)$ to the edges $\{u, v\}$ and try to find the shortest path from v_0 to every other point v_k, i.e. the path to v_k, along which the sum of the edge weights is smallest possible. This sum is the *distance* of the points v_0, v_k. If every edge has unity weight, this reduces to the previous problem.

Suppose the points $v_0, v_1, v_2, \ldots, v_{k-1}, v_k$ determine (in this order) a shortest path between v_0 and v_k. Then the initial segment $v_0, v_1, v_2, \ldots, v_{k-1}$ of this path must also be the shortest between v_0 and v_{k-1} (since if another path were shorter, that could be extended by the edge $\{v_{k-1}, v_k\}$ to a shorter path between v_0 and v_k).

Let $V(G) = V_1 \cup V_2$ and $v_0 \in V_1$. Suppose we already know the shortest paths from v_0 to every other point $x, y, \ldots$ of V_1 and denote their lengths by $d_x, d_y, \ldots$, respectively.

Imagine a list of the shortest paths between v_0 and the points $u, v, \ldots$ of V_2 as well, with respective lengths $d_u, d_v, \ldots$. Let d_u be the minimum. Then the shortest path $v_0, v_a, v_b, \ldots, v_t, v_u$ must have every point (except u) from V_1 and, by the above observation, d_u must equal $d_t + w(t, u)$.

Hence if we compare the weights $w(p, q)$ for every adjacent pair $p \in V_1$ and $q \in V_2$ and find that $w(p_0, q_0)$ is the minimum then $d_{q_0} = d_{p_0} + w(p_0, q_0)$ and we can place q_0 from V_2 to V_1 (and continue the process).

This is already a BFS algorithm but not very effective (see Exercise 3.4.4). Comparing all the $w(p, q)$ weights takes too much time if done in every step. If $q \in V_2$ then the new minimum $w(p_0, q)$ can be better than before only if p_0 moved to V_1 just in the previous step. Hence the following quicker algorithm is obtained [Dijkstra, 1959].

Algorithm 3.2 *Shortest distances from a given point in an edge-weighted graph.*

Input A graph G with a given point v_0 and a nonnegative weight function w on the set of edges.

Output The distance d_k of every point $v_k \in V(G)$ from v_0.

Step 1 We prepare two files V_1 and V_2 as above, plus we store the "head" z of the BFS and the distance D of the head from v_0. Initially let $d_0 = 0, d_k = \infty$ for every other point of G, furthermore let $V_1 = \{v_0\}$; $V_2 = V(G) - V_1$; $z = v_0$ and $D = 0$.

Step 2 Check all those neighbours of z which are in V_2. If no such point exists, go to **End**. If v_k is such a point, calculate $D + w(z, v_k)$, and if the result is smaller than d_k then put this sum instead of d_k.

Step 3 Among those neighbours of z which are in V_2, choose the point v_i with d_i minimum. This will be the new "head", so put $z = v_i$, $D = d_i$; and replace v_i from V_2 to V_1. Go to **Step 2**.

End The algorithm terminates. If $V_2 \neq \emptyset$, G was disconnected and V_1 contains only a component. The values d_k are the distances of v_k from v_0 (and $d_k = \infty$ if $v_k \notin V_1$).

Of course, a large number (e.g., the sum of all the edge weights plus one) can be used instead of ∞. Furthermore, one might record the edges $\{z, v_k\}$ in Step 2 (i.e. the "father" of v_k) in order to reconstruct the shortest paths themselves, as well, not only their values.

This algorithm is explained in Fig. 3.25. It shows the whole graph in every step; solid circles denote the elements of V_1 and the actual value of D is shown separately. The heavy edges show the "father–son relations" and together always give a spanning tree. This tree has the property that the (unique) path from v_0 to any given point v in the tree is just a shortest path from v_0 to v.

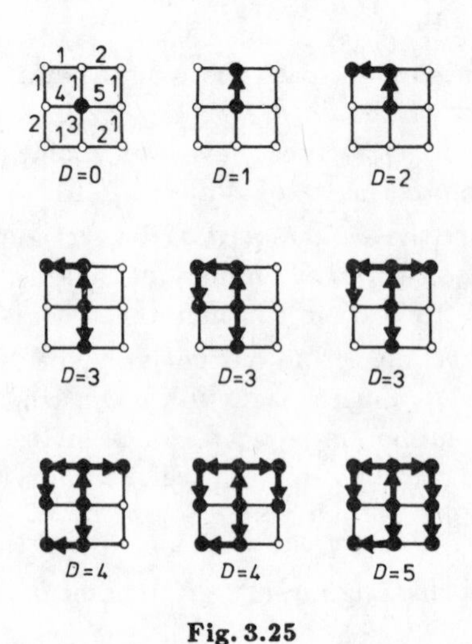

Fig. 3.25

Exercises for Section 3.4

3.4.1 Choose an arbitrary spanning tree of K_4. Show that its points can be numbered so that this tree be just the result of either the DFS or the BFS tree search. Is this also true for K_5?

3.4.2 Define the depth-first-search for directed graphs (forward move is permitted in accordance with the direction of the edge only). Characterize those edges (x, y) which are not used in the DFS. Is $f(x) < f(y)$ or $f(x) > f(y)$ possible?

3.4.3 Prove that a directed graph G has no directed circuit if and only if one can assign numbers $t(p)$ to every point p of G so that if (x, y) is a directed edge then $t(x) < t(y)$.

3.4.4* Give an upper bound to the number of operations required for Algorithm 3.2.

3.4.5 Modify Algorithm 3.2 to obtain shortest directed paths from a given point in an edge-weighted directed graph.

Problems for Section 3.4

3.4.6 Applying Exercises 3.4.2–3 give an algorithm to decide whether a given directed graph has at least one directed circuit.

3.4.7 Prove that if a directed graph G has directed circuits then, for any DFS spanning forest F of G, there is at least one so that all but one of its edges belong to F.

3.4.8 Give an algorithm to assign orientation to the edges of a 2-edge-connected graph so that the result is strongly connected.

Chapter 4

Applications

Section 4.1 Duality in electric network theory I: The classical results

Certain concepts in graph theory were seen to arise "in pairs", see **Box 3.1.** This phenomenon can be observed in electric network theory as well.

Recall that loops correspond to separating edges and parallel edges correspond to series edges (**Box 3.1**). A device "corresponding" to a loop has obviously zero voltage while that, "corresponding" to a separating edge, has zero current. Similarly, if devices were in parallel, their voltages are identical and their currents should be added and *vice versa* for series devices.

This leads to the straightforward idea that voltages and currents should correspond to each other in this symmetry. Then voltage and current sources, capacitors and inductors also correspond to each other.

What does correspond to a resistor? Interchange the role of voltage and current in Ohm's Law. The result, formally, is $i = Ru$; this also defines a resistor whose resistance is R^{-1}. Hence a resistor with resistance R corresponds to another resistor with admittance R.

Statement 4.1.1 *Suppose that a network N has RLC elements, voltage and current sources, and that its network graph G is planar. Let G^* be a dual of G. Prepare a new network N^* so that its network graph be G^* and its elements are obtained from those of N by replacing voltage sources by current sources (of the same numerical values) and vice versa, capacitors by inductors (of the same numerical values) and vice versa, while resistors by other resistors whose resistances are the reciprocal values of the original ones. Then N^* is uniquely solvable if and only if N is, and then the current of any device of N^* is numerically the same as the voltage of the corresponding device of N and vice versa.*

For example, the network graphs of the two networks of Fig. 4.1 are duals of each other and the correspondence among the element values is as follows: $I_1' = U_1$; $R_2' = R_2^{-1}$; $C_3' = L_3$; $L_4' = C_4$; $U_5' = I_5$. If we choose $\begin{pmatrix} i_3 \\ i_4 \end{pmatrix}$ and $\begin{pmatrix} u_3' \\ i_4' \end{pmatrix}$ as the state vectors of the two networks, respectively, then the state equations will be

$$\frac{d}{dt}\begin{pmatrix} i_3 \\ u_4 \end{pmatrix} = \begin{pmatrix} -R_2/L_3 & 0 \\ 0 & 0 \end{pmatrix}\begin{pmatrix} i_3 \\ u_4 \end{pmatrix} + \begin{pmatrix} 1/L_3 & -R_2/L_3 \\ 0 & 1/C_4 \end{pmatrix}\begin{pmatrix} u_1 \\ i_5 \end{pmatrix}$$

$$\frac{d}{dt}\begin{pmatrix} u_3' \\ i_4' \end{pmatrix} = \begin{pmatrix} -1/R_2'C_3' & 0 \\ 0 & 0 \end{pmatrix}\begin{pmatrix} u_3' \\ i_4' \end{pmatrix} + \begin{pmatrix} 1/C_3' & -1/R_2'C_3' \\ 0 & 1/L_4 \end{pmatrix}\begin{pmatrix} i_1' \\ u_5' \end{pmatrix}$$

and the outputs are described by

$$\begin{pmatrix} i_1 \\ u_5 \end{pmatrix} = \begin{pmatrix} -1 & 0 \\ -R_2 & -1 \end{pmatrix}\begin{pmatrix} i_3 \\ u_4 \end{pmatrix} + \begin{pmatrix} 0 & -1 \\ 1 & -R_2 \end{pmatrix}\begin{pmatrix} u_1 \\ i_5 \end{pmatrix}$$

$$\begin{pmatrix} u_1' \\ i_5' \end{pmatrix} = \begin{pmatrix} -1 & 0 \\ -1/R_2' & -1 \end{pmatrix}\begin{pmatrix} u_3' \\ i_4' \end{pmatrix} + \begin{pmatrix} 0 & -1 \\ 1 & -1/R_2' \end{pmatrix}\begin{pmatrix} i_1' \\ u_5' \end{pmatrix}.$$

The *proof* of the statement is clear from this example; the numerical equations are simply the same for N and N^*. The voltage equations of N were obtained from $\mathbf{C}(G)$ which is identical to $\mathbf{Q}(G^*)$ hence leading to current equations of N^* (and *vice versa*). Only the signs must be still checked e.g. the reference directions for U_1 and I_5 were identical but for I_1' and U_5' were opposite in Fig. 4.1. This is clarified as Problem 4.1.12.

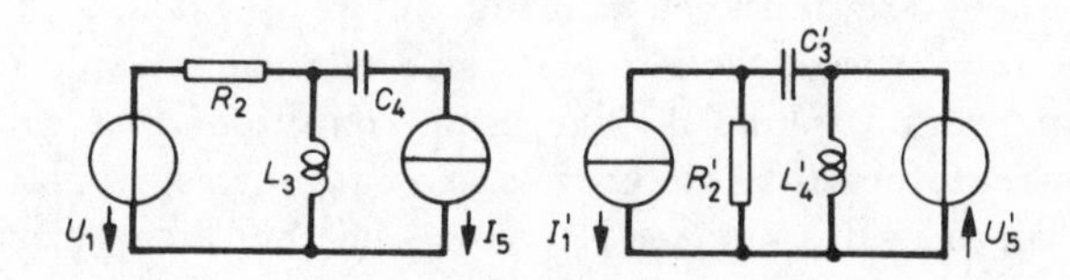

Fig. 4.1

The network N^* of Statement 4.1.1 is usually called the *dual* of N by most electric network theorists. In this book we shall call it the *inverse* of N (in order to reserve the notion "dual" for another network, related to duality of matroid theory, see Section 10.1).

This symmetry is summarized as **Box 4.1.** In the rest of the section the last two concepts of this box will be studied; all the others are already well known to the reader.

As shown in Chapter 2, the fundamental systems of circuits and cut sets with respect to a spanning forest give just the right number of linearly independent voltage and current equations for network analysis. However, in practical applications, current equations are obtained not from such cut sets. Since the network graph of every practical network is 2-connected, simply the "stars" (i.e., the set of edges, incident to a single point) are cut sets. If the stars of all but one point are considered, they serve as a maximal system of linearly independent cut sets.

Occasionally even the fundamental system of circuits can be avoided, if small networks (with planar network graphs) are analyzed. Simply consider a planar drawing of the network graph, and take those circuits which bound the regions. The reader should verify (Exercise 4.1.6) that this method always works.

Box 4.1. Correspondence among network concepts, related to the classical duality principle of electric networks

Voltage	Current
Resistance (impedance)	Admittance
Capacitance	Inductance
Short circuit	Open circuit
Parallel connection	Series connection
Voltage source	Current source
Capacitor	Inductor
Resistor	Resistor
Kirchhoff's Voltage Law	Kirchhoff's Current Law
Method of the "loop currents"	Method of the "node potentials"

These methods are described and illustrated by examples in **Box 4.2.** (It should be emphasized that the left hand side method is applicable only if the graph of the network is planar.) The usual conditions (the subgraphs of the voltage and current sources be circuit free and cut set free, respectively) are only implicitly contained in **Box 4.2** (see Exercise 4.1.5) but are obviously necessary.

Exercises for Section 4.1

4.1.1 What is the inverse of the "devices" in Figs 2.6–2.7?

4.1.2 What is the "inverse" of Problem 2.1.10?

4.1.3 Formulate the inverse of Millman's theorem (Problem 2.1.13).

4.1.4 Formulate the inverse of Foster's theorem (Problem 2.5.8).

4.1.5 What happens during the methods, described in **Box 4.2,** if the set of voltage sources contains a circuit or the set of current sources contains a cut set?

4.1.6 Let G be a planar graph and let $C_1, C_2, \ldots, C_k$ denote the circuits, determined by the bounded regions of a planar representation of G. Prove that they form a maximum set of linearly independent circuits over $\mathbb{B}$.

4.1.7 (continued) Show that these circuits determine a totally unimodular submatrix of the circuit matrix $\mathbf{C}(G)$.

Problems for Section 4.1

4.1.8 Several previous problems considered interconnected network elements as a single network element (keeping contact with the rest of the world over the specified two points only). For example, Problem 2.3.6 could

Box 4.2. The method of the "loop–currents" and the "node–potentials"

The method of the "loop–currents"

Step 1 Consider a planar representation of the network graph G and assign fictitious currents to the circuits determined by the bounded regions of the plane. For example, the "loop–currents" j_1, j_2 and j_3 in the figure belong to the circuits $\{1,2,3\}$, $\{3,4,5\}$ and $\{5,6\}$, respectively.

The method of the "node–potentials"

Step 1 Choose one of the points of the network graph G as a "reference" and consider the voltages, measured between this point and the others, as unknown quantities. For example, if p_0 is the reference in the figure, then the voltage v_1 is measured between p_1 and p_0, the voltage v_2 between p_2 and p_0 etc.

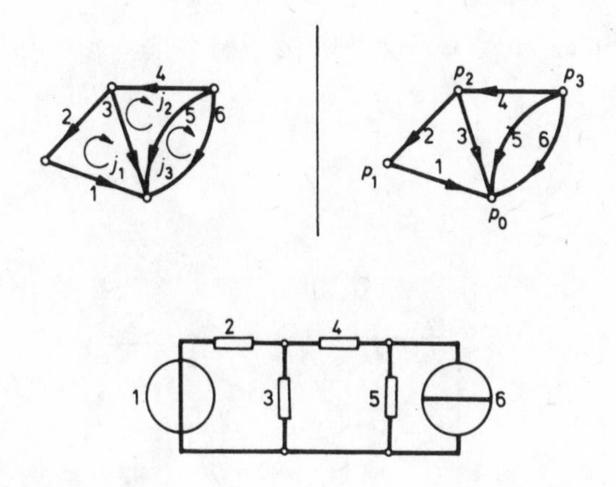

The method of the "loop–currents"

Step 2 If the network contains current sources, they reduce the number of unknown quantities. In our example j_3 is not unknown since $j_3 = i_6$.

Step 3 Apply Ohm's Law to express every voltage as functions of the "loop currents", like $u_3 = = R_3(j_1 - j_2)$; $u_5 = R_5(j_2 - j_3)$ etc. in our example.

Step 4 Apply Kirchhoff's Voltage Law to the obtained voltages, for example $R_3(j_1 - j_2) - R_4 j_2 = = R_5(j_2 - j_3)$.

Step 5 Solve the resulting system of $n(G) - e_I$ equations.

The method of the "node–potentials"

Step 2 If the network contains voltage sources, they reduce the number of unknown quantities. In our example v_1 is not unknown since $v_1 = u_1$.

Step 3 Apply Ohm's Law to express every current as functions of the "node–potentials", like $i_3 = = R_3^{-1} v_2$; $i_4 = R_4^{-1}(v_3 - v_2)$ etc. in our example.

Step 4 Apply Kirchhoff's Current Law to the obtained currents, for example $R_4^{-1}(v_3 - v_2) = = R_2^{-1} v_1 + R_3^{-1} v_2$.

Step 5 Solve the resulting system of $r(G) - e_U$ equations.

equivalently be formulated as follows: Prove that the "network element" in the broken-line frame of Fig. 4.2 is just like a resistor (if the value of all the four interconnected elements was unity). Define the inverse of such "network elements" as well. Is Statement 4.1.1 still valid?

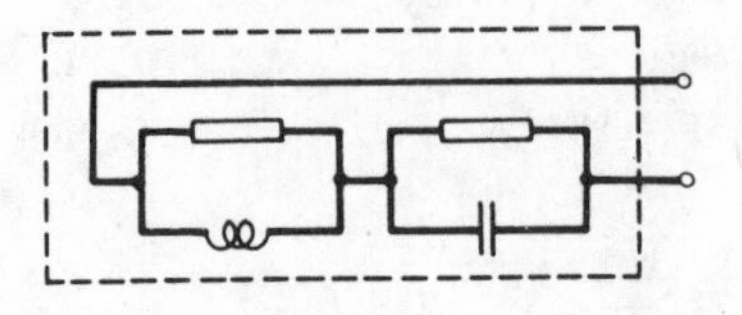

Fig. 4.2

4.1.9 Determine in the easiest possible way the inverse of the network shown in Fig. 4.3.

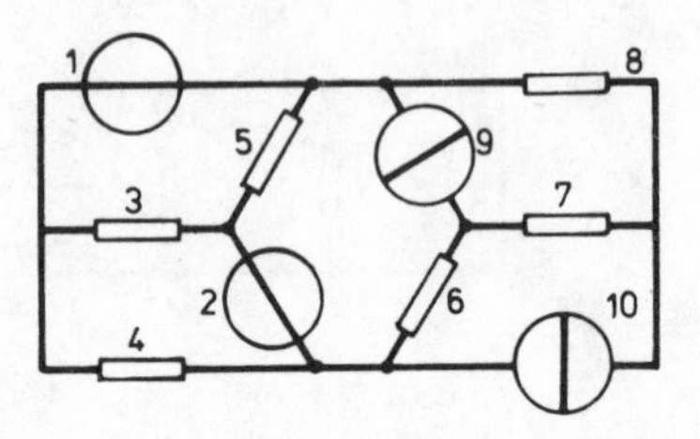

Fig. 4.3

4.1.10 The two "network elements" of Fig. 4.4 are called *filters*. If the voltage sources input different signals then the first network has a large "resistance" against the low-frequency signals and small one against the high-frequency signals while the second network has *vice versa*. (Hence they are called highpass and lowpass filters, respectively.) One can immediately see, for example, that the second filter has no resistance in case of direct current. Determine the inverse of these filters.

4.1.11 (continued) Similarly to Problem 4.1.8, define the inverse of those "network elements" as well which keep contact with the rest of the world over two pairs of points only.

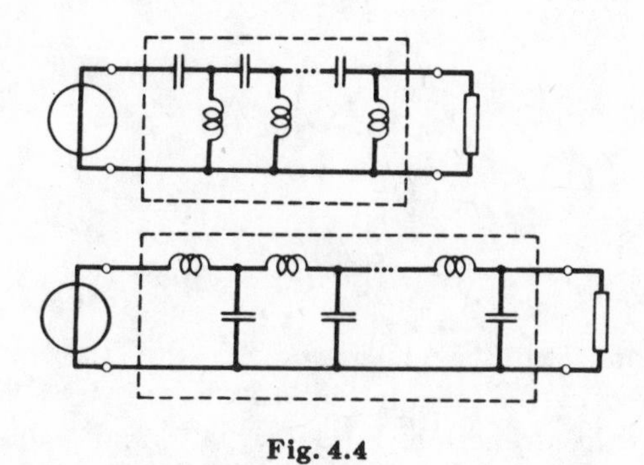

Fig. 4.4

4.1.12 Finish the proof of Statement 4.1.1 by introducing the dual of planar *directed* graphs.

4.1.13 Apply the result of Problem 3.1.13 to give a simpler solution to Problem 2.4.3.

Section 4.2 Maxwell–Cremona diagrams and the "reciprocity" of planar frameworks

All the other engineering applications in this book are on qualitative problems of linear systems (solvability of electric networks, rigidity of frameworks, etc); the one we present in this section is mostly quantitative. The reason why it is included is that these questions are historically related to combinatorics and are believed to be interesting for mathematicians as well.

For simplicity, only planar frameworks are considered, the rods are completely rigid, the joints are point-like and completely flexible, and all sorts of external forces attack on the joints only (i.e., not on the "middle" of a rod).

Let us recall that a system is in equilibrium if and only if the vectorial sum of its attacking forces is zero. Since the joints are completely flexible, this condition must hold at every joint: The vectorial sum of the forces in the rods, incident to this joint and the outside forces (if any), attacking at this joint, must be zero. These constraints (somewhat similar to Kirchhoff's Current Laws) enables us to calculate all the forces in the rods.

Consider, for example, the system of Fig. 4.5. It is fixed to the plane at points C, D and only a force $\mathbf{F}$ attacks at point B. The only way to avoid a motion of B is if there is a force in the rod AB (pulling joint B with a force, identical in strength to $\mathbf{F}$, but in opposite direction). The rod BC must not have any effect whatsoever.

The tension of strength $\mathbf{F}$ in rod AB should be balanced by the rods AD and AC since joint A is also in equilibrium. Being horizontal, it must be balanced by

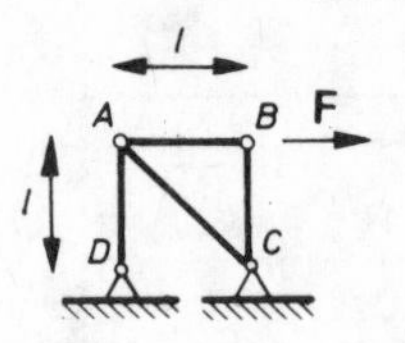

Fig. 4.5

the horizontal component of the force in rod AC while the vertical component of the latter is balanced by the force in rod AD. Hence AC is under compression by a force $\sqrt{2}$ times the strength of $\mathbf{F}$ (see Fig. 4.6) and AD is under tension by a force whose strength is identical to that of $\mathbf{F}$. Finally Fig. 4.7 also shows the effect of $\mathbf{F}$ to the "earth" to which the system is fixed.

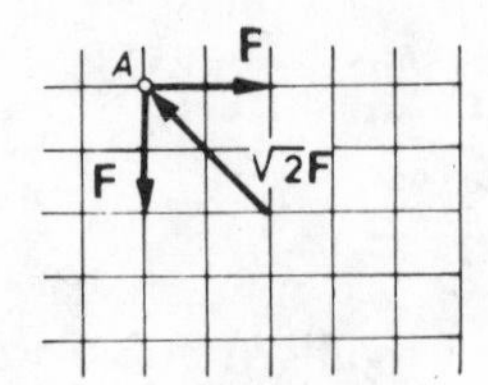

Fig. 4.6

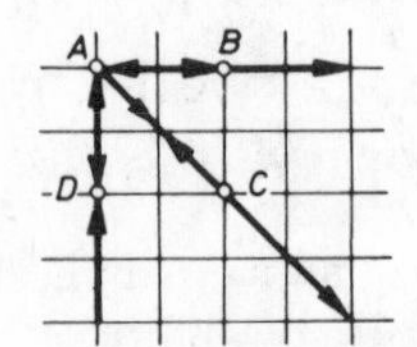

Fig. 4.7

Before proceeding further, an important remark is in order. We already saw in Section 2.6 that by removal of rod AC from the framework of Fig. 4.5 the result was not rigid. Should we, instead, add a new rod BD as well, the rigidity were "even more assured" but the stresses in the rods could not be uniquely determined any more. In the first case (Fig. 4.5 minus rod AC) we speak about a *dynamically underdetermined* system, in the second case (Fig. 4.5 plus rod BD) we have a *statically underdetermined* (and *dynamically overdetermined*) system. Only systems which are neither statically nor dynamically underdetermined, will be considered in the rest of this section.

Even seemingly more complicated systems can conveniently be analyzed in this way, see Exercises 4.2.1–3. However, if we are not interested in the stress in every rod, only in some of them, then another method can be suggested (whose relation to the above one is quite similar to the relation of Kirchhoff's Current Law to that in the original form, referring to "stars" of points only).

Consider, for example, the system of Fig. 4.8, fixed at the joints D and H with a vertical force $\mathbf{F}$ attacking at joint A. If we are interested in the stress of rod CD only, we simply consider the "cut set" determined by the rods CD, CG and FG. The vectorial sum of the forces in these three rods must balance $\mathbf{F}$, i.e. the sum of the four forces must be zero. Now the vectorial sum of the forces in FG and CG (whose line obviously passes through joint G) must balance the sum of $\mathbf{F}$ and the force in CD (whose line obviously passes through joint A). This is possible only if both sums of forces has just the line $\overline{AG}$. Thus Fig. 4.9 immediately tells us that the force in CD has twice the strength of $\mathbf{F}$. (The other two forces can also be seen in Fig. 4.10 which also shows that rods CD and CG are under compression and FG under tension.)

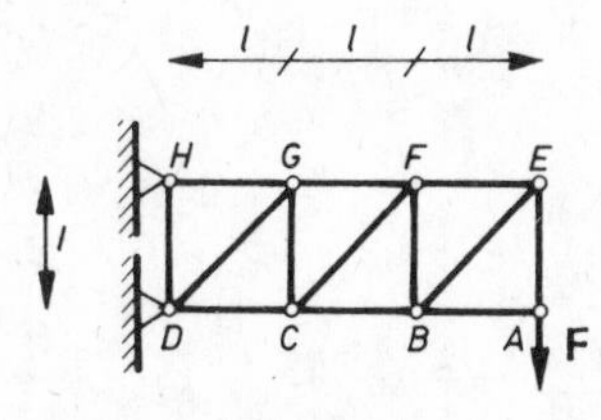

Fig. 4.8

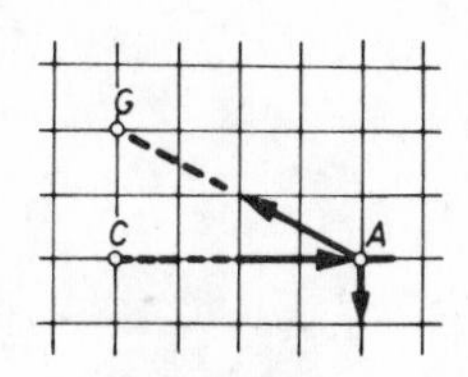

Fig. 4.9

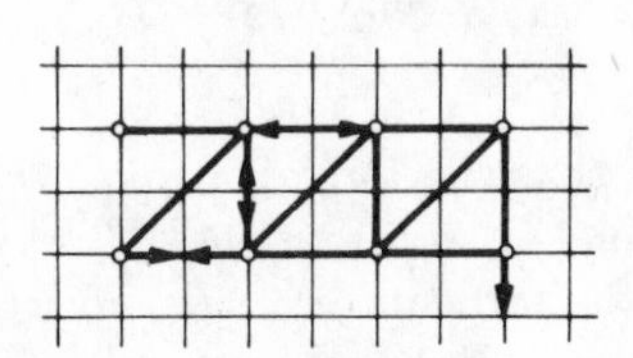

Fig. 4.10

This method (both for graphical and numerical calculations) was already known in the middle of the last century (Culmann and Ritter, respectively). The other method (Cremona and Maxwell) is somewhat more recent (though also more than 100 years old); this will be presented in the rest of this section.

Consider the system of Fig. 4.11 loaded by a vertical force $\mathbf{F}_a$ at joint A, and fixed at joints B, C. (The way how joint B is fixed implies that equilibrium is possible only if the force attacking at B has no component, parallel to the "wall".) Even if we do not know the stresses in the rods, we can easily find the forces $\mathbf{F}_b$ and $\mathbf{F}_c$, attacking at joints B and C to balance $\mathbf{F}_a$ (see Fig. 4.12 where we determined the direction of $\mathbf{F}_c$ first).

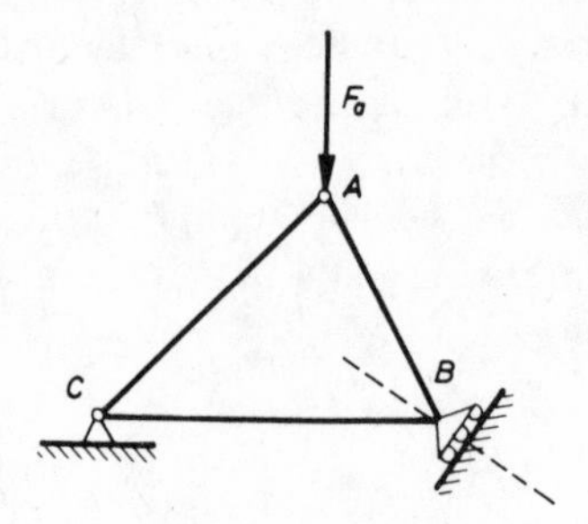

Fig. 4.11

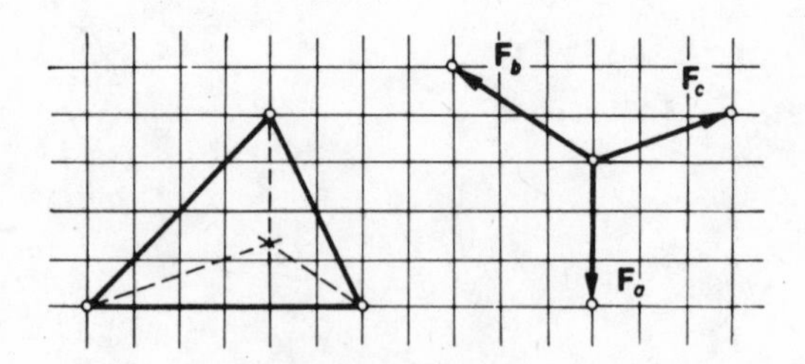

Fig. 4.12

Another possibility is that the forces in rods AB and AC are determined first (to balance $\mathbf{F}_a$ in joint A), see Fig. 4.13. Then $\mathbf{F}_b$ and the force in rod BC are determined from the force in rod AB and from the direction of $\mathbf{F}_b$, as in Fig. 4.14. Finally, we can check whether the obtained forces in rods AC and BC really balance $\mathbf{F}_c$ (Fig. 4.15).

Look at Figs 4.13–15 and observe that the force in every rod is drawn twice. Should we arrange the forces for the joints not in "stars" but in "force-polygons" (see Figs 4.16–18 instead of Figs 4.13–15, respectively), we could draw everything into a single diagram, see Fig. 4.19. This is called the *(Maxwell-) Cremona diagram* of this framework with respect to this loading.

If we put the geometric drawing of the framework beside its Cremona diagram, we shall soon realize that there is a one–one correspondence between the set of rods and external forces on the one hand and the set of "edges" of the diagram on the other hand. Corresponding lines are parallel; length-ratios reflect ratios of real geometric lengths of the rods on the one hand and ratios of strengths of the forces in the corresponding rods on the other hand. All

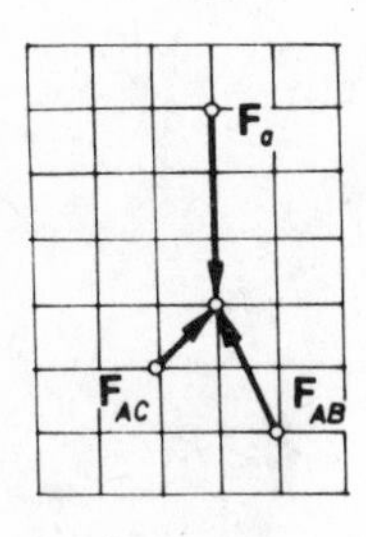

Fig. 4.13

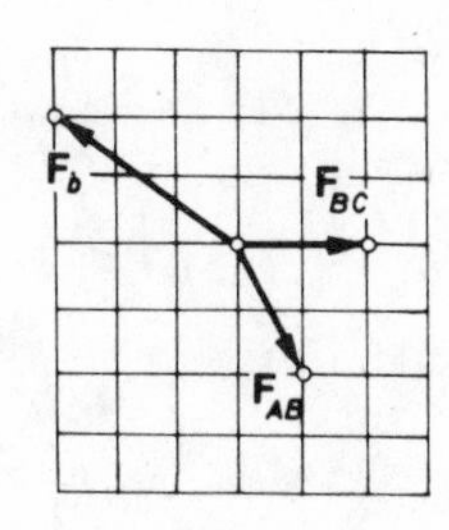

Fig. 4.14

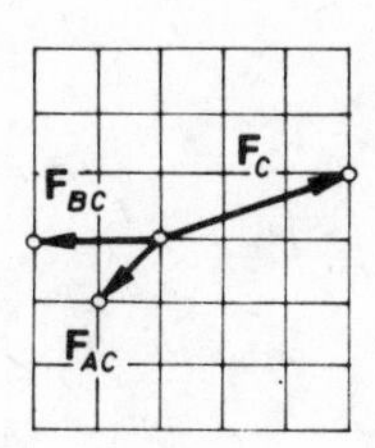

Fig. 4.15

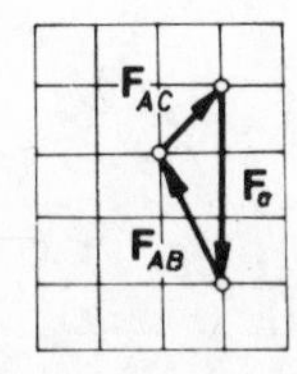

Fig. 4.16

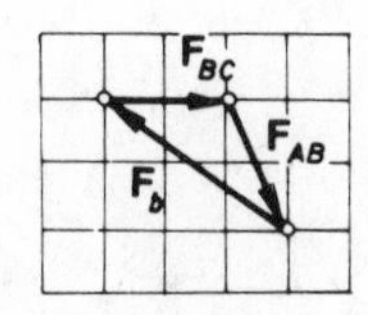

Fig. 4.17

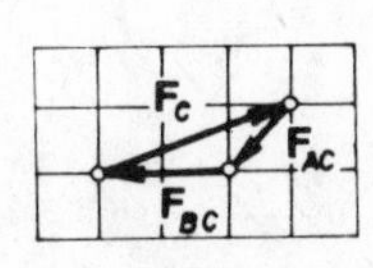

Fig. 4.18

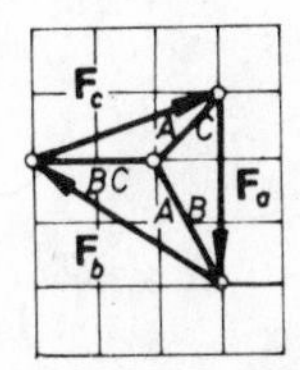

Fig. 4.19

the forces, attacking a joint (and balancing one another), form a "star" on the geometric drawing and a "polygon" on the Cremona diagram.

This last sentence (roughly speaking points correspond to circuits) shows the relation of this theory to the duality of planar graphs. Even the way of constructing the Cremona diagram is similar to the process to construct duals of planar graphs as shown in Fig. 3.6. However, while each region of the planar representation corresponded to a point of the dual graph (including one point for the "outer" region), here we need as many points for the "outer" region as the number of attacking forces.

For example, consider the system in Fig. 4.20. The vertical force $\mathbf{F}$ attacking at joint C is balanced by the forces $\mathbf{F}_a$ and $\mathbf{F}_b$. The drawing cuts the plane into six regions, the bounded ones a, b, c and the unbounded ones d, e, f. (This way of drawing is usually called *Bow's notation.*) Hence the "graph" of the Cremona diagram will have 6 points. For example, edge 5 (joining points a and b in the Cremona diagram) corresponds to the rod 5 (separating the regions a, b in the original drawing). Of course, the two drawings cannot be presented on the same picture (like a planar graph and its dual could in Fig. 3.6) since the corresponding "edges" are now parallel instead of crossing.

This relation between the geometry of the framework and the Cremona

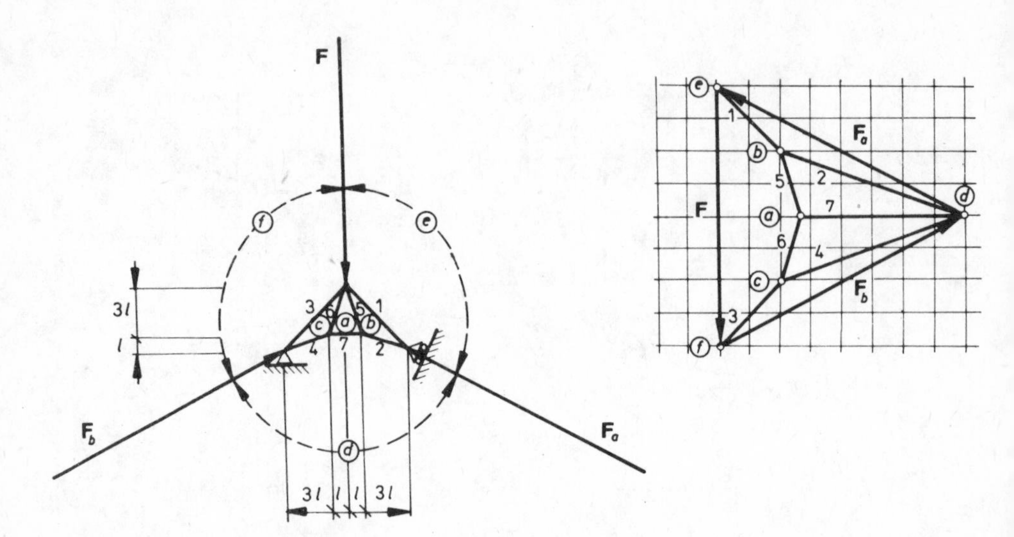

Fig. 4.20

diagram of the forces is sometimes called *reciprocity* in statics. In practical cases, however, some of these beautiful one–one correspondences are not met. For example, if a rod happens to be neither under tension nor under compression for a particular loading, then the corresponding "line" in the Cremona diagram is contracted to a point only, see Exercise 4.2.5 for example. Two different "points" might be in the same position in the Cremona diagram in other cases as well, see Problem 4.2.8. It can also happen that the forces in the rods, incident to a given joint, form a self-intersecting polygon (like in Problem 4.2.8) or that a point of the polygon is inside another edge (like in Exercise 4.2.5).

Let us emphasize again that Cremona diagrams refer not to frameworks but to frameworks with a particular loading. One and the same framework with different loadings can have very different Cremona diagrams (see especially Problem 4.2.9).

Exercises for Section 4.2

4.2.1 Determine the stresses in the rods of both frameworks of Fig. 4.21 if the attacking forces are as indicated in the figure (pulling the joints B and C).

4.2.2 Determine the stresses in the rods of the "tower" of Fig. 4.22 if a single horizontal force attacks it at joint A.

4.2.3 Determine the stresses in the rods of the "crane" of Fig. 4.23 if a single vertical force (a loading at joint A) attacks it.

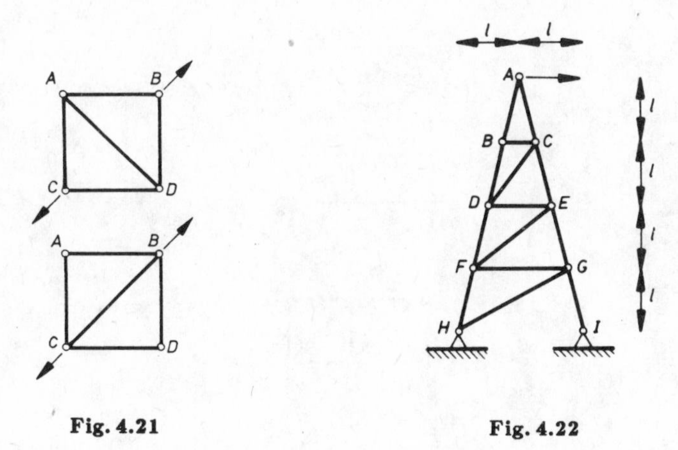

Fig. 4.21

Fig. 4.22

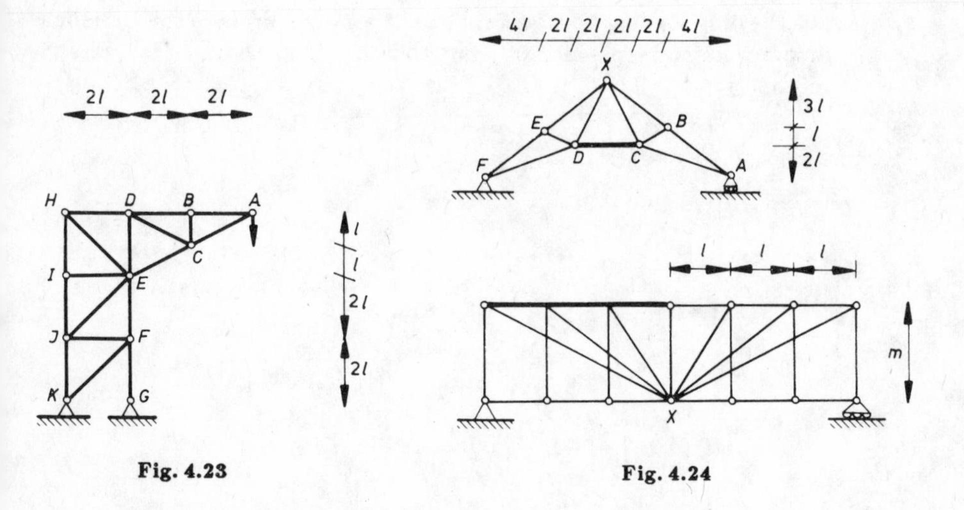

Fig. 4.23

Fig. 4.24

4.2.4 Determine the stresses in the heavy horizontal rods of the frameworks of Fig. 4.24 if a single vertical force (a loading at joint X) attacks them.

4.2.5 Determine the Cremona diagrams of the frameworks of Fig. 4.25 with the given loadings.

4.2.6 Determine the Cremona diagram of the first framework of Fig. 4.21. Apply Bow's notation.

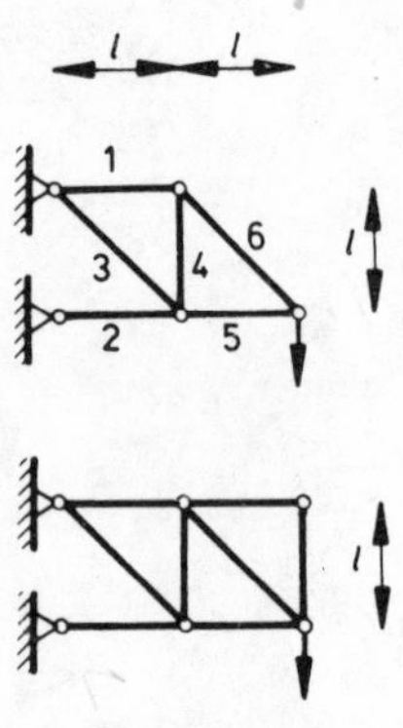

Fig. 4.25

Problems for Section 4.2

4.2.7 Consider the framework of Fig. 4.26. It is fixed to the wall at joints C, D; the position of joint B is also given but the position of joint A is changed along the dotted line (by changing the lengths of rods 1 and 2). Let the only loading be a vertical one at joint A. Prepare the Cremona diagram for every position of joint A. In particular, study the stress in rod 3.

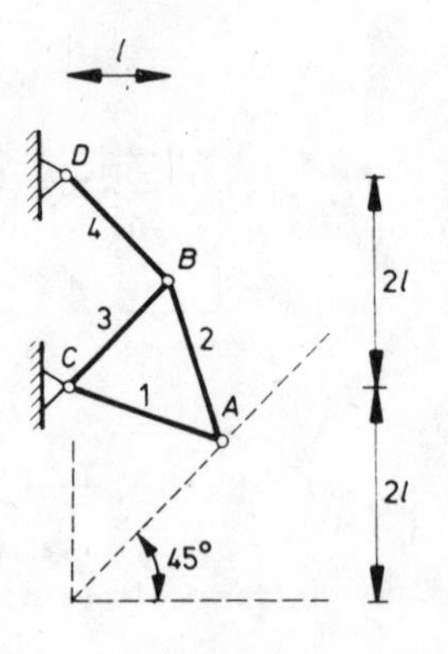

Fig. 4.26

4.2.8 The "bridge" of Fig. 4.27 is a symmetric framework with asymmetric loading. Prepare its Cremona diagram. Observe that it will contain many similar (but not congruent) polygons.

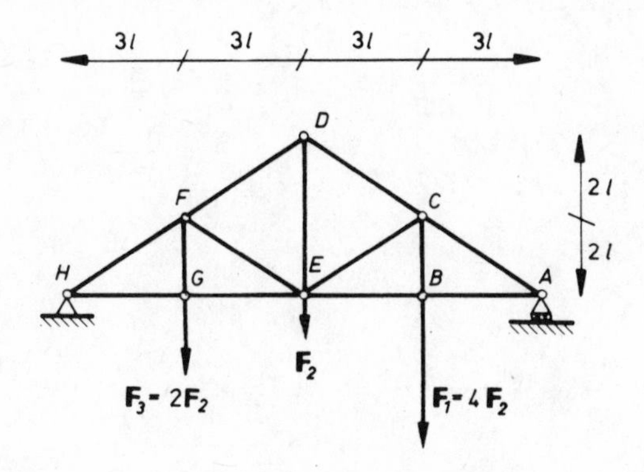

Fig. 4.27

4.2.9 Consider the asymmetric framework of Fig. 4.28. It is loaded in three different ways as follows (weights measured in a suitable unit):
(a) $Q_1 = 3;\ Q_2 = 9;\ Q_3 = 0$
(b) $Q_1 = 0;\ Q_2 = 8;\ Q_3 = 4$
(c) $Q_1 = 0;\ Q_2 = 0;\ Q_3 = 12$

The loadings are balanced by vertical forces $\mathbf{F}_a$ and $\mathbf{F}_b$ in each case and it is easy to check that the strength of $\mathbf{F}_a$ is 8, 6 and 4, respectively, in the above three cases. Prepare the Cremona diagrams and observe how this "shift of the centre of gravity" is reflected on them.

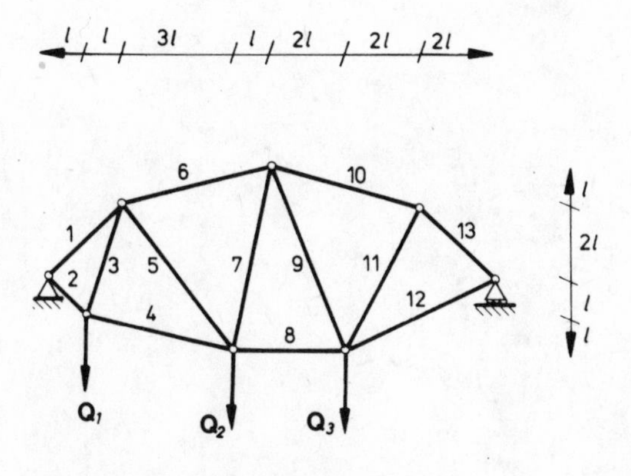

Fig. 4.28

4.2.10 (continued) Study the stress of the two "almost vertical" rods in the middle of the framework at all the three loadings above. Are they under compression or under tension?

Chapter 5

The theorems of König and Menger

Section 5.1 Bipartite graphs, 1-factors, König's theorem

Many of our graphs in the previous chapters had the property that their point set $V(G)$ decomposed into two disjoint sets X_1 and X_2 and every edge connected one point of X_1 to one of X_2. Such graphs are called *bipartite* graphs with *bipartition* X_1, X_2. A simple bipartite graph with bipartition X_1, X_2 where $|X_1| = k$, $|X_2| = \ell$ is called a *complete bipartite graph* and is denoted by $K_{k,\ell}$ if every point of X_1 is adjacent to every point of X_2. For example, $K_{3,3}$ is the second Kuratowski graph.

Lemma 5.1.1 *A graph G is bipartite if and only if every circuit of G has even length.*

Proof. Let G be a bipartite graph with bipartition X_1, X_2. If C is a circuit of G then its points are alternating in X_1 and X_2, hence $|V(C)|$ is even. On the other hand, if $v \in V(G)$ is an arbitrary point, put it in X_1, its neighbours in X_2, the unscanned neighbours of these latter points in X_1 again etc. This will lead to a bipartition since if two adjacent points were put in the same subset, it would lead to a circuit of odd length.

□

Let X_1 and X_2 be the sets of boys and girls and $v_1 \in X_1$ and $v_2 \in X_2$ be adjacent if the boy, corresponding to v_1 and the girl, corresponding to v_2, know each other. How many couples can then be formed? In a graph a collection of point-disjoint edges is called a *matching*. It is not difficult to see that the second bipartite graph of Fig. 5.1 has a matching with three edges, see Fig. 5.2, but the first one has not; at most two disjoint edges can be found in it.

The reason why the first graph of Fig. 5.1 has no matching with three edges is clear: Although the three boys together know all the three girls, the second and third boys together know only one girl (namely the third one). In general, a matching containing every boy can exist only if every k boys together know at least k girls. This simple necessary condition will be proved to be sufficient as well.

Let $N(X)$ denote the set of neighbours of the points of X for a set $X \subseteq V(G)$, i.e. $y \in N(X)$ if there is at least one point $x \in X$ so that $\{x, y\} \in E(G)$.

Theorem 5.1.2 [Hall, 1935]. *A bipartite graph G with bipartition X_1, X_2 has a matching which contains every point of X_1 if and only if $|N(X)| \geq |X|$ for every $X \subseteq X_1$.*

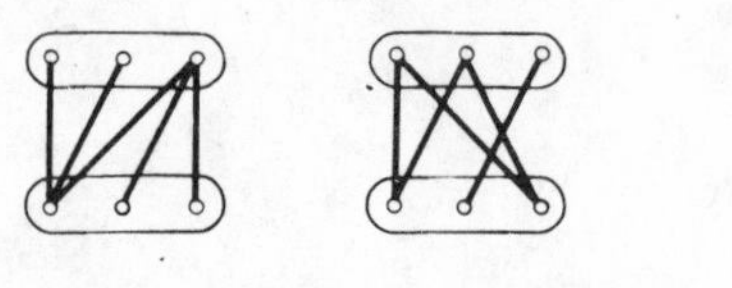

Fig. 5.1

Fig. 5.2

Proof. The necessity is obvious. Suppose the condition is met and try to find a matching containing every point of X_1. Suppose we have already a matching M which contains $X \subset X_1$ but there exists some element $u \in X_1 - X$ (see Fig. 5.3). For every $v \in X$ let v' denote its neighbour in M; similarly let X' be the set of points from X_2, contained by M. If u is adjacent to at least one point of $X_2 - X'$ then simply add this edge to M.

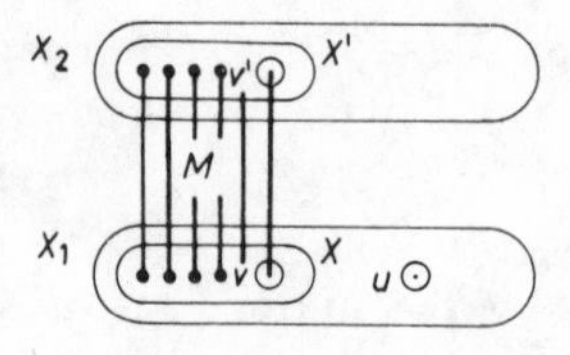

Fig. 5.3

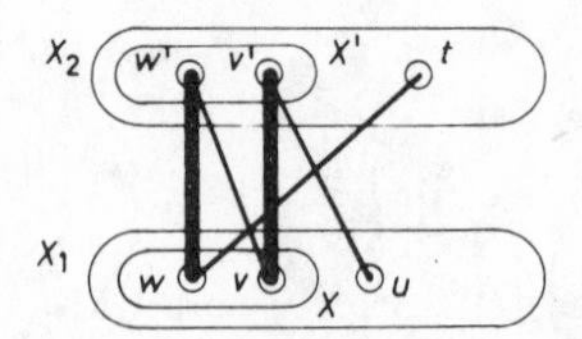

Fig. 5.4

But it can happen that every neighbour of u is in X'. Even in this case we may extend our matching. For example, if M consisted of the edges $\{w, w'\}$ and $\{v, v'\}$ and u was incident to v' only, the three thin edges in Fig. 5.4 are appropriate. More generally, if there is a path P of odd length joining a point from $X_1 - X$ to a point from $X_2 - X'$ so that every second edge of P belongs to M then such a so-called *alternating path* can be used to find a matching M' with one more edge; simply put $M' = [M - (M \cap P)] \cup (P - M)$.

Suppose finally that even this method fails, i.e. if T denotes the set of those points of X_2 which can be reached from u along alternating paths then T is contained in X' (see Fig. 5.5). Let $S \subseteq X$ be the set of points, matched by M to those of T. We prove that $N(S) = T$. The relation $\subseteq$ is obvious by the edges of M. Suppose $x \in S$ and there is an edge $\{x, y\}$ with $y \notin T$. Let P be an alternating path from u to x'. This path cannot contain x (since the edge $\{x, x'\}$ cannot be the last edge of an alternating path and if it were an internal edge then P would contain x' twice). But then P plus $\{x', x\}$ and $\{x, y\}$ is an alternating path from u to y, a contradiction to $y \notin T$.

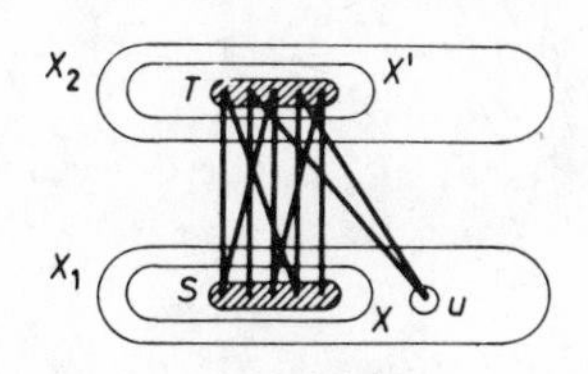

Fig. 5.5

Hence, by $N(S) = T$ and $N(\{u\}) \subseteq T$, we have found a set $S \cup \{u\}$ with $|N(S \cup \{u\})| = |T|$ which, by $|T| = |S| < |S \cup \{u\}|$, contradicts the assumption that $|N(X)| \geq |X|$ for every $X \subseteq X_1$.

□

A set of edges is called a *perfect matching* if it is a matching and contains every point of the graph.

Corollary 5.1.3 [Frobenius, 1875]. *A bipartite graph G with bipartition X_1, X_2 has a perfect matching if and only if $|X_1| = |X_2|$ and $|N(X)| \geq |X|$ for every $X \subseteq X_1$.*

Proof. The necessity is trivial. The second condition assures the existence of a matching which contains X_1 and then, by the first condition, X_2 as well.

□

Let $\nu(G)$ denote the maximum number of point-disjoint edges in a graph G. This is the size of the largest matching in G and G has a perfect matching if and only if $\nu(G) = v(G)/2$. The point-disjoint edges are also called *independent* edges.

A set $X \subseteq V(G)$ is called a *covering set of points* if X contains at least one end point from every edge of G. The minimum number of covering points is denoted by $\tau(G)$.

For example, $\tau = 3$ for both graphs of Fig. 5.6 but $\nu = 2$ for the first graph and $\nu = 3$ for the second.

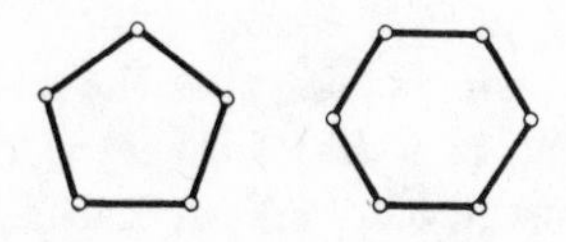

Fig. 5.6

Statement 5.1.4 *$\nu(G) \leq \tau(G)$ for every graph G.*

Proof. Let $M \subseteq E(G)$ be an arbitrary set of independent edges and $X \subseteq V(G)$ be an arbitrary set of covering points. Since different edges of M can be covered by different points of X only, obviously $|M| \leq |X|$. Since $\nu(G)$ is the size of the maximum such M, we have $\nu(G) \leq |X|$ for every set X of covering points, in particular for a minimum size such set X too, which has $\tau(G)$ points.

□

The first graph of Fig. 5.6 shows that $\nu < \tau$ is also possible. However, equality always holds for bipartite graphs:

Theorem 5.1.5 [König, 1931]. *If G is bipartite then $\nu(G) = \tau(G)$.*

Proof. Let M be a maximum matching, containing $\nu(G)$ edges. Let the end points of the M-edges form the sets X, X' (see Fig. 5.7). Let $U = X_1 - X$ and T be the set of those points of X_2 which can be reached from U-points along alternating paths. We know already that $T \subseteq X'$.

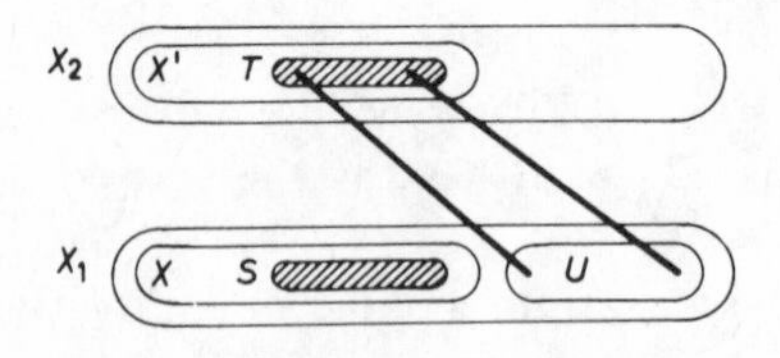

Fig. 5.7

Let $Y = T \cup (X - S)$. This set has $\nu(G)$ points, hence if we can prove that Y covers every edge then $\tau(G) \leq |Y|$ which, by Statement 5.1.4, gives the assertion.

Suppose there is an edge $\{x_1, x_2\}$ with $x_1 \in X_1$ and $x_2 \in X_2$ which is not covered by Y. Then $x_1 \in S \cup U$ and $x_2 \in X' - T$. Now $x_1 \in U$ is clearly impossible by the definition of T. If $x_1 \in S$ then, in the same way as in the proof of Theorem 5.1.2, the edges $\{x_1', x_1\}$ and $\{x_1, x_2\}$ would extend an alternating path from a point of U to x_1', which is also a contradiction to the definition of T.

□

This was a *minimax-theorem*, stating that the minimum of a quantity is equal to the maximum of another. In most of these results the relation max $\leq$ min is the easy one (like Statement 5.1.4 in our case) and proving max $\geq$ min is more difficult. A large number of such minimax theorems will be presented throughout this book, especially in the next two sections and in Chapters 13 and 15. Some remarks on the algorithmic aspects of minimax theorems will be given in Section 5.4.

A perfect matching is also called a *1-factor*. In general, a subset $X \subseteq E(G)$ is called a *k-factor* of G if for every point $v \in V(G)$ the number of edges from X, incident to v, is k.

A necessary and sufficient condition for the existence of a 1-factor in a bipartite graph was given in Corollary 5.1.3. Its proof is essentially algorithmic. A similar condition for arbitrary (nonbipartite) graphs is due to Tutte [1947], see Problems 5.1.19–22. Efficient algorithms are also known for finding such 1-factors [Edmonds, 1965b], see also [Lovász, 1979a] and [Lawler, 1976].

Exercises for Section 5.1

5.1.1 Characterize those graphs where the length of every circuit is odd.

5.1.2 Prove that if G is simple and bipartite then $e(G) \leq [v(G)]^2/4$.

5.1.3 Give a quick algorithm which decides whether a given graph is bipartite.

5.1.4 Describe the adjacency matrix $\mathbf{A}(G)$ of a bipartite graph.

5.1.5 Remove two opposite corners from an 8×8 chessboard. Show that the remaining 62 squares cannot be covered by 31 rectangles of size 2×1.

5.1.6* A graph is *regular* if every point has the same degree. Prove that every regular bipartite graph G has a 1-factor unless $e(G) = 0$.

5.1.7 A set of cardinality nk $(n, k \geq 2)$ can be decomposed in two different ways into the disjoint union of n subsets of cardinality k each. Prove that there exist n elements so that they are all in different subsets in both decompositions. Is a similar statement true for three decompositions as well?

5.1.8 Show that a tree cannot have two different 1-factors.

5.1.9 Prove that the Petersen graph (Fig. 1.8) cannot be decomposed to the union of three 1-factors. Give another 3-regular graph which can be so decomposed.

5.1.10 Does the graph of Fig. 5.8 have a 1-factor?

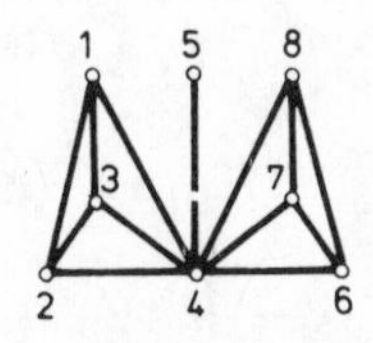

Fig. 5.8

5.1.11 Is it possible that a graph has a closed edge sequence, covering each edge exactly once, yet it has no 2-factor?

5.1.12 The *line graph* $L(G)$ of a graph G is defined so that the points of $L(G)$ are the edges of G and $e_1, e_2 \in V(L(G))$ are adjacent if and only if e_1 and e_2 were adjacent edges in G. Show that $\nu(G) = \alpha(L(G))$, see Problem 5.1.18.

Problems for Section 5.1

5.1.13 Let G be nonbipartite, hence it has odd circuits. Prove that in every strongly connected orientation of G there exists an odd directed circuit.

5.1.14 Let Q be a cut set in a bipartite graph G. Prove that G/Q is also bipartite.

5.1.15 A graph is called *k-dimensional cube* if its points are the 0–1 sequences of length k and two points are adjacent if and only if the two sequences differ in exactly one position, see Fig. 5.9 if $k = 3$.

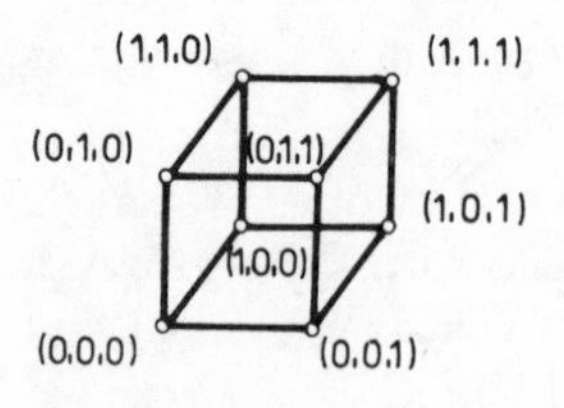

Fig. 5.9

(a) Draw the 4-dimensional cube.
(b) Prove that the k-dimensional cube has 2^k points and $k \cdot 2^{k-1}$ edges.
(c) Prove that the k-dimensional cube is bipartite and possesses a 1-factor.

5.1.16 Prove [Hetyei, 1964] that the following three statements are equivalent for a bipartite graph G with bipartition X_1, X_2:
(1) G is connected and every edge of G is contained in a 1-factor.
(2) For every $x_1 \in X_1$ and $x_2 \in X_2$ the graph $G - \{x_1, x_2\}$ contains a 1-factor.
(3) $|X_1| = |X_2|$ and for every $X \subset X_1$ $(X \neq \emptyset; X \neq X_1)$ we have $|N(X)| > |X|$.

5.1.17 Let the degree of every point in a bipartite graph G be at most d. Show that $E(G)$ can be decomposed into the union of d matchings.

5.1.18 Let $\alpha(G)$ denote the maximum number of *independent* (i.e. pairwise nonadjacent) points in the graph G and $\rho(G)$ denote the minimum number of *covering edges* (i.e., every point is incident to at least one of the edges). Prove that

(a) $\alpha(G) \le \rho(G)$ for every graph G.
(b) $\tau(G) + \alpha(G) = v(G)$ for every graph G without loops.
(c) $\nu(G)+\rho(G) = v(G)$ for every graph G without isolated points [Gallai, 1959].
(d) $\alpha(G) = \rho(G)$ for every bipartite graph G without isolated points.

5.1.19 An arbitrary (not necessarily bipartite) graph G has a 1-factor if and only if the number of components with an odd number of points in $G-X$ is at most $|X|$ for every $X \subseteq V(G)$ [Tutte, 1947]. Show the necessity of the condition and apply to the graph of Fig. 5.8.

5.1.20 (continued) Let G have a 1-factor, hence the number of the odd components in $G - X$ is at most $|X|$. Can this number be exactly $|X| - 1$?

5.1.21 (continued) In order to prove the sufficiency of Tutte's condition, indirectly suppose that G meets this condition, has no 1-factor and has a minimum number of points among the counterexamples. Let $Y \subseteq V(G)$ be a maximal point set with the property that the number of odd components of $G - Y$ is exactly $|Y|$. Prove [Gallai, 1963] that
(a) there exists such a Y;
(b) the even components of $G - Y$ have perfect matchings;
(c) whenever we delete an arbitrary point from an arbitrary odd component of $G - Y$ the result has a perfect matching.

5.1.22 (continued) Prove the sufficiency of Tutte's condition.

5.1.23 Let M be a matching in an arbitrary (not necessarily bipartite) graph G. A path P is *alternating* (with respect to M) if its edges alternatively belong to M and to $E(G) - M$. Obviously, the length of P is odd if its terminal points are not covered by M. Prove that M is a maximal matching if and only if G has no alternating path with odd length [Berge, 1957].

5.1.24 Show that every connected, $(2d)$-regular graph with an even number of edges can be decomposed to the union of two d-factors.

Section 5.2 Flows in networks, Menger's theorem

Let G be a directed graph with two distinguished points $s, t \in V(G)$. Let the edges of G be weighted by nonnegative numbers and let $c(e)$ denote the weight of e. G can be taken to be a road net along which as much produce should be transported from s to t as possible; however, the roads (i.e., the edges) may be used according to their orientation only, and no more than $c(e)$ units of produce can be transported along the edge e. Such a system $(G; s, t; c)$ is called a *network*, s and t are the *producer* and the *consumer*, respectively, and $c(e)$ is the *capacity* of the edge e.

Let us denote by $f(e)$ the amount of produce transported along the edge e. A function $f(e)$, associating nonnegative values to the edges, is legitimate if $f(e) \le c(e)$ holds for every $e \in E(G)$ and the quantity

$$m(v) = \sum\{f(e) \mid v \text{ is the head of } e\} - \sum\{f(e) \mid v \text{ is the tail of } e\}$$

is zero for every $v \in V(G)$ except for $v = s, t$. Such a legitimate function is called a *flow*. Then it is easy to show that $m(s) = -m(t)$ and this number (the amount of produce transported from s to t) is denoted by $m(f)$ and is called the *value of the flow*. Every network has flows (e.g., $f \equiv 0$ for every edge, in the worst case).

Consider such a network in Fig. 5.10. The first number on every edge e is the flow $f(e)$ and the second number (in the bracket) is the capacity $c(e)$. This flow has value 2 and is certainly not optimal. The heavy edges form a path, along which the capacities are higher than the actual $f(e)$-values. Should we increase f by 1 on these edges, we get a flow of value 3 and in this example this is optimal (since $f(e) = c(e)$ holds, that is e is *saturated* for every edge e).

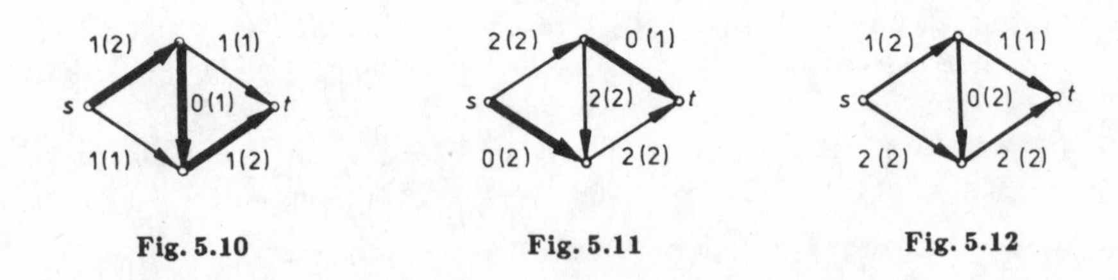

Fig. 5.10 Fig. 5.11 Fig. 5.12

However, one cannot hope to have such easy flow-increases in every network. For example, the value of the flow is 2 in the network of Fig. 5.11 as well. If we denote by heavy lines those edges e where $f(e) < c(e)$, the resulting subgraph does not contain any directed path from s to t. Yet, there is a flow of value 3 in this network as well, see Fig. 5.12.

Suppose we have a path with points $(s = v_0, v_1, v_2, \ldots, v_{k-1}, v_k = t)$ in the underlying undirected graph of the network. We can increase the value of the flow in the network if, for every $i = 0, 1, \ldots, k-1$, *either* there is a directed edge $e_i = (v_i, v_{i+1})$ with $f(e_i) < c(e_i)$ *or* there is a directed edge $e_i = (v_{i+1}, v_i)$ with $f(e_i) > 0$. (Intuitively, in the first case we still have unused capacity on the edge, while in the second case we have some flow in the opposite direction to be reduced.) Paths with these properties are called *augmenting paths* from s to t.

Theorem 5.2.1 *The value of a flow is maximum if and only if there is no augmenting path from s to t.*

Proof. Suppose P is an augmenting path. For every edge of P we have a strictly positive value, namely $c(e_i) - f(e_i)$ in the first case and $f(e_i)$ in the second. Let d be the minimum of these values, then clearly $d > 0$. Now increase $f(e_1)$ by d for the first type of edges, decrease $f(e_i)$ by d for the second type, and do not modify its value for those edges which are not in P. Then $m(v)$ does not change for $v \in V(G) - \{s, t\}$ but the value of the flow increases by d, hence the original flow was not maximum.

Suppose now that the network has no augmenting path from s to t. Still there might be some points which can be reached from s along an augmenting-like path (where all the above conditions except $v_k = t$ are met). Let us call such points accessible and denote by $X \subset V(G)$ the set of such points. Neither X nor $V(G) - X$ is empty (since $s \in X$ and $t \notin X$). Consider the edges, joining an accessible and an inaccessible point (Fig. 5.13). If such an edge e_1 points "forward" (i.e., if its tail is in X) then $f(e_1) = c(e_1)$ since its head u would otherwise be in X. On the other hand, if such an edge e_2 points "backward" (i.e., if its head is in X) then $f(e_2) = 0$ since its tail w would otherwise be in X. Then all the edges connecting X to $V(G) - X$ are "saturated", that is, no more produce can be transported along this "cut". Hence the value of f is maximum. □

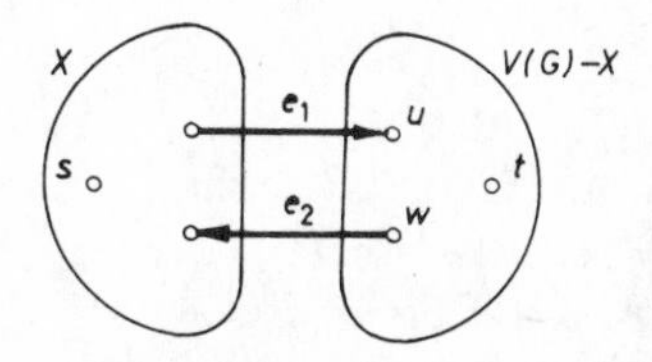

Fig. 5.13

This theorem has a better-known alternative formulation. Let a set C of edges be called an *(s,t)-cut* if it consists of those edges which join a point from X and a point from $V(G) - X$, for some $X \subset V(G), s \in X, t \notin X$. If the tail of such an edge is in X, we call it a forward edge (with respect to C). Otherwise it is called a backward edge. The sum of the flows on the forward edges minus the sum of the flows on the backward edges clearly equals the value of the flow. The *capacity* $c(C)$ of an (s,t)-cut is defined as the sum of the capacities of the forward edges.

Corollary 5.2.2 [Ford and Fulkerson, 1956]. *Max flow equals min cut, i.e.*

$$\max\{m(f) | f \text{ is a flow from } s \text{ to } t\} = \min\{c(C) | \ C \text{ is an } (s,t)\text{-cut}\}.$$

Proof. Max flow $\leq$ min cut is obvious since if every "forward-edge" of C is saturated and no "backward-edge" of C carries any produce then obviously we cannot expect any flow with higher value. On the other hand, if f is a maximum flow, the proof of Theorem 5.2.1 gives a cut with the same value. □

In order to have an algorithm to find a maximum flow, we must find augmenting paths. If we have an actual flow f in the network $(G; s, t; c)$ then define an auxiliary directed graph H_f with the same point set as G so that a directed edge (v_i, v_j) is in $E(H_f)$ if and only if

$$\text{either } e = (v_i, v_j) \in E(G) \text{ and } f(e) < c(e)$$

$$\text{or } e' = (v_j, v_i) \in E(G) \text{ and } f(e') > 0.$$

The augmenting paths from s to t are just the directed paths in H_f from s to t. Hence the following algorithm can be formulated:

Algorithm 5.1 *Maximum flow in a single producer single consumer network with edge capacities.*

Input The network $(G; s, t; c)$.
Output A flow f with maximum value.
Step 1 Start with the initial flow $f \equiv 0$ for every edge.
Step 2 Construct the auxiliary directed graph H_f. Apply Algorithm 3.4 to find the shortest path (i.e., that with a minimum number of edges) from s to t. If no such path exists, go to **End**. Otherwise augment the flow along this path and go to **Step 2**.
End The algorithm terminates, f is a flow with maximum value.

Observe that the algorithm provides the min cut as well, hence the user can easily check the result.

A very important feature of the algorithm is that we need a shortest path in Step 2. Although any path could serve for augmenting, without this extra requirement the algorithm could be very slow (see Exercise 5.2.2) or even non-terminating (Problem 5.2.9). However, using shortest paths one can show that the total number of steps is at most proportional to the cube of the number of points [Dinic, 1970], [Edmonds and Karp, 1972], see also [Lovász, 1979a].

If there are several producers and/or several consumers, this does not raise any significantly new questions, see Exercise 5.2.4 and Problem 5.2.10, as long as a single type of produce is considered. (The so called multicommodity flow problem, see Problem 5.2.16, is much more complicated.)

Another possible generalization appears if not only the edges but the points also have capacities. One can easily imagine that all the roads entering a city have 2 units of capacity each but due to downtown street conditions the total traffic entering the city must not be larger than 6 units.

A simple idea helps here: cut every point v into two new points v_1 and v_2 so that if an edge had head in v it gets moved to v_1 and it if had tail in v it gets moved to v_2 (see Fig. 5.14). Keep the old capacities of the edges and transfer the capacity of point v to edge (v_1, v_2).

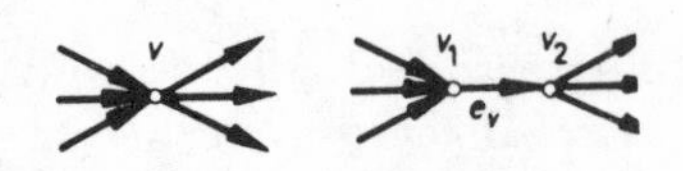

Fig. 5.14

Algorithm 5.2 *Maximum flow in a single producer single consumer network with edge and point capacities.*

Input The network $(G; s, t; c)$ and point capacities.
Output A flow f with maximum value.
Step 1 For every point with finite capacity prepare the transformation as shown in Fig. 5.14.
Step 2 Apply Algorithm 5.1.

Suppose that the capacity of every edge is an integer. If the actual flow f is also integer-valued then the augmenting value d (see the first part of the proof of Theorem 5.2.1) is also integer. Hence by starting from the $f \equiv 0$ flow the result is also integer-valued.

Corollary 5.2.3 *In the case of integer capacities the maximum flow can be obtained with integer values on every edge as well.*

In particular, if the capacity of every edge is 1 then there exists a maximal flow which either saturates an edge or does not use it at all. If those edges with $f(e) = 0$ are simply deleted, the result is a collection of edge-disjoint directed paths from s to t (and possibly some circuits).

Theorem 5.2.4 [Menger, 1927]. *If after deleting any set of $k-1$ edges from a directed graph G we still have a directed path from s to t then G contains k edge-disjoint directed paths from s to t.*

Proof. By the condition, every (s,t)-cut has capacity greater than $k-1$. Since everything was integer, the min cut capacity is at least k, hence the statement follows from Corollary 5.2.2.

□

Theorem 5.2.5 [Menger, 1927]. *If s and t are nonadjacent points in the directed graph G and after deleting any set of $k-1$ points, not containing s and t, we still have a directed path from s to t then G contains k internally disjoint directed paths from s to t.*

Proof. Apply the idea depicted in Fig. 5.14, and the previous theorem.

□

These are also minimax theorems, like König's theorem in the previous section. (In fact, most of the results of the previous section can easily be deduced from these, see Exercise 5.2.5.) The following theorem will also be useful, it is the undirected version of Menger's theorems.

Theorem 5.2.6 [Menger, 1927]. *If a graph is k-connected or k-edge-connected then one can find k internally disjoint paths (edge-disjoint paths, respectively) between any two points of it.*

The proof is left to the reader as Problem 5.2.17.

Exercises for Section 5.2

5.2.1 Most books use source and sink instead of producer and consumer. Discuss the differences among their meanings and our definitions of source and sink in Section 1.3.

5.2.2 Four among the five edges in the network of Fig. 5.15 has a large capacity n. Suppose that in every step we use the augmenting path containing the "vertical" edge of unit capacity. What is the total number of steps to find a maximum flow and how does this result depend on the size of the input of the problem?

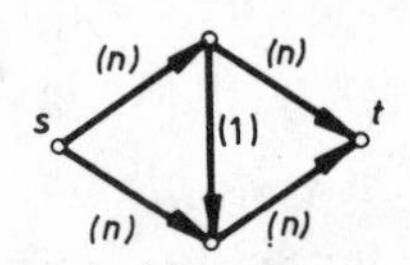

Fig. 5.15

5.2.3 Determine the value of the maximal flow in the network of Fig. 5.16 (capacities are given in brackets) as functions of x and y.

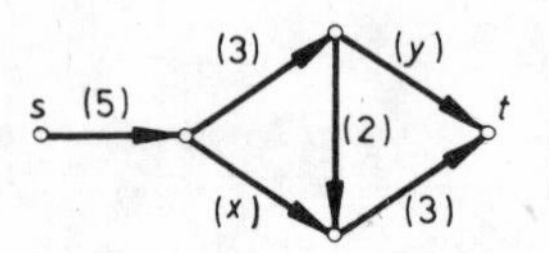

Fig. 5.16

5.2.4 If several producers $s_1, s_2, \ldots, s_k$ and several consumers $t_1, t_2, \ldots, t_\ell$ are in the network, give an algorithm to maximize the value of the total flow. (Observe that there is only one type of produce; for example the demand of t_1 can be obtained partly from s_i, partly from s_j etc. Do not confuse it with multicommodity flows, see Problem 5.2.16.)

5.2.5 Deduce Hall's and König's theorem from Theorem 5.2.5.

5.2.6 Let $X \subseteq V(G)$ and denote $\delta(X)$ the number of directed edges with tail in X, head not in X. Prove

$$\delta(X) + \delta(Y) \geq \delta(X \cup Y) + \delta(X \cap Y)$$

5.2.7 Let $X \subseteq V(G)$ be a minimum size point set, containing at least one point from every path between s and t. Let G_s and G_t denote the components of $G - X$, containing s and t, respectively. Prove that
(a) for every point $x \in X$ there exists a path P between s and t so that $V(P) \cap X = \{x\}$
(b) every point $x \in X$ has neighbours both in G_s and G_t.

5.2.8 Let $A, B \subseteq V(G)$ be two disjoint sets of points. Suppose that after deleting any set of $k-1$ points from G, one still has a path between A and B (a directed path from A to B, respectively, if G is directed). Prove that there are k point-disjoint paths (directed paths, respectively) between points of A and points of B.

Problems for Section 5.2

5.2.9 Consider the network in Fig. 5.17 [Lovász 1979a]. The capacity of the edges (z, w), (w, v), (v, u) and (u, z) are unity, those of the other edges are 2 units each. A flow is shown, where x is the unique real solution of the equation $x^3 + x - 1 = 0$ (about 0.682). Suppose we cyclically use the following four augmenting paths: (s, z, w, v, u, t); (s, z, u, v, w, t); (s, v, u, z, w, t) and (s, u, w, z, u, t). Prove that in this way our augmenting process will never terminate.

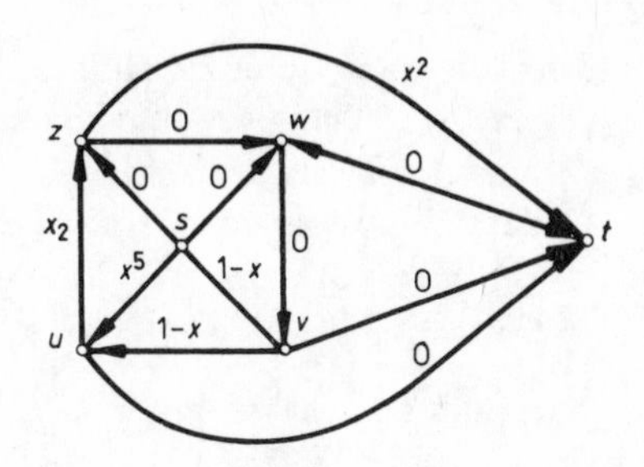

Fig. 5.17

5.2.10 Rain can enter the sewage system of a city at several points, and there is just one "output". The system turns out to be insufficient; in case of big thunderstorms part of the water comes to the streets. If the capacity of each section of the system is known, how can you find the "critical" parts (whose extension would extend the global system as well)?

5.2.11 Law of Lilliput requires that marriages be performed *via* matchmakers only. This law also prescribes that a matchmaker must not organize more than m marriages per year. Boy-dwarfs may have several girl-dwarf-friends to choose among; and most boy-dwarfs contacted several matchmakers (by informing them about the full list of their girl-dwarf-friends). Give an algorithm which maximizes the number of marriages for this year. Solve the South–Lilliputian version of the problem as well (where polygamy is still allowed, i.e. a man may have several wives).

5.2.12 Let C be an (s,t)-cut and denote by $\overline{C}$ the component of $G - C$, containing s. If C_1 and C_2 are such cuts then let $C_1 \cup C_2$ denote the cut D for which $\overline{C}_1 \cup \overline{C}_2 = \overline{D}$ and define $C_1 \cap C_2$ in an analogous way.
(a) Prove that $c(C_1) + c(C_2) \geq c(C_1 \cup C_2) + c(C_1 \cap C_2)$.
(b) Prove that among the minimum value cuts of the network there is a unique C_0 where $\overline{C}_0$ is minimum (as a set). Characterize this set.

5.2.13 Let $(G; s, t; c)$ be a network. Add some planarity-type condition and define its dual. What will be the "dual concepts" of flow and (s,t)-cut?

5.2.14 Suppose that the value $f(e)$ of the flow on the edges is not only upper bounded (by the capacities) but also bounded from below.
(a) Give an example that the flow problem may have no solution at all.
(b) Extending the network by additional points and edges reduce the problem to usual network flow problem. Check your solution to (a) and explain how the possible unsolvability can be recognized.

5.2.15 Let G be an *undirected* graph with two distinguished points s, t. Maximize the total number of products transported from s to t. Transportation can be realized on any edge in either direction; edge capacities give upper bounds for the total activity (if the same edge is used in both directions then the sum of the transportations is bounded). Reduce this problem to the usual one.

5.2.16 Let G be an undirected graph again, with k producers $s_1, s_2, \ldots, s_k$ and k consumers $t_1, t_2, \ldots, t_k$. There are k *different* products; s_1 produces for t_1 only, s_2 produces for t_2 only etc. (This is the *multicommodity flow problem*). Prove that the statement of Corollary 5.2.3 does not remain in general true, even for $k = 2$.

5.2.17 Prove Theorem 5.2.6.

5.2.18 Let $k \geq 2$. Prove that any set of k points in a k-connected graph is contained in a circuit [Dirac, 1960].

Section 5.3 Term rank, algebraic dependence

Let $\mathbf{A} = (a_{ij})$ be an $n \times n$ matrix. Recall that det $\mathbf{A}$ is the signed sum of all those products $a_{1,k_1} a_{2,k_2} \cdots a_{n,k_n}$ where $k_1, k_2, \ldots, k_n$ is a permutation of the numbers $1, 2, \ldots, n$. These products are called the *expansion members* of det $\mathbf{A}$.

The maximum number of nonzero entries of a matrix $\mathbf{A}$, no two in the same row or in the same column, is called the *term rank* of $\mathbf{A}$ and is denoted by $r_t(\mathbf{A})$.

If the "usual" rank $r(\mathbf{A})$ of $\mathbf{A}$ is k then $\mathbf{A}$ has a $k \times k$ submatrix $\mathbf{A}_0$ with det $\mathbf{A}_0 \neq 0$. Since det $\mathbf{A}_0$ is the sum of the expansion members, at least one of them must be nonzero, determining k nonzero entries in k different rows and k different columns. Hence $r_t(\mathbf{A}) \geq k$ and we obtain the following observation:

Lemma 5.3.1 *$r_t(\mathbf{A}) \geq r(\mathbf{A})$ for every matrix* $\mathbf{A}$.

One cannot hope for equality in general. Let, for example, $\mathbf{A} = \begin{pmatrix} 1 & 2 \\ 2 & 4 \end{pmatrix}$. Although $r_t(\mathbf{A}) = 2$, the nonzero expansion members $1 \cdot 4$ and $2 \cdot 2$ "cancel out" each other. If $r_t(\mathbf{A}) > r(\mathbf{A})$, there must exist some algebraic dependence among the entries of $\mathbf{A}$. This idea will now be studied in some detail.

Let L be a field and K a subfield (that is, $K \subset L$). If $x \in L - K$ then $K(x)$ denotes the "smallest" field containing K and x. There are such fields, for example L itself. If $y_1, y_2 \in L - K$ are another elements then $y_1 \in K(x)$ and $y_2 \notin K(x)$ is possible.

For example, if $K = \mathbb{Q}$ (i.e., the field of rationals), $L = \mathbb{R}$ (i.e., the field of reals) and $x = \sqrt{2}$ then $\sqrt{8} \in K(x)$ but $\sqrt{3} \notin K(x)$. None of $\sqrt{2}, \sqrt{3}$ and $\sqrt{8}$ is rational; $\sqrt{8}$ is contained in every field containing $\mathbb{Q}$ and $\sqrt{2}$ (since $\sqrt{8} = 2\sqrt{2}$); but $\sqrt{3} \notin K(x)$ since all the numbers of form $a + b\sqrt{2}$ (with $a, b \in \mathbb{Q}$) form a field (this is the smallest field containing $\mathbb{Q}$ and $\sqrt{2}$) and $\sqrt{3}$ does not arise in this form, see Exercise 5.3.3.

Similarly, if $x = \pi$ then $\mathbb{Q}(\pi)$ certainly contains every member of the form $\left(\sum_{i=0}^{k} a_i \pi^i\right) / \left(\sum_{j=0}^{\ell} b_j \pi^j\right)$, where k, ℓ are nonnegative integers, every a_i and b_j is in $\mathbb{Q}$ and at least one of the b_j's is nonzero. Since all these members form a field, this is the smallest field containing $\mathbb{Q}$ and π.

These fields $\mathbb{Q}(\sqrt{2})$ and $\mathbb{Q}(\pi)$ are very different. The reason for the difference is that $\sqrt{2}$ is an *algebraic* number over $\mathbb{Q}$, that is, there exists a nonidentically zero polynomial with coefficients from $\mathbb{Q}$ so that $\sqrt{2}$ is a root of it. (The simplest such polynomial is $x^2 - 2$.) The nonalgebraic numbers are *transcendental* over $\mathbb{Q}$.

π is transcendental, hence all of its powers have to be taken into consideration, while in the case of an algebraic number its powers after a certain exponent do not play any role.

Observe that $\sqrt{\pi}$ is transcendental over $\mathbb{Q}$ but, although it is not contained in $\mathbb{Q}(\pi)$, it is algebraic over it since it is the root of the polynomial $x^2 - \pi$, where both coefficients (1 and $-\pi$) belong to $\mathbb{Q}(\pi)$.

Two elements $x, y \in L$ are called algebraically independent over K if x is transcendental over $K(y)$ and y is transcendental over $K(x)$. More generally, the elements $x_1, x_2, \ldots \in L$ are *algebraically independent* over K if any one of them is transcendental over the smallest field containing K and all the other x_i's. The following alternative characterization will be useful:

Theorem 5.3.2 *The elements $x_1, x_2, \ldots$ are algebraically independent over K if and only if there is no nonidentically zero multivariable polynomial with coefficients from K, whose root-system is just $(x_1, x_2, \ldots)$.*

Its *proof* is left to the reader as Problem 5.3.7.

Corollary 5.3.3 [Edmonds, 1967]. *If $K \subset L$ and the nonzero entries of the matrix $\mathbf{A}$ belong to $L - K$ and are algebraically independent over K then $r(\mathbf{A}) =$ $= r_t(\mathbf{A})$.*

Its *proof* is now trivial, since the determinant is a special multivariable polynomial in the entries of the matrix.

Observe that the rank in Corollary 5.3.3 is over the field L, see also Problem 5.3.8.

Exercises for Section 5.3

5.3.1 Let us call both the rows and the columns of a matrix *lines*. Determine the minimum number of lines which cover all the nonzero entries of the matrices $\mathbf{A}_1 = \begin{pmatrix} x & 0 & 0 \\ y & 0 & 0 \\ z & u & v \end{pmatrix}$ and $\mathbf{A}_2 = \begin{pmatrix} x & y & 0 \\ 0 & 0 & z \\ u & v & 0 \end{pmatrix}$. Prove that this number is always greater than or equal to the term rank.

5.3.2 Show that if all the nonzero entries of an $n \times n$ matrix $\mathbf{A}$ can be covered by less than n lines then $\det \mathbf{A} = 0$.

5.3.3 Prove that the numbers $a + b\sqrt{2}$ $(a, b \in \mathbb{Q})$ form a field and that $\sqrt{3}$ does not belong to this field.

Problems for Section 5.3

5.3.4 Prove that the minimum number of lines (see Exercise 5.3.1) which cover all the nonzero entries of a matrix $\mathbf{A}$ is equal to the term rank $r_t(\mathbf{A})$ [Egervári, 1931].

5.3.5 Let the degree d_i of every point v_i of a connected graph G be even. Write down its circuit matrix $\mathbf{B}$ so that the row, corresponding to v_1 be written in $d_1/2$ copies one under the other, the row, corresponding to v_2, be written in $d_2/2$ copies etc. The resulting matrix has $e(G)$ rows and $e(G)$ columns. What will be its term rank? What do its expansion members correspond to?

5.3.6 A matrix is called *doubly stochastic* if its entries are nonnegative and its row-sums and column-sums equal to one.
(a) Prove that every doubly stochastic matrix is square.
(b) Prove that the term rank of a doubly stochastic matrix is n.

5.3.7 Prove Theorem 5.3.2.

5.3.8 Consider the matrices $\mathbf{A}_1 = \begin{pmatrix} 1 & 2 \\ 2 & 4 \end{pmatrix}$, $\mathbf{A}_2 = \begin{pmatrix} 1 & \sqrt{2} \\ \sqrt{8} & 4 \end{pmatrix}$ and $\mathbf{A}_3 = \begin{pmatrix} 1 & \sqrt{27} \\ \sqrt{3} & 9 \end{pmatrix}$
(a) Determine their term rank over $\mathbb{Q}, \mathbb{Q}(\sqrt{2})$ and over $\mathbb{R}$.
(b) Determine their rank over $\mathbb{R}$ and find out why the problems "determine their rank over $\mathbb{Q}$ and over $\mathbb{Q}(\sqrt{2})$" are not posed to you.

5.3.9 Let $\mathbf{x}_1, \mathbf{x}_2, \ldots, \mathbf{x}_k$ be k-dimensional vectors with integer entries. Suppose that none of them can be expressed as a linear combination of the others with rational coefficients. Prove that none of them can be expressed by the others with real coefficients either.

5.3.10 Let $\mathbf{M} = (m_{ij})$ be a $k \times (k+1)$ matrix with real entries and with $r(\mathbf{M}) = k$ over $\mathbf{R}$. Then one of its column vectors $\mathbf{m}_1, \mathbf{m}_2, \ldots, \mathbf{m}_{k+1}$ can be expressed as a linear combination $\sum \lambda_i \mathbf{m}_i$ of the others. Prove that the coefficients λ_i are in the smallest field, containing $\mathbb{Q}$ and the m_{ij} entries.

5.3.11 Using Corollary 5.3.3, give a linear algebra proof for the Theorem of Frobenius (Corollary 5.1.3), see [Edmonds, 1967].

Section 5.4 Graph theory algorithms III: NP-problems

When speaking about algorithms, we always considered the total number of operations required for them. Although, for example, a multiplication requires more time than an addition, we considered all operations alike (which is an acceptable approach since we only say that the total number of operations is proportional to some function $f(n)$ of the size of the input and do not study the coefficients of $f(n)$ at all). Furthermore, all of our estimates referred to the "worst case" and we do not consider other important evaluations, like the average number of operations.

Let n denote the larger of the two numbers giving the input and the output sizes. Then both Algorithms 1.3 and 3.1 were in a sense the best possible since the total number of operations was proportional to n. Note that Algorithm 3.1 requires up to a constant times v^3 steps because here $n = v^3$, while on the other hand Algorithm 3.2 requires only a constant times v^2 steps but is not necessarily the best possible since here $n = \max(v, e)$. (It is an open question whether a max (v, e) algorithm exists for the shortest path problem.)

Needless to say, Algorithm 3.2 is still very applicable for practical purposes. Sometimes we are even satisfied with any algorithm whose total number of operations is proportional to a polynomial of n.

Suppose for a moment that a problem has two algorithmic solutions. One of them simply checks all the possible cases and requires 2^n steps. The other applies some highly sophisticated ideas and requires $3000n^5$ steps. If we have one minute of time on a high-speed computer with 10^9 steps/s then the first algorithm can handle problems with size at most $\lfloor {}^2\log 6 \cdot 10^{10} \rfloor = 35$ while the second at most $\lfloor \sqrt[5]{6 \cdot 10^{10}/3000} \rfloor = 28$. But if we have one hour of time on this computer, the limit of the first algorithm jumps from 35 to 41 only, while that of the second from 28 to 65. This difference is even greater if the speed of the computer is increased (see **Box 5.1**).

Hence we will now consider only such graph theory algorithms whose total number of operations is polynomial of n. For some time we shall consider *decision problems* only, where the input is a question (like "Is G connected?") for which the output is "yes" or "no". Let us call *class* $\mathbf{P}$ the collection of those decision problems which can be solved in polynomial time. For example, checking connectedness or planarity are in the class $\mathbf{P}$.

Box 5.1. Comparison of polynomial and exponential algorithms

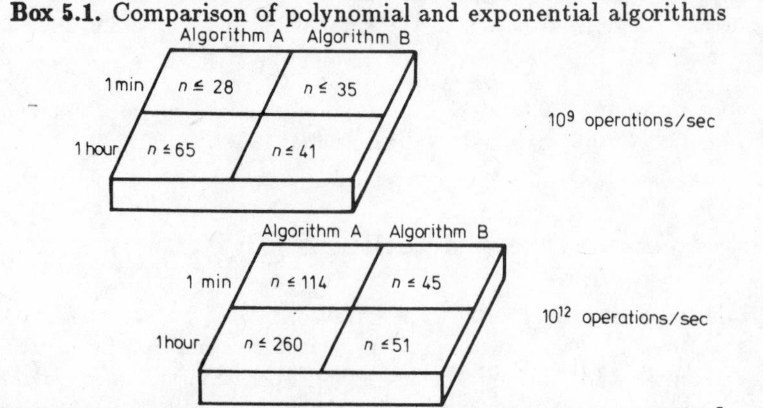

What is the maximal n so that Algorithm A (which requires $3000n^5$ operations) or Algorithm B (which requires 2^n operations) can be used by computers of speed $10^9 \ldots 10^{12}$ operations/sec, if 1 minute or 1 hour computing time is available?

Consider the following two problems:

(P1) Does the given graph have a circuit of length at most k?

(P2) Does the given graph have a circuit of length at least k?

We show (Exercise 5.4.1) that (P1) is in **P**; in fact we give a polynomial algorithm for the "more difficult" problem: "determine the *girth* (the length of the shortest circuit) of a given graph". On the other hand, (P2) is not known to be in **P** (most people believe it is not, see below), even the "easier" problem: "determine whether a given graph has a *Hamiltonian circuit*" (i.e. a circuit which passes through every point of the graph) is open.

Now we know that

(P3) Is the girth of a given graph equal to k?

is a decision problem in class **P**. Obviously, its *negation*.

(P3′) Is the girth of a given graph not equal to k?

is also in class **P**. On the other hand, as long as we do not know whether the decision problem

(P4) Does the given graph have a Hamiltonian circuit?

is in **P** or not, we do not know this about its negation either:

(P4′) Does the given graph have no Hamiltonian circuit?

Yet, while (P3) and (P3′) are both "easy", the logical structure in the case of (P4) and (P4′) is, in a sense, not the same. If a wizard (with some mystical abilities) could decide the answer to (P4) by giving us a Hamiltonian circuit if the answer is yes, we could at least *prove* in polynomial time that he is right; once a Hamiltonian circuit is given, it is trivial to find a polynomial (in fact, linear) algorithm which verifies that it is really a Hamiltonian circuit. But if he says no, all we can do is trust him since no polynomial algorithm is known for verifying the nonexistence of a Hamiltonian circuit.

On the other hand, an "affirmative" answer to (P4′), that is that the graph has no Hamiltonian circuit, cannot be verified in polynomial time, while a "negative" answer can, by simply asking our wizard to show a Hamiltonian circuit and then checking that it is really Hamiltonian.

Those decision problems for which a given affirmative answer can be proved in polynomial time, using the help of the wizard, form the *class* **NP**. Hence (P4) is in **NP** while (P4′) is not known to be.

Those decision problems, where the negative answer can be proved in polynomial time, using the help of the wizard, form the *class* co-**NP**. Of course, a problem is in co-**NP** if and only if its negation is in **NP**; for example (P4′) is in co-**NP**.

Obviously, $\mathbf{P} \subseteq \mathbf{NP}$ and the question whether this inclusion is proper, is one of the most interesting open problems in the theory of computational complexity. If somebody proved $\mathbf{P} = \mathbf{NP}$, this whole Section 5.4 would be of no interest.

Let us mention some problems from **NP**$\cap$ co-**NP**:

(P5) Does a given bipartite graph have a perfect matching?

(P6) Does a given network have a flow with value of least k?

(P7) Is a given graph planar?

Obviously, they are in **NP**; if our wizard gives a perfect matching or a flow with value at most k or a planar representation then we can quickly verify them. But they are in co-**NP** as well; in the case of negative answers we can ask our wizard to give a subset X of a bipartition-class so that $|N(X)| < |X|$ in case of (P5); or give an (s,t)-cut with capacity less than k in case of (P6); or a subgraph which is isomorphic to one of the Kuratowski graphs or to their series extensions, in case of (P7). This "counterexample-type" information can also be quickly verified.

The above three problems are not only in **NP**$\cap$ co-**NP**; they are actually in **P**. Some people believe that $\mathbf{P} = \mathbf{NP} \cap$ co-**NP**.

In fact, the history of discrete mathematics suggests that if a problem is obviously in **NP** and one can prove that it is in co-**NP** as well then sooner or later one will find a polynomial algorithm for it. For example, the existence of a perfect matching is trivially in **NP** (even for nonbipartite graphs); it has been known to be in co-**NP** [Tutte, 1947] for some time and now it is known to be in **P** [Edmonds, 1965b]. Similarly, since 1956 [Ford and Fulkerson, 1956] a proof that (P6) is in co-**NP** is known and we saw (Exercise 5.2.2 and Problem 5.2.9) that the first algorithms were not yet polynomial. The proof of (P6)$\in$ **P** followed in 1972 [Edmonds and Karp]. Planarity is also obviously in **NP** and while it is known to be in co-**NP** since 1930 [Kuratowski, 1930], the first polynomial algorithms came much later [Auslander and Parter, 1961]; [Goldstein, 1963]; [Lempel *et al.* 1967].

The logical structure of the theorems of Kuratowski, Hall and Ford and Fulkerson is the same:

$$\text{There exists a} \left\{ \begin{array}{c} \text{planar representation} \\ \text{complete matching} \\ \text{flow with value} \geq k \end{array} \right\} \text{if and only if}$$

$$\text{there exists no} \left\{ \begin{array}{c} \text{Kuratowski subgraph} \\ \text{large set with few neighbours} \\ \text{cut with value } < k \end{array} \right\}.$$

Such theorems are called *good characterizations* [Edmonds, 1965a], see also [Chvátal, 1973]. They just state that the problems are in **NP**$\cap$ co-**NP**. On the other hand, a theorem with the logical structure "something exists if and only if something else exists" is not a good characterization (although, of course, it might express a very deep mathematical idea).

Suppose we could solve problem P_1 in polynomial time if, in addition to the usual steps in a computer, we could also ask for the solution of another problem P_2. Then we say that P_1 is polynomially *reducible* to P_2 [Cook, 1971]. One can imagine the situation where the computer program to solve problem P_1 has only a polynomial number of steps but among the steps there may be some calls of a subroutine "Solve P_2".

If $P_2 \in \mathbf{P}$ then so is P_1 since if $z = f(y)$ and $y = g(x)$ are both polynomials then $z = f(g(x))$ is also a polynomial in x. Hence P_2 is "at least as hard" as P_1.

A problem is called **NP**-*hard* [Cook,1971] if every problem of **NP** is polynomially reducible to it. Such a "very hard" problem may or may not itself be in **NP** and need not even be a decision problem. If it is in **NP**, it is called **NP**-complete.

These concepts are introduced not for the sake of abstraction since Cook proved that there are **NP**-complete problems. In fact, exactly the most important open problems turn out to be **NP**-complete, see for example [Karp, 1972]. Only a few of these are listed in **Box 5.2**; the reader should consult the Appendix of Garey and Johnson [1979] and the Ongoing Guide in the Journal of Algorithms.

All these concepts are summarized in **Box 5.3**.

Why should we know about these concepts? If we are faced with a problem which is obviously in **NP** but for which we have not found any quick algorithm (apart from methods which check essentially every possible case in an exponential number of steps) then we should try to reduce an **NP**-complete problem to ours. If this succeeds, we may tell our boss that the problem is "essentially unsolvable", that is, if we had a solution to it, we would be able to solve at the same time hundreds of very difficult problems which have resisted the attacks of very bright people for decades.

Box 5.2. Some **NP**-complete problems

(1) Does a given graph contain a path of length at least k between two given points?
(2) Does a given graph contain a circuit of length at least k?
(3) Are there at least k independent points in a given graph (i.e. does $\alpha(G) \geq k$ hold)?
(4) Is there a set of at most k points in a given graph which covers every edge (i.e. does $\tau(G) \leq k$ hold)?
(5) Can all the directed circuits of a directed graph be covered by at most k points (or edges)?
(6) Does the multicommodity flow problem (Problem 5.2.16) have an integer solution?
(7) Does the given graph G have a subgraph which is isomorphic to a given graph H?

Remarks. These problems are strongly related (for example, (2) is a special case of (7)). They have many polynomially solvable special cases (for example the edge-version of (5) if the directed graph is planar; or (7) for many special graphs H).

The following important problem has no known polynomial solution. However, it is not known to be **NP**-complete either.

(8) Decide whether two given graphs are isomorphic?

Such news does not make our boss happy but at least he/she realizes that the job is too difficult (and not that we are too stupid). Since the job has to be performed anyhow, the fact that it turns out to be **NP**-hard means the start, rather than the end, of the story. It means that at least one of the five suggestions in **Box 5.4** should be seriously considered.

Exercises for Section 5.4

5.4.1 Show that the girth of the graph can be determined by the following algorithm:

Algorithm 5.3 *Minimum length circuit in a simple graph.*

Input A simple graph G.
Output A minimum length circuit C of G.
Step 1 For every edge $e = \{x, y\}$ of G construct the graph $G - e$ and determine the distance d_e between x and y in $G - e$, by Algorithm 3.2. This gives a path P_e which together with e gives a circuit C_e with length $d_e + 1$.
Step 2 The circuit C_e of minimum length is the required circuit C.
End

Box 5.3. Survey of some concepts of computational complexity

(1) Class **P** (problems solvable in polynomial time).
(2) Class **NP** (problems verifiable in polynomial time with the wizard's help).
(3) Class co-**NP** (problems whose negation is in **NP**).
(4) **NP**-hard problems (every **NP**-problem is reducible to them).

$(2) \cap (3)$ These problems have good characterization.
$(2) \cap (4)$ These are the **NP**-complete problems.

The diagram of Fig. 5.18 shows the most probable relation of these concepts. Even the suggested "statement" $(3) \cap (4) = \emptyset$ is conjectured only. Some people believe the truth of the following conjectures as well:

$(1) = (2) \cap (3)$
$(1) \neq (2)$
$(2) \neq (3)$

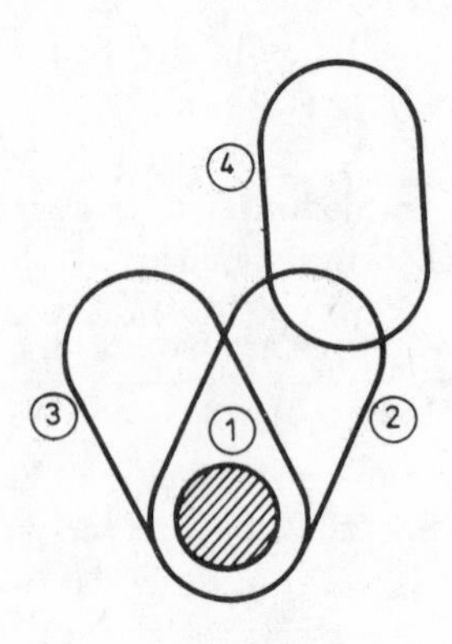

Fig. 5.18

Check that this is polynomial and thus conclude that (P1) and (P3) are in **P**.

5.4.2 Give a polynomial algorithm to decide whether a given graph with v points is k-connected. (Here v is part of the input, k is not, hence an algorithm requiring, say, v^k steps, is acceptable. However, see Problem 5.4.7 as well.)

5.4.3 The problem "Are there at least k disjoint edges in a graph (i.e. does $\nu(G) \geq k$ hold)?" can be shown to be in **P** while the similar question about $\alpha(G)$ is **NP**-complete (see (3) in **Box 5.2**). Is there no contradiction to the result of Exercise 5.1.12?

5.4.4 Real life computers as well as their mathematical models (Turing Machines, Random Access Machines) are *deterministic*; if there are several

Box 5.4. What should you do if your problem turns out to be **NP**-hard?

(1) Check the size of the input, required for the actual application. It can happen that an exponential algorithm is still acceptable. For example, an algorithm whose running time is proportional to 1.01^n can solve relatively large problems in a reasonable amount of time.
(2) Check your original modelling assumptions which lead you to this problem; perhaps some of your assumptions were too general. If you have some stronger assumptions which take more specific features of your original problem into consideration, you might be led to a polynomially tractable subcase of an otherwise **NP**-hard problem.
(3) There are many algorithms which are exponential in the worst case but polynomial on the average.
(4) There are many polynomial algorithms which find an "almost optimal" solution. For example, consider the *chromatic index* $q(G)$ of a simple graph G, i.e. the minimum number of colours required to paint the edges of G so that adjacent edges have different colours. If D is the maximum degree of G then $D \le q(G)$ is trivial and $q(G) \le D+1$ also holds [Vizing, 1964]. Finding $q(G)$ is **NP**-complete but there is a polynomial algorithm using not more than $D+1$ colours.
(5) There are many polynomial algorithms using random numbers at certain steps. Their output is optimal with high probability. We shall see an example in Chapter 17.

choices in a step of an algorithm, they choose exactly one of them (perhaps they later return and continue with another choice etc). Imagine, instead, a so-called *nondeterministic* machine [Rabin and Scott, 1959] which, in case of a choice among k possibilities, replicates itself into k identical copies and each copy continues the algorithm with a different choice. (In the next step they may replicate themselves again, etc.) If any one copy finds a solution, all copies stop.
Prove that a problem can be solved on such a nondeterministic machine in polynomial time if and only if it is in **NP**. (Hence the abbreviation: Nondeterministic Polynomial.)

5.4.5 Prove that if someone could prove $\mathbf{P} = \mathbf{NP}$, he/she could also conclude that every problem of **NP** is **NP**-complete as well.

Problems for Section 5.4

5.4.6 Let $\mathbf{A}$ be an $n \times n$ matrix. How many steps are needed to decide whether $\det \mathbf{A} = 0$ or not, if
(a) the entries of $\mathbf{A}$ are integers;
(b) the entries of $\mathbf{A}$ are partly zeros, partly real numbers which are algebraically independent over $\mathbb{Q}$;

(c) the entries of $\mathbf{A}$ are of form $p_i(x_1, x_2, \ldots)$ where $x_1, x_2, \ldots$ are reals, algebraically independent over $\mathbb{Q}$ and the p_i's are multivariable polynomials with coefficients from $\mathbb{Q}$?

5.4.7 (a) Give a polynomial algorithm which decides about a given graph with n points, whether it is $n/3$-connected?

(b) Give a polynomial algorithm which finds the largest k for which the given graph is k-connected.

5.4.8 Let (P) be the problem whether a given graph has Hamiltonian circuit and let (Q) be the problem of actually finding one. (This (Q) is not a decision problem.) Show that (Q) is polynomially reducible to (P).

5.4.9 A path p between the points a, b are called a *Hamiltonian path* if it contains all the other points of the graph as well. Prove that the problems, whether a graph has a Hamiltonian circuit or a Hamiltonian path, are polynomially reducible to each other.

5.4.10 Prove that the problems of the existence of Hamiltonian circuits and paths in the undirected graphs, and their analogues in case of directed graphs, are polynomially reducible to one another.

5.4.11 Prove that if a single problem of co-**NP** were **NP**-complete then **NP** = co-**NP** would follow.

Chapter 6

Applications

Section 6.1 Unique solvability of RLC networks containing negative resistors

Theorems 2.2.1 and 2.3.2 gave necessary and sufficient conditions for the unique solvability of networks, containing voltage and current sources and positive resistors (positive RLC elements, respectively).

If we allow negative resistors as well, these necessary conditions are not sufficient any more. The simplest counterexample, though a very artificial one, is a circuit of length three, formed by a voltage source U_1, and two resistors R_2 and R_3. If, say $R_2 = 500$ Ohm and $R_3 = -500$ Ohm then the network has no unique solution, although the subgraph, formed by the voltage sources, is circuit-free.

Since negative resistors are useful to model certain physical devices, we do not wish to exclude them. Yet, such a counterexample can be excluded since it does not reflect reality. The parameters of the devices (like 500 Ohm above) are not exact numerical data, one can argue, but approximate values, subjects to technological constraints. The resistance of such a resistor is not exactly 500 Ohm, hence its series connection to the other does not lead to a short circuit, in fact no exact algebraic relation ($R_2 + R_3 = 0$ or anything similar) can *a priori* be prescribed among them.

Statement 6.1.1 *If a network N consists of voltage and current sources, and such resistors, positive and negative alike, whose values are algebraically independent over the field $\mathbb{Q}$ of the rationals, then N is uniquely solvable if and only if the subgraphs, formed by the voltage and current sources, are circuit-free and cut set free, respectively.*

Proof. We have seen the necessity in Section 2.1. For the sufficiency recall that if these conditions are met, the network graph has spanning forests, containing every voltage source and none of the current sources. These spanning forests were shown in Section 2.5 to be in one–one correspondence with the nonzero expansion members of the network determinant. Hence there are nonzero expansion members and if their signed sum were zero, it were just an algebraic relation among the resistances.

□

Observe that the condition implied that the term rank of the network matrix is full. Yet we could not finish the proof simply by referring to Corollary 5.3.3

since only the resistances are algebraically independent but some other entries of the matrix (those expressing Kirchhoff's Laws) are ± 1.

Statement 6.1.2 *The statements of Theorem 2.3.2 remain true if, instead of the positivity of the values of the* RLC *devices, we prescribe their algebraic independence over* $\mathbb{Q}$.

Its *proof* is omitted, as that of Theorem 2.3.2 was already. The same proof should be analyzed again, and whenever the positivity of the element values was applied (for the nonsingularity of a square matrix), the algebraic independence should be applied instead.

Exercise for Section 6.1

6.1.1 Give a network where an algebraic relation among resistances, although does not lead to unsolvability, reduces the number of those capacitor voltages, which can independently be prescribed as initial values.

Problem for Section 6.1

6.1.2 Find a network where such an algebraic relation among the RLC values leads to unsolvability (or to a decrease of the number of the independent initial values), which is not purely among the resistances or among the capacitances or inductances.

Section 6.2 Unique solvability of networks containing ideal transformers or gyrators

By now we are familiar with networks, consisting of voltage and current sources and RLC elements. However, most of the practical electric circuits contain such devices, too, which have more than two terminals (like transistors, integrated circuits etc.). Hence, our approach of assigning one edge of a graph to each network element, cannot directly be used any more.

A large class of such "multiterminal" devices will be modelled by a new concept (to be introduced in Chapter 8). The unique solvability of the networks, containing such elements, is much more difficult and some questions of this type will be studied still in Chapter 18. But there are two multiterminal devices which can still be handled by graph theoretical methods.

One of these devices is the *ideal transformer* (Fig. 6.1). Its four terminals are grouped in two pairs (hence it should be called 2-port rather than 4-terminal device, see Section 8.1) and its property is that it transforms the voltage of one of its ports to a constant multiple on the other port; and a similar statement holds to its port currents in the other direction. Hence its describing system of equations is

$$u_2 = ku_1; \quad i_1 = -ki_2. \tag{6.1}$$

The negative sign is meant in accordance with the sign conventions as shown in the figure; later we shall see (Exercise 8.1.6) the physical content of these signs. k is called the *transfer ratio* of the transformer.

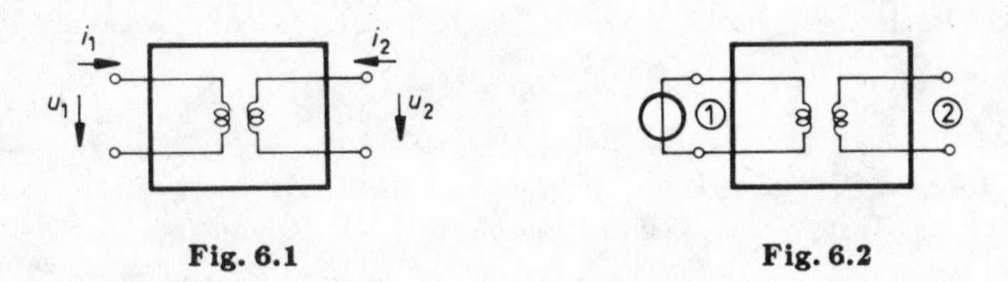

Fig. 6.1 **Fig. 6.2**

The drawing of Fig. 6.1 suggests some relations between transformers and inductances. These are studied in Problem 6.2.8.

The most important feature of an ideal transformer is that if a voltage source of value U is put to its first port (Fig. 6.2), its second port looks like a voltage source of value $k \cdot U$. This also means that putting voltage sources to both ports of an ideal transformer is as illegitimate as putting two voltage sources in parallel.

Suppose that our network consists of voltage and current sources, RLC elements and ideal transformers. The interconnection will be modelled so that sources and RLC elements will correspond to one edge each, as usual, but the ideal transformers correspond to two edges each (one for each port of each transformer), see Fig. 6.3 for example. If we put a voltage source on place 4, the network would have no unique solution. Yet, the network graph had a spanning forest (e.g. $\{1,4,5\}$) containing all the voltage sources and none of the current sources. This shows that the usual necessary condition (Statement 2.1.1) is in general not sufficient.

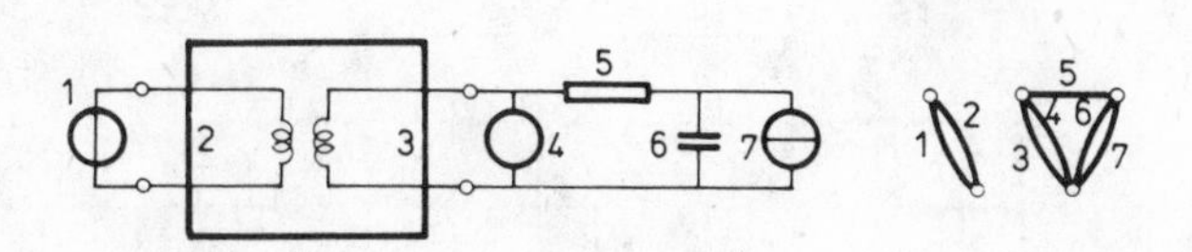

Fig. 6.3

Let us try to find such a spanning forest of the network graph which, in addition to containing all the voltage source edges and none of the current source edges, contains exactly one edge from every pair, corresponding to ideal transformers. Such spanning forests are called *normal*. For example, if the network of Fig. 6.3 is considered then $\{1,3,5\}$ or $\{1,3,6\}$ are normal spanning forests if we put a current source to place 4, and there is no normal spanning forest in the case of a voltage source.

The advantage of normal spanning forest is, intuitively speaking, that if both edges were in the complement of a spanning forest then both ports of the transformer could be terminated by voltage sources; or if both edges were in the forest then both ports could be terminated by current sources which is also illegitimate.

Theorem 6.2.1 *If the graph of a network consisting of voltage and current sources, resistors and ideal transformers, has normal spanning forest then the determinant of its system of equations has nonzero expansion members.*

Proof. Let F be a normal spanning forest. Let $E(F) = U \cup X \cup S$ and $E(G) - E(F) = I \cup Y \cup T$, where U and I are the sets of voltage sources and current sources, respectively, $X \cup Y$ is the set of resistors, and each ideal transformer has one edge from S and one from T. As an analogy of Fig. 2.22, the coefficient matrix of network equations can now be written in the symbolic form of Fig. 6.4, where $(\mathbf{C}_1|\mathbf{E})$ and $(\mathbf{E}|\mathbf{Q}_1)$ are the circuit and cut set matrices, determined by F. Every empty position (except the shaded areas) means a zero entry. One can immediately see that the diagonal elements of the submatrices with heavy lines determine a nonzero expansion member. □

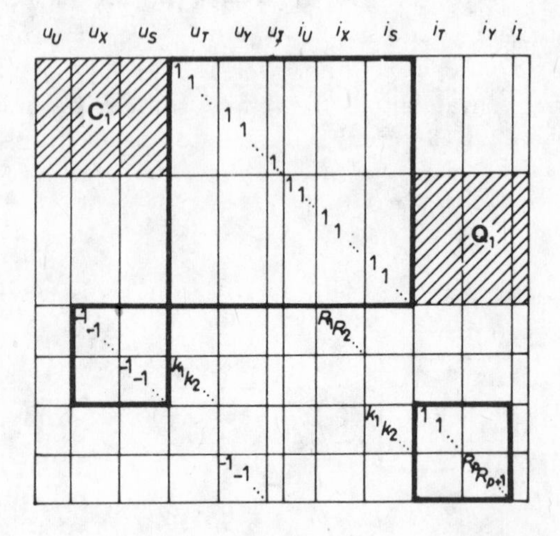

Fig. 6.4

Although this statement is very similar to the results of Section 2.5, the analogy is fairly limited. Not every nonzero expansion member corresponds to normal spanning forests, see Problem 6.2.9. Furthermore, if there exists a normal spanning forest (and hence the network matrix has full term rank), the unique solvability of the network need not follow (Exercise 6.2.6). However, the following two theorems are true:

Theorem 6.2.2 [Abdullah, 1972], [Bryant and Tow, 1972]. *If a network consisting of voltage and current sources, resistors and ideal transformers, has unique solvability then it has normal spanning forests.*

Theorem 6.2.3 *If the resistances and the transfer ratios of the above network are algebraically independent over* $\mathbb{Q}$ *then the existence of normal spanning forests is already sufficient to unique solvability.*

The *proof* of Theorem 6.2.2 is left to the reader as Problem 6.2.11. In order to prove Theorem 6.2.3 let us consider Fig. 6.4 again. For proving Theorem 6.2.1 we assigned numbers to the transformer-port-edges according to the normal spanning forest. If this numbering is given at first and the normal spanning forest is given thereafter then some of the submatrices (framed by heavy lines on the figure) contain not only ones but some k_j's as well in their diagonals. Anyhow, each expansion member will be the product of some resistances R_i and of the square of some transfer ratios k_j, but one and the same product will not arise several times, hence cannot be cancelled out, by the algebraic independence.

These results are summarized in the second column of **Box 6.1**.

The other interesting multiterminal device to be studied in this section is the *gyrator*. It is defined by the system

$$u_1 = -R \cdot i_2; \quad u_2 = R \cdot i_1 \tag{6.2}$$

of equations, and is denoted by the sign shown in Fig. 6.5. Later we shall see some important features of this device (Exercises 6.2.3, 8.1.5 and 8.1.7). Should we put a voltage source U to its first port (Fig. 6.6), it looks like a current source of value $-U/R$ at the second port. Hence it is obvious that, unlike in case of ideal transformers, the illegitimate situation arises here if one of its ports is terminated by a voltage source and the other by a current source.

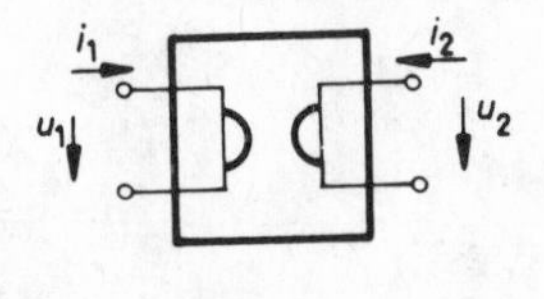

Fig. 6.5

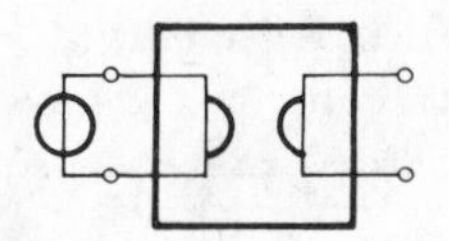

Fig. 6.6

Accordingly, let us call a spanning forest (of a network, consisting of voltage and current sources, RLC elements and gyrators) *normal* if, in addition to containing all the voltage sources and none of the current sources, it contains either none or both edges, corresponding to each gyrator. In case of several gyrators, such a normal spanning forest may contain both edges of one gyrator and none of the other (see **Box 6.2**).

Box 6.1. Conditions for the unique solvability of networks containing resistors, ideal transformers and gyrators, see [Recski, 1984]

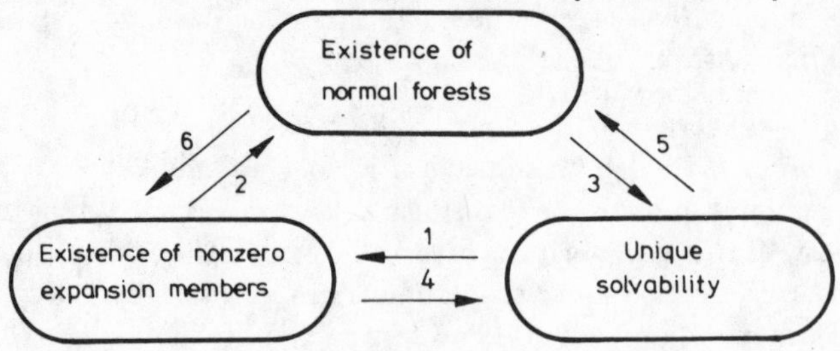

Implication#	Resistors only	Resistors and ideal transformers	Resistors and gyrators
1	yes (1)	yes (1)	yes (1)
2	yes Theorem 2.5.1	yes Theorem 6.2.2	no Problem 6.2.13
3	no (3)	no (4)	no (5)
4	no (3)	no (4)	no (6)
5	yes (2)	yes (2)	yes Theorem 6.2.5
6	yes Theorem 2.5.1	yes Theorem 6.2.1	yes Theorem 6.2.4

Remarks

(1) Trivial from linear algebra
(2) Trivial by implications 1 and 2
(3) Becomes "yes" if the resistances are positive (Theorem 2.2.1) or if their values are algebraically independent (Theorem 6.1.1)
(4) Exercise 6.2.6, but becomes "yes" if resistances and transfer ratios are algebraically independent (Theorem 6.2.3)
(5) Exercise 6.2.7, but becomes "yes" if resistances and gyrator constants are algebraically independent (Theorem 6.2.6)
(6) Problem 6.2.13; does not become "yes" with additional conditions on algebraic independence
(7) A fourth column (for networks, containing resistors, ideal transformers and gyrators) can also be given, see Problem 6.2.16

Theorem 6.2.4 *If the graph of a network, consisting of voltage and current sources, resistors and gyrators, has normal spanning forests then the determinant of its system of equations has nonzero expansion members.*

Proof. Same as for Theorem 6.2.1 with the only difference that S and T will denote the pairs of gyrator edges in the spanning forest and in its complement, respectively, and the symbolic form of the coefficient matrix will be in Fig. 6.7. □

Box 6.2. Interconnection of an ideal transformer or a gyrator with an RLC element, voltage or current source

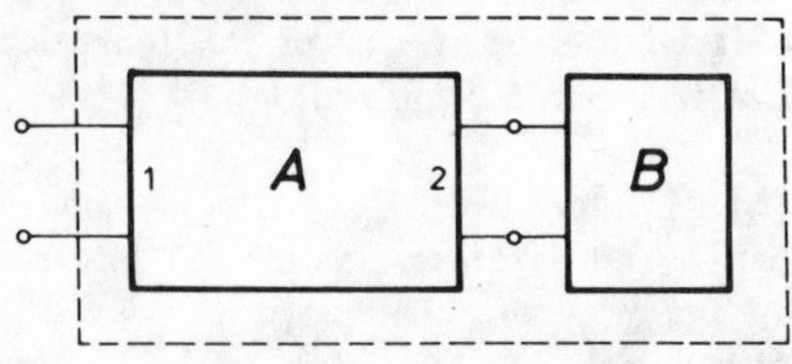

A \ B	R	L	C	U	I
$u_2 = k u_1$ $i_1 = -k i_2$ (ideal transf.)	$\frac{R}{k^2}$	$\frac{L}{k^2}$	$k^2 \cdot C$	$\frac{U}{k}$	$k \cdot I$
$u_1 = -\varrho i_2$ $u_2 = \varrho i_1$ (gyrator)	$\frac{\varrho^2}{R}$	$\frac{L}{\varrho^2}$	$\varrho^2 \cdot C$	$\frac{U}{\varrho}$	$\varrho \cdot I$

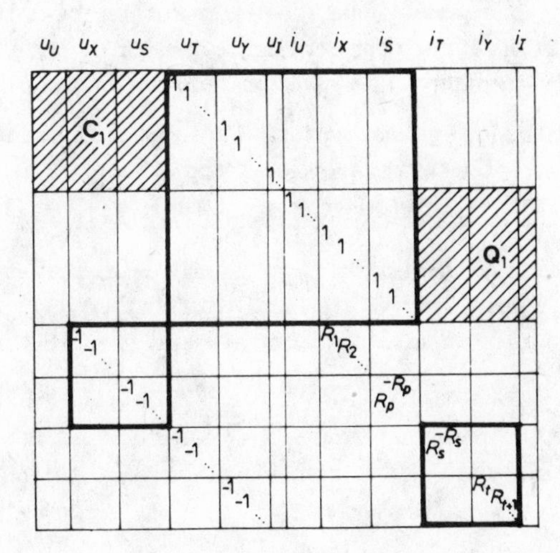

Fig. 6.7

Just like in the case of transformer networks, not all nonzero expansion members correspond to normal spanning forests (Problem 6.2.12), and unique solvability is not implied by the existence of a normal spanning forest (Exercise 6.2.7). However, even the existence of the nonzero expansion members does not imply the existence of normal spanning forests (see Problem 6.2.13), although this was true for transformer networks. Hence it is somewhat surprising that the analogy of Theorem 6.2.2 is still true.

Theorem 6.2.5 [Milić, 1974]. *If a network, consisting of voltage and current sources, resistors and gyrators, has unique solvability then it has normal spanning forests.*

Its *proof* is difficult, even a sketch is postponed until Section 18.1. On the other hand, the analogy of Theorem 6.2.3 is not only true but even the proof is about the same.

Theorem 6.2.6 *If the resistances and the gyrator constants of the above network are algebraically independent over $\mathbb{Q}$ then the existence of normal spanning forests is already sufficient to unique solvability.*

The results of this section together with the previous results on resistive networks are summarized in **Box 6.1.** Observe that all the three columns are different. Also note that while in the case of resistive networks we have algorithms (Section 2.4) to check the existence of a spanning forest with the required properties, in the other two cases we do not have yet (see Sections 12.2 and 18.1, respectively).

Exercises for Section 6.2

6.2.1 Terminate one port of the ideal transformer by a resistor of value R. Prove that the other port acts like a resistor of value R/k^2. What is the analogous statement for capacitors and inductors?

6.2.2 (continued) More generally, terminate one port of the ideal transformer by an RLC subnetwork. How does the other port act like?

6.2.3 What is the analogue of Exercises 6.2.1–2 in case of a gyrator?

6.2.4 The *cascade-connection* of two 2-ports is defined by Fig. 6.8. What happens if both 2-ports are ideal transformers or both are gyrators or one of them is an ideal transformer and the other is a gyrator? (See **Box 6.3**.)

6.2.5 Put voltage or current source to position A in the network shown in Fig. 6.9 [Belevitch, 1968; Fischer, 1975]. Will the network have unique solution?

6.2.6 Show a network, containing ideal transformers, which has normal spanning forest but no unique solution.

6.2.7 Show the same example with gyrators.

Box 6.3. Cascade interconnection of some important 2-ports
(see the reference directions in Fig. 6.1 and the illustration on p. 140)

A \ B	(1) $u_2 = ku_1$ $i_1 = -ki_2$	(2) $u_1 = -ri_2$ $u_2 = ri_1$	(3) $i_1 = 0$ $i_2 = gu_1$	(4) $i_1 = 0$ $u_2 = cu_1$	(5) $u_1 = 0$ $i_2 = di_1$	(6) $u_1 = 0$ $u_2 = ri_1$	(7) $u_1 = 0$ $i_1 = 0$
(1) $u_2 = lu_1$ $i_1 = -li_2$	(1) $u_2 = klu_1$ $i_1 = -kli_2$	(2) $u_1 = -(r/l)i_2$ $u_2 = (r/l)i_1$	(3) $i_1 = 0$ $i_2 = glu_1$	(4) $i_1 = 0$ $u_2 = clu_1$	(5) $u_1 = 0$ $i_2 = (d/l)i_1$	(6) $u_1 = 0$ $u_2 = (r/l)i_1$	(7) $u_1 = 0$ $i_1 = 0$
(2) $u_1 = -si_2$ $u_2 = si_1$	(2) $u_1 = -ski_2$ $u_2 = ski_1$	(1) $u_2 = (r/s)u_1$ $i_1 = -(r/s)i_2$	(5) $u_1 = 0$ $i_2 = gsi_1$	(6) $u_1 = 0$ $u_2 = csi_1$	(3) $i_1 = 0$ $i_2 = (d/s)u_1$	(4) $i_1 = 0$ $u_2 = (r/s)u_1$	(7) $u_1 = 0$ $i_1 = 0$
(3) $i_1 = 0$ $i_2 = hu_1$	(3) $i_1 = 0$ $i_2 = (h/k)u_1$	(4) $i_1 = 0$ $u_2 = -rhu_1$	(7) $i_1 = 0$ $u_1 = 0$	(7) $i_1 = 0$ $u_1 = 0$	(3) $i_1 = 0$ $i_2 = -hdu_1$	(4) $i_1 = 0$ $u_2 = -rhu_1$	(7) $u_1 = 0$ $i_1 = 0$
(4) $i_1 = 0$ $u_2 = bu_1$	(4) $i_1 = 0$ $u_2 = kbu_1$	(3) $i_1 = 0$ $i_2 = -(b/r)u_1$	(3) $i_1 = 0$ $i_2 = gbu_1$	(4) $i_1 = 0$ $u_2 = bcu_1$	(7) $i_1 = 0$ $u_1 = 0$	(7) $i_1 = 0$ $u_1 = 0$	(7) $u_1 = 0$ $i_1 = 0$
(5) $u_1 = 0$ $i_2 = ei_1$	(5) $u_1 = 0$ $i_2 = (e/k)i_1$	(6) $u_1 = 0$ $u_2 = -rei_1$	(7) $u_1 = 0$ $i_1 = 0$	(7) $u_1 = 0$ $i_1 = 0$	(5) $u_1 = 0$ $i_2 = -dei_1$	(6) $u_1 = 0$ $u_2 = -rei_1$	(7) $u_1 = 0$ $i_1 = 0$
(6) $u_1 = 0$ $u_2 = si_1$	(6) $u_1 = 0$ $u_2 = ksi_1$	(5) $u_1 = 0$ $i_2 = -(s/r)i_1$	(5) $u_1 = 0$ $i_2 = gsi_1$	(6) $u_1 = 0$ $u_2 = csi_1$	(7) $u_1 = 0$ $i_1 = 0$	(7) $u_1 = 0$ $i_1 = 0$	(7) $u_1 = 0$ $i_1 = 0$
(7) $u_1 = 0$ $i_1 = 0$	(7) $u_1 = 0$ $i_1 = 0$	(7) $u_1 = 0$ $i_1 = 0$	(7) $u_1 = 0$ $i_1 = 0$	(7) $u_1 = 0$ $i_1 = 0$	(7) $u_1 = 0$ $i_1 = 0$	(7) $u_1 = 0$ $i_1 = 0$	(7) $u_1 = 0$ $i_1 = 0$

(1) Ideal transformer (2) Gyrator (3) Voltage controlled current source (4) Voltage controlled voltage source
(5) Current controlled current source (6) Current controlled voltage source (7) Pair of nullator and norator

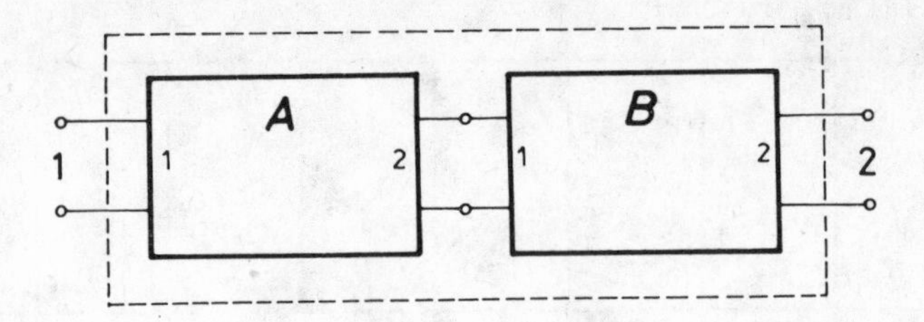

Illustration for Box 6.3

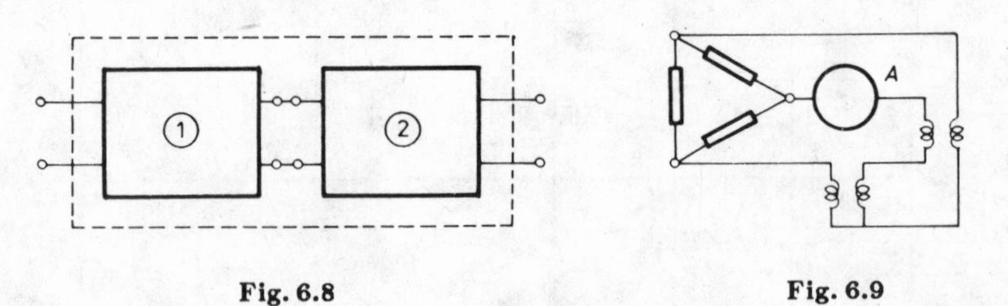

Fig. 6.8 **Fig. 6.9**

Problems for Section 6.2

6.2.8 Figure 6.10 shows a "real" transformer with two inductances on the same core. Then current i in one of the inductances induces a voltage, proportional to $\frac{di}{dt}$, in the other inductance as well. What sort of further assumptions are required for the device to meet Eqs (6.1)?

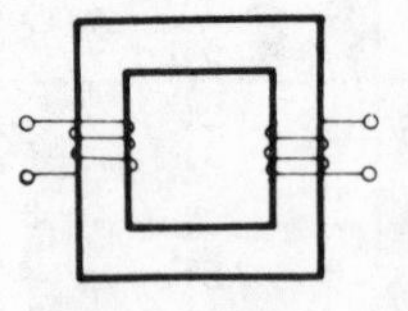

Fig. 6.10

6.2.9 Give an example to a network, containing ideal transformers, where not all the nonzero expansion members correspond to normal spanning forests.

6.2.10 The "yes" and "no" answers in the first and second columns of **Box 6.1** are the same. Why is there no contradiction to the result of the previous problem?

6.2.11 Prove Theorem 6.2.2.

6.2.12 Show a network, containing gyrators, where not all the nonzero expansion members correspond to normal spanning forests.

6.2.13 Give a gyrator network with nonzero expansion members but without normal spanning forests.

6.2.14 Let G be the graph of a network, containing gyrators. Let G' be another graph, obtained from G by interchanging the numbers for each pair of gyrator edges. Show that nonzero expansion members are in one–one correspondence with those spanning forests of G which are spanning forests in G' as well [Tow, 1970]. What is the relation between these "common forests" and the normal forests?

6.2.15 Let a network with a connected network graph G consist of capacitors (n_c in number) and gyrators. Prove that if $n_c - v(G)$ is even, then it is impossible to independently assign the voltage of all the capacitors as initial values [Milić, 1974].

6.2.16 Add a fourth column to **Box 6.1** for networks, containing resistors, ideal transformers *and* gyrators.

Section 6.3 Rigidity of trusses I: Graph model, the definition of rigidity

In the statics applications described in the rest of the book (except in Section 10.2) the frameworks are of arbitrary shape, i.e. not necessarily planar grids (like in Section 2.6). Like before, we always suppose that the rods are perfectly rigid and the joints are perfectly rotatable.

Let us define the *graph* $G(T)$ of the framework T in the normal way; the joints correspond to the points of $G(T)$ and two points are adjacent if and only if there is a rod in T between the corresponding joints. For example, all the three frameworks of Fig. 6.11 have the same graph, namely the circuit of length 4.

In certain cases (for example just in Fig. 6.11) the graph itself immediately determines that the framework is not rigid (that is, it is dynamically underdetermined, see Section 4.2). In some other cases the graph does not carry enough information; the length of the rods must also be known. For example, both frameworks of Fig. 6.12 have the same graph yet only the first is rigid (Exercise 6.3.4). Sometimes this difference is quite hard to be found, see Problems 6.3.7–8, for example.

Hence we have two goals. Firstly, to find a mathematical model for deciding rigidity about every framework — for this purpose graph theory (or even combinatorics) is certainly not enough — , secondly, to find purely combinatorial answers whenever possible, since the complete solution involves numerical

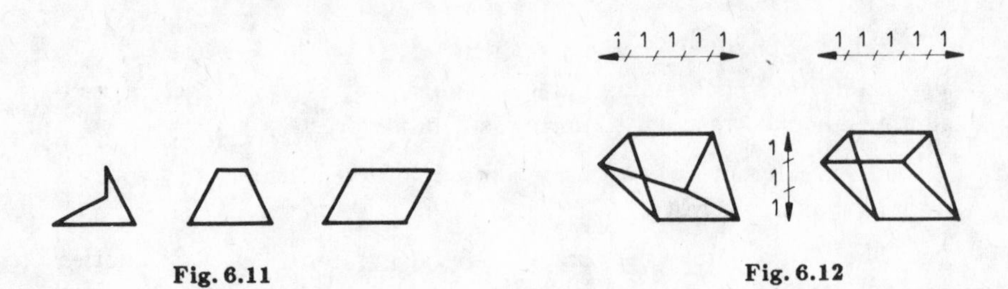

Fig. 6.11 **Fig. 6.12**

calculations with roundoff errors etc. Our first goal will be the subject of the present section while the second is considered later.

Two more problems should yet be clarified. The problem setting is incorrect without knowing the dimension of the space; for example the framework in Fig. 6.13 is rigid in the plane but nonrigid in the 3-dimensional space. The other problem is explained by Fig. 6.14. The graphs of the two frameworks are isomorphic and the first framework is obviously rigid. What about the second one? Sometimes it is considered rigid since fixing, say, two adjacent points of it to the plane, the position of every other joint is uniquely determined. Some other people consider it nonrigid since a small, "infinitesimal" motion can be obtained by a force, attacking the joint according to the arrow. (This can be made precise, see Problem 6.3.5). In what follows, we wish to consider such frameworks nonrigid.

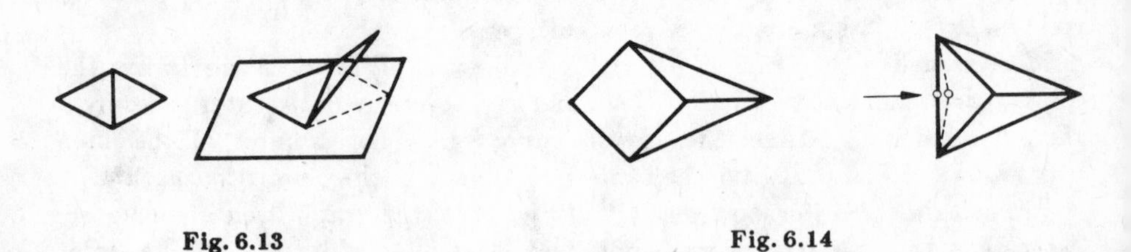

Fig. 6.13 **Fig. 6.14**

Let F be a framework consisting of e rods and v joints. Let the coordinates of the joints $P_1, P_2, \ldots$ be $(x_1, y_1, z_1), (x_2, y_2, z_2), \ldots$, respectively (if we are in the 3-dimensional space) or $(x_1, y_1), (x_2, y_2), \ldots$, respectively, if we are on the plane. A rod R_{ij} between the joints P_i, P_j means that the distances between P_i and P_j is constant

$$\sqrt{(x_i - x_j)^2 + (y_i - y_j)^2 + (z_i - z_j)^2} = \text{constant} \tag{6.3}$$

or $\sqrt{(x_i - x_j)^2 + (y_i - y_j)^2}$ = constant in case of the plane. It is not clear what can we do with e quadratic equations and $3v$ or $2v$ unknowns. We show that the problem can be solved by a linear system of equations as well.

Imagine our framework moving. (Even if it is rigid, it can have, say, a translation.) Hence (x_i, y_i, z_i) is really $(x_i(t), y_i(t), z_i(t))$, a triple of functions of time. The square of Eq. (6.3) is

$$(x_i(t) - x_j(t))^2 + (y_i(t) - y_j(t))^2 + (z_i(t) - z_j(t))^2 = c_{ij} \tag{6.4}$$

for every rod R_{ij}. Differentiating these equations and dividing by two, we obtain

$$\begin{aligned}(x_i(t) - x_j(t))(\dot{x}_i(t) - \dot{x}_j(t)) + (y_i(t) - y_j(t))(\dot{y}_i(t) - \dot{y}_j(t)) + \\ +(z_i(t) - z_j(t))(\dot{z}_i(t) - \dot{z}_j(t)) = 0,\end{aligned} \tag{6.5}$$

where $\dot{x}_i(t)$ stands for $\frac{d}{dt}x_i(t)$ etc.

What does Eq. (6.5) mean? If (x_i, y_i, z_i) and $(\dot{x}_i, \dot{y}_i, \dot{z}_i)$ are denoted by $\mathbf{h}_i$ and by $\mathbf{v}_i$ respectively, (let us call them position and velocity, respectively) then Eq. (6.5) can shortly be written as

$$(\mathbf{h}_i - \mathbf{h}_j)(\mathbf{v}_i - \mathbf{v}_j) = 0 \text{ for every rod } R_{ij}$$

i.e. the difference of the position vectors and that of the velocity vectors are perpendicular. (Recall that the zero vector is perpendicular to everything, by definition.)

Restricting our attention for a moment to the plane, for simplicity, any motion of a rigid rod is

- either a translation (Fig. 6.15) when $\mathbf{v}_i = \mathbf{v}_j$, i.e. their difference is $\mathbf{0}$;
- or a rotation around the joint P_i (Fig. 6.16) when $\mathbf{v}_i = \mathbf{0}$ and $\mathbf{v}_j$ is always perpendicular to the direction of the rod, i.e. to $\mathbf{h}_i - \mathbf{h}_j$;
- or a combination of these two motions.

Fig. 6.15 **Fig. 6.16**

Equation (6.5) can obviously be rewritten as

$$\begin{aligned}(x_i(t) - x_j(t))\dot{x}_i(t) + (x_j(t) - x_i(t))\dot{x}_j(t) + \ldots \\ \ldots + (z_i(t) - z_j(t))\dot{z}_i(t) + (z_j(t) - z_i(t))\dot{z}_j(t) = 0.\end{aligned} \tag{6.6}$$

The e equations can be collected into a single one

$$\mathbf{A}\mathbf{u} = 0 \tag{6.7}$$

where $\mathbf{u} = (\dot{x}_1, \dot{x}_2, \ldots, \dot{x}_v, \dot{y}_1, \dot{y}_2, \ldots, \dot{y}_v, \dot{z}_1, \dot{z}_2, \ldots, \dot{z}_v)^T$ and $A = (a_{k\ell})$ is an $e \times 3v$ matrix with

$$a_{k\ell} = \begin{cases} x_i - x_j & \text{if rod } k \text{ is between joints } i \text{ and } j, \text{ and } \ell = i \\ y_i - y_j & \text{if rod } k \text{ is between joints } i \text{ and } j, \text{ and } \ell = i + v \\ z_i - z_j & \text{if rod } k \text{ is between joints } i \text{ and } j, \text{ and } \ell = i + 2v \\ 0 & \text{otherwise} \end{cases} \tag{6.8}$$

For example, in the case of the framework in Fig. 6.17, we have

$$\mathbf{A} = \begin{pmatrix} x_1 - x_2 & x_2 - x_1 & 0 & y_1 - y_2 & y_2 - y_1 & 0 & z_1 - z_2 & z_2 - z_1 & 0 \\ x_1 - x_3 & 0 & x_3 - x_1 & y_1 - y_3 & 0 & y_3 - y_1 & z_1 - z_3 & 0 & z_3 - z_1 \\ 0 & x_2 - x_3 & x_3 - x_2 & 0 & y_2 - y_3 & y_3 - y_2 & 0 & z_2 - z_3 & z_3 - z_2 \end{pmatrix}.$$

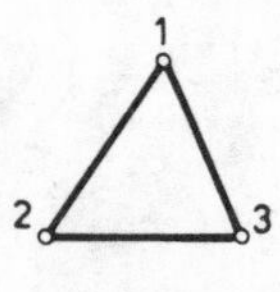

Fig. 6.17

Observe that Eq. (6.7) is linear. The entries of $\mathbf{A}$ come from the positions of the joints, hence the framework is described by a real matrix (in a given instant). Intuitively one might expect that "rigidity" means that no motion is possible, hence $\mathbf{u} = \mathbf{0}$ is the only solution of Eq. (6.7).

This is not true. Imagine a translation of the rigid framework in the direction determined by a vector (ξ, η, ς). Then we can define a vector

$$\mathbf{u} = \{\dot{x}_i = \xi, \dot{y}_i = \eta, \dot{z}_i = \varsigma \text{ for every } i = 1, 2, \ldots v\}$$

and $\mathbf{Au} = \mathbf{0}$ will hold. This is trivial for physical reasons ($\mathbf{v}_i - \mathbf{v}_j = \mathbf{0}$ for every pair i, j in Eq. (6.6)), but for mathematical reasons as well (the sum of the first (or second or third) group of v columns of $\mathbf{A}$ is $\mathbf{0}$).

More generally, Eq. (6.7) does have nontrivial solutions, even for rigid frameworks, hence $r(\mathbf{A}) < 3v$. Those nontrivial solutions which come from the congruent transformations of the whole 3-dimensional space like translations, rotations etc. (also called *rigid body like motions*), can be shown to form a 6-dimensional vector space (Exercise 6.3.1), hence $r(\mathbf{A}) \le 3v - 6$. (We always suppose $v \ge 2$, i.e. that there is at least one rod in the framework.)

Lemma 6.3.1 *If a framework F has at least 3 joints and $r(\mathbf{A}) < 3v - 6$ holds, then F is dynamically underdetermined (i.e., not rigid).*

Its *proof* is now trivial. The solutions of $\mathbf{Au} = \mathbf{0}$ form a subspace of dimension at least 7. Were F rigid, every motion of F were the congruent motion of the whole space, but in this way at most dimension 6 could be obtained. □

Thus we shall define a *3-dimensional* framework to be *rigid* if $r(\mathbf{A}) = 3v - 6$ holds for its matrix $\mathbf{A}$. If a framework is nonrigid, the nontrivial solutions of Eq. (6.7) determine the not rigid body like motions of the framework. They may be physical motions of a mechanism or "infinitesimal motions" (like the one shown on the right hand side of Fig. 6.14) as well (cf., Exercise 6.3.3).

Corollary 6.3.2 [Maxwell, 1864]. *If a 3-dimensional framework with v joints and e rods is rigid then $e \geq 3v - 6$.*

Proof. $r(\mathbf{A}) = 3v - 6$ by the rigidity. Since matrix $\mathbf{A}$ has e rows, obviously $r(\mathbf{A}) \leq e$.

□

Similarly, a *2-dimensional* framework is defined to be *rigid* if $r(\mathbf{A}) = 2v - 3$.

Corollary 6.3.3 [Maxwell, 1864]. *If a 2-dimensional framework with v joints and e rods is rigid then $e \geq 2v - 3$.*

These conditions are necessary but not sufficient (see Exercise 6.3.2). Stronger necessary conditions will be presented in Section 12.3, they will be essentially sufficient as well in the 2-dimensional case.

If the position of the joints of a framework F is known then rigidity of F can be determined by checking the condition $r(\mathbf{A}) = 3v - 6$ (or $2v - 3$, respectively). Exercise 6.3.4 and Problems 6.3.7–8 can be solved in this way. Since the rank of a real matrix of size $e \times 3v$ should be determined and the coordinates of the positions of the joints are — in the practical cases — rationals (even integers if the unit length is small enough), this rigidity can be checked in polynomial time (see Part (a) of Problem 5.4.1). This is what really happens in engineering, see e.g. [Szabó and Roller, 1978].

However, if only the graph of a framework F is known then the matrix $\mathbf{A}$ of F cannot be determined (not even $r(\mathbf{A})$, see Exercise 6.3.4). We may pose the following problem: Suppose we know the graph of F and the additional information that there is no algebraic relation among the coordinates (for the positions of the joints), i.e. that these coordinates be algebraically independent over the field $\mathbb{Q}$ of the rationals. For example, such an independence excludes the second system in Figs 6.12 and 6.14.

Unfortunately, such a task can require a very large (exponential or even larger) number of operations, see part (c) of Problem 5.4.1. Applying the special structure of $\mathbf{A}$, one still can give a polynomial algorithm at least in the 2-dimensional case (Section 12.3) but for this matroids will be needed, graph theory is not enough.

Exercises for Section 6.3

6.3.1 Prove that the trivial solutions of $\mathbf{Au} = \mathbf{0}$ (determined by the congruent transformations of the space or the plane) form a subspace of dimension 6 and 3, respectively.

6.3.2 Give examples to show that Maxwell's conditions are not sufficient for rigidity.

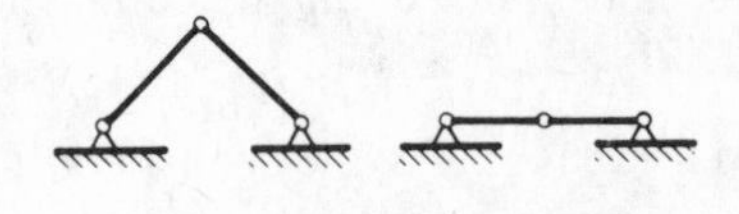

Fig. 6.18

6.3.3 Prove that the second framework in Fig. 6.18 is not rigid and generally, that our definition of rigidity prohibits the "infinitesimal" motions.

6.3.4 Prove by explicitly constructing matrix $\mathbf{A}$ that among the two frameworks in Fig. 6.12 the first is rigid and the second is not.

Problems for Section 6.3

6.3.5 *Hooke's Law* states that a small force F leads to a change of the length of a rigid body, proportional to the strength of F. Using this determine the strength of the vertical force, required for a small vertical motion of the middle joint in the two frameworks of Fig. 6.18.

6.3.6 Prove that the trivial solutions of $\mathbf{Au} = \mathbf{0}$ in the n-dimensional space form a subspace of dimension $n(n+1)/2$.

6.3.7 Are the plane frameworks of Fig. 6.19 rigid?

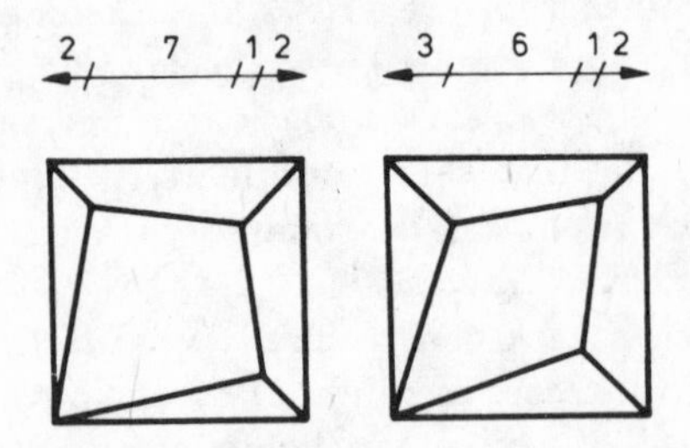

Fig. 6.19

6.3.8 Two 3-dimensional frameworks are shown in Fig. 6.20 Their general study is complicated [Foeppl, 1888; Tarnai, 1979 and 1980] but a simplified analysis, taking one "ring" instead of 6 and square shape instead of hexagonal one, still preserves some of its features. Hence determine whether the 3-dimensional frameworks of Fig. 6.21 are rigid. (The second framework is shown from above in Fig. 6.22.)

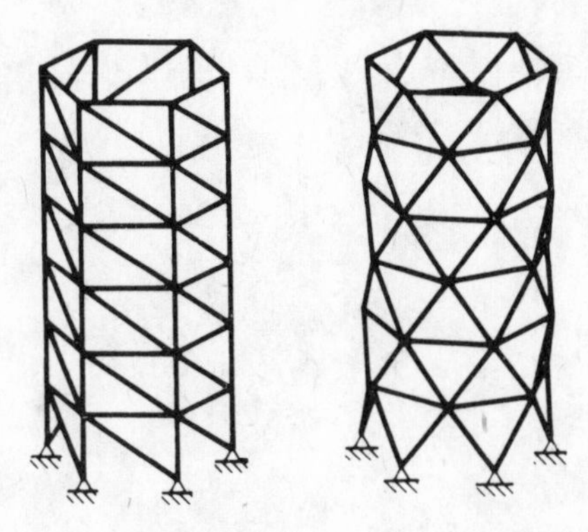

Fig. 6.20

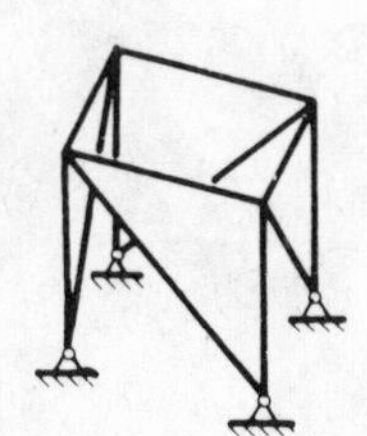

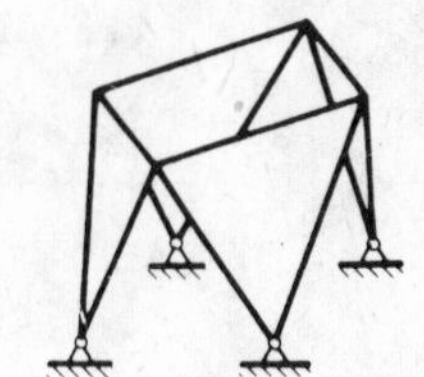

Fig. 6.21

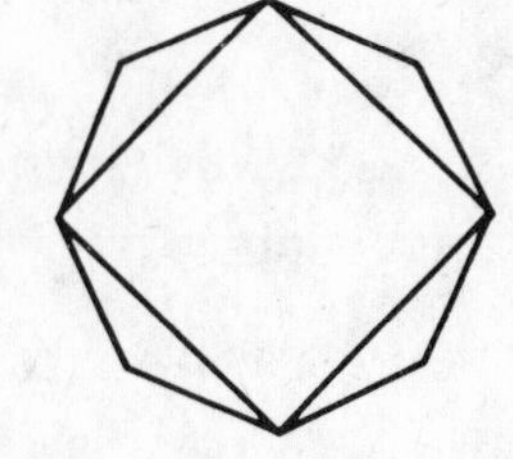

Fig. 6.22

6.3.9 Show that Corollaries 6.3.2–6.3.3 would be false if an alternative definition of rigidity excluded mechanisms only but permitted infinitesimal motions.

PART TWO

Chapter 7

Basic concepts in matroid theory

Section 7.1 Matroids; independence, base, rank

Since the introduction of 2-isomorphism we saw several examples, both in the theory and its applications, where not all the properties of a graph play a role; in most cases only subgraphs with or without circuits need to be distinguished. Similarly, sign conventions or the underlying field for the matrices of the graphs were not important; we did not care about the numerical values of the entries of the matrices, only the linear dependence or independence of the columns of the matrices.

For example, the three graphs in Fig. 7.1 are pairwise 2-isomorphic (although pairwise nonisomorphic); their edge set is $E = \{1,2,3,4,5,6\}$ and, for any of them, a subset $X \subseteq E$ of edges is circuit-free if and only if

(a) X does not contain edge 6;

(b) X contains at most two from the edges $1,2,3,4$;

(c) X contains at most one of the edges $1,2$.

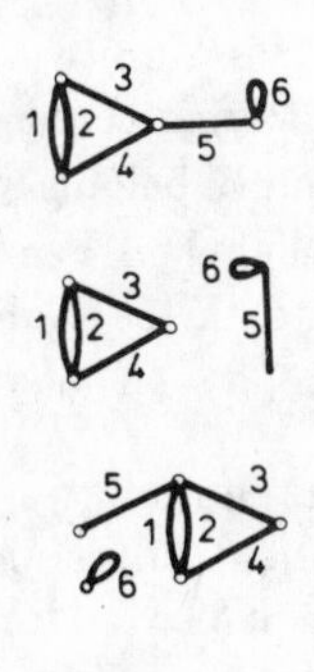

Fig. 7.1

Now consider the following four matrices

$$\mathbf{M}_1 = \begin{pmatrix} 1 & 1 & 0 & 1 & 0 & 0 \\ 0 & 0 & 1 & 1 & 0 & 0 \\ 0 & 0 & 0 & 0 & 1 & 0 \end{pmatrix}; \quad \mathbf{M}_2 = \begin{pmatrix} 1 & -1 & 0 & 1 & 0 & 0 \\ 0 & 0 & 1 & -1 & 0 & 0 \\ 0 & 0 & 0 & 0 & -1 & 0 \end{pmatrix};$$

$$\mathbf{M}_3 = \begin{pmatrix} 5 & 3 & 0 & 1 & 0 & 0 \\ 0 & 0 & -2 & -1 & 0 & 0 \\ 0 & 0 & 0 & 0 & 7 & 0 \end{pmatrix}; \quad \mathbf{M}_4 = \begin{pmatrix} 1 & 2 & 1 & 0 & 1 & 0 \\ 2 & 4 & 0 & 2 & 1 & 0 \\ 3 & 6 & 0 & 3 & 1 & 0 \end{pmatrix}$$

All of them are different but have several common properties. The last column is $\mathbf{0}$ in each, the first column is a nonzero constant times the second one in each, and $\mathbf{0}$ can be obtained in each as a linear combination of the third, fourth and either the first or the second columns. More formally, if $\mathbf{a}, \mathbf{b}, \mathbf{c}, \mathbf{d}, \mathbf{e}, \mathbf{f}$ denote their columns (in this order) then the column set is $T = \{\mathbf{a}, \mathbf{b}, \mathbf{c}, \mathbf{d}, \mathbf{e}, \mathbf{f}\}$ in each, and a subset $X \subseteq T$ is linearly independent in each if and only if

(a) $\mathbf{f} \notin X$

(b) $\{\mathbf{a}, \mathbf{b}, \mathbf{c}, \mathbf{d}\} \cap X$ has at most two elements

(c) $\{\mathbf{a}, \mathbf{b}\} \cap X$ has at most one element.

Our abstraction is by now far away from the special features of graphs or matrices. If we speak about a set $S = \{s_1, s_2, \ldots, s_6\}$ and about a collection $\mathbf{F}$ of subsets of S so that, for $X \subseteq S, X \in \mathbf{F}$ holds if and only if

(a) $s_6 \notin X$

(b) $|\{s_1, s_2, s_3, s_4\} \cap X| \leq 2$

(c) $|\{s_1, s_2\} \cap X| \leq 1$

then we cannot really say whether this pair $(S, \mathbf{F})$ came from graphs or from matrices. Such a pair $(S, \mathbf{F})$ will be called matroid and this concept will thus become a common generalization of graphs and matrices.

Should we consider an arbitrary pair $(S, \mathbf{F})$, where $\mathbf{F}$ is a collection of subsets of S, it would be such a general concept that only trivialities could be said about it. Hence we prescribe some "reasonable" properties for $\mathbf{F}$; not too many properties (since we wish a general concept) but enough to prove nontrivial statements.

The pair $(S, \mathbf{F})$ is a *matroid* if S is a finite set and $\mathbf{F}$ is a collection of certain subsets of S which satisfies the following three properties:

(F1) $\emptyset \in \mathbf{F}$ (the empty set must belong to $\mathbf{F}$);

(F2) If $X \in \mathbf{F}$ and $Y \subseteq X$ then $Y \in \mathbf{F}$ must also hold;

(F3) If $X \in \mathbf{F}$ and $Y \in \mathbf{F}$ and $|X| > |Y|$ then there must exist an element $x \in X - Y$ so that $Y \cup \{x\} \in \mathbf{F}$ also holds.

Let us return to column vectors of matrices. For any set S of vectors, $(S, \mathbf{F})$ is a matroid if the linearly independent subsets form $\mathbf{F}$. (F1) and (F2) are trivial since the empty set of vectors is by definition linearly independent, and so is any subset of any linearly independent set. (F3) can easily be verified too: The vectors of X span a subspace A_1 of dimension $|X|$, those in Y span a subspace A_2 of lower dimension, hence X must contain at least one vector not in A_2.

Similarly, if S is the edge set $E(G)$ of any graph G and $\mathbf{F}$ consists of the edge sets of the circuit free subgraphs of G then $(S, \mathbf{F})$ is a matroid. (F1) and (F2) are trivial again since a subgraph without edges has no circuit and a subgraph of a circuit free graph is circuit free again. (F3) can also be verified, see Exercise 7.1.2.

If $\mathbf{M} = (S, \mathbf{F})$ is a matroid then S is called the *underlying set* of $\mathbf{M}$. Among the subsets of S, those which belong to $\mathbf{F}$ are called *independent*, the others are called *dependent*. Two matroids $\mathbf{M}_i = (S_i, \mathbf{F}_i)$ for $i = 1, 2$ are *isomorphic* if there exists a bijection between S_1 and S_2 so that independent subsets of S_1 correspond to independent ones of S_2 and dependent subsets of S_1 correspond to dependent ones of S_2.

Let $|S| = n$ and let $\mathbf{F}$ contain those subsets of S whose cardinality is at most k (for some $0 \leq k \leq n$). Then the pair $(S, \mathbf{F})$ is obviously a matroid. This is called *uniform matroid* and denoted by $\mathbf{U}_{n,k}$. The matroid $\mathbf{U}_{n,n}$ is also called *free-matroid*; here every subset is independent. The matroid $\mathbf{U}_{n,0}$ is also called *trivial matroid*; here no subset except $\emptyset$ is independent.

Let us consider $\mathbf{U}_{3,2}$. Every proper subset of the 3-element underlying set is independent. If we consider a graph G with $e(G) = 3$ where the three edges just form a circuit then the matroid, corresponding to this graph, will be isomorphic to $\mathbf{U}_{3,2}$. On the other hand, no graph corresponds to $\mathbf{U}_{4,2}$ since no graph with four edges has the property that any three of these edges form a circuit.

A matroid $\mathbf{M}$ is *graphic* if there is a graph, corresponding to $\mathbf{M}$. Thus $\mathbf{U}_{3,2}$ is graphic while $\mathbf{U}_{4,2}$ is not. This also shows that the concept of matroids is a proper generalization of that of graphs. The following observation is straightforward:

Statement 7.1.1 *Let G_1 and G_2 be two graphs. The corresponding matroids are isomorphic if and only if G_1 and G_2 are 2-isomorphic.*

The maximal independent subsets of a matroid are called the *bases*. (Recall that maximal only means that they are not properly contained in any other independent subset. However, we shall see that they are maximum as well, all having the same cardinality.)

Theorem 7.1.2 *Let $\mathbf{B}$ denote the collection of bases of the matroid $\mathbf{M} = (S, \mathbf{F})$. Then the following properties hold:*

(B1) $\mathbf{B} \neq \emptyset$ *(i.e., $\mathbf{M}$ does have bases).*

(B2) $|X_1| = |X_2|$ *for every $X_1, X_2 \in \mathbf{B}$ (i.e., bases are equicardinal).*

(B3) *If $X_1, X_2 \in \mathbf{B}$ and $x \in X_1$ then there exists an $y \in X_2$ so that $(X_1 - \{x\}) \cup \{y\} \in \mathbf{B}$.*

Proof. Since $\mathbf{F}$ is nonempty (by (F1) in the definition of the matroids), there are independent subsets, hence maximals among them do exist as well. Thus (B1) follows. (B2) is also obvious since in the case of, say, $|X_1| > |X_2|$ we could extend X_2, by (F3). Finally (B3) is a special case of (F3) since, by (B2), $|X_1 - \{x\}| < |X_2|$ and both are independent.

□

If the matroid was obtained from a graph, $\mathbf{B}$ is the set of its spanning forests. If the matroid is $\mathbf{U}_{n,k}$ then $\mathbf{B}$ is the set of the k-element subsets.

We have seen ((B2) of Theorem 7.1.2) that all the maximal independent subsets of S are equicardinal. This is obviously true for the subsets of S as well.

Statement 7.1.3 *Let $X \subseteq S$ be an arbitrary subset of the underlying set S of a matroid $\mathbf{M} = (S, \mathbf{F})$. All the maximal independent subsets of X are equicardinal.*

This common cardinality is denoted by $r(X)$ and is called the *rank* of X. In particular, $r(S)$ is also called the *rank* of the *matroid*.

For example, the rank in the uniform matroid $\mathbf{U}_{n,k}$ is $r(X) = \min(|X|, k)$. If a matroid is determined by a graph then its rank function is the same as that defined in Section 1.2 for graphs. If a set of vectors is considered as a matroid then its rank is the same as that in linear algebra (i.e., the dimension of the subspace spanned by the vectors of the subset).

Theorem 7.1.4 *Let $r(X)$ be the rank function of the matroid $\mathbf{M} = (S, \mathbf{F})$. Then*

(R1) *r associates nonnegative integers to every subset of S.*

(R2) $0 \le r(X) \le |X|$ *for every* $X \subseteq S$.

(R3) $X \subseteq Y$ *implies* $r(X) \le r(Y)$.

(R4) $r(X) + r(Y) \ge r(X \cup Y) + r(X \cap Y)$ *for every* $X, Y \subseteq S$.

Corollary 7.1.5 *Let $r(X)$ be the rank function of the matroid $\mathbf{M} = (S, \mathbf{F})$. Then*

(R5) $r(\emptyset) = 0$.

(R6) $r(X) \le r(X \cup \{y\}) \le r(X) + 1$ *for every* $X \subseteq S$ *and* $y \in S$.

(R7) *If* $X \subseteq S$; $y, z \in S$; *and* $r(X) = r(X \cup \{y\}) = r(X \cup \{z\})$ *then* $r(X) = r(X \cup \{y, z\})$.

The property (R4), also called *submodularity* will play an important role later on.

Proofs. (R1), (R2) and (R3) are trivial consequences of the definition of rank. For proving (R4), let $A \subseteq X \cap Y$ be a maximal independent subset of $X \cap Y$. Extending A by some further element we can obtain a subset $B \subseteq X \cup Y$ so that $A \subseteq B$ and B is a maximal independent subset of $X \cup Y$, see Fig. 7.2. Obviously, $|B \cap X| + |B \cap Y| = |B| + |A|$ since exactly those elements of B are counted twice on the left hand side, which are in A. But B is independent hence so are $B \cap X$ and $B \cap Y$. Then $|B \cap X| \le r(X)$ and $|B \cap Y| \le r(Y)$ leads to $r(X) + r(Y) \ge |B| + |A| = r(X \cup Y) + r(X \cap Y)$.

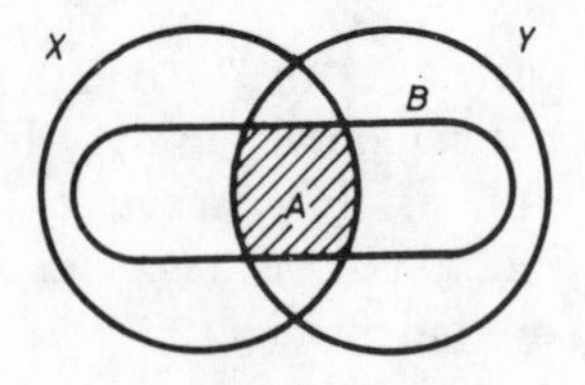

Fig. 7.2

Statements (R5), (R6) and (R7) will be deduced from (R1) through (R4). First of all, (R5) is trivial, (apply (R2) for $X = \emptyset$). $r(X) \le r(X \cup \{y\})$ is a special case of (R3), and $r(X \cup \{y\}) \le r(X) + 1$ is obtained from (R4) with the choice $Y = \{y\}$, and by using $r(\{y\}) \le 1$ from (R2). Hence (R6) also follows.

Finally, apply (R3) and also (R4) to the sets $X \cup \{y\}$ and $X \cup \{z\}$ to obtain (R7).

□

Exercises for Section 7.1

7.1.1 Let $x, y \in S$ and let $\mathbf{F}$ contain those subsets T of S for which $|T \cap \cap \{x, y\}| < 2$ hold. Is the pair $(S, \mathbf{F})$ a matroid?

7.1.2 We associated a matroid to a graph G by choosing $S = E(G)$ and $\mathbf{F}$ be the set of circuit-free subgraphs. Show that $(S, \mathbf{F})$ satisfies (F3).

7.1.3 Prove that the uniform matroid $\mathbf{U}_{n,k}$ is graphic if $k = 0, 1, n-1$ or n. (We shall see in Section 9.1 that it is nongraphic if $2 \leq k \leq n-2$.)

7.1.4 Let $\mathbf{M} = (S, \mathbf{F})$ be a rank p matroid ($p \geq 1$) and define $\mathbf{F}' = \{X | X \subset S, X \in \mathbf{F}$ and $|X| < p\}$. Prove that $(S, \mathbf{F}')$ is also a matroid. (This is called the *truncation* of $\mathbf{M}$.) Determine the truncation of the matroid, corresponding to the complete graph of 4 points.

7.1.5 Suppose that only S and the bases are known about a matroid $\mathbf{M} = = (S, \mathbf{F})$. How can $\mathbf{F}$ be obtained? And how can it be obtained if S and the rank function of $\mathbf{M}$ are known?

7.1.6 Can one of the properties below be used instead of (F2) in the definition of the matroids?

(F2′) If $X \in \mathbf{F}, Y \subset X$ and $|Y| = |X| - 1$ then $Y \in \mathbf{F}$ must also hold.

(F2″) If $X \in \mathbf{F}$ for $|X| \geq 1$ then there must exist an $x \in X$ so that $X - \{x\} \in \mathbf{F}$ also holds.

7.1.7 Can one of the properties below be used instead of (F3) in the definition of the matroids?

(F3′) If $X \in \mathbf{F}$ and $Y \in \mathbf{F}$ and $|X| = |Y| + 1$ then there must exist an element $x \in X - Y$ so that $Y \cup \{x\} \in \mathbf{F}$ also holds.

(F3″) If $X \in \mathbf{F}$ and $Y \in \mathbf{F}$ and $|X| = |Y|$ then for every $y \in Y$ there must exist an element $x \in X$ so that $(Y - \{y\}) \cup \{x\} \in \mathbf{F}$ also holds.

7.1.8 List all the nonisomorphic matroids with underlying sets of at most four elements. (Hint: There are 2,4,8 and 17 ones, respectively, if the underlying set has cardinality 1,2,3 or 4.)

Problems for Section 7.1

7.1.9* Give upper bound to the number of nonisomorphic matroids on an n-element set.

7.1.10 "Modify" $\mathbf{U}_{n,k}$ so that one of its k-element subsets be dependent (and keep the other $\binom{n}{k} - 1$ subsets independent). Prove that the result is a matroid again. Show that it becomes graphic if $n = 4$, $k = 2$.

7.1.11 Let $n = 2p$ and S_1, S_2 be two disjoint p-element subsets of the n-element set $S (p \geq 2)$. Denote by $\mathbf{F}$ the set of those subsets of S which have cardinality at most $n - 3$ and do not contain S_1 or S_2. Prove that $\mathbf{V}_p = (S, \mathbf{F})$ is a matroid (cf., Exercise 7.3.6(c) as well).

7.1.12 Give nonisomorphic graphic matroids whose truncations (see Exercise 7.1.4) are isomorphic.

7.1.13 Can the truncation of a graphic matroid be nongraphic? And *vice versa*?

7.1.14 Let S be a finite set and $\mathbf{B}$ be the set of certain subsets of S, satisfying properties (B1), (B2) and (B3) of Theorem 7.1.2. Prove that there exists a matroid $\mathbf{M} = (S, \mathbf{F})$ so that its bases are exactly the members of $\mathbf{B}$.

7.1.15 Let $\mathbf{M} = (S, \mathbf{F})$ be a rank k matroid and let $k < p < |S|$. Define $\mathbf{M}^p$ on the same set S so that those p-element subsets be its bases which contain at least one base of $\mathbf{M}$. Show that $\mathbf{M}^p$ is a matroid. Give examples that $\mathbf{M}$ is graphic and $\mathbf{M}^p$ is not, or *vice versa*, or both are graphic.

7.1.16 Problems 7.2.26 and 13.2.6 later will show that the properties below, though stronger than (B3) in Theorem 7.1.2, are still true:

(B3′) If $X_1, X_2 \in \mathbf{B}$ and $x \in X_1$ then there exists an $y \in X_2$ so that $(X_1 - \{x\}) \cup \{y\} \in \mathbf{B}$ and $(X_2 - \{y\}) \cup \{x\} \in \mathbf{B}$ both hold.

(B3″) If $X_1, X_2 \in \mathbf{B}$ and $A_1 \subseteq X_1$ then there exists an $A_2 \subseteq X_2$ so that $(X_1 - A_1) \cup A_2 \in \mathbf{B}$ and $(X_2 - A_2) \cup A_1 \in \mathbf{B}$ both hold.

Show, however, that one cannot usually give a bijection between two arbitrary bases X_1, X_2 so that, denoting the pair of x by x', $(X_1 - \{x\}) \cup \{x'\} \in \mathbf{B}$ and $(X_2 - \{x'\}) \cup \{x\} \in \mathbf{B}$ both hold. Why is it in no contradiction to (B3′).

7.1.17 Let r_1 be the rank function of $\mathbf{M}_1 = (S, \mathbf{F})$ and r_2 be that of $\mathbf{M}_2$, obtained by the truncation of $\mathbf{M}_1$. Show that $r_1(X) + r_2(Y) \geq r_1(X \cap Y) + r_2(X \cup Y)$ holds for every $X, Y \subseteq S$.

7.1.18 Let S be a finite set and r be a function satisfying (R1) through (R4) in Theorem 7.1.4. Prove that there exists a matroid $\mathbf{M} = (S, \mathbf{F})$ so that r is its rank function.

7.1.19 Let S be a finite set and r be an integer valued function on it, satisfying (R5), (R6) and (R7) in Corollary 7.1.5. Prove that there exists a matroid $\mathbf{M} = (S, \mathbf{F})$ so that r is its rank function.

7.1.20 Show that (R3) and (R4) together are equivalent to the following property:

(R′) $r(X) + r(X \cup Y \cup Z) \leq r(X \cup Y) + r(X \cup Z)$ for every $X, Y, Z \subseteq S$.

7.1.21 Consider the following property:

(R″) $r(X) + r(X \cup Y \cup Z) + r(X \cup Y \cup W) + r(Z \cup W) \leq r(X \cup Y) + r(X \cup Z) + r(X \cup W) + r(Y \cup Z) + r(Y \cup W)$ for every $X, Y, Z, W \subseteq S$.

Show that this, too, implies (R3) and (R4), though not *vice versa*, see Problem 9.2.14.

7.1.22 Let $r(X)$ be the rank function of the matroid $\mathbf{M} = (S, \mathbf{F})$ and let c be a nonnegative integer. Prove that $f(X) = \min(|X|, r(X) + c)$ is also the rank function of a matroid on S.

Section 7.2 Circuits, duality, cut sets

Property (F2) in the definition of matroids shows that subsets of independent sets are also independent. Similarly, if $X \subseteq Y$ and X is dependent then so is Y. Just like in the case of independent sets it was enough to consider the maximal ones (i.e., the bases), in the case of dependent sets it will be enough to consider the minimal ones. They will be called the circuits.

The subset $X \subseteq S$ in a matroid $\mathbf{M} = (S, \mathbf{F})$ is called a *circuit* if $X \notin \mathbf{F}$ but $Y \in \mathbf{F}$ for every proper subset $Y \subset X$. Obviously, $r(X) = |X| - 1$ if X is a circuit.

If $\mathbf{M}$ is graphic then this concept coincides with that of graph theory. However, if $\mathbf{M}$ was obtained from a set of vectors then its circuits cannot be recognized so easily. Of course, if the zero vector arises in the set then it is a single element circuit and if there are parallel vectors (i.e., $\mathbf{a} \neq \mathbf{0}$ and $\mathbf{b} \neq \mathbf{0}$ and there is a number λ so that $\mathbf{a} = \lambda\mathbf{b}$) then $\{\mathbf{a}, \mathbf{b}\}$ is a 2-element circuit. More generally, the set $\{\mathbf{a}_1, \mathbf{a}_2, \ldots, \mathbf{a}_k\}$ of vectors is a circuit in $\mathbf{M}$ if and only if they together span a $(k-1)$-dimensional subspace only but this subspace is spanned by any $k-1$ of these vectors as well.

An element $x \in S$ is called a *loop* if $\{x\} \notin \mathbf{F}$; the elements $x_1, x_2, \ldots, x_k \in S$ are called *parallel* if none of them is a loop (i.e., $\{x_i\} \in \mathbf{F}$ for every i) but for any pair of them $\{x_i, x_j\} \notin \mathbf{F}$; see Exercise 7.2.1 as well.

Theorem 7.2.1 *Let* $\mathbf{C}$ *denote the collection of circuits of the matroid* $\mathbf{M} = (S, \mathbf{F})$. *Then the following properties hold:*

(C1) *If* $X \in \mathbf{C}, Y \in \mathbf{C}$ *and* $X \subseteq Y$ *then* $X = Y$ *(i.e., no proper subset of a circuit is a circuit).*

(C2) *If* $X \in \mathbf{C}, Y \in \mathbf{C}, X \neq Y$ *and* $u \in X \cap Y$ *then there exists a* $Z \in \mathbf{C}$ *so that* $Z \subseteq X \cup Y$ *and* $u \notin Z$ *(see Fig. 7.3).*

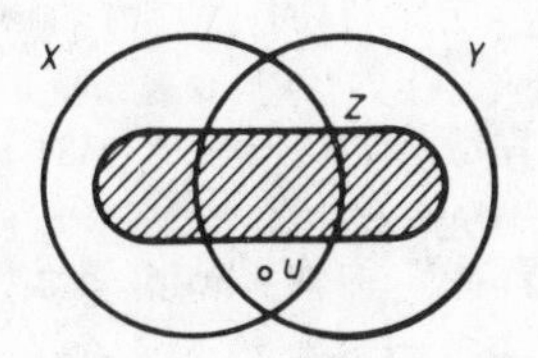

Fig. 7.3

Proof. (C1) trivially follows from the definition. In order to obtain (C2) we have to prove that the set $T = (X \cup Y) - \{u\}$ is dependent (and hence contains a circuit). Suppose that $T \in \mathbf{F}$, i.e. $r(T) = |T|$ would hold. Then

$$r(T) \leq r(X \cup Y) \leq r(X) + r(Y) - r(X \cap Y) =$$

$$= |X| - 1 + |Y| - 1 - |X \cap Y| = |X \cup Y| - 2 = |T| - 1,$$

a contradiction. (We used properties (R3) and (R4) for the rank function, and that $X \cap Y \in \mathbf{F}$, which is trivial by (C1).)

□

Corollary 7.2.2 *The following property also holds for the collection* $\mathbf{C}$ *of circuits:*

(C3) *If* $X \in \mathbf{C}$, $Y \in \mathbf{C}$, $u \in X \cap Y$ *and* $x \in X - Y$ *then there exists a* $Z \in \mathbf{C}$ *so that* $Z \subseteq X \cup Y$, $u \notin Z$ *and* $x \in Z$.

Its *proof* is more complicated, see Problem 7.2.19. It states that $T = (X \cup \cup Y) - \{u\}$ is not only dependent but even if we prescribe an element x contained in exactly one of X and Y, there is a circuit in T containing x.

Theorem 7.2.3 *If* X *is independent in the matroid* $\mathbf{M} = (S, \mathbf{F})$ *but* $X \cup \{x\}$ *contains a circuit* C, *for some* $x \in S - X$, *then* $(X \cup \{x\}) - \{y\} \in \mathbf{F}$ *for every* $y \in C$.

This important statement expresses that, roughly, if joining x leads to dependence then exactly one circuit arises.

Proof. Should $(X \cup \{x\}) - \{y\}$ contain a circuit C' then $x \in C'$ by $X \in \mathbf{F}$. Hence, by property (C2) of Theorem 7.2.1, there was a circuit in $(C \cup C') - \{x\}$, contradicting $X \in \mathbf{F}$ since $(C \cup C') - \{x\} \subseteq X$.

□

During the study of the dual of planar graphs we saw that spanning forests of the graph are the complements of the spanning forests of its dual. This can be generalized for matroids.

Theorem 7.2.4 *Let* $\mathbf{B}$ *be the collection of bases in the matroid* $\mathbf{M} = (S, \mathbf{F})$. *Put* $\mathbf{B}' = \{X | \, S - X \in \mathbf{B}\}$, *i.e. exchange every set with its complement on the "list" of the bases. Then there exists a matroid* $\mathbf{M}' = (S, \mathbf{F}')$, *so that* $\mathbf{B}'$ *is the collection of its bases.*

Proof. By Problem 7.1.14, only (B1), (B2) and (B3) of Theorem 7.1.2 must be shown. (B1) and (B2) are trivial. (If $\mathbf{B}$ was nonempty and its elements were equicardinal then so is $\mathbf{B}'$.) In order to prove (B3) let $X_1, X_2 \in \mathbf{B}'$ and $x \in X_1$. We have to prove that there always exists an $y \in X_2$ so that $(X_1 - \{x\}) \cup \{y\} \in \mathbf{B}'$, see Fig. 7.4. (On the figure $x \in X_1 - X_2$ holds but we shall not use this.) Since $S - X_1$ was a base of $\mathbf{M}$, the set $(S - X_1) \cup \{x\}$ contains a circuit C with $x \in C$. If C contains at least one element of X_2 then this is chosen to y; the set $[(S - X_1) \cup \{x\}] - \{y\}$ is a base in $\mathbf{M}$ by Theorem 7.2.3 and its complement is in $\mathbf{B}'$. The only trouble was in the case of $C \cap X_2 = \emptyset$. However, $X_2 \in \mathbf{B}'$ implies that $S - X_2$ is a base of $\mathbf{M}$, thus it cannot contain the circuit C.

□

Hence, this matroid $\mathbf{M}'$ (with base set $\mathbf{B}'$) exists by Problem 7.1.14 and is obviously unique. It will be called the *dual* of $\mathbf{M} = (S, \mathbf{F})$ and will be denoted by $\mathbf{M}^* = (S, \mathbf{F}^*)$. Unlike the case of graphs, every matroid has a dual (see Exercise 7.2.7).

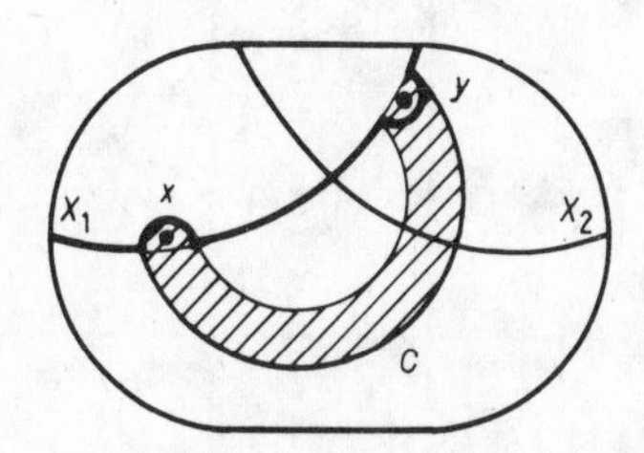

Fig. 7.4

Theorem 7.2.5 *If* $\mathbf{M}^* = (S, \mathbf{F}^*)$ *is the dual of* $\mathbf{M} = (S, \mathbf{F})$ *then* $(\mathbf{M}^*)^* = \mathbf{M}$. *Furthermore, for the rank function* r^* *of* $\mathbf{M}^*$ *we have*

$$r^*(X) = |X| + r(S - X) - r(S). \tag{7.1}$$

The first statement is trivial, the second is *proved* in Problem 7.2.24.

□

The circuits of $\mathbf{M}^*$ are called the *cut sets* of $\mathbf{M}$. Since every matroid has a dual, we immediately obtain the following results.

Statement 7.2.6 *Let* X, Y *be cut sets in the matroid* $\mathbf{M} = (S, \mathbf{F})$. *Then* $X \subseteq Y$ *implies* $X = Y$. *If* $X \neq Y$ *and* $u \in X \cap Y$ *then* $\mathbf{M}$ *has a cut set* Z *with* $Z \subseteq (X \cup Y) - \{u\}$. *Furthermore, if* T *is a base and* X *is a cut set then* $X \cap T \neq \emptyset$.

The element $x \in S$ is a *bridge* of the matroid $\mathbf{M} = (S, \mathbf{F})$ if $\{x\}$ is a cut set of $\mathbf{M}$. This happens if and only if x is a loop in $\mathbf{M}^*$. If none of $x, y \in S$ is a bridge and $\{x, y\}$ is a cut set of $\mathbf{M}$ then x, y are *series elements* in $\mathbf{M}$. This happens if and only if x, y are parallel in $\mathbf{M}^*$.

Theorem 7.2.7 *Let* $\mathbf{C}$ *be the collection of circuits in the matroid* $\mathbf{M} = (S, \mathbf{F})$. *Then a subset* $X \subseteq S$ *is a cut set of* $\mathbf{M}$ *if and only if*
(a) $X \neq \emptyset$.
(b) $|X \cap Y| \neq 1$ *for every* $Y \in \mathbf{C}$.
(c) X *is minimal with respect to these properties.*

By this theorem circuits and cut sets of a matroid can be characterized by each other, with no reference to bases, duality etc. Its *proof* is Problem 7.2.28. Such characterizations are listed in **Box 7.1.**

Graphic matroids were introduced in Section 7.1. If matroid $\mathbf{M}$ "belongs" to the graph G then the circuits of $\mathbf{M}$ are just the circuits of G (in the old sense of graph theory). Now we can associate another matroid $\mathbf{N}$ to G as well; the circuits of $\mathbf{N}$ are just cut sets of G. Obviously, $\mathbf{N} = \mathbf{M}^*$ and if G happens to be planar and G^* is a dual of it (in the sense of graph theory) then $\mathbf{N}$ just "belongs" to G^*.

Box 7.1 Basic concepts of matroid theory characterized by each other

The matroid $\mathbf{M} = (S, \mathbf{F})$ is given by	A subset $X \subseteq S$ is independent if and only if	a base if and only if	a circuit if and only if	a cut set if and only if	The rank of subset $X \subseteq S$ equals to
the system $\mathbf{F}$ of independent sets	$X \in \mathbf{F}$	X is maximal with $X \in \mathbf{F}$	X is minimal with $X \notin \mathbf{F}$	X is minimal with $(X \cap Y \neq \emptyset$ for every maximal $Y \in \mathbf{F})$	$\max_{Y \in \mathbf{F}} \lvert X \cap Y \rvert$
the system $\mathbf{B}$ of bases	X is a subset of an element of $\mathbf{B}$	$X \in \mathbf{B}$	X is minimal with $(X \not\subseteq Y$ for every $Y \in \mathbf{B})$	X is minimal with $(X \cap Y \neq \emptyset$ for every $Y \in \mathbf{B})$	$\max_{Y \in \mathbf{B}} \lvert X \cap Y \rvert$
the system $\mathbf{C}$ of circuits	$Y \subseteq X$ for no $Y \in \mathbf{C}$	X is maximal with $(Y \subseteq X$ for no $Y \in \mathbf{C})$	$X \in \mathbf{C}$	X is min. nonempty with $(\lvert X \cap Y \rvert \neq 1$ for every $Y \in \mathbf{C})$	$\max\{\lvert Y \rvert; Y \subseteq X$ and Y contains no element of $\mathbf{C}\}$
the system $\mathbf{Q}$ of cutsets	X is contained in a min. Z with $(Z \cap Y \neq$ $\neq \emptyset$ for every $Y \in \mathbf{Q})$	X is minimal with $(X \cap Y \neq \emptyset$ for every $Y \in \mathbf{Q})$	X is min. nonempty with $(\lvert X \cap Y \rvert \neq 1$ for every $Y \in \mathbf{Q})$	$X \in \mathbf{Q}$	
its rank function r	$\lvert X \rvert = r(X)$	$\lvert X \rvert = r(X) = r(S)$	$\lvert X \rvert > r(X)$ and $\lvert Y \rvert = r(Y)$ for every $Y \subset X$	X is minimal with $r(S - X) < r(S)$	$r(X)$

See [Brylawski, 1986b] for a more detailed table

In what follows, $\mathbf{M}$ and $\mathbf{N}$ will be called the *circuit matroid* and the *cut set matroid* of G, respectively, and they will be denoted by $\mathbf{M}(G)$ and $\mathbf{M}^*(G)$, respectively. That is, both $\mathbf{M}(G)$ and $\mathbf{M}^*(G)$ are defined on the edge set $E(G)$ of the graph G and $X \subseteq E$ is independent in $\mathbf{M}(G)$ or in $\mathbf{M}^*(G)$ if and only if X, as a subgraph of G, is circuit-free or cut set free, respectively, in G. A matroid was called graphic if it arises as the circuit matroid of some graph. A matroid will be called *cographic* if it arises as the cut set matroid of some graph. Of course, a matroid is cographic if and only if its dual is graphic, and a matroid is graphic and cographic at the same time if and only if it is the circuit matroid of a planar graph, by Theorem 3.2.7.

Exercises for Section 7.2

7.2.1 Show that if x is parallel to y and y to z then either $x = z$ or they are parallel.

7.2.2 Characterize the subset $X \subseteq S$ if it intersects every circuit of the matroid $\mathbf{M} = (S, \mathbf{F})$.

7.2.3 Let C be a circuit and $x \in C$. Prove that there exists a base B so that C is the unique circuit contained in $B \cup \{x\}$.

7.2.4 Prove that x belongs to every base if and only if it belongs to no circuit.

7.2.5 Let $X, Y \subseteq S$ and define their *symmetric difference* $X \triangle Y = (X - Y) \cup \cup (Y - X)$. If X, Y are circuits of a graphic matroid then $X \triangle Y$ is either a circuit or the union of disjoint circuits. Give examples that this may or may not be true for nongraphic matroids.

7.2.6 A matroid is *simple* if every circuit of it has at least three elements. Show that $\mathbf{M}(G)$ is simple if and only if G is a simple graph. Give an example that $\mathbf{M}$ is simple and $\mathbf{M}^*$ is not.

7.2.7 What is wrong with the following reasoning: Every graph is a matroid and every matroid has a dual hence every graph must have a dual, not the planar ones only, as claimed in Theorem 3.2.7.

7.2.8 Can $X \subseteq S$ be circuit and cut set, at the same time, in a matroid $\mathbf{M} = (S, \mathbf{F})$?

7.2.9 What is the graph theory meaning of the rank function of the cut set matroid of a graph?

7.2.10 Let $A, B \subseteq S$, $A \cap B = \emptyset$ and $A \in \mathbf{F}, B \in \mathbf{F}^*$ in the matroid $\mathbf{M} = (S, \mathbf{F})$. Prove that $\mathbf{M}$ has a base K with $A \subseteq K$ and $B \cap K = \emptyset$.

7.2.11 Let C be a circuit and D be a cut set in a matroid $\mathbf{M}$. Prove $|C \cap D| \neq 1$. Is $|C \cap D| = 3$ possible?

7.2.12 Define the fundamental system of circuits and cut sets of a matroid, with respect to a base.

7.2.13 Let the cut sets $D_1, D_2, \ldots$ be disjoint and choose an arbitrary element $x_i \in D_i$ from each. Show that $\{x_1, x_2, \ldots\}$ is independent.

7.2.14 Prove that x is a loop if and only if $x \notin B$ for every base of the matroid.

7.2.15 Suppose that only S and the circuits are known about a matroid $\mathbf{M} = (S, \mathbf{F})$. How can $\mathbf{F}$ be obtained? And how can it be obtained if S and the cut sets are known?

7.2.16 Define the series and the parallel extensions of an element in a matroid.

Problems for Section 7.2

7.2.17 Let $\mathbf{M}_1 = (S, \mathbf{F}_1)$ and $\mathbf{M}_2 = (S, \mathbf{F}_2)$, let $X \subseteq S$ be a common base of them. Give an example that if their systems of fundamental circuits with respect to X are identical, they still may be nonisomorphic.

7.2.18 Let S be a finite set and C be the set of certain nonempty subsets of S, satisfying properties (C1) and (C2) in Theorem 7.2.1. Prove that there exists a matroid $\mathbf{M} = (S, \mathbf{F})$ so that its circuits are exactly the members of $\mathbf{C}$.

7.2.19 Prove Corollary 7.2.2.

7.2.20 Prove that the set $\mathbf{C}$ of circuits of the matroid $\mathbf{M} = (S, \mathbf{F})$ satisfies the following property:

(C4) If $X \in \mathbf{C}$ and $Y \in \mathbf{C}$ and $X \cap Y \neq \emptyset$ then for every $x \in X$ and $y \in Y$ there exists a circuit $Z \in \mathbf{C}$ so that $x, y \in Z$ and $Z \subseteq X \cup Y$.

7.2.21 Show by an example that, in (C4) of the previous problem, we cannot have an additional requirement of $u \notin Z$ for an *a priori* given $u \in X \cap Y$.

7.2.22 The matroid $\mathbf{M} = (S, \mathbf{F}_1)$ is a *refinement* of the matroid $\mathbf{N} = (S, \mathbf{F}_2)$ if every circuit of $\mathbf{N}$ is a circuit of $\mathbf{M}$ as well [Bean, 1972]. Prove that $\mathbf{N}$ has a proper refinement if and only if it does not have any circuit of cardinality $r(S) + 1$.

7.2.23 Let x, y be distinct elements of a circuit X of a matroid. Prove that the matroid has a cut set Y so that $X \cap Y = \{x, y\}$.

7.2.24 Deduce the formula $r^*(X) = |X| + r(S - X) - r(S)$ for the rank function of the dual matroid (Theorem 7.2.5).

7.2.25 Let $z \in S$ be an arbitrary element in the matroid $\mathbf{M} = (S, \mathbf{F})$. Consider an arbitrary partition of $S - \{z\}$ into $A \cup B$. Show that either there exists a circuit X in $\mathbf{M}$ with $z \in X$ and $B \cap X = \emptyset$ or there exists a cut set Y in $\mathbf{M}$ with $z \in Y$ and $A \cap Y = \emptyset$.

7.2.26 Prove property (B3′) of Problem 7.1.16.

7.2.27 Let $\mathbf{M}$ be a matroid. Prepare the truncation of its dual $\mathbf{M}^*$ and denote the dual of the result by $\mathbf{N}$. Determine the rank function of $\mathbf{N}$ (cf., Problem 7.1.22).

7.2.28 Prove Theorem 7.2.7.

Section 7.3 Minors, direct sum, connectivity

At first, we generalize two well known graph theory concepts, the deletion and the contraction of edges.

If $\mathbf{M} = (S, \mathbf{F})$ is an arbitrary matroid and $X \subset S$ then define $\mathbf{F}'$ so that for $Y \subseteq S - X$ let $Y \in \mathbf{F}'$ if and only if $Y \in \mathbf{F}$. (That is, consider those independent subsets of $\mathbf{M}$ which are disjoint from X.) Then $\mathbf{F}'$ is the collection of independent subsets of a matroid $\mathbf{M}'$ on $S - X$. This matroid will be denoted by $\mathbf{M}\backslash X$ and called the *deletion* of X from $\mathbf{M}$ or the *restriction* of $\mathbf{M}$ to $S - X$.

The generalization of the contraction is more complicated. Even if a single element $x \in S$ is to be contracted, we must distinguish between two cases. If x is a loop then its contraction and deletion are just the same (as it was in the case of graphs as well). If x is not a loop then $Y \subseteq S - \{x\}$ will be considered independent (after the contraction) if and only if $Y \cup \{x\}$ was independent before the contraction.

Hence contraction of a subset could be defined as successive contractions of its elements but it is not clear whether this definition is independent of the order of the single element contractions. Another possible way to define contractions of a set X is to choose a maximum independent subset X_0 of X and to say that $Y \subseteq S - X$ will be independent if and only if $Y \cup X_0$ was independent in the original matroid. But in this case one should prove that this definition is independent of the actual choice of X_0, which is also not trivial.

In order to avoid these problems, contraction will be defined *via* the rank function rather than by describing the independent sets. Let r be the rank function of the matroid $\mathbf{M} = (S, \mathbf{F})$; let $X \subset S$ be arbitrary and define the function

$$R(Y) = r(X \cup Y) - r(X) \tag{7.2}$$

for every subset Y of the set $S - X$.

Theorem 7.3.1 *R is the rank function of a matroid on the set $S - X$.*

Proof. By Problem 7.1.18 all we have to prove are (R1),...,(R4) from Theorem 7.1.4. (R1) and (R3) are trivial. For (R2) we must show $R(Y) \geq 0$ (this is obvious) and $R(Y) \leq |Y|$. This is also clear if we rewrite it into $r(X \cup Y) \leq$ $\leq r(X) + |Y|$, which follows by property (R6) of r. Finally, we wish to prove $R(Y_1) + R(Y_2) \geq R(Y_1 \cup Y_2) + R(Y_1 \cap Y_2)$. First let us rewrite it into

$$r(X \cup Y_1) + r(X \cup Y_2) \geq r(X \cup (Y_1 \cup Y_2) + r(X \cup (Y_1 \cap Y_2))$$

Then observe that $X \cup (Y_1 \cap Y_2) = (X \cup Y_1) \cap (X \cup Y_2)$ and apply (R4) to r. □

The matroid with rank function $R(X)$ on the set $S - X$ is denoted by $\mathbf{M}/X$ and is called the *contraction* of $\mathbf{M}$ to $S - X$ (or we shall speak about the contraction of X from $\mathbf{M}$ as well).

Theorem 7.3.2 *Let X_0 be a maximum independent subset of X in a matroid $\mathbf{M}$. Then $Y \subseteq S - X$ is independent in $\mathbf{M}/X$ if and only if $Y \cup X_0$ is independent in $\mathbf{M}$.*

Its *proof* is left to the reader as Problem 7.3.12. Hence all the properties, mentioned (and expected) before the definition of the contraction, are really true.

By deleting or contracting subsets of S, many new matroids can be obtained from an original one $\mathbf{M} = (S, \mathbf{F})$. The result of a sequence of deletions and contractions is called a *minor* of $\mathbf{M}$.

Statement 7.3.3 *If* $\mathbf{N}$ *is a minor of* $\mathbf{M} = (S, \mathbf{F})$ *then there are two disjoint subsets* $A, B \subset S$ *so that* $\mathbf{N} = (\mathbf{M}\backslash A)/B$. *(That is, every minor can be obtained by just one suitable deletion and just one suitable contraction.)*

Proof. The relations $(\mathbf{M}\backslash A_1)\backslash A_2 = \mathbf{M}\backslash(A_1 \cup A_2)$ and $(\mathbf{M}/B_1)/B_2 = \mathbf{M}/(B_1 \cup$ $\cup B_2)$ are obvious, hence our statement follows if we prove

$$(\mathbf{M}\backslash A)/B = (\mathbf{M}/B)\backslash A. \tag{7.3}$$

Let B_0 a maximum independent subset of B in $\mathbf{M}$. Since $A \cap B = \emptyset$, B_0 is also a maximum independent subset of B in $\mathbf{M}\backslash A$. Hence, in either matroid of Eq. (7.3), a subset $Y \subseteq S - (A \cup B)$ is independent if and only if $Y \cup B_0 \in \mathbf{F}$. Thus the two matroids are the same.

□

The importance of these two operations is further emphasized by the fact that they are the "dual" of each other.

Theorem 7.3.4 *Let* $X \subseteq S$ *be an arbitrary subset in the matroid* $\mathbf{M} = (S, \mathbf{F})$. *Then*

$$(\mathbf{M}/X)^* = \mathbf{M}^*\backslash X \text{ and } (\mathbf{M}\backslash X)^* = \mathbf{M}^*/X. \tag{7.4}$$

Proof. We prove the second statement only, since if it is true then its dual $\mathbf{N}\backslash X = (\mathbf{N}^*/X)^*$ is also true for every $\mathbf{N}$, including $\mathbf{N} = \mathbf{M}^*$. Applying Eq. (7.1) for the rank function of $\mathbf{M}^*$, we obtain that the rank function r_1 of $\mathbf{M}^*/X$ is

$$\begin{aligned} r_1(Y) &= r^*(X \cup Y) - r^*(X) = \\ &= [|X \cup Y| + r(S - X - Y) - r(S)] - [|X| + r(S - X) - r(S)] = \\ &= |Y| + r(S - X - Y) - r(S - X). \end{aligned}$$

On the other hand, the rank function r_2 of $(\mathbf{M}\backslash X)^*$ is $r_2(Y) = |Y| + r(T - Y) -$ $- r(T)$, where T is the underlying set of $\mathbf{M}\backslash X$, hence $T = S - X$. Thus $r_1(Y) = r_2(Y)$ for every Y which proves the assertion.

□

Another useful operation on matroids can "increase" the size of the underlying set. If $\mathbf{M}_1 = (S_1, \mathbf{F}_1)$ and $\mathbf{M}_2 = (S_2, \mathbf{F}_2)$ are two matroids with disjoint underlying sets, i.e. if $S_1 \cap S_2 = \emptyset$ then a new matroid $\mathbf{N}$ can be defined on $S_1 \cup S_2$ so that a subset $X \subseteq S_1 \cup S_2$ be independent in $\mathbf{N}$ if and only if $S_1 \cap X \in \mathbf{F}_1$ and $S_2 \cap X \in \mathbf{F}_2$. This matroid $\mathbf{N}$ is called the *direct sum* of the matroids $\mathbf{M}_1$ and $\mathbf{M}_2$, and is denoted by $\mathbf{M}_1 \oplus \mathbf{M}_2$. It satisfies the matroid axioms (Exercise 7.3.5) and can be generalized to more than two summands in a straightforward way.

For example, if the 2-connected components of a graph G are $G_1, G_2, \ldots$ then $\mathbf{M}(G) = \mathbf{M}(G_1) \oplus \mathbf{M}(G_2) \oplus \ldots$.

If a matroid arises as the direct sum of simpler matroids then most problems concerning the original matroid can be reduced to the study of the summands. Let us call a matroid *connected* if it does not arise as direct sum.

Statement 7.3.5 *The matroid $\mathbf{M} = (S, \mathbf{F})$ is connected if and only if, for any decomposition $S = S_1 \cup S_2$ of its underlying set into two disjoint nonempty subsets, there is a circuit C with $C \cap S_1 \neq \emptyset$ and $C \cap S_2 \neq \emptyset$.*

Proof. If $\mathbf{M} = \mathbf{M}_1 \oplus \mathbf{M}_2$ then no such C exists for the decomposition of S into the underlying sets S_1 and S_2 of $\mathbf{M}_1$ and $\mathbf{M}_2$, respectively, since $C \cap S_1$ and $C \cap S_2$ were independent in $\mathbf{M}_1$ and in $\mathbf{M}_2$, respectively, hence their union would not be in a circuit in $\mathbf{M}$.

On the other hand, if $S = S_1 \cup S_2$ is a decomposition with the required properties then let $\mathbf{M} \backslash S_2$ and $\mathbf{M} \backslash S_1$ be denoted by $\mathbf{M}_1$ and $\mathbf{M}_2$, respectively. We show that $\mathbf{M} = \mathbf{M}_1 \oplus \mathbf{M}_2$. If X is independent in $\mathbf{M}$ then $X \cap S_i$ is independent in $\mathbf{M}_i$ by the definition of the deletion. If X is dependent in $\mathbf{M}$ then it contains a circuit C which must be entirely contained in one of the S_i's, but then $X \cap S_i$ is also dependent.

□

The following — more difficult — theorem is obviously stronger.

Theorem 7.3.6 *$\mathbf{M} = (S, \mathbf{F})$ is connected if and only if for every pair x, y of elements of S there exists a circuit of $\mathbf{M}$ containing x and y.*

Proof. The sufficiency of the condition is trivial, the necessity is left to the reader as Problem 7.3.16.

□

Observe that this is just Theorem 3.3.4 if $\mathbf{M}$ was graphic. Since the terminology is somewhat misleading, let us emphasize again:

Statement 7.3.7 *Let G be a graph with $v(G) \geq 3$ and without isolated points. Then $\mathbf{M}(G)$ is connected if and only if G is 2-connected.*

See Cunningham [1981] and Oxley [1981] for higher connectivity numbers of matroids.

Exercises for Section 7.3

7.3.1 Are the minors of the uniform matroids uniform?

7.3.2 Show that $X \subseteq S - A$ is a circuit in $\mathbf{M}/A$ if and only if it is nonempty, arises as $X = Y \cap (S - A)$ from a circuit Y of $\mathbf{M}$, and is minimal with respect to these two properties.

7.3.3 Let C be a circuit in $\mathbf{M} = (S, \mathbf{F})$ and $x \in S - C$. Suppose that C is not a circuit in $\mathbf{M}/\{x\}$. Then, for $\mathbf{M}$ graphic, C is the disjoint union of two circuits of $\mathbf{M}/\{x\}$. Is this true for an arbitrary $\mathbf{M}$ as well?

7.3.4 Define series contraction. Describe the bases and the circuits of the result.

7.3.5 Prove that the direct sum of some matroids is a matroid again.

7.3.6 (a) Let $S_1, S_2, \ldots, S_p$ be the *partition* of S, i.e. let S be the union of the disjoint, nonempty subsets S_i. Consider, furthermore, p nonnegative integers $a_1, a_2, \ldots, a_p$ so that $a_i \leq |S_i|$. Let $\mathbf{F}$ consist of those subsets X of S which meet $|X \cap S_i| \leq a_i$ for every i. Prove that $(S, \mathbf{F})$ is a matroid. (Matroids, arising in this way, are called *partitional matroids*).
(b) Which is the smallest nonpartitional matroid?
(c) Give a new proof for Problem 7.1.11.

7.3.7 Let $\mathbf{M} = \mathbf{M}_1 \oplus \mathbf{M}_2 \oplus \ldots \oplus \mathbf{M}_k$. Characterize the bases, circuits and the rank function of $\mathbf{M}$ in terms of those of the $\mathbf{M}_i$'s.

7.3.8 (continued) Describe $\mathbf{M}^*$; characterize the cut sets of $\mathbf{M}$.

7.3.9 The subset $T \subseteq S$ is a *separator* of the matroid $\mathbf{M} = (S, \mathbf{F})$ if $\mathbf{M}$ arises as the direct sum of $\mathbf{M} \backslash T$ and $\mathbf{M} \backslash (S - T)$. Prove that the following statements are equivalent:
(1) T is a separator.
(2) Either $C \subset T$ or $C \cap T = \emptyset$ for every circuit C of $\mathbf{M}$.
(3) $\mathbf{M} \backslash T = \mathbf{M}/T$.
(4) $r(T) + r(S - T) = r(S)$.

7.3.10 Prove that $\mathbf{M} = (S, \mathbf{F})$ is connected if and only if $r(X) + r(S - X) > r(S)$ for every proper subset X of S.

7.3.11 Show that $\mathbf{M}$ is connected if and only if $\mathbf{M}^*$ is connected.

Problems for Section 7.3

7.3.12 Prove Theorem 7.3.2.

7.3.13 Show that the decomposition of a matroid into connected direct summands is unique.

7.3.14 Let $\mathbf{M} = (S, \mathbf{F})$ be connected and $x \in S$. Prove that at least one of $\mathbf{M}/\{x\}$ and $\mathbf{M} \backslash \{x\}$ is connected.

7.3.15 (continued) Show that both $\mathbf{M}/A$ and $\mathbf{M} \backslash A$ may be disconnected if $\mathbf{M}$ was connected and $A \subseteq S$.

7.3.16 Prove Theorem 7.3.6.

7.3.17 Two circuits X_1, X_2 of a matroid are neighbours if the union of any two of the three subsets $X_1 \cap X_2, X_1 - X_2$ and $X_2 - X_1$ is a circuit. A matroid is connected if and only if it has no bridges and any circuit can be reached from any other along a sequence of neighbouring circuits, see [Tutte, 1965]. Prove the sufficiency of this condition (the necessity is much more complicated).

7.3.18 Let $X \cup Y \neq S$ and $\mathbf{M} \backslash X$ and $\mathbf{M} \backslash Y$ be connected. Show that $\mathbf{M} \backslash (X \cap Y)$ is also connected. Reformulate the statement by using the expression "restriction".

Section 7.4 Greedy algorithms II: Matroids

In this section we prove at first that the greedy algorithm (see Section 1.5) is optimal for matroids as well as for graphs. Then we show that it is not always optimal for more general structures than matroids.

Let $\mathbf{M} = (S, \mathbf{F})$ be an arbitrary matroid with nonnegative weights $w(s)$ associated to each element $s \in S$. We extend the definition of weight for subsets X of S as well by $w(X) = \sum_{x \in X} w(x)$. Let us find a subset $X \subseteq S$ which is independent in $\mathbf{M}$ and has maximum possible weight, i.e. determine $\max_{X \in \mathbf{F}} w(X)$. By the nonnegativity of the weights we may suppose that the optimal subset is not only independent but a base as well.

Algorithm 7.1 *Maximum weight base in a matroid*

Input A matroid $\mathbf{M} = (S, \mathbf{F})$ and nonnegative weights $w(s)$ for every $s \in S$.

Output A base X of $\mathbf{M}$ so that $w(X) \geq w(B)$ for every base B of $\mathbf{M}$.

Step 1 Prepare File 1 for the elements which will form the required maximum weight base. Initially let File 1 be empty.

Step 2 Prepare File 2 for those elements which can be applied to extend the already existing independent set. Formally, let X be the set of elements in File 1 and put into File 2 the set $Y = \{x | x \in S - X$ and $X \cup \{x\} \in \mathbf{F}\}$.

Step 3 If $Y = \emptyset$, go to **End**. Otherwise choose an element $x \in Y$ with maximum weight (if there are several such elements, choose randomly among them), add x to File 1 and go to **Step 2**.

End The algorithm terminates, X is the set of the elements in File 1.

That is, we extend X with the maximum possible amount in each step. This will be called the *greedy algorithm.*

Theorem 7.4.1 *If $\mathbf{M} = (S, \mathbf{F})$ is an arbitrary matroid and w is an arbitrary weight function then the greedy algorithm determines a maximum weight base.*

Proof. Suppose that the output of Algorithm 7.1 is $B_m = \{a_1, a_2, \ldots, a_n\}$ while a maximum weight base $B_0 = \{b_1, b_2, \ldots, b_n\}$ would exist with $\sum_{i=1}^n w(a_i) <$ $< \sum_{i=1}^n w(b_i)$. Suppose that the elements of B_m were obtained in this order, i.e.

$$w(a_1) \geq w(a_2) \geq \ldots \geq w(a_n).$$

Furthermore, we may suppose that

$$w(b_1) \geq w(b_2) \geq \ldots \geq w(b_n).$$

also holds (since the subscripts can be permuted otherwise). The relation $w(a_i) \geq w(b_i)$ cannot hold for every $i = 1, 2, \ldots, n$ since then $\sum w(a_i) \geq \sum w(b_i)$ would follow. Let k be the first subscript with $w(a_k) < w(b_k)$ and consider the subsets $A = \{a_1, a_2, \ldots, a_{k-1}\}$ and $B = \{b_1, b_2, \ldots, b_k\}$. (If $k = 1$ then let $A = \emptyset$.) Obviously $A, B \in \mathbf{F}$ (since they are subsets of bases) and $|A| < |B|$, hence there exists a $b_j \in B$ so that $A \cup \{b_j\} \in \mathbf{F}$. But then

$w(b_j) \geq w(b_k) > w(a_k)$, hence the greedy algorithm could not choose a_k at the kth instance of Step 3, a contradiction.

□

We saw in Section 7.1 that every graph leads to a matroid. Hence Algorithm 7.1 is a generalization of Algorithm 1.4. One might wonder if a further generalization is possible. Although matroids can further be generalized (see Problems 7.4.10–11), we show that if the axioms for $(S, \mathbf{F})$ are weakened, the greedy algorithm can fail.

Theorem 7.4.2 [Edmonds, 1971]. *Let $\mathbf{M} = (S, \mathbf{F})$ be a pair containing a finite set S and a collection $\mathbf{F}$ of certain subsets of S. Suppose $\emptyset \in \mathbf{F}$ and $X \in \mathbf{F}, Y \subseteq X$ implies $Y \in \mathbf{F}$ but $\mathbf{M}$ is not a matroid (since (F3) is violated). Then one can associate weights $w(s)$ to the elements s of S so that the greedy algorithm gives a wrong answer.*

Proof. If (F3) is violated, there are subsets $X, Y \in \mathbf{F}$ with $|X| > |Y|$ so that $Y \cup \{x\} \notin \mathbf{F}$ for every $x \in X - Y$. Let the cardinalities of $X \cap Y$, $X - Y$ and $Y - X$ be denoted by ℓ, k and m, respectively (Fig. 7.5); then $k > m$. Let us associate weight a to each element of Y, weight b to those of $X - Y$ and zero to the rest. If $a > b > 0$ then the greedy algorithm chooses the elements of Y at first, and then cannot use the elements of $X - Y$ any more, hence it stops (perhaps choosing some more elements from $S - (X \cup Y)$ as well). Its solution has total weight $(\ell + m)a$. On the other hand, the total weight of X is $\ell a + kb$ which is greater if $kb > ma$, i.e. if $b > \frac{m}{k}a$. Since $k > m$, the conditions $a > b > \frac{m}{k}a$ can simultaneously be satisfied, hence a and b can be chosen so that the answer of the greedy algorithm be wrong.

□

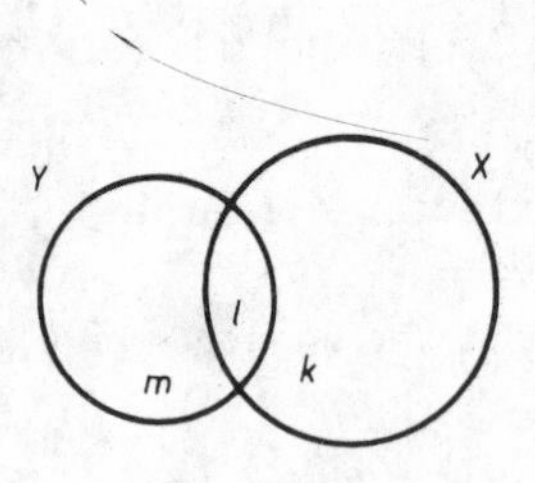

Fig. 7.5

Exercises for Section 7.4

7.4.1 Let the weight of every element of S be different in the matroid $\mathbf{M} = (S, \mathbf{F})$. Let B_m denote the base given by the greedy algorithm. Let $x \in B_m$. Prove that
(a) there exists a cut set in $\mathbf{M}$ where x is the maximum weight element;
(b) there exists no circuit in $\mathbf{M}$ where x is the minimum weight element.

7.4.2 Modify the statements of the previous exercise if equal weights are also permitted.

7.4.3 Prove that the following algorithm is also correct:

Algorithm 7.2 *Maximum weight base in a matroid.*

Input and **Output** as in Algorithm 7.1.
Step 1 as in Algorithm 7.1.
Step 2 Prepare File 2 for those elements which possibly can be applied to extend the already existing independent set. Initially let File 2 contain every element of S.
Step 3 If File 2 is empty, go to **End**. Otherwise choose an element x from File 2, with maximum value (if there are several such elements, choose randomly among them).
Step 4 If x is not a loop, add x to File 1.
Step 5 Delete x from File 2, contract x in the matroid, and go to **Step 3**.
End The algorithm terminates, File 1 contains the elements of the required base X.

7.4.4 Give new solution to Exercise 1.5.7.

7.4.5 Let G be a directed graph with positive weights associated to its edges. Consider those subgraphs of G where the indegree of every point is at most one. Give an algorithm which finds such a subgraph with maximum weight.

Problems for Section 7.4

7.4.6 Prove the following strengthening of the statements of Exercise 7.4.1. If B is an arbitrary maximum weight base of $\mathbf{M} = (S, \mathbf{F})$ then
(a) from every cut set of $\mathbf{M}$ at least one of the maximum weight elements is contained in B
(b) from every circuit of $\mathbf{M}$ at least one of the minimum weight elements is not contained in B.

7.4.7 While the greedy algorithm chooses a maximum weight base X from the matroid $\mathbf{M} = (S, \mathbf{F})$, it also finds a minimum weight base $X' = S - X$ from the dual matroid $\mathbf{M}^*$. This is one of the ideas in the following algorithm [Lawler, 1976].

Algorithm 7.3 *Maximum weight base in a matroid.*

Input and **Output** as in Algorithm 7.1.
Step 1 We prepare three files. File 1 will contain the requested base, File 2 will contain its complement, File 3 has those elements which are not yet assigned to File 1 or 2. Initially let Files 1 and 2 be empty and let File 3 contain every element of S.
Step 2 If File 3 is empty, go to **End**.
Step 3 If the set of elements in Files 1 and 3 contains a circuit C of $\mathbf{M}$ then choose a minimum weight element of C and put this element from File 3 to File 2. Alternatively, if the set of elements in Files 2 and 3

contains a cut set Q of $\mathbf{M}$ then choose a maximum weight element of Q and put this element from File 3 to File 1. (If both conditions are met, perform any of the two operations.) Go to **Step 2.**

End The algorithm terminates, File 1 contains the elements of the requested base X.

(a) Prove that the element to be replaced in Step 3 is really in File 3.
(b) Prove that the algorithm terminates after a finite number of steps.
(c) Prove that the output is really a maximum weight base.

7.4.8 Give new solution to Problem 1.5.15.

7.4.9 Generalize the greedy algorithm for arbitrary (not necessarily nonnegative) weight functions.

7.4.10 The pair $(S, \mathbf{G})$ is called a *greedoid* [Korte and Lovász, 1981] if S is a finite set and $\mathbf{G}$ is a collection of certain subsets of S satisfying the following three conditions:

(G1) $\emptyset \in \mathbf{G}$.

(G2) If $X \in \mathbf{G}$ and $|X| \geq 1$ then there exists an $x \in X$ so that $X - \{x\} \in \mathbf{G}$.

(G3) If $X \in \mathbf{G}, Y \in \mathbf{G}$ and $|X| > |Y|$ then there exists an $x \in$ $\in X - Y$ so that $Y \cup \{x\} \in \mathbf{G}$.

Prove that the greedy algorithm may be wrong for greedoids.

7.4.11 (continued) However, the greedy algorithm does give an optimal solution for greedoids as well, if the weight function has an additional property: Roughly, if joining an element x to a set A is optimal then joining it to a superset of A should also be optimal, provided x can be added to this superset. Formally, if $A \subseteq B, A \in \mathbf{G}, B \in \mathbf{G}, A \cup \{x\} \in \mathbf{G}, B \cup \{x\} \in \mathbf{G}$ and $w(A \cup \{x\}) \geq w(A \cup \{y\})$ for any y, subject to $A \cup \{y\} \in \mathbf{G}$ then $w(B \cup \{x\}) \geq w(B \cup \{y\})$ also holds for any y, subject to $B \cup \{y\} \in \mathbf{G}$, see [Korte and Lovász, 1984]. Check whether your solution to the previous problem violates this condition.

Chapter 8

Applications

Section 8.1 The concept of multiports and their various descriptions

We have seen some network elements in Section 6.2, which had more than two terminals. Recall that both the ideal transformer and the gyrator (Fig. 8.1) have two pairs of terminals; among the voltages and currents, measured on these pairs, there are two linear equations.

More generally, we can speak about 2-*ports*; they are abstract network elements with two pairs of terminals (i.e., two *ports*) and the two voltages and two currents of such an element are related via two linear equations. 2-ports will be denoted as in Fig. 8.2, the arrows indicate the reference directions for the voltages and currents. The two equations can be written as $\mathbf{M}\begin{pmatrix} u_1 \\ u_2 \\ i_1 \\ i_2 \end{pmatrix} = \mathbf{0}$ but in most cases the first two and the second two columns of $\mathbf{M}$ are written separately as $\mathbf{A}\begin{pmatrix} u_1 \\ u_2 \end{pmatrix} + \mathbf{B}\begin{pmatrix} i_1 \\ i_2 \end{pmatrix} = \mathbf{0}$. The right hand side $\mathbf{0}$ is a 2-dimensional vector in both cases and the rank of $\mathbf{M} = (\mathbf{A}|\mathbf{B})$ is 2 (over the reals).

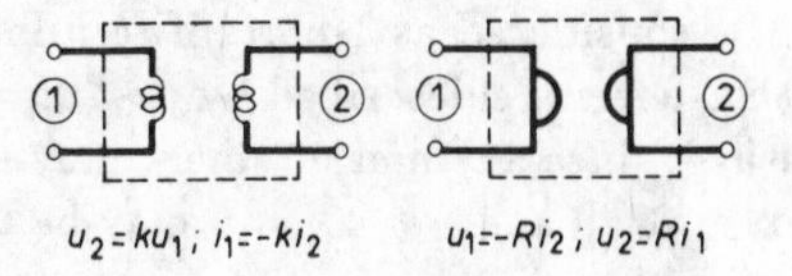

Fig. 8.1

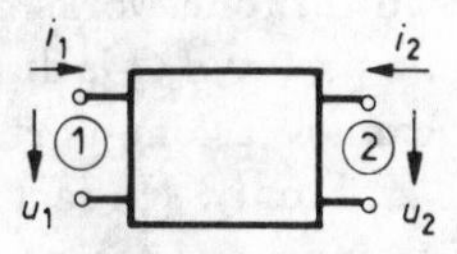

Fig. 8.2

We can further generalize and introduce the concept of *n-ports*. They are abstract network elements with n pairs of terminals (Fig. 8.3) and with n linearly independent equations among the voltages and currents of the ports. The system of equations can be written as $\mathbf{Au} + \mathbf{Bi} = \mathbf{0}$ with $\mathbf{u} = \begin{pmatrix} u_1 \\ u_2 \\ \cdot \\ \cdot \\ \cdot \\ u_n \end{pmatrix}$, $\mathbf{i} = \begin{pmatrix} i_1 \\ i_2 \\ \cdot \\ \cdot \\ \cdot \\ i_n \end{pmatrix}$,

where $\mathbf{A}, \mathbf{B}$ are real $n \times n$ matrices with $r(\mathbf{A}|\mathbf{B}) = n$. In what follows, the arrows for the reference directions will not be drawn at each time but the convention of Fig. 8.3 will be valid throughout. If the number n of ports in an n-port does not play any specific role, we shall speak about *multiports* as well.

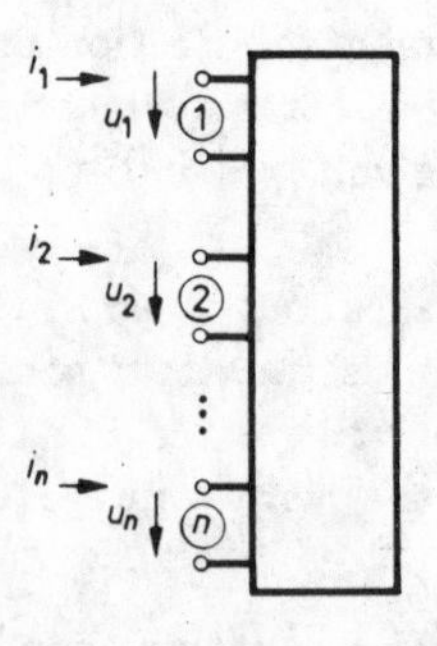

Fig. 8.3

Once the matrices $\mathbf{A}, \mathbf{B}$ are given, they uniquely determine the relationships between $\mathbf{u}$ and $\mathbf{i}$ but not *vice versa*; the same multiport has several different matrix descriptions. More about this in Section 10.1. We also mention that n-ports with $r(\mathbf{A}|\mathbf{B}) \neq n$ might be of some interest, too, see Section 14.2.

Unlike resistors, transformers etc, multiports (as abstract systems of equations) are not real devices, just a possibility to model certain devices. We should use these models if only those properties of the devices are interesting for us which are reflected by the relations among the port voltages and currents. For example, both subnetworks of Fig. 8.4 can be considered as 1-ports if terminals A, B form a port and C is disregarded. Both 1-ports are described by $u - 20i = 0$ and they cannot, as one-ports, be distinguished, since any measurement may use terminals A, B only. (In fact, C may not exist at all and $u - 20i = 0$ may be the model of a single resistor as well.)

If we wish to use terminal C, too, then this subnetwork should be modelled as a 2-port, like in the way, shown in Fig. 8.5. If the pairs of terminals A, C and B, C are numbered by 1 and by 2, respectively, then the description of the 2-port is $\begin{pmatrix} -1 & 0 \\ 0 & -1 \end{pmatrix} \begin{pmatrix} u_1 \\ u_2 \end{pmatrix} + \begin{pmatrix} R_1 & 0 \\ 0 & R_2 \end{pmatrix} \begin{pmatrix} i_1 \\ i_2 \end{pmatrix} = \begin{pmatrix} 0 \\ 0 \end{pmatrix}$. The same subnetwork can also be modelled in the way, shown in Fig. 8.6. Then let us number the pairs

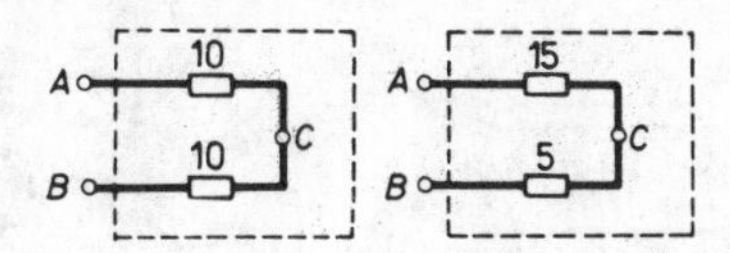

Fig. 8.4

of terminals A, B and B, C by 1 and 2, respectively, and $\begin{pmatrix} -1 & 0 \\ 0 & -1 \end{pmatrix} \begin{pmatrix} u_1 \\ u_2 \end{pmatrix} +$
$+ \begin{pmatrix} R_1 + R_2 & 0 \\ 0 & R_2 \end{pmatrix} \begin{pmatrix} i_1 \\ i_2 \end{pmatrix} = \begin{pmatrix} 0 \\ 0 \end{pmatrix}$ will be the 2-port description.

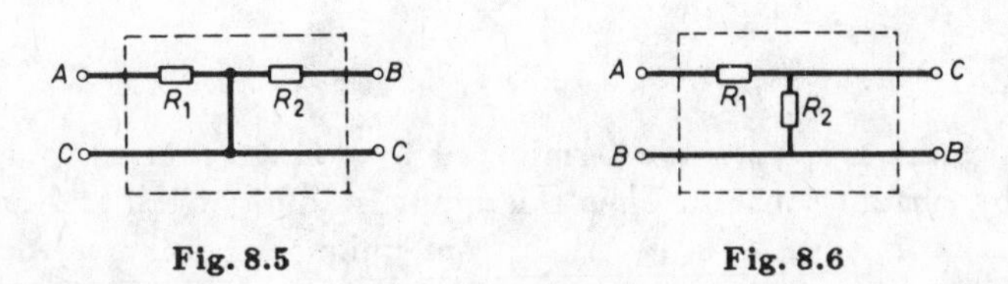

Fig. 8.5 Fig. 8.6

More generally, a device with k terminals can always be modelled by a $(k-1)$-port; one of its terminals is chosen as a reference point and all the other terminals are in pair with this one. Of course, one and the same device leads to different $(k-1)$-ports if the reference is chosen in different ways (for example, Figs 8.5 and 8.6 have C and B, respectively, as reference points). Let us mention that k-terminal devices can be modelled by $(k-1)$-ports in other ways as well, see Exercise 8.1.3 and Problem 10.1.11.

We saw that the two ports of an ideal transformer could not be terminated by two voltage sources or by two current sources, but it could by a voltage source at port 1 and by a current source at port 2 (or *vice versa*). In case of gyrators the situation was just the other way around. One can easily see that any combination of sources can be used for the 2-ports in Figs 8.5–8.6. Similarly, the first 2-port of Fig. 8.7 does not allow current sources at both ports (neither does the second 2-port allow two voltage sources) but any other combination is legitimate. On the other hand, one can hardly tell at the first look the legitimate combinations of sources for the 3-port *circulator* of Fig. 8.8. This network element is defined by $a_1 = b_2,\ a_2 = b_3,\ a_3 = b_1$ where $a_j = u_j + i_j,\ b_j = u_j - i_j$ (for $j = 1, 2, 3$).

An arbitrary combination of voltage and current sources, joining the ports of a multiport, will be called an *input*. Such an input is called *admissible* if the

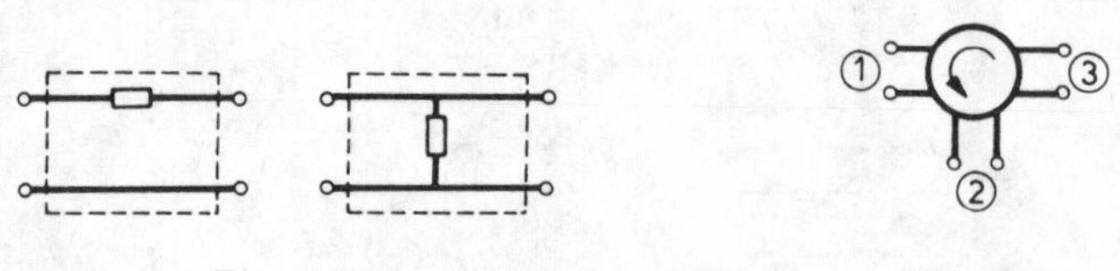

Fig. 8.7 Fig. 8.8

network, formed by this n-port and by the n sources, is uniquely solvable. For example, voltage source at port 1 and current source at port 2 is an admissible input for an ideal transformer but not for a gyrator. An n-port clearly has 2^n different inputs (since we have independent choices between voltage and current sources at each port). However, it depends on the actual n-port, how many of these inputs are admissible.

The n-port is described by $\mathbf{Au} + \mathbf{Bi} = \mathbf{0}$. If $\mathbf{a}_1, \mathbf{a}_2, \ldots, \mathbf{a}_n$ and $\mathbf{b}_1, \mathbf{b}_2, \ldots, \mathbf{b}_n$ denote the columns of the matrices $\mathbf{A}$ and $\mathbf{B}$, respectively, then this description becomes

$$\mathbf{a}_1 u_1 + \mathbf{a}_2 u_2 + \ldots + \mathbf{a}_n u_n + \mathbf{b}_1 i_1 + \mathbf{b}_2 i_2 \ldots + \mathbf{b}_n i_n = \mathbf{0}. \tag{8.1}$$

Suppose that ports $1, 2, \ldots, k$ are terminated by voltage sources, ports $k+1$, $k+2, \ldots, n$ by current sources. Then this input is admissible if and only if the quantities $i_1, i_2, \ldots, i_k, u_{k+1}, u_{k+2}, \ldots, u_n$ can uniquely be expressed as functions of the others. Since Eq. (8.1) can easily be rewritten as

$$\begin{aligned} &\mathbf{a}_{k+1} u_{k+1} + \mathbf{a}_{k+2} u_{k+2} + \ldots + \mathbf{a}_n u_n + \mathbf{b}_1 i_1 + \ldots + \mathbf{b}_k i_k = \\ &= -[\mathbf{a}_1 u_1 + \mathbf{a}_2 u_2 + \ldots + \mathbf{a}_k u_k + \mathbf{b}_{k+1} i_{k+1} + \mathbf{b}_{k+2} i_{k+2} + \ldots + \mathbf{b}_n i_n], \end{aligned} \tag{8.2}$$

admissibility means that the matrix, formed by the column vectors $\mathbf{a}_{k+1}$, $\mathbf{a}_{k+2}, \ldots, \mathbf{a}_n, \mathbf{b}_1, \mathbf{b}_2, \ldots, \mathbf{b}_k$, is nonsingular.

More generally, let us consider an arbitrary input. If port j is terminated by a voltage source, delete the jth column of $\mathbf{A}$, if it is terminated by a current source, delete the jth column of $\mathbf{B}$. Thus we delete n columns from $(\mathbf{A}|\mathbf{B})$. If the resulting $n \times n$ matrix is invertible, then the input is admissible; if the matrix is singular, it is not.

Recall that a matrix $\mathbf{K}$ defined a matroid $\mathbf{M} = (S, \mathbf{F})$ on the set S of columns of $\mathbf{K}$ so that a subset $X \subseteq S$ of columns was independent ($X \in \mathbf{F}$) if and only if the column vectors were linearly independent. Hence if $(\mathbf{A}|\mathbf{B})$ is the matrix of an n-port N then a matroid $\mathbf{M}_N = (S_N, \mathbf{F}_N)$ can be defined in this way. Obviously, $|S_N| = 2n$ and the rank of $\mathbf{M}_N$ is just n.

For example, let T and G be an ideal transformer and gyrator, respectively, with the describing systems of equations as given in Fig. 8.1. Let us describe their matroids $\mathbf{M}_T$ and $\mathbf{M}_G$, respectively. Every pair is base in $\mathbf{M}_T$ except $\{u_1, u_2\}$ and $\{i_1, i_2\}$; this can be best seen from the matrix $\begin{pmatrix} k & -1 & 0 & 0 \\ 0 & 0 & 1 & k \end{pmatrix}$ of the transformer. Similarly, $\{u_1, i_2\}$ and $\{u_2, i_1\}$ are circuits in $\mathbf{M}_G$ and all the other pairs are bases. Hence, if we consider the circuit matroids of the graphs of Fig. 8.9, we obtain $\mathbf{M}_T \cong \mathbf{M}(G_1)$ and $\mathbf{M}_G \cong \mathbf{M}(G_2)$. These figures are for illustration only, one can easily construct n-ports whose matroids are nongraphic (Exercise 8.1.12).

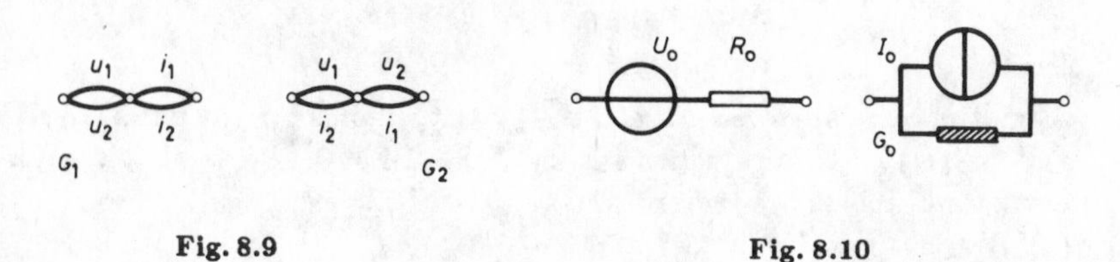

Fig. 8.9 **Fig. 8.10**

The above observation can be summarized as follows.

Statement 8.1.1 *Consider an input of the n-port N, terminating ports $j_1, j_2, .., j_k$ with voltage sources and the others with current sources. This input is admissible if and only if $\{u_{j_1}, u_{j_2}, \ldots, u_{j_k}\} \cup [\{i_1, i_2, \ldots, i_n\} - \{i_{j_1}, i_{j_2}, \ldots, i_{j_k}\}]$ is the complement of a base in the matroid $\mathbf{M}_N$.*

Two further remarks are in order. First, the matroid $\mathbf{M}_N$ contains significantly less information than the $(\mathbf{A}|\mathbf{B})$ description of N, see Problem 8.1.22, for example. Furthermore, observe that Statement 8.1.1 does not state a one–one correspondence between the set of admissible inputs of N and the set of bases of $\mathbf{M}_N$, see Exercise 8.1.13. We shall see more about this in Sections 12.1 and 14.2.

Exercises for Section 8.1

8.1.1 One could introduce the more general concept of linear memoryless (but not necessarily homogeneous) n-ports by $\mathbf{Au} + \mathbf{Bi} = \mathbf{c}$, where $r(\mathbf{A}|\mathbf{B}) = r(\mathbf{A}|\mathbf{B}|\mathbf{c}) = n$. Prove the *theorem of Helmholtz*: Any such 1-port can be substituted by one of the two 1-ports of Fig. 8.10. (They are called *Thévenin* and *Norton-equivalents*, respectively; R_0 is an arbitrary resistance, G_0 is an arbitrary admittance, they might be zero as well.)

8.1.2 The four 2-ports of Fig. 8.11 are called *controlled sources*. (The first one is called voltage-controlled current source, the second one is voltage controlled voltage source etc.) What does it mean if their only parameter (g, c, d and r, respectively) becomes zero?

8.1.3 The 4-terminal device, consisting of three resistors, was modelled as a 3-port in two different ways, see Fig. 8.12. Determine their descriptions in form $\mathbf{Au} + \mathbf{Bi} = \mathbf{0}$.

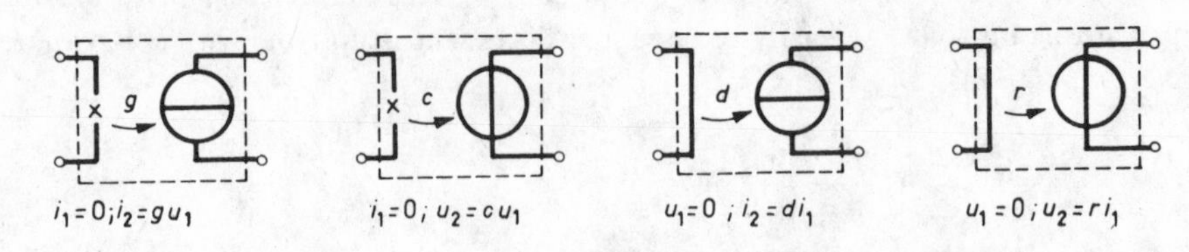

Fig. 8.11

8.1.4 A multiport is called *reciprocal* if $\mathbf{u}_1^T\mathbf{i}_2 = \mathbf{u}_2^T\mathbf{i}_1$ holds for any pairs $(\mathbf{u}_1, \mathbf{i}_1)$ and $(\mathbf{u}_2, \mathbf{i}_2)$ of vectors which satisfy $\mathbf{Au}+\mathbf{Bi}=\mathbf{0}$. Prove that the following multiports are reciprocal:
(a) Every 1-port.
(b) The ideal transformer.
(c) The 3-ports of the previous exercise.

8.1.5 (continued) Prove that the following multiports are not reciprocal:
(a) The gyrator.
(b) The controlled sources, introduced in Exercise 8.1.2.
(c) The 3-port circulator.

Box 8.1. Further important concepts of network theory

Let the system of equations for a multiport N be

$$\mathbf{Au}+\mathbf{Bi}=\mathbf{0} \qquad (^*)$$

N is reciprocal if $\mathbf{u}_1^T\mathbf{i}_2 = \mathbf{u}_2^T\mathbf{i}_1$ holds for any pairs $(\mathbf{u}_1,\ \mathbf{i}_1)$ and $(\mathbf{u}_2,\ \mathbf{i}_2)$ of vectors, satisfying $(^*)$.
N is passive if $\mathbf{u}^T\mathbf{i} \geq 0$ holds for any pair $(\mathbf{u},\mathbf{i})$ of vectors, satisfying $(^*)$.
N is lossless if $\mathbf{u}^T\mathbf{i} = 0$ holds for any pair $(\mathbf{u},\mathbf{i})$ of vectors, satisfying $(^*)$.
N has impedance description if $\det \mathbf{A} \neq 0$.
N has admittance description if $\det \mathbf{B} \neq 0$.

8.1.6 An n-port is called *passive* if $\mathbf{u}^T\mathbf{i} = \sum_{j=1}^{n} u_j i_j \geq 0$ holds for any pair $(\mathbf{u},\mathbf{i})$ of vectors, which satisfies $\mathbf{Au}+\mathbf{Bi}=\mathbf{0}$. If the stronger relation $\mathbf{u}^T\mathbf{i}=0$ holds then the n-port is *lossless*. If the n-port is not passive, it is *active*. Characterize the passive and the lossless 1-ports.

8.1.7 (continued) Prove that the ideal transformer and the gyrator are passive, in fact, they are even lossless. What can be said about the controlled sources and about the 3-port circulator?

Box 8.2. The properties of some important multiports

	Is it reciprocal?	Is it passive?	Is it lossless?	Does it have **Z** description?	Does it have **Y** description?	Does it have at least one hybrid descr.?
Positive resistor	+	+	—	+	+	+
Negative resistor	+	—	—	+	+	+
Open circuit	+	+	+	—	+	+
Short circuit	+	+	+	+	—	+
Nullator	(*)	+	+	—	—	—
Norator	—	—	—	—	—	—
Ideal transformer	+	+	+	—	—	+
Gyrator	—	+	+	+	+	+
Voltage controlled c.source	—	—	—	—	+	+
Voltage controlled v.source	—	—	—	—	—	+
Current controlled c.source	—	—	—	—	—	+
Current controlled v.source	—	—	—	+	—	+
Nullator-norator pair	—	—	—	—	—	—
3-port circulator	—	+	+	—	+	+

(*) The nullator satisfies the definition given in Exercise 8.1.4 hence most authors consider it reciprocal, see p. 912 footnote 4 in [Carlin–Youla, 1961]. However, it does not satisfy the definition, given in the solution of Problem 14.2.10 for the reciprocity of generalized multiports.

8.1.8 More generally, a (not necessarily memoryless) multiport is called passive if $\int_{t_1}^{t_2} \mathbf{u}^T(t)\mathbf{i}(t)dt \geq 0$ holds for every interval $[t_1, t_2]$ of time. What additional constraints are required for the system of equations of the "real" transformer (Problem 6.2.8) to be passive?

8.1.9 If $\mathbf{A}$ or $\mathbf{B}$ is nonsingular then the multiport equation $\mathbf{Au} + \mathbf{Bi} = \mathbf{0}$ can be written as $\mathbf{u} = -\mathbf{A}^{-1}\mathbf{Bi} = \mathbf{Zi}$ or as $\mathbf{i} = -\mathbf{B}^{-1}\mathbf{Au} = \mathbf{Yu}$, respectively. In this case $\mathbf{Z}$ and $\mathbf{Y}$ are called the *impedance-* and the *admittance-matrix* of the multiport, respectively. Determine whether the following n-ports have impedance or admittance matrices (or both): All the 1-ports, the ideal transformer, the gyrator, the controlled sources and the 3-port circulator.

8.1.10 Does the 2-port with equations $u_1 = Ri_1$, $i_2 = ai_1$ have a $\mathbf{Z}$-or a $\mathbf{Y}$-matrix description? Determine the admissible inputs of this 2-port.

8.1.11 Describe the matroid of the 3-port circulator and determine its admissible inputs.

8.1.12 Construct an n-port whose matroid is nongraphic.

8.1.13 Give an example to a base in the matroid $\mathbf{M}_N$ of a multiport N, whose complement does not correspond to an admissible input.

Problems for Section 8.1

8.1.14 Show that the 3-terminal subnetworks in Fig. 8.13 have the same multiport-equations if $a = BC/(A + B + C)$, $b = AC/(A + B + C)$ and $c = AB/(A + B + C)$; or, with a transformation in the opposite direction, $A = d/a$, $B = d/b$, $C = d/c$ with $d = ab + bc + ca$. (This transformation is usually called $Y - \Delta$ *transformation*).

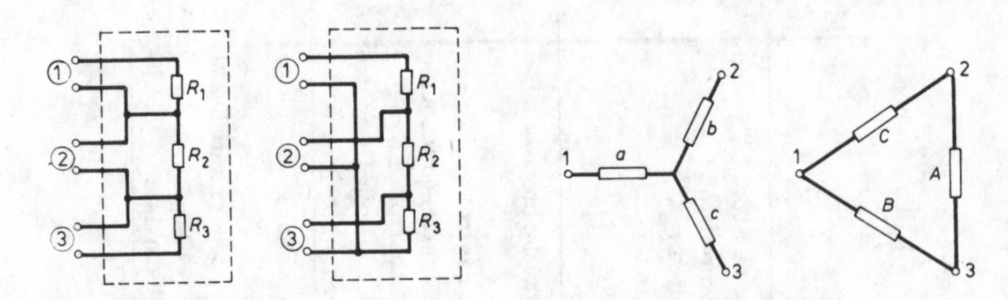

Fig. 8.12 **Fig. 8.13**

8.1.15* Show that the Δ-connection (right hand side in Fig. 8.13) cannot always be substituted by the Y-connection if not only resistors but capacitors and inductors are also permitted in positions A, B and C, and equivalence is required at every frequency.

8.1.16 Suppose that a multiport has a **Z** or a **Y** description. How can one recognize from these matrices if the multiport is reciprocal, passive or lossless?

8.1.17 Characterize the lossless 2-ports.

8.1.18 Realize the 2-port of Exercise 8.1.10 with controlled sources. How many solutions are there?

8.1.19 Resistors can be imagined as current controlled voltage sources (controlled by their own currents). Using this observation give an absurd network where "normal" and controlled voltage sources form a circuit. (To emphasize the difference from controlled sources, the "normal" voltage and current sources, introduced in Section 2.1, are called *independent sources.*) Is this network a counterexample to Statement 2.1.1?

8.1.20 (continued) Prove Purslow's theorem [1970]: A necessary condition of the unique solvability of a network is that those independent and controlled voltage sources whose currents do not control other sources, form a circuit free subgraph; and those independent and controlled current sources whose voltages do not control other sources, form a cut set free subgraph. (cf., Problem 10.1.5(c) as well.)

8.1.21 Change the formula in the definition of reciprocity to $\mathbf{u}_1^T\mathbf{i}_2 = -\mathbf{u}_2^T\mathbf{i}_1$. Thus another concept is defined, usually called *anti-reciprocity.* (Do not confuse with *nonreciprocity,* which is simply the negation of reciprocity.) Prove that a multiport is antireciprocal if and only if it is lossless. (Hence, in particular, both the ideal transformer and the gyrator are antireciprocal.)

8.1.22 Suppose that a multiport N is given by its matroid $\mathbf{M}_N$. Can you decide from $\mathbf{M}_N$ only whether N is passive, or reciprocal, or lossless?

Section 8.2 Rigidity of trusses II: Various tie-down structures

Consider a framework F in the 2- or 3-dimensional space. Its rigidity can be determined by the rank of its matrix $\mathbf{A}(F)$, introduced in Section 6.3. However, practical frameworks are not "floating" in the air but fixed at some joints to the plane (or to some points of the space). Then their rigidity depends not on their own structures only, but on the way of their fixings as well.

For example, the triangular framework of Fig. 6.17 was rigid (both in the plane and in the 3-dimensional space). Yet, if one of its joints is fixed only then it can move, even in the plane (see Fig. 8.14). The frameworks forming a quadrangle (like those in Fig. 6.11) were nonrigid. If we fix two of their joints to the plane, they might become "rigid", depending on the choice of the two joints (see Fig. 8.15).

Recall that the describing matrix $\mathbf{A}(F)$ of a framework F with v joints and e rods had e rows and $2v$ or $3v$ columns (depending on the dimension of the

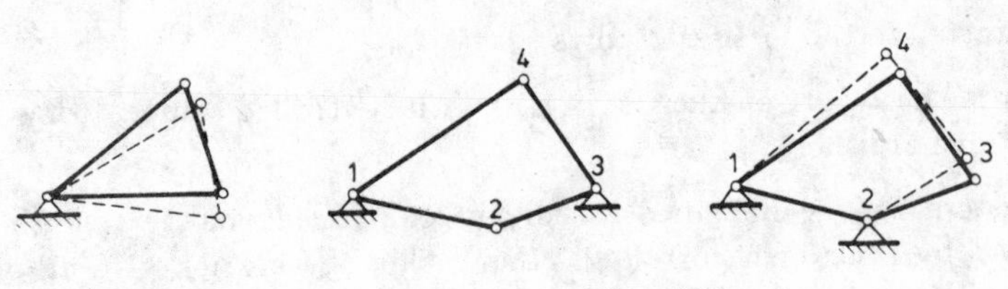

Fig. 8.14

Fig. 8.15

space). The first, the $(v+1)^{th}$ (and, in case of 3 dimensions, the $(2v+1)^{th}$) columns correspond to the velocity coordinates of the first joint etc.

If joint k is fixed, its velocity vector $(\dot{x}_k, \dot{y}_k)$ or $(\dot{x}_k, \dot{y}_k, \dot{z}_k)$ becomes zero; the kth, $(v+k)^{th}$ and $(2v+k)^{th}$ columns of $\mathbf{A}(F)$ do not correspond to unknown quantities anymore.

For example, the matrix $\mathbf{A}(F)$ of the 2-dimensional "quadrangle"-framework of Fig. 8.15 is

$$\mathbf{A}(F) = \begin{pmatrix} x_1-x_2 & x_2-x_1 & 0 & 0 & y_1-y_2 & y_2-y_1 & 0 & 0 \\ 0 & x_2-x_3 & x_3-x_2 & 0 & 0 & y_2-y_3 & y_3-y_2 & 0 \\ 0 & 0 & x_3-x_4 & x_4-x_3 & 0 & 0 & y_3-y_4 & y_4-y_3 \\ x_1-x_4 & 0 & 0 & x_4-x_1 & y_1-y_4 & 0 & 0 & y_4-y_1 \end{pmatrix}$$

and since $r(\mathbf{A}) = 4 < 5 = 2v-3$, F is nonrigid by the planar analogue of Lemma 6.3.1. Really, $\mathbf{Au} = \mathbf{0}$ (see Eq. (6.7)) has many nontrivial solutions. If we fix joints 1 and 3, as shown on the left hand side of Fig. 8.15, then $\dot{x}_1 = \dot{x}_3 = \dot{y}_1 =$ $= \dot{y}_3 = 0$ leads to the system

$$\begin{pmatrix} x_2-x_1 & 0 & y_2-y_1 & 0 \\ x_2-x_3 & 0 & y_2-y_3 & 0 \\ 0 & x_4-x_3 & 0 & y_4-y_3 \\ 0 & x_4-x_1 & 0 & y_4-y_1 \end{pmatrix} \begin{pmatrix} \dot{x}_2 \\ \dot{x}_4 \\ \dot{y}_2 \\ \dot{y}_4 \end{pmatrix} = \begin{pmatrix} 0 \\ 0 \\ 0 \\ 0 \end{pmatrix}$$

of equations. Its determinant is nonzero (hence it has only the trivial solution $\dot{x}_2 = \dot{x}_4 = \dot{y}_2 = \dot{y}_4 = 0$) provided that neither joint 2 nor joint 4 is on the line connecting joints 1 and 3 (because in case of such a collinearity an "infinitesimal" motion would be still possible). On the other hand, if the first and second joints are fixed, see the right hand side of Fig. 8.15, we obtain

$$\begin{pmatrix} 0 & 0 & 0 & 0 \\ x_3-x_2 & 0 & y_3-y_2 & 0 \\ x_3-x_4 & x_4-x_3 & y_3-y_4 & y_4-y_3 \\ 0 & x_4-x_1 & 0 & y_4-y_1 \end{pmatrix} \begin{pmatrix} \dot{x}_3 \\ \dot{x}_4 \\ \dot{y}_3 \\ \dot{y}_4 \end{pmatrix} = \begin{pmatrix} 0 \\ 0 \\ 0 \\ 0 \end{pmatrix}$$

which has an infinite number of solutions, independently of the actual numerical values, since we have 3 equations only for the 4 unknown quantities.

Let us formulate our observation in general:

Statement 8.2.1 *Delete from the matrix* $\mathbf{A}(F)$ *of the framework* F *those columns which correspond to the velocity coordinates of the fixed points. Then* F *(with this system of fixed joints) is rigid if and only if the resulting matrix* $\mathbf{W}$ *has linearly independent columns only.*

Since we can associate a matroid $\mathbf{M}_F$ to the framework (whose elements are the column vectors of $\mathbf{A}(F)$ and independence is simply linear independence over the field $\mathbb{R}$ of the reals), this statement can be reformulated.

Statement 8.2.2 *After fixing some joints of framework* F*, the resulting system is rigid if and only if, deleting the 2-2 columns, corresponding to each of these joints, the remaining subset is independent in* $\mathbf{M}_F$.

Figure 8.14 showed that the triangular framework still can move in the plane if one of its joints is fixed. It obviously cannot move any more if we fix two of its joints. Our intuitive feeling might be that, although fixing one joint was not enough, fixing two joints is a bit "too much". The matrix $\mathbf{A}(F)$ of this framework F had size 3×6 and rank 3. Hence if we delete two columns, the remaining four will surely remain linearly dependent, but why should we delete 4 columns? Deleting three suitable columns must also lead to a linearly independent system.

What corresponds physically to fixing "one and a half" joints? If we fix joint 1, the columns corresponding to $\dot{x}_1$ and $\dot{y}_1$ should be deleted. If we wish to delete the column, corresponding to $\dot{x}_2$ as well, we must introduce a constraint ensuring $\dot{x}_2 = 0$ without affecting the value of $\dot{y}_2$. This means that the possible motion of joint 2 must not have any horizontal component (but can have an arbitrary vertical one). Such a constraint is shown in Fig. 8.16. We shall say that joint 2 is on a vertical *track*. (The drawing is slightly misleading; the "coach" carrying joint 2 must stay on the track all the time.)

Hence we can generalize Statement 8.2.2 as follows:

Theorem 8.2.3 *Let* F *be a 2-dimensional framework. Fix joints* $i_1, i_2, \ldots, i_\ell$ *to the plane; put joints* $j_1, j_2, \ldots, j_m$ *and* $k_1, k_2, \ldots, k_p$ *to vertical and to horizontal tracks, respectively. (Among these sets* $\{i_1, i_2, \ldots, i_\ell\}, \{j_1, j_2, \ldots, j_m\}$ *and* $\{k_1, k_2, \ldots, k_p\}$ *there might be empty ones; but they must be disjoint.) Then the system becomes rigid (i.e., dynamically determined) if and only if the complement of the set*

$$\{\dot{x}_{i_1}, \dot{x}_{i_2}, \ldots, \dot{x}_{i_\ell}, \dot{x}_{j_1}, \dot{x}_{j_2}, \ldots, \dot{x}_{j_m}, \dot{y}_{i_1}, \dot{y}_{i_2}, \ldots, \dot{y}_{i_\ell}, \dot{y}_{k_1}, \dot{y}_{k_2}, \ldots, \dot{y}_{k_p}\}$$

is independent in the matroid $\mathbf{M}_F$.

□

Let us consider again the planar triangular framework F. Recall that

$$\mathbf{A}(F) = \begin{pmatrix} x_1 - x_2 & x_2 - x_1 & 0 & y_1 - y_2 & y_2 - y_1 & 0 \\ 0 & x_2 - x_3 & x_3 - x_2 & 0 & y_2 - y_3 & y_3 - y_2 \\ x_1 - x_3 & 0 & x_3 - x_1 & y_1 - y_3 & 0 & y_3 - y_1 \end{pmatrix}$$

If there is no algebraic relation among the 6 coordinates of the 3 joints then the matroid $\mathbf{M}_F$ defined by the columns of $\mathbf{A}(F)$ is similar to the uniform matroid $\mathbf{U}_{6,3}$ except that the triples $\{\dot{x}_1, \dot{x}_2, \dot{x}_3\}$ and $\{\dot{y}_1, \dot{y}_2, \dot{y}_3\}$ are not bases but circuits. Hence we obtain the matroid $\mathbf{V}_3$ of Problem 7.1.11.

The "improper" fixings of F, corresponding to these two circuits arise if all the three joints are on horizontal tracks or all are on vertical tracks. Obviously, in these cases F can move along the tracks. All the other fixings, $\binom{6}{3} - 2 = 18$ in number, are good; there are essentially two sorts of them. Either we fix one joint and put another to a track (Fig. 8.16) or we put all three to tracks so that only two of the tracks are parallel (Fig. 8.17). (There are 12 and 6 such cases, respectively.)

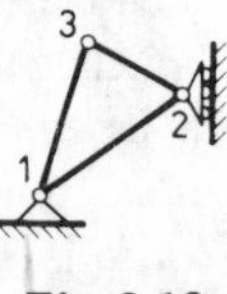

Fig. 8.16

The situation completely changes if algebraic relations are allowed among the coordinates of the joints. If, for example, $y_1 = y_2$ (i.e., if rod 12 is horizontal) then both fixings of Fig. 8.18 are improper. The first one still allows an infinitesimal motion where joint 1 remains at its original position, joint 2 moves vertically. The second fixing still allows a motion where joints 1 and 2 move vertically (but in opposite direction), leading to a horizontal move of joint 3. But, of course, in this case $\mathbf{M}_F$ will be different from $\mathbf{V}_3$, see Problem 8.2.4.

Instead of speaking about tracks, another constraint on the joints may be if we introduce a new rod (which does not belong to the framework), fix one end of it to our joint and fix the other end of it to the plane. Hence a "proper" fixing is obtained in Fig. 8.19 instead of that in Fig. 8.17. It is intuitively clear that the "degree of freedom" of a joint is decreased by one in this way (just like in the case of the tracks).

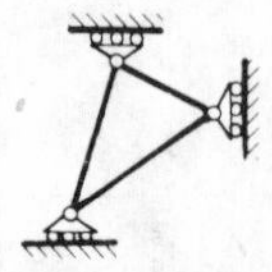

Fig. 8.17

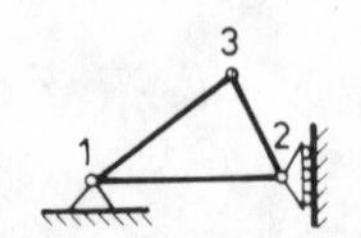

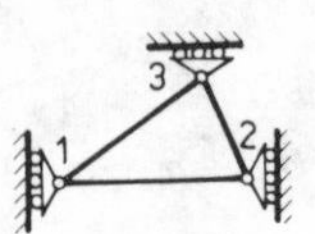

Fig. 8.18

Thus we can even avoid the fixing of a joint, simply by using two such new rods for the joint (see Fig. 8.20 instead of Fig. 8.16). Of course, the problems (like those mentioned in connection with Fig. 8.18) with the algebraic relations among the coordinates remain the same, see Exercise 8.2.1.

Finally, let us mention that all the results in this section are valid in case of 3-dimensional frameworks as well, see Exercises 8.2.2–3 and Problems 8.2.6–7.

Exercises for Section 8.2

8.2.1 Redraw the improper fixings of Fig. 8.18 using the alternative constraints introduced in Figs 8.19 and 8.20.

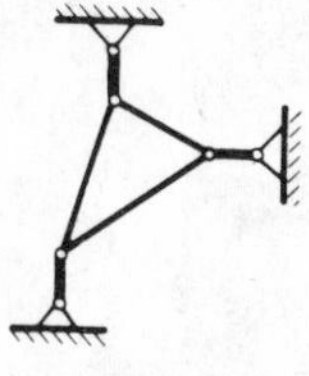

Fig. 8.19

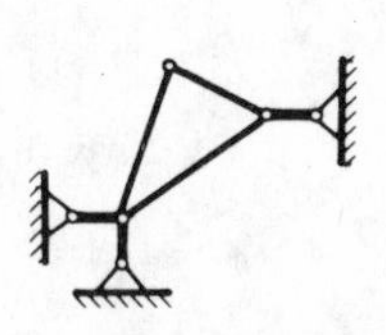

Fig. 8.20

8.2.2 What is the 3-dimensional analogue of a constraint $\dot{x}_i = 0$?

8.2.3 What is the 3-dimensional analogue of a "track"?

Problems for Section 8.2

8.2.4 Determine the matroid $\mathbf{M}_F$ of the framework F of Fig. 8.18 (where rod 12 is horizontal but there are no other algebraic relations among the coordinates). Also determine the matroid $\mathbf{M}_G$ of the framework G of Fig. 8.21 (where $y_1 = y_2$ and $x_1 = x_3$ are met).

8.2.5 (continued) Prepare an arrangement which fixes F but not G.

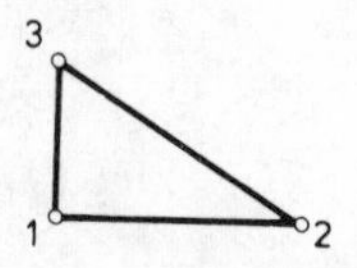

Fig. 8.21

8.2.6 What is the 3-dimensional analogue of the constraints introduced in Figs 8.19 and 8.20?

8.2.7 Consider now the triangular framework F in the 3-dimensional space (and suppose there are no algebraic relations among the coordinates). Give all the possible fixings of F (using tracks etc of the joints). Draw the results also by using the constraints, introduced in the previous problem. (Hint: There are 4 essentially different solutions.)

8.2.8 Compare matroid $\mathbf{V}_3$ with the matroids $\mathbf{M}_F$ and $\mathbf{M}_G$ of Problem 8.2.4. Find a relation among them. Are there graphic ones among them?

Chapter 9

Algebraic and geometric representation of matroids

Section 9.1 Representation over various fields

One of our first examples at the introduction of matroids was the following. Let T be an arbitrary field, $V[T]$ be a vector space over T and S be a set of vectors from this vector space. This set leads to a matroid $\mathbf{M} = (S, \mathbf{F})$ as follows: $X \subseteq S$ is independent (denoted by $X \in \mathbf{F}$) if and only if the vectors, belonging to X are linearly independent over T.

For example, the vectors (1,1,0), (1,0,1) and (0,1,1) are linearly dependent over $\mathbb{B}$, they form a uniform matroid $\mathbf{U}_{3,2}$. The same vectors are linearly independent over $\mathbb{Q}$ or $\mathbb{R}$, they form a free matroid $\mathbf{U}_{3,3}$. Independently of the choice of the field, the zero vector corresponds to a loop, parallel (specially equal) vectors correspond to parallel elements.

A matroid $\mathbf{M} = (S, \mathbf{F})$ is called *coordinatizable* (or *representable*) over a field T if suitable vectors from a vector space over T can play the role of S in the above construction. For example, $\mathbf{U}_{4,2}$ is coordinatizable over $\mathbb{R}$ (consider the vectors (1,0), (0,1), (1,1) and (1,2), for example) but not over $\mathbb{B}$, see below.

In what follows, vectors will always be column vectors and instead of listing them we write a matrix. For example, the above coordinatization of $\mathbf{U}_{4,2}$ is then $\begin{pmatrix} 1 & 0 & 1 & 1 \\ 0 & 1 & 1 & 2 \end{pmatrix}$. The reason for the name of coordinatization is clear: all elements of S correspond to (not necessarily distinct) vectors and we use the coordinates of this vector. If we use another base in the same vector space, the coordinates would be different:

Statement 9.1.1 *Let* $\mathbf{A}$ *be an arbitrary* $k \times \ell$ *matrix and* $\mathbf{B}$ *be a nonsingular* $k \times k$ *matrix. Then the column vectors of* $\mathbf{A}$ *and* $\mathbf{BA}$ *determine isomorphic matroids.*

Proof. For any $k \times p$ submatrix $\mathbf{A}'$ of $\mathbf{A}$ we have $r(\mathbf{A}') = r(\mathbf{BA}')$.

□

This trivial statement is very important. For example, if $\mathbf{A}$ represents a matroid $\mathbf{M} = (S, \mathbf{F})$ and $X \subseteq S$ is a base of $\mathbf{M}$ then we may suppose that the columns of $\mathbf{A}$, corresponding to X, form a unity matrix. $|X| = t$ is the rank of the matroid then we may suppose that $\mathbf{A}$ has exactly t rows. (Less than t is impossible since the columns, corresponding to X, must be linearly independent. More than t is not necessary, since if we choose t linearly independent rows, the others are their linear combinations only, hence they can be deleted without changing the matroid.) Next, if the columns of $\mathbf{A}$ corresponding to X, form a submatrix $\mathbf{A}_0$ then $\mathbf{A}_0$ is nonsingular (since $r(X) = t$). Hence the above statement can be applied with $\mathbf{B} = \mathbf{A}_0^{-1}$.

We can easily prove now that the matroid $\mathbf{U}_{4,2}$ defined over the set $S = \{a, b, c, d\}$ cannot be represented over $\mathbb{B}$. Since $\{a, b\}$ is a base, a representation could look like $\begin{pmatrix} 1 & 0 & p & q \\ 0 & 1 & r & s \end{pmatrix}$. None of the numbers p, q, r, s can be zero (if $p = 0$, for example, then $\{b, c\}$ cannot be independent). Then all of them must equal the other element of $\mathbb{B}$, i.e. 1. However, then $\{c, d\}$ is dependent, a contradiction.

As a byproduct, we obtained that $\mathbf{U}_{4,2}$ can be coordinatized over any field T different from $\mathbb{B}$. Really, T has at least one element t, different than 0 and 1, and then $\begin{pmatrix} 1 & 0 & 1 & 1 \\ 0 & 1 & 1 & t \end{pmatrix}$ is a suitable coordinatization.

Theorem 9.1.2 *Every graphic matroid can be coordinatized over every field.*

Proof. Let G be an arbitrary graph, T be an arbitrary field. We construct a representation of $\mathbf{M}(G)$ over T. Assign, for a moment, the first unity vector $\mathbf{e}_1 = (1, 0, 0, \ldots, 0)^T$ to the first point v_1 of G, the second unity vector $\mathbf{e}_2 = (0, 1, 0, \ldots, 0)^T$ to the second point v_2 etc. Then assign the difference $\mathbf{v}_e = \mathbf{e}_i - \mathbf{e}_j$ to the edge $e = \{v_i, v_j\}$. (Should T be just $\mathbb{B}$, we may write $\mathbf{v}_e = \mathbf{e}_i + \mathbf{e}_j$ as well.)

If the edges $e_1, e_2, \ldots$ form a circuit then the corresponding vectors $\mathbf{v}_{e_1}, \mathbf{v}_{e_2}, \ldots$ are linearly dependent over T (their linear combination with coefficients ± 1 leads to $\mathbf{0}$). We only have to prove the converse, that linear dependence of some vectors implies the existence of a circuit in the subgraph, determined by them.

Suppose that $\mathbf{v}_{e_1}$ is the linear combination of the vectors $\mathbf{v}_{e_2}, \mathbf{v}_{e_3}, \ldots$ and denote by H the subgraph determined by $e_1, e_2, \ldots$. Let $e_1 = \{x, y\}$. Since the 1 or -1 in the positions x and y of $\mathbf{v}_{e_1}$ must be obtained, at least one more edge of H must be incident to x and at least one to y. More generally, the degree of every point must be 0 or at least 2 in H, hence H contains a circuit by Problem 1.1.20.

□

Observe that all the $\mathbf{v}_e$ vectors (for every $e \in E(G)$) together just give the incidence matrix $\mathbf{B}(G)$ of an orientation of G, apart from the difference that loops correspond to zero columns. Similarly, the cut set matrix $\mathbf{Q}(G)$ of the graph G also coordinatizes the matroid $\mathbf{M}(G)$, this was Theorem 1.4.7 in the case that $T = \mathbb{Q}$.

A matroid is called *regular* if it is representable over every field. Hence, graphic matroids are regular, by Theorem 9.1.2. So are cographic matroids as well, as it will follow from the following basic result:

Theorem 9.1.3 *If a matroid is representable over a field then so is its dual.*

Proof. Let $|S| = n$ and let the rank of the matroid $\mathbf{M} = (S, \mathbf{F})$ be t. We may suppose that the matrix $\mathbf{A}$, coordinatizing $\mathbf{M}$, has t linearly independent rows. $\mathbf{A}$ can be transformed into $(\mathbf{E}_t | \mathbf{A}_0)$ where the columns of the $t \times t$ unity matrix $\mathbf{E}_t$ correspond to the elements of a base (see Statement 9.1.1) and the columns of the submatrix $\mathbf{A}_0$ correspond to the rest. We shall prove that $\mathbf{B} = (-\mathbf{A}_0^T|\mathbf{E}_{n-t})$ is a coordinatization of $\mathbf{M}^*$.

In fact, we shall only need that $\mathbf{A}$ and $\mathbf{B}$ are *orthogonal complements* of each other, which means that both of them have n columns, $r(\mathbf{A}) + r(\mathbf{B}) = n$ and $\mathbf{AB}^T = \mathbf{O}$. Let us prove that their matroids $\mathbf{M}$ and $\mathbf{N}$ are then duals of each other. A well known theorem of linear algebra states that exactly those row vectors $\mathbf{x}^T$ arise as linear combinations of the rows of $\mathbf{A}$ for which $\mathbf{Bx} = \mathbf{0}$.

Let $U \subseteq S$ contain a base of the matroid $\mathbf{M}$ and consider the submatrix $\mathbf{A}'$ of $\mathbf{A}$, corresponding to the U columns. The rows of $\mathbf{A}'$ are linearly independent (since $r(\mathbf{A}') = t$). Thus, if a row vector $\mathbf{x}^T$ arises as a nontrivial linear combination of the rows of $\mathbf{A}$, some coordinates belonging to the U columns must be nonzero. Accordingly, by the above theorem, if $\mathbf{Bx} = \mathbf{0}$ and $\mathbf{x} \neq \mathbf{0}$ then some of the nonzero elements of $\mathbf{x}$ must belong to U. Hence $S - U$ is independent in $\mathbf{N}$.

Therefore, if U is a base of $\mathbf{M}$ then the subset $S - U$ with cardinality $n - r(\mathbf{A}) = r(\mathbf{B})$ must be a base of $\mathbf{N}$. Conversely, bases of $\mathbf{N}$ correspond to base complements of $\mathbf{M}$, by the symmetry of $\mathbf{A}$ and $\mathbf{B}$ in the definition of orthogonal complements. Hence $\mathbf{N} = \mathbf{M}^*$.

□

This theorem has still another important consequence. Let $\mathbf{M} = (S, \mathbf{F})$ be a matroid and $x \in S$. If $\mathbf{A}$ coordinatizes $\mathbf{M}$ over T and we delete the column, corresponding to x, then the remaining matrix coordinatizes $\mathbf{M}\backslash\{x\}$. Now applying $(\mathbf{M}^*\backslash X)^* = \mathbf{M}/X$, see Theorem 7.3.4, we obtain

Corollary 9.1.4 *If* $\mathbf{M}$ *is representable over a field then its minors are as well.*

For example, if a matroid $\mathbf{M}$ contains $\mathbf{U}_{4,2}$ as a minor then $\mathbf{M}$ cannot be represented over $\mathbb{B}$. The converse of this statement is a much deeper theorem. Let us call a matroid *binary* if it is representable over $\mathbb{B}$.

Theorem 9.1.5 [Tutte, 1965]. *A matroid is binary if and only if it does not contain* $\mathbf{U}_{4,2}$ *as a minor.*

Its *proof* will be obtained by Exercise 9.1.12 and Problem 9.1.17.

Another important feature of binarity is related to Exercise 7.2.11:

Theorem 9.1.6 *A matroid is binary if and only if, for every circuit* C *and every cut set* Q *of it,* $|C \cap Q|$ *is even.*

Its *proof* is also left to the reader as Exercise 9.1.10 and Problem 9.1.18.

Several further characterizations of binarity can be found in **Box 9.2.**

Representability over other fields can also be studied in detail, see especially Section 9.2. Similarly, further properties of regular matroids will also be studied in Sections 9.2 and 17.2. Relations among these concepts are summarized in **Box 9.1.**

Exercises for Section 9.1

9.1.1 Suppose that representability of $\mathbf{M} = (S, \mathbf{F})$ over T had been defined in the following way: The vector space $V[T]$ over T has a subset V_0 so

Box 9.1. Relations among various concepts in matroid coordinatization

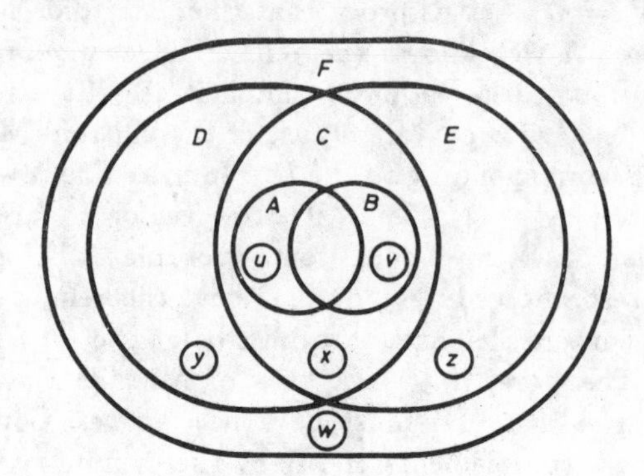

A graphic matroids
B cographic matroids
C regular matroids
D binary matroids
E ternary matroids
F all the matroids

$A \cap B$ circuit matroids of planar graphs
$D \cap E = C$ (Problem 17.2.9)

Examples for some special matroids:
x $\mathbf{R}_{10}$ (Section 17.2)
y $\mathbf{F}_7$ (Section 9.2)
z $\mathbf{U}_{4,2}$
u circuit matroids of the Kuratowski graphs
v cut set matroids of the Kuratowski graphs
w $\mathbf{K}_8$ (Section 9.2)

that V_0 and S are in one–one correspondence and for $X \subseteq S$, let $X \in \mathbf{F}$ if and only if the vectors in the corresponding subset of V_0 are linearly independent.

(a) What is the difference between this definition and our one?
(b) Show that neither Theorem 9.1.2 nor Theorem 9.1.3 were true with this definition.

9.1.2 Let $\mathbf{A}$ coordinatize the matroid $\mathbf{M} = (S, \mathbf{F})$ over a field T and let $x \in S$ be an arbitrary nonloop element in $\mathbf{M}$. We saw that the matrix which coordinatizes $\mathbf{M}\backslash\{x\}$ can be obtained from $\mathbf{A}$ by deleting a column. Show that constructing a matrix which coordinatizes $\mathbf{M}/\{x\}$ corresponds to that process in linear algebra when we have a (fictitious) system of equations with $\mathbf{A}$ as the matrix of coefficients, and the unknown quantity, corresponding to x, is expressed from one equation and is substituted into the others.

Box 9.2. Various characterizations of binary matroids

The following 8 statements are equivalent:

(1) $\mathbf{M}$ is binary, i.e. coordinatizable over the field $\mathbb{B}$.

(2) $\mathbf{M}$ has no minor, isomorphic to $\mathbf{U}_{4,2}$.

(3) The intersection of any circuit and any cut set of $\mathbf{M}$ is of even cardinality.

(4) There are no circuit and cut set containing exactly 3 elements in common.

(5) If C_1, C_2 are distinct, intersecting circuits of $\mathbf{M}$ then $C_1 \triangle C_2$ is dependent.

(6) If $C_1, C_2, \ldots$ are circuits of $\mathbf{M}$ then $C_1 \triangle C_2 \triangle \ldots$ is the disjoint union of circuits of $\mathbf{M}$.

(7) If B is an arbitrary base and C is an arbitrary circuit in $\mathbf{M}$ with $C - B = \{x, y, \ldots\}$ then $C = C_x \triangle C_y \triangle \ldots$ where C_u is the unique circuit of $\mathbf{M}$, contained in $B \cup \{u\}$.

(8) If C_1, C_2 are distinct circuits of $\mathbf{M}$ and $|C_1 \cap C_2| \geq 2$ then $(C_1 \cup C_2) - \{a, b\}$ is dependent for any pair $a, b \in C_1 \cap C_2$.

a Exercise 9.1.10
b trivial
c Problem 9.1.18
d Problem 9.1.17
e trivial
f Problem 9.1.19
g Problem 9.1.17
h Exercise 9.1.12
j Problem 9.1.17
k Exercise 9.1.11

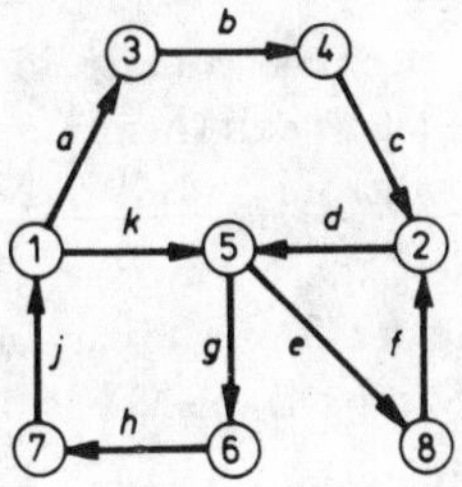

9.1.3 Give an example that $\mathbf{M}$ is representable over T but its truncation is not (cf., Problem 9.2.18).

9.1.4 Let G be a graph and T an arbitrary field, other than $\mathbb{B}$. The proof of Theorem 9.1.2 shows that $\mathbf{M}(G)$ can be coordinatized over T by a matrix with entries $0, \pm 1$ only. Prove that 0 and $+1$ are enough if G is bipartite.

9.1.5 (continued) Does this property characterize bipartite graphs?

9.1.6 Let GF(p) denote the field of the residue classes mod p. (p is a prime, for $p = 2$ we shall use $\mathbb{B}$ further on.) A matroid is *ternary* if it is representable over GF(3). Find the smallest nonternary matroid.

9.1.7 Let $1 < k < n - 1$. Prove that $\mathbf{U}_{n,k}$ is nongraphic (cf., Exercise 7.1.3).

9.1.8 Let C be a circuit in the matroid $\mathbf{M} = (S, \mathbf{F})$ and let $x \in S - C$. We have seen (Exercise 7.3.3) that if C is not a circuit in $\mathbf{M}/\{x\}$ then it need not be the union of two disjoint circuits. Prove that it is if **M** is binary.

9.1.9 Let $k \geq 2$, $S = \{a, b, c_1, c_2, \ldots, c_k\}$ and define $\mathbf{M} = (S, \mathbf{F})$ so that $S - \{a\}$, $S - \{b\}$ and all the triples of form $\{a, b, c_i\}$ are the circuits of $\mathbf{M}$ $(i = 1, 2, \ldots, k)$. Prove that **M** is a matroid for $k = 2$ only.

9.1.10 Prove the necessity of the condition of Theorem 9.1.6.

9.1.11 Let the *symmetric difference* $X \triangle Y$ of two sets X, Y be defined as $(X - -Y) \cup (Y - X)$. Prove that if C_1, C_2 are two intersecting circuits of a binary matroid **M** then $C_1 \triangle C_2$ is dependent.

9.1.12 Show implication $6 \to 7$ in the figure of **Box 9.2.**

Problems for Section 9.1

9.1.13 Prove that every rank 2 matroid is representable over a suitable field.

9.1.14 The *characteristic* of a field T is the smallest positive integer k (if it exists) for which $x + x + \ldots + x$ (k summands) equals zero for every $x \in T$. For example, GF(p) has characteristic p. If no such k exists (like in cases $T = \mathbf{Q}, \mathbf{R}$ or $\mathbf{C}$) then the field is said to be characteristic 0. If k is a nonzero field-characteristic then k is prime.
Give a field of characteristic 2, different from **B**, and represent $\mathbf{U}_{4,2}$ over it.

9.1.15 Let every circuit of a matroid **M** be of even cardinality. (By an analogue of graphs, let us call them *bipartite matroids.*) Show that if **M** is a binary bipartite matroid and Q is a cut set of **M** then $\mathbf{M}/Q$ is also bipartite.

9.1.16 Generalize Euler's theorem (Problem 1.1.28) for binary matroids: The underlying set S of a binary matroid $\mathbf{M} = (S, \mathbf{F})$ is the union of disjoint circuits if and only if $\mathbf{M}^*$ is bipartite [Welsh, 1969b].

9.1.17 Prove implications $2 \to 5, 5 \to 6$ and $7 \to 1$ of **Box 9.2.**

9.1.18 Prove the sufficiency of the condition of Theorem 9.1.6.

9.1.19 We have seen in Section 7.2 that the system **C** of circuits of a matroid satisfy (C1) and (C2) and we could replace (C2) with (C3) or (C4). Prove that the following — stronger — condition (C5) [Fournier, 1981] is true if and only if the matroid is binary.
(C5) If $X \in \mathbf{C}, Y \in \mathbf{C}, X \neq Y$ and $a, b \in X \cap Y$ then there exists a $Z \in \mathbf{C}$ so that $Z \subseteq X \cup Y, a \notin Z$ and $b \notin Z$.

9.1.20 Let P be a finite set, **H** be a collection of certain subsets of P. A nonempty subcollection $\mathbf{H}_0 \subseteq \mathbf{H}$ is called a circuit if every element of P is contained in an even number of subsets from $\mathbf{H}_0$ and $\mathbf{H}_0$ is minimal with respect to this property. If **F** denotes the circuit free subcollections of **H** then $(\mathbf{H}, \mathbf{F})$ forms a binary matroid.
(a) What does this construction mean if every member of **H** is of cardinality two?
(b) Let $P = \{a, b, c\}, \mathbf{H} = \{\{a, b\}, \{b, c\}, \{c, a\}, \{a, b, c\}\}$. What is the resulting matroid?

(c) Prove [Ünver and Ceyhun, 1983] that every binary matroid arises in this way.

9.1.21 Let $\mathbf{M}_1 = (S_1, \mathbf{F}_1)$ and $\mathbf{M}_2 = (S_2, \mathbf{F}_2)$ be two arbitrary matroids, representable over the same field F and satisfying $r_1(S_1) = r_2(S_2)$. Prove that there exists a matrix whose row space matroid is isomorphic to $\mathbf{M}_1$ and column space matroid is isomorphic to $\mathbf{M}_2$.

9.1.22 Prove that if a matroid is representable over a field F then it can also be represented by a symmetric matrix with entries from F.

Section 9.2 Geometric representation

The "representation" of matroids, to be described in this section, is a visualization. Not all matroids can be "drawn" in this way, yet it is a very useful tool.

Consider two matroids; $\mathbf{M}_1$ will be the uniform matroid $\mathbf{U}_{4,2}$ and $\mathbf{M}_2$ will be the circuit matroid of the graph of Fig. 9.1. Both matroids can be coordinatized over the field $\mathbb{R}$, in fact, in an infinite number of ways. For example, $\mathbf{A}_1 = \begin{pmatrix} 1 & 0 & 1 & 1 \\ 0 & 1 & 1 & 2 \end{pmatrix}$ and $\mathbf{A}'_1 = \begin{pmatrix} 0 & 1 & 1 & 2 \\ 1 & -1 & 2 & 1 \end{pmatrix}$ both represent $\mathbf{M}_1$ and $\mathbf{A}_2 = \begin{pmatrix} 1 & 0 & 1 & 0 \\ 0 & 1 & 1 & 0 \\ 0 & 0 & 0 & 1 \end{pmatrix}$ and $\mathbf{A}'_2 = \begin{pmatrix} 1 & 2 & 0 & 1 \\ 0 & 1 & 1 & 0 \\ 0 & 0 & 0 & 1 \end{pmatrix}$ both represent $\mathbf{M}_2$.

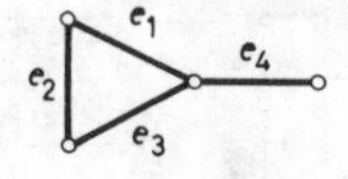

Fig. 9.1

The four column vectors of $\mathbf{A}_1$ and those of $\mathbf{A}'_1$ are very different (see Fig. 9.2). From the point of view of matroid theory they carry the same information: We have four vectors, any two are linearly independent and the other two can be expressed as their linear combination.

Let us consider a line e "in general position" in the plane (i.e., e does not pass through the origin and is not parallel to any of our vectors). Instead of our vectors, consider the intersection of their lines with e (Fig. 9.3). The two results are essentially the same: four distinct points on a line, see Fig. 9.4. This drawing still carries all the information about $\mathbf{M}_1$.

Similarly, the four column vectors of $\mathbf{A}_2$ and $\mathbf{A}'_2$ are different (see Fig. 9.5) but if we take a plane S "in general position" and instead of the vectors we consider the intersections of their line with S (Fig. 9.6) then the results are

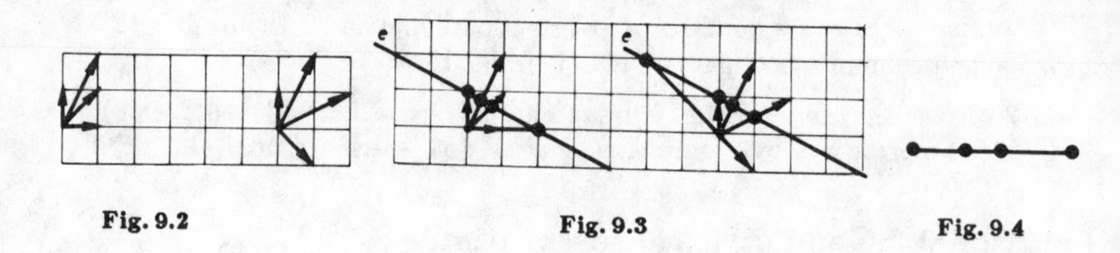

Fig. 9.2 Fig. 9.3 Fig. 9.4

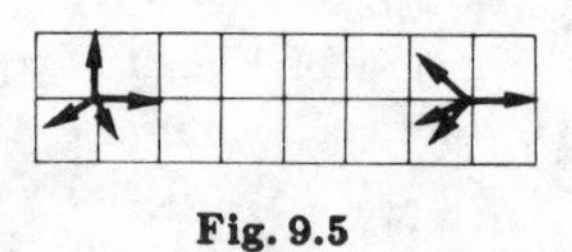

Fig. 9.5

essentially the same: four distinct points on a plane, the first three are collinear, see Fig. 9.7. This drawing also contains all the information about $\mathbf{M}_2$.

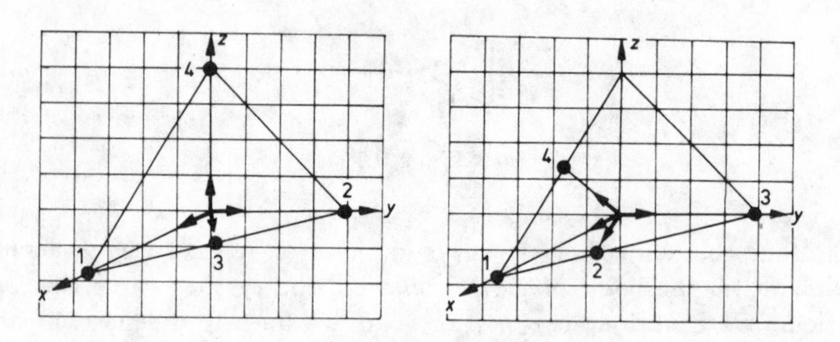

Fig. 9.6

We try to generalize this way of "visualization". If the matroid has loops, we do not associate points to them (since the zero vector does not intersect the line or plane in general position); if the matroid has parallel elements, we associate the same points to them (since their vectors had the same line); if three elements form a circuit in the matroid, we associate colinear points to them (since their vectors were coplanar) etc.

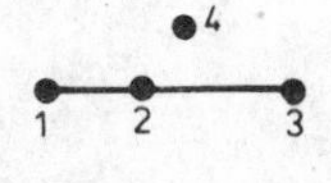

Fig. 9.7

Try now to visualize $\mathbf{M}(K_4)$. If we choose the basis $\{1,2,3\}$, see Fig. 9.8 then it can be coordinatized by $\begin{pmatrix} 1 & 0 & 0 & a & 0 & e \\ 0 & 1 & 0 & b & c & 0 \\ 0 & 0 & 1 & 0 & d & f \end{pmatrix}$ where a, b, c, d, e, f are suitable nonzero numbers. (For example, $\{1,2,4\}$ is a circuit hence the fourth column is the linear combination of the first two columns.) Should we choose $a = b = c = d = e = f = 1$, the six vectors of Fig. 9.9 (or, after the intersection with a plane in general position, the six points of Fig. 9.10) are obtained. This is obviously wrong since $\{4,5,6\}$ must be a circuit in $\mathbf{M}(K_4)$; hence the corresponding three points must be collinear.

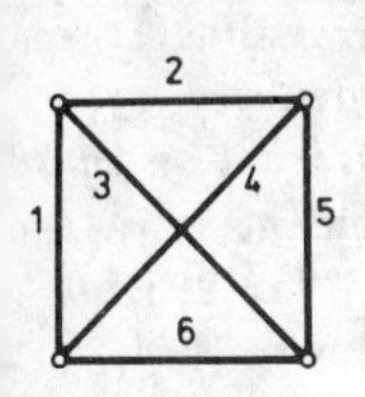

Fig. 9.8

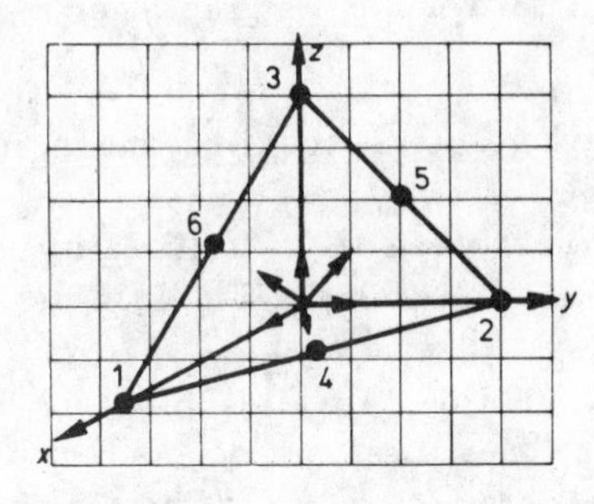

Fig. 9.9

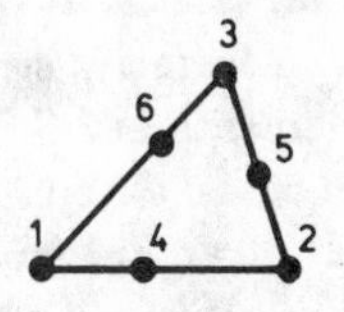

Fig. 9.10

Our process failed since we choose a, b, c, d, e, f arbitrarily, despite the requirement of $\begin{vmatrix} a & 0 & e \\ b & c & 0 \\ 0 & d & f \end{vmatrix} = 0$ (since $\{4,5,6\}$ is dependent in the matroid). Should we choose, say, $a = b = c = 1, d = f = 2, e = -1$ instead, then the vectors 4, 5, 6 become coplanar (Fig. 9.11) and the corresponding points collinear (Fig. 9.12).

This drawing of Fig. 9.12 could have been found much more quickly if we had not considered coordinatization and 3-dimensional space. Should we simply say that $\{1,2,4\}, \{1,3,6\}, \{2,3,5\}$, and $\{4,5,6\}$ are the 3-element circuits of $\mathbf{M}(K_4)$, which can trivially be seen from Fig. 9.8, it were a few seconds' job to find the drawing of Fig. 9.12. This quickness is one of the advantages of this visualization. Another advantage is that if we need a matroid with certain

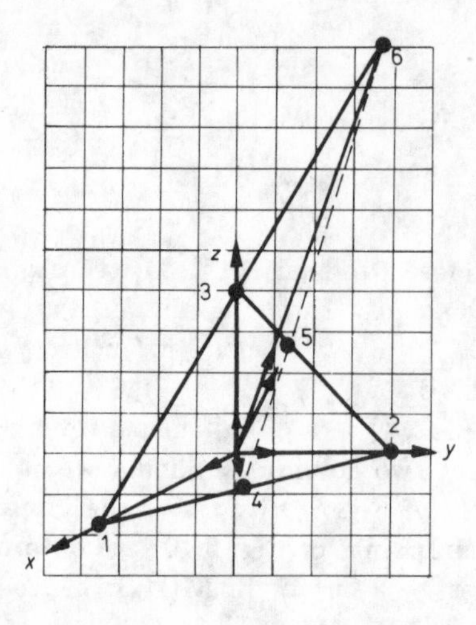

Fig. 9.11

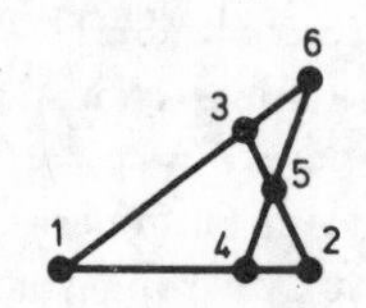

Fig. 9.12

properties, then such a drawing can be prepared independently of whether the matroid is graphic (and if so, what is its graph), or representable (and if so, what is its matrix) etc.

As in graph theory (where points could be drawn anywhere and the lines could be drawn as straight or not), here we also need not worry about the actual positions of the points, even their "collinearity" can be denoted by a not necessarily straight line. For example, Fig. 9.13 denotes a matroid on the 7-elements set $S = \{1, 2, \ldots, 7\}$ with rank 3, where all the 3-element subsets are bases except $\{1,2,4\}, \{1,3,6\}, \{1,5,7\}, \{2,3,5\}, \{2,6,7\}, \{3,4,7\}, \{4,5,6\}$. Only 6 among these 7 "collinearities" are denoted by straight lines, $\{3,4,7\}$ is by a circle. (In fact, it is easy to prove that no drawing of these 7 points in the plane can ensure these seven collinearities in the geometrical sense.) Yet this is a matroid, in fact a very interesting one that we shall call *Fano-matroid* and denote by $\mathbf{F}_7$.

Try now to determine $\mathbf{F}_7 \backslash \{7\}$. Although the "lines" $\{2,6\}, \{1,5\}$ and $\{3,4\}$ still could be drawn they carry no information any more since any 2 distinct points are on a "line". Hence these lines need not be drawn at all and the result (Fig. 9.14) is obviously $\mathbf{M}(K_4)$.

We may conclude that once a matroid is represented in this way, the deletion of some elements of it can be obtained by deleting the corresponding points of the drawing. If an element x should be contracted, the whole drawing is to be projected from the point, corresponding to x, to a subspace of general position and of dimension one less (see Exercise 9.2.4). For example, contractions of element 6 from $\mathbf{F}_7$ and from $\mathbf{M}(K_4)$ are shown in Fig. 9.15. The resulting matroids are graphic (the circuit matroids of the two graphs of Fig. 9.16).

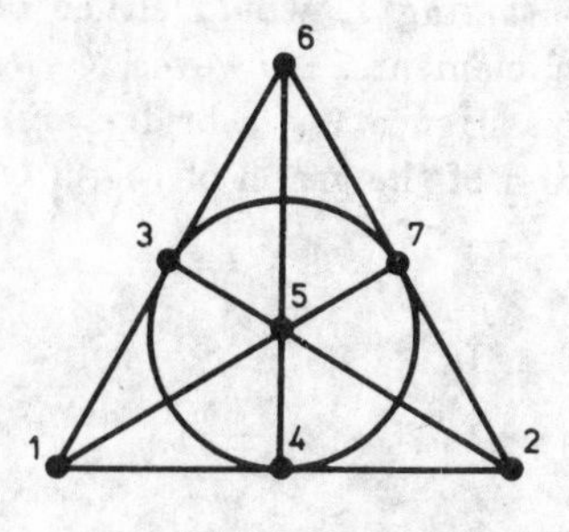

Fig. 9.13

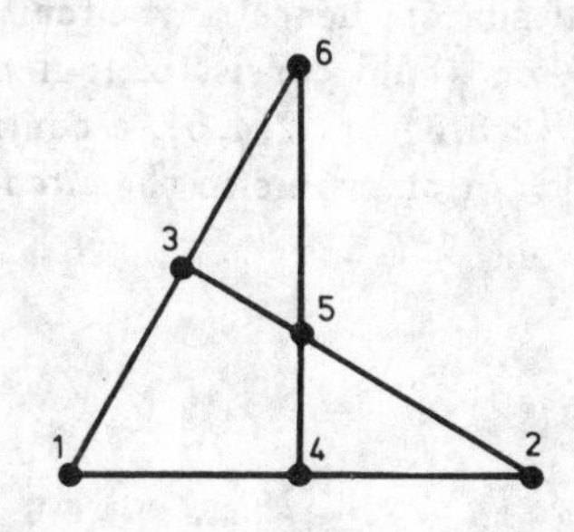

Fig. 9.14

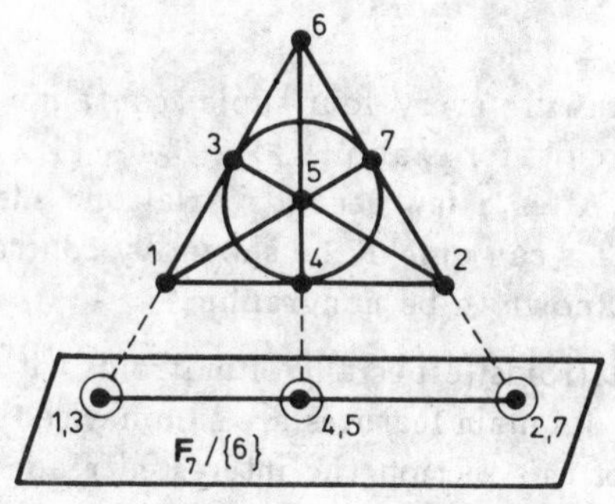

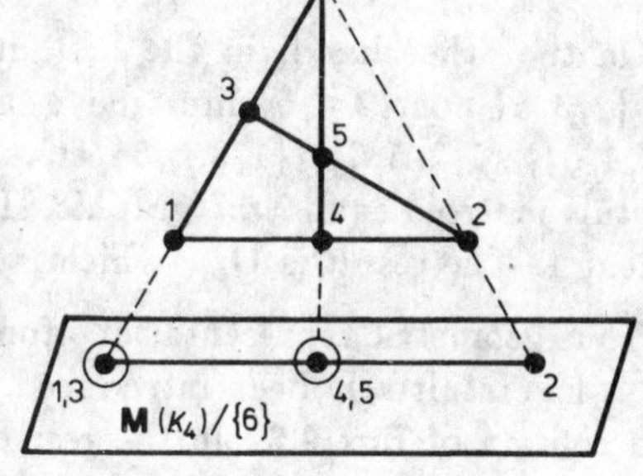

Fig. 9.15

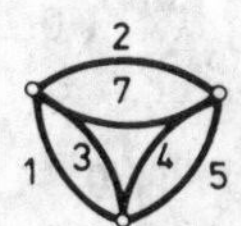

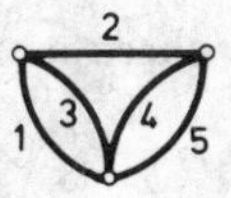

Fig. 9.16

As another example, consider the two matroids, represented by the drawings of Fig. 9.17. The sets $\{2,3,5\}, \{2,4,6\}, \{3,4,5,6\}$ are obviously circuits in both. In the second drawing the matroid has rank 4 (since the drawing is in the 3-dimensional space), hence any circuit C, containing 1, would either be of cardinality 5, or would consist of four coplanar elements. However, C would then contain $\{2,3,5\}$ or $\{2,4,6\}$, a contradiction. Hence 1 is a bridge and the whole matroid is isomorphic to the circuit matroid of the graph of Fig. 9.18.

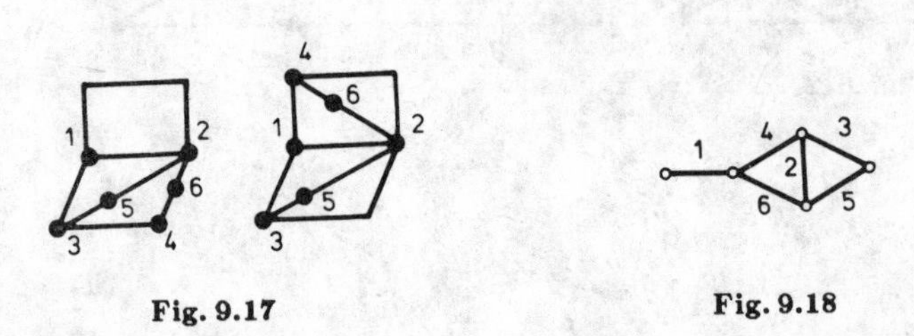

Fig. 9.17 **Fig. 9.18**

On the other hand, in the first drawing every four-tuple, containing neither $\{2,3,5\}$ nor $\{2,4,6\}$ must be a circuit, for example $\{1,2,3,4\}$, $\{1,2,3,6\}$, $\{1,2,4,5\}$, $\{1,2,5,6\}$, $\{1,3,4,5\}$ etc. After a few seconds' trial one suspects that this matroid cannot be graphic. This can quickly be shown by contracting element 1. The result is $\mathbf{U}_{5,2}$ which is known to be nongraphic.

Thus geometric representation of matroids (especially of matroids with rank 3 or 4) has intuitively been introduced. Its main features are summarized in the third column of **Box 9.3**. In the rest of this section the interested reader may find some further remarks related to this representation. They are, however, not essential for the rest of this book.

The drawings (e.g., in Figs 9.4, 9.7, 9.10, 9.12) were obtained by a projection of the 2- or 3-dimensional space to a general position subspace of dimension 1 or 2, respectively. If such a drawing corresponds to a representable matroid, one way of constructing such a matroid representation is the following.

Suppose the matroids $\mathbf{M}_1$ and $\mathbf{M}_2$ are given by the drawings of Figs 9.19 and 9.20, respectively. Imagine their points, for a moment, in the 1- and 2-dimensional space, respectively, and write their vectors in a suitable system of coordinates. The results are, say, $(0,1,2,3)$ and $\begin{pmatrix} 0 & 1 & 2 & 0 & 0 \\ 0 & 0 & 0 & 1 & 2 \end{pmatrix}$, respectively. Then add an all-one row to these matrices:

$$\mathbf{A}_1 = \begin{pmatrix} 0 & 1 & 2 & 3 \\ 1 & 1 & 1 & 1 \end{pmatrix} \quad \mathbf{A}_2 = \begin{pmatrix} 0 & 1 & 2 & 0 & 0 \\ 0 & 0 & 0 & 1 & 2 \\ 1 & 1 & 1 & 1 & 1 \end{pmatrix}$$

These are just the right coordinatizations of $\mathbf{M}_1$ and $\mathbf{M}_2$, respectively.

Box 9.3. How can we imagine certain concepts of matroid theory?

Matroid theory concept	Suppose that the matroid can be given by		
	a graph	vectors	geometrical drawing
element	edge	vector	point
parallel elements	parallel edges	parallel vectors	coinciding points
loop	loop	zero vector	<does not belong to the model>
rank	rank	dimension of subspace	dimension of subspace +1
deletion	deletion	deletion	deletion
contraction	contraction	elimination of an unknown	projection from a point
dual	dual for planar graphs, nothing for the others	orthogonal complement	<has no visual meaning>
sum of matroids	see Section 11.2		
transversal matroids	see Section 15.1		

Fig. 9.19

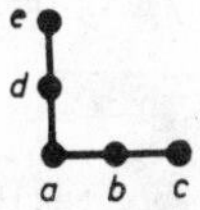

Fig. 9.20

The idea behind these examples is that deleting the kth row of the matrix corresponds to the projection of the vectors to a subspace of dimension $k-1$, orthogonal to the kth axis. The fact that the kth row contains no zero elements, assures that the remaining $(k-1)$-dimensional subspace will be in "general position", i.e. not parallel to any of the vectors.

Consider now the plane $z = 1$ in the 3-dimensional space (Fig. 9.21). If P_1, P_2 are points in this plane, let $\mathbf{u}_1$ and $\mathbf{u}_2$ denote the vectors OP_1 and OP_2, respectively. If $\mathbf{v} = \lambda_1\mathbf{u}_1 + \lambda_2\mathbf{u}_2$ is an arbitrary linear combination then the end point of $\mathbf{v}$ will usually not be on the plane. However, it will be if $\lambda_1 + \lambda_2 = 1$.

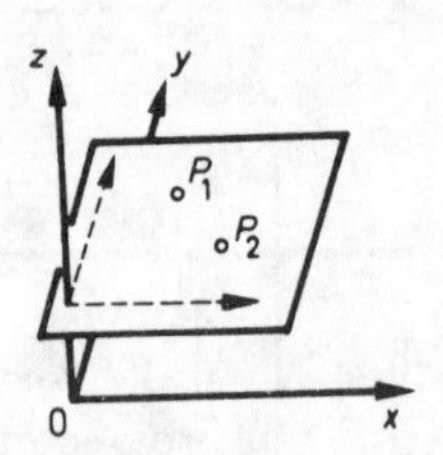

Fig. 9.21

More generally, the linear combination $\lambda_1\mathbf{u}_1 + \lambda_2\mathbf{u}_2 + \ldots + \lambda_n\mathbf{u}_n$ is called an *affine combination* if $\lambda_1 + \lambda_2 + \ldots + \lambda_n = 1$. Vectors are *affinely independent* if $\lambda_1\mathbf{u}_1 + \lambda_2\mathbf{u}_2 + \ldots + \lambda_n\mathbf{u}_n = \mathbf{0}$ with the additional requirement that $\lambda_1 + \lambda_2 + \ldots + \lambda_n = 0$ is possible if and only if $\lambda_i = 0$ for every i. One can easily see that vectors are affinely independent if and only if none of them can be expressed as the affine combination of the others. Hence affine independence is implied by linear independence but not *vice versa*.

Statement 9.2.1 *The column vectors of a matrix are affinely independent if and only if the column vectors of another matrix, obtained from the first one by adding an all-one row, are linearly independent.*

Its *proof* is left to the reader as Problem 9.2.22.

Exercises for Section 9.2

9.2.1 Determine the graphic one(s) among the three matroids drawn in Fig. 9.22.

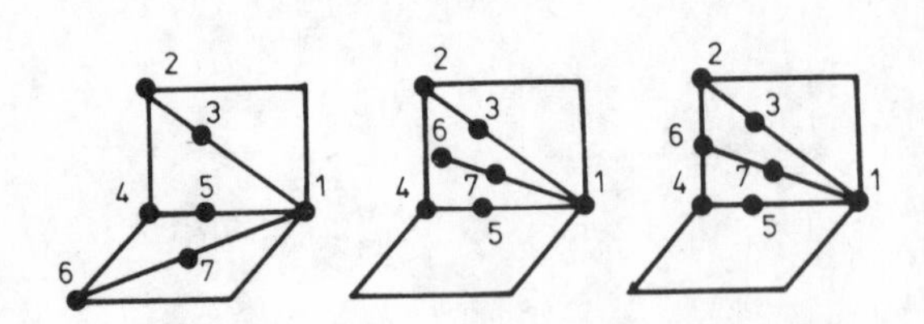

Fig. 9.22

9.2.2 Prove that $\mathbf{F}_7$ is neither graphic nor cographic.

9.2.3 Coordinatize $\mathbf{F}_7$ and its dual $\mathbf{F}_7^*$. Over which fields is this possible?

9.2.4 Prove that the contraction of x corresponds to a projection of the whole k-dimensional drawing from x to a general position subspace of dimension $k-1$.

9.2.5 Delete the "circle" from $\mathbf{F}_7$ (see Fig. 9.23). The resulting matroid is sometimes called *anti-Fano matroid* and is denoted by $\mathbf{F}_7^-$. Over which fields is $\mathbf{F}_7^-$ coordinatizable?

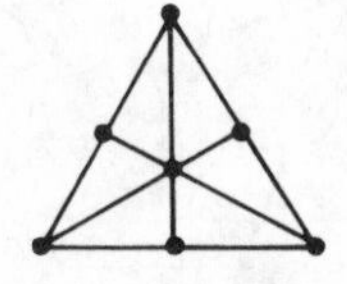

Fig. 9.23

9.2.6 Find a matroid which cannot be coordinatized over any field.

9.2.7 The rank four matroid $\mathbf{M}$ is represented by the 3-dimensional drawing, consisting of the 8 vertices of a cube.
(a) Draw the representation of $\mathbf{M}/\{x\}$ for an element x of $\mathbf{M}$.
(b) Show that $\mathbf{M}$ is nonbinary.

Problems for Section 9.2

9.2.8 Generalize Exercise 9.2.3. Let p be a prime and consider the matroid $\mathbf{M}_p$ which is coordinatized over GF(p) by the following matrix [Lazarson, 1958]:

$$\overbrace{\qquad\qquad}^{p+1}\qquad\overbrace{\qquad\qquad}^{p+1}$$
$$\left[\begin{array}{cccccc|cccccc|c}
1 & & & & & & 0 & & & & & & 1\\
 & 1 & & \mathbf{0} & & & & 0 & & \mathbf{1} & & & 1\\
 & & 1 & & & & & & 0 & & & & 1\\
 & & & \ddots & & & & & & \ddots & & & \vdots\\
 & & & & 1 & & & & & & 0 & & 1\\
 & \mathbf{0} & & & & 1 & & \mathbf{1} & & & & 0 & 1\\
 & & & & & 1 & & & & & & 0 & 1
\end{array}\right]$$

Prove that $\mathbf{M}_p$ can be coordinatized over fields of characteristic p only.

9.2.9 *Desargues' theorem* states that if e_1, e_2, e_3 are three straight lines, having a common point P (see Fig. 9.24) and A_i, B_i are arbitrary points on e_i ($i = 1, 2, 3$) then the three points Q_{ij} of intersections of the lines A_iA_j and B_iB_j are collinear. Consider the figure as the 3-dimensional drawing of a rank 4 matroid $\mathbf{M}$ on the 10-element set. Prove that $\mathbf{M}$ is graphic.

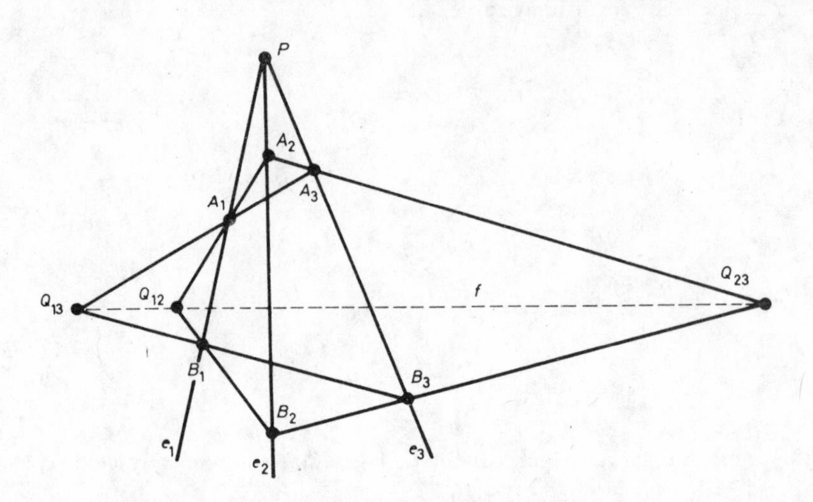

Fig. 9.24

9.2.10 (continued) Consider the 10 points of Fig. 9.24 as points of a graph and call two points adjacent if their interconnecting straight line is not shown on the drawing. What sort of graph do we obtain?

9.2.11 (continued) Consider Fig. 9.24 as a 2-dimensional drawing and delete line f (i.e., decide that $\{Q_{12}, Q_{13}, Q_{23}\}$ be independent). Prove that the new drawing also represents a rank 3 matroid (the so called *non-Desarguesian matroid*). This is, of course, not representable over any field.

9.2.12* *Pappos' theorem* states the following: Let A_i, B_i, C_i be three–three collinear points on the lines e_1 and e_2, respectively (see Fig. 9.25). Denote the intersection of the lines A_iB_{3-i} by $C_3 (i = 1,2)$; that of the lines A_iC_{3-i} by B_3 and B_iC_{3-i} by A_3. Then the points A_3, B_3 and C_3 are collinear.
If we delete line f (i.e., decide that $\{A_3, B_3, C_3\}$ be independent), a new rank 3 matroid is obtained, the so called *non-Papposian matroid.* This is not representable over any field either.

9.2.13 Let $S = \{a, a', b, b', c, c', d, d'\}$ and let $A = \{a, a'\}$, $B = \{b, b'\}$ etc. Define a matroid $\mathbf{K}_8$ on S so that all of its 4-element subsets be bases except $A \cup B, A \cup C, A \cup D, B \cup C$ and $B \cup D$. (This is not the matroid represented by the drawing of Fig. 9.26 since $C \cup D$ is a base in $\mathbf{K}_8$.) Prove that $\mathbf{K}_8$ cannot be represented over any field [Higgs, Vámos].

9.2.14 (continued) Prove that $\mathbf{K}_8$ does not satisfy property (R'') of the rank-functions, see Problem 7.1.20 [Ingleton, 1971].

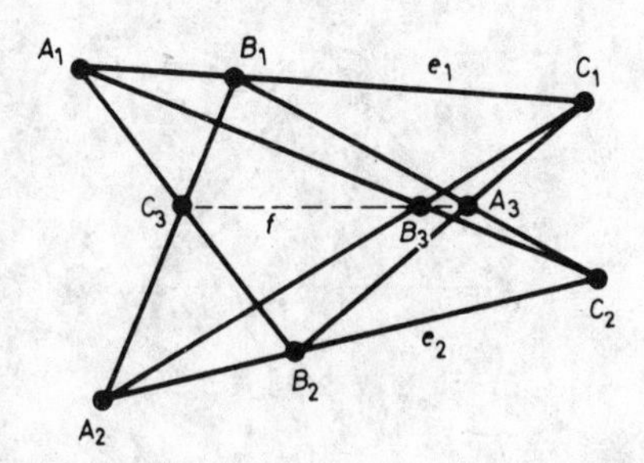

Fig. 9.25

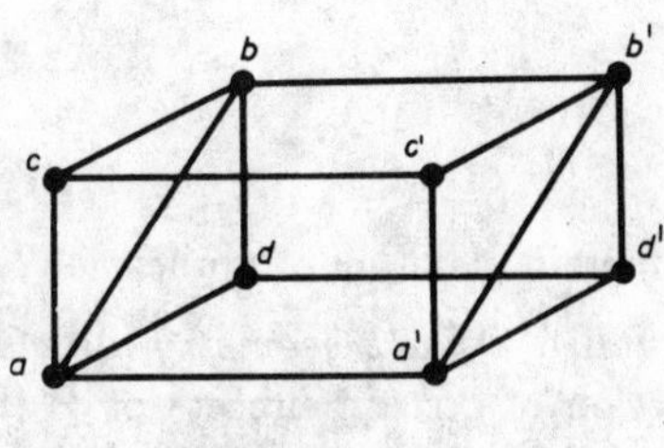

Fig. 9.26

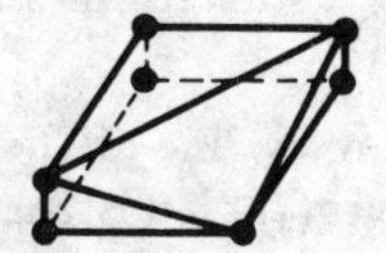

Fig. 9.27

9.2.15 Which matroid is shown in Fig. 9.27?

9.2.16 Over which fields is the matroid of Fig. 9.28 coordinatizable [Aigner, 1979]? In particular, prove that it is representable over the field $\mathbb{C}$ of the complex numbers but not over the field $\mathbb{R}$ of the reals [MacLane, 1936].

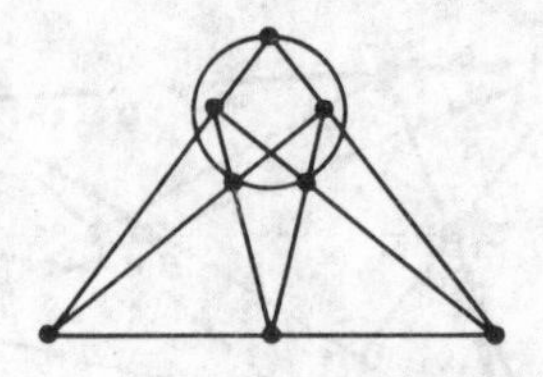

Fig. 9.28

9.2.17 The set $P = \{0, 2, 3, 5, 7, \ldots\}$ consists of zero and the primes. A subset $K \subseteq P$ is the *characteristic set* of a matroid $\mathbf{M}$ if $\mathbf{M}$ is representable over some extension of a field if and only if the characteristic of the field is an element of K. Every finite subset of P not containing zero, and the complement of every such set, arise as characteristic sets of suitable matroids [Kahn, 1982]. Give examples for matroids with characteristic sets
(a) $K = P$
(b) $K = \emptyset$
(c) $|K| = 1$
(d) $K = P - \{2\}$.

9.2.18 What is the geometric meaning of truncation?

9.2.19 A pair (P, L) is called a *finite geometry* (or a *projective plane*) if P is a set, L is a collection of certain subsets of P (the elements of P and L are called points and lines, respectively) and
(1) for every two points there is exactly one line containing both of them;
(2) for every two lines there is exactly one point contained in both of them;
(3) there are four noncollinear points.
Show that the drawing of $\mathbf{F}_7$ determines a finite geometry (the so called *Fano geometry*).

9.2.20 (continued) Prove that every line of the finite geometry has the same number of points and every point is contained in the same number of lines.

9.2.21 (continued) Prove that the number of points in a finite geometry is $k^2 + k + 1$ for a suitable k.

9.2.22 Prove Statement 9.2.1.

9.2.23 Is the matroid of Fig. 9.29 representable over some field?

Section 9.3 The closed sets of a matroid

One of the most important concepts of linear algebra is the linear combination of vectors. This will be generalized for matroids.

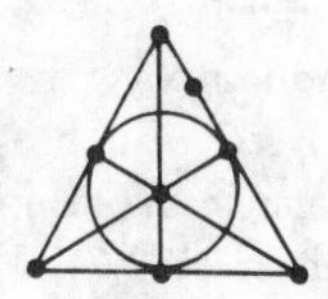

Fig. 9.29

For the matroid $\mathbf{M} = (S, \mathbf{F})$ if $X \subset S$ then x will be said to belong to the closure of X if $r(X \cup \{x\}) = r(X)$. If $\mathbf{M}$ was a set of certain vectors from a vector space then this exactly means that x arises as a linear combination of the vectors in X. By definition, the elements of X are also considered as members of the closure. Hence the *closure* $\sigma(X)$ of the subset X is defined by

$$\sigma(X) = \{x | r(X \cup \{x\}) = r(X)\}$$

In particular, if $\mathbf{M}$ is the circuit matroid of a graph G then $x \in \sigma(X)$ holds for $x \in S - X$ if and only if either x is a loop or its two end points can be connected by such a path of G which contains edges of X only.

Obviously, $X \subseteq \sigma(X)$. It is important to see that a subset can be extended in this way only once:

Statement 9.3.1 *$\sigma(\sigma(X)) = \sigma(X)$ for every $X \subseteq S$.*

Proof. $\sigma(\sigma(X)) \supseteq \sigma(X)$ is obvious. Suppose there were an element $x \in \in \sigma(\sigma(X)) - \sigma(X)$. Then $r(X \cup \{x\}) > r(X)$ but $r(\sigma(X) \cup \{x\}) = r(\sigma(X))$ would hold. However, by property (R7) of the rank functions (Section 7.1) we have $r(\sigma(X)) = r(X)$, thus

$$r(\sigma(X)) = r(X) < r(X \cup \{x\}) \leq r(\sigma(X) \cup \{x\}) = r(\sigma(X)),$$

a contradiction.

□

That is, if $x \notin X$ but $x \in \sigma(X)$ then the matroid has a circuit C with $x \in C$ and $C - \{x\} \subseteq X$. This could be an alternative definition of closure, see Exercise 9.3.3. Sometimes this property can be applied more easily than the original definition.

A subset $X \subseteq S$ of a matroid $\mathbf{M} = (S, \mathbf{F})$ is called a *closed set* if $\sigma(X) = X$. The underlying set S itself is always closed. The empty set is closed if and only if $\mathbf{M}$ has no loops. The single element set $\{x\}$ is closed if and only if there exists no element parallel to x. Hence a matroid is simple (see Exercise 7.2.6) if and only if $\emptyset$ and all singletons are closed sets.

Theorem 9.3.2 *Let $\mathbf{M} = (S, \mathbf{F})$ be a matroid, consisting not only of loops (i.e., $r(S) \geq 1$). Then a subset $X \subset S$ with rank $r(S) - 1$ is closed if and only if $S - X$ is a cut set.*

Proof.

(a) Let $S - X$ be a cut set. If X were not closed, there would be an element $x \notin X$ with $x \in \sigma(X)$. Then $X \cup \{x\}$ would contain a circuit C with $(S-X) \cap C = \{x\}$, a contradiction.

(b) Let X be closed. $S - X$ is nonempty for $r(X) < r(S)$. No circuit C of $\mathbf{M}$ can satisfy $|C \cap (S - X)| = 1$ since such a single element is also in $\sigma(X)$ if all the other elements of C are in X. Hence $S - X$ is either a cut set, or contains a cut set Q (by Theorem 7.2.7). However, in the latter case $S - Q$ is closed by part (a) of the proof, but its rank could not be between $r(X)$ and $r(S)$.

□

Theorem 9.3.3 *$X \subseteq S$ is closed in $\mathbf{M} = (S, \mathbf{F})$ if and only if $S - X$ is either empty or is the union of cut sets.*

Proof. If $S - X$ is empty or is the union of cut sets then the statement follows as in part (a) of the previous proof. On the other hand, let $X \subseteq S$ be closed. If $X = S$, we are done, otherwise let $x \in S - X$. We construct a cut set of $S - X$ containing X. Let Y be a maximal independent subset of X. $Y \cup \{x\}$ is independent and is therefore contained in a base B_x. Then $S - \sigma(B_x - \{x\})$ is a cut set by the previous theorem and has the required properties.

□

Exercises for Section 9.3

9.3.1 Prove that the following statements are equivalent.
(1) $x \in S$ is a loop in the matroid $\mathbf{M} = (S, \mathbf{F})$
(2) $x \in \sigma(\emptyset)$
(3) $x \in \sigma(X)$ for every $X \subseteq S$.
(4) $x \in X$ for every closed subset $X \subseteq S$.

9.3.2 Prove that if $x \in \sigma(X)$ and x, y are parallel elements, then $y \in \sigma(X)$ also holds.

9.3.3 Prove that $X \subset S$ is closed if and only if $|C - X| = 1$ holds for no circuit C.

9.3.4 Show that $X \subset S$ is closed in $\mathbf{M} = (S, \mathbf{F})$ if and only if $\mathbf{M}/X$ has no loops.

9.3.5 Prove that if $r(X_1) + r(X_2) = r(X_1 \cup X_2)$ then $\sigma(X_1) \cap \sigma(X_2)$ is empty or contains loops only.

9.3.6 Show that every matroid of rank r has at least 2^r closed subsets. Is this bound sharp?

9.3.7 How can one determine the matroid if only the closure function $X \rightarrow \sigma(X)$ is known?

9.3.8 Let X, Y be closed sets. Prove that $X \cap Y$ is always closed while $X \cup Y$ not.

9.3.9 Give an example which shows that the cut sets of Theorem 9.3.3 cannot be prescribed to be disjoint.

9.3.10 Let $\mathbf{A}$ be an $r \times n$ matrix with entries from a field F. Let the set of its columns be denoted by $S = \{s_1, s_2, \ldots, s_n\}$. The *support* of a row vector $(a_{k,1}, a_{k,2}, \ldots, a_{k,n})$ is the set $\{s_i | a_{k,i} \neq 0\}$. List the supports of all the rows of $\mathbf{A}$ and those of all the linear combinations of the rows (with coefficients from F). Finally, consider the set Q of those nonempty elements of this list of supports, which are minimal (i.e., which do not contain any other one). Prove that they are just the circuits of a matroid $\mathbf{M}$ on S.

Problems for Section 9.3

9.3.11 One can prove that the closed sets of a matroid $\mathbf{M}$ form a lattice $\mathbf{L}(\mathbf{M})$ with the operations $X \wedge Y = X \cap Y; X \vee Y = \sigma(X \cup Y)$.
(a) What is the lattice of the uniform matroids?
(b) Determine $\mathbf{L}(\mathbf{M}(K_4))$.
(c)* Give a lattice which does not arise as $\mathbf{L}(\mathbf{M})$ for any $\mathbf{M}$.

9.3.12* Consider only those closed sets of $\mathbf{M}$ which arise as unions of suitable circuits of $\mathbf{M}$. These closed sets also form a lattice. What is the corresponding lattice for $\mathbf{M}(K_4)$?

9.3.13 List all the closed sets of a matroid $\mathbf{M}$ of rank k. Then delete the rank $k-1$ closed sets from the list. Prove that the new list is the list of closed sets of another matroid $\mathbf{M}'$.

9.3.14 Prove that the matroid $\mathbf{M}$ of Exercise 9.3.10 is just the dual of the matroid $\mathbf{N}$, coordinatized by $\mathbf{A}$.

9.3.15 (continued) How is the closure operation of $\mathbf{N}$ reflected in the matrix $\mathbf{A}$?

9.3.16 Give a new proof for Exercise 9.1.10, using the concept of the support.

9.3.17 Let X be a circuit and $S - X$ be a cut set in the matroid $\mathbf{M} = (S, \mathbf{F})$. Prove that $\mathbf{M}' = (S, \mathbf{F}')$ with $\mathbf{F}' = \mathbf{F} \cup \{X\}$ is also a matroid [Seymour, 1981]. See also [Kahn, 1985].

Chapter 10

Applications

Section 10.1 Duality in electric network theory II: Inverse, dual and adjoint multiports; the full symmetry

The "classical" duality of electric network theory was introduced in Section 4.1. There we used the word "inverse" rather than the more usual word "dual". Our reason to do this was that we wish to use "dual" for a concept obtained from matroidal duality. Hence we shall also emphasize that there are several symmetries in electric network theory.

Recall that n-ports, as introduced in Section 8.1, were described as $\mathbf{Au} + \mathbf{Bi} = \mathbf{0}$ with $r(\mathbf{A}|\mathbf{B}) = n$. We mentioned already then that one and the same multiport can have several matrix descriptions. If the pairs $\mathbf{A}, \mathbf{B}$ and $\mathbf{A}', \mathbf{B}'$ determine two multiports N and N' respectively, then N and N' will be considered equivalent (denoted by $N \cong N'$) if, for any pair $\mathbf{u}, \mathbf{i}$ of vectors, either both $\mathbf{Au} + \mathbf{Bi} = \mathbf{0}$ and $\mathbf{A}'\mathbf{u} + \mathbf{B}'\mathbf{i} = \mathbf{0}$ hold or none of them. (In other words, they describe the same voltage-current relations, although by different pairs of matrices.)

Statement 10.1.1 *$N \cong N'$ holds if and only if there exist matrices $\mathbf{S}, \mathbf{T}$ so that $\mathbf{S}(\mathbf{A}|\mathbf{B}) = (\mathbf{A}'|\mathbf{B}')$ and $\mathbf{T}(\mathbf{A}'|\mathbf{B}') = (\mathbf{A}|\mathbf{B})$.*

Proof. Trivial since equivalence means that $\mathbf{Au} + \mathbf{Bi} = \mathbf{0}$ and $\mathbf{A}'\mathbf{u} + \mathbf{B}'\mathbf{i} = \mathbf{0}$ determine the same (n-dimensional) subspace of the $2n$-dimensional space, and then $\mathbf{S}$ and $\mathbf{T}$ are just transformations of the systems of coordinates.

□

Of course, in order to have equivalence, $r(\mathbf{A}|\mathbf{B}) = r(\mathbf{A}'|\mathbf{B}')$ must hold and if this common rank is, at the same time, the number of rows in $\mathbf{A}, \mathbf{B}, \mathbf{A}', \mathbf{B}'$ as well, then $\mathbf{S}, \mathbf{T}$ are nonsingular square matrices.

If $N \cong N'$ then their matroids $\mathbf{M}_N$ and $\mathbf{M}_{N'}$ are identical. (But this can happen for nonequivalent multiports as well, see Exercise 10.1.3.)

In Section 4.1 a correspondence among network elements was obtained by the voltage–current symmetry. Let us extend this to multiports at first. If $\mathbf{Au} + \mathbf{Bi} = \mathbf{0}$ describes an n-port N then the *inverse* of N (denoted by inv N) is the n-port N' satisfying $\mathbf{Bu} + \mathbf{Ai} = \mathbf{0}$. This is simply the interchange of the roles of voltage and current.

Obviously if $N' =$inv N then their matroids $\mathbf{M}_N$ and $\mathbf{M}_{N'}$ are isomorphic; the one–one correspondence ϕ simply gives $\phi(u_j) = i_j$ and $\phi(i_j) = u_j$ for every $j = 1, 2, \dots, n$.

Once again, observe that $N' =$inv N implies $\mathbf{M}_{N'} = \phi(\mathbf{M}_N)$ but not *vice versa* (see Exercise 10.1.3).

Matroid theory suggests an entirely different duality concept among electric network elements. If $\mathbf{Au} + \mathbf{Bi} = \mathbf{0}$ describes an n-port N then the *dual* of N (denoted by dual N) is an n-port N' with matrices $\mathbf{A}'$, $\mathbf{B}'$ if, for every pair $\mathbf{u}, \mathbf{i}$ satisfying $\mathbf{Au} + \mathbf{Bi} = \mathbf{0}$ and for every pair $\mathbf{u}', \mathbf{i}'$ satisfying $\mathbf{A}'\mathbf{u}' + \mathbf{B}'\mathbf{i}' = \mathbf{0}$, the relation $\mathbf{u}^T\mathbf{u}' + \mathbf{i}^T\mathbf{i}' = 0$ holds.

Since $r(\mathbf{A}|\mathbf{B}) = r(\mathbf{A}'|\mathbf{B}') = n$, this condition means that $(\mathbf{A}|\mathbf{B})$ and $(\mathbf{A}'|\mathbf{B}')$ are orthogonal complements. Hence if N and N' are dual n-ports then their matroids $\mathbf{M}_N$ and $\mathbf{M}_{N'}$ are also dual to each other (and this implication is also not true *vice versa*, see Exercise 10.1.3).

For example, let N be a 2-port with $i_1 = 0,\ u_2 = ku_1$. One directly obtains $u_1 = 0,\ i_2 = ki_1$ for inv N. In order to find dual N, let $\mathbf{u} = \begin{pmatrix} u_1 \\ u_2 \end{pmatrix}$; $\mathbf{i} = \begin{pmatrix} i_1 \\ i_2 \end{pmatrix}$; $\mathbf{u}' = \begin{pmatrix} u_1' \\ u_2' \end{pmatrix}$; $\mathbf{i}' = \begin{pmatrix} i_1' \\ i_2' \end{pmatrix}$. Then $\mathbf{u}^T\mathbf{u}' + \mathbf{i}^T\mathbf{i}' = 0$ becomes $u_1u_1' + u_2u_2' + i_1i_1' + $ $+ i_2i_2' = 0$. Taking $i_1 = 0$ and $u_2 = ku_1$ into consideration, we obtain that $u_1(u_1' + ku_2') + i_2i_2' = 0$ must hold. Since u_1 and i_2 are arbitrary (i_1 and u_2 were eliminated), this is possible only if $u_1' = -ku_2',\ i_2' = 0$ hold for dual N.

Our results are shown in Fig. 10.1. The dual of a voltage controlled voltage source is *not* the current controlled current source (that is the inverse) but another voltage controlled voltage source, with opposite direction of control.

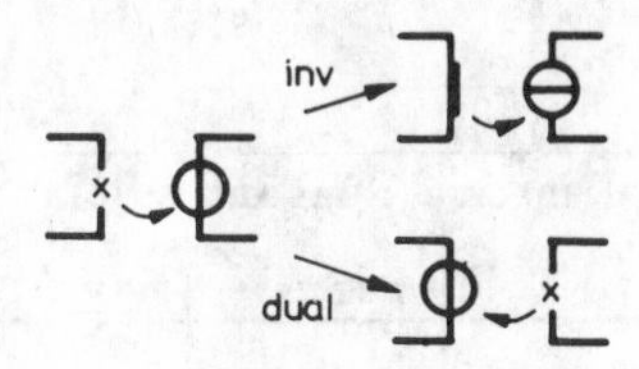

Fig. 10.1

Another important concept of network theory is the *adjoint* of multiports. The n-ports N, N' with matrices $\mathbf{A}, \mathbf{B}$ and $\mathbf{A}', \mathbf{B}'$, respectively, are called adjoint to each other if, for any pair $\mathbf{u}, \mathbf{i}$ of vectors, subject to $\mathbf{Au} + \mathbf{Bi} = \mathbf{0}$, and for any pair $\mathbf{u}', \mathbf{i}'$ of vectors, subject to $\mathbf{A}'\mathbf{u}' + \mathbf{B}'\mathbf{i}' = \mathbf{0}$, the relation $\mathbf{u}^T\mathbf{i}' = \mathbf{u}'^T\mathbf{i}$ holds.

Since this relation can be written as $(\mathbf{u}, \mathbf{i})^T(\mathbf{i}', -\mathbf{u}') = 0$ as well, and since $r(\mathbf{A}|\mathbf{B}) = r(\mathbf{A}'|\mathbf{B}') = n$, this condition means that $(\mathbf{A}|\mathbf{B})$ and $(\mathbf{B}'|-\mathbf{A}')$ are orthogonal complements.

For example, the adjoint of the 2-port N of the above example (adj N, for short) is $i_1' = -ki_2',\ u_2' = 0$, i.e. a current controlled current source.

Let us introduce one more concept, for completeness. If the n-ports N, N' with matrices $\mathbf{A}, \mathbf{B}$ and $\mathbf{A}', \mathbf{B}'$, respectively, satisfy $(\mathbf{A}|\mathbf{B}) \cong (\mathbf{A}'|-\mathbf{B}')$ then N

and N' are called *negative* to each other (N =neg N', for short). In this case their matroids $\mathbf{M}_N$ and $\mathbf{M}_{N'}$ are isomorphic. (In fact, the matroid does not change if some of the columns of its representing matrix are multiplied by -1, or by any other nonzero number.)

Theorem 10.1.2 *adj N = neg dual inv N for any multiport N.*

Proof. If N was given by the pair $\mathbf{A}, \mathbf{B}$ of matrices then adj N should be given by such $\mathbf{C}, \mathbf{D}$ that $\mathbf{AD}^T = \mathbf{BC}^T$. On the other hand, inv N is given by $\mathbf{B}, \mathbf{A}$, hence dual inv N should be given by such $\mathbf{P}, \mathbf{Q}$ that $\mathbf{BP}^T + \mathbf{AQ}^T = \mathbf{O}$. Immediately follows that $(\mathbf{C}|\mathbf{D}) \cong (\mathbf{P}|-\mathbf{Q})$, that is adj N and dual inv N are negative to each other.

□

The relation of these four "operations" for an example is shown in Fig. 10.2. Obviously

$$\text{neg neg } N = \text{inv inv } N = \text{dual dual } N = \text{adj adj } N = N \tag{10.1}$$

and one can very easily verify that they are commutative. Taking this result and Theorem 10.1.2 into consideration, we obtain

$$\begin{aligned} &\text{neg inv } N = \text{inv neg } N = \text{dual adj } N = \text{adj dual } N \\ &\text{neg dual } N = \text{dual neg } N = \text{inv adj } N = \text{adj inv } N \\ &\text{neg adj } N = \text{adj neg } N = \text{dual inv } N = \text{inv dual } N \end{aligned} \tag{10.2}$$

The negative, inverse, dual and adjoint of the most important 2-ports is summarized in **Box 10.1.**

Box 10.1. The negative, inverse, dual and adjoint of some important 2-ports

N	neg N	inv N	dual N	adj N
Ideal transformer	ideal tr.	ideal tr.	ideal tr.	ideal tr.
Gyrator	gyrator	gyrator	gyrator	gyrator
Voltage controlled current source (VCCS)	VCCS	CCVS	CCVS*	VCCS*
Voltage controlled voltage source (VCVS)	VCVS	CCCS	VCVS*	CCCS*
Current controlled current source (CCCS)	CCCS	VCVS	CCCS*	VCVS*
Current controlled voltage source (CCVS)	CCVS	VCCS	VCCS*	CCVS*

* In the dual and adjoint of the controlled sources the direction of the control is reversed.

See **Box 14.1** as well.

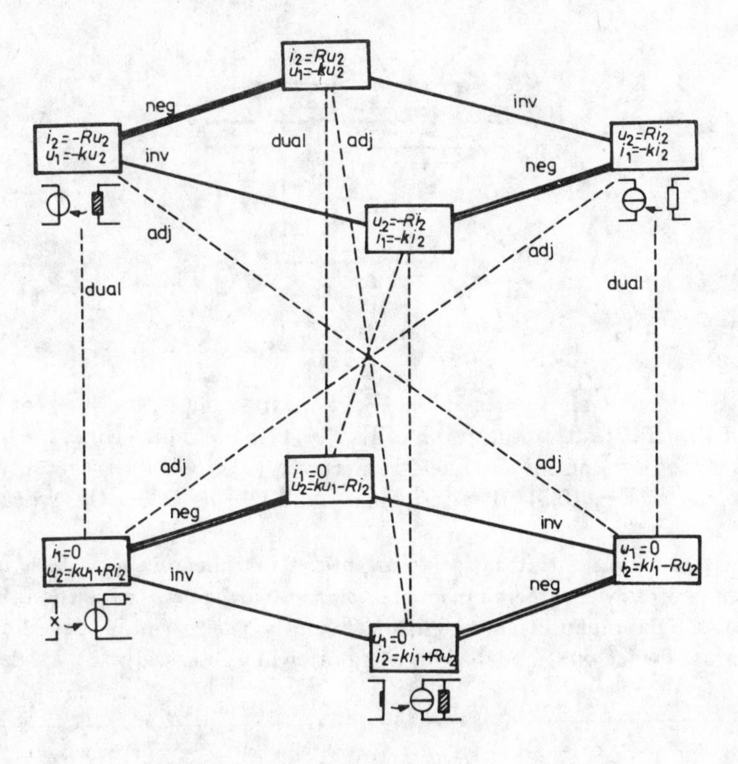

Fig. 10.2

When discussing the classical results of electric network duality in Section 4.1, finding the "pairs" of the elements (actually their inverses) was only the first step. The main result was Statement 4.1.1, stating that if devices were interconnected along a planar graph G and their "pairs" along G^* then the results will also be the "pairs" of each other. Since our more recent devices have more than two terminals each, we should reconsider what the graph of the interconnection means.

Theoretically, arbitrary multiports can be interconnected arbitrarily, and if we distinguish k pairs of points in the result, we can call it a new k-port (see Fig. 10.3, for example). However, the number of linearly independent equations among the voltages and currents of these k ports may then be less or more than k as well, see Problem 10.1.8. Although the results of this section can be generalized to such "generalized multiports" as well, their study follows in Section 14.2 only.

Suppose we have some interconnected multiports forming a k-port (i.e. k pairs of points are distinguished; the $2k$ points need not be all distinct). Let us join the pairs of points by red edges and imagine the "interconnecting wires"

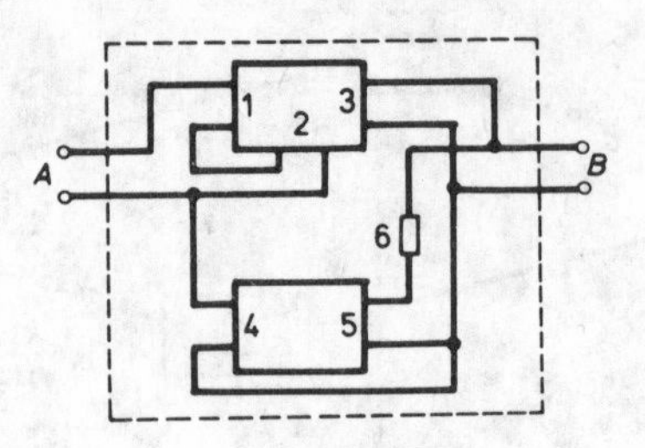

Fig. 10.3

among the different multiports as blue edges. For example, the interconnection shown in Fig. 10.3 leads to the graph of Fig. 10.4 (red and blue edges are denoted by broken and by continuous lines, respectively). Finally, let us contract all the blue edges (Fig. 10.5). The resulting graph will be called the *graph of the interconnection.*

Figure 10.6 calls our attention to an interesting phenomenon. Although the ways how sources were interconnected with the two 2-ports, were the same, the graphs of the interconnections are not. Obviously, the reason is that in case (a) a 3-terminal device was modelled by a 2-port while in case (b) by a 4-terminal one.

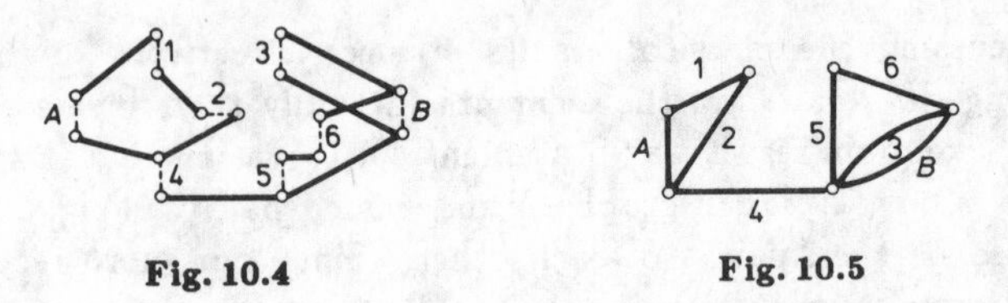

Fig. 10.4

Fig. 10.5

This suggests that if certain terminals are short-circuited inside the device, then this fact should be reflected by the symbolic drawing, already before the interconnection. For example, the symbols for the two 2-ports must be different (see the upper part of Fig. 10.7) and in case of (a) the "red" port-edges must be adjacent already before the interconnection (lower part of Fig. 10.7).

Incidentally, among the two networks of Fig. 10.6, (b) is uniquely solvable (and, of course, trivial) while (a) is not. This might be somewhat surprising since not only the ways are identical how they were interconnected by sources, but also their matrix descriptions $\begin{pmatrix} u_1 \\ u_2 \end{pmatrix} = \begin{pmatrix} R_1 & 0 \\ 0 & R_2 \end{pmatrix} \begin{pmatrix} i_1 \\ i_2 \end{pmatrix}$ are the same.

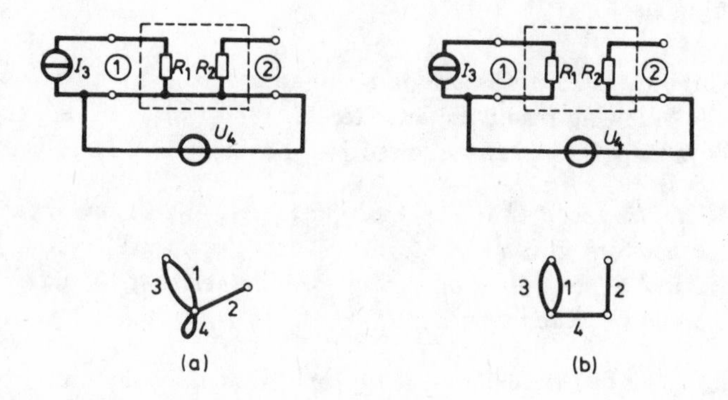

Fig. 10.6

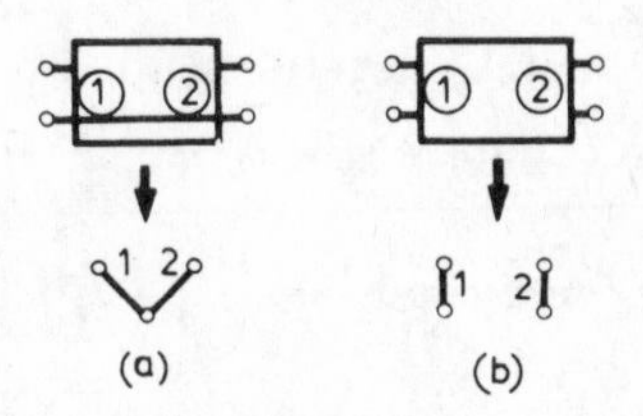

Fig. 10.7

In summary, if a device is modelled by a multiport and certain terminals belong to several ports then these "internal short circuits" must be reflected in the network graph. The importance of this convention is also shown by Exercise 10.1.2.

Statement 10.1.3 *Suppose that the multiports $N_1, N_2, \ldots$ are interconnected along a network graph G to form a k-port N.*

(a) If $N_1 \cong N_1', N_2 \cong N_2', \ldots$ and G and G' are 2-isomorphic then the interconnection of the multiports $N_1', N_2', \ldots$ along G' results in a k-port N' with $N \cong N'$.

(b) If G is planar, G^* is a dual of G and $N_1 \cong \text{inv } N_1', N_2 \cong \text{inv } N_2', \ldots$ then the interconnection of the multiports $N_1', N_2', \ldots$ along G^* results in a k-port N' with $N \cong \text{inv } N'$.

Proof. List the multiport-equations for $N_1, N_2, \ldots$ and add the Kirchhoff Laws for G. Eliminate the port voltages and currents of the original multiports. The resulting system has k linearly independent equations by the condition that N

is a k-port. In case (a) the systems of equations are equivalent for N and N', hence the statement is trivial. In case (b) all we have to observe is that not only $N_j \to N'_j$ but $G \to G^*$ also interchanges the roles of voltages and currents (and leave everything else).

□

While Statement 10.1.3 was only a straightforward generalization of Statement 4.1.1, the following result [Iri and Recski, 1980] is less trivial, especially if "generalized" multiports are also allowed (see Problem 14.2.11 as well).

Theorem 10.1.4 *Suppose that the multiports $N_1, N_2, \ldots$ are interconnected along a planar network graph G to form a k-port N. Let N'_1 =dual N_1, $N'_2 =$ = dual $N_2, \ldots$ and G^* be a dual of G. Suppose further that the interconnection of $N'_1, N'_2 \ldots$ along G^* leads to a k-port N'. Then $N \cong$dual N'.*

Proof. Let $\mathbf{u}_k, \mathbf{i}_k$ and $\mathbf{u}'_k, \mathbf{i}'_k$ denote the vectors, formed by the port voltages and currents, respectively, of the k-ports N and N', respectively. The multiports $N_1, N_2, \ldots$ also have port voltages and currents, they together form the vectors $\mathbf{u}_b, \mathbf{i}_b$, respectively. Similarly, $N'_1, N'_2, \ldots$ gives $\mathbf{u}'_b$ and $\mathbf{i}'_b$.

Since the multiports N_j and N'_j are dual to each other, we have

$$\mathbf{u}_b^T \mathbf{u}'_b + \mathbf{i}_b^T \mathbf{i}'_b = 0 \tag{10.3}$$

On the other hand, if $\mathbf{C}$ and $\mathbf{Q}$ denote the circuit and cut set matrices of G then $\mathbf{Cu} = \mathbf{0}$ and $\mathbf{Qi} = \mathbf{0}$ hold where $\mathbf{u} = \begin{pmatrix} \mathbf{u}_b \\ \mathbf{u}_k \end{pmatrix}$ and $\mathbf{i} = \begin{pmatrix} \mathbf{i}_b \\ \mathbf{i}_k \end{pmatrix}$. Hence Kirchhoff's equations can be written as $\begin{pmatrix} \mathbf{C} & \mathbf{O} \\ \mathbf{O} & \mathbf{Q} \end{pmatrix} \begin{pmatrix} \mathbf{u} \\ \mathbf{i} \end{pmatrix} = \mathbf{0}$ for N and as $\begin{pmatrix} \mathbf{Q} & \mathbf{O} \\ \mathbf{O} & \mathbf{C} \end{pmatrix} \begin{pmatrix} \mathbf{u}' \\ \mathbf{i}' \end{pmatrix} = \mathbf{0}$ for N'. These coefficient matrices are also orthogonal complements to each other, hence $\mathbf{u}^T \mathbf{u}' + \mathbf{i}^T \mathbf{i}' = 0$ also holds, which is

$$\mathbf{u}_b^T \mathbf{u}'_b + \mathbf{u}_k^T \mathbf{u}'_k + \mathbf{i}_b^T \mathbf{i}'_b + \mathbf{i}_k^T \mathbf{i}'_k = 0 \tag{10.4}$$

The difference of Eqs (10.3) and (10.4) gives $\mathbf{u}_k^T \mathbf{u}'_k + \mathbf{i}_k^T \mathbf{i}'_k = 0$. Hence the multiports N and N' — if they exist at all — are dual to each other.

□

The reason why the theorem and the last sentence of the proof is formulated in such a complicated way is that we do not know yet whether the multiports can be interconnected at all (i.e., whether the number of the linearly independent equations will be equal to the number of the ports). We shall return to this question in Section 14.2.

A combination of Theorems 10.1.2 and 10.1.4 leads to the following result.

Corollary 10.1.5 *Suppose that the multiports $N_1, N_2, \ldots$ are interconnected along a network graph G to form a k-port N. Let $N'_1 = $ adj N_1, $N'_2 = $ adj $N_2, \ldots$ and G' be 2-isomorphic to G. If $N'_1, N'_2, \ldots$ can be interconnected along G' then the resulting k-port N' is the adjoint of N.*

Observe that the planarity of G was not prescribed any more, since in the proof of Theorem 10.1.4 we used the matroid, coordinatized by $\begin{pmatrix} \mathbf{Q} & \mathbf{O} \\ \mathbf{O} & \mathbf{C} \end{pmatrix}$ only (that is, the direct sum of the circuit and cut set matroids of G), but did not require that it be graphic.

Exercises for Section 10.1

10.1.1 Cascade connection was introduced in Exercise 6.2.4. What is the graph of this interconnection? What is the result if transformers, gyrators and controlled sources (in different combinations) are interconnected in cascade?

10.1.2 Figure 10.8 shows the *series-parallel* interconnection of 2-ports N_1 and N_2. (The series-series, parallel-series and parallel-parallel interconnections can be defined analogously.)

(a) Determine the resulting 2-port if, using the notations of Fig. 10.9,

(1) $N_1 = N_1'$ and $N_2 = N_2'$

(2) $N_1 = N_1'$ and $N_2 = N_2''$

(3) $N_1 = N_1''$ and $N_2 = N_2'$.

(b) What is the difference between N_k' and N_k''?

(c) Determine the graph of the interconnection based on Fig. 10.8 at first (supposing that there is no short circuit inside the 2-ports), then for the above three cases as well.

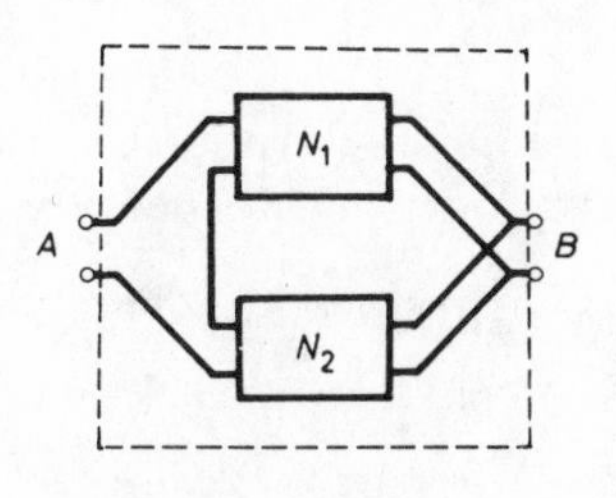

Fig. 10.8

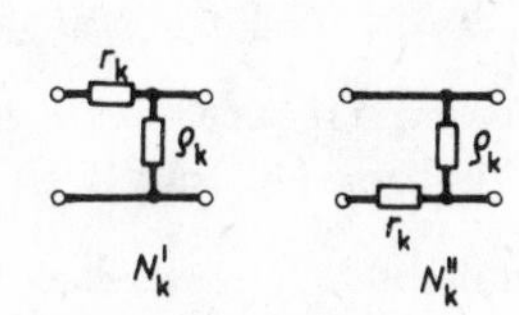

Fig. 10.9

10.1.3 Find multiports N, N' so that

(a) $N \not\cong N'$ but $\mathbf{M}_N = \mathbf{M}_{N'}$;

(b) $N \not\cong$inv N' but $\mathbf{M}_N = \phi(\mathbf{M}_{N'})$;

(c) $N \not\cong$dual N' but $\mathbf{M}_N = \mathbf{M}^*_{N'}$.

10.1.4 Prove that a multiport N is reciprocal if and only if $N =$adj N. What does Corollary 10.1.5 mean for reciprocal multiports?

Problems for Section 10.1

10.1.5 (a) Prove that the interconnection of two controlled sources as in Fig. 10.10 leads to an ideal transformer.

(b) Give a similar realization of the gyrator by controlled sources.

(c) Terminate port A of Fig. 10.10 by a voltage source and port B by a current source. This network must have unique solution. Is there no contradiction to Purslow's theorem (Problem 8.1.20)?

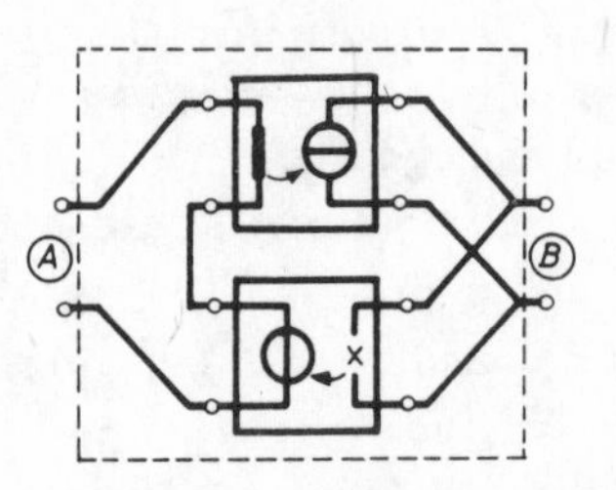

Fig. 10.10

10.1.6 Give examples to multiports with **Z**-descriptions so that their suitable interconnection has no **Z**-description.

10.1.7 Prove that if two 2-ports have **Z**-description then so has their series-series interconnection as well.

10.1.8 Give an example where multiports are interconnected and the resulting "k-port" has more (or less) than k linearly independent equations.

10.1.9 Find a matroidal necessary condition for the reciprocity of a multiport. Is this condition also sufficient?

10.1.10 Call a multiport N *qualitatively reciprocal* if $\mathbf{M}_N = \phi(\mathbf{M}_N^*)$ holds.

(a) Prove that both the reciprocal and the antireciprocal multiports are qualitatively reciprocal.

(b) Is it possible that N is qualitatively reciprocal, yet there exists no reciprocal multiport N' with $\mathbf{M}_N = \mathbf{M}_{N'}$?

10.1.11 k-terminal devices are usually modelled by a $(k-1)$-port as discussed in Section 8.1. (Choose a reference point and all the other points are in pair with this one.) Then the $k-1$ port edges form a star before the interconnection. Prove that any other three with $k-1$ edges could be used for modelling these devices. What about the other graphs with $k-1$ edges?

Section 10.2 How to brace a one-story building

In Section 2.6 we considered the question how to brace a square grid using diagonal rods. That was only preliminary to the study of the "real" one-story buildings. For example, the building with 3×3 rooms (Fig. 10.11) can be made rigid by suitably chosen 8 rods and we shall see that one of the two solutions in Fig. 10.12 is good.

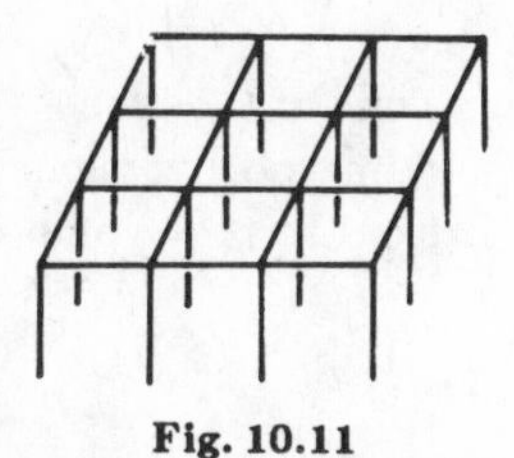

Fig. 10.11

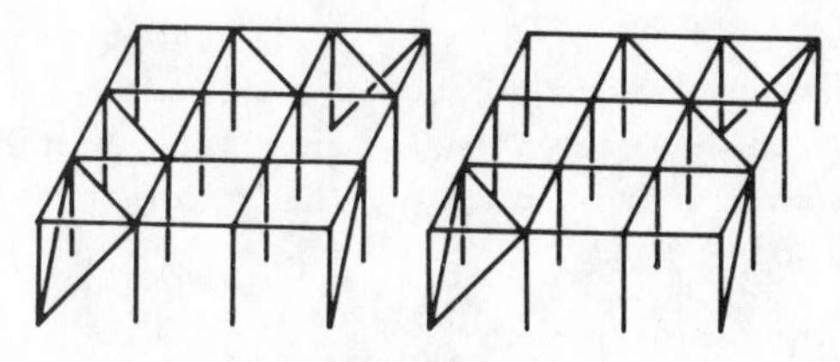

Fig. 10.12

First of all, let us consider the vertical rods, fixed to the earth with joints. If such a rod AB has a force $\mathbf{F}$ at point B then the vertical component $\mathbf{F}_v$ of $\mathbf{F}$ leads to a compression or to a tension of the rod (with no effects since the rod is supposed to be rigid) while the horizontal component $\mathbf{F}_h$ of $\mathbf{F}$ leads to a horizontal motion (Fig. 10.13). Strictly speaking, point B could move along a sphere (with centre at A) but any small motion is essentially horizontal. Even if every horizontal square were braced by a diagonal, the whole building would not be rigid; a translation or a rotation (Figs 10.14–15) were still possible. Hence some vertical squares should also be braced by diagonals.

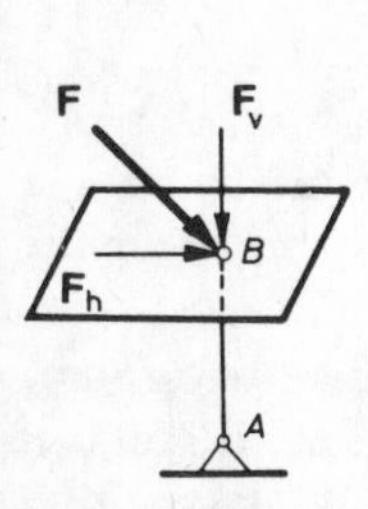

Fig. 10.13

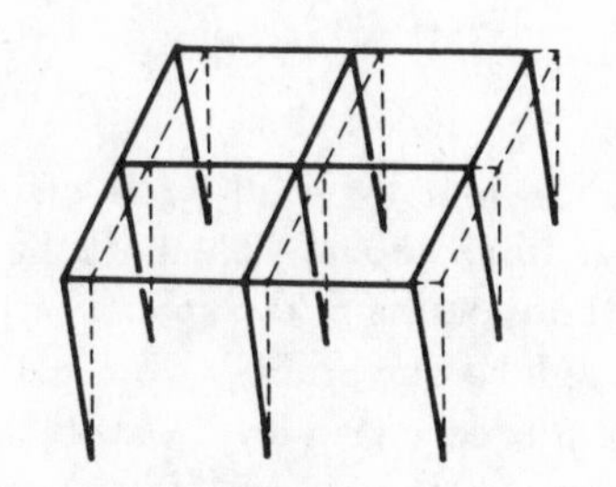

Fig. 10.14

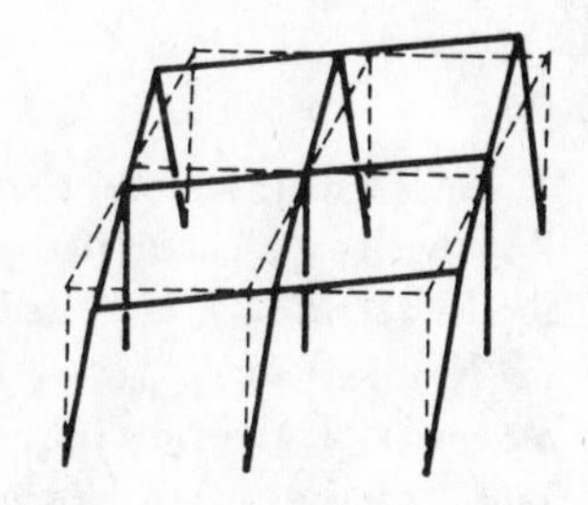

Fig. 10.15

First of all, what is the effect of such a brace in a vertical wall? Figure 10.16 shows the only possible type of deformation of a vertical wall *along itself* and this can obviously be blocked by any one diagonal.

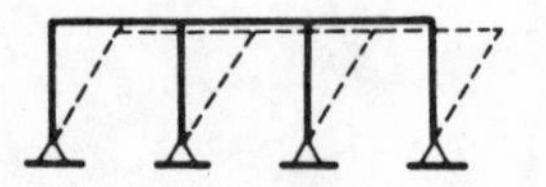

Fig. 10.16

If we use two such diagonals x and y (see Fig. 10.17) then none of the planes S_x, S_y can move along itself. Hence, their intersection has no admissible motion whatsoever. Thus x and y together means that the joint P is fixed to that point of the space.

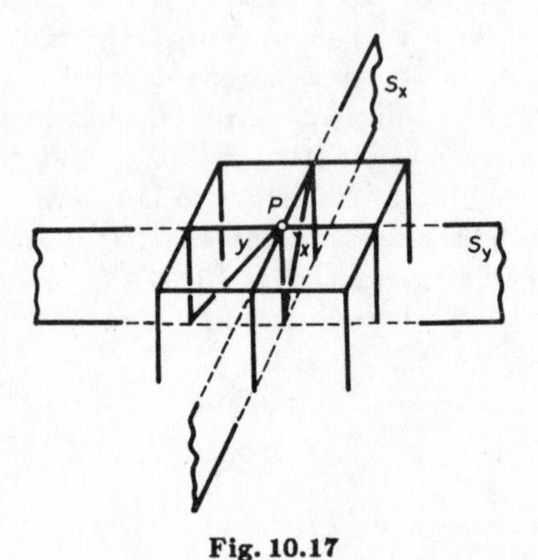

Fig. 10.17

Suppose therefore that all the four external walls of our building have a diagonal brace each (like in Fig. 10.12 above). Then all the four corners of the quadrilateral roof are fixed to those points of the space.

Needless to say, if we thus reduce our problem to a planar one, we are not yet ready. If the four corners of a square grid are fixed to the plane, it still can have (infinitesimal) deformations, see Fig. 10.18, for example. We know from Theorem 2.6.2 that such a $k \times \ell$ grid can be made rigid by a suitable set of

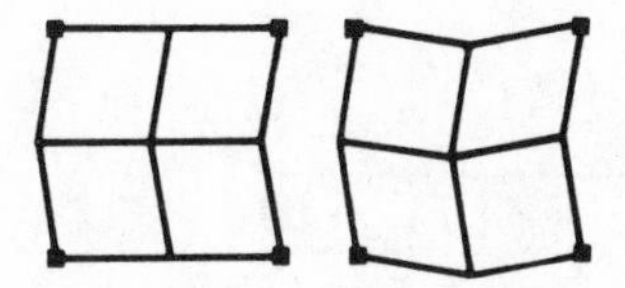

Fig. 10.18

$k + \ell - 1$ diagonal rods, but since the corners are now fixed, we expect that somewhat less is also enough.

Recall that $x_1, x_2, \ldots$ and $y_a, y_b, \ldots$ denoted the deformations of the rows and the columns, respectively, of the square grid in Section 2.6. A diagonal rod in, say, room $a2$ means an equation $y_a = x_2$.

If the four corners of the grid is fixed to the plane then the sum of all the vertical deformations and the sum of all the horizontal deformations must separately be zero

$$\sum_{i=1}^{k} x_i = 0; \qquad \sum_{j=1}^{\ell} y_j = 0. \tag{10.5}$$

This system should be extended by equations of form $x_i = y_j$ to a "large" system whose only solutions are $x_1 = x_2 = \ldots = y_1 = y_2 = \ldots = 0$. Since this is a homogeneous linear system with $k + \ell$ unknown quantities, we need at least $k + \ell - 2$ equations of form $x_i = y_j$.

Recall that a bipartite graph G was associated to our square grid in Section 2.6, with edges corresponding to diagonal braces. We have also seen that a circuit of G indicates that one of the braces is redundant. Hence, if a suitable set of $k + \ell - 2$ diagonal rods gives a solution, the corresponding subgraph must be a 2-component forest of G.

Let, for example, $k = 4, \ell = 6$ and consider the two arrangements of Fig. 10.19. The second one is rigid, see below, while the first one has infinitesimal deformations, see Fig. 10.20. Both arrangements have 8 diagonals and both bipartite graphs (Fig. 10.21) are 2-component forests. What is the difference?

The ratio of the cardinalities of the two bipartition classes of G is 4:6. In the first case both components have this same ratio while in the second case one of the components has 2:2, the other has 2:4.

In general, if $G = (V, E)$ is a bipartite graph with bipartition $V_1 \cup V_2$ then consider the ratio $|V_1| : |V_2|$. Call a 2-component forest *symmetric* if both components have this same ratio, and *asymmetric* otherwise.

Theorem 10.2.1 *If a $k \times \ell$ square grid is fixed to the plane at its four corners then at least $k + \ell - 2$ diagonal braces are required to make it rigid. A set of*

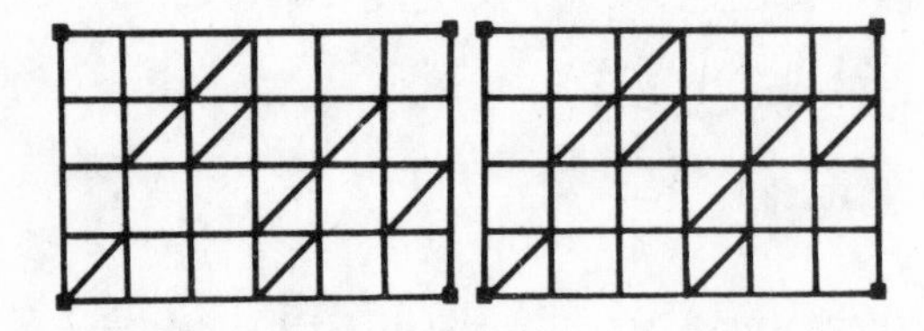

Fig. 10.19

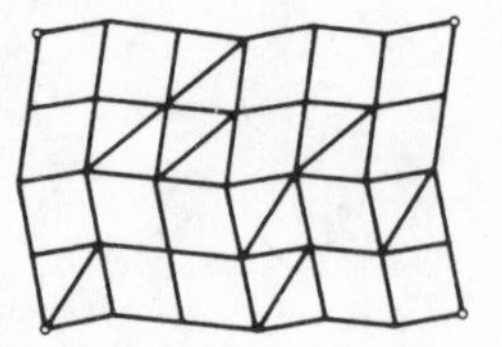

Fig. 10.20

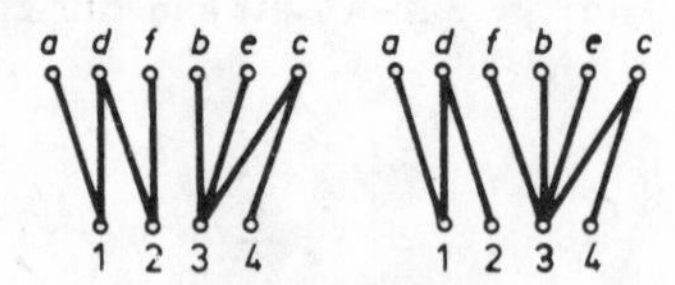

Fig. 10.21

exactly $k+\ell-2$ braces is good if and only if they determine an asymmetric 2-component forest.

Proof. The necessity of at least $k+\ell-2$ braces have already been shown above. Suppose at first that a system of $k+\ell-2$ braces is "wrong" (i.e., it leads to a nonrigid system). We shall prove the symmetry of the corresponding forest.

Consider the example of Fig. 10.20. The equations of this system can be written as follows (we permute the columns for some future purposes):

$$\begin{array}{ccccccccccl}
x_1 & & & & -y_a & & & & & & =0\\
x_1 & & & & & -y_d & & & & & =0\\
 & x_2 & & & & -y_d & & & & & =0\\
 & x_2 & & & & & -y_f & & & & =0\\
 & & x_3 & & & & & -y_b & & & =0\\
 & & x_3 & & & & & & -y_e & & =0\\
 & & x_3 & & & & & & & -y_c & =0\\
 & & & x_4 & & & & & & -y_c & =0\\
x_1 & +x_2 & +x_3 & +x_4 & & & & & & & =0\\
 & & & & y_a & +y_d & +y_f & +y_b & +y_e & +y_c & =0
\end{array} \qquad (10.6)$$

This is a linearly dependent system of equations having nontrivial solutions as well, for example $x_1 = x_2 = y_a = y_d = y_f = 1$, $x_3 = x_4 = y_b = y_c = y_e = -1$. Thus a nontrivial this linear combination of the rows of the coefficient matrix of Eq. (10.6) can lead to the zero vector. Let us write down this matrix again, together with the coefficients of this linear combination.

coef.										
2	1				−1					
1	1					−1				
1		1				−1				
2		1					−1			
2			1					−1		
2			1						−1	
−1			1							−1
3				1						−1
−3	1	1	1	1						
2					−1	−1	−1	−1	−1	−1

(10.7)

The ratio of the last two coefficients is $(-3) : 2$, while the ratio of the bipartite graph was $2 : 3$. We show that in general, if the bipartite graph has ratio $k : \ell$ then such a linear combination necessarily has the ratio $(-\ell) : k$ for the last two coefficients. First of all, these coefficients must be rationals (see Problem 5.3.9) and then can be supposed to be integers as well. Above the thick horizontal line there are $k + \ell - 2$ occurrences of $+1$ and -1 each, hence no matter what the coefficients of the first $k + \ell - 2$ rows are, the sum of the $+1$'s (left from the vertical thick line) and that of the -1's (right from that line) must be the same (apart from the sign). Since all these must be eliminated by the k and ℓ ones, respectively, the ratio of the coefficients of the two last rows must be $(-\ell) : k$.

Furthermore, since the graph G has $k+\ell-2$ edges, it is disconnected. Hence, after some permutation of the points, a block-diagonal arrangement at both sides of the vertical thick line can be obtained (like in Eq. (10.7)). Then the above reasoning should be repeated for each component, leading to the symmetry of the 2-component forest.

We still have to prove that symmetric 2-component forests imply nonrigidity. Let G_1 and G_2 denote the two components of G with ratios $k_1 : \ell_1$ and $k_2 : \ell_2$, respectively. Consider the "solution" of the above system of equations, where all the x_i's and y_j's equal c_1 if they belong to G_1 and equal c_2 if they belong to G_2. The first $k + \ell - 2$ equations are automatically satisfied and so are the last two as well, if the relations $k_1 c_1 + k_2 c_2 = \ell_1 c_1 + l_2 c_2 = 0$ are met. By the symmetricity of the forest, c_1 and c_2 can be chosen so that both equalities are simultaneously met.

□

Although our result and its proof could be told without using matroids, we intentionally postponed them until this chapter. If S denotes the set of all the $k \cdot \ell$ squares and a minimal system X of $k + \ell - 2$ diagonal braces to make our building rigid is considered as a subset of S and is called a "base", then all these "bases" give just a system of bases of a matroid $\mathbf{M}_{k,\ell}$.

When the four corners were not fixed (in Section 2.6), the corresponding matroid was just the circuit matroid of the bipartite graph $K_{k,\ell}$ having $k + \ell$

points and all the possible $k \cdot \ell$ edges. However, if the corners are fixed, the matroids are not graphic in general, see Exercise 10.2.3 and Problem 10.2.6.

Exercises for Section 10.2

10.2.1 Find the rigid ones among the four systems of Fig. 10.22. (Their corners are fixed to the plane.) Draw a deformation each for the others.

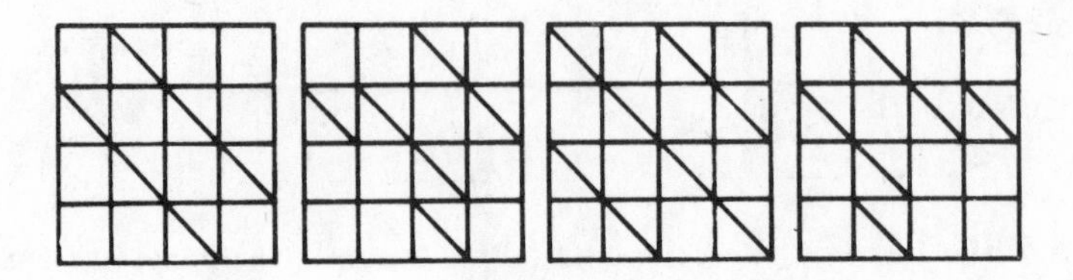

Fig. 10.22

10.2.2 We have seen that a one-story building cannot be braced by horizontal diagonals only. What about diagonals in the vertical walls only?

10.2.3 Describe the matroid $\mathbf{M}_{k,\ell}$ if $k = \ell = 2$ and if $k = 2,\ \ell = 3$.

10.2.4 What is the physical meaning of the minors of this matroid $\mathbf{M}_{k,\ell}$?

10.2.5 Suppose we use $k + \ell - 1$ diagonal rods (plus 4 ones in the vertical walls, as usual). If the system is rigid, which are the critical rods?

Problems for Section 10.2

10.2.6 Repeat Exercise 10.2.3 if $k = \ell = 3$ and if $k = 2,\ \ell = 4$.

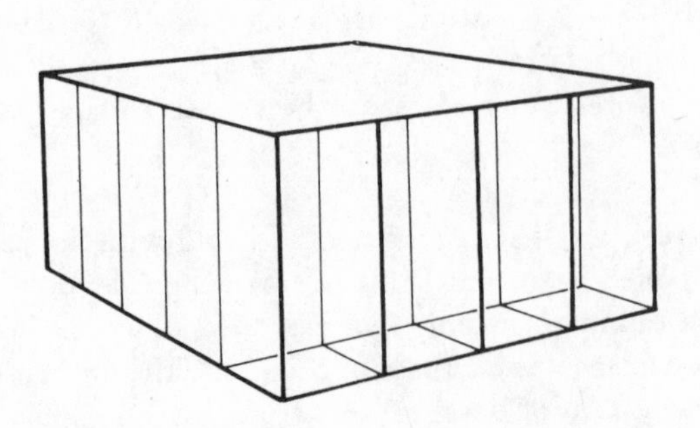

Fig. 10.23

10.2.7 Show by an example that even the graphic minors of $\mathbf{M}_{k,\ell}$ need not correspond to bipartite graphs.

10.2.8 The "cottage" of Fig. 10.23 is of size 4×5 units and one fifth part of it (its "front") is a terrace where diagonal rods in the vertical "wall" would not be beautiful. Give a system of acceptable diagonal rods to make it rigid.

10.2.9 Can we use less than 4 diagonals in the vertical walls?

10.2.10 Suppose we use $k + \ell - 1$ diagonal rods, as in Exercise 10.2.5. Can all of them be critical? Can none of them be critical?

10.2.11 Let $k = \ell$ and we use $2k - 1$ diagonal rods (plus 4 ones in the vertical walls, as usual). Suppose that the $2k - 1$ rods correspond to a spanning tree of the bipartite graph and that there are critical among them. Prove that there is a row or a column where the number of diagonal rods is even.

Chapter 11

The sum of matroids I

Section 11.1 The concept of the sum

Let $\mathbf{M}_1 = (S, \mathbf{F}_1)$, $\mathbf{M}_2 = (S, \mathbf{F}_2), \ldots, \mathbf{M}_k = (S, \mathbf{F}_k)$ be k matroids on the same underlying set S. We define the *sum* (or *union*) of these matroids to be the matroid on the same set S with $X \subseteq S$ independent if and only if X can be decomposed as $X_1 \cup X_2 \cup \ldots \cup X_k$ where $X_1 \in \mathbf{F}_1$, $X_2 \in \mathbf{F}_2, \ldots, X_k \in \mathbf{F}_k$.

For example, if $\mathbf{M}_1, \mathbf{M}_2$ and $\mathbf{M}_3$ are the circuit matroids of the three graphs of Fig. 11.1 then their sum is $\mathbf{U}_{4,3}$: every 3-element subset is independent in the sum (for example, $\{1,2,3\}$ can be decomposed as $\{3\} \cup \{2\} \cup \{1\}$); the full set $\{1,2,3,4\}$ is dependent since all the three matroids have rank 1.

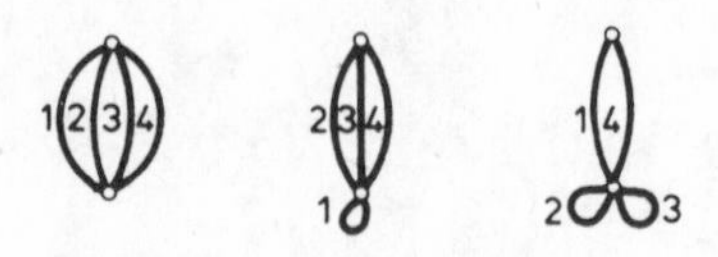

Fig. 11.1

This sum will be denoted by $\mathbf{M}_1 \vee \mathbf{M}_2 \vee \ldots \vee \mathbf{M}_k$ or briefly by $\vee_{i=1}^{k} \mathbf{M}_i$. The notation $(S, \vee_{i=1}^{k} \mathbf{F}_i)$ will also be used.

The sum should not be confused with direct sum; the latter was defined on matroids with disjoint underlying sets. The two concepts are related, however, see Exercise 11.1.5 and the construction after Theorem 13.2.3.

Theorem 11.1.1 *The sum of matroids is indeed a matroid, i.e. the axioms* (F1)–(F3) *are satisfied.*

Before giving the proof we shall explain why verifying property (F3) is non-trivial. However, the reader may omit the whole proof since the assertion will later be a trivial consequence of Theorem 13.3.2.

Let $\mathbf{M}_1 \vee \mathbf{M}_2 = (S, \mathbf{F})$; $X, Y \in \mathbf{F}$ and $|X| > |Y|$. Then X and Y can be decomposed as $X_1 \cup X_2$ and $Y_1 \cup Y_2$, respectively, with $X_1, Y_1 \in \mathbf{F}_1$ and X_2, $Y_2 \in \mathbf{F}_2$. We may even suppose that $X_1 \cap X_2 = Y_1 \cap Y_2 = \emptyset$ (see Exercise 11.1.7) and then at least one of $|X_1| > |Y_1|$ and $|X_2| > |Y_2|$ holds, since otherwise $|X| \leq |Y|$ would follow, a contradiction. Suppose $|X_1| > |Y_1|$.

Applying property (F3) of $\mathbf{M}_1$ there is an element $x \in X_1 - Y_1$ so that $Y_1 \cup \{x\} \in \mathbf{F}_1$. However, we cannot be sure whether $x \in Y_2$ (case (a) in Fig. 11.2) or $x \notin Y_2$ (case (b)) holds. In the former case x is not an appropriate choice.

Proof. The relation $\mathbf{M}_1 \vee \mathbf{M}_2 \vee \mathbf{M}_3 = (\mathbf{M}_1 \vee \mathbf{M}_2) \vee \mathbf{M}_3$ is obvious, hence the theorem need be proved for two summands only. Let $\mathbf{M}_1 \vee \mathbf{M}_2 = (S, \mathbf{F})$. $\emptyset \in \mathbf{F}$ is obvious since $\emptyset$ was independent both in $\mathbf{M}_1$ and in $\mathbf{M}_2$. Similarly, if $X \in \mathbf{F}$, then $X = X_1 \cup X_2$ for some $X_1 \in \mathbf{F}_1$ and $X_2 \in F_2$, and thus $Y \subseteq X$ implies $Y \in \mathbf{F}$ by $(Y \cap X_1) \cup (Y \cap X_2) = Y$.

For (F3) we give an indirect proof. Suppose (F3) does not always hold. Imagine all the possible counterexamples X, Y and for these pairs all the possible decompositions $X = X_1 \cup X_2$, $Y = Y_1 \cup Y_2$ with $X_1, Y_1 \in \mathbf{F}_1$ and $X_2, Y_2 \in \mathbf{F}_2$. Now choose such a counterexample for which $|X_1 \cap Y_2| + |X_2 \cap Y_1|$ is minimum (see Fig. 11.3).

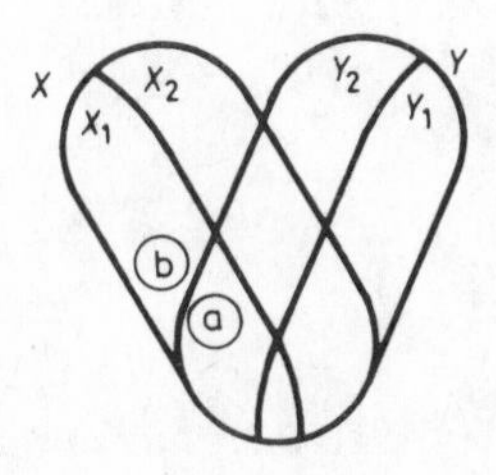

Fig. 11.2

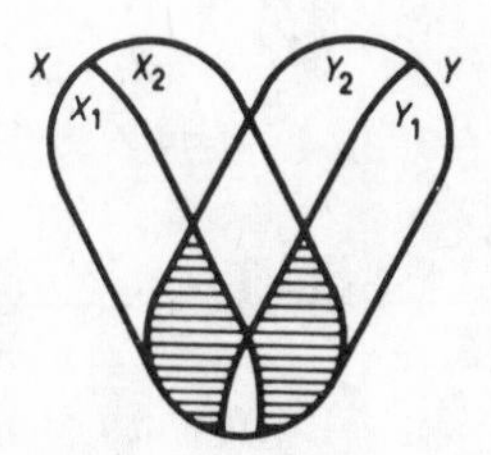

Fig. 11.3

Suppose $|X_1| > |Y_1|$ and that the element $x \in X_1 - Y_1$ with $Y_1 \cup \{x\} \in \mathbf{F}_1$ is in Y_2. Then, instead of the decomposition $Y = Y_1 \cup Y_2$ we could take $Y = (Y_1 \cup \{x\}) \cup (Y_2 - \{x\})$. But the above minimum quantity would then be decreased by one, a contradiction. Hence $x \notin Y_2$, thus $x \in X - Y$ and $Y \cup \{x\} \in \mathbf{F}$ by the decomposition $(Y_1 \cup \{x\}) \cup Y_2$. Therefore X, Y is not a counterexample. This contradiction proves (F3).

□

We have seen that $X \in \vee_{i=1}^{k} \mathbf{F}_i$ also implies a decomposition into *disjoint* subsets $X_1 \cup X_2 \cup \ldots \cup X_k$. This is a partition (apart from the possible empty subsets X_i). Thus, the problem of whether a given subset $X \subseteq S$ is independent in the sum is usually called the *matroid partition problem.* This has a "quick" solution, see Chapter 13.

Suppose $\mathbf{M}_1 = (S, \mathbf{F}_1)$ and $\mathbf{M}_2 = (S, \mathbf{F}_2)$ have the same rank r. We shall frequently encounter the problem of whether they have a common base, i.e. an r-element subset from $\mathbf{F}_1 \cap \mathbf{F}_2$. This is called the *matroid intersection problem* (see Chapter 13). One should recognize immediately that this "intersection set"

$\mathbf{F}_1 \cap \mathbf{F}_2$ is not necessarily the collection of independent sets of a matroid, see Problem 11.1.14.

The matroid partition problem and the matroid intersection problem are essentially equivalent:

Statement 11.1.2 *If $\mathbf{M}_1 = (S, \mathbf{F}_1)$ and $\mathbf{M}_2 = (S, \mathbf{F}_2)$ are two matroids with the same rank r then they have a common base if and only if $\mathbf{M}_1 \vee \mathbf{M}_2^*$ is the free matroid $(S, 2^S)$.*

Proof. If X is a common base then $S - X$ is a base of $\mathbf{M}_2^*$. Hence $S = X \cup (S - X)$ with $X \in \mathbf{F}_1$ and $S - X \in \mathbf{F}_2^*$, thus $S \in \mathbf{F}_1 \vee \mathbf{F}_2^*$.

On the other hand, if $S = S_1 \cup S_2$ with $S_1 \in \mathbf{F}_1$, $S_2 \in \mathbf{F}_2^*$ then, using the notation r_1 and r_2^* for the respective rank functions,

$$|S| \leq |S_1| + |S_2| = r_1(S_1) + r_2^*(S_2) \leq r_1(S) + r_2^*(S) = r + (|S| - r) = |S|,$$

hence S_1 must be a base of $\mathbf{M}_1$, S_2 a base of $\mathbf{M}_2^*$ and they must be disjoint. Therefore S_1 is a common base of $\mathbf{M}_1$ and $\mathbf{M}_2$.

□

We close the section with a very important theorem:

Theorem 11.1.3 [Nash-Williams, 1967]. *Let $r_1, r_2, \ldots, r_k$ denote the rank functions of the matroids $\mathbf{M}_1 = (S, \mathbf{F}_1)$, $\mathbf{M}_2 = (S, \mathbf{F}_2), \ldots, \mathbf{M}_k = (S, \mathbf{F}_k)$, respectively. If $\mathbf{N} = \vee_{i=1}^k \mathbf{M}_i$ with rank function R then*

$$R(X) = \min_{Y \subseteq X} \left[\sum_{i=1}^{k} r_i(Y) + |X - Y| \right].$$

Proof. The relation $\leq$ is left to the reader as Problem 11.1.12. The more difficult relation $\geq$ will be proved later in Section 13.2

□

Exercises for Section 11.1

11.1.1 Determine the sum of the circuit matroids of the graphs of Fig. 11.4.

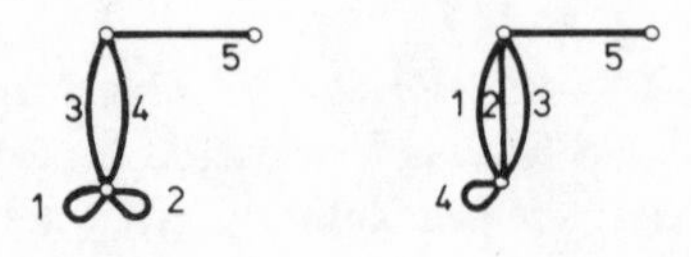

Fig. 11.4

11.1.2 Let x be a bridge in $\mathbf{M}_1 = (S, \mathbf{F}_1)$. Prove that it is a bridge in $\mathbf{M}_1 \vee \mathbf{M}_2$ as well, for every $\mathbf{M}_2$.

11.1.3 Prove the following equality for every $X \subseteq S$

$$(\vee_{i=1}^{k}\mathbf{M}_i)\backslash X = \vee_{i=1}^{k}(\mathbf{M}_i\backslash X).$$

Is the analogous statement true for contraction as well?

11.1.4 Let x, y be two series or parallel elements in $\mathbf{M}_1$. Do they remain series or parallel, respectively, in $\mathbf{M}_1 \vee \mathbf{M}_2$?

11.1.5 Show that the direct sum can be considered as a special case of sum.

11.1.6 With the notation of Theorem 11.1.3 show that $R(S) \leq \sum_{i=1}^{k} r_i(S)$. Give an example where the relation is not an equality.

11.1.7 Let X be independent in $\mathbf{M}_1 \vee \mathbf{M}_2$. Prove that X can be decomposed into the *disjoint* union of two subsets, each independent in the respective matroid.

11.1.8 Let X be a base in $\mathbf{M}_1 \vee \mathbf{M}_2$. Prove that X can be decomposed into the disjoint union of two subsets, one independent in $\mathbf{M}_1$ and one base in $\mathbf{M}_2$.

11.1.9 However, give a counterexample to the following statement: If X is a base of $\mathbf{M}_2$, one can extend it by an independent set of $\mathbf{M}_1$ to form a base of $\mathbf{M}_1 \vee \mathbf{M}_2$.

11.1.10 Determine $\mathbf{M} \vee \mathbf{M}^*$.

11.1.11 $\mathbf{M}_1$ is given. Characterize those matroids $\mathbf{M}_2$ which satisfy $\mathbf{M}_1 \vee \mathbf{M}_2 = \mathbf{M}_1$.

Problems for Section 11.1

11.1.12 Prove the $\leq$ relation in Theorem 11.1.3.

11.1.13 Prove the following generalization of Statement 11.1.2: Let r_1 and r_2 denote the rank functions of $\mathbf{M}_1 = (S, \mathbf{F}_1)$ and $\mathbf{M}_2 = (S, \mathbf{F}_2)$, respectively. $\mathbf{M}_1$ and $\mathbf{M}_2$ have a common independent set of cardinality t if and only if the rank of $\mathbf{M}_1 \vee \mathbf{M}_2^*$ is at least $t + r_2^*(S)$.

11.1.14 Consider the intersection $\mathbf{F}_1 \cap \mathbf{F}_2$ of the system of independent sets of two matroids. Show that this system does not necessarily satisfy property (F3) of matroids. (Hint: Give a counterexample with $|S| = 3$.)

11.1.15 Statement 11.1.2 "reduced" the problem of the intersection of two matroids to the matroid partition problem with $k = 2$ summands. Show that the matroid partition problem with any number $k \geq 2$ of summands can also be "reduced" to the problem of the intersection of two appropriate matroids.

Section 11.2 Algebraic and geometric meaning of the sum

Suppose that the matrices $\mathbf{A}_1, \mathbf{A}_2, \ldots, \mathbf{A}_k$ coordinatize the matroids $\mathbf{M}_1 = (S, \mathbf{F}_1)$, $\mathbf{M}_2 = (S, \mathbf{F}_2), \ldots, \mathbf{M}_k = (S, \mathbf{F}_k)$, respectively, over the same field T. The common underlying set S means that a one–one correspondence exists among the column sets of the matrices. In what follows, we shall simply suppose that if $S = \{s_1, s_2, \ldots, s_n\}$ then, for every $i = 1, 2, \ldots, k$, the first columns of the matrices $\mathbf{A}_i$ correspond to s_1, the second columns of these matrices to s_2 etc. We shall see later (Problems 13.2.7–8) that $\vee_{i=1}^{k}\mathbf{M}_i$ will also be coordinatizable over a suitable field.

Consider the circuit matroids $\mathbf{M}_i = \mathbf{M}(G_i)$ of the graphs of Fig. 11.5. The first two can certainly be coordinatized by the matrices $\mathbf{A}_1 = \begin{pmatrix} 0 & 0 & 1 & 1 & 0 \\ 0 & 0 & 0 & 0 & 1 \end{pmatrix}$ and $\mathbf{A}_2 = (1\ \ 1\ \ 1\ \ 0\ \ 0)$, respectively. One can directly see that $\mathbf{M}_1 \vee \mathbf{M}_2 = \mathbf{M}_3$ and that $\mathbf{M}_3$ can be represented by $\mathbf{A}_3 = \begin{pmatrix} 0 & 0 & 1 & 1 & 0 \\ 0 & 0 & 0 & 0 & 1 \\ 1 & 1 & 1 & 0 & 0 \end{pmatrix}$. This may give the idea to consider the matrix $\mathbf{A} = \begin{pmatrix} \mathbf{A}_1 \\ \mathbf{A}_2 \\ \cdot \\ \cdot \\ \cdot \\ \mathbf{A}_k \end{pmatrix}$ and to check whether it coordinatizes the matroid $\mathbf{N} = \vee_{i=1}^{k}\mathbf{M}_i$.

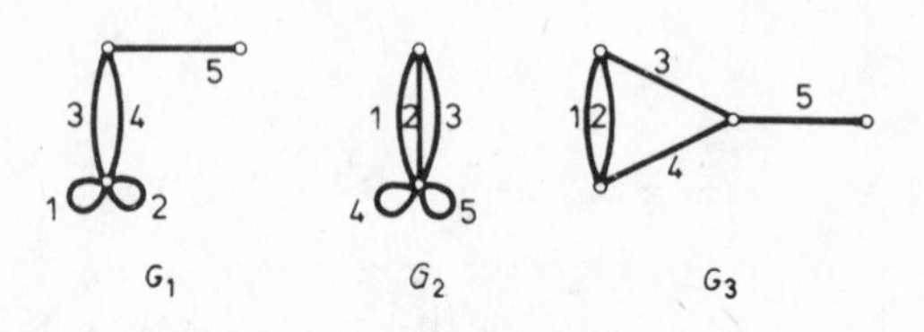

Fig. 11.5

Statement 11.2.1 *Let $X \subseteq S$ be independent in the matroid determined by* $\mathbf{A}$. *Then $X \in \vee_{i=1}^{k}\mathbf{M}_i$, i.e. X is independent in* $\mathbf{N}$ *as well.*

Proof. Suppose $k = 2$ (for $k \geq 3$ see Problem 11.2.4.). Let $|X| = p$. Since the X-columns of $\mathbf{A}$ are linearly independent, we can find p rows in $\mathbf{A}$ so that the $p \times p$ submatrix $\mathbf{W}$ formed by these rows and by the X-columns be nonsingular. Let us number the rows of $\mathbf{W}$ so that the first p_1 rows belong to $\mathbf{A}_1$ and the remaining p_2 belong to $\mathbf{A}_2$. If one of them, say p_1, is zero, we are done, since then X is independent in $\mathbf{M}_2$ (and "even more independent" in $\mathbf{M}_1 \vee \mathbf{M}_2$).

If neither p_1, nor p_2 is zero, consider the Laplace-expansion of det $\mathbf{W}$ along the heavy horizontal line of Fig. 11.6. We choose a p_1-element subset X_1 of X; these columns and the first p_1 rows form submatrix $\mathbf{W}_1$, while the other columns form $\mathbf{W}_2$ with the other rows. det $\mathbf{W}$ is the sum of products of the form det $\mathbf{W}_1 \cdot$ det $\mathbf{W}_2$. By the nonsingularity of $\mathbf{W}$, at least one of these products must be nonzero, hence there must exist a decomposition $X = X_1 \cup X_2$ so that the X_1-columns are linearly independent in $\mathbf{A}_1$ and the X_2-columns are independent in $\mathbf{A}_2$. This means $X_1 \in \mathbf{F}_1$ and $X_2 \in \mathbf{F}_2$, that is, $X \in \mathbf{F}_1 \vee \mathbf{F}_2$.

□

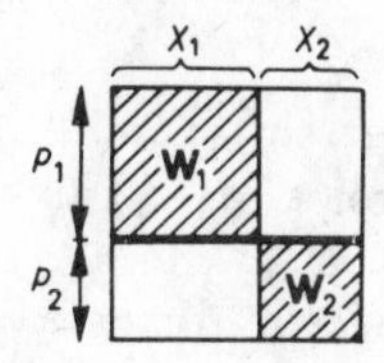

Fig. 11.6

The converse statement is in general false, i.e. $X \in \mathbf{F}_1 \vee \mathbf{F}_2$ does not imply that X is independent in the matroid determined by $\mathbf{A}$. The reason for this is that det $\mathbf{W}$ is the signed sum of several products of the form det $\mathbf{W}_1 \cdot$ det $\mathbf{W}_2$ — in fact $\binom{p}{p_1}$ in number — and even if some of them are nonzero, their signed sum can vanish. For example, the "matrix" (1 1) coordinatizes the matroid $\mathbf{U}_{2,1}$ but $\mathbf{A} = \begin{pmatrix} 1 & 1 \\ 1 & 1 \end{pmatrix}$ does not coordinatize $\mathbf{U}_{2,1} \vee \mathbf{U}_{2,1} = \mathbf{U}_{2,2}$ since det $\mathbf{A} = 1 \cdot 1 - 1 \cdot 1 = 0$. However, if such or similar cancellations could be avoided, the converse statement would also be true.

Theorem 11.2.2 [Edmonds, 1967]. *If the real matrices $\mathbf{A}_1, \mathbf{A}_2, \ldots, \mathbf{A}_k$ coordinatize the respective matroids $\mathbf{M}_1, \mathbf{M}_2, \ldots, \mathbf{M}_k$ and all the nonzero entries of all these matrices are algebraically independent over the field $\mathbb{Q}$ of the rationals then the matrix* $\mathbf{A} = \begin{pmatrix} \mathbf{A}_1 \\ \mathbf{A}_2 \\ \cdot \\ \cdot \\ \cdot \\ \mathbf{A}_k \end{pmatrix}$ *coordinatizes the matroid* $\vee_{i=1}^{k} \mathbf{M}_i$.

□

This theorem will have several applications since its hypothesis arises naturally.

In order to visualize the sum of two matroids, the following result is more useful.

Suppose that $\mathbf{M}_1 = (S, \mathbf{F}_1)$ and $\mathbf{M}_2 = (S, \mathbf{F}_2)$ are two matroids with ranks r_1 and r_2, respectively, and they are visualized in the $r_1 - 1$ and in the $r_2 - 1$ dimensional spaces, as in Section 9.2, by the point sets $S' = \{s'_1, s'_2, \ldots, s'_n\}$ in the case of $\mathbf{M}_1$ and $S'' = \{s''_1, s''_2, \ldots, s''_n\}$ in the case of $\mathbf{M}_2$, where $|S| = n$. Put

S' and S'' in the (r_1+r_2-1)- dimensional space so that the subspaces $<S'>$ and $<S''>$, spanned by S' and by S'', respectively are in skew position. Now, for every pair of elements (s_i', s_i''), prepare a line $e_i = s_i's_i''$ and consider a point s_i on this line in general position. If an element, say the ith is a loop in one of the matroids, say in $\mathbf{M}_1$, then s_i' did not exist by definition (see Section 9.2). Then if s_i'' exists, we set $s_i = s_i''$. If s_i'' also does not exist (because the ith element is a loop in both matroids) then s_i does not exist either.

Theorem 11.2.3 [Mason, 1977]. *The points* $s_1, s_2, \ldots, s_n$ *give the affine representation of* $\mathbf{M}_1 \vee \mathbf{M}_2$.

Its proof will follow after Theorem 13.2.4.

For example, the circuit matroids $\mathbf{M}(G_1)$ and $\mathbf{M}(G_2)$ of the first two graphs of Fig. 11.5 are visualized in Fig. 11.7. (Recall that loops are not shown.) The sum should be visualized in a space of dimension $r_1 + r_2 - 1 = 2 + 1 - 1 = 2$, i.e. in the plane. Being skew means here only that the 0-dimensional subspace $<S''>$ is not on the 1-dimensional subspace $<S'>$, i.e. on the line f determined by s_3', s_4' and by s_5'. Figure 11.8 shows e_3 as well (the other lines e_i do not exist since every other element is a loop in one of the two matroids), while the final result is shown in Fig. 11.9, which is of course the circuit matroid of the third graph of Fig. 11.5.

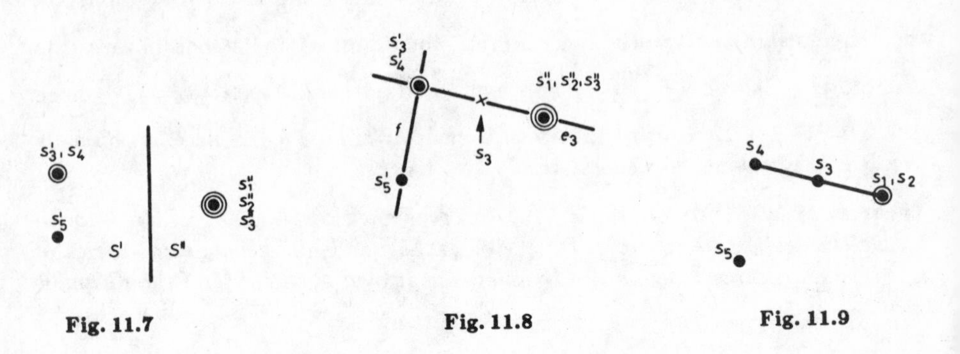

Fig. 11.7 Fig. 11.8 Fig. 11.9

The following example is somewhat less trivial. Determine the sum of the circuit matroids of the two graphs of Fig. 11.10. The first matroid is visualized in Fig. 11.11, the second has just three points s_1'', s_3'' and s_7'' in the same position (since the other 4 elements are loops). Being skew means here that these 3 points (in the same position) must be outside the plane $<S'>$. Thus the visualization of the sum immediately follows (Fig. 11.12); one can readily see that the result is a rank 4 matroid where all the $\binom{7}{4}$ four-tuples are bases except $\{s_1, s_2, s_3, s_4\}, \{s_1, s_5, s_6, s_7\}, \{s_2, s_4, s_5, s_6\}$ and $\{s_3, s_4, s_6, s_7\}$.

Exercises for Section 11.2

11.2.1 Prove that every partitional matroid is representable over the field $\mathbf{R}$ of the reals.

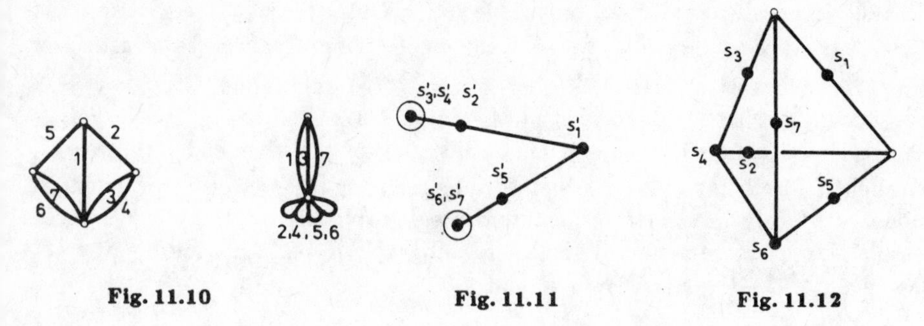

Fig. 11.10 Fig. 11.11 Fig. 11.12

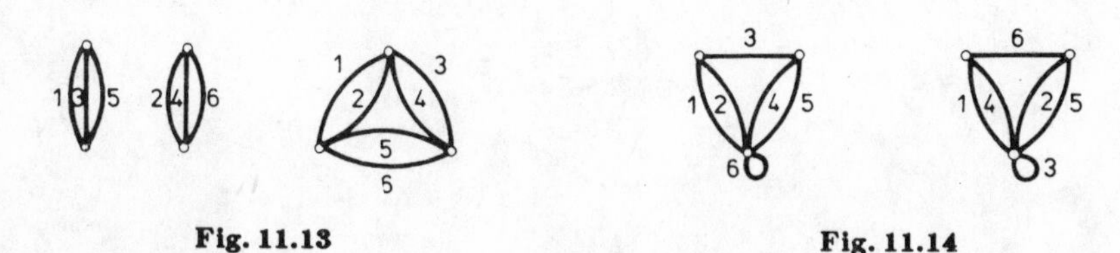

Fig. 11.13 Fig. 11.14

11.2.2 Determine the sum of the circuit matroids of the two graphs of Fig. 11.13.

11.2.3 The same exercise for the two graphs of Fig. 11.14.

Problems for Section 11.2

11.2.4 Finish the proof of Statement 11.2.1 for $k \geq 3$.

11.2.5 Try to find a weaker hypothesis than that of Theorem 11.2.2 so that the statement is still true. (Hint: Not all the nonzero entries should be algebraically independent.)

11.2.6 Consider the rank 4 matroid **M** which can be visualized by the four vertices of a tetrahedron, plus by 6 further points, one from each edge of the tetrahedron, all in general position. Prove that **M** arises as the sum of two rank 2 matroids.

11.2.7 Show by an example that the points s_i on the lines e_i (see the construction preceeding Theorem 11.2.3) must be in general position, i.e. it is not enough that s_i differs from s_i' and s_i'' (if none of them were a loop).

Section 11.3 Applications in graph theory

We close this chapter with two examples showing that very simple, short proofs can be given to fairly difficult problems of graph theory, if we are familiar with the sum of matroids.

Let $G = (V, E)$ be a graph. How many circuit-free subgraphs $G_1, G_2, \ldots, G_k$ are required to cover E, i.e. to meet $\cup_{i=1}^{k} E(G_i) = E$? We suppose that G has no loops (since otherwise the problem is unsolvable) and that $E \neq \emptyset$.

If G_i is circuit-free, we certainly have $e(G_i) \leq v(G_i) - c(G_i)$, see Section 1.2. Hence we need at least $\lceil \frac{e(G)}{v(G)-c(G)} \rceil$ subgraphs. For example, the graph of Fig. 11.15 requires at least $\lceil \frac{6}{5-1} \rceil = 2$ circuit-free subgraphs. In this case 2 are enough, e.g. let G_1 be formed by the edges 1, 3, 4, and G_2 by the others.

However, this quantity $\lceil \frac{e}{v-c} \rceil$ is not always enough. If we add a third edge, parallel to 1 and 2, to the graph of Fig. 11.15, then $\lceil \frac{7}{5-1} \rceil$ would still equal 2, though obviously three distinct circuit-free subgraphs are needed to cover three parallel edges. More generally, for every subgraph H of G, containing at least one edge, $\lceil \frac{e(H)}{v(H)-c(H)} \rceil$ is a lower bound.

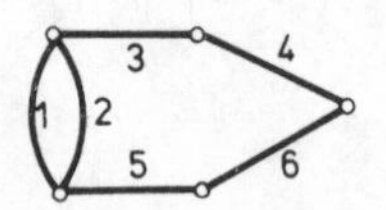

Fig. 11.15

Theorem 11.3.1 [Nash-Williams, 1961, 1964]. *Let G be a loop free graph, containing at least one edge. $E(G)$ can be covered by k circuit-free subgraphs of G if and only if*

$$k \geq \max_{\substack{H \subseteq G \\ c(H) > 0}} \left\lceil \frac{e(H)}{v(H) - c(H)} \right\rceil .$$

Proof. We have seen the "only if" direction above. Let now the above maximum equal k and let us show that $E(G)$ can be covered by k circuit free subgraphs. Let $\mathbf{M} = \mathbf{M}(G)$ be the circuit matroid of G. Our condition is $|X| \leq kr(X)$ for every $X \subseteq E$; i.e. $\min_{X \subseteq E}[kr(X) - |X|] \geq 0$ Thus we can write

$$|E| \leq \min_{X \subseteq E} [kr(X) + |E - X|].$$

The right hand side of this inequality is just $R(E)$ where R is the rank function of $\mathbf{N} = \mathbf{M} \vee \mathbf{M} \vee \ldots \vee \mathbf{M}$ (k times), see Theorem 11.1.3. Since $R(E) \leq |E|$ holds for every matroid, we obtain equality. This means that $\mathbf{N} = (E, 2^E)$, i.e. that E can be decomposed into the union of k independent (i.e., circuit-free) subsets. □

The original proof of this theorem (without using matroids) was much longer. Similarly, the concept of the sum significantly shortens the proof of the following result as well.

Theorem 11.3.2 [Tutte, 1961]. *A connected graph G has k edge-disjoint spanning trees if and only if, for an arbitrary decomposition of the points of G into m nonempty classes, there are at least $k(m-1)$ edges connecting points from different classes.*

For example, the graph of Fig. 11.16 has 10 edges but does not contain two edge-disjoint spanning trees (of 5 edges each), since considering the decomposition $\{1\} \cup \{2\} \cup \{3,4,5,6\}$ of the point set into $m = 3$ classes, only the three bold edges connect points of different classes, and $3 < 2(3-1) = 4$.

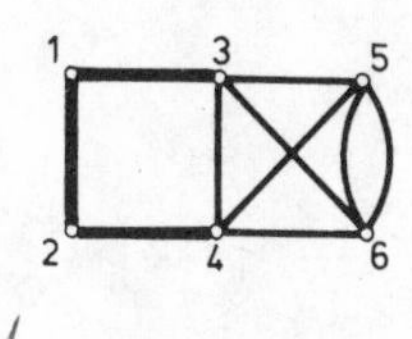

Fig. 11.16

As before, in the *proof* of this theorem the necessity of the condition is easy (Exercise 11.3.2) but the sufficiency is not (Problem 11.3.7).

Exercises for Section 11.3

11.3.1 Prove that, in Theorem 11.3.1, it is enough to take the maximum for every induced subgraph H only. (This was the original form of Nash-Williams' theorem.)

11.3.2 Prove the necessity of the condition of Theorem 11.3.2.

11.3.3 Applying Theorem 11.1.3 prove that the matroid $\mathbf{M} = (S, \mathbf{F})$ has k disjoint bases if and only if $|S - A| \geq k[r(S) - r(A)]$ holds for every subset $A \subseteq S$.

11.3.4 At least how many colours are required to colour the edges of a loop free graph so that it contains no unicoloured circuits?

11.3.5 We shall give a quick algorithm for the two matroid intersection problem in Section 13.1. Which of the following problems of graph theory have quick solutions, using this result?

(a) Find the maximum number of point-disjoint edges in a bipartite graph.

(b) Find a spanning forest in a directed graph where the outdegree of every point is at most one.

(c) Find a spanning forest in an undirected graph where the degree of every point is at most k.

(d) Let X be a set of pairwise nonadjacent points in an undirected graph. Find a spanning forest where the degree of every point in X is at most k.

11.3.6 Let S be a subset of vectors in a vector space. Prove that S can be covered by k subsets of linearly independent vectors if and only if $k \dim A \geq |A|$ for every subset $A \subseteq S$ [Horn,1955].

Problems for Section 11.3

11.3.7 Prove the sufficiency of the condition of Theorem 11.3.2.

11.3.8 Prove that the edge set of the complete graph K_n of n points can be covered by $\lceil n/2 \rceil$ forests. (Similarly one can prove that the edge set of the bipartite graph $K_{m,n}$ with $m+n$ points and all the possible $m \cdot n$ edges can be covered by $\left\lceil \frac{mn}{m+n-1} \right\rceil$ forests.)

Chapter 12

Applications

Section 12.1 The existence of hybrid immitance descriptions of multiports

Recall that n-ports are described by $\mathbf{Au} + \mathbf{Bi} = \mathbf{0}$ with $\mathbf{u} = (u_1, u_2, \ldots, u_n)^T, \mathbf{i} = (i_1, i_2, \ldots, i_n)^T$ with $r(\mathbf{A}|\mathbf{B}) = n$. This condition for the rank means that the set $S = \{u_1, u_2, \ldots, u_n, i_1, i_2, \ldots, i_n\}$ contains at least one n-element subset so that the corresponding columns of $(\mathbf{A}|\mathbf{B})$ are linearly independent. For example, if the first n or the last n columns can be chosen so then our n-port can be described as $\mathbf{u} = \mathbf{Zi}$ or as $\mathbf{i} = \mathbf{Yu}$, respectively.

Some multiports have neither **Z**-, nor **Y**- descriptions but have the property that the voltage of some ports and the current of the other ports can be chosen independently. For example, the voltage controlled voltage source $(u_2 = ku_1, i_1 = 0)$ can be written as $\begin{pmatrix} k & -1 \\ 0 & 0 \end{pmatrix} \begin{pmatrix} u_1 \\ u_2 \end{pmatrix} + \begin{pmatrix} 0 & 0 \\ 1 & 0 \end{pmatrix} \begin{pmatrix} i_1 \\ i_2 \end{pmatrix} = \begin{pmatrix} 0 \\ 0 \end{pmatrix}$. Here **A** and **B** are both singular but the quantities $\{u_2, i_1\}$ determine a nonsingular submatrix $\begin{pmatrix} -1 & 0 \\ 0 & 1 \end{pmatrix}$.

Such descriptions — perhaps after re-numbering the ports — can always be written in form $\begin{pmatrix} \mathbf{u}_1 \\ \mathbf{i}_2 \end{pmatrix} = \mathbf{H} \begin{pmatrix} \mathbf{i}_1 \\ \mathbf{u}_2 \end{pmatrix}$. These are called *hybrid immitance descriptions*, the name indicates that this is a common generalization of the impedance and the admittance descriptions. There are multiports with no hybrid immitance descriptions whatsoever, e.g. the 2-port with equations $u_1 = 0, i_1 = 0$, see Exercise 12.1.2. Generally speaking, $r(\mathbf{A}|\mathbf{B}) = n$ only means that at least one of the $\binom{2n}{n}$ different column-n-tuples must be linearly independent, while the existence of a hybrid immitance description means that at least one of 2^n special n-tuples is independent. (Recall that $\binom{2n}{n}$ is nearly 4^n if n is large.)

It is very important to know whether a multiport has at least one hybrid immitance description. (More about its importance in Section 14.2). While checking the rank of a matrix requires a polynomial number of operations only (use the Gauss elimination, for example), the existence of at least one hybrid description may appear to be a much more complicated problem since, by the definition, one may be faced with checking the rank of 2^n different matrices of size $n \times n$.

Yet we can give a polynomial order algorithm for deciding whether a given n-port has at least one hybrid immitance description. Let $S = \{u_1, u_2, \ldots, u_n, i_1, i_2, \ldots, i_n\}$ and define a matroid $\mathbf{B}_n = (S, \mathbf{F})$ so that $X \subseteq S$ be independent if and only if $|X \cap \{u_k, i_k\}| \leq 1$ holds for every $k = 1, 2, \ldots, n$.

Statement 12.1.1 [Iri and Tomizawa, 1975]. *An n-port N, given by* $(\mathbf{A}|\mathbf{B})$ *or by its matroid* $\mathbf{M}_N$ *has at least one hybrid immitance description if and only if the rank of* $\mathbf{M}_N$ *is n and* $\mathbf{M}_N \vee \mathbf{B}_n = (S, 2^S)$.

Proof. If N has a hybrid description then $\mathbf{M}_N$ has a base X meeting $|X \cap \cap \{u_k, i_k\}| = 1$ for every k, and then obviously $S - X$ is a base of $\mathbf{B}_n$. On the other hand, if $\mathbf{M}_N \vee \mathbf{B}_n$ is the free matroid then the decomposition of S gives an n-element subset which determines a hybrid immitance description.

□

Thus, using the matroid partition algorithm to be presented in Section 13.1, one can decide in polynomial order whether N has at least one hybrid immitance description. In case of an answer in the affirmative, one such description is actually given by the algorithm.

Exercises for Section 12.1

12.1.1 Suppose the multiport N has a hybrid immitance description. Prove that so have neg N, inv N, dual N and adj N as well. (See **Box 12.1**).

Box 12.1. The negative, inverse, dual and adjoint of a network, possessing a hybrid immittance description

N	$\begin{pmatrix}\mathbf{u}'\\ \mathbf{i}''\end{pmatrix} = \begin{pmatrix}\mathbf{Z} & \mathbf{U}\\ \mathbf{V} & \mathbf{Y}\end{pmatrix}\begin{pmatrix}\mathbf{i}'\\ \mathbf{u}''\end{pmatrix}$
neg N	$\begin{pmatrix}\mathbf{u}'\\ \mathbf{i}''\end{pmatrix} = \begin{pmatrix}-\mathbf{Z} & \mathbf{U}\\ \mathbf{V} & -\mathbf{Y}\end{pmatrix}\begin{pmatrix}\mathbf{i}'\\ \mathbf{u}''\end{pmatrix}$
inv N	$\begin{pmatrix}\mathbf{i}'\\ \mathbf{u}''\end{pmatrix} = \begin{pmatrix}\mathbf{Z} & \mathbf{U}\\ \mathbf{V} & \mathbf{Y}\end{pmatrix}\begin{pmatrix}\mathbf{u}'\\ \mathbf{i}''\end{pmatrix}$
dual N	$\begin{pmatrix}\mathbf{i}'\\ \mathbf{u}''\end{pmatrix} = \begin{pmatrix}-\mathbf{Z}^T & -\mathbf{V}^T\\ -\mathbf{U}^T & -\mathbf{Y}^T\end{pmatrix}\begin{pmatrix}\mathbf{u}'\\ \mathbf{i}''\end{pmatrix}$
adj N	$\begin{pmatrix}\mathbf{u}'\\ \mathbf{i}''\end{pmatrix} = \begin{pmatrix}\mathbf{Z}^T & -\mathbf{V}^T\\ -\mathbf{U}^T & \mathbf{Y}^T\end{pmatrix}\begin{pmatrix}\mathbf{i}'\\ \mathbf{u}''\end{pmatrix}$

12.1.2 The "1-port" with the equation $u = i = 0$ is called a *nullator*, the "1-port" with no equation (u and i are arbitrary) is called a *norator*. Their symbol is shown in Fig. 12.1. They are *not* 1-ports according to the definition of multiports in Section 8.1. However, if we use one of each, we have a 2-port with equations $u_1 = i_1 = 0, u_2$ and i_2 are arbitrary.

This 2-port is called *a pair of nullator and norator*. Show that it has no hybrid immitance description and that this is essentially the only 2-port with no hybrid description.

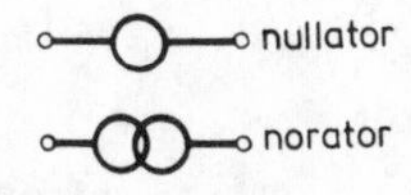

Fig. 12.1

12.1.3 Give examples for 2-ports with $0, 1, 2, 3$ or 4 hybrid immitance descriptions, respectively.

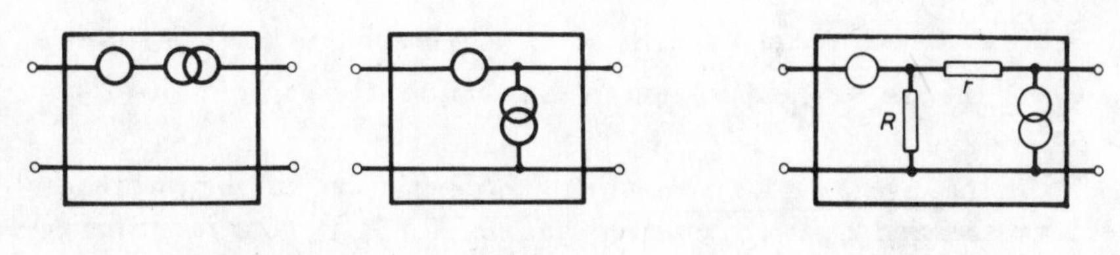

Fig. 12.2 **Fig. 12.3**

12.1.4 Do the 2-ports of Fig. 12.2 have hybrid immitance descriptions?

12.1.5 Prove that the 2-port of Fig. 12.3 is a voltage controlled voltage source (also called *operational amplifier*).

12.1.6 Give a similar model by nullator and norator to the current controlled voltage source as well.

12.1.7 Determine the equations of the 2-port of Fig. 12.4 [Carlin and Youla, 1961].

12.1.8 Determine the equation of the 1-port of Fig. 12.5 [Carlin and Youla, 1961].

Problems for Section 12.1

12.1.9 Prove that every multiport, possessing at least one hybrid immitance description, can be realized by resistors and controlled sources.

12.1.10 The 3-port $u_1 = 0,\ i_1 = 0,\ u_3 = R_3 i_3$ consists of a nullator, a norator and a usual 1-port. Give a significantly different 3-port, also having no hybrid immitance description.

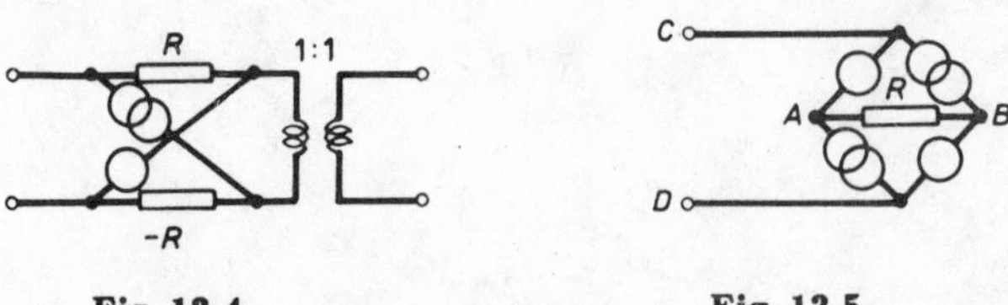

Fig. 12.4 **Fig. 12.5**

12.1.11 Suppose that an n-port N has no hybrid immitance description. Prove that one can choose k ports $(1 \leq k \leq n-1)$ of N so that the voltages and currents of these k ports form a submatrix $n \times 2k$ with rank less than k. (One may define *generalized nullators* as k-ports with more than k equations for the k voltages and k currents. Then a multiport either has at least one hybrid immitance description or contains a generalized nullator [Oono, 1960].)

12.1.12 Give a 4-port, having no hybrid immitance description but containing neither nullators nor norators. Are similar 3-ports also possible?

12.1.13 Prove that every reciprocal multiport has hybrid immitance descriptions.

12.1.14 Check the last row and column for **Box 6.3**, for the pair of nullator and norator.

12.1.15 Prove that every 2-port, possessing a **Z** description, can be realized by 4 resistors and 2 pairs of nullators and norators. Try your result for gyrators [Bendik, 1967]. Can you give also a simpler solution for gyrators?

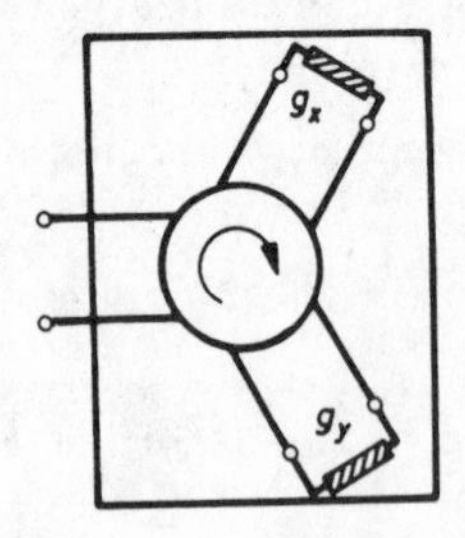

Fig. 12.6

12.1.16 Consider the 1-port of Fig. 12.6, consisting of two resistors of admittance g_x, g_y, respectively, and a 3-port circulator. Determine the equation of the 1-port. Try to choose g_x and g_y so that the result be a nullator or a norator [Carlin and Youla, 1961].

12.1.17 Analogously to the 3-port circulator, define the 4-port circulator and prepare a 2-port as in Fig. 12.7. What happens if $R_2 = 1$ Ohm and $R_4 = -1$ Ohm or *vice versa*? [Carlin and Youla, 1961].

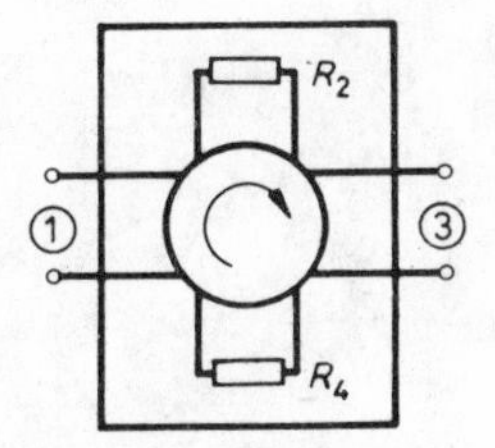

Fig. 12.7

12.1.18 Prove that the converse of the statement of Problem 12.1.9 is not true, i.e. realize a pair of nullator and norator by controlled sources.

Section 12.2 Unique solvability of linear active networks I: Necessary conditions

The necessary and sufficient condition for the unique solvability of a network consisting of positive RLC elements and voltage and current sources was fairly simple (Theorem 2.3.2). If the network also contained ideal transformers and/or gyrators then this condition could still be given by graph theory tools (Section 6.3), although deciding whether the conditions are met could not be made in polynomial time, unless we use matroid theory algorithms. In this section unique solvability of arbitrary networks (consisting of sources and multiports) is considered.

Statement 12.2.1 *Suppose that the subgraph of the voltage sources is circuit free and that of the current sources is cut set free in a network N. Prepare a new network N_0 by putting short circuits instead of the voltage sources and open circuits instead of the current sources. N is uniquely solvable if and only if N_0 is.*

Proof. Since Kirchhoff's laws and the describing equations of the multiports are linear, the total systems of equations can be written as $\mathbf{Wx} = \mathbf{b}$ and as $\mathbf{Wx} = \mathbf{0}$ in case of N and N_0, respectively, where $\mathbf{x}$ contains the unknown quantities and $\mathbf{b}$ contains the voltages of the voltage sources and the currents of the current sources. This latter vector becomes zero if we form N_0. Since both systems of equations have unique solution if and only if det $\mathbf{W} \neq 0$, the statement follows. □

Therefore, we suppose from now on that our network consists of the multiports $M_1, M_2, \ldots$ only. Let G be the network graph with edge set

$\{e_1, e_2, \ldots, e_n\}$. Let the voltages and currents of these elements form the sets $E_u = \{u_1, u_2, \ldots, u_n\}$ and $E_i = \{i_1, i_2, \ldots, i_n\}$, respectively. Let $\mathbf{A}_i$ denote the matroid $\mathbf{M}_{N_i}$ of the multiport N_i (for $i = 1, 2, \ldots$) and let $\mathbf{A} = \mathbf{A}_1 \oplus \mathbf{A}_2 \oplus \ldots$ denote their direct sum. The underlying set of $\mathbf{A}$ is the set $S = E_u \cup E_i$.

For example, the network of Fig. 12.8 contains a 2-port and two 1-ports (namely resistors), hence the above matroid $\mathbf{A}$ is determined by the column vectors of the matrix

$$\begin{pmatrix} \multicolumn{2}{|c|}{\mathbf{A}'} & 0 & 0 & \multicolumn{2}{|c|}{\mathbf{A}''} & 0 & 0 \\ & & 0 & 0 & & & 0 & 0 \\ 0 & 0 & -1 & 0 & 0 & 0 & R_3 & 0 \\ 0 & 0 & 0 & -1 & 0 & 0 & 0 & R_4 \end{pmatrix},$$

where the 2-port is described by $\mathbf{A}' \begin{pmatrix} u_1 \\ u_2 \end{pmatrix} + \mathbf{A}'' \begin{pmatrix} i_1 \\ i_2 \end{pmatrix} = \mathbf{0}$. The graph of the network is shown in Fig. 12.9 (observe that edge 5 was contracted and edge 6 was deleted). If we write down all the Kirchhoff's laws as well, then the matrix of the coefficients of these equations will be $\left(\begin{array}{cccc|cccc} -1 & 0 & 1 & 0 & 0 & 0 & 0 & 0 \\ 0 & -1 & 0 & 1 & 0 & 0 & 0 & 0 \\ \hline 0 & 0 & 0 & 0 & 1 & 0 & 1 & 0 \\ 0 & 0 & 0 & 0 & 0 & 1 & 0 & 1 \end{array}\right)$ in this example and $\begin{pmatrix} \mathbf{C} & \mathbf{O} \\ \mathbf{O} & \mathbf{Q} \end{pmatrix}$ in general, where $\mathbf{C}$ and $\mathbf{Q}$ are the circuit and cut set matrices of G, respectively.

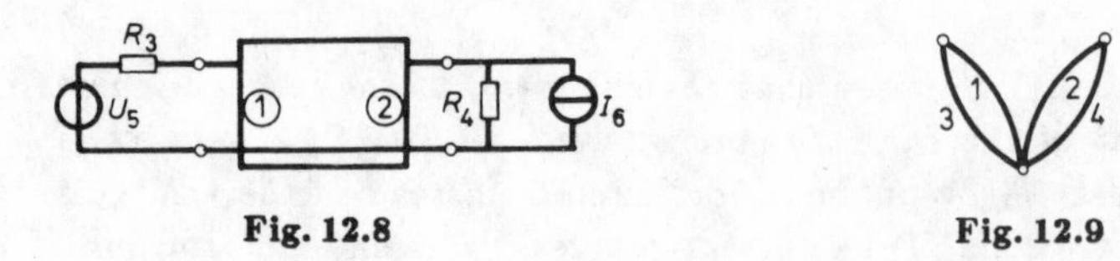

Fig. 12.8 Fig. 12.9

Since the column vectors of $\mathbf{C}$ and $\mathbf{Q}$ determine just $\mathbf{M}^*(G)$ and $\mathbf{M}(G)$, respectively, if we define them over the sets E_u and E_i, respectively, we get a matroid

$$\mathbf{G} = (E_u, \mathbf{M}^*(G)) \oplus (E_i, \mathbf{M}(G)) \tag{12.1}$$

which is just the matroid over S, coordinatized by $\begin{pmatrix} \mathbf{C} & \mathbf{O} \\ \mathbf{O} & \mathbf{Q} \end{pmatrix}$.

Theorem 12.2.2 *A necessary condition of the unique solvability of a network, consisting of multiports, is that* $\mathbf{G} \vee \mathbf{A}$ *equals the free matroid* $(S, 2^S)$.

Proof. Let $\mathbf{W}$ denote the matrix of coefficients of the total system of equations (Kirchhoff's laws and multiport equations alike). This could symbolically be written as $\mathbf{W} = \begin{pmatrix} \mathbf{C} & \mathbf{O} \\ \mathbf{O} & \mathbf{Q} \\ \hline \multicolumn{2}{c}{\mathbf{T}} \end{pmatrix}$ where the columns of $\mathbf{T}$ coordinatize the matroid $\mathbf{A}$, and the columns of the submatrix above the line coordinatize $\mathbf{G}$. The necessary *and sufficient* condition of the unique solvability is $\det \mathbf{W} \neq 0$. In order to have $\det \mathbf{W} \neq 0$, it is necessary (but not sufficient any more) that at least one nonzero member must exist in the Laplace expansion of $\det \mathbf{W}$ along the horizontal line. This means that one half of the set S of column vectors of $\mathbf{W}$ is linearly independent in $\mathbf{T}$ while the other half in the other submatrix. That is $S = S_1 \cup S_2$ with S_1 independent in $\mathbf{A}$ and S_2 in $\mathbf{G}$, hence S is independent in $\mathbf{G} \vee \mathbf{A}$.

□

Observe that Statement 11.2.1 was now proved again, for a special case.

Let us consider a very simple example at first. If both ports of a 2-port are terminated by voltage sources (Fig. 12.10) then G consists of two loops, and the matroid $\mathbf{G}$ can be visualized as the circuit matroid of the graph of Fig. 12.11. Should the 2-port be an ideal transformer, a voltage controlled source, or a gyrator, then matroid $\mathbf{A}$ can be visualized as the circuit matroid of the three graphs of Fig. 12.12, respectively.

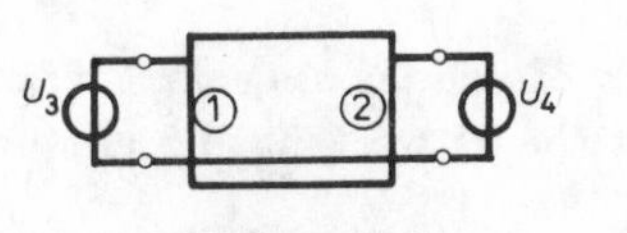

Fig. 12.10

Fig. 12.11

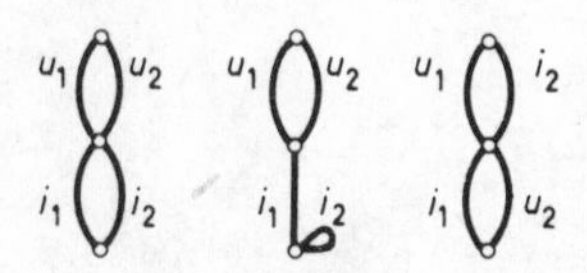

Fig. 12.12

$\mathbf{G} \vee \mathbf{A}$ can immediately be seen to be not the free matroid in the first two cases, since the rank of $\{i_1, i_2\}$ is 0 in $\mathbf{G}$ and 1 in $\mathbf{A}$. In the third case $\mathbf{G} \vee \mathbf{A} = (S, 2^S)$, since $\{u_1, u_2\}$ and $\{i_1, i_2\}$ are independent in $\mathbf{G}$ and in $\mathbf{A}$, respectively. Really, putting voltage sources to both ports is an admissible input for gyrators but not for ideal transformers or voltage controlled voltage sources.

A somewhat less trivial example is shown in Fig. 12.13. The reader should find $\mathbf{G}$ and $\mathbf{A}$ first (Exercise 12.2.3) to check that $\mathbf{G} \vee \mathbf{A}$ is the free matroid (for example, $\{u_1, u_3, u_4, u_5, u_6, i_3, i_6\}$ is independent in $\mathbf{A}$, $\{u_2, u_7, i_1, i_2, i_4, i_5, i_7\}$ is independent in $\mathbf{G}$).

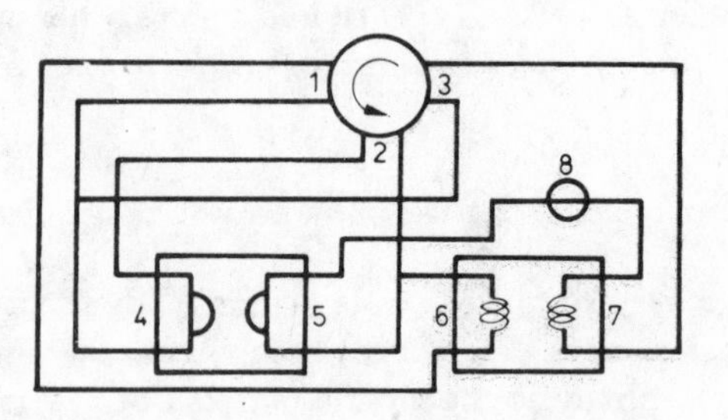

Fig. 12.13

However, if we change the source 8 to a current source then the rank of $\{u_1, u_2, u_3, u_6, u_7\}$ becomes 1 in $\mathbf{G}$ and 3 in $\mathbf{A}$, hence the network cannot have unique solution.

Exercises for Section 12.2

12.2.1 Assign a suitable combination of voltage and current sources to places A and B of Fig. 12.14 so that the network has no unique solution. The 2-port is given by

(a) $i_2 = ki_1, \quad u_2 = u_1 + ri_2$.

(b) $u_1 = Ri_1, \quad u_2 = r_1 i_1 + r_2 i_2$.

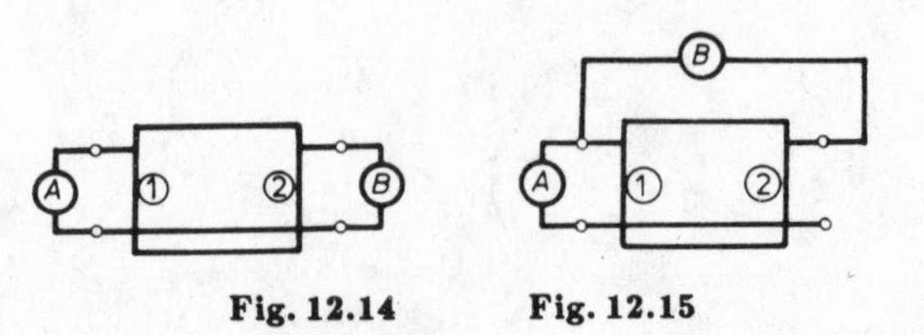

Fig. 12.14 **Fig. 12.15**

12.2.2 Assign all possible combinations of voltage and current sources to places A and B of Fig. 12.15. Determine whether the network has unique solution. The 2-port is given by
(a) $i_1 = 0,\ u_2 = 0.$
(b) $i_2 = ki_1,\ u_1 = Ri_1.$

12.2.3 Describe the matroids **G** and **A** for the network shown in Fig. 12.13.

12.2.4 Consider the necessary condition of Theorem 12.2.2 for the network of Fig. 12.16. Prove that this condition is met if and only if the 2-port has at least one hybrid immitance description.

12.2.5 (continued) Give a suitable 2-port to show that this condition is not always sufficient.

12.2.6 Give new proofs for Theorems 6.2.1 and 6.2.4.

12.2.7 Figure 12.17 shows a "generalization" of the ideal transformer [Belevitch, 1968], where three inductors are on a common core. Its system of equations is $\begin{pmatrix} i_1 \\ u_2 \\ u_3 \end{pmatrix} = \begin{pmatrix} 0 & -k & -\ell \\ k & 0 & 0 \\ \ell & 0 & 0 \end{pmatrix} \begin{pmatrix} u_1 \\ i_2 \\ i_3 \end{pmatrix}$. Modify the statement of Theorem 6.2.1 (by the help of Theorem 12.2.2) so that such transformers could also be included in the network.

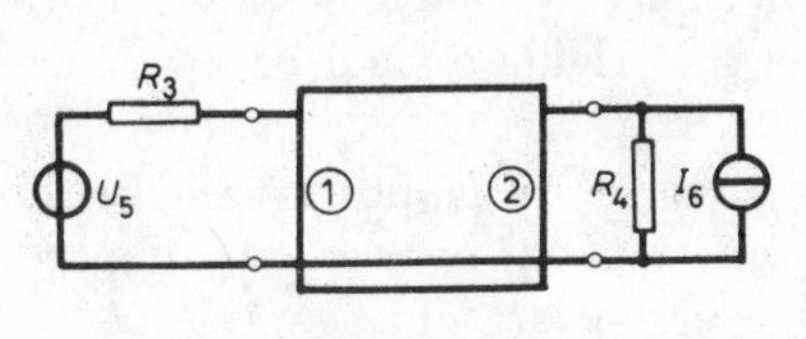

Fig. 12.16

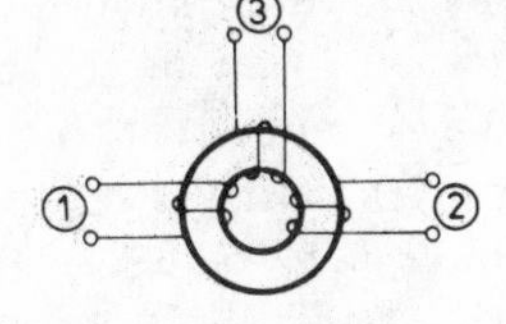

Fig. 12.17

Problems for Section 12.2

12.2.8 Determine the necessary and sufficient condition for the unique solvability of the network of Fig. 12.16. Compare your result with Exercises 12.2.4–5.

12.2.9 What does Theorem 12.2.2 state for networks, consisting of resistors and pairs of nullators and norators only? (See [Grimbleby, 1981] but implicitly [Coates, 1958] and [Mayeda, 1958] as well.)

12.2.10 Let N be a network, consisting of resistors and controlled sources only. Define (and denote by N') the *passive counterpart* of N so that controls are disregarded (i.e., short circuits are contracted, open circuits are deleted and controlled sources are considered as independent sources).

Prove that if the sets of voltage and current sources of N' are circuit free and cut set free, respectively, then N satisfies the necessary condition of Theorem 12.2.2 (see [Ozawa, 1976; Petersen, 1979]).

12.2.11 Generalize Theorem 12.2.2 for networks, containing multiports plus L and C elements (see [Recski, 1978; Petersen, 1979]).

12.2.12 (continued) Give an algorithm to check the obtained condition.

12.2.13 Let N be a network consisting of RLC elements and controlled sources only. Let σ denote the maximum number of capacitor voltages and inductor currents whose initial values can independently be prescribed. Let N' be the passive counterpart of N and σ' be the corresponding number for N'. Give examples that $\sigma > \sigma'$ is possible [Purslow and Spence, 1967; Hirano *et al.*, 1974].

12.2.14* The *topological degree of freedom* of a resistive network is the minimum number of unknown quantities (voltages and currents) so that every other quantity can be obtained as their linear combination. (For example, this number was 2 for the network of Fig. S.2.4, see the solution of Problem 2.2.5.) Find this number using the matroid partition problem.

Section 12.3 Rigidity of trusses III: Laman's theorem

Consider a framework of e rods and $v \geq 3$ joints in the 2- or 3-dimensional space. In Section 6.4 we assigned a matrix $\mathbf{A}$ of e rows and $2v$ or $3v$ columns to this framework which was called rigid if $r(\mathbf{A})$ was $2v - 3$ and $3v - 6$, respectively. Recall that the so called "infinitesimal" deformations were also excluded by this definition. We have seen several examples that rigidity could be decided from this numerically given matrix.

If only the graph of the framework is given, $r(\mathbf{A})$ cannot usually be determined (Figs 6.12, 6.14, 6.18–20) since the length of the rods are also important. If we "overcome" this difficulty by considering *generic* frameworks, i.e. if we suppose that the coordinates of the joints are algebraically independent over the field $\mathbb{Q}$ of the rationals, then $r(\mathbf{A})$ becomes uniquely determined. However, in this case it is not clear whether $r(\mathbf{A})$ can be computed in polynomial time (see Problem 5.4.6(c)). So far from the graph we could only deduce necessary conditions for the rigidity, see Corollaries 6.3.2–3, the classical results of Maxwell. In this section — at least for the 2-dimensional case — we give a necessary *and sufficient* condition for the rigidity (of generic frameworks) in terms of the graph of the framework only, which can be checked by polynomial order algorithm.

A few words about the practical applicability of the genericity assumption is in order. We gave an intuitive "justification" of a similar assumption in electric network theory at the beginning of Section 6.1. That idea does not work here since buildings have a lot of prefabricated elements whose lengths etc. are clearly not independent. Yet, the question is not only of theoretical interest, since frameworks also serve as models for other problems. For example, in geodesical networks the position of an object should uniquely be determined from several measurements [Snay, 1978], [Spriggs and Snay, 1982]. Another type of example follows in Section 14.3.

Restrict ourselves to the 2-dimensional case and recall Corollary 6.3.3. The condition $e \geq 2v-3$ is obviously not sufficient for the rigidity. If a framework has v joints and exactly $2v-3$ rods then it can still be nonrigid (dynamically underdetermined), see Fig. 12.18 for example, since it may have a "subsystem" with v' joints and e' rods satisfying $e' > 2v'-3$ and then the rest can have a motion (i.e., $v'=4$ and $e'=6$ in Fig. 12.18). Hence let us "generalize" the old necessary condition:

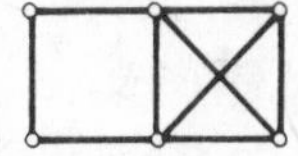

Fig. 12.18

Statement 12.3.1 *If a planar framework R has v joints and $e = 2v-3$ rods and is rigid then $e' \leq 2v'-3$ holds for every subsystem of R, having v' joints and e' rods.*

The "converse" of this theorem is about 100 years more recent:

Theorem 12.3.2 [Laman, 1970]. *If a planar framework R with v joints and $e = 2v-3$ rods is generic then the stronger necessary condition of Statement 12.3.1 is already sufficient for the rigidity of R.*

We shall only sketch the *proof* of this theorem by Exercise 12.3.1 and Problem 12.3.3.

Even now, it is not clear how can generic rigidity in the plane be quickly decided, since Laman's theorem still requires the check of an exponential number of subsystems.

Theorem 12.3.3 [Lovász and Yemini, 1982]. *If a planar framework R with v joints and $e = 2v-3$ rods is generic and G denotes the graph of R then R is rigid if and only if the edge set of G_x, obtained from G by doubling an edge x of G, can be covered by two edge disjoint spanning trees for every x.*

Observe that this theorem requires to check $\mathbf{M}(G_x) \vee \mathbf{M}(G_x) = (S, 2^S)$ for all the possible e choices of x. Hence if the matroid partition problem is polynomially solvable (Section 13.1) then this condition can also be checked in polynomial time.

Proof. If $e' \leq 2v'-3$ holds for every subgraph G' of G (with v' points and e' edges), as prescribed in Statement 12.3.1, then $e'' \leq 2v''-2$ follows for every subgraph G'' of G_x (with v'' points and e'' edges), since even $e'' \leq 2v''-3$ holds if $x \notin E(G'')$ and otherwise the bound is just one greater. Hence G_x can be covered by two trees, by Theorem 11.3.1 (and these trees must be disjoint by $e(G_x) = 2v(G_x) - 2$).

On the other hand, if R is not rigid then, by Laman's theorem, it has a subgraph G' with v' points, e' edges and $e' > 2v' - 3$. If we choose x from $E(G')$ and consider $G' + x$ as a subgraph of G_x, we obtain $e'' > 2v'' - 2$, hence G_x cannot be covered by two trees.

□

Observe that genericity was applied in the second part of the proof only:

Corollary 12.3.4 *The condition of Theorem 12.3.3 is necessary for the rigidity of any framework with $e = 2v - 3$ rods, even if the framework is not generic.*

By the analogy of Theorem 12.3.2 one could expect that a 3-dimensional generic framework with v joints and $e = 3v - 6$ rods is rigid if and only if, adding three further edges to its graph G, the result can be covered by three spanning trees. This is obviously not true (e.g., if an edge becomes of multiplicity four, it certainly cannot be covered by three circuit-free graphs) but can slightly be modified (Problem 12.3.5) to obtain the 3-dimensional analogue of Corollary 12.3.4.

However, the 3-dimensional analogues of Theorem 12.3.2–3 are simply not true. Figure 12.19 [Asimow and Roth, 1978] shows a 3-dimensional framework with $v = 8$ joints and $e = 18$ rods. One can check that $e = 3v - 6$ and that $e' \leq 3v' - 6$ holds for every subsystem with v' joints and e' rods. Yet it is dynamically underdetermined; imagine a vertical line connecting the uppermost and the lowermost joints of it and observe that fixing one half, the other "turns around" this line, like around a hinge.

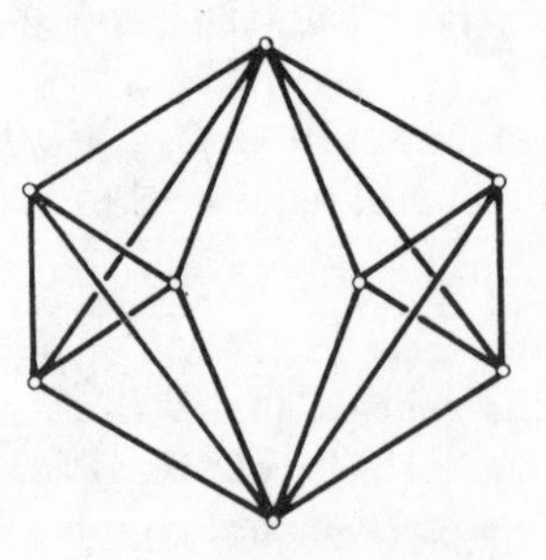

Fig. 12.19

In fact, it is an outstanding open problem whether 3-dimensional generic rigidity can be determined from the graph of the framework in polynomial time.

Exercises for Section 12.3

12.3.1 A framework has v joints and $e = 2v - 3$ rods. Prove that its graph has points with degree at most 3. At least how many such points exist?

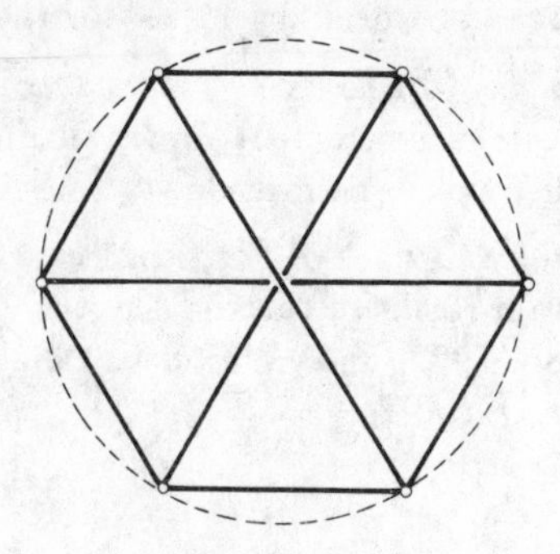

Fig. 12.20

12.3.2 Consider the complete bipartite graph $K_{3,3}$ (with two times three points and all the possible 9 edges). Show that the corresponding generic framework is rigid.

Problems for Section 12.3

12.3.3* Suppose that Laman's theorem were false and G were the graph of a counterexample with a minimum number of points. Prove that G cannot have points of degree 2.

12.3.4 Prove the following strengthening of Theorem 12.3.3. If a planar framework R with v points and $e = 2v - 3$ rods is generic and G denotes the graph of R then R is rigid if and only if, adding an edge between any two (not necessarily adjacent) points of G the edges of the resulting graph can be covered by two edge disjoint spanning trees.

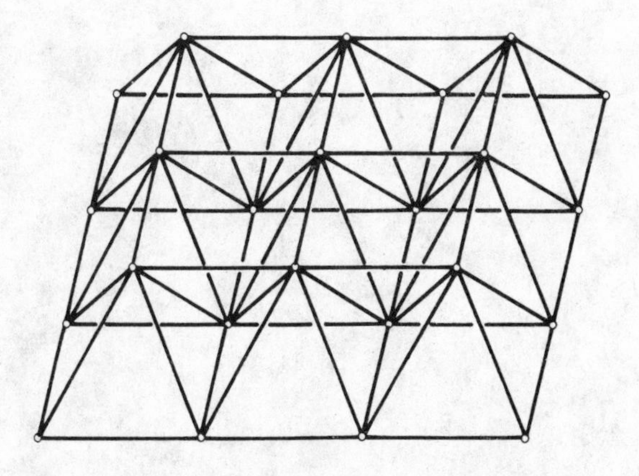

Fig. 12.21

12.3.5 What is the analogue of Corollary 12.3.4 for the 3-dimensional case?

12.3.6 Suppose the graph $K_{3,3}$ of Exercise 12.3.2 as a framework is not generic but has a very special arrangement shown in Fig. 12.20. Show that it is nonrigid. Generalize your observation.

12.3.7 The 3-dimensional framework of Fig. 12.21 can be proved to be nonrigid in the case of certain ratios of the rod lengths. Its detailed analysis can be found in [Crapo, 1982] and [Whiteley, 1982b]. How can it be made rigid by adding extra rods?

Chapter 13

The sum of matroids II

Section 13.1 Matroid theory algorithms I: Partition and intersection

For all the applications in Chapter 12 we needed quick algorithm for the matroid partition problem: Given the matroids $\mathbf{M}_1 = (S, \mathbf{F}_1)$, $\mathbf{M}_2 = (S, \mathbf{F}_2), \ldots$, $\mathbf{M}_k = (S, \mathbf{F}_k)$ with respective rank functions $r_1, r_2, \ldots, r_k$, and given a subset $X \subseteq S$. One should decide whether X is independent in the sum $\mathbf{N} = \vee_{i=1}^k \mathbf{M}_i$ of the matroids.

In case of an affirmative answer the algorithm should present a partition $\bigcup_{i=1}^k X_i$ of X with $X_i \in \mathbf{F}_i$ for every i; in the case of a negative answer the algorithm should present a subset $Y \subseteq X$ with $|Y| > \sum_{i=1}^k r_i(Y)$. The former answer proves that X is independent in $\mathbf{N}$, the latter proves that even Y is dependent in $\mathbf{N}$.

What does "quick" algorithm mean in matroid theory? Recall that in graph theory, although graphs could be stored in a number of different ways, all the data structures presented in Chapter 1 could be converted into all the others in polynomial time. Hence the statement that a graph theory algorithm was polynomial was independent of the way of the storage of the input graph.

This will not be the case in matroid theory. Some of the "natural" descriptions of matroids cannot be converted into some others in polynomial time (see Section 17.1 for details). However, the matroids in our applications are given by graphs or by matrices. Hence if a matroid is defined over a set S and $|S| = n$ then an input, proportional to, say, n^2, is possible, and both the linear independence of some columns of a matrix, or the question whether a subgraph is circuit free, can be answered in time, polynomial in n.

Thus, until a more detailed study of matroid algorithms in Section 17.1, we shall suppose that our matroids $\mathbf{M} = (S, \mathbf{F})$ are given so that the question "Is $X \subseteq S$ independent in $\mathbf{M}$?" can be answered in polynomial time.

Saying it in another way, for each matroid $\mathbf{M}_i$ we have a "subroutine" R_i which decides independence of given subsets in $\mathbf{M}_i$; and then a matroid algorithm is *quick* if the total number of operations is proportional to a polynomial of n, *provided* that callings of such subroutines are also considered as single steps.

Hence, on the one hand, the "quickness" of this algorithm is independent of the actual storage of the matroid (we do not care the real complexity of the subroutines R_i); on the other hand, if these R_i's are polynomial (which is the case in the presented applications) then the total algorithm is polynomial.

Let us return to the matroid partition problem. If we already have some disjoint subsets $I_1, I_2, \dots, I_k \subseteq S$ with $I_j \in \mathbf{F}_j$ for every $j = 1, 2, \dots, k$, but $\bigcup_{j=1}^k I_j$ is not equal to X yet, then consider an element $x \in X - \bigcup_{j=1}^k I_j$ and try to join it to some of the I_j's.

If there is a j so that $I_j \cup \{x\} \in \mathbf{F}_j$ then we join x to this I_j. If this is impossible for every j, it does not mean that $\{x\} \cup \bigcup_{j=1}^k I_j$ is dependent in $\mathbf{N}$; can happen that it is independent, just our partition is inappropriate.

For example, if $\mathbf{M}_1$ and $\mathbf{M}_2$ are the circuit matroids of the graphs of Fig. 13.1 and $I_1 = \{1, 2\}$, $I_2 = \{3, 4\}$ then 5 cannot be joined to either of them, although $\{1, 2, 3, 4, 5\}$ is independent in the sum, with a partition $I_1' = \{2, 5\}$, $I_2' = \{1, 3, 4\}$. Obviously, the point is that 5 could not be joined to I_1, only "exchanged" to 1 which could be joined to I_2. Sometimes quite a sequence of such exchanges is required.

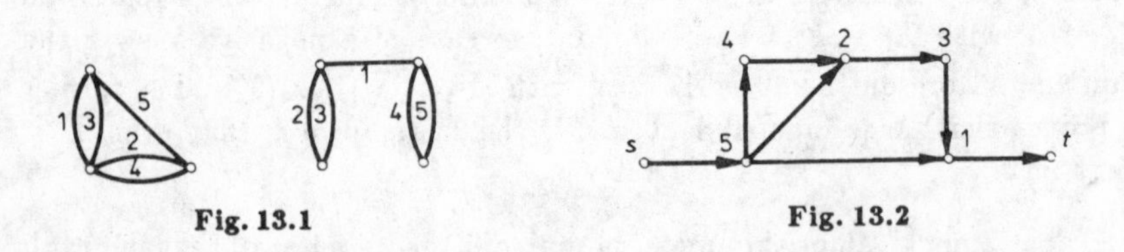

Fig. 13.1 **Fig. 13.2**

In order to see all these exchange possibilities, we prepare a graph model. Let the directed graph $G = (V, E)$ have a point set V, containing all the elements of S plus two further points s, t. Let us draw an edge from s to $x \in S$ if and only if $x \notin \bigcup_{j=1}^k I_j$, i.e. if x *should be added* to the present partition; draw an edge from $y \in S$ to t if and only if there exists a j so that $y \notin I_j$ and $I_j \cup \{y\} \in \mathbf{F}_j$, i.e. if y *can be added* to I_j; finally if $a, b \in S$ then draw an edge from a to b if there exists a j so that $b \in I_j$, $I_j \cup \{a\} \notin \mathbf{F}_j$ and $(I_j \cup \{a\}) - \{b\} \in \mathbf{F}_j$, i.e. if b *can be exchanged* to a in $\mathbf{M}_j$.

For example, if $\mathbf{M}_1$ and $\mathbf{M}_2$ are still the circuit matroids of the graphs of Fig. 13.1 and $I_1 = \{1, 2\}$, $I_2 = \{3, 4\}$ then our graph G is shown in Fig. 13.2. The edge $(s, 5)$ shows that $5 \notin I_1 \cup I_2$. The edge $(1, t)$ shows that $I_2 \cup \{1\} \in \mathbf{F}_2$. Among the other edges, $(5, 4)$ is present due to the circuit $\{4, 5\}$ of the set $I_2 \cup \{5\}$ in $\mathbf{M}_2$; $(5, 1)$ and $(5, 2)$ are present due to the circuit $\{1, 2, 5\}$ of the set $I_1 \cup \{5\}$ in $\mathbf{M}_1$ etc.

One can readily see that the extension $I_1' = (I_1 \cup \{5\}) - \{1\}$ and $I_2' = I_2 \cup \{1\}$ are prescribed by the directed path $(s, 5, 1, t)$. More generally, we shall prove that the extension can be done along any *minimal* directed path $(s, x_1, x_2, \dots, x_p, t)$, where minimality means that there is no edge (s, x_j) for $j > 1$, no edge (x_i, x_j) with $j > i + 1$, and no edge (x_j, t) for $j < p$.

Similarly to Algorithm 5.1 (maximum flow), either our solution is optimal or there exists an augmenting path. However, augmentation is not necessarily possible here, if the directed path is not minimal (see Problem 13.1.11).

Algorithm 13.1 *Matroid partition.*

Input The matroids $\mathbf{M}_j = (S, \mathbf{F}_j)$ for $j = 1, 2, \dots, k$ are given by subroutines R_j which check the independence of any subset of S in $\mathbf{M}_j$ in unit time.

Output Either a partition $\bigcup_{j=1}^k I_j$ of S with $I_j \in \mathbf{F}_j$ for every j (if $\mathbf{N} = \vee_{j=1}^k \mathbf{M}_j$ is the free matroid), or a subset $Y \subseteq S$ with $|Y| > \sum_{j=1}^k r_j(Y)$ (if $\mathbf{N}$ is not the free matroid).

Step 1 Initially let $I_1 = I_2 = \ldots = I_k = \emptyset$.

Step 2 If $S = \bigcup_{j=1}^k I_j$ then go to **End 1**.

Step 3 Prepare the above graph model. If t can be reached from s along directed paths then choose a minimal one and augment the I_j sets accordingly, then go to **Step 2**.

Step 4 If T denotes the set of those points which can be reached from s along directed paths and $t \notin T$ then go to **End 2**.

End 1 The algorithm terminates, $\mathbf{N} = \vee_{j=1}^k \mathbf{M}_j$ is the free matroid and $\bigcup_{j=1}^k I_j$ is a partition of S.

End 2 The algorithm terminates, $\mathbf{N}$ is not the free matroid and $Y = T - \{s\}$ is dependent in $\mathbf{N}$ since $|Y| > \sum_{j=1}^k r_j(Y)$.

Recall that a minimal path for Step 3 can be found by Algorithm 3.4.

In order to see the correctness of this algorithm we have to prove that the augmentation always works (Problems 13.1.9–11) and that Y is dependent in $\mathbf{N}$ at End 2.

Theorem 13.1.1 *If Algorithm 13.1 terminates at End 2 then* $|Y| > \sum_{j=1}^k r_j(Y)$.

Proof. $Y = \emptyset$ is impossible since $T = \{s\}$ would mean that the algorithm terminated at End 1. Hence if A denotes the neighbours of s in the graph (i.e., $A = Y - \bigcup_{j=1}^k I_j$) then $A \neq \emptyset$. Let B denote the set of other elements of Y (if any); by any means, $|B| < |Y|$.

We prove $r_j(Y) = |Y \cap I_j|$ for $j = 1, 2, \dots, k$. Obviously, $Y \cap I_j \in \mathbf{F}_j$ (since $I_j \in \mathbf{F}_j$) hence $r_j(Y) \geq |Y \cap I_j|$ holds. Suppose there were an element $x \in Y - I_j$ with $(Y \cap I_j) \cup \{x\} \in \mathbf{F}_j$. Then either $I_j \cup \{x\} \in \mathbf{F}_j$ and then an edge (x, t) exists; or the circuit of $\mathbf{M}_j$, contained in $I_j \cup \{x\}$ were not entirely contained in Y but would have an element $y \in I_j - Y$ as well, and then an edge (x, y) exists. Both possibilities contradict to the definition of Y.

Then, finally, $\sum_{j=1}^k r_j(Y) = \sum_{j=1}^k |Y \cap I_j| = |B| < |Y|$, as requested. □

For the intersection problem of two matroids the following very similar algorithm is suggested.

Algorithm 13.2 *2-matroid intersection.*

Input The matroids $\mathbf{M}_1 = (S, \mathbf{F}_1)$ and $\mathbf{M}_2 = (S, \mathbf{F}_2)$ are given by subroutines R_1, R_2, respectively, which check the independence of any subset of S in the respective matroid in unit time.

Output A maximum cardinality subset $I \subseteq S$, subject to $I \in \mathbf{F}_1 \cap \mathbf{F}_2$.

Step 1 Initially let $I = \emptyset$.

Step 2 Prepare a directed graph G with point set $S \cup \{s, t\}$ so that, for some $a, b \in S$,
(s, a) is an edge if $a \notin I$ and $I \cup \{a\} \in \mathbf{F}_2$;
(b, t) is an edge if $b \notin I$ and $I \cup \{b\} \in \mathbf{F}_1$;
(a, b) is an edge if either $a \notin I$, $I \cup \{a\} \notin \mathbf{F}_1$ and $(I \cup \{a\}) - \{b\} \in \mathbf{F}_1$ or $b \notin I$, $I \cup \{b\} \notin \mathbf{F}_2$ and $(I \cup \{b\}) - \{a\} \in \mathbf{F}_2$.

Step 3 If t can be reached from s along directed paths then let $(s, a_1, a_2, \ldots$ $\ldots, a_p, t)$ be a *minimal* such path and let us augment along it, i.e. let $I = (I \cup \{a_1, a_3, a_5, \ldots\}) - \{a_2, a_4, a_6, \ldots\}$; then go to **Step 2**.

Step 4 If T denotes the set of those points which can be reached from s along directed paths and $t \notin T$ then go to **End**.

End The algorithm terminates, I is the solution; furthermore $|I| = r_1(T) +$ $+ r_2(S - T)$ holds.

In order to see the correctness of this algorithm, we have to prove that the augmentation always works (this happens in the same way, as for Algorithm 13.1) and the relation $|I| = r_1(T) + r_2(S - T)$, see Exercise 13.1.2. This relation will automatically imply that I is maximum, see Theorem 13.2.2 below.

We have seen (Statement 11.1.2, Problems 11.1.13 and 11.1.15) that the matroid partition problem (for $k \geq 2$) and the intersection problem for 2 matroids are equivalent. However, Algorithm 13.1 is suggested for the former and Algorithm 13.2 for the latter. One can namely prove (Problems 13.1.13–14) that the reduction of any of them to the other may significantly increase the number of requested operations.

In general, the *k-matroid intersection problem* means that a subset $X \subseteq S$ with maximum cardinality must be found, subject to $X \in \cap_{j=1}^{k} \mathbf{F}_j$ (i.e., X must be independent in all the matroids $\mathbf{M}_j = (S, \mathbf{F}_j)$, $j = 1, 2, \ldots, k$).

Theorem 13.1.2 *The k-matroid intersection problem is polynomially solvable for $k = 2$ and is* **NP**-*hard for $k \geq 3$.*

Proof. Algorithm 13.2 above is polynomial for $k = 2$, see Exercise 13.1.3. Let $k \geq 3$. If $k > 3$ then the case $k = 3$ is reducible to this by choosing $\mathbf{M}_t = (S, 2^S)$ for $t = 4, 5, \ldots, k$; hence it is enough to prove that the case $k = 3$ is **NP**-hard.

Recall that the following problem is **NP**-complete. Decide whether there exists a directed Hamiltonian path from a given point u to a given point v in a given directed graph G.

Let $S = E(G)$, $\mathbf{M}_1 = \mathbf{M}(G)$. Define $\mathbf{M}_2 = (S, \mathbf{F}_2)$ and $\mathbf{M}_3 = (S, \mathbf{F}_3)$ as follows. An edge set $X \subseteq S$ belongs to $\mathbf{F}_2$ if the indegree of u is zero and the indegree of every other point is at most one in the subgraph, determined by X; while $X \in \mathbf{F}_3$ if the outdegree of v is zero and the outdegree of every other point is at most one in this subgraph. (For example, if G is the graph of Fig. 13.3 then the three matroids are just the circuit matroids of the graph of Fig. 13.4).

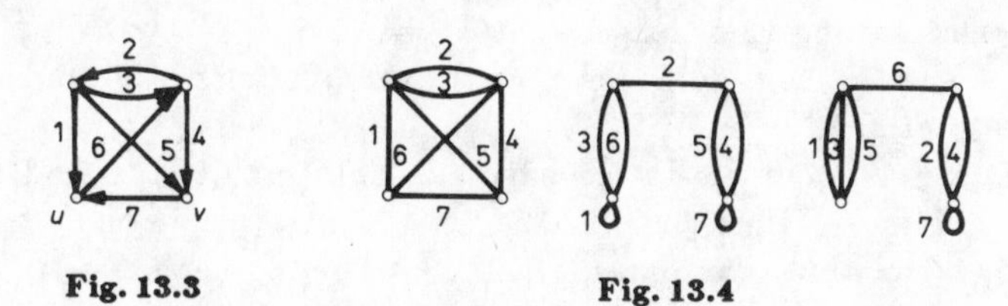

Fig. 13.3

Fig. 13.4

G has a directed Hamiltonian path from u to v if and only if $\mathbf{M}_1, \mathbf{M}_2$ and $\mathbf{M}_3$ have a common base. Hence the above **NP**-complete problem is reducible to the 3-matroid intersection problem.

□

We close this section with the *weighted matroid intersection algorithm.* If the elements of the underlying set S of the matroids $\mathbf{M}_1 = (S, \mathbf{F}_1)$ and $\mathbf{M}_2 = (S, \mathbf{F}_2)$ have weights w and they are extended to subsets $X \subseteq S$ by $w(X) = \sum_{x \in X} w(x)$ then we may ask

$$W = \max\{w(X);\ X \text{ is a maximum cardinality independent set in } \mathbf{M}_1 \text{ and } \mathbf{M}_2\}. \tag{13.1}$$

This problem was first solved by Edmonds [1969], here we present a more recent algorithm [Frank, 1981]. The weights may be negative as well.

Algorithm 13.3 *Weighted matroid intersection.*

Input Two matroids, as in Algorithm 13.2; and a weight function w.

Output A subset $I \subseteq S$ where the maximum W of Eq. (13.1) is obtained.

Step 1 We introduce two n-dimensional vectors $\mathbf{s}_1 = (s_1^1, s_2^1, \ldots, s_n^1)$ and $\mathbf{s}_2 = (s_1^2, s_2^2, \ldots, s_n^2)$, where $n = |S|$. Initially, let $s_j^1 = 0$ for every j and let s_j^2 contain the weight of the jth element of S. Furthermore, let $I = \emptyset$ and $k = 0$.

Step 2 Let $m_i = \max\{s_j^i | j \notin I \text{ and } I \cup \{j\} \in \mathbf{F}_i\}$. Prepare a directed graph G with point set $S \cup \{s, t\}$ so that, for some $a, b \in S$,
(s, a) is an edge if $a \notin I,\ \ I \cup \{a\} \notin \mathbf{F}_2$ and $s_a^2 = m_2$;
(b, t) is an edge if $b \notin I,\ \ I \cup \{b\} \notin \mathbf{F}_1$ and $s_b^1 = m_1$;
(a, b) is an edge if either $a \notin I,\ \ I \cup \{a\} \notin \mathbf{F}_1,\ \ (I \cup \{a\}) - \{b\} \in \mathbf{F}_1$ and $s_a^1 = s_b^1$ or $b \notin I,\ \ I \cup \{b\} \notin \mathbf{F}_2,\ \ (I \cup \{b\}) - \{a\} \in \mathbf{F}_2$ and $s_a^2 = s_b^2$.
If a pair of points satisfies all the above conditions except the last ones (referring to the vectors $\mathbf{s}_i$) then the corresponding directed edge is drawn dotted only; and then the difference $m_2 - s_a^2$ or $m_1 - s_b^1$ or $s_b^1 - s_a^1$ or $s_a^2 - s_b^2$ is the weight of this dotted edge.

Step 3 If t can be reached from s along a directed path of continuous (not dotted) edges then let $(s, a_1, a_2, \ldots, a_\ell, t)$ be a *minimal* such path and let us augment along it, i.e. let $I = (I \cup \{a_1, a_3, a_5, \ldots\}) - \{a_2, a_4, a_6, \ldots\}$; then let $k = k + 1$ and go to **Step 2**.

Step 4 Let T denote the set of those points which can be reached from s along a directed path of continuous edges. If there are dotted edges with tail in T and head in $(S \cup \{t\}) - T$ then let δ be the minimum weight among these edges; put $s_x^1 = s_x^1 + \delta$ and $s_x^2 = s_x^2 - \delta$ for $x \in T - \{s\}$, and do not change the other entries of $\mathbf{s}_1$ and $\mathbf{s}_2$; then go to **Step 2**.

Step 5 If $t \notin T$ and no dotted edge exists with tail in T and head in $(S \cup \{t\}) - T$ then go to **End**.

End The algorithm terminates, I is the required k-element set.

Exercises for Section 13.1

13.1.1 Let $\mathbf{M} = (S, \mathbf{F})$ be a rank r matroid on the set S of cardinality n. The question "Is $X \subseteq S$ independent in $\mathbf{M}$?" can be answered in one step.
(a) How many steps are needed to find a base of $\mathbf{M}$?
(b) How many steps are needed to answer the question "Is $X \subseteq S$ independent in $\mathbf{M}^*$?"

13.1.2 Prove that $|I| = r_1(T) + r_2(S - T)$ holds at the end of Algorithm 13.2.

13.1.3 Estimate the number of operations in Algorithm 13.2.

13.1.4 Modify Algorithm 13.1 so that in the case of $\mathbf{N} \neq (S, 2^S)$ the output be not only a base $B = \bigcup_{j=1}^k I_j$ of $\mathbf{N}$ but the fundamental system of circuits with respect to B as well.

13.1.5 Let $W' = \max\{w(X) | X \in \mathbf{F}_1 \cap \mathbf{F}_2\}$. What is the relation of this number to W, defined by Eq. (13.1)?

13.1.6 (continued) Applying Algorithm 13.3 for W, give a polynomial algorithm for W' as well.

13.1.7 Perform step by step Algorithm 13.3 if $\mathbf{M}_1$ and $\mathbf{M}_2$ are the circuit matroids of the graphs of Fig. 13.5 and $w_1 = 5, w_2 = 1, w_3 = 2$.

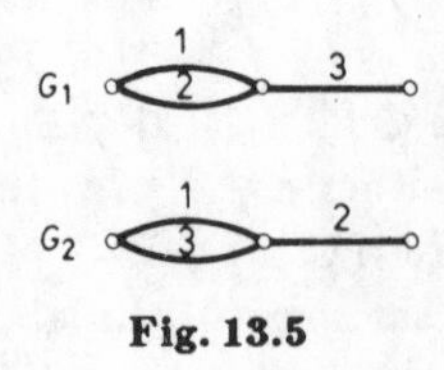

Fig. 13.5

13.1.8 Estimate the number of operations in Algorithm 13.3.

Problems for Section 13.1

13.1.9 Prove that if a minimal directed path is of length at most 4 then the augmentation (Step 3 in Algorithm 13.1) is always possible.

13.1.10 (continued) Prove that the augmentation is always possible along minimal directed paths.

13.1.11 Give an example that the minimality is necessary for the augmentation.

13.1.12 Estimate the number of operations in Algorithm 13.1.

13.1.13 How many operations are required if the 2-matroid intersection problem is solved by Algorithm 13.1, making use of Problem 11.1.13?

13.1.14 How many operations are required if the k-matroid partition problem is solved by Algorithm 13.2, making use of Problem 11.1.5?

13.1.15 Let $s_k = \max\{w(X); |X| = k \text{ and } X \in \mathbf{F}_1 \cap \mathbf{F}_2\}$. We have seen (Exercise 13.1.5) that the sequence $s_1, s_2, \ldots$ need not be monotone increasing. However, prove that this is "concave", i.e. $s_{k+1} - s_k \leq s_k - s_{k-1}$ [Krogdahl, 1975; Lawler, 1976].

Section 13.2 Important minimax theorems

First of all, we prove Theorem 11.1.3, what we have applied several times already.

Theorem 13.2.1 [Nash-Williams, 1967]. *If $r_1, r_2, \ldots, r_k$ and R denote the rank functions of the matroids $\mathbf{M}_1 = (S, \mathbf{F}_1)$, $\mathbf{M}_2 = (S, \mathbf{F}_2), \ldots, \mathbf{M}_k = (S, \mathbf{F}_k)$ and $\mathbf{N} = \vee_{j=1}^k \mathbf{M}_j$, respectively, then, for an arbitrary subset $X \subseteq S$,*

$$R(X) = \min_{T \subseteq X} \left[\sum_{j=1}^{k} r_j(T) + |X - T| \right] \tag{13.2}$$

Proof. Applying the relation $\mathbf{N} \backslash Z = \vee_{i=1}^k (\mathbf{M}_i \backslash Z)$ of Exercise 11.1.3 with $Z = S - X$ we may suppose that the underlying set of $\mathbf{M}_1, \mathbf{M}_2, \ldots, \mathbf{M}_k$ and $\mathbf{N}$ is just X. Then applying Algorithm 13.1 we obtain a partition $I = I_1 \cup I_2 \cup \ldots \cup I_k$ with $|I| = R(X)$, and a subset $Y \subseteq X$ meeting $r_j(Y) = |Y \cap I_j|$ for every j, see the proof of Theorem 13.1.1. (If $I = X$ then $Y = \emptyset$). Then

$$R(X) = |I| = |I \cap Y| + |I - Y| = \sum_{j=1}^{k} |I_j \cap Y| + |I - Y| = \sum_{j=1}^{k} r_j(Y) + |I - Y| =$$

$$= \sum_{j=1}^{k} r_j(Y) + |X - Y| \geq \min_{T \subseteq X} \left[\sum_{j=1}^{k} r_j(T) + |X - T| \right],$$

since $I - Y = X - Y$ (by $X = I \cup Y$). The other inequality $\leq$ was seen as Problem 11.1.12 already.

□

The following minimax theorem, considering intersection of matroids rather than partition, is also of basic importance.

Theorem 13.2.2 [Edmonds, 1968]. *Let r_1 and r_2 be the rank functions of $\mathbf{M}_1 = (S, \mathbf{F}_1)$ and $\mathbf{M}_2 = (S, \mathbf{F}_2)$, respectively. Then*

$$\max_{Y \in \mathbf{F}_1 \cap \mathbf{F}_2} |Y| = \min_{X \subseteq S} [r_1(X) + r_2(S - X)]. \tag{13.3}$$

Proof. Let the value on the left hand side be t and let $Y_0 \in \mathbf{F}_1 \cap \mathbf{F}_2$ with $|Y_0| = t$. Then for every bipartition $X \cup (S - X)$ of S, we have $t = |Y_0 \cap X| + |Y_0 \cap (S - X)|$, and since $Y_0 \cap X \in \mathbf{F}_1$ and $Y_0 \cap (S - X) \in \mathbf{F}_2$, we have $t \leq r_1(X) + r_2(S - X)$ for every X. Hence the relation max $\leq$ min is trivial, as usually.

We know that the rank of $\mathbf{M}_1 \vee \mathbf{M}_2^*$ is $t + r_2^*(S)$, see Problem 11.1.13. Thus, by the previous theorem, there exists a subset $T \subseteq S$ with $r_1(T) + r_2^*(T) + |S - T| = t + r_2^*(S)$. Finally, $r_2^*(T) - r_2^*(S) = r_2(S - T) - |S - T|$, by Theorem 7.2.5, leading to $r_1(T) + r_2(S - T) = t$. Hence $t \geq \min_{X \subseteq S}[r_1(X) + r_2(S - X)]$ also holds.

□

As one can expect by Theorem 13.1.2, this theorem cannot be extended for more than 2 matroids (see Exercise 13.2.3).

Now we introduce an important concept. Let $\mathbf{M} = (S, \mathbf{F})$ be a matroid with rank function r and ρ be a function from S onto $\rho(S)$. We define the *homomorphic image* $\mathbf{M}_\rho = (\rho(S), \mathbf{F}_\rho)$ of $\mathbf{M}$ as a matroid where $X \subseteq \rho(S)$ is an element of $\mathbf{F}_\rho$ if and only if there exists a subset $T \subseteq S$ so that $T \in \mathbf{F}$ and $X = \rho(T)$. For example, if ρ is defined by Fig. 13.6 and $\mathbf{M}$ is the circuit matroid of the graph G_1 of Fig. 13.7 then $\mathbf{M}_\rho$ is the circuit matroid of G_2 in the same figure. Of course, one should prove that this definition leads to a matroid. We directly give its rank function.

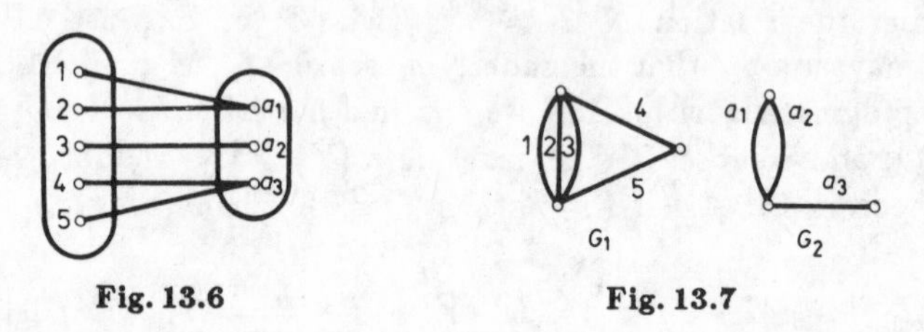

Fig. 13.6 **Fig. 13.7**

Theorem 13.2.3 [Nash-Williams, 1967].

$$r_\rho(X) = \min_{Y \subseteq X}[r(\rho^{-1}(Y)) + |X - Y|]. \tag{13.4}$$

Proof. Let $\rho(S) = \{a_1, a_2, \ldots\}$ and $A_i = \rho^{-1}(\{a_i\})$. For any bipartition $Y \cup (X - Y)$ of any subset $X \subseteq \rho(S)$ we have $r_\rho(Y) \leq r\left(\bigcup_{a_i \in Y} A_i\right) = r(\rho^{-1}(Y))$ and $r_\rho(X - Y) \leq |X - Y|$, hence the relation $r_\rho(X) \leq$ min is obvious.

Define now a partitional matroid $\mathbf{P}$ on S so that $T \subseteq S$ be independent in $\mathbf{P}$ if and only if $|T \cap A_i| \leq 1$ holds for every i. Observe that the elements of $\mathbf{F}_\rho$ are just the images of the common independent subsets of $\mathbf{M}$ and $\mathbf{P}$, and that $r_\rho(T) = |\rho(T)|$. We may suppose that $\mathbf{M}$ and $\mathbf{P}$ are restricted to the set $S_0 = \rho^{-1}(X)$.

Let us denote $r_\rho(X)$ by t. There must be a t-element subset, independent both in $\mathbf{M}$ and $\mathbf{P}$, hence $\min_{T\subseteq S_0}[r(T)+|\rho(S_0-T)|]=t$, by Theorem 13.2.2. Should such a T (where the minimum is attained) arise in form $\rho^{-1}(Y)$ for some $Y\subseteq X$, we were ready, but it can happen that T contains only a subset of some A_i.

Let, therefore, $T_0=\bigcup_{A_i\subseteq T}A_i$, that is, let T_0 be the union of only those subsets A_i which are entirely in T. Then $\rho(S_0-T)=\rho(S_0-T_0)$ and $r(T_0)\le r(T)$ by $T_0\subseteq T$. Hence the relation

$$r(T_0)+|\rho(S_0-T_0)|\le t=\min_{T\subseteq S_0}[r(T)+|\rho(S_0-T)|]$$

must be met by equality and Eq. (13.4) follows by $Y=X-\rho(S_0-T_0)$. □

The importance of the concept of homomorphic image is best seen by the following construction. Let $S=\{a,b,\dots,z\}$ and prepare this set in k copies $S_1,S_2,\dots,S_k$. Consider the mapping ρ of the union of these k sets to S, as shown in Fig. 13.8, i.e. let $\rho(t_i)=t$ for every $i=1,2,\dots,k$. Define a matroid $\mathbf{M}_i=(S_i,\mathbf{F}_i)$ on every set S_i and consider $\mathbf{N}=\oplus_{i=1}^k\mathbf{M}_i$.

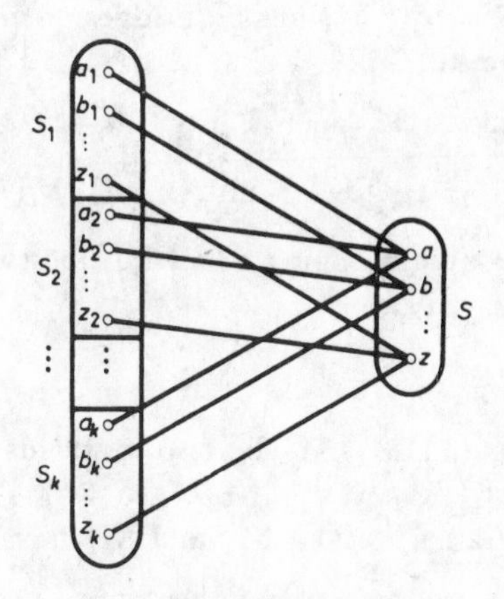

Fig. 13.8

One can easily see that $\mathbf{N}_\rho$ is just $\vee_{i=1}^k\mathbf{M}_i$, since $X\subseteq S$ was defined to be independent in N_ρ if and only if $X=\rho(Y)$ for some $Y\subseteq\bigcup_{i=1}^k S_i$, which is independent in $\mathbf{N}$, and this latter holds if and only if $Y\cap S_i\in\mathbf{F}_i$ for every i. Hence the sum of matroids arises as the homomorphic image of the direct sum.

Using Theorem 13.2.3 we immediately obtain that the rank of a subset $X\subseteq S$ in $\mathbf{N}_\rho$ is just $\min_{Y\subseteq S}[r(\rho^{-1}(Y))+|X-Y|]$, where r is the rank function of $\oplus_{i=1}^k\mathbf{M}_i$. However, if r_i denotes the rank function of $\mathbf{M}_i$ (for $i=1,2,\dots,k$) then $r(Z)=\sum_{i=1}^k r_i(Z\cap S_i)$, hence we obtain Theorem 13.2.1. This means, all the 3 theorems of this section imply the other two.

We close this section with another interesting property of homomorphic image.

Theorem 13.2.4 [Mason, 1977]. *Let a set P of points be the affine representation of the rank r matroid $\mathbf{M} = (S, \mathbf{F})$ in the $(r-1)$-dimensional space. Let $\rho : S \longrightarrow \rho(S)$ be a mapping and the inverse images $\rho^{-1}(\{a_i\})$ of the elements $a_1, a_2, \ldots$ of $\rho(S)$ be denoted by A_i. Each subset A_i corresponds to a point set $P_i \subseteq P$ and these point sets span the subspaces $<P_i>$ of dimension $r(A_i) - 1$. If we choose a point p_i from each $<P_i>$ in general position then this set $\{p_1, p_2, \ldots\}$ is the affine representation of $\mathbf{M}_\rho$.*

Observe that this theorem specializes to Theorem 11.2.3 if ρ is the mapping shown in Fig. 13.8 and if $k = 2$. A somewhat more general form will be proved in Problem 13.2.9.

Exercises for Section 13.2

13.2.1 Suppose that the minimum in $R(S) = \min_{T \subseteq S} \left[\sum_{i=1}^{k} r_i(T) + |S - T|\right]$ is attained at a subset $T_0 \subseteq S$.
(a) Prove that the elements of $S - T_0$ are bridges in $\mathbf{N} = \vee_{i=1}^{k} \mathbf{M}_i$.
(b) Prove that choosing such a bridge $x \in S - T_0$ and writing Eq. (13.2) for $R(S - \{x\})$ the minimum is attained at T_0 again.

13.2.2 Prove that if $\mathbf{N} = \vee_{i=1}^{k} \mathbf{M}_i$ has no bridges then $R(S) = \sum_{i=1}^{k} r_i(S)$, see [Lovász and Recski, 1973], [Kelmans *et al.*, 1976].

13.2.3 Give three matroids $\mathbf{M}_i = (S, \mathbf{F}_i)$ with $i = 1, 2, 3$ so that

$$\max\{|Y|;\ Y \in \mathbf{F}_1 \cap \mathbf{F}_2 \cap \mathbf{F}_3\} \neq \min\{r_1(A) + r_2(B) + r_3(C);\ A \cup B \cup C = S\}.$$

13.2.4 Let $S \longrightarrow \rho(S)$ be a mapping and $\mathbf{M}, \mathbf{N}$ be two matroids on S. Prove $(\mathbf{M} \vee \mathbf{N})_\rho = \mathbf{M}_\rho \vee \mathbf{N}_\rho$.

Problems for Section 13.2

13.2.5 Let $|S| = k \cdot \ell$ and $\mathbf{M}_1, \mathbf{M}_2$ be two matroids on S with rank ℓ each. Prove that if $\vee_{i=1}^{k} \mathbf{M}_1 = \vee_{i=1}^{k} \mathbf{M}_2 = (S, 2^S)$, i.e. if S partitions into k bases in both matroids, then $\mathbf{M}_1$ and $\mathbf{M}_2$ have a common base as well.

13.2.6 Prove the following strengthening of property (B3) of the bases of a matroid (cf., Problem 7.1.16).
(B3$''$) If $X_1, X_2 \in \mathbf{B}$ and $A_1 \subseteq X_1$ then there exists an $A_2 \subseteq X_2$ so that $(X_1 - A_1) \cup A_2 \in \mathbf{B}$ and $(X_2 - A_2) \cup A_1 \in \mathbf{B}$ both hold.

13.2.7 Let two matroids be coordinatizable over the field $\mathbb{R}$ of the reals. Prove [Piff and Welsh, 1970] that so is their sum.

13.2.8 Using the previous problem what can you say about the coordinatizability of the sum of two matroids if both are coordinatizable over some finite fields? (Hint: The only property of $\mathbb{R}$ we needed was that its cardinality is sufficiently large.)

13.2.9 Let a set P of points be the affine representation of the matroid $\mathbf{M} = (S, \mathbf{F})$ in the $(r(S) - 1)$-dimensional space. Let $P_i \subseteq P$ be arbitrary subsets and let F_i denote the subspace spanned by P_i (for $i = 1, 2, \ldots$). Choose a point x_i in general position from every F_i. Prove [Lovász, 1977] that the rank function R of the matroid whose affine representation is the set $X = \{x_1, x_2, \ldots\}$ is

$$R(X) = \min_{Y \subseteq X} \left[r(\bigcup_{x_i \in X} F_i) + |X - Y| \right].$$

Section 13.3 Submodular functions

In this section we study functions which satisfy (R1),(R3) and (R4) — but not necessarily (R2) — of the properties of the rank functions of matroids. So let $f(X)$ be a function which assigns nonnegative integers to the subsets X of S so that

(a) f is normalized i.e. $f(\emptyset) = 0$;
(b) f is monotone, i.e. $A \subseteq B$ implies $f(A) \leq f(B)$; and
(c) f is submodular, i.e. $f(A) + f(B) \geq f(A \cup B) + f(A \cap B)$ holds for every pair $A, B \subseteq S$.

Such functions will be called *polymatroid functions* for reasons we are not studying in this book. We shall see that they are "almost like matroid rank functions" and that they have very strong relations with the results of the previous two sections.

Observe, for example, that if $\mathbf{M}_i = (S, \mathbf{F}_i)$ are matroids with rank functions r_i on the same set S $(i = 1, 2, \ldots, k)$ then the function $\rho(X) = \sum_{i=1}^{k} r_i(X)$ is a polymatroid function. Of course, $\rho(X) \leq |X|$ is not true in general. Recall that $R(X) = \min_{Y \subseteq X}[\rho(Y) + |X - Y|]$ was the rank function of a matroid, by Theorem 13.2.1. This observation is more generally true:

Theorem 13.3.1 [Edmonds, 1970]. *If $f(X)$ is a polymatroid function then $\tilde{f}(A) = \min_{A_0 \subseteq A}[f(A_0) + |A - A_0|]$ is the rank function of a matroid.*

Proof. $\tilde{f}$ is trivially normalized. For proving that $\tilde{f}$ is monotone, let $A \subseteq B$ and let $\tilde{f}(B) = f(X) + |B - X|$ for some X. Then, choosing $A_0 = X \cap A$, we have $\tilde{f}(A) = \min_{A_0 \subseteq A}[f(A_0) + |A - A_0|] \leq f(X \cap A) + |A - X|$ and since $f(X \cap A) \leq f(X)$ and $|A - X| \leq |B - X|$, the statement follows.

Proving submodularity is somewhat longer (Problem 13.3.6). Finally, we must show $\tilde{f}(A) \leq |A|$ for every A. But this is trivial with the choice $A_0 = \emptyset$ in $\tilde{f}(A) = \min[f(A_0) + |A - A_0|]$.

□

Theorem 13.3.2 [Edmonds, 1970]. *Consider the matroid determined by $\tilde{f}$ of the previous theorem. A subset $X \subseteq S$ is independent in it if and only if $|Y| \leq f(Y)$ holds for every $Y \subseteq X$.*

Proof. X is independent if and only if $|X| = \tilde{f}(X) = \min_{Y \subseteq X}[f(Y) + |X - Y|]$, i.e. if and only if $f(Y) + |X - Y| - |X| \geq 0$ for every $Y \subseteq X$.

□

Observe that in the above example $\rho(X) = \sum_{i=1}^{k} r_i(X)$ and $\tilde{\rho}(X)$ is the rank function of $\vee_{i=1}^{k}\mathbf{M}_i$. Hence we obtained a new proof for Theorem 13.2.1 as well.

Submodular functions play an important role in discrete optimization (more about this in Section 17.1), somewhat similarly to the role of convex functions in continuous optimization. This similarity is illustrated by a recent result. It is intuitively clear (see Fig. 13.9) that if $f_1(x)$ is a convex function and $f_2(x)$ is a concave one and $f_2(x) \leq f_1(x)$ holds for every x then there is a linear function $f(x)$ meeting $f_2(x) \leq f(x) \leq f_1(x)$.

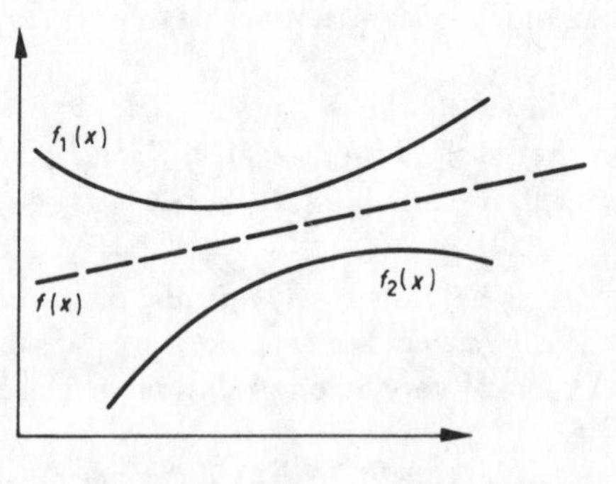

Fig. 13.9

Let us call a set function $f(x)$ *supermodular* if $-f(X)$ is submodular. A function being sub- and supermodular at the same time is *modular*; i.e. f is modular if $f(X) + f(Y) = f(X \cup Y) + f(X \cap Y)$ holds for every $X, Y \subseteq S$. Modular functions have a very simple structure see Exercise 13.3.5.

Theorem 13.3.3 [Frank, 1982]. *Let f_1 and f_2 be integer-valued functions on the same set; with $f_1(\emptyset) = f_2(\emptyset) = 0$. Let f_1 be submodular and f_2 be supermodular, and let $f_2(X) \leq f_1(X)$ for every $X \subseteq S$. Then there exists an integer-valued modular function $f(X)$ satisfying $f_2(X) \leq f(X) \leq f_1(X)$ for every X.*

Its *proof* is not included here. The original proof by the author is lengthy but constructive. A shorter but nonconstructive proof can be found in [Lovász and Plummer, 1986]. This theorem — also called *sandwich-theorem* — implies many results of the previous section (see Problem 13.3.9).

Exercises for Section 13.3

13.3.1 Let f be the rank function of a matroid $\mathbf{M}$. What matroid results by Theorem 13.3.1 from $\tilde{f}$?

13.3.2 Let r be the rank function of a matroid $\mathbf{M} = (S, \mathbf{F})$. Is the function $f(X) = r(S - X)$ submodular?

13.3.3 Let r be monotone and submodular. Prove that $f_k(X) = \min[k, r(X)]$ is also submodular if k is fixed.

13.3.4 Let $A_1, A_2, \ldots, A_t$ be a collection of subsets of a set S. Define a function $f(X) = r(\cup_{i \in X} A_i)$ on the set $T = \{1, 2, \ldots, t\}$, where r was the rank function of a matroid $\mathbf{M} = (S, \mathbf{F})$. Prove that f is submodular. (Every polymatroid function arises in this way from a suitable matroid [Lovász, 1973; Helgason, 1974; McDiarmid, 1975].)

13.3.5 Let $\rho : S \longrightarrow \mathbf{R}$ be an arbitrary real-valued function and let $f(X) = c + \sum_{x \in X} \rho(x)$. Prove that f is modular and that every modular function arises in this way.

Problems for Section 13.3

13.3.6 Prove that $\tilde{f}$, as defined in Theorem 13.3.1, is submodular.

13.3.7 Give examples to normalized monotone but not submodular functions f so that the corresponding functions $\tilde{f}$
(a) do not arise as the rank function of a matroid,
(b) do arise as the rank function of a matroid.

13.3.8 Let f be a nonnegative function on the subsets of S, satisfying (a) and (c).
(a) Prove that $\overline{f}(A) = \min_{A \subseteq X} f(X)$ is a polymatroid function.
(b) Prove that $f'(A) = \min_{Y \subseteq S}[f(Y) + |A - Y|]$ is the rank function of a matroid.

13.3.9 Obtain the nontrivial part of Theorem 13.2.2 from the sandwich-theorem.

13.3.10 Prove that f is submodular if and only if $f(X) - f(X - \{a\})$ is monotone decreasing, i.e. if

$$f(X) - f(X - \{a\}) \geq f(X \cup Y) - f((X \cup Y) - \{a\})$$

holds for every $X, Y \subseteq S$ and $a \in X$.

13.3.11 Let $|S| \geq n + 2$ and define the function $f_n(X)$ for the subsets X of S as follows.
If $|X| \leq n$ then let $f_n(X) = 2 \cdot |X|$; if $|X| \geq n+2$ then let $f_n(X) = 2n+2$; and for the $(n+1)$-element subsets X choose the values $f(X)$ randomly from $\{2n+1, 2n+2\}$.
Prove that all these functions f_n are submodular [Lovász, 1981].

13.3.12 Let f_1, f_2 be two submodular functions on the same set and let $f_1 - f_2$ be monotone. Prove that $f(X) = \min(f_1(X), f_2(X))$ is submodular.

Chapter 14

Applications

Section 14.1
Unique solvability of linear active networks II: Necessary and sufficient conditions for "general" networks

A necessary condition for the unique solvability of multiport networks was given in Theorem 12.2.2. This condition was not always sufficient, see Exercise 12.2.5 and Problem 12.2.8. Prescribing some "generality"-type condition (i.e., ensuring that the nonzero expansion members of det $\mathbf{W}$ cannot cancel out each other) we may hope to make this condition sufficient as well.

When Kirchhoff's classical results were extended to networks, containing negative resistors as well (see Section 6.2), we simply prescribed that the values of the resistances (as real numbers) be algebraically independent over the field $\mathbb{Q}$ of the rationals. Analogously, we may now formulate the so called *strong generality assumption* [Iri and Tomizawa, 1975]: Suppose that all the nonzero entries in the matrices of the interconnected multiports are algebraically independent over $\mathbb{Q}$. This means that algebraic relations are forbidden both within the matrices of single multiports, and among entries of matrices of different multiports.

This assumption does not take actual numerical values of the matrix entries into consideration, it concentrates on the zero-nonzero pattern of the matrices only. This assumption obviously ensures the sufficiency of the condition of Theorem 12.2.2.

How realistic is this assumption? The original generality assumption in Section 6.1 gave a restriction to the set of devices to be interconnected. (For example, two resistors with the same resistance — or with any other exact numerical relation between their resistances — were excluded.) However, the above strong generality assumption refers not to the devices but to their describing equations. One can easily construct two, mathematically equivalent matrix descriptions for one and the same device so that this assumption is met in one of the two cases only (Exercise 14.1.1). Another problem is that this assumption is too restrictive; such important devices like ideal transformers or gyrators were also excluded since their transfer ratios (gyrator constants, respectively) arise in 2 equations each.

Some of these difficulties can be avoided by using the so called *weak generality assumption* [Recski, 1978]: Suppose that among the nonzero entries of the multiport matrices such algebraic relations are possible only, which are reflected by the structure of the matroid $\mathbf{A}$, modelling the multiports. This means that certain — not arbitrary — relations within the matrices of single multiports are allowed while other algebraic relations among entries of matrices of different multiports are still prohibited.

This assumption just corresponds to the weakening of the condition of Theorem 11.2.2 as described in the solution of Problem 11.2.5. Hence we obtain the following result.

Statement 14.1.1 *The necessary condition for the unique solvability of multiport networks as given in Theorem 12.2.2, is also sufficient if the weak generality assumption holds.*

One has to prove that this weak generality assumption is really weaker than the "strong" one (Exercise 14.1.2), and also that this assumption depends only on the actual device and not on the actual matrix description of it (Problem 14.1.6). It is also possible that a device satisfies the weak generality assumption but does not satisfy the strong one by any matrix description (see Problem 14.1.7). This problem will be further studied in Section 16.1.

Although the weak generality assumption excludes less devices than the strong one, some difficulties still remain. One can find such pathological situations where algebraic relations among nonzero entries within the matrix of a single multiport do exist but are not reflected in the structure of the matroid of the multiport; yet they may cause singularity in the network (see Problem 14.1.8). We return to this problem in Section 18.1.

Exercises for Section 14.1

14.1.1 Show that the matrices

$$\mathbf{A}' = \begin{pmatrix} -1 & 2 & 0 & 0 \\ 0 & -1 & 2 & 1 \end{pmatrix} \text{ and } \mathbf{A}'' = \begin{pmatrix} -1 & 1 & 2 & 1 \\ 0 & -1 & 2 & 1 \end{pmatrix}$$

both describe the 2-port of Fig. 14.1. Replace the nonzero entries of these matrices by algebraically independent numbers over $\mathbb{Q}$ and determine the matroids $\mathbf{M}'$ and $\mathbf{M}''$, respectively, of the resulting matrices. Is this 2-port legitimate in case of the strong generality assumption?

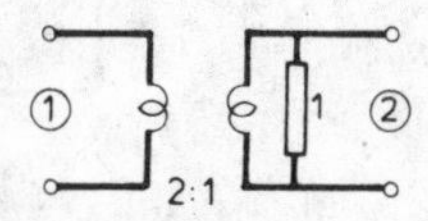

Fig. 14.1

14.1.2 Explain the reason of calling the two generality assumptions as "weak" and "strong".

14.1.3 Suppose the weak generality assumption holds. Determine the maximum number σ of capacitor voltages and inductor currents whose initial values can independently be prescribed, in a network containing multiports, capacitors and inductors. (Hint: Use the solution of Problem 12.2.11.)

14.1.4 (continued) Give an algorithm to find such a maximum system.

14.1.5 Apply the result of Exercise 14.1.3 for the network of Fig. 14.2, where S_5 is a short circuit.

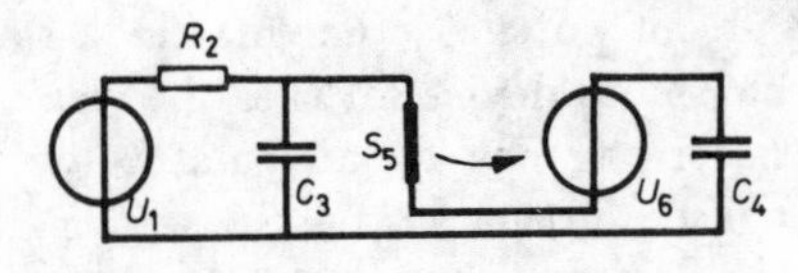

Fig. 14.2

Problems for Section 14.1

14.1.6 Prove that the weak generality assumption refers to the devices and not to their matrix descriptions.

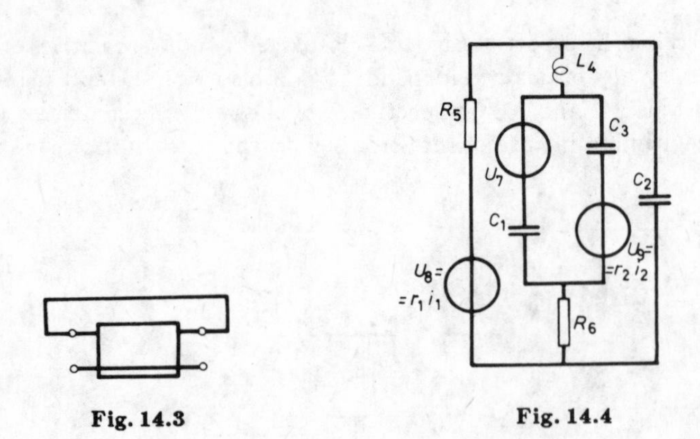

Fig. 14.3 **Fig. 14.4**

14.1.7 Prove that considering any matrix description of the 3-port circulator, there will be some algebraic relations among the nonzero entries of the matrix.

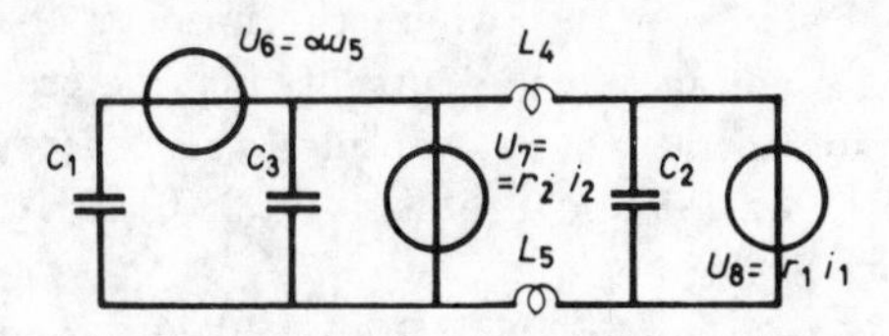

Fig. 14.5

14.1.8 Consider the "network" of Fig. 14.3 where the 2-port has impedance matrix $\mathbf{Z} = \begin{pmatrix} 0 & R_1 \\ R_2 & 0 \end{pmatrix}$ with $R_1 \neq 0, R_2 \neq 0$. Does the network have unique solution?

14.1.9 Show that the analogue of the statement of Exercise 2.4.2 is not true for networks, containing multiports as well.

14.1.10 Determine σ (see Exercise 14.1.3) for the network of Fig. 14.4 [Tow, 1968], under the weak generality assumption.

14.1.11 The same problem for the network of Fig. 14.5 [Mark and Swamy, 1976].

Section 14.2 Interconnection of multiports

We have seen that multiports, modelling really existing physical devices (like transformers, gyrators, controlled sources, circulators) have at least one hybrid immitance description (see Section 12.1). If, on the other hand, a multiport is obtained by mathematically considering a matrix — rather than modelling a device — then multiports with no hybrid descriptions can also arise, like the pair of nullator and norator, for example. They are also interesting (e.g., for modelling other devices, see Figs 12.2–12.5) but do not really exist.

A "really existing" linear device always has at least one hybrid immitance description, due to some physical reasons. However, this property is not closed with respect to interconnection (see Problem 12.1.18, for example).

This indicates a significant lack of our knowledge in circuit theory. We use multiports for modelling devices; and if the interconnection of these devices gives an electric circuit, it would be a reasonable requirement that the circuit be modelled by the electric network, interconnected from these multiports (see also the diagram at the beginning of Section 2.1).

Unfortunately, it does not always work. There are some classes of devices (like passive ones, reciprocal ones) which are known — for physical reasons — to be closed with respect to interconnection; however, these classes are not broad enough from the point of view of practical applications. On the other hand, the class of all multiports, having at least one hybrid immitance description, is "too broad"; the interconnection of such devices may not have this property.

(Incidentally, if the even broader class of all multiports is considered, it is still not closed with respect to interconnection, see Problem 14.2.9, for example. The still broader class of *generalized multiports* is closed with respect to interconnection; they are "n-ports" where the number of linearly independent equations among the n voltages and n currents may be greater or less than n as well. Nullators and norators are such examples, their theory could be developed further, see Exercises 14.2.4–5 and Problems 14.2.10–11, but they do not model real devices any more.)

In the case of practical applications this problem has to be solved somehow. In the process of computer-aided network design, the multiport models of the devices are stored, and the modelling of the circuit is tried at first by interconnecting these multiports. If these multiports cannot be interconnected (i.e., if the result is either a generalized multiport only, or a "normal" one but without possessing any hybrid immitance description) then the computer program should indicate this.

Such an indication might have several meanings. *Either* the real devices cannot be interconnected (for example if there was an input error in the specification of the network) *or* the devices can be interconnected but — at least for this interconnection — the modelling assumptions were wrong, the devices should be modelled in a different way. Evaluating such an indication is the task of the engineer who uses this program (and has some insight on the physical behaviour of the circuit). The task of the mathematician is only to develop a quick algorithm which gives this indication if necessary.

Statement 14.2.1 *Let the multiport N be obtained by the interconnection of the multiports $N_1, N_2, \ldots$ along the network graph G. If the weak generality assumption holds then $\mathbf{M}_N = (\mathbf{G} \vee \mathbf{A})/B$, where $\mathbf{G}$ was defined by Eq. (12.1); $\mathbf{A} = \mathbf{M}_{N_1} \oplus \mathbf{M}_{N_2} \oplus \ldots$ and B is the set of port voltages and port currents of the multiports $N_1, N_2, \ldots$.*

Recall that different edges correspond to the ports of the multiports $N_1, N_2, ..$ and to those of N in the graph G (see Section 10.1). Hence if we contract B from the set S (i.e., from the set of voltage and current copies of $E(G)$) then the result is the set K of port voltages and currents of the resulting multiport N.

For example, if N is the 2-port of Fig. 14.6 then $B = \{u_1, u_2, u_3, i_1, i_2, i_3\}$; $K = \{u_A, u_B, i_A, i_B\}$; and the matroids $\mathbf{M}_{N_1}, \mathbf{M}_{N_2}$ and $\mathbf{G}$ are visualized as the circuit matroids of the three graphs of Fig. 14.7. In this example $\mathbf{G} \vee \mathbf{A}$ happens to be graphic (see the circuit matroid of the graph of Fig. 14.8). Finally if we contract the elements, corresponding to internal port voltages and currents, we obtain the circuit matroid of the graph of Fig. 14.9, in agreement with the solution of Exercise 14.1.1.

The *proof* of Statement 14.2.1 is trivial. $\mathbf{G} \vee \mathbf{A}$ describes the interconnected network, by Statement 14.1.1, and by Theorem 11.2.2 (we apply the weak generality assumption at this point). Exercise 9.1.2 gives then that contraction of B means the elimination of the internal variables. □

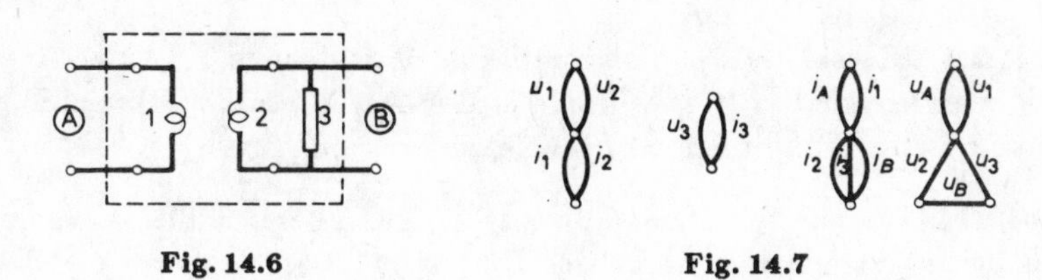

Fig. 14.6 **Fig. 14.7**

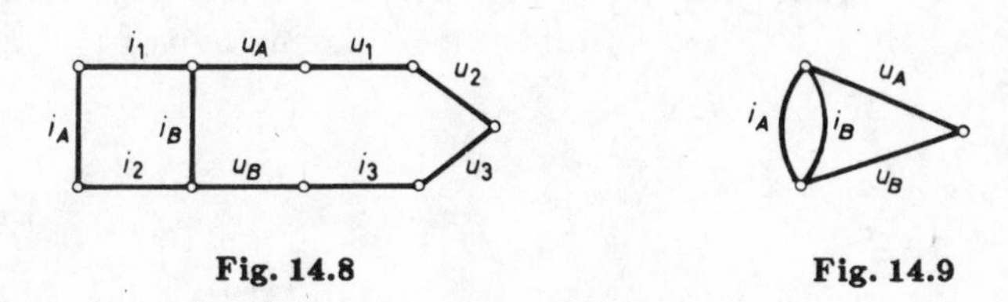

Fig. 14.8 **Fig. 14.9**

Now we can start preparing a quick algorithm to check whether certain multiports can be interconnected. Using the previous statement and Statement 12.1.1 we obviously obtain

Statement 14.2.2 *Let us interconnect the multiports $N_1, N_2, \ldots$ along the network graph G. If the weak generality assumption holds then the result will be an n-port with at least one hybrid immitance description if and only if the rank of $(\mathbf{G} \vee \mathbf{A})/B$ is n and $[(\mathbf{G} \vee \mathbf{A})/B] \vee \mathbf{B}_n$ is the free matroid. ($\mathbf{B}_n$ was defined at Statement 12.1.1, everything else in Statement 14.2.1.)*

The real question is, how can we quickly check this condition. Recall that the underlying set $S = E_i \cup E_u$ of the matroid $\mathbf{G}$ decomposes as $B \cup K$, where B corresponds to the internal ports (to be eliminated) and K to the external ports (n in number). Observe that the underlying set of $\mathbf{A}$ is essentially B only (since every element of K is a loop in $\mathbf{A}$) and that the underlying set of $\mathbf{B}_n$ is K only. We may extend the definition of $\mathbf{B}_n$ to S (by adding loops for each element of B). Then the following observation is correct:

Theorem 14.2.3 *If $\mathbf{G} \vee \mathbf{A} \vee \mathbf{B}_n = (S, 2^S)$ then the second condition of Statement 14.2.2 necessarily holds.*

Proof. If S decomposes as $S_1 \cup S_2 \cup S_3$ where these sets are independent in $\mathbf{G}$, in $\mathbf{A}$ and in $\mathbf{B}_n$, respectively, then $S_2 \subseteq B$ and $S_3 \subseteq K$. We further decompose S_1 into $(S_1 \cap B) \cup (S_1 \cap K)$. Then $(S_1 \cap B) \cup S_2$, i.e. B itself, is independent in $\mathbf{G} \vee \mathbf{A}$ and $S_1 \cap K$ is independent in $(G \vee A)/B$. Since $K = S_3 \cup (S_1 \cap K)$, the statement follows.

□

This condition can easily be checked by Algorithm 13.1. However, this condition $\mathbf{G} \vee \mathbf{A} \vee \mathbf{B}_n = (S, 2^S)$ is sufficient but not necessary [Nakamura, 1979], see Problem 14.2.12.

Theorem 14.2.4 [Recski, 1979]. *The conditions of Statement 14.2.2 are met if and only if every element of K is a bridge in $\mathbf{G} \vee \mathbf{A} \vee \mathbf{B}_n$ and $r(B) = r(S) - n$ holds, where r is the rank function of $\mathbf{G} \vee \mathbf{A}$ and $n = |K|/2$.*

Its *proof* is left to the reader as Problem 14.2.13. These conditions can also be checked by polynomial algorithms (see Exercise 14.2.8).

Exercises for Section 14.2

14.2.1 Let the 2-port N in box D of Fig. 14.10 be a pair of nullator and norator ($i_2 = u_2 = 0$). Then the "resulting" 1-port is a norator. Can we obtain a norator also if N is a suitable 2-port having hybrid immitance description?

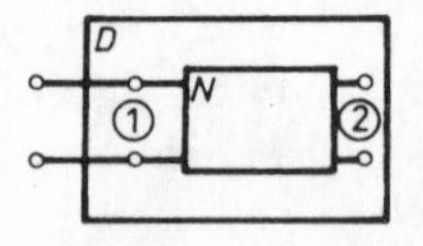

Fig. 14.10

14.2.2 (a) Show by an example that Statement 14.2.1 is not always true if the generality assumption is not postulated.
(b) Can we still say something (instead of equality) in this case?

14.2.3 Let $i_A = 0, u_B = ku_A$ and $i_C = 0, u_D = u_C$ be the equations of two 2-ports. They are interconnected in cascade in two different ways in Fig. 14.11. Determine whether the two resulting 2-ports have hybrid immitance descriptions.

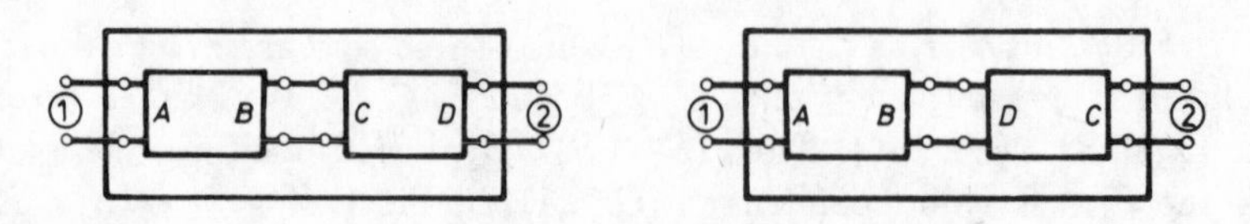

Fig. 14.11

Box 14.1. Extension of **Box 10.1**

N	neg N	inv N	dual N	adj N
Nullator	Nullator	Nullator	Nullator	Norator
Norator	Norator	Norator	Norator	Nullator
Pair of nullator & norator	Pair of nullator & norator	Pair of nullator & norator	Pair of nullator & norator*	Pair of nullator & norator*

* The role of the two ports is interchanged.

14.2.4 Check the statements in **Box 8.2** referring to nullators, norators, and pairs of nullator and norator.

14.2.5 Extend **Box 10.1** by nullators, norators, and pairs of nullator and norator. (See **Box 14.1**).

14.2.6 "Generalize" the last row of **Box 6.3** by characterizing those 2-ports whose cascade connection does not modify a pair of nullator and norator.

14.2.7 How can you determine the bridges in the sum of two matroids?

14.2.8 Give a polynomial algorithm to check the conditions of Theorem 14.2.4.

Problems for Section 14.2

14.2.9 Give an example that the interconnection of 2-ports leads to a generalized multiport only.

14.2.10 Extend the concepts of negative, inverse, dual and adjoint to generalized multiports as well.

14.2.11 Are the statements of Section 10.1 also valid for generalized multiports?

14.2.12 Show by an example that the condition of Theorem 14.2.3 is not necessary for the conditions of Statement 14.2.2 to hold.

14.2.13 Prove Theorem 14.2.4.

14.2.14 Prove by examples that the two conditions in Theorem 14.2.4 are independent from each other.

Section 14.3 Reconstruction of polyhedra from projected images

How can we reconstruct a polyhedron if only its projection (say, from above) is known? (We suppose that the sets of vertices, edges and faces, with a list of their incidences is known.) In the case of a tetrahedron (Fig. 14.12) if its image is given (lower part of the figure) then the height of all the four vertices must be given. On the other hand, in the case of a pyramid (Fig. 14.13) the vertices

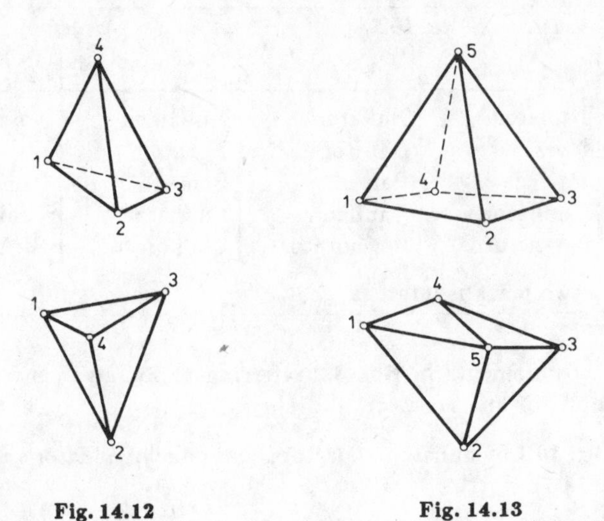

Fig. 14.12 Fig. 14.13

1, 2, 3, 4 are coplanar, hence the height of only three of them must be given (plus that of vertex 5, of course).

This problem has several applications. When preparing or checking an engineering drawing, the question directly arises in case of complicated polyhedra. Another example is computer aided design. The user may specify a polyhedron by drawing its projection to a graphical display and prescribing the height of some vertices separately. In this case he/she must know how many further data can yet be prescribed without causing a contradiction.

In this latter case (if we define the polyhedron by a projection), even the question whether such polyhedron exists at all, can be nontrivial. The reader might have seen funny drawings like those in Fig. 14.14 (see [Sugihara, 1979] for a larger choice). Their mistakes can be detected by an analysis of the incidences of vertices, edges and faces. However, a picture might look correct at first. For example, the two drawings of Fig. 14.15 are wrong; the extensions of the three bold edges should intersect each other in one and the same point. We return to this problem at the end of this section.

Let $F_1, F_2, \ldots, F_\ell$ be the faces and $V_1, V_2, \ldots, V_p$ be the vertices of a polyhedron. The coordinates of the vertex V_i are (x_i, y_i, z_i) and the equation of the plane of face F_j is $\alpha_j x + \beta_j y + \gamma_j z + \delta_j = 0$. We shall suppose throughout that projection is performed parallel to the z-axis (to the xy-plane) and that the polyhedron is in general position with respect to the system of coordinates, hence, in particular, none of the faces is perpendicular to the xy-plane. Hence, in the above equation of the plane of face F_j we have $\gamma_j \neq 0$. Introducing the notation $\alpha_j/\gamma_j = a_j; \beta_j/\gamma_j = b_j$ and $\delta_j/\gamma_j = d_j$, this equation can be written as $a_j x + b_j y + z + d_j = 0$. The vertex V_i is incident to the face F_j if and only if

Fig. 14.14

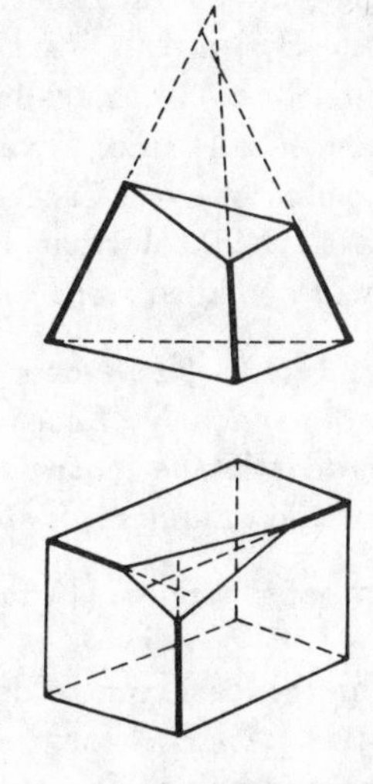

Fig. 14.15

$$a_j x_i + b_j y_i + z_i + d_j = 0 \tag{14.1}$$

If the polyhedron is to be reconstructed from its projection then the pairs (x_i, y_i) are known (they are contained in the projection). The quantities a_j, b_j, d_j are unknown (just as z_i's are). For example, our system of equations for the tetrahedron of Fig. 14.12 will be

$$\begin{array}{ccc|ccc|ccc|ccc|cccc}
\multicolumn{3}{c|}{F_1} & \multicolumn{3}{c|}{F_2} & \multicolumn{3}{c|}{F_3} & \multicolumn{3}{c|}{F_4} & z_1 & z_2 & z_3 & z_4
\end{array}$$

$$\left(\begin{array}{ccc|ccc|ccc|ccc|cccc}
x_1 & y_1 & 1 & & & & & & & & & & 1 & & & \\
x_2 & y_2 & 1 & & & & & & & & & & & 1 & & \\
x_3 & y_3 & 1 & & & & & & & & & & & & 1 & \\
 & & & x_1 & y_1 & 1 & & & & & & & 1 & & & \\
 & & & x_2 & y_2 & 1 & & & & & & & & 1 & & \\
 & & & x_4 & y_4 & 1 & & & & & & & & & & 1 \\
 & & & & & & x_1 & y_1 & 1 & & & & 1 & & & \\
 & & & & & & x_3 & y_3 & 1 & & & & & & 1 & \\
 & & & & & & x_4 & y_4 & 1 & & & & & & & 1 \\
 & & & & & & & & & x_2 & y_2 & 1 & & 1 & & \\
 & & & & & & & & & x_3 & y_3 & 1 & & & 1 & \\
 & & & & & & & & & x_4 & y_4 & 1 & & & & 1
\end{array}\right)
\begin{pmatrix} a_1 \\ b_1 \\ d_1 \\ a_2 \\ b_2 \\ d_2 \\ a_3 \\ b_3 \\ d_3 \\ a_4 \\ b_4 \\ d_4 \\ z_1 \\ z_2 \\ z_3 \\ z_4 \end{pmatrix} = \mathbf{0}$$

where the zero entries of the matrix are not shown. In general, the matrix can be partitioned by a vertical line into two parts, consisting of 3ℓ and p columns, respectively. (The number of rows need not be 3ℓ since the faces need not be all triangles, in general.) For every projection P of polyhedra we can assign a matrix $(\mathbf{A}_P|\mathbf{B}_P)$ in this way; or a matroid $\mathbf{M}_P$, formed by its column vectors.

The first 3ℓ columns are linearly independent (one could directly prove this, see Exercise 14.3.1) since if we knew the z_i's (in addition to all the x_i's and y_i's) then the plane of every face F_j would be known. Such a system of equations always have trivial solutions, for example where all the vertices are coplanar. A solution will be called *nontrivial* if no two faces F_j are in the same plane.

Statement 14.3.1 *The z-coordinates of the vertices $V_{i_1}, V_{i_2}, \ldots, V_{i_k}$ can be prescribed independently of each other if and only if the set $X = \{z_{i_1}, z_{i_2}, \ldots, z_{i_k}\}$ is independent in the matroid $\mathbf{M}_P$. If X is a base then these data are already enough for the unique reconstruction of the projection P.*

The proof is trivial. (In fact, we did not need matroids for this so far.) □

Even in this case, matroids are useful to visualize the situation. For example, if the matrix $(\mathbf{A}_P|\mathbf{B}_P)$ for the projection P of Fig. 14.13 is considered then the part, corresponding to $\mathbf{B}_P$ can be visualized as the circuit matroid of the graph of Fig. 14.16, since z_5 and three further z_i's are to be prescribed. A somewhat less trivial example is the projection shown in Fig. 14.17. For its reconstruction the z-coordinate of suitable four among the six points should be prescribed; the remaining two then follow. The corresponding matroid of rank 4 on the 6-element set is shown both by a graph and by a geometric representation in Fig. 14.18 (see Exercise 14.3.2).

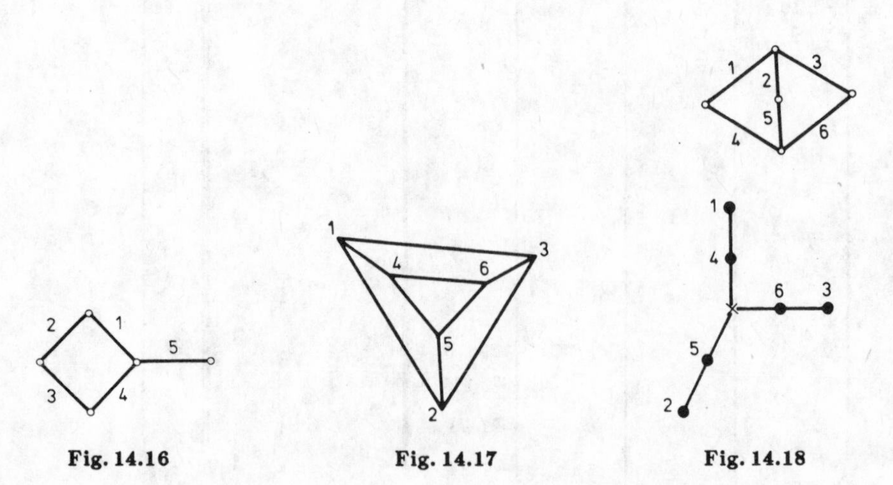

Fig. 14.16 **Fig. 14.17** **Fig. 14.18**

Now we are in the position of considering the problem, mentioned in connection with Fig. 14.15. A projection (actually a list of incidences of the vertices, edges and faces) is called *regular* if the system of equations (14.1), in which $x_1, x_2, \ldots, x_p, y_1, y_2, \ldots, y_p$ are algebraically independent over the field $\mathbb{Q}$ of the rationals, has a nontrivial solution. In this case, it is not only the projection of an appropriate polyhedron but even if we slightly perturbate its vertices, the result still corresponds to a polyhedron.

For example, the lower parts of Figs 14.12–13 are regular projections, Fig. 14.17 is not since perturbation of vertex 6 would lead to the upper part of Fig. 14.15.

One can prove [Sugihara, 1979] that a projection P of a convex polyhedron is regular if and only if the rows of the matrix $(\mathbf{A}_P|\mathbf{B}_P)$ are linearly independent over $\mathbb{Q}$ (while $x_1, x_2, \ldots, x_p, y_1, y_2, \ldots, y_p$ are algebraically independent over $\mathbb{Q}$). However, this condition cannot be checked directly in polynomial time, see Problem 5.4.6(c).

In the examples so far, we could prescribe the z-coordinate of 4 vertices. In general *at least* four z-coordinates are required (Problem 14.3.4). Hence, if X is an arbitrary subset (of cardinality at least two) of the set $F = \{F_1, F_2, \ldots, F_\ell\}$ of faces, and equations like (14.1) are considered for the X-faces then the number of linearly independent equations (i.e., the rank of the submatrix) must be smaller than the number of columns by at least four. The number of columns is obviously $3|X| + v_X$ where v_X is the number of vertices, incident to the X-faces.

In the case of a regular projection the above rank is just the number r_X of incidences to the X-faces, i.e. just the number of equations referring to the X-faces, since no linear dependence is possible among the rows. For example, in the case of the lower part of Fig. 14.13 we have $|F| = 5$, and in case of $X = F$ we obtain $r_X = 16$ (since we have a quadrangle and four triangles) and $v_X = 5$. Hence the relation $r_X \leq 3|X| + v_X - 4$ is met with equality. On the other hand this relation does not hold at all, with choice $X = F$, for the nonregular projection of Fig. 14.17, since $r_X = 18, v_X = 6$ and $|X| = 5$. We obtained:

Statement 14.3.2 *If F is the set of faces of a regular polyhedron then*

$$r_X \leq 3|X| + v_X - 4 \tag{14.2}$$

holds for every $X \subseteq F$ provided $|X| \geq 2$.

It is far less trivial that the converse of this theorem is also true for certain classes of polyhedra, e.g. for convex ones.

Theorem 14.3.3 [Sugihara, 1979]. *A convex polyhedron is regular if and only if (14.2) holds for every $X \subseteq F, |X| \geq 2$.*

This theorem (whose *proof* is not included) is a good characterization only, but does not directly give a polynomial algorithm (the number of conditions to be checked is nearly 2^ℓ). But such an algorithm is also available:

Theorem 14.3.4 [Sugihara, 1979]. *The necessary and sufficient condition for the regularity of the projection of a convex polyhedron P with ℓ faces is that, for every choice of $i, j \leq \ell$, $i \neq j$, the deletion of the columns corresponding to the* *unknown quantities a_i, b_i, a_j, b_j from the matrix $(\mathbf{A}_P|\mathbf{B}_P)$ does not decrease the term rank r_L.*

Observe that this theorem requires $\binom{\ell}{2}$ checks of term ranks (searches for 1-factors in bipartite graphs) hence it gives a polynomial algorithm.

Proof. The sufficiency is trivial: Let $X \subseteq F$; $|X| \geq 2$ and $F_i, F_j \in X$. If the term rank r_L does not decrease by deleting the columns a_i, b_i, a_j, b_j from the matrix $(\mathbf{A}_P|\mathbf{B}_P)$ then, restricting us to the equations, corresponding to the X-faces, the term rank r_X does not decrease either, by deleting these columns. Hence (14.2) follows. The proof of the necessity is more complicated (Problem 14.3.6).

□

The reader can easily verify (using Exercise 13.3.5) that the function $3|X| + v_X - r_X$ is submodular. Hence this algorithm decided whether the minimum of this submodular function is less than four or not. More about minimizing a submodular function will be mentioned in Section 17.1.

One can find some similarity between Section 12.3 and the present one. We had a rather natural necessary condition (for rigidity and for reconstructibility, respectively) which turned out to be sufficient if genericity (resp. regularity) was postulated. The theorems gave good characterizations and could be used for polynomial algorithms as well.

However, not only the form but the content of the two topics are also related:

Theorem 14.3.5 [Maxwell, 1864]. *A planar framework with v joints and $2v - 3$ rods is rigid if and only if no part of it is the projection of a 3-dimensional polyhedron.*

For example, among the two frameworks of Fig. 6.12 the first one is rigid, the second is not, since this second is the projection of a triangular prism.

Exercises for Section 14.3

14.3.1 Give a direct proof that the column vectors of $\mathbf{A}_P$ are linearly independent.

14.3.2 Check that the matroid of Fig. 14.18 corresponds to the projection shown in Fig. 14.17.

14.3.3 Give a new solution to Problem 6.3.7.

Problems for Section 14.3

14.3.4 Prove that during the reconstruction of a polyhedron from its projection at least four z-coordinates can independently be prescribed.

14.3.5 Give two polyhedra with regular projections so that, in the first case, the relation (14.2) be met by equality for every $X \subseteq F$; $|X| \geq 2$ and in the second case not. Prove that in the first case the projection contains enough information to decide about any four points whether they are coplanar while in the second case it is not necessarily the case.

14.3.6 Prove the necessity of the condition of Theorem 14.3.4.

Chapter 15

Matroids induced by graphs

Section 15.1 Transversal matroids

Consider a bipartite graph G with bipartition $X_1 \cup X_2$. Recall Hall's theorem: G has a matching which covers every point of X_1 if and only if $|N(X)| \geq |X|$ holds for every $X \subseteq X_1$, where $N(X)$ denotes the set of neighbours of X.

Even if such a matching does not exist, one can study those subsets of X_1 which can be covered by point-disjoint edges of the bipartite graph. For example, in the bipartite graph of Fig. 15.1, exactly those subsets of X_1 are contained in matchings which contain at most three elements from the set $\{1, 2, 3, 4\}$. The "good" subsets of this example are just the independent sets of the circuit matroid of the graph of Fig. 15.2. This observation is true generally.

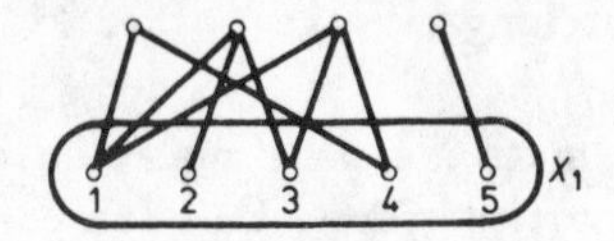

Fig. 15.1

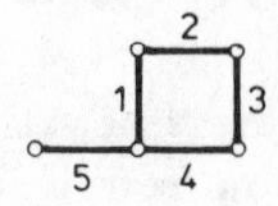

Fig. 15.2

Theorem 15.1.1 [Rado, 1942]. *If S is one of the bipartition classes of a bipartite graph then those subsets of S which are contained in matchings, are the independent sets of a suitable matroid.*

Matroids arising in this way are called *transversal* matroids.

We present two *proofs* for Theorem 15.1.1. First, try to prove directly properties (F1)–(F3) of the system of independent sets of a matroid for our system of subsets (contained in matchings). (F1) and (F2) are trivial. For (F3) let $X, Y \subseteq S$, $|X| = |Y| + 1$ and let E_X and E_Y be the edge sets of matchings containing X and Y respectively. Consider their symmetric difference $E_X \triangle E_Y$. It determines a subgraph H consisting of point-disjoint paths and circuits. By $|E_X| = |E_Y| + 1$ this H has an odd number of edges. On the other hand, its circuits are of even length since it is a subgraph of a bipartite graph. Hence at least one of its paths is of odd length. These odd paths are alternating and in at least one of them E_X has more edges than E_Y. One of the terminal points of this does not belong to Y but can be joined to it.

□

This proof is visualized in Fig. 15.3. Let $X = \{1,2,3,4,5,6\}$ and $Y = \{1,2,3,4,7\}$. Only the edges of E_X and E_Y are shown in the figure (by continuous and by dotted lines, respectively). H consists of one circuit and two paths, the odd path shows that point 5 can be added to Y.

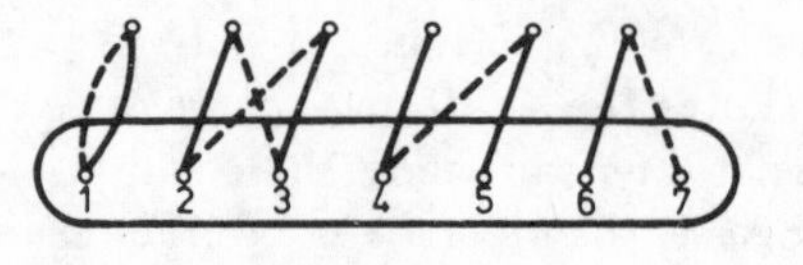

Fig. 15.3

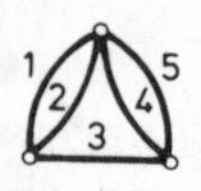

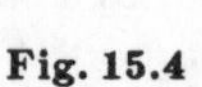

Fig. 15.4

The second proof is shorter but requires many previous results. Consider the function $f(X) = |N(X)|$ for every $X \subseteq S$. This is trivially normalized, monotone, and, by Exercise 5.2.6, submodular. It defines a matroid with rank function $\hat{f}$, where, by Statement 13.3.2, exactly those subsets $X \subseteq S$ are independent which satisfy $f(Y) \geq |Y|$ for every $Y \subseteq X$; but these are, by Hall's theorem, just the subsets contained in matchings.

□

In this section we study transversal matroids, i.e. matroids which arise with this construction from a suitable bipartite graph. One can easily see, for example, that the circuit matroid of the first graph in Fig. 15.4 is transversal (see the bipartite graph in Fig. 15.5) while the second is not (see Exercise 15.1.1).

While the "lower" set of bipartition of the point set of the bipartite graph is considered as the underlying set S of our matroid, the "upper" points could be considered as subsets of S (see Fig. 15.6); an element $x \in S$ belongs to a subset $X \subseteq S$ if and only if the "lower" point x and the "upper" point X are adjacent in the bipartite graph. Hence transversal matroids can be defined by a system $\mathbf{A} = (A_1, A_2, \ldots, A_k)$ of not necessarily different subsets A_i of S so that a subset $T = \{x_1, x_2, \ldots, x_p\}$ of S is independent in the matroid defined by $\mathbf{A}$, if and only if the elements of T can be assigned to different subsets $A_{i_1}, A_{i_2}, \ldots, A_{i_p}$ satisfying $x_j \in A_{i_j}$ for every j. Such a subset T is usually called the *transversal* of the subsystem $(A_{i_1}, A_{i_2}, \ldots, A_{i_p})$; hence the name of transversal matroids.

Consider the matroid shown in Figs 15.5 and 15.6. It can obviously be represented by the matrix $\mathbf{M} = \begin{pmatrix} x & x & x & 0 & 0 \\ 0 & 0 & x & x & x \end{pmatrix}$ over the field $\mathbb{R}$ of the reals, where the x's are arbitrary nonzero entries. The more general statement also holds.

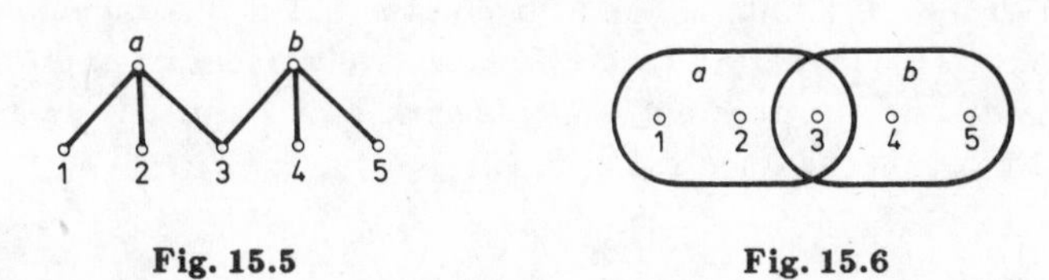

Fig. 15.5

Fig. 15.6

Theorem 15.1.2 *Every transversal matroid can be coordinatized over* $\mathbb{R}$.

Proof. Let the transversal matroid **T** be presented by the bipartite graph, consisting of the "lower" point set $S = \{s_1, s_2, \ldots, s_n\}$ and the "upper" points $a_1, a_2, \ldots, a_k$ (or, alternatively, by the set S and the set system $\mathbf{A} = (A_1, A_2, \ldots, A_k)$. Let $\mathbf{M} = (m_{ij})$ be a $k \times n$ matrix so that its columns correspond to the elements of S and its rows to the "upper" points (or to the subsets A_i in $\mathbf{A}$, respectively), and $m_{ij} = 0$ if $s_j \notin A_i$, while the nonzero entries be algebraically independent over $\mathbb{Q}$.

If $X \subseteq S$ with $|X| = t$ corresponds to a linearly independent set of columns of **M** then any nonzero expansion member of any $t \times t$ submatrix formed by these columns determines a matching in the bipartite graph (or a transversal in the set-system). Hence X is independent in **T**. On the other hand, an independent subset in **T** determines a nonzero expansion member of the corresponding square submatrix of **M**. By the algebraic independence this ensures nonsingularity of this submatrix.

□

Instead of generating algebraically independent entries over $\mathbb{Q}$, one can simply choose integers randomly from the interval $1 \ldots N$. This will also lead to a suitable representation with large probability, if N is large enough, see Problem 15.1.18.

As a byproduct, we obtained

Corollary 15.1.3 *A matroid is transversal if and only if it arises as sum of rank one matroids.*

Proof. The above construction and Theorem 11.2.2 together ensures that every transversal matroid arises as sum of these matroids which are coordinatized by the rows of **M**. On the other hand, if $\mathbf{M}_1, \mathbf{M}_2, \ldots$ are rank one matroids then let A_i denote the set of nonloop elements of $\mathbf{M}_i$ (for $i = 1, 2, \ldots, k$). One can easily see that $\vee_{i=1}^{k} \mathbf{M}_i$ is just the transversal matroid, determined by $\mathbf{A} = (A_1, A_2, \ldots, A_k)$.

□

Using these ideas we can visualize transversal matroids, given by "lower" points $S = \{s_1, s_2, \ldots, s_n\}$ and "upper" points $a_1, a_2, \ldots, a_k$ in a geometric way as well. Fix some points $a_1, a_2, \ldots, a_k$ in the $(k-1)$-dimensional space so that they form a *simplex* (i.e. that none of them be contained in the subspace, spanned by the other $k-1$ points). Then, for each s_i, consider its neighbours in the bipartite graph. These are "upper" points which span a subspace. Put s_i into this subspace but otherwise in general position.

For example, the transversal matroid given by the bipartite graph of Fig. 15.7, is visualized in Fig. 15.8. Points 1, 3 and 2 are in the 0, 1 and 2-dimensional subspaces, spanned by $\{a\}$, $\{a, b\}$ and $\{a, b, c\}$, respectively. General position means that 3 is on the ab line but does not coincide with a or b and 2 is in the plane, but for example, not on the lines 1, 4 or 3, 4.

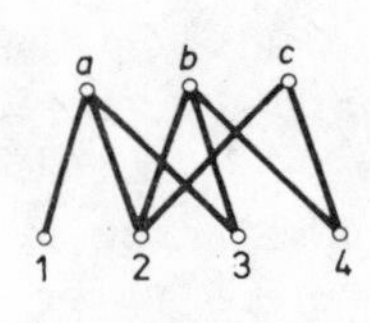

Fig. 15.7

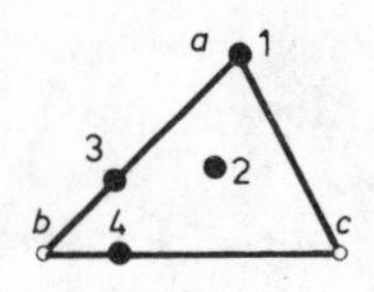

Fig. 15.8

This geometric approach leads to further information as well. For example, the circuit matroid of the complete graph K_4 of four vertices cannot be transversal. Figure 15.9 cannot be obtained by the above construction, whichever three noncollinear points are considered first. If, say, points 1, 2, 3 are considered first (as vertices of the simplex) then Fig. 15.10 cannot be finished.

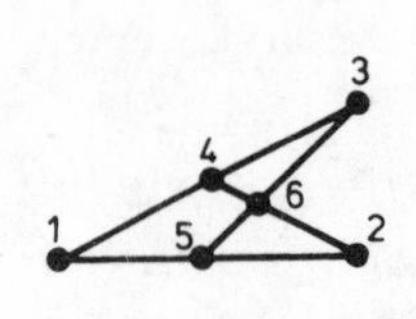

Fig. 15.9

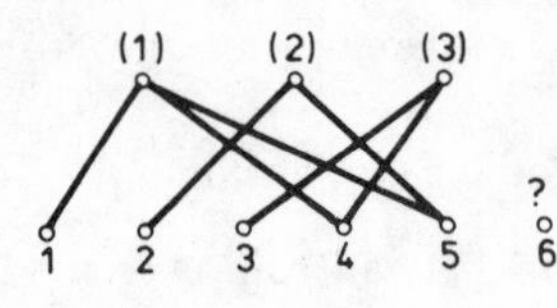

Fig. 15.10

More general results can also be obtained in this way.

Theorem 15.1.4 [Las Vergnas, 1970, Bondy and Welsh, 1971]. *A transversal matroid* $\mathbf{T}$ *of rank* r *can be presented by a set-system* $\mathbf{A} = (A_1, A_2, \ldots, A_k)$ *with* $k = r$. *Furthermore, the subsets* A_i *can be cut sets of* $\mathbf{T}$.

Before sketching the proof observe that one and the same transversal matroid can have several different presentations. In order to obtain the free matroid $\mathbf{U}_{n,n}$, for example, just n disjoint edges are enough but even $K_{n,n}$ is appropriate.

$k \geq r$ is trivial since the bases of $\mathbf{T}$ could have no matchings otherwise. $k > r$ is not necessary since $\mathbf{T}$ can be visualized in the space of dimension $r - 1$.

All the subsets A_i, corresponding to the vertices of the simplex, contain cut sets, since their deletion decreases the rank of the matroid. (Recall that cut sets are just the complements of the rank $r-1$ closed subsets, see Section 9.3.) Finally, if the system $\mathbf{A} = (A_1, A_2, \ldots, A_k)$ is minimal, that is, if for any subscript $j \in \{1, 2, \ldots, k\}$ and for any element $x \in A_j$, the new system

$$\mathbf{A}' = (A_1, A_2, \ldots, A_{j-1}, A_j - \{x\}, A_{j+1}, \ldots, A_k)$$

determines a matroid, different from $\mathbf{T}$, then the elements of $\mathbf{A}$ are cut sets.

Finally, we mention that although every rank r transversal matroid $\mathbf{T}$ can be presented by a set system $\mathbf{A} = (A_1, A_2, \ldots, A_r)$ consisting of r cut sets of $\mathbf{T}$, one cannot expect that these cut sets form just a fundamental system of cutsets with respect to a base of $\mathbf{T}$. If this is the case, $\mathbf{T}$ is called *fundamental transversal matroid* [Bondy and Welsh, 1971].

Exercises for Section 15.1

15.1.1 Prove that the circuit matroid of the second graph of Fig. 15.4 is not transversal.

15.1.2 Prove that the class of transversal matroids is closed with respect to deletion but not for contraction. (One can prove that it is for series contraction [Bondy, 1972].)

15.1.3 Is the Fano matroid $\mathbf{F}_7$ transversal?

15.1.4 Show that neither the dual nor the truncation of a transversal matroid is necessarily transversal.

15.1.5 Prove that a matroid is transversal if and only if it is the sum of partitional matroids.

15.1.6 Prove that the sum of transversal matroids is also transversal.

15.1.7 Show that the class of transversal matroids is closed with respect to series extension but not for parallel extension.

15.1.8 Prove that the circuit matroid of the graph in Fig. 15.11 is transversal but not a fundamental transversal matroid.

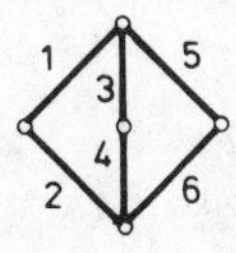

Fig. 15.11

15.1.9 (continued) Give a matrix representation of this matroid (with algebraically independent elements) and visualize it geometrically as well.

15.1.10 Prove that every transversal matroid can be obtained from a suitable fundamental transversal matroid by deleting a suitable subset.

15.1.11 Define a matroid on the *edge set* of a bipartite graph G with bipartition $A \cup B$ so that a subset X of edges be independent if and only if the end points of the X-edges in A are all different. Prove that this is a matroid.

Problems for Section 15.1

15.1.12 Fig. 15.12 is the geometric representation of a rank 3 matroid. Prove that it is not transversal.

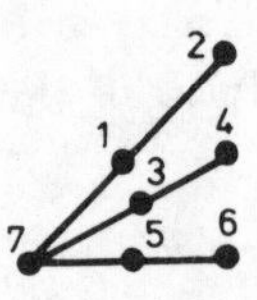

Fig. 15.12

15.1.13 Let r denote the rank function in the circuit matroid of the second graph of Fig. 15.4. Let us define a function f on the set $S = \{1,2,3,4,5,6,7\}$ so that $f(X) = r(X)$ if $7 \notin X$ and $f(X) = 2$ in every other case. Prove [Murty and Simon, 1976] that f is monotone, normalized, submodular, but does not arise as the sum of matroid rank functions.

15.1.14 Let $X \subseteq V(G)$ be an independent subset of the point set $V(G)$ of an arbitrary graph G if all the points in X can be covered by some suitable disjoint edges of G. (Of course, these edges may cover further points as well.) Prove [Edmonds and Fulkerson, 1965] that this is a matroid. (Applying some deep results of [Edmonds, 1965b] one can also show that exactly the transversal matroids arise in this way, see [Edmonds and Fulkerson, 1965].)

15.1.15 Prove that the homomorphic image of a transversal matroid is transversal.

15.1.16 Characterize fundamental transversal matroids in algebraic and in geometric way as well.

15.1.17 Prove [Las Vergnas, 1970] that the class of fundamental transversal matroids is closed with respect to duality.

15.1.18 Let $\mathbf{A}$ be a $k \times n$ matrix whose nonzero entries were randomly chosen from the set of integers $\{1,2,\ldots,N\}$. Let $\mathbf{M}$ be the matroid represented by **A**. Changing the nonzero entries to algebraically independent transcendentals over $\mathbb{Q}$, the new matrix represents a matroid $\mathbf{M}'$. Prove that the probability of the event $\mathbf{M} \neq \mathbf{M}'$ is less than $2^n/N$.

Section 15.2 Matroids induced by bipartite graphs

We saw in the previous section that the edge set of a bipartite graph determines a matroid in one of the bipartition classes of the point set of the graph. This construction can be generalized; if G is a bipartite graph with bipartition $V(G) = A \cup B$ then $E(G)$ and a matroid defined on A together determine a matroid on B.

Theorem 15.2.1 [Perfect, 1969]. *Let G be a bipartite graph with bipartition $V(G) = A \cup B$ and let **A** be a matroid on A. Consider those subsets of B which, via disjoint edges of G, can be matched with a subset of A which is independent in **A**. Then the independent subsets of a matroid on B are obtained.*

For example, if G is the bipartite graph in Fig. 15.13 and **A** is just the free matroid $\mathbf{U}_{2,2}$ on A then the circuit matroid of the first graph of Fig. 15.4 is obtained on the set B. If $\mathbf{A} = \mathbf{U}_{2,1}$ then the new matroid will be $\mathbf{U}_{5,1}$. If **A** has a as loop, b as bridge then 1 and 2 become loops in the new matroid while $3, 4, 5$ become parallel elements.

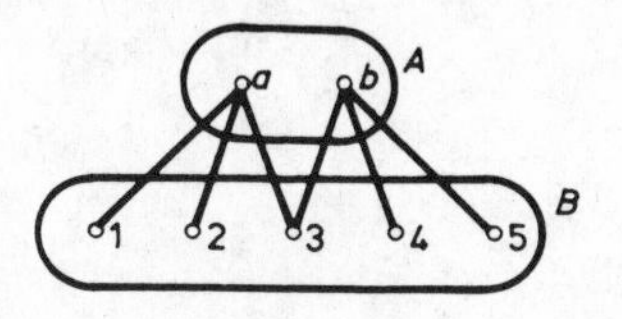

Fig. 15.13

The *proof* will be given at the end of this section. Observe that if **A** is the free matroid then just the transversal matroids arise in this way, hence this theorem is a generalization of Theorem 15.1.1.

Let us call a subset of edges **A**-*independent* if the end points of these edges in A form a subset which is independent in **A**. One can easily prove (Exercise 15.2.1) that they are the independent subsets of a matroid on $E(G)$. We have seen a special case of this result as Exercise 15.1.11 (where **A** was the free matroid).

The subsets of $E(G)$, applied in Theorem 15.2.1, were matchings (i.e. consisted of point-disjoint edges) and were **A**-independent as well. As a possible further generalization one might define two matroids **A** and **B**, respectively, on the two subsets A, B of the bipartite graph G with bipartition $V(G) = A \cup B$. Then those subsets of $E(G)$ which are simultaneously **A**- and **B**-independent, do not form a matroid in general (Exercise 15.2.2), yet they have a beautiful property.

Theorem 15.2.2 [Aigner and Dowling, 1970,1971], [Brualdi, 1970a]. *Let G be a bipartite graph with bipartition $V(G) = A \cup B$ and let $\mathbf{A}, \mathbf{B}$ be two matroids on A and B, respectively. Then the cardinality of the maximum size subsets of $E(G)$ which are simultaneously **A**- and **B**-independent, is $\min_T[r_A(T \cap A) + r_B(T \cap B)]$ where r_A and r_B are the rank functions of the matroids **A** and **B**, respectively, and the minimum is taken over all subsets T of $V(G)$ which cover every edge of G.*

Proof. Let $\mathbf{M}_A$ and $\mathbf{M}_B$ be the matroids of the **A**-independent and **B**-independent subsets on $E(G)$, respectively. Applying Edmonds' theorem (Theorem 13.2.2) the requested maximum is $k = \min_{F \subseteq E}[r_{\mathbf{M}_A}(F) + r_{\mathbf{M}_B}(E - F)]$. We have to prove $k = \ell$ where $\ell = \min_T[r_A(T \cap A) + r_B(T \cap B)]$.

Obviously, $r_{\mathbf{M}_A}(X) = r_A(V(X) \cap A)$ where $V(X)$ denotes the set of end points of the edges in X. In order to show $k \leq \ell$ consider a subset T of points, covering every edge (Fig. 15.14). Choose a subset F_T of edges, containing all the type 3 edges (joining $T \cap A$ and $B - T$) and none of the type 1 edges (joining $T \cap B$ and $A - T$). We have no prescription for the type 2 edges (joining $T \cap A$ and $T \cap B$), and no edge is possible between $A - T$ and $B - T$ since T covers everything. Now

$$r_A(T \cap A) + r_B(T \cap B) \geq r_{\mathbf{M}_A}(F_T) + r_{\mathbf{M}_B}(E - F_T) \geq k$$

holds for every such T, hence for the minimum of the left hand side expressions as well.

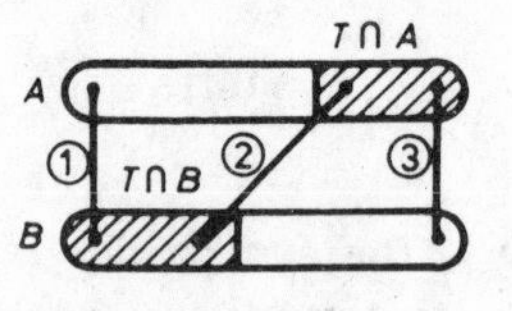

Fig. 15.14

On the other hand, let $k = r_{\mathbf{M}_A}(F_0) + r_{\mathbf{M}_B}(E - F_0)$. Since $V(F_0) \cap A$ and $V(E - F_0) \cap B$ together cover every edge, we have $k = r_A(V(F_0) \cap A) + r_B(V(E - F_0) \cap B)$. Hence k cannot be smaller than the minimum of the right hand side expressions.

□

Several known results follow as special cases of this theorem. If **A** and **B** are arbitrary matroids and G consists of n disjoint edges only then Theorem 13.2.2

is obtained. If $\mathbf{A}, \mathbf{B}$ are the free matroids and G is an arbitrary bipartite graph then $\nu(G) = \tau(G)$ follows (Theorem 5.11.5 by König).

Let now $\mathbf{A}$ and G be arbitrary and $\mathbf{B}$ be the free matroid. Then we obtain

Theorem 15.2.3 [Rado, 1942]. *Let G be a bipartite graph with bipartition $V(G) = A \cup B$ and let $\mathbf{A}$ be a matroid on A. Then all the points of B can be covered by an $\mathbf{A}$-independent matching of G if and only if $r_A(N(X)) \geq |X|$ holds for every $X \subseteq B$, where $N(X)$ is the set of neighbours of the points of X.*

Observe that this reduces to Hall's theorem (Theorem 5.1.2) if $\mathbf{A}$ is the free matroid. The necessity of the condition is trivial. For the sufficiency suppose the contrary, i.e. let $|B| = k$ and suppose that only a proper subset of B can be covered by an $\mathbf{A}$-independent matching. Then, by Theorem 15.2.2, there exists a subset T of $V(G)$ which covers every edge of G, yet $r_A(T \cap A) + |T \cap B| < k$. Now choose $X = B - T$. Obviously $N(X) \subseteq T \cap A$, see Fig. 15.14, hence $r_A(N(X)) \leq r_A(T \cap A) < k - |T \cap B| = |X|$, a contradiction.

□

Now we can prove Theorem 15.2.1 as well. Consider the function $f(X) = r_A(N(X))$. It is trivially normalized and monotone, and it is submodular by Exercise 15.2.6. Hence it defines a matroid with rank function $\tilde{f}$. In this matroid exactly those subsets $X \subseteq B$ are independent (Statement 13.3.2) which meet $f(Y) \geq |Y|$ for every $Y \subseteq X$. But these are just the subsets possessing an $\mathbf{A}$-independent matching, by Theorem 15.2.3.

Exercises for Section 15.2

15.2.1 Let G be a bipartite graph with bipartition $V(G) = A \cup B$ and let $\mathbf{A}$ be a matroid on A. Define a matroid on $E(G)$ so that X be independent if and only if those end points of the X-edges which are in A form an independent subset of $\mathbf{A}$. Show that this is a matroid.

15.2.2 Give a bipartite graph G with bipartition $V(G) = A \cup B$ and two matroids $\mathbf{A}, \mathbf{B}$ on A and B, respectively, so that those subsets of $E(G)$ which are simultaneously $\mathbf{A}$- and $\mathbf{B}$-independent matchings, do not form a matroid.

15.2.3 The property (B2) for the bases of a matroid was shown (Problem 7.1.16) to be weaker than the following property.

(BP) If $X_1, X_2 \in \mathbf{B}$ then there exists a bijection ϕ between X_1 and X_2 so that for every $x \in X_1$ both $(X_1 - \{x\}) \cup \{\phi(x)\}$ and $(X_2 - \{\phi(x)\}) \cup \{x\}$ belong to $\mathbf{B}$.

If (BP) is met, call the matroid *base orderable* [Brualdi, 1969]. Show that the class of base orderable matroids is closed with respect to deletion, contraction and forming dual.

15.2.4 Let C_n^2 denote (for $n \geq 3$) the graph, obtained from the length n circuit by doubling every edge. Applying the remark at the end of Exercise 15.1.2 show that the circuit matroid $\mathbf{M}(G)$ cannot be transversal if G contains K_4 or C_n^2 or their series extensions. (One can prove [Bondy, 1972], [Las Vergnas, 1970] that all the other graphic matroids are transversal. The fundamental transversal matroids among the graphic ones were also characterized [Nakamura, 1978].)

15.2.5 (continued) Prove that $\mathbf{M}(G)$ cannot be base orderable if G contains K_4 or its series extension. (This condition is also sufficient [Bondy, 1972].)

15.2.6 Prove that the function $f(X) = r_A(N(X))$ at the end of this section is submodular.

Problems for Section 15.2

15.2.7 Prove that the class of base orderable matroids is closed with respect to forming sum.

15.2.8 (continued) Prove that every transversal matroid is base orderable. (We mention that the statements of these two problems are valid for the *strongly base orderable matroids* as well, which are defined by property (BP′) of **Box 15.1.**

Box 15.1. Base exchange and base orderability properties

(B3) If $X_1, X_2 \in \mathbf{B}$ and $x \in X_1$ then there exists an $y \in X_2$ so that $(X_1 - \{x\}) \cup \{y\} \in \mathbf{B}$ (Theorem 7.1.2).

(B3′) If $X_1, X_2 \in \mathbf{B}$ and $x \in X_1$ then there exists an $y \in X_2$ so that $(X_1 - \{x\}) \cup \{y\} \in \mathbf{B}$ and $(X_2 - \{y\}) \cup \{x\} \in \mathbf{B}$ both hold (Problem 7.2.26).

(B3″) If $X_1, X_2 \in \mathbf{B}$ and $A_1 \subseteq X_1$ then there exists an $A_2 \subseteq X_2$ so that $(X_1 - A_1) \cup A_2 \in \mathbf{B}$ and $(X_2 - A_2) \cup A_1 \in \mathbf{B}$ both hold (Problem 13.2.6).

(BP) If $X_1, X_2 \in \mathbf{B}$ then there exists a bijection ϕ between X_1 and X_2 so that for every $x \in X_1$ both $(X_1 - \{x\}) \cup \{\phi(x)\} \in \mathbf{B}$ and $(X_2 - \{\phi(x)\}) \cup \{x\} \in \mathbf{B}$ hold. This is not always true, see Problem 7.1.16. Matroids satisfying (BP) are the base orderable matroids, see Exercises 15.2.3, 15.2.5 and Problems 15.2.7, 15.3.9.

(BP′) If $X_1, X_2 \in \mathbf{B}$ then there exists a bijection ϕ between X_1 and X_2 so that for every $A \subseteq X_1$ both $(X_1 - A) \cup \phi(A) \in \mathbf{B}$ and $(X_2 - \phi(A)) \cup A \in \mathbf{B}$ hold. This is even less frequently true, see **Box 15.3.** Matroids satisfying (BP′) are the strongly base orderable matroids.

See [Kung, 1986 b] for further results.

Section 15.3 Matroids induced by directed graphs

In this section we use directed graphs instead of undirected bipartite graphs. Just like matroids, defined on one subset of points of a bipartite graph induced a matroid on the other subset (*via* the edges of the graph), see Theorem 15.2.1, we shall see that if X, Y are subsets of the points of a directed graph G and there is a matroid on X then it induces another matroid on Y (*via* the edges of G). This construction — though reducible to the previous one — leads to a more general class of matroids, which is already closed with respect to the main operations.

Let $X, Y \subseteq V$ be two, *not necessarily disjoint* subsets of the directed graph $G = (V, E)$. We say we can *link* X onto Y if $|X| = |Y|$ and there are $|X|$ point-disjoint directed paths in G with initial points in X and terminal points in Y. A path may consist of a single point, hence, in particular, every subset can be linked onto itself.

For example, if G is the directed graph in Fig. 15.15 then $X = \{1, 2\}$ can be linked onto every 2-element subset except $\{1, 3\}$. If X should be linked to, say, $\{1, 4\}$ or $\{3, 4\}$, the respective pairs of paths are (1), (2,4) and (1,3),(2,4).

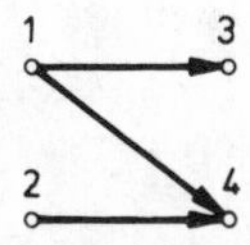

Fig. 15.15

Theorem 15.3.1 [Perfect, 1969], [Pym, 1969b]. *If $G = (V, E)$ is a directed graph and $X \subseteq V$ is a fixed set of points then those subsets of V which can be linked onto X form the bases of a matroid.*

Matroids arising in this way will be called *strict gammoids* [Mason, 1972]. This class is not closed with respect to deletion (see Exercise 15.3.2). A matroid is called *gammoid* if it can be obtained from a strict gammoid by deleting a suitable subset.

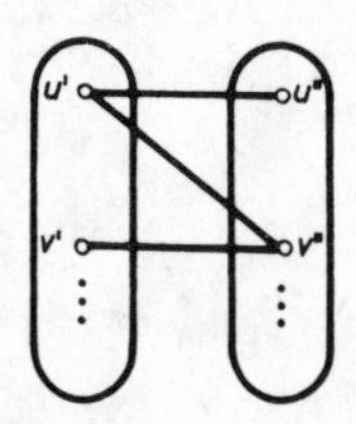

Fig. 15.16

The *proof* of Theorem 15.3.1 is given by a construction of Ingleton and Piff [1973]. Let us draw the point set V of G in two copies, and connect the corresponding pairs by edges (that is, take all the edges of form $\{u', u''\}, \{v', v''\}$, see Fig. 15.16); furthermore, connect a point u' of the first copy with a point v'' of the second copy if and only if there was a directed edge (v, u) in the directed graph G. Hence a directed path $(a, b, c, \ldots)$ of G will become a path $(a', a'', b', b'', c', c'', \ldots)$ in the resulting (undirected) bipartite graph H.

We claim that $A \subseteq V$ can be linked onto $B \subseteq V$ if and only if $V' - A'$ and $V'' - B''$ can be connected by point-disjoint paths in H. For example, the pair $\{4, 5\}$ can be linked onto $\{1, 5\}$ in the directed graph of Fig. 15.17. Accordingly, the point sets $\{1', 2', 3'\}$ and $\{2'', 3'', 4''\}$ can be joined by a matching, see Fig. 15.18. The figure essentially contains the general proof: if an edge (like (4,2)) was used for the linking then consider the corresponding edge of H (like $\{2', 4''\}$ in our case); if a point (like 3) was not used for the linking then consider the corresponding "horizontal" edge of H (like $\{3', 3''\}$ in our case). Thus the theorem is reduced to Theorem 15.1.1. □

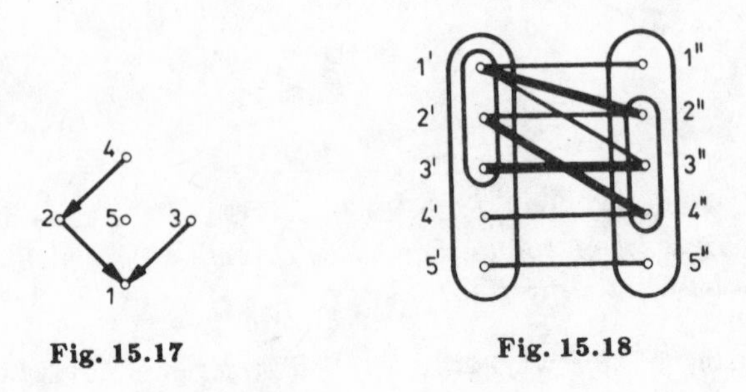

Fig. 15.17 **Fig. 15.18**

An obvious byproduct of this proof is that the bases of these matroids are just the base complements of a transversal matroid, that is

Theorem 15.3.2 [Ingleton and Piff, 1973]. **M** *is a strict gammoid if and only if* $\mathbf{M}^*$ *is transversal.*

Hence several further problems can easily be solved (Exercises 15.3.2–3). The relations among the four basic concepts of this chapter are shown on the diagram of **Box 15.2** [Ingleton, 1977], see also **Box 15.3**.

Theorem 15.3.3 [Brualdi, 1971b], [Mason, 1972]. *If* $G = (V, E)$ *is a directed graph,* $X \subseteq V$ *is a fixed set of points and* **M** *is a matroid on* X *then those subsets of* V *which can be linked onto independent subsets of* **M** *form the independent subsets of a new matroid* **N**.

Proof. Using the same construction as for the proof of Theorem 15.3.1, it is enough to apply Theorem 15.2.1. In fact, we obtain somewhat more: If **N** is the

Box 15.2. The classes of matroids induced by graphs [Ingleton, 1977]

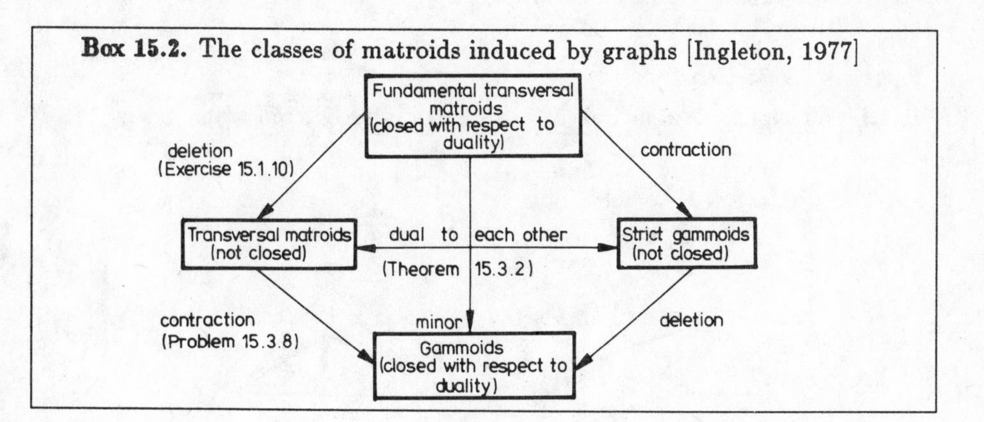

Box 15.3. Relations among the matroid classes induced by graphs, and the (strongly) base orderable matroids

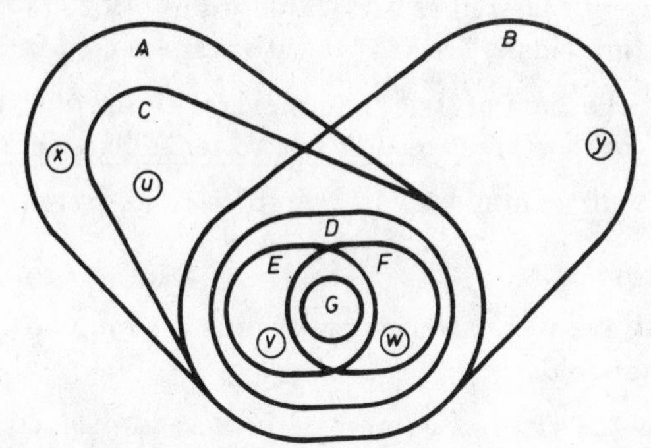

A	Base orderable matroids
B	Matroids, representable over the field $\mathbb{R}$ of the reals
C	Strongly base orderable matroids
D	Gammoids
E	Strict gammoids
F	Transversal matroids
G	Fundamental transversal matroids

Examples for some special matroids:

x	See [Ingleton, 1976]
y	$\mathbf{M}(K_4)$
u	The non-Papposian matroid (Problem 9.2.12)
v	$\mathbf{M}(C_3^2)$ (see the second graph of Fig.15.4)
w	$\mathbf{M}(K_{3,2})$ (see Fig. 15.11)

new matroid then $\mathbf{N}^*$ could be obtained by inducing $\mathbf{M}^*$ by a bipartite graph and then contracting a suitable subset.

$\square$

Exercises for Section 15.3

15.3.1 Which pairs can be linked onto $\{4, 6\}$ in the directed graph of Fig. 15.19?

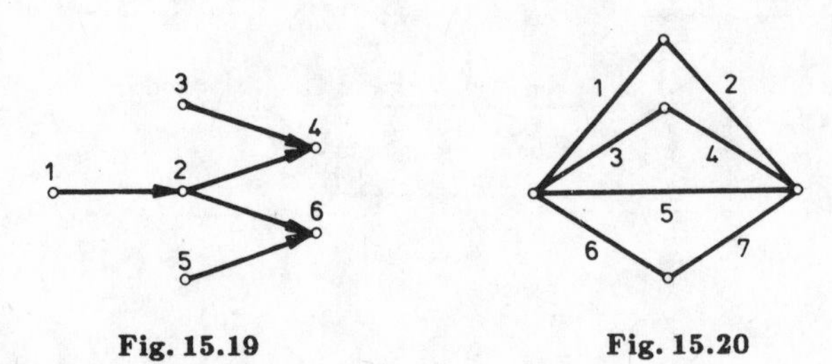

Fig. 15.19 **Fig. 15.20**

15.3.2 Use the circuit matroid of the graph of Fig. 15.20 to show that the class of strict gammoids is not closed with respect to deletion.

15.3.3 Prove that the class of strict gammoids is closed with respect to contraction and parallel extension but not to series extension and dualization.

15.3.4 Prove that the gammoids can be represented over the field of the reals.

Problems for Section 15.3

15.3.5 Prove that every transversal matroid is a gammoid but not necessarily a strict gammoid.

15.3.6 Show that the class of gammoids is closed with respect to dualization and forming minors.

15.3.7 Prove that the class of gammoids is closed with respect to forming sums [Narayanan and Vartak, 1973].

15.3.8 Show that a matroid is a gammoid if and only if it can be obtained from a transversal matroid by contracting a suitable subset.

15.3.9 Prove that every gammoid is base orderable.

15.3.10 A graph G is called *series-parallel* if it can be obtained from one of the single edge graphs with series and parallel extensions. Show that in this case $\mathbf{M}(G)$ is a gammoid.

15.3.11 Let G be a series-parallel graph. Prove that $\mathbf{M}(G)$ cannot have $\mathbf{M}(K_4)$ as a minor. (One can prove [Dirac, 1952], [Duffin, 1965] that this condition is sufficient as well for 2-connected graphs.) .

15.3.12 Prove that every series-parallel graph is planar.

Chapter 16

Applications

Section 16.1 On the data structure of network analysis programmes

The strong and weak generality assumptions were introduced in Section 14.1. The weak assumption gave less restriction on the applicable devices. On the other hand we have now to consider the methods of describing (storing) the multiports, since our model should be independent from the describing matrix.

Theoretically, an ordinary n-port, described by $\mathbf{Au} + \mathbf{Bi} = \mathbf{0}$, can be given by the $n \times 2n$ matrix $(\mathbf{A}|\mathbf{B})$. This requires $2n^2$ storage and for an arbitrary subset of columns, we can check linear independence in time proportional to n^3. Since $r(\mathbf{A}|\mathbf{B}) = n$, we can even use a nonsingular transformation which transforms $(\mathbf{A}|\mathbf{B})$ into the form $(\mathbf{E}|\mathbf{C})$ (after a suitable permutation of the columns) and then one has to store $\mathbf{C}$ (and the permutation) only.

This supposes that the entries of $\mathbf{A}, \mathbf{B}$ and $\mathbf{C}$, as real numbers, can be stored with high precision and that no roundoff-error arises during the rank checks. On the other hand, we saw in Section 14.1 that although the matrix descriptions $\begin{pmatrix} -1 & 2 & 0 & 0 \\ 0 & -1 & 2 & 1 \end{pmatrix}$ and $\begin{pmatrix} -1 & 1 & 2 & 1 \\ 0 & -1 & 2 & 1 \end{pmatrix}$ are equivalent, the former one is more suitable. If its nonzero entries are considered as algebraically independent numbers over the field $\mathbb{Q}$ of the rationals, the new matrix determines the same matroid and we can use term-ranks instead of ranks.

Let $\mathbf{T}(\mathbf{A})$ denote the matroid which is obtained so that the nonzero entries of the matrix $\mathbf{A}$ are changed to algebraically independent numbers over $\mathbb{Q}$ at first, and the matroid, determined by the columns of this new matrix is considered thereafter.

We would like to describe the n-port N with such an $n \times n$ matrix $(\mathbf{A}|\mathbf{B})$ which meets $\mathbf{T}(\mathbf{A}|\mathbf{B}) = \mathbf{M}_N$. Even better if such an $n \times n$ matrix $\mathbf{C}$ exists that $\mathbf{T}(\mathbf{E}|\mathbf{C}) = \mathbf{M}_N$ holds. Such matrices need not necessarily exist. We have shown as Problem 14.1.7 that no such $\mathbf{C}$ exists in case of a 3-port circulator.

Theorem 16.1.1 [Recski and Iri, 1980]. *There exists a matrix* $(\mathbf{A}|\mathbf{B})$ *satisfying* $\mathbf{T}(\mathbf{A}|\mathbf{B}) = \mathbf{M}_N$ *if and only if* $\mathbf{M}_N$ *is transversal. Moreover, there exists a matrix* $\mathbf{C}$ *satisfying* $\mathbf{T}(\mathbf{E}|\mathbf{C}) = \mathbf{M}_N$ *if and only if* $\mathbf{M}_N$ *is fundamental transversal.*

Its *proof* is almost trivial; the first statement is a reformulation of Corollary 15.1.3, the second is that of the solution of Problem 15.1.16.

□

Hence, using the result of the previous chapter we immediately obtain that 3-port circulators cannot be stored at all with matrices, consisting of algebraically independent entries. One can find such multiport as well (see Exercise 16.1.1 and Problem 16.1.2) which can be stored in such a way by an $n \times 2n$ matrix but not by an $n \times n$ matrix.

Network analysis has quantitative problems as well (like explicitly finding the solution), not only qualitative ones (like solvability, independence of the state variables etc.) Hence, when developing a program for network analysis, the n-ports — which model the devices — must be stored by a matrix anyhow. If the matroid of the multiport is transversal then the above theorem ensures that — in the case of appropriately chosen matrices — the zero-nonzero pattern of the matrix will be enough for the qualitative problems.

Let, finally, D be a device whose matroid is nontransversal. Then — in addition to a matrix description — one might find useful to store the matroid as well (e.g., by a $0-1$ vector of length 2^{2n} which tells about every subset of the port voltages and currents, whether it is independent). Preparing this vector might require much time but it should be done only once and can be used each time when D is used in any network.

Exercise for Section 16.1

16.1.1 Find a matrix description $(\mathbf{A}|\mathbf{B})$ to the 4-port N of Fig. 16.1 which satisfies $\mathbf{T}(\mathbf{A}|\mathbf{B}) = \mathbf{M}_N$.

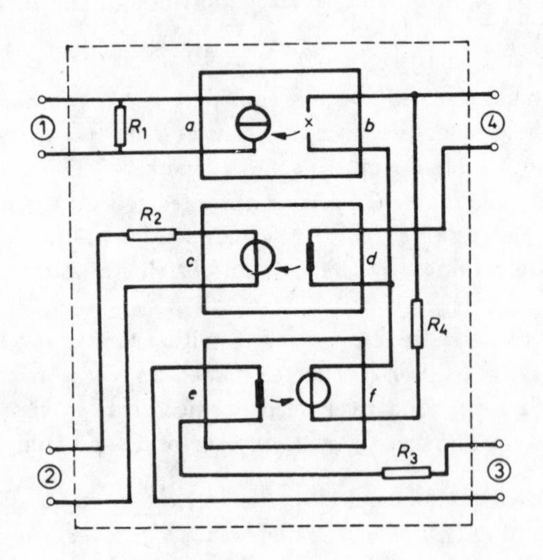

Fig. 16.1

Problem for Section 16.1

16.1.2 Show that the 4-port N of Fig. 16.1 has no such matrix description $(\mathbf{E}|\mathbf{C})$ which satisfies $\mathbf{T}(\mathbf{E}|\mathbf{C}) = \mathbf{M}_N$.

Section 16.2 "Unrealizability" of certain multiports

All the electric applications so far were related to network analysis. We supposed that the descriptions of the devices and their interconnection are given and wanted to know certain properties of the resulting circuit. The real engineering problem is rather the inverse one: Given the properties of the requested circuit, find the way to realize it.

More exactly, using a given class of network elements try to obtain a network with *a priori* given properties. Such problems belong to *network synthesis*. For example, the problem

realize a 2-port with impedance matrix $\mathbf{Z} = \begin{pmatrix} a & b \\ c & d \end{pmatrix}$ using positive resistors

is solvable if and only if a, b, c, d are real numbers, $b = c$ and $a \geq |b|, d \geq |b|$. (If $b, c \geq 0$ then the network if Fig. 16.2 is a possible solution. If $b, c < 0$ then the solution is more complicated, see also Exercise 16.2.2.) If $b \neq c$ then reciprocity is violated, hence the problem has no solution, not even with positive and negative resistors. If $b = c$ then — using possibly negative resistors as well — the problem is always solvable.

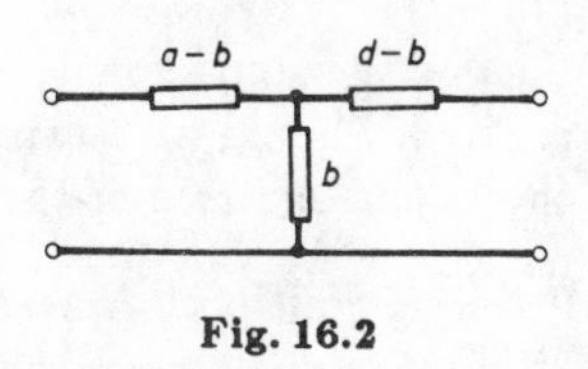

Fig. 16.2

(Although we do not go into details, let us mention that the above problem — the so called *resistive n-port synthesis* — is open even for 4×4 matrices. The case 3×3 is solved [Cederbaum, 1959, 1963] but fairly difficult.)

Several synthesis problems were presented in this book (though the expression was not used), like Problems 12.1.9, 12.1.15 and 12.1.18. Similarly, Fig. 12.4 showed that a gyrator can be realized by nullator, norator, ideal transformer and a pair of resistors with values $\pm R$ (see Exercise 12.1.7). Should the sum of the resistances of the positive and the negative resistor be not exactly zero, then another 2-port were obtained instead of the gyrator.

Obviously, it is more advantageous — if possible at all — to realize a network so that if we slightly change the properties of the elements then the result has no qualitative change either. Such a realization will be called *qualitatively reliable*

[Recski, 1980b]. For example, both networks of Fig. 16.3 realize the 2-port, given by $\begin{pmatrix} -1 & 2 \\ 0 & -1 \end{pmatrix}\begin{pmatrix} u_1 \\ u_2 \end{pmatrix} + \begin{pmatrix} 0 & 0 \\ 2 & 1 \end{pmatrix}\begin{pmatrix} i_1 \\ i_2 \end{pmatrix} = \begin{pmatrix} 0 \\ 0 \end{pmatrix}$, but only the second is qualitatively reliable, since if the resistances of the second network are slightly changed, the two port voltages shall not remain proportional.

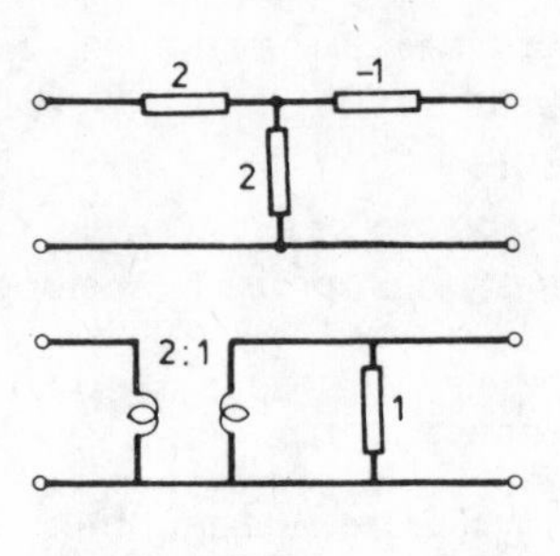

Fig. 16.3

Lemma 16.2.1 *Let us interconnect the multiports $N_1, N_2, \ldots$ along the network graph G. This realizes the resulting multiport N in a qualitatively reliable way if and only if $\mathbf{M}_N = (\mathbf{G} \vee \mathbf{A})/B$ holds (see the notations at Statement 14.2.1).*

The *proof* is trivial since qualitative reliability is just a reformulation of the weak generality assumption from the point of view of synthesis.

□

For example, Figs 14.7–9 show the steps how this condition can be checked for the second network of Fig. 16.3. Should we start from the first network, the result were $\mathbf{U}_{4,2}$ instead of the circuit matroid of the graph of Fig. 14.9.

Recall (Problem 10.1.10) that a multiport N was called qualitatively reciprocal if $\mathbf{M}_N = \phi(\mathbf{M}_N^*)$ holds for its matroid, where ϕ is the transformation which interchanges the roles of voltage and current. In particular, all the reciprocal and antireciprocal multiports are qualitatively reciprocal.

Theorem 16.2.2 [Iri and Recski, 1980]. *If weak generality is assumed then the class of qualitatively reciprocal multiports is closed with respect to interconnection.*

Its *proof* is left to the reader as Problem 16.2.4.

Corollary 16.2.3 *If $\mathbf{M} = \phi(\mathbf{M}^*)$ is not met for the matroid $\mathbf{M}$ of a multiport N then N cannot be realized in a qualitatively reliable way from reciprocal devices and gyrators.*

Proof. Trivial since the gyrator is also qualitatively reciprocal.

□

In particular, the controlled sources of Fig. 8.11 cannot be realized in this way. Should we forget about the qualitative reliability, this realization were possible, see Fig. 16.4, for example, for the current controlled current source. (Negative resistors are obviously necessary since the controlled sources are active.)

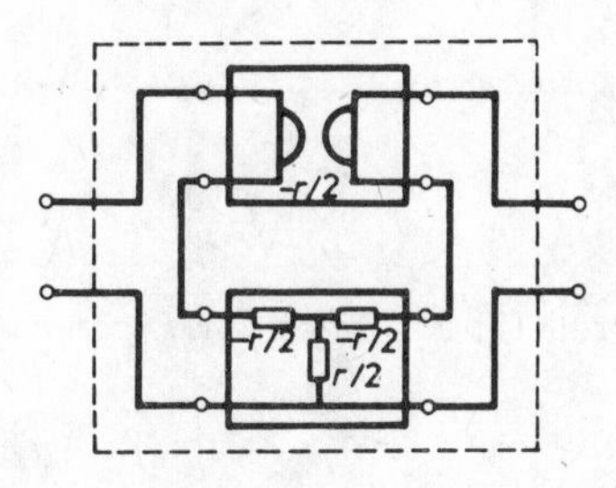

Fig. 16.4

More generally, if reciprocal devices and gyrators are available, every multiport, possessing at least one hybrid immitance description, can be realized [Oono, 1972], see Problem 16.2.6 in a special case. Hence Corollary 16.2.3 suggests that qualitative reliability is quite restrictive.

Obviously, if a class C of matroids is closed with respect to contraction and sum, and contains the matroid $\mathbf{G}$ defined for the network graph, then the class of those network elements whose matroids belong to C, is closed with respect to qualitatively reliable interconnection. Hence we obtain, for example

Theorem 16.2.4 [Recski, 1980b]. *The 3-port circulator cannot be realized in a qualitatively reliable way by series-parallel connection of 1- and 2-ports.*

Proof. If the graph of the interconnection is series-parallel then the corresponding matroid $\mathbf{G}$ is a gammoid, by Problem 15.3.6, 15.3.10 and 15.3.12. Similarly, the matroid of every 1- or 2-port is a gammoid (in fact, every matroid over a set of cardinality at most 5 is a gammoid). Finally, since the class of gammoids is closed with respect to contraction and sum (Problem 15.3.6–7), we cannot obtain a circulator since its matroid is $\mathbf{M}(K_4)$.

□

Exercises for Section 16.2

16.2.1 Let $\mathbf{Y} = \begin{pmatrix} y_{11} & y_{12} \\ y_{12} & y_{22} \end{pmatrix}$. Show that $\mathbf{Y}$ can be realized as the admittance matrix of a 2-port consisting of positive resistors, if $y_{12} \leq 0$ and $y_{11}, y_{22} \geq -y_{12}$.

16.2.2 Determine the $\mathbf{Z}$ matrix of the 2-port of Fig.16.5 and then show by an example that among the realizability conditions for $\mathbf{Z}$, mentioned at the beginning of the section, the relation $b, c \geq 0$ is not necessary.

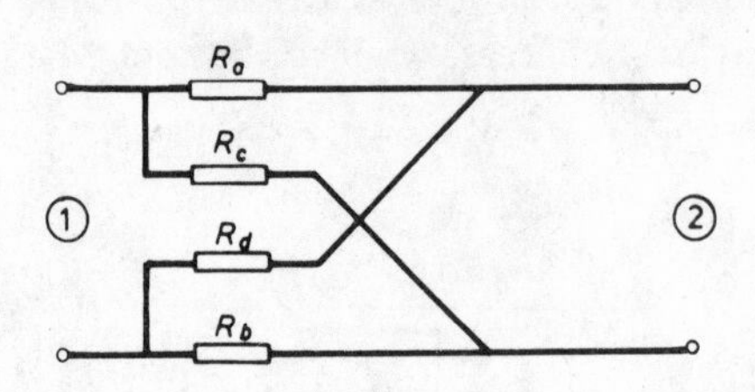

Fig. 16.5

16.2.3 Prove that Theorem 16.2.4 holds for k-port circulators with $k \geq 4$ as well.

Problems for Section 16.2

16.2.4 Prove Theorem 16.2.2.

16.2.5 Prove that every $n \times n$ antisymmetric matrix can be realized as the impedance matrix of an n-port, consisting of gyrators only.

16.2.6 Prove that every multiport with an impedance description $\mathbf{Z}$ can be realized with a reciprocal multiport and with gyrators.

11.3.6 Let S be a subset of vectors in a vector space. Prove that S can be covered by k subsets of linearly independent vectors if and only if $k \dim A \geq |A|$ for every subset $A \subseteq S$ [Horn,1955].

Problems for Section 11.3

11.3.7 Prove the sufficiency of the condition of Theorem 11.3.2.

11.3.8 Prove that the edge set of the complete graph K_n of n points can be covered by $\lceil n/2 \rceil$ forests. (Similarly one can prove that the edge set of the bipartite graph $K_{m,n}$ with $m+n$ points and all the possible $m \cdot n$ edges can be covered by $\left\lceil \frac{mn}{m+n-1} \right\rceil$ forests.)

Chapter 17

Some recent results in matroid theory

Section 17.1 Matroid theory algorithms II: Oracles

All the matroid algorithms presented so far (in Sections 7.4 and 13.1) were polynomial in the sense that the total number of operations was proportional to a polynomial of the size n of the input (usually n was the cardinality of the underlying set S of the matroids $\mathbf{M}_1, \mathbf{M}_2, \ldots$ in question), supposing that questions like "Is $X \subseteq S$ independent in $\mathbf{M}_i$?" could be answered by just one step.

That means, these matroids $\mathbf{M}_1, \mathbf{M}_2, \ldots$ were given by some "subroutines" $R_1, R_2, \ldots$, respectively, and when speaking about operations in the algorithm, any call of these subroutines was considered as a single operation.

Since our applications usually require graphic or representable matroids, which can be given by a graph or by a matrix, in both cases with storage requirement proportional to n^2 at most; and since checking independence in such matroids can be realized by at most about n^3 steps; the above assumption leads to "really" polynomial algorithms. But what is the case in general?

As we have mentioned in the solution of Problem 7.1.9, the number of non-isomorphic matroids on the n-element set S is nearly 2^{2^n}. Even if we wish to write down this number (say, in binary representation), it requires a storage, proportional to 2^n. Hence the storage of a "general" matroid requires exponential space (as function of n).

Thus, it has not much sense to require that an algorithm be polynomial in the size of such an input (namely the description of the matroid $\mathbf{M} = (S, \mathbf{F})$) since this input itself is already an exponential function of $n = |S|$. That is the reason why we require the algorithms to be polynomial in n instead.

Of course, in this case $\mathbf{M}$ is not really described; all we have is a subroutine R which answers the question "Is $X \subseteq S$ independent in $\mathbf{M}$?" in one step. For certain decision problems (like "Is the matroid loopless?") we need then only a few steps (a few call of this subroutine R) to obtain the answer, but we cannot "identify" the matroid by these questions.

Hence we can be sure that such a subroutine R for a "general" matroid cannot be realized so that its number of operations is polynomial in n. Rather, we imagine this subroutine like an "oracle" (a medium by which god reveals hidden knowledge).

We could have chosen another subroutine as well, which answers the question "Is $X \subseteq S$ a circuit in $\mathbf{M}$?" Therefore our above subroutine will be called *independence oracle* and this one *circuit oracle*. Similarly, we can define *rank oracle, base oracle* and *girth oracle*. For all these subroutines, the input is a subset X of S. In case of the independence, base and circuit oracles, the output is yes or no (whether X is independent (base, circuit, respectively) or not). In case of the rank oracle, the output is $r(X)$. In case of the girth oracle, the output is the cardinality of the minimum size circuit contained in X (or ∞ if X is independent).

Recall that all the data structures for graphs presented in Section 1.4, could be converted to all the others in polynomial time (as function of the size of the input). For certain oracles we have similar results.

Statement 17.1.1 *The independence oracle and the rank oracle are polynomially equivalent.*

Proof. Let O_1 be the rank oracle and O_2 be the independence oracle. If O_1 is available, we can apply it for producing O_2 as follows. If X is the input of O_2, we input X to O_1 first. If the answer of O_1 is $|X|$, we output "yes" for O_2; if the answer of O_1 is less than $|X|$, we output "no" for O_2.

On the other hand, if O_2 is available, we can use our greedy algorithm (with weight 1 for elements of X and weight 0 for the elements of $S - X$) to obtain $r(X)$. Since O_2 was called in a polynomial number of times only, we reduced O_1 to it.

□

However, in some other cases oracles are not equivalent polynomially.

Statement 17.1.2 *The circuit oracle is less powerful than the independence oracle.*

Proof. Let O_3 be the circuit oracle. We can produce it as follows if the independence oracle O_2 is available. Let X be the input for O_3. Input X to O_2. If the answer is "yes" (i.e., if X is independent), the output of O_3 is "no". If the answer of O_2 is "no" then input $X - \{x\}$ to O_2 for every $x \in X$. If all the answers of O_2 is "yes" then the output of O_3 is "yes"; otherwise the output of O_3 is "no".

On the other hand, if O_3 is available only, we cannot produce O_2 with only a polynomial number of calls of O_3. For suppose the contrary, i.e. that someone produced an algorithm which realizes O_2 with calling O_3 in a polynomial number of times only. Let $\mathbf{M}$ be the free matroid $(S, 2^S)$ and $\mathbf{M}_X$ be another matroid having a single circuit $X \subseteq S$ (and all the elements of $S-X$ are bridges). Clearly, O_2 should produce different output for the input S in the two cases ("yes" for $\mathbf{M}$ and "no" for $\mathbf{M}_X$). However, if the algorithm called O_3 in a polynomial number of times only, it could not ask about every subset of S whether it was a circuit (there are 2^n subsets of S). Hence if for each call of O_3 the algorithm received a "no" output, it still cannot decide whether it has $\mathbf{M}$ or $\mathbf{M}_X$ (with an X, which has never been asked from O_3), a contradiction.

□

Further examples for oracles are contained in **Box 17.1** and their relations are summarized in **Box 17.2**.

Box 17.1. Some important matroid oracles

Name	Input	Output
Independence oracle	$X \subseteq S$	Yes, if X is independent, no otherwise
Base oracle	$X \subseteq S$	Yes, if X is a base, no otherwise
Circuit oracle	$X \subseteq S$	Yes, if X is a circuit, no otherwise
Closed set oracle	$X \subseteq S$	Yes, if X is a closed set, no otherwise
Rank oracle	$X \subseteq S$	$r(X)$
Girth oracle	$X \subseteq S$	Cardinality of the minimum size circuit, contained in X (or ∞ if X is independent)
Closure oracle	$X \subseteq S$	$\sigma(X)$

Box 17.2. Relations among matroid oracles

Let $O_1 \to O_2$ denote that O_1 is less powerfull than O_2 and let $O_1 \leftrightarrow O_2$ denote that they are polynomially equivalent [Haussmann–Korte, 1981, see also Robinson–Welsh, 1980].

a	Exercise 17.1.3	Base oracle $\xrightarrow{a}$ Independence oracle
b	Statement 17.1.2	Independence oracle $\overset{d}{\updownarrow}$ Rank oracle
c	Problem 17.1.7	Circuit oracle $\xrightarrow{b}$ Rank oracle $\xrightarrow{f}$ Girth oracle
d	Statement 17.1.1	Rank oracle $\overset{e}{\updownarrow}$ Closure oracle
e	Exercise 17.1.1	Closed set oracle $\xrightarrow{c}$ Closure oracle
f	Exercise 17.1.2	

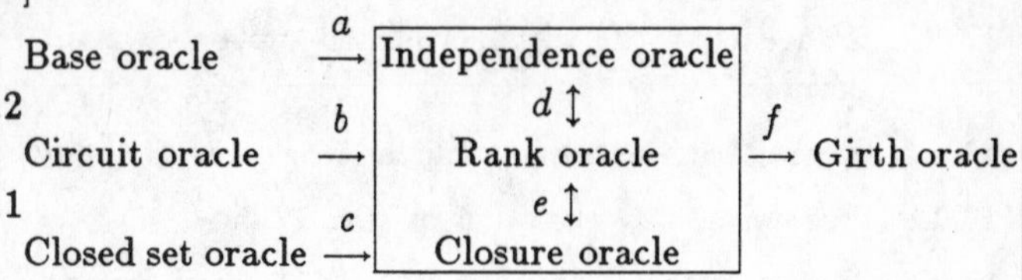

Let us emphasize again that all the important algorithms we have seen so far (like the greedy algorithm, the matroid partition and intersection algorithms) supposed that the matroids are given by an independence oracle.

However, quite a lot of properties of matroids cannot be decided by polynomial algorithms, if the matroid is given by an independence oracle. Such properties are listed in **Box 17.3**. We shall return to related questions in the next section.

We close this section by a list of some polynomially solvable, by some **NP**-hard and by some polynomially unsolvable problems of matroid theory (**Box 17.4**). Some items were contained in **Box 17.3** as well, some others will be explained in the rest of this chapter only.

A last remark refers to submodular functions. Many important algorithms are special cases of submodular function minimization, see Exercise 17.1.5. The general problem is also polynomial [Grötschel *et al.*, 1981] but the only known

Box 17.3. Properties which cannot be recognized in polynomial time if the matroid $\mathbf{M}$ is given by independence oracle

(1) Is $\mathbf{M}$ uniform?
(2) Is $\mathbf{M}$ isomorphic to its dual?
(3) Is $\mathbf{M}$ transversal?
(4) (If yes, is $\mathbf{M}$ fundamental transversal?)
(5) Is $\mathbf{M}$ representable?
(6) Is $\mathbf{M}$ orientable (see Section 17.4)?
(7) Is $\mathbf{M}$ bipartite?
(8) Is $\mathbf{M}$ binary?

See Exercise 17.1.4 for (1) and (7); Problem 17.1.8 for (2) and (5); Problem 17.1.10 for (8) and [Jensen–Korte, 1982] for (3), (4) and (6).

solution uses the famous ellipsoid method [Khachiyan, 1979; Shor, 1970]. A combinatorial solution (without real arithmetic etc.) would be a major result.

Exercises for Section 17.1

17.1.1 Prove that the closure oracle is polynomially equivalent to the independence and rank oracles.

17.1.2 Prove that the girth oracle is more powerful than the independence oracle.

17.1.3 Prove that the base oracle is less powerful than the independence oracle.

17.1.4 Prove items (1) and (7) in **Box 17.3.**

17.1.5 Show that the problems "Is $X \subseteq S$ independent in the sum of some matroids?" and "Does there exist a common base in two matroids?" can be formulated as the minimization of suitable submodular functions.

Problems for Section 17.1

17.1.6 Suppose you know one base in the matroid. Prove that this makes the base oracle polynomially equivalent to the independence oracle.

17.1.7 Prove that the closed set oracle is less powerful than the closure oracle.

17.1.8 (a) Prove item (2) in **Box 17.3.** (Hint: Let $\mathbf{M}_1 = \mathbf{U}_{4k,2k}$ and let $\mathbf{M}_2$ have $\binom{4k}{2k} - 3$ bases in a suitable way.)
(b) Prove item (5) in **Box 17.3.** (Hint: Let $\mathbf{M}_1 = \mathbf{U}_{2r,r}$ and let $\mathbf{M}_2$ be a suitable matroid containing $\mathbf{K}_8$ (see Problem 9.2.13) as a minor.)

17.1.9 Let $S = \{x_1, x_2, \ldots, x_k, y_1, y_2, \ldots, y_k\}$ with $k \geq 3$ and consider the subsets of S of form $\{x_i, x_j, y_i, y_j\}$; furthermore those subsets T which satisfy $|T \cap \{x_i, y_i\}| = 1$ for every i and $|T \cap \{y_1, y_2, \ldots, y_k\}|$ is even. Prove that these subsets of S are just the circuits of a binary matroid $\mathbf{M}$ [Seymour, 1981].

Box 17.4. Some polynomially solvable, some **NP**-hard and some polynomially unsolvable problems of matroid theory

A Polynomially solvable problems
1 Base of maximal weight in a matroid (Algorithms 7.1–7.3)
2 Rank of a subset in the sum of k matroids (Algorithm 13.1)
3 Maximum size subset, independent in two matroids (Algorithm 13.2)
4 Maximum weight subset, independent in two matroids (Algorithm 13.3)
5 Submodular function minimization [Grötschel–Lovász–Schrijver, 1981]
6 Graphicness–test for a binary matroid [Tutte, 1960; Iri, 1968a; Bixby–Cunningham, 1980]
7 Graphicness–test for an arbitrary matroid [Seymour, 1981]
8 Regularity–test for an arbitrary matroid [Truemper, 1982b]
9 2-polymatroid matching for linearly represented matroids [Lovász, 1980b]

B **NP**-hard problems
1 Maximum size subset, independent in k ($k \geq 3$) linearly represented matroids (Theorem 13.1.2)
2 k-polymatroid grouping ($k \geq 3$) for linearly represented matroids (Corollary 17.3.2)
3 Length of a shortest/longest circuit in a linearly represented matroid

C Polynomially unsolvable problems
1 Length of a shortest/longest circuit in a matroid (Exercise 17.1.2 for the shortest; [Jensen–Korte, 1982] for the longest circuit)
2 Submodular function maximization (Problem 17.3.5)
3 The general 2-polymatroid matching problem (Problem 17.3.3)
4 Binary–test for an arbitrary matroid (Problem 17.1.10)
5 Representability–test for an arbitrary matroid (Problem 17.1.8)

17.1.10 Prove [Seymour, 1981] that binarity cannot be recognized in polynomial time if the matroid is given by independence oracle. (Hint: apply Problems 9.3.17 and 17.1.9.)

Section 17.2 The characterization of regular matroids

We recall that a matroid is called regular if it is coordinatizable over every field. Examples for regular matroids are the graphic and the cographic matroids; another interesting regular matroid is $\mathbf{R}_{10}$, see [Bixby, 1977a]; this is a matroid whose representation over $\mathbb{B}$ is given by the following matrix:

$$\begin{pmatrix} 1 & 1 & 1 & 1 & 1 & 1 & 0 & 0 & 0 & 0 \\ 1 & 1 & 1 & 0 & 0 & 0 & 1 & 1 & 1 & 0 \\ 1 & 0 & 0 & 1 & 1 & 0 & 1 & 1 & 0 & 1 \\ 0 & 1 & 0 & 1 & 0 & 1 & 1 & 0 & 1 & 1 \\ 0 & 0 & 1 & 0 & 1 & 1 & 0 & 1 & 1 & 1 \end{pmatrix}$$

On the other hand, examples for nonregular matroids include all the uniform matroids $\mathbf{U}_{n,k}$ with $n \geq 4$, $1 < k < n-1$ (they are nonbinary, see Exercise 9.1.7), or the Fano matroid $\mathbf{F}_7$ (which is binary but nonrepresentable over fields with characteristic $\neq 2$).

By Theorem 9.1.4 every minor of a regular matroid is regular. Hence the necessity of the condition in the next deep theorem immediately follows. (The sufficiency is difficult, its proof is not included here.)

Theorem 17.2.1 [Tutte, 1965]. *A matroid is regular if and only if it is binary and has no minor isomorphic to $\mathbf{F}_7$ or $\mathbf{F}_7^*$.*

Recall that a matroid is binary if and only if it has no minor, isomorphic to $\mathbf{U}_{4,2}$. Hence binarity is characterized with a single *excluded minor* (namely $\mathbf{U}_{4,2}$), regularity by three excluded minors ($\mathbf{U}_{4,2}$, $\mathbf{F}_7$ and $\mathbf{F}_7^*$), and graphic matroids can be characterized by five excluded minors:

Theorem 17.2.2 [Tutte, 1965]. *A matroid is graphic if and only if it is regular and has no minor isomorphic to the cut set matroids of the two Kuratowski graphs, i.e. to $\mathbf{M}^*(K_5)$ or to $\mathbf{M}^*(K_{3,3})$.*

The necessity of these conditions is again obvious while their sufficiency is not proved here.

These important theorems enable us to prove if a matroid $\mathbf{M}$ is *not* binary or *not* regular or *not* graphic. (If someone has found an excluded minor, we can verify it in a polynomial number of steps provided that $\mathbf{M}$ is given by an independence oracle.)

How can we prove that $\mathbf{M}$ *is*, say, graphic? If someone has found a graph G and claims that $\mathbf{M} \cong \mathbf{M}(G)$ then a single call of an *isomorphism oracle* is needed. (The input of such an oracle is a pair of matroids, the output is "yes" if the two matroids are isomorphic and "no" otherwise.) Similarly, binarity can be verified in the same sense by presenting a $0-1$ matrix.

But how can we prove that $\mathbf{M}$ is regular? If someone presents a matrix $\mathbf{A}$ and claims that its column space matroid $\mathbf{M}_\mathbf{A}$ (over a given field) is isomorphic to $\mathbf{M}$, we can verify it but shall still not be convinced that $\mathbf{M}$ is regular. We could ask for matrix representations over every field but then a "proof" would require an infinite number of calls of our isomorphism oracle. Alternatively, if we know Tutte's theorem (Problem 17.2.8 below), we can ask for a totally unimodular matrix $\mathbf{A}$ and require a single call of the isomorphism oracle, but verifying that $\mathbf{A}$ is totally unimodular would require an exponential number of operations. Hence we need an entirely different result.

Let us call direct sum as *1-sum* for a moment. In the case of graphic matroids 1-sum can be visualized by Fig. 17.1. The drawings of Fig. 17.2–3 suggest two further operations, called *2-sum* and *3-sum*; the former "identifies" two edges

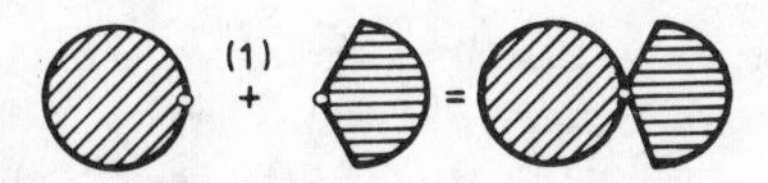

Fig. 17.1

(one element from each matroid) and then deletes it; the latter "identifies" a circuit of length 3 from each matroid and then deletes these 3 elements. (The formal definitions are left to the reader as Exercise 17.2.5.) It is not difficult to show (Exercise 17.2.6) that the 1, 2 and 3-sums of regular matroids are also regular.

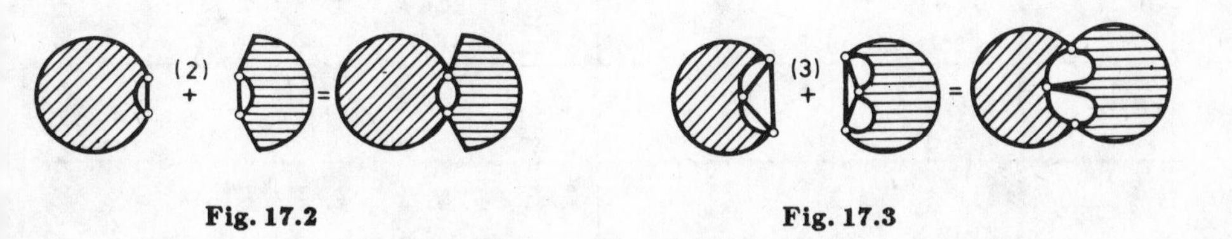

Fig. 17.2 Fig. 17.3

Theorem 17.2.3 [Seymour, 1980]. *Every regular matroid can be obtained by 1-, 2- and 3-sums from graphic matroids, from cographic matroids, and from several copies of* $\mathbf{R}_{10}$.

This theorem (whose proof is difficult and is not included here) can help to verify regularity by explicitly showing how the matroid was built up.

Finally, let a matroid $\mathbf{M}$ be given by an independence oracle. How can we decide whether it is graphic, regular or binary?

There exist several polynomial algorithms for testing graphicness of a binary matroid [Tutte, 1960; Iri, 1968a; Bixby and Cunningham, 1980], even of an arbitrary matroid [Seymour, 1981]. Regularity can also be tested in polynomial time [Truemper, 1982b]. However, binarity cannot be decided in polynomial time, see Problem 17.1.10.

Exercises for Section 17.2

17.2.1 Prove that deletion of an element from $\mathbf{R}_{10}$ leads to $\mathbf{M}(K_{3,3})$.

17.2.2 Prove that $\mathbf{R}_{10}$ is isomorphic to its dual but is not identically self-dual.

17.2.3 Prove that $\mathbf{R}_{10}$ is neither graphic nor cographic.

17.2.4 Prove that $\mathbf{R}_{10}$ is regular.

17.2.5 Describe the circuits of the 2-sum and 3-sum of two matroids in terms of the circuits of the two matroids.

17.2.6 Prove that the 1-, 2- and 3-sums of regular matroids are also regular.

17.2.7 The results of successive deletions and series contractions are called *series minors* of a matroid. **Box 17.5** summarizes both the excluded minor and the excluded series minor characterizations of some important matroidal properties. Why is $\mathbf{R}_{10}$ contained in the last column for graphic matroids (but not for cographic ones)?

Box 17.5. Characterization of some matroidal properties by excluded minors or by excluded series minors

A matroid is	if and only if it does not contain any of the following matroids as minors	as series minors
binary	$\mathbf{U}_{4,2}$ [Tutte, 1965]	$\mathbf{U}_{k,k-2}$ for $k \geq 4$ [Bixby, 1976]
ternary	$\mathbf{U}_{5,2}$, $\mathbf{U}_{5,3}$, $\mathbf{F}_7$, $\mathbf{F}_7^*$ [Reid*]	$\mathbf{U}_{k,k-2}$ for $k \geq 5$, $\mathbf{U}_{5,2}$, $\mathbf{F}_7$, $\mathbf{F}_7^*$ [Bixby, 1979]
regular	$\mathbf{U}_{4,2}$, $\mathbf{F}_7$, $\mathbf{F}_7^*$ [Tutte, 1965]	$\mathbf{U}_{k,k-2}$ for $k \geq 4$, $\mathbf{F}_7$, $\mathbf{F}_7^*$ [Bixby, 1976]
graphic	$\mathbf{U}_{4,2}$, $\mathbf{F}_7$, $\mathbf{F}_7^*$, $\mathbf{M}^*(K_5)$, $\mathbf{M}^*(K_{3,3})$ [Tutte, 1965]	$\mathbf{U}_{k,k-2}$ for $k \geq 4$, $\mathbf{F}_7$, $\mathbf{F}_7^*$, R_{10}, $\mathbf{M}^*(K_5)$, $\mathbf{M}^*(K_{3,3})$, $\mathbf{M}^*(G_A)$, $\mathbf{M}^*(G_B)$ [Bixby, 1977a]
cographic	$\mathbf{U}_{4,2}$, $\mathbf{F}_7$, $\mathbf{F}_7^*$, $\mathbf{M}(K_5)$, $\mathbf{M}(K_{3,3})$ [Tutte, 1965]	$\mathbf{U}_{k,k-2}$ for $k \geq 4$, $\mathbf{F}_7$, $\mathbf{F}_7^*$, $\mathbf{M}(K_5)$, $\mathbf{M}(K_{3,3})$ [Bixby, 1977a]

For G_A and G_B see Fig. 17.4.

* Reid's result is unpublished. See [Bixby, 1979], [Kahn, 1984], [Seymour, 1979] and [Truemper, 1982a].

Problems for Section 17.2

17.2.8 A matrix is totally unimodular (Exercise 1.4.14) if every square submatrix of it has determinant 0 or ± 1. Call a matroid *totally unimodular* if it can be coordinatized by such a matrix over the field $\mathbb{Q}$ of the rationals. Prove the "only if" part of the following basic theorem [Tutte, 1965]: A matroid is totally unimodular if and only if it is regular.

17.2.9 Prove that, if $\mathbf{M}$ is binary and ternary, then it is regular (cf., **Box 9.1**).

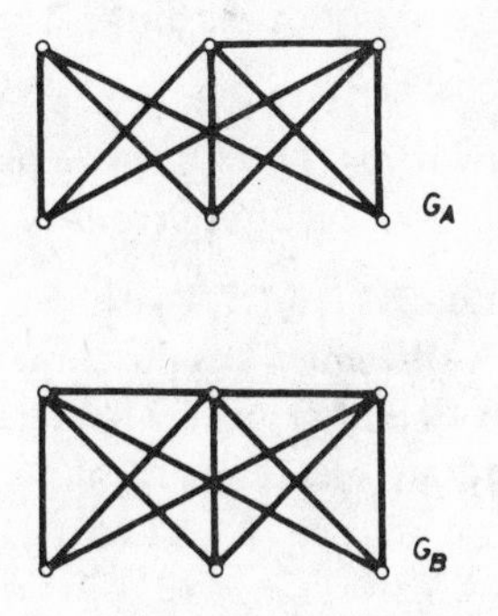

Fig. 17.4

Section 17.3 The 2-polymatroid matching problem

Theorems 5.1.2 and 5.1.5 (Hall, König) gave good characterization for the existence of a maximum matching in a bipartite graph. On the one hand, it was generalized [Tutte, 1947] for arbitrary graphs (Problem 5.1.19). On the other hand, it could also be obtained as special case of the intersection theorem for two matroids [Edmonds, 1968], see part (a) of Exercise 11.3.5. In this section we show that all these are special cases of the 2-polymatroid matching problem, see Fig. 17.5.

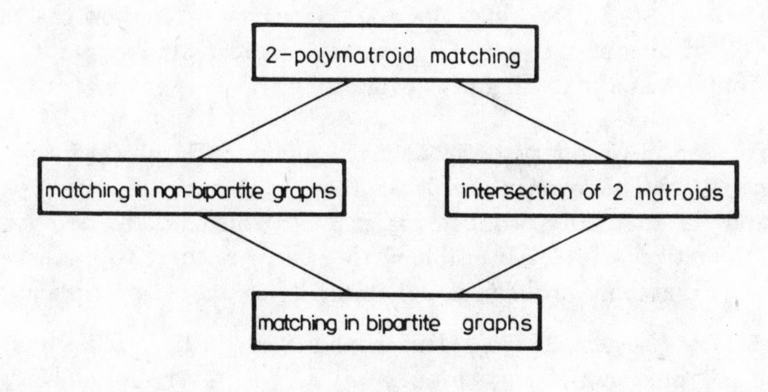

Fig. 17.5

Let f be a normalized, monotone, submodular function on the set S. If $f(\{x\}) \leq 1$ holds for every $x \in S$, we have a matroid on S with rank function f; otherwise f was called a polymatroid function (Section 13.3). We shall call it a

2-polymatroid if $f(\{x\}) = 2$ holds for every $x \in S$. Then $f(X) \leq 2|X|$ obviously follows for every $X \subseteq S$, by the submodularity. A subset $X \subseteq S$ is called a *matching* in a 2-polymatroid f if $f(X) = 2|X|$. The decision problem "Does the 2-polymatroid f have a matching of cardinality at least p?" will be called the *2-polymatroid matching problem.*

Statement 17.3.1 *The problem "Does there exist a common base in two matroids?" is a special case of the 2-polymatroid matching problem.*

Proof. Let $\mathbf{M}_1 = (S, \mathbf{F}_1)$ and $\mathbf{M}_2 = (S, \mathbf{F}_2)$ be the two matroids with respective rank functions r_1 and r_2. We may suppose that they are loop-free (possible loops can be deleted from S and from both matroids). Then $f = r_1 + r_2$ is a 2-polymatroid function and a p-element subset X belongs to $\mathbf{F}_1 \cap \mathbf{F}_2$ if and only if X is a matching of f.

□

One can generalize this statement for *k-polymatroid functions* (i.e. where $f(\{x\}) = k$ holds for every $x \in S$). A subset $X \subseteq S$ is called a *k-grouping* in f if $f(X) = k|X|$. Then the intersection problem of k matroids is a special case of the problem "Does the k-polymatroid f have a k-grouping of cardinality at least p?". Applying Theorem 13.1.2 we immediately obtain

Corollary 17.3.2 *The problem "Does the k-polymatroid f have a k-grouping of cardinality at least p?" is* **NP**-*hard for* $k \geq 3$.

Statement 17.3.3 *The problem of the maximum matching in a graph is a special case of the 2-polymatroid matching problem.*

Proof. Recall (Exercise 13.3.5) that if $A_1, A_2, \ldots, A_t$ are subsets of the underlying set S of a matroid $\mathbf{M}$ (with rank function r) then $f(X) = r\left(\bigcup_{r \in X} A_i\right)$ is a polymatroid on the set $T = \{1, 2, \ldots, t\}$. Obviously f is a k-polymatroid if $r(A_i) = k$ for every $i \in T$.

Let now G be a graph and let $\mathbf{M}$ be the free matroid on its point set $V(G)$. If the subsets $A_1, A_2, \ldots$ are just the nonloop edges of G then the maximum number $\nu(G)$ of disjoint edges in G is just the largest number p for which the 2-polymatroid f has a matching of cardinality p.

□

We have seen in Corollary 17.3.2 that the analogous problem for k-polymatroids for $k \geq 3$ is **NP**-hard. On the other hand, both the 2-matroid intersection problem and the matching problem for graphs turned out to be special cases of the 2-polymatroid matching problem. It is very natural to ask whether the 2-polymatroid matching problem is polynomially solvable in general.

Theorem 17.3.4 [Lovász, 1980a], [Jensen and Korte, 1982]. *If a 2-polymatroid is given by a "rank oracle" (which outputs $f(X)$ for every input $X \subseteq S$) then no polynomial algorithm can solve the matching problem.*

Its proof is left to the reader as Problem 17.3.3.

□

On the other hand, there is a polynomial algorithm for a very important special case. Let $\mathbf{M} = (S, \mathbf{F})$ be a loop-free matroid with rank function r

and let $A_1, A_2, \ldots$ be pairs of nonparallel elements of S. The function $f(X) = r\left(\bigcup_{i \in X} A_i\right)$ gives a 2-polymatroid and every 2-polymatroid arises in this way.

Suppose that **M** is represented over the field $\mathbb{R}$ of the reals. Then the elements of this polymatroid can be imagined as planes spanned by pairs of column vectors of the coordinatizing matrix.

Theorem 17.3.5 [Lovász, 1980a]. *If the 2-polymatroid arises in this way from a matroid* **M** *coordinatized over* **R** *then the 2-polymatroid matching problem can be solved in polynomial time.*

The proof (and the algorithm) are very complicated and not included here. Note that **M** should be coordinatized, not just coordinatizable. (If somebody assures us that **M** is coordinatizable over **R**, we still cannot find an actual coordinatization in polynomial time.) Accordingly, unlike in case of Theorem 17.3.4, here the 2-polymatroid is given by this coordinatization of **M**, not just by a rank-oracle. However, as we shall see in Sections 18.1–2, this condition is met in the applications.

The original algorithm of Lovász [1980a] was very complicated. More recently some simpler algorithms are also available [Orlin *et al.*, 1986; Gabow and Stallmann, 1986].

Let us return to the 2-polymatroid function $f(X) = r\left(\bigcup_{i \in X} A_i\right)$ where A_i are pairs $\{\mathbf{a}_i, \mathbf{b}_i\}$ of nonparallel vectors with rational entries. Define the *wedge product* $\mathbf{a}_i \wedge \mathbf{b}_i$ of these vectors as the matrix $\mathbf{W}_i = \mathbf{a}_i \mathbf{b}_i^T - \mathbf{b}_i \mathbf{a}_i^T$.

Theorem 17.3.6 [Lovász, 1979b]. *The maximum rank of the matrix* $c_1\mathbf{W}_1 + c_2\mathbf{W}_2 + \ldots$*, where the maximum is taken over all reals* $c_1, c_2, \ldots$*, is twice the cardinality of the maximum matching of the 2-polymatroid* f*. This maximum is attained when* $c_1, c_2, \ldots$ *are algebraically independent over* $\mathbf{Q}$.

The *proof* is not presented here. This theorem can be applied for a probabilistic algorithm (in the same sense as the matrix **A** of Problem 15.1.18 was used to coordinatize **M'**).

Exercises for Section 17.3

17.3.1 Consider the matroid **M** coordinatized by the matrix

$$\begin{pmatrix} 0 & 0 & 0 & 1 & 0 & 1 \\ 0 & 0 & 0 & 0 & 1 & 1 \\ 1 & 2 & 0 & 0 & 0 & 0 \\ 1 & 1 & 1 & 1 & 1 & 1 \end{pmatrix}.$$

Let A_1, A_2, A_3 be the pairs, formed by the first two, second two and last two columns of this matrix. Describe the corresponding 2-polymatroid, and find a maximum matching $Y \subseteq \{1, 2, 3\}$.

17.3.2 (continued) Determine the wedge products $\mathbf{W}_1, \mathbf{W}_2$ and $\mathbf{W}_3$, and the rank of $c_1\mathbf{W}_1 + c_2\mathbf{W}_2 + c_3\mathbf{W}_3$ for the previous exercise.

Problems for Section 17.3

17.3.3 Prove Theorem 17.3.4. (Hint: Apply Problem 13.3.11.)

17.3.4 A submodular function f on T is called a $(\leq k)$-*polymatroid* if $f(\{x\}) \leq k$ holds for every $x \in T$. Show that the analogue of Theorem 17.3.5 also holds for (≤ 2)-polymatroids, represented over $\mathbf{R}$.

17.3.5 Show that the 2-polymatroid matching problem can be formulated as the maximization problem of a submodular function [Lovász, 1983].

17.3.6 Let $\mathbf{M}_1$ and $\mathbf{M}_2$ be two matroids on the same set, both represented over $\mathbf{Q}$. The question whether the 2-polymatroid matching problem is polynomial for $\mathbf{M}_1 \vee \mathbf{M}_2$ is open, in spite of the result of Piff and Welsh (see Problem 13.2.7). Devise a probabilistic algorithm with a polynomial number of steps.

Section 17.4 Oriented matroids

Recall (Problem 7.2.18) that matroids could be defined as collections $\mathbf{C}$ of certain subsets of S, satisfying

(C1) If $X \in \mathbf{C}$, $Y \in \mathbf{C}$ and $X \subseteq Y$ then $X = Y$.

(C2) If $X \in \mathbf{C}$, $Y \in \mathbf{C}$, $X \neq Y$ and $u \in X \cap Y$ then there exists a $Z \in \mathbf{C}$ so that $Z \subseteq (X \cup Y) - \{u\}$.

We have also shown (Corollary 7.2.2) that (C2) could be replaced by the following, seemingly stronger, property:

(C3) If $X \in \mathbf{C}, Y \in \mathbf{C}, u \in X \cap Y$ and $x \in X - Y$ then there exists a $Z \in \mathbf{C}$ so that $Z \subseteq (X \cup Y) - \{u\}$ and $x \in Z$.

We try to generalize the orientation of graphs to matroids. We assigned orientations to the circuits of the directed graph (at the end of Section 1.4), but this orientation (clockwise or anticlockwise in a planar drawing) was of no real importance (its change multiplied a row of the circuit matrix by -1); its significance was a partitioning of the edge set of the circuit into two parts (namely to those edges whose orientation agreed with that of the circuit, and to the rest). For this reason, both possible orientations of the circuits will be considered.

Accordingly, a set S and a collection $\mathbf{D}$ of signed sets will be called an *oriented matroid* if it satisfies the three properties below. A *signed set* means a set X with a partition $X = X^+ \cup X^-$.

(OC1) If $X \in \mathbf{D}$ then $X \neq \emptyset$, and another signed set Y with $Y^+ = X^-$ and $Y^- = X^+$ also belongs to $\mathbf{D}$. This Y is denoted by $-X$.

(OC2) If $X \in \mathbf{D}$, $Y \in \mathbf{D}$ and $X \subseteq Y$ then either $X^+ = Y^+$ and $X^- = Y^-$ or $X^+ = Y^-$ and $X^- = Y^+$, that is, $X = \pm Y$.

(OC3) If $X \in \mathbf{D}, Y \in \mathbf{D}, X \neq \pm Y$ and $u \in (X^+ \cap Y^-) \cup (X^- \cap Y^+)$ then there exists a $Z \in \mathbf{D}$ so that $Z^+ \subseteq (X^+ \cup Y^+) - \{u\}$ and $Z^- \subseteq (X^- \cup Y^-) - \{u\}$.

Property (OC3) can be illustrated in Fig. 17.6(a). If edge u is oriented upwards then $u \in X^- \cap Y^+$; if it is oriented downwards then $u \in X^+ \cap Y^-$. Should u belong to $X^+ \cap Y^+$ or to $X^- \cap Y^-$ (like in Fig. 17.6(b)), the statement were false, even for directed graphs.

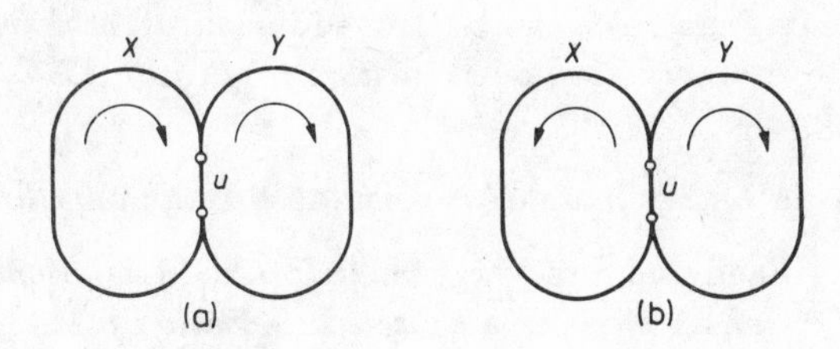

Fig. 17.6

The seemingly stronger (but really equivalent) property (C3) can also be generalized for oriented matroids, in fact in two, seemingly nonequivalent ways.

Theorem 17.4.1 [Bland and Las Vergnas, 1978]. *If* $\mathbf{M}$ *is an oriented matroid then*

(OC4) *If* $X \in \mathbf{D}$, $Y \in \mathbf{D}$, $X \neq \pm Y$, $u \in (X^+ \cap Y^-) \cup (X^- \cap Y^+)$ *and* $x \in X - Y$ *then there exists a* $Z \in \mathbf{D}$ *so that* $Z^+ \subseteq (X^+ \cup Y^+) - \{u\}$, $Z^- \subseteq \subseteq (X^- \cup Y^-) - \{u\}$, *and* $x \in Z$.

(OC5) *If* $X \in \mathbf{D}$, $Y \in \mathbf{D}$, $X \neq \pm Y$, $u \in (X^+ \cap Y^-) \cup (X^- \cap Y^+)$ *and* $x \in (X^+ - -Y^-) \cup (X^- - Y^+)$ *then there exists a* $Z \in \mathbf{D}$ *so that* $Z^+ \subseteq (X^+ \cup \cup Y^+) - \{u\}$, $Z^- \subseteq (X^- \cup Y^-) - \{u\}$ *and* $x \in Z$.

The *proof* of this theorem is long and is not included here.

If $\mathbf{M}$ is a matroid and its collection $\mathbf{C}$ of circuits can be partitioned into two subsets so that the resulting collection $\mathbf{D}$ of signed sets forms an oriented matroid then $\mathbf{M}$ is called *orientable*. Obviously every graphic matroid is orientable (just assign arbitrary orientation to the edges). The following observation is also very easy.

Proposition 17.4.2 *If* $\mathbf{M}$ *is coordinatizable over the field* $\mathbb{R}$ *of the reals then it is orientable.*

Proof. Let the matrix $\mathbf{A}$ coordinatize $\mathbf{M}$ over $\mathbb{R}$. If C is a circuit of $\mathbf{M}$, the corresponding column vectors $\mathbf{a}_1, \mathbf{a}_2, \ldots, \mathbf{a}_k$ of $\mathbf{A}$ are linearly dependent, hence there exists a linear combination $\sum_{i=1}^{k} \lambda_i \mathbf{a}_i = \mathbf{0}$ where some coefficients are nonzero. (In fact, every λ_i is nonzero, by the minimality of the circuits.) Furthermore, if $\sum_{i=1}^{k} \mu_i \mathbf{a}_i = \mathbf{0}$ is another linear combination then there exists a real constant α that $\mu_i = \alpha\lambda_i$ holds for every $i = 1, 2, \ldots, k$ (since one of the $\mathbf{a}_i$'s could be eliminated otherwise, contradicting to the minimality, again). Thus C

can be partitioned into two parts according to the positive and the negative coefficients (without depending on the actual linear combination). These partitions can easily be proved to satisfy (OC3) as well.

□

However, nonrepresentable matroids may be orientable as well. For example, the Higgs–Vámos matroid $\mathbf{K}_8$ is orientable, see [Bland and Las Vergnas, 1978]. They also prove that the matroids $\mathbf{M}_p$ of Problem 9.2.8 (in particular, the Fano matroid $\mathbf{F}_7$) are not orientable.

Minors of oriented matroids can be defined in a straightforward way. Hence, combining the above observation on $\mathbf{F}_7$ with Proposition 17.4.2 and with Theorem 17.2.1, we obtain

Statement 17.4.3 *A binary matroid is orientable if and only if it is regular.*

Theorem 17.4.4 [Bland and Las Vergnas, 1978]. *If* $\mathbf{M}$ *is orientable, so is* $\mathbf{M}^*$. *For any orientation of* $\mathbf{M}$ *there is a unique orientation of* $\mathbf{M}^*$ *so that if* X *is a circuit of* $\mathbf{M}$ *and* Y *is a circuit of* $\mathbf{M}^*$ *then* $X \cap Y \neq \emptyset$ *implies that neither* $(X^+ \cap Y^+) \cup (X^- \cap Y^-)$, *nor* $(X^+ \cap Y^-) \cup (X^- \cap Y^+)$ *is empty.*

Its *proof* is not included here (however, see Problem 17.4.4).

For other approaches to oriented matroids the reader is referred to Folkman and Lawrence [1978]; Edmonds, Fukuda [1982] and Mandel [1982].

The importance of oriented matroids lies among others in the fact that they serve for the generalization of linear programming in the same way as matroids do for that of linear algebra (see especially [Bland, 1977]).

Exercises for Section 17.4

17.4.1 Given an oriented planar graph, find an orientation of its dual, satisfying the property of Theorem 17.4.4.

17.4.2 An orientation of a matroid is *acyclic* if, for every circuit X of it, neither X^+, nor X^- is empty. It is called *totally cyclic* if every element of its underlying set is contained in a circuit X with $X^- = \emptyset$. Prove that if $\mathbf{M}$ has a totally cyclic orientation then the unique orientation of $\mathbf{M}^*$, as described in Theorem 17.4.4, is acyclic. (The converse is also true, see Las Vergnas, [1977].) What does this mean for graphic matroids?

Problems for Section 17.4

17.4.3 Where did we apply the special properties of the field $\mathbf{R}$ during the proof of Proposition 17.4.2? What about other fields?

17.4.4 Prove that if the orientation of $\mathbf{M}$ is given then the orientation of $\mathbf{M}^*$, as described in Theorem 17.4.4, is unique.

Chapter 18

Applications

Section 18.1 Unique solvability of linear active networks III: Sufficient conditions

Let N be a network, consisting of linear multiports only. We have presented a necessary condition for its unique solvability (Theorem 12.2.2). Under the weak generality assumption this condition became sufficient as well (Statement 14.1.1).

Recall that the weak generality assumption excluded algebraic relations among the nonzero entries of the describing matrices of different devices; but allowed certain algebraic relations within a single device D if they are reflected by the structure of its matroid $\mathbf{M}_D$.

For example, the matrix $\mathbf{A}'' = \begin{pmatrix} -1 & 1 & 2 & 1 \\ 0 & -1 & 2 & 1 \end{pmatrix}$ of Exercise 14.1.1 has the feature that the last two columns are parallel vectors; this fact is reflected by the dependence of the set $\{i_1, i_2\}$ in its matroid $\mathbf{M}''$. Similarly, the matrix of the 3-port circulator has quite a lot of numerical relations among its entries but all of them are reflected by the $\mathbf{M}(K_4)$ structure of its matroid (see Exercise 8.1.11). On the other hand, the matrix $\begin{pmatrix} -1 & 0 & 0 & R_1 \\ 0 & -1 & R_2 & 0 \end{pmatrix}$ of the 2-port in Problem 14.1.8 may or may not describe the gyrator — depending on whether $-R_1 = R_2$ holds or not — but this difference is not reflected by its matroid. Similarly, in the matrix $\begin{pmatrix} -1 & k & 0 & 0 \\ 0 & 0 & k & 1 \end{pmatrix}$ of the ideal transformer there are two identical entries (the transfer ratio k) but the matroid would not change if the value of one of them were modified.

This means that the necessary condition of Theorem 12.2.2 for the unique solvability of the network N need not be sufficient if N contains gyrators or ideal transformers. Really, we have seen a pathological example (Problem 14.1.8), where the "network" of Fig. 14.3 is singular if the 2-port is a gyrator and is uniquely solvable if $R_1 + R_2 \neq 0$; while it always satisfies the necessary condition of Theorem 12.2.2.

Theorem 18.1.1 *Let the network N contain resistors and gyrators only, and let the resistances and gyrator constants be algebraically independent over $\mathbb{Q}$. (Observe that this requirement is weaker than the weak generality assumption.) Then N is uniquely solvable if and only if the graph of N has a normal spanning forest.*

Recall that a spanning forest was called normal if it contained either none or both edges of each gyrator. We proved the sufficiency (Theorem 6.2.6). Now we are in the position to sketch a proof [Lovász and Plummer, 1986] for the necessity, which is significantly different from the original proof of [Milić, 1974].

Proof. (sketch). Suppose that N consists of n gyrators only and let $\mathbf{B}$ denote the $v \times 2n$ incidence matrix of the directed graph of N. Then Kirchhoff's Current Laws can be written as $\mathbf{Bi} = \mathbf{0}$. Let the entries of a vector $\mathbf{p}$ be the "node-potentials" (see **Box 4.2**). Then the voltages can simply be expressed as $\mathbf{u} = \mathbf{B}^T\mathbf{p}$. Finally, if the reciprocal values of the gyrator constants are denoted by $G_1, G_2, \ldots, G_n$, respectively, then $\mathbf{i} = \mathbf{Gu}$, where

$$\mathbf{G} = \begin{pmatrix} \boxed{\begin{matrix} 0 & G_1 \\ -G_1 & 0 \end{matrix}} & & \mathbf{0} \\ & \boxed{\begin{matrix} 0 & G_2 \\ -G_2 & 0 \end{matrix}} & \\ \mathbf{0} & & \ddots \end{pmatrix}$$

Thus $\mathbf{0} = \mathbf{Bi} = \mathbf{BGu} = (\mathbf{BGB}^T)\mathbf{p}$. The network is uniquely solvable if and only if $r(\mathbf{BGB}^T) = v - c$, since one of the potentials can always be initially prescribed in each connected component of the network graph.

Let $\mathbf{b}_1, \mathbf{b}_2, \ldots, \mathbf{b}_{2n}$ be the column vectors of $\mathbf{B}$. A routine calculation shows that $\mathbf{BG} = G_1(-\mathbf{b}_2, \mathbf{b}_1, 0, 0, \ldots) + G_2(0, 0, -\mathbf{b}_4, \mathbf{b}_3, 0, 0, \ldots) + \ldots$ and finally

$$\mathbf{BGB}^T = \sum_{i=1}^{n} G_i(\mathbf{b}_{2i-1} \wedge \mathbf{b}_{2i}).$$

We can now apply Theorem 17.3.6, that is, $r(\mathbf{BGB}^T) = v - c$ holds if the G_i's are algebraically independent over $\mathbb{Q}$ and there exists a maximum matching of cardinality $(v - c)/2$. This latter is just a normal spanning forest.

□

Theorem 18.1.2 *The condition of the above theorem can be checked in polynomial time.*

Proof. This is a trivial special case of Theorem 17.3.5 if N consists of gyrators only. If N contains resistors as well, a small modification is needed which is left to the reader as Problem 18.1.4.

□

If a simple device like the gyrator leads to problems which require matroid matching algorithm, what can we expect in general? As long as the network consists of resistors and 3-terminal 2-ports only, essentially this is the only situation which can lead to singularity, yet is not reflected in the structure of the matroids of the 2-ports.

In order to formulate this claim in a precise way, let us call a 3-terminal 2-port D with describing equations $\begin{pmatrix} a & b \\ c & d \end{pmatrix}\begin{pmatrix} u_1 \\ u_2 \end{pmatrix} + \begin{pmatrix} \alpha & \beta \\ \gamma & \delta \end{pmatrix}\begin{pmatrix} i_1 \\ i_2 \end{pmatrix} = \begin{pmatrix} 0 \\ 0 \end{pmatrix}$ to be *potentially singular* if at least one of the following relations holds:

$$ad = bc; \quad a\delta = \beta c; \quad \alpha d = b\gamma; \quad \alpha\delta = \beta\gamma; \tag{18.1}$$

$$(a+b)(\gamma-\delta) = (c+d)(\alpha-\beta) \tag{18.2}$$

Observe that (18.2), which is also satisfied by gyrators, is the only one among these five relations which is not reflected by the matroid $\mathbf{M}_D$ of the above 2-port D.

Theorem 18.1.3 *Instead of the weak generality assumption let us suppose only that no algebraic relation exists among nonzero entries of the matrices of different 2-ports. Let the network N consist of 3-terminal 2-ports and resistors only. If the necessary condition of Theorem 12.2.2 is met but N is singular then at least one of its 2-ports is potentially singular.*

Its *proof* is not given here, see [Recski, 1980a]. □

Hence if the above network N contains no potentially singular 2-ports then the necessary condition of Theorem 12.2.2 is also sufficient for the unique solvability, even if the above requirement is postulated only, instead of the weak generality assumption.

Exercises for Section 18.1

18.1.1 Show that the definition of potential singularity does not depend on the actual matrix description of the 2-port.

18.1.2 Check that the gyrator meets Eq. (18.2). Give an example where the potentially singular 2-port, satisfying Eq. (18.2), is reciprocal.

18.1.3 Apply Theorem 18.1.3 to the network of Fig. 18.1 [Recski and Takács, 1981.] The equation of the 2-ports are

$u_1 = R_1 i_1 + R_0 i_2 \qquad i_3 = 0 \qquad i_5 = 0$

$u_2 = R_2 i_1 + R_0 i_2 \qquad i_4 = G_1 u_3 + G_2 u_4 \qquad u_5 = r i_6$

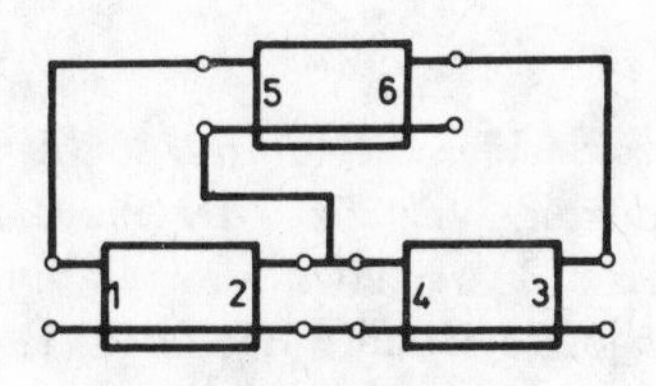

Fig. 18.1

Problems for Section 18.1

18.1.4 Finish the proof of Theorem 18.1.2.

18.1.5 Generalize Theorems 18.1.1–18.1.2 to networks, containing resistors, gyrators and ideal transformers.

18.1.6 (continued) Formulate (in terms of field extensions, as in Section 5.3) the algebraic observation required to prove your statement.

Section 18.2 Rigidity of trusses IV: How many joints must be pinned down?

Let F be a planar framework composed of v joints and e rods. We described it by a matrix $\mathbf{A}(F)$ of e rows and $2v$ columns (Section 6.3). If certain joints were fixed to the plane ("pinned down") then we deleted from $\mathbf{A}(F)$ the 2–2 columns, corresponding to the velocity coordinates of these joints, and concluded (Statement 8.2.1) that F became fixed if and only if the columns of the remaining matrix $\mathbf{W}$ are linearly independent.

For example, if F is the framework of Fig. 18.2 then

$$\mathbf{A}(F) = \begin{pmatrix} x_1 - x_2 & x_2 - x_1 & 0 & 0 & y_1 - y_2 & y_2 - y_1 & 0 & 0 \\ 0 & x_2 - x_3 & x_3 - x_2 & 0 & 0 & y_2 - y_3 & y_3 - y_2 & 0 \\ 0 & 0 & x_3 - x_4 & x_4 - x_3 & 0 & 0 & y_3 - y_4 & y_4 - y_3 \\ x_1 - x_4 & 0 & 0 & x_4 - x_1 & y_1 - y_4 & 0 & 0 & y_4 - y_1 \end{pmatrix}$$

We have seen (Fig. 8.15) that fixing joints 1, 3 makes the whole system rigid while fixing 1 and 2 does not.

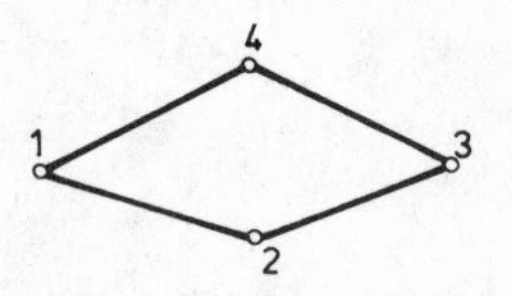

Fig. 18.2

Alternatively, we may consider the column space matroid $\mathbf{M}_F$ of the matrix $\mathbf{A}(F)$. It is a rank 4 matroid, visualized by the drawing of Fig. 18.3 (Exercise 18.2.2). If joints 1 and 3 are fixed, the remaining subset $\{\dot{x}_2, \dot{x}_4, \dot{y}_2, \dot{y}_4\}$ is independent in $\mathbf{M}_F$; if joints 1 and 2 are fixed, the subset $\{\dot{x}_3, \dot{x}_4, \dot{y}_3, \dot{y}_4\}$ is dependent.

In this section the general question is considered; determine the minimal number of joints whose fixing makes the whole system rigid.

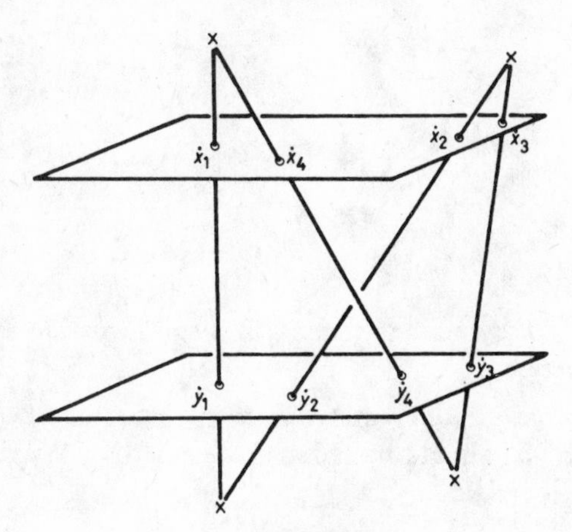

Fig. 18.3

Theorem 18.2.1 [Lovász, 1980b]. *The minimum number of joints whose fixing makes the whole 2-dimensional system F rigid, can be found in polynomial time.*

Proof. Consider the matroid $\mathbf{M}_F$ of the system and define the pairs $P_1 = \{\dot{x}_1, \dot{y}_1\}, P_2 = \{\dot{x}_2, \dot{y}_2\}, \ldots, P_v = \{\dot{x}_v, \dot{y}_v\}$. $\mathbf{M}_F$ determines a 2-polymatroid on the set $T = \{P_1, P_2, \ldots, P_v\}$, see Problem 18.2.4 as well. Apply Theorem 17.3.5. The complement of a maximum matching is just a minimum system of joints to be fixed.

□

What is the analogue of this result in case of 3-dimensional frameworks? Using the same argument the problem of finding the minimum number of joints to be fixed could be reduced to finding a maximum matching in a (≤ 3)-polymatroid. This latter is known to be **NP**-hard, see Corollary 17.3.2. This fact does not imply directly that our original problem is **NP**-hard as well, since the special structure of the matroids of the frameworks was not taken into consideration. Nevertheless, one can prove — using an entirely different method, see [Mansfield, 1981] — that this problem is also very difficult:

Theorem 18.2.2 *The problem to decide whether a 3-dimensional framework can be made rigid by fixing a given number of joints is* **NP**-*complete.*

Exercises for Section 18.2

18.2.1 Find a minimum system of joints whose fixing makes the whole system rigid, for the planar frameworks of Fig. 18.4.

18.2.2 Check that the drawing of Fig. 18.3 correctly represents the matroid $\mathbf{M}_F$ of the framework F of Fig. 18.2.

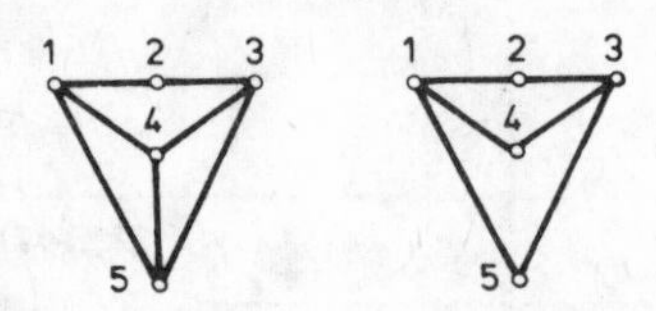

Fig. 18.4

Problems for Section 18.2

18.2.3 Change Fig. 18.2 by making joints 1, 2, 3 collinear. Describe the matroid of the new system.

18.2.4 In order to obtain a 2-polymatroid in the proof of Theorem 18.2.1, the pairs $P_i = \{\dot{x}_i, \dot{y}_i\}$ should be independent in $\mathbf{M}_F$. What does it mean if a pair P_i is dependent?

Section 18.3 Tensegrity frameworks

In case of a certain loading some of the rods of a framework are under compression, some others under tension. (Some of them may have no stress whatsoever, they could theoretically be removed without effecting anything as long as the loading does not change.) The rods under tension could be changed to cables. Even if one does not really change these rods to cables, it is important to know this distinction since real life rods with given size, made from a given material, might be more reliable under tension than under compression.

Therefore one might study *tensegrity frameworks* where certain pairs of joints are interconnected by one of the following three types of elements:
(a) rods (which are rigid both under tension and compressions).
(b) cables (which are used under tension only).
(c) struts (which are used under compression only).
In what follows, these elements will be denoted by the symbols of Fig. 18.5.

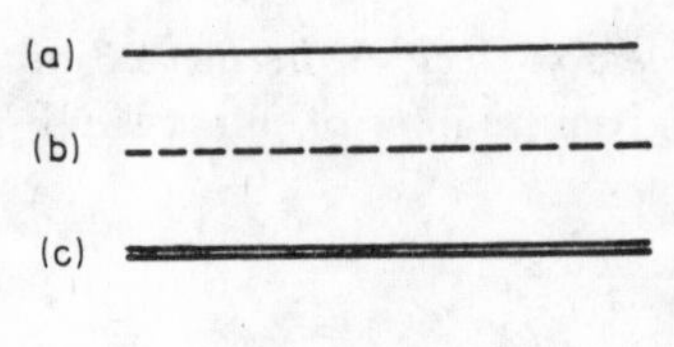

Fig. 18.5

For example, a square-shaped framework consists of four rods. We have seen in Section 2.6 that it can be made rigid using one diagonal rod or two diagonal cables. It is easy to see (Exercise 18.3.1) that in the latter case the original four rods can be changed to struts without effecting rigidity.

In order to study tensegrity frameworks we return to "normal" frameworks at first. Unless the system F or a part of it is statically underdetermined, the rows of its matrix $\mathbf{A}(F)$ are linearly independent. Then changing any rod to cable or strut would increase the degrees of freedom of this system.

Let us consider, therefore, statically underdetermined systems, like the one in Fig. 18.6. Its matrix $\mathbf{A}$, after deleting the columns corresponding to $\dot{x}_1, \dot{y}_1$ and $\dot{y}_2$ becomes

$$\mathbf{W} = \begin{pmatrix} x_2 - x_1 & 0 & 0 & 0 & 0 \\ 0 & x_3 - x_1 & 0 & y_3 - y_1 & 0 \\ 0 & 0 & x_4 - x_1 & 0 & y_4 - y_1 \\ x_2 - x_3 & x_3 - x_2 & 0 & y_3 - y_2 & 0 \\ x_2 - x_4 & 0 & x_4 - x_2 & 0 & y_4 - y_2 \\ 0 & x_3 - x_4 & x_4 - x_3 & y_3 - y_4 & y_4 - y_3 \end{pmatrix}$$

If the numerical values of the coordinates x_2, x_4, y_3, y_4 are chosen to 1 and that of the others to zero, as on the figure, then we obtain

$$\mathbf{W} = \begin{pmatrix} 1 & 0 & 0 & 0 & 0 \\ 0 & 0 & 0 & 1 & 0 \\ 0 & 0 & 1 & 0 & 1 \\ 1 & -1 & 0 & 1 & 0 \\ 0 & 0 & 0 & 0 & 1 \\ 0 & -1 & 1 & 0 & 0 \end{pmatrix}$$

The rows of $\mathbf{W}$ are obviously linearly dependent. If a linear combination of them with coefficients $\alpha, \beta, \gamma, \delta, \varepsilon, \varsigma$, respectively, leads to the zero vector then we have $\alpha + \beta = 0,\ \delta + \varsigma = 0, \gamma + \varsigma = 0, \beta + \delta = 0$ and $\gamma + \epsilon = 0$. Solving this system of equations we obtain that any solution $(\alpha, \beta, \gamma, \delta, \epsilon, \varsigma)$ is a constant multiple of $(1, 1, -1, -1, 1, 1)$. This indicates that rods c and d have one type of stress and all the others have the opposite type of stress, under any loading.

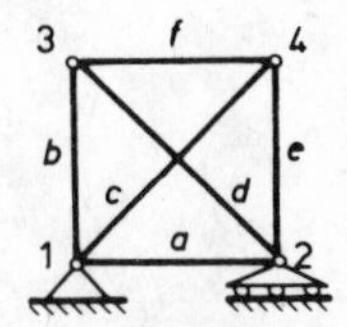

Fig. 18.6

Let us formulate our observation in general:

Proposition 18.3.1 *Suppose that a system S of k rods is rigid. Delete those columns from its matrix $\mathbf{A}$ which correspond to the fixation of S (two columns for each fixed joint, one for each "track"). Suppose that the resulting matrix $\mathbf{W}$ has size $k \times (k-1)$, hence S is statically underdetermined. Let $\mathbf{x}^T$ be a nonzero row vector satisfying $\mathbf{x}^T\mathbf{W} = \mathbf{0}^T$. Then either the rods of S, corresponding to the positive entries of $\mathbf{x}$, can be changed to cables and those, corresponding to the negatives ones, can be changed to struts, or vice versa, without effecting rigidity. If an entry of $\mathbf{x}$ is zero, the corresponding rod cannot be changed to cable or strut.*

Let us generalize this observation for an arbitrary tensegrity framework F. We define the *underlying system F' of rods* by simply replacing cables and struts by rods. (If F contained rods as well, they remain unchanged.) Then consider the oriented matroid $\mathbf{M}(F)$, coordinatized by the *row vectors* of the matrix $\mathbf{A}(F)$. Its underlying set S has a tripartition $S_C \cup S_S \cup S_R$, corresponding to cables, struts and rods, respectively, in the original tensegrity framework F.

For example, the underlying system of rods for both tensegrity frameworks F_1, F_2 of Fig. 18.7 is the same. After deleting the $\dot{x}_1, \dot{y}_1, \ldots, \dot{x}_4, \dot{y}_4$ columns of matrix $\mathbf{A}$ of either F_1' or F_2', we obtain $\mathbf{W}(F_1) = \mathbf{W}(F_2) =$
$$= \begin{pmatrix} x_5 - x_1 & y_5 - y_1 \\ x_5 - x_2 & y_5 - y_2 \\ x_5 - x_3 & y_5 - y_3 \\ x_5 - x_4 & y_5 - y_4 \end{pmatrix} = \begin{pmatrix} 1 & 1/2 \\ 0 & 1/2 \\ 1 & -1/2 \\ 0 & -1/2 \end{pmatrix},$$
and the corresponding oriented matroids $\mathbf{M}(F_1) = \mathbf{M}(F_2)$ can be described as the circuit matroid of the directed graph of Fig. 18.8a.

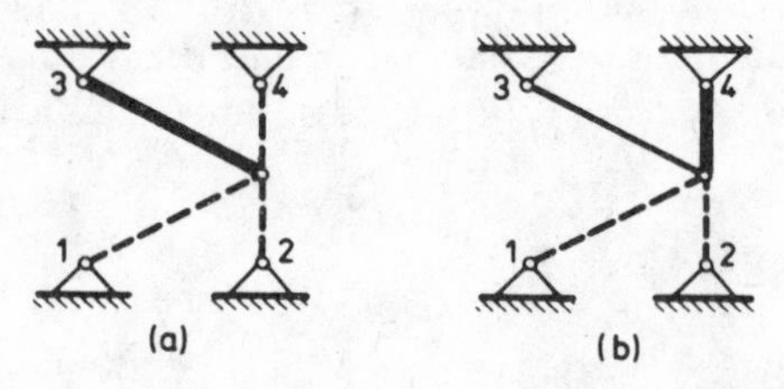

Fig. 18.7

Suppose for a while that the oriented matroid $\mathbf{M}(F)$ of the underlying system of rods F' is graphic and is described by a directed graph G. We define the *tensegrity transformation* of G (with respect to the tripartition $S_C \cup S_S \cup S_R$) as follows: If $e \in E(G)$ corresponds to a cable, leave the edge unaltered. If it corresponds to a strut, reverse its orientation. If it corresponds to a rod, replace it with a pair of oppositely oriented edges. For example, the tensegrity transformation of the directed graph of Fig. 18.8a with respect to the tripartitions

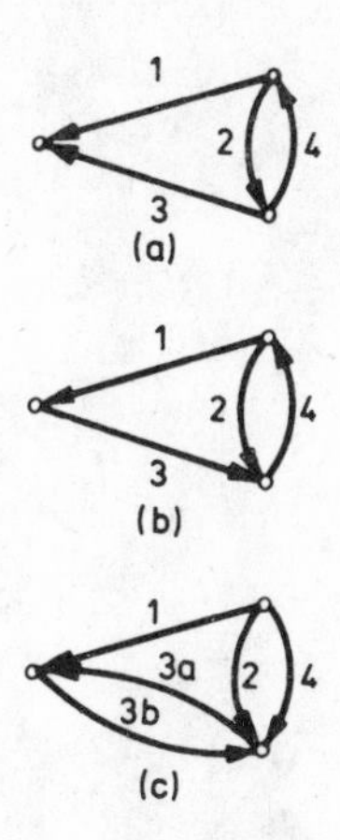

Fig. 18.8

$\{1,2,4\} \cup \{3\} \cup \emptyset$ and $\{1,2\} \cup \{4\} \cup \{3\}$ are shown in Figs 18.8b and 18.8c, respectively.

Theorem 18.3.2 *Let F be a tensegrity framework and suppose that the underlying system F' of rods is rigid (i.e. dynamically determined). Suppose that the oriented matroid $\mathbf{M}(F)$ is graphic and is described by a directed graph G. Then F is rigid if and only if the tensegrity transformation of G is strongly connected.*

For example, the graph of Fig. 18.8b is strongly connected while that of Fig. 18.8c is not. Accordingly, the first tensegrity framework of Fig. 18.7 is rigid and the second one is not.

The graphicness of $\mathbf{M}(F)$ is not essential. The tensegrity transformation can clearly be generalized for any oriented matroid and strong connectedness is replaced by the requirement that the oriented matroid be totally cyclic.

Applying the results of Exercises 9.3.10, 17.4.2 and Problem 9.3.14, one can reformulate Theorem 18.3.2 as follows:

Theorem 18.3.3 *Let F be a tensegrity framework and suppose that the underlying system F' of rods is rigid. Then F is rigid if and only if the support (see Exercise 9.3.10) of every linear combination of the columns of $\mathbf{W}(F)$*

(a) either contains a rod of F;

(b) or contains two cables (or two struts) of F so that the corresponding entries of this column have different signs;

(c) or contains a cable and a strut of F so that the corresponding entries of this column have identical signs.

Exercises for Section 18.3

18.3.1 Show directly that the tensegrity framework of Fig. 18.9 is rigid. Starting from a square and its two diagonals, is there any other rigid configuration of cables and struts?

18.3.2 Consider the system of Fig. 18.10 with coordinates $x_1 = y_1 = y_2 = 0$, $x_3 = \frac{1}{2}$, $x_2 = y_3 = 1$. Replace the rods by cables or struts if the coordinates of joint 4 are
(a) $x_4 = y_4 = \frac{1}{2}$
(b) $x_4 = 2$; $y_4 = 1$
(c) $x_4 = \frac{1}{2}$; $y_4 = 0$

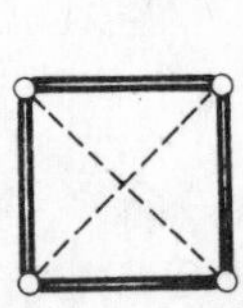

Fig. 18.9

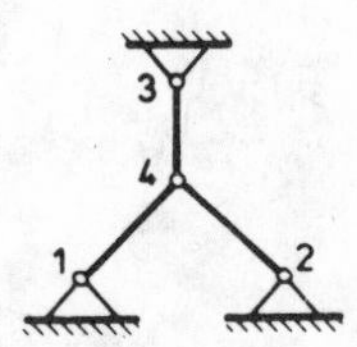

Fig. 18.10

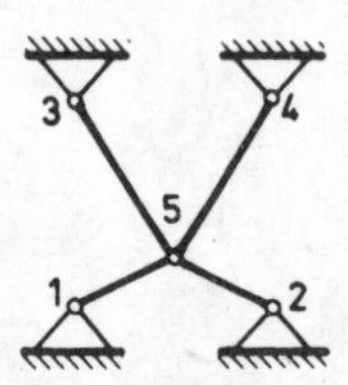

Fig. 18.11

18.3.3 (continued) Describe the oriented matroids, determined by the above three systems.

18.3.4 Consider the system of Fig. 18.11 with coordinates $x_1 = x_3 = y_1 = {}$ $= y_2 = 0$, $x_2 = x_4 = y_3 = y_4 = 1$. Replace the rods by cables or struts if
(a) $x_5 = y_5 = \frac{1}{2}$
(b) joint 2 and 5 coincide.

18.3.5 (continued) Describe the oriented matroids, determined by the above two systems.

18.3.6 Let T_1, T_2 be two tensegrity frameworks with isomorphic graphs and with identical tripartitions of the edge sets to rods, cables and struts. Let T_1 be rigid and T_2 not. Is it true that a small perturbation of the joints of T_2 leads to a rigid framework?

Problems for Section 18.3

18.3.7 Modify the system of Fig. 18.6 so that $x_1 = x_3 = y_1 = y_2 = 0$, $x_2 = y_3 = 1$ but x_4 and y_4 become $\frac{1}{2}$. Describe its rigidity properties.

18.3.8 Study the system of Fig. 18.11 if $x = \frac{1}{2}$, $y = \frac{2}{3}$.

18.3.9 The results of this section suggests a "symmetry" between cables and struts. Show a tensegrity framework where interchanging their role changes infinitesimal motion to a "real" one.

18.3.10 Replace the edges of a regular hexagon by struts and add a suitable set of four diagonals as cables. Decide whether the resulting planar tensegrity frameworks are rigid.

Appendix 1

Some important results in chronological order

1736 Euler	first paper on graph theory
1812 Lhuilier	applications of duality of planar graphs
1847 Kirchhoff	first application of graph theory in electricity
1852 De Morgan	formulation of the 4-colour-conjecture which motivated much of the research on graphs and matroids in the next 125 years
1864 Maxwell	first application of graph theory in statics
1875 Frobenius	first minimax theorem
1891 Petersen	first result in matching theory
1926 Borůvka	greedy algorithm
1930 Kuratowski	characterization of planar graphs
1932 Whitney	relation between planarity and duality
1933 Whitney	2-isomorphism of graphs
1935 Whitney	definition of matroids
1935 van der Waerden	unified study of linear and algebraic dependence
1936 Birkhoff	relation between matroids and lattices
1936 König	first book on graph theory
1942 Rado	first minimax theorem in matroid theory
1959 Tutte	characterization of regular and graphic matroids
1965 Edmonds	matroid partition
1970 Laman	characterization of generic rigidity in the plane
1971 Cook	**NP**-completeness
1972 Karp	many famous problems are **NP**-complete
1976 Welsh	first monography on matroid theory
1979 Garey–Johnson	first monography on **NP**-theory
1980 Lovász	matroid matching
1980 Seymour	decomposition of regular matroids

Appendix 2

List of the Boxes

Appendix 3

List of the Algorithms

Appendix 4

Solutions to the Exercises and Problems

Chapter 1

1.1.1 Yes, the points $1, 2, 3, 4, 5, 6, 7$ correspond to the points a, d, g, c, f, b, e, respectively. (Due to symmetry one may suppose that 1 corresponds to a. Among the four neighbours 2, 3, 6 and 7 there are exactly two (2 and 7) which are adjacent. Hence they should correspond to d and e (or to e and d), respectively. From now on, everything is straightforward.)

1.1.2 The two graphs of Fig. S.1.1 determine π uniquely. If G and H are complete graphs with the same number of points then any π will do.

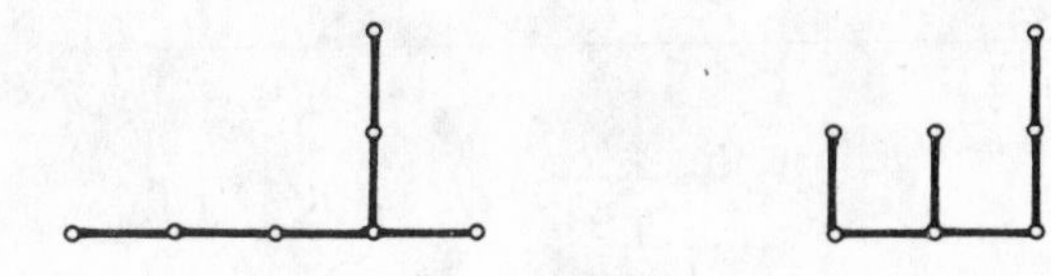

Fig. S.1.1

1.1.3 All the three graphs are isomorphic, see Fig. S.1.2.

1.1.4 (a) Let G be the graph of Fig. 1.8.

(b) This is impossible. If G is disconnected, let A be a proper subset of $V(G)$ so that no point of A is adjacent to any point of $B = V(G) - A$. We show that any two distinct points x, y of H can be connected in H by a path of length 1 or 2. If $x \in A$ and $y \in B$ or *vice versa*, they are adjacent in H. Otherwise let $x, y \in A$, for example, and let $z \in B$. Both $\{x, z\}$ and $\{z, y\}$ are edges in H.

1.1.5 (a) Take a circuit of length 5.

(b) This is impossible. If G has v points and e edges then its complement has $\frac{v(v-1)}{2} - e$ edges. In order to have an isomorphic complement, $\frac{v(v-1)}{2}$ must be even which is not the case if v is of form $4k + 2$ or $4k + 3$.

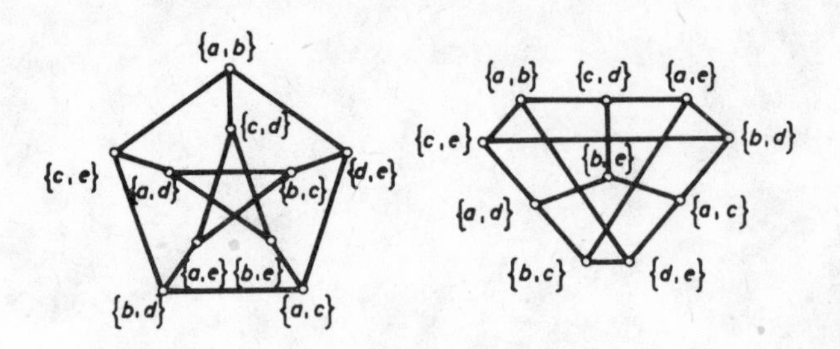

Fig. S.1.2

1.1.6 If all the points of the edge sequence are different, we are done. Otherwise let v_i be the first point of the sequence which arises more than once. Delete the subsequence between the first and last appearance of v_i. Hence we obtain a shorter edge sequence between v_0 and v_k. Repeat this process, if necessary, until receiving a path.

The answer is negative to the second question, see Fig. S.1.3.

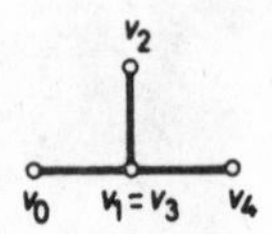

Fig. S.1.3

1.1.7 If G is not simple, there is a circuit of length 1 or 2. If G is simple, the paths P_1, P_2 between v_1 and v_2 can be given by listing their points:

$$P_1 = [v_1 = u_0, u_1, u_2, \ldots, u_k = v_2]; \; P_2 = [v_1 = w_0, w_1, w_2, \ldots, w_l = v_2].$$

Let $i > 0$ be the first subscript with $u_i \neq w_i$. (The case $i = 1$ is also possible but the two paths may start in the same way, see Figure S.1.4.) Let P_1' be the subpath $[u_i, u_{i+1}, \ldots, u_k]$ of P_1 and P_2' be the subpath $[w_i, w_{i+1}, \ldots, w_l]$ of P_2. Let u_j be the first point of P_1' which also appears on P_2'. (The case $j = 1$ is also possible, provided that u_i is not the first but a "later" point of P_2'.) Then the subpath of P_1' from u_i to u_j and the subpath of P_2' back from u_j to w_i and the edges $\{w_i, u_{i-1}\}$ and $\{u_{i-1}, u_i\}$ form a circuit. Figure S.1.4 shows that the circuit need not contain v_1 or w_2.

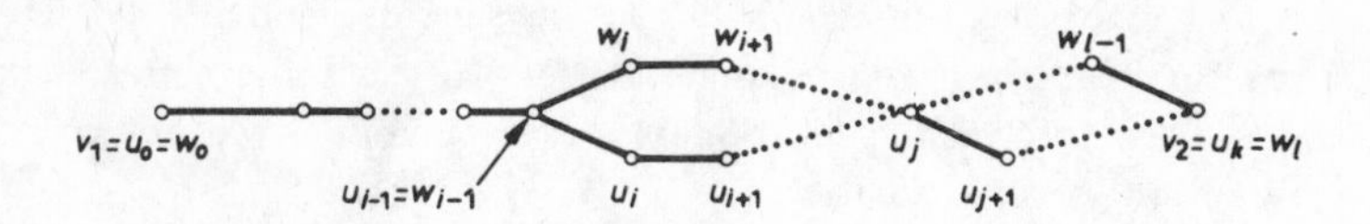

Fig. S.1.4

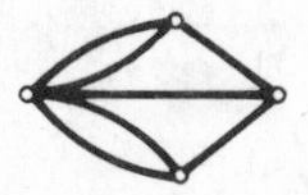

Fig. S.1.5

The statement need not be true for distinct edge sequences, see Fig. S.1.3 with $P_1 = [v_0, v_1, v_2, v_1, v_4]$ and $P_2 = [v_0, v_1, v_4]$.

1.1.8 One must leave each point of the graph (Fig. S.1.5) of the bridges as many times as one reaches that point. This is impossible since each point is of odd degree.

1.1.9 If e is a separating edge, there exist two points a, b so that every path P from a to b contains e. Should e be contained in a circuit C, the edges in $(P \cup C) - \{e\}$ would form an edge sequence from a to b. Then Exercise 1.1.6 leads to a contradiction.

On the other hand, if $e = \{u, v\}$ is not contained in any circuit then it forms the only possible path between u and v.

1.1.10 No. On the one hand, a separating point can be contained in a circuit, see v_4 in Fig. 1.6. On the other hand, if a point is not contained in any circuit, it need not be separating, it may be of degree one.

Nevertheless, if a point v is of degree at least two and is not contained in any circuit, it is separating. Let a, b be two distinct neighbours of v. If there were a path P from a to b, not containing v, then this path and the edges $\{a, v\}$ and $\{v, b\}$ form a circuit, containing v.

1.1.11 If $e = \{v_1, v_2\}$ is a separating edge and v_1 is of degree at least 2 then let v_3 be a neighbour of it with $v_2 \neq v_3$. Every path P from v_2 to v_3 must contain v_1 (since otherwise P with e and $\{v_1, v_3\}$ would form a circuit, containing e). Hence v_1 is a separating point.

The answer to the second question is negative, see v_4 in Fig. 1.6.

1.1.12 Let v be a separating point with neighbours u_1, u_2, u_3. If none of the edges $e_i = \{u_i, v\}$ are separating, each of them is contained in circuits. Let, for example, the circuit containing e_1 also contain e_2. Deleting v from the graph, u_1 and u_2 must belong to the same component, hence u_3 belongs to another component. But then e_3 were not contained in any circuit, a contradiction.

1.1.13 $c(G) \leq c(G-e)$ is trivial. Suppose $c(G-e) > c(G)+1$, that is, suppose that the number of components increases by at least 2 if e is deleted. Let x, y, z be three distinct points so that they belong to the same component of G but no two of them belong to the same component of $G-e$. Since e is separating, it is not contained in any circuit (Exercise 1.1.9). Hence we may say that a point of G is on one "side" or on the other "side" of e. Now, at least two of the above three points are on the same "side", hence they can be connected by a path, not containing e, a contradiction.

If a point v is deleted from G, the quantity $c(G)$ may even decrease (if v was an isolated point) or arbitrarily increase (if $E(G)$ consists of edges of form $\{v, u_i\}$ for $i = 1, 2, \dots$ only).

1.1.14 If we add all the degrees, each edge is counted exactly twice.

1.1.15 Let points correspond to people, edges to mutual acquaintances. By the previous problem, the sum of all the degrees is even, hence the number of odd degree points must also be even.

1.1.16 Let $e = \{u_1, u_2\}$ be the separating edge and G_1, G_2 be the two components of $G-e$. G had an even number of points (by the previous problem), hence $v(G_1)$ and $v(G_2)$ are both odd or both even. The latter is impossible since u_1 and u_2 are the only even degree points of G_1 and G_2, respectively.

1.1.17 No, unless the graph has one point and no edges. If G is simple, $0 \leq$ $\leq d(v) \leq n-1$ holds for every $v \in V(G)$. Should all these degrees be different, all the numbers $0, 1, \dots, n-1$ would arise. But 0 indicates an isolated point and $n-1$ indicates a point which is adjacent to all the others, a contradiction if $n > 1$.

1.1.18 By the previous problem, all the degrees are either
(a) between 0 and $n-2$ (and then there exists an isolated point), or
(b) between 1 and $n-1$ (and then there exists a point which is adjacent to all the others).

Deleting this distinguished point a new graph with the same property arises (and case (a) turns to case (b) and *vice versa*). Thus all the graphs with this property can be obtained from the graphs with two points, by alternatingly using the following two constructions: If there is no isolated point, add one. If there is one, add a point and join to all the others.

1.1.19 (a) There is an even number $2k$ of points with odd degrees; join them by k edges in an arbitrary way. The remaining degree requirements are even at every point; realize them by loops.

(b) $n = 2,\ d_1 = 0,\ d_2 = 2$.

(c) $n = 4,\ d_1 = 0,\ d_2 = 1,\ d_3 = 2,\ d_4 = 3$.

1.1.20 G may be supposed to be simple (since there is a circuit of length 1 or 2 in G otherwise). Let $P = [v_1, e_1, v_2, e_2, \ldots v_{t-1}, e_{t-1}, v_t]$ be a maximal path in G. Since $d(v_t) > 1$, there is an edge $e = \{v_t, u\}$, different from e_{t-1}. The point u cannot be different from all the points $v_1, v_2, \ldots, v_{t-2}$ of P (since P was maximal), hence the subpath from u to v_t plus e form a circuit.

1.1.21 The graph G has a circuit C_1, by the previous problem. Every point of $G - E(C_1)$ has even degree again, hence this graph has a circuit C_2 etc. The process terminates when all the edges are covered by $C_1 \cup C_2 \cup \ldots$.

1.1.22 The statement is true if $v(G) = 1$. Suppose that the statement is not always true and let G_0 be a counterexample with $v(G_0)$ minimum. G_0 has a point p with degree at most 1 (by Problem 1.1.20). Then $G_0 - p$ is also a counterexample, which is impossible.

1.1.23 Consider the maximal path P as introduced in the solution of Problem 1.1.20. v has at least $k - 1$ neighbours among $\{v_0, v_1, \ldots, v_{t-2}\}$ (since it cannot have any neighbour outside P, by the maximality). Hence the number of these points, that is $t - 1$, must be greater than or equal to $k - 1$, hence the length of P is at least k.

1.1.24 Suppose that G is disconnected, its components $G_1, G_2, \ldots$ have $k_1, k_2, \ldots$ points, respectively. Since the degree of any point of G_i is at most $k_i - 1$, we obtain $k_i > \lfloor n/2 \rfloor$, hence $k_i \geq \lfloor n/2 \rfloor + 1$. Since $c(G) \geq 2$ it leads to $n \geq k_1 + k_2 \geq \lfloor n/2 \rfloor + \lfloor n/2 \rfloor + 2 > n$, a contradiction.

If the degrees may equal $\lfloor n/2 \rfloor - 1$, G may be disconnected (put $v(G) = 2$ or 3 and $e(G) = 0$, for example).

1.1.25 Solution 1. While Algorithm 1.2 is running, stop it when the kth point is just removed from File 3. (This is the $(k-1)^{th}$ application of Step 2 since the first point was "removed" in Step 1.) These k points span a connected subgraph (since the algorithm would run in the same way for this subgraph as well).

Solution 2. Delete as many edges of G as possible without disconnecting it. The result has no circuits, hence, by Problem 1.1.20, has a point of degree one. This point can be deleted from G without disconnecting it. Repeat this procedure $n - k$ times.

1.1.26 $n - 2$ separating points are possible (see a path of length $n - 1$). This is best possible since the end points of a maximal path can easily be proved to be nonseparating.

1.1.27 Suppose there were two paths P_1, P_2 with maximum length ℓ so that they have no point in common. Let P be an arbitrary path connecting a point of P_1 with a point of P_2. (Such P exists by the connectivity of G.) Let x be the last point along P which still belongs to P_1 and then let y be the first point after x which already belongs to P_2. Let a_1, b_1 be the lengths of the two subpaths of P_1, determined by x and let a_2, b_2 be the length of the two subpaths of P_2, determined by y. We may suppose that $a_1 \geq b_1$ and $a_2 \geq b_2$. Then $a_1+b_1 = a_2+b_2 = \ell$ implies $a_1+a_2 \geq \ell$. If the distance between x and y on P is c, we obtain a path of length $a_1 + a_2 + c$, a contradiction.

1.1.28 Suppose the contrary and let G be a counterexample with a minimum number of edges. G contains closed edge sequences (e.g. a circuit by Problem 1.1.20), let C be such a closed edge sequence with a maximum number of edges. $E(C)$ is a proper subset of $E(G)$, hence, by the connectivity, there exists an edge $e = \{x, y\}$ so that $e \notin E(C)$ yet $x \in V(C)$. Consider the component of $G - C$ which contains x. The degrees are still even, hence the edges of this component can be covered by a closed edge sequence C' (since G was a minimum counterexample). Start from x to form an edge sequence along C first and along C' thereafter. This edge sequence is also closed, contradicting the maximality of C.

1.1.29 Let a, b be the two exceptional points of G and add a new edge $e = \{a, b\}$ to G. Apply the previous problem and delete e thereafter.

1.1.30 Replace each edge by two parallel edges and apply Problem 1.1.28.

1.1.31 (a) All we change is that, when File 3 becomes empty, the modified algorithm still checks the elements of File 2 whether they have neighbours from File 3. The estimation max (n, e) of the number of steps, applying the sum of the degrees, is still valid.

(b) The total number of steps in Algorithm 1.2 was proportional to n for complete graphs with n points, but becomes proportional to n^2 after this change.

1.1.32 Step 3 joins a new edge and a new point to the graph each time. Step 2 checks if such a "forward" step is possible. If not, we "backtrack" to try other directions for further unmarked points. The result H is a maximum connected subgraph (for otherwise there were two adjacent points x, y so that only x is marked; a violation of Step 2). The total number of steps is proportional to max (n, e), as in Algorithm 1.2, this can be proved by the same argument again.

1.2.1 G satisfies $e(G) = v(G) - 1$ by Proposition 1.2.1, hence the definition of a tree implies (1). The implication (1)$\rightarrow$(2) also follows from Proposition 1.2.1 (if G were connected but not circuit-free, deleting an edge of a circuit would lead to a proper subgraph with at least $v(G) - 1$ edges). (2) implies the definition since if G were disconnected, at least one component G_i of G would have $e(G_i) \geq v(G_i)$ and then G_i would contain a circuit by Problem 1.1.22.

Exercise 1.1.7 shows that (3) is implied by the definition while the other implication is trivial (a circuit of G were just the union of two distinct paths between the points).

(3) and (4) are equivalent. For (3)→(4) add a new edge $e = \{x, y\}$ to G. The resulting graph has circuit(-s) since there was a path from x to y, but cannot have two distinct circuits C_1, C_2 because then $C_1 - \{e\}$ and $C_2 - \{e\}$ were two distinct paths from x to y. For (4)→(3) let x, y be two distinct points in G. There must be a path from x to y (since adding $e = \{x, y\}$ must lead to a circuit) but several distinct paths would lead to distinct circuits.

1.2.2 Should G contain a circuit, its edges were not separating. For the other direction apply statement (3) of the previous exercise.

1.2.3 Every graph has a spanning forest with exactly $e(G) - v(G) + c(G) \geq 0$ edges.

1.2.4 This number is at least 2 since, considering a path of maximal length, its end points cannot have degree greater than one. This bound is sharp, since a path has exactly two such points. On the other hand, all the points cannot have degree one (unless $v(T) = 2$), by Problem 1.1.14. However, $v(T) - 1$ of them can have degree one, if all of them are adjacent to the only exceptional point.

If the tree T has a point with degree k and t points with degree 1 each, then the sum of the degrees of the remaining points is at least $2(v-t-1)$ and, at the same time, equals $2(v-1) - k - t$, by Problem 1.1.14. Hence $t \geq k$.

1.2.5 $r(G) = 0$ if and only if G has either no edges or has loops only; $n(G) = 0$ if and only if G has no circuits.

1.2.6 An edge is contained in every spanning forest if and only if it is a separating edge. An edge is contained in no spanning forest if and only if it is a loop.

1.2.7 A graph G has exactly $n(G)$ circuits if and only if no two distinct circuits of G have any edge in common.

1.2.8 No, there need not be any circuit whatsoever, containing both e and f. But if C is a circuit which contains e and f then $C - \{e\}$ can be extended to a spanning forest T and then the unique circuit of $T \cup \{e\}$ is C.

1.2.9

Algorithm 1.8 *Spanning forest construction.*

Input A graph G with at least one point.

Output A spanning tree of the connected component H of G, containing the first point of G. (If $G = H$, this is a spanning tree of G. Otherwise repeating the algorithm with input $G - H$ one can stepwise obtain every component of the spanning forest.)

Description Apply Algorithm 1.2 with the following difference in **Step 2**. If a point w is replaced from File 3 to File 2, mark the edge $\{v, w\}$. The marked edges form a spanning tree of H.

Algorithm 1.9 *Spanning forest construction.*

Input and **Output** as in Algorithm 1.8.

Description Apply Algorithm 1.3 and mark the "forward" edges $\{x, y\}$ each time when **Step 3** is performed. These form a spanning tree of H.

The total number of operations is increased by $v(G) - c(G)$ only, hence the previous estimations (proportional to max (n, e)) are still valid.

1.2.10 No, the loops are not. Any other edge e is contained in at least one cut set, since $\{e\}$ can be extended to a spanning forest T and then consider the fundamental cut set Q_e with respect to T.

1.2.11 No. If the given point p was separating, the deletion of its star increases the number of components by more than two. On the other extreme, if all the edges, incident to p, are loops then the deletion of its star decreases the number of components by one.

1.2.12 If X was the edge set of a cut set free subgraph but the subgraph H with edge set $E(G) - X$ is not a spanning forest then H must contain a circuit C. (The other possible reason why H is not a spanning forest could be $c(H) > c(G)$ but then X would contain a cut set.) Remove an edge of C and join it to X. Apply this process, possibly several times, until X becomes the complement of a spanning forest.

1.2.13 Consider the edge set $\{1, 2, 3\}$ of the graph of Fig. 1.12.

1.2.14 Extend $C - \{e\}$ to a spanning forest F and consider the fundamental cut set Q_f with respect to F.

1.2.15 $y \in E(C_x)$ if and only if the only path of F, containing the two end points of x, contains y. This is obviously equivalent to $x \in E(Q_y)$. Moreover, (2)$\longleftrightarrow$(3) is also obvious since $(F - \{y\}) \cup \{x\}$ has the same number of edges as F and it is also circuit-free (since the only circuit of $F \cup \{x\}$ is destroyed by deleting y).

1.2.16 Let F denote the spanning forest, formed by the "forward" edges (see the solution of Exercise 1.2.9). Let $e \in E(G) - E(F)$ be an edge $\{v_1, v_2\}$ and suppose that the algorithm marked v_1 earlier than v_2. Then the sequence $a_{v_2} = v_3, a_{v_3} = v_4, a_{v_4} = v_5, \ldots$ leads to v_1 and all the edges $\{v_i, v_{i+1}\}$ with $i = 2, 3, \ldots$ belong to F. They form a path in F between v_1 and v_2. This path and e gives the fundamental circuit [Paton, 1969; Gibbs, 1975].

(Using the expression "father of i" for a_i, as in Algorithm 1.3, one may say that this process works since either v_1 or v_2 is a "descendant" of the other for any edge $\{v_1, v_2\}$ of the graph.)

The total number of steps required is clearly at least proportional to the sum S of the lengths of the fundamental circuits with respect to F. If S is proportional to max (e, n) then this modified algorithm is as effective as Algorithm 1.3 was. But S may be proportional even to n^3, see the next problem.

1.2.17 F will be a path of length $n-1$. The fundamental system will consist of 1 circuit of length n, 2 circuits of length $n-1$ etc. Hence $S=\sum_{k=1}^{n-2} k(n-k+1)=$
$= n\sum_{k=1}^{n-2} k - \sum_{k=1}^{n-2} k^2 + \sum_{k=1}^{n-2} k = (n-1)(n-2)(n+6)/6$, which is proportional to n^3. This is "worst possible" since the number $n(G)$ of fundamental circuits is proportional to at most n^2 and the length of each circuit is at most n.

1.2.18 If X is a circuit, it clearly intersects the complement of every spanning forest F (since X were contained in F otherwise) and no proper subset Y of X has this property (since Y can be extended to a spanning forest).

If X intersects the complement of every spanning forest, it must contain a circuit. The minimal sets, containing circuits, are clearly just the circuits.

1.2.19 If X is the complement of a spanning forest, it must intersect every circuit C (since C were contained in the complement of X otherwise) and no proper subset Y of X has this property (let $e \in X-Y$ and consider C_e).

If X intersects every circuit, its complement must be circuit-free, and also spanning, due to the minimality of X.

1.2.20 Let P_1 and P_2 be two maximum paths of G. Walking along P_1, let x be the first and y be the last point which also belongs to P_2. The subpaths of P_1 and P_2 between x and y must be the same (and must be identical to the whole intersection $P_1 \cap P_2$), since G is a tree. Let a_1, a_2 denote the length of the subpaths of P_1 and P_2, respectively, until x; and let b_1, b_2 denote the lengths of those from y.

We claim that $a_1 = a_2$ and $b_1 = b_2$. Since $a_1+b_1 = a_2+b_2$, an inequality, say $a_1 > a_2$, would imply $b_1 < b_2$. Then walking along P_1 until y and continuing thereafter along P_2 would lead to a longer path.

Similarly, if the common length of P_1 and P_2 is ℓ, all the lengths a_1, a_2, b_1, b_2 are at most $\ell/2$. (For example, $a_1 = a_2 > \ell/2$ would imply a path of length $a_1 + a_2$.)

Hence, the central point of every longest path (or the central edge if ℓ is odd) is the same.

1.2.21 We may suppose that $d_1 \geq d_2 \geq d_3 \geq \ldots \geq d_n$. Join the first point to the second, third,..., $(d_1+1)^{st}$ points, then join the second point to those with numbers $d_1+2, d_1+3, \ldots, d_1+d_2$. (The total number of these latter points is d_2-1 only, but the second point is adjacent to the first one as well.) Next join the third point to those with numbers $d_1+d_2+1, d_1+d_2+2, \ldots, d_1+d_2+d_3-1$. Continue this process until all the points with degree at least 3 were considered.

If this process works, the result is clearly a tree and clearly with the required degree sequence. The process fails if, in Step k, the required edges (up to the point with number $d_1+d_2+\ldots d_k-(k-2)$) cannot be realized since this number is greater than n. But this is impossible since, by $d_1+d_2+\ldots+d_n = 2n-2$, it would imply $d_{k+1}+d_{k+2}+\ldots+d_n < n-k$ while the sum of $n-k$ positive integers is at least $n-k$.

1.2.22 (a) If the root has degree one, all the points (except the other end point) will appear in the Prüfer-code in the same order as they form the path. If the root has degree two, it will appear twice while all the other points of degree two appear once.

(b) If the root has degree $n-1$ then it will stand alone in the Prüfer-code, $n-1$ times. If it has degree 1 and the point with number k is its neighbour then k will appear $n-2$ times, followed by n.

1.2.23 For a number v_k in the code let w_k denote the point of degree one which was deleted (with the incident edge $\{w_k, v_k\}$) when v_k was recorded. The function $v_k \to w_k$ for each k is clearly enough for the reconstruction.

Obviously, w_1 is the smallest number, not included in the code. More generally, w_k is the smallest number, not included among the numbers $w_1, w_2, \ldots, w_{k-1}, v_k, v_{k+1}, \ldots, v_{n-1}$. The forbidden numbers form a set of cardinality at most $n-1$ hence its complement is nonempty.

In this way we obtain a graph with n points and $n-1$ edges of form $\{v_k, w_k\}$. It must be a tree. Otherwise it would contain a circuit C and every point of C would appear once among the v_i's and once among the w_j's. Then applying the function for the point of C in leftmost position among the v_i's leads to a contradiction.

Hence we proved that the Prüfer-code gives a one–one correspondence between the rooted trees and the number sequences. The last number in the code is always n (i.e., the number of the root), and we obtain n^{n-2} variations for the first $n-2$ numbers.

1.2.24 Consider the Prüfer-code which uses each point v exactly $d(v)-1$ times.

1.2.25 If X is a cut set, it intersects every spanning forest since $G-X$ cannot be connected. The minimality also follows since the complement of any proper subset of X can be extended to a spanning forest. If X intersects every spanning forest then $c(G-X) > c(G)$ follows, hence X contains a cut set. By the minimality, X is a cut set.

1.2.26 If X intersects every cut set then $c(X) = c(G)$, hence X contains a spanning forest. The other part is a trivial consequence of the definition of the cut sets.

1.2.27 The "only if" part follows from Theorem 1.2.4.

Let X be a minimal edge set intersecting every cut set of G by an even number of edges. Let G_0 be the graph, consisting of the edges of X (and of the necessary points) and let $v \in V(G_0)$. If v is nonseparating in G, the edges, incident to v, form a cut set, hence the degree of v is even in G_0. If v is separating, consider the components of $G-v$. The edges of G, connecting v with any one of these components, form a cut set which also intersects X by an even number of edges. Hence the degree of v in

G_0 is even in this case as well. Since every point of G_0 is of even degree, Problem 1.1.21 can be applied.

1.2.28 Theorem 1.2.4 implies that the cut sets do possess this property. Let X intersect each circuit by an even number of edges. Then $c(G-X) > c(G)$ will prove the assertion. Otherwise the complement of X would contain a spanning forest. Adding an edge of X to this forest, the resulting fundamental circuit would intersect X by just one edge.

1.2.29 Consider the graph $H = G - Y$. X is circuit free in H as well, hence there exists a spanning forest F of H with $X \subseteq E(F)$. Since Y had no cut set, $c(H) = c(G)$. Hence F is a spanning forest of G as well.

1.2.30 X intersects every circuit by an even number of edges. Hence X contains a cut set Q of G (Problem 1.2.28). If $X \neq Q$ then $X' = X - Q$ intersects every circuit by an even number of edges, hence it contains a cut set Q' etc.

1.3.1 If the directed edges are "arrows" then the sum of the indegrees is just the number of heads of these arrows while the sum of the outdegrees equals the number of tails. This common value is just the number of arrows (i.e., the number of the directed edges).

1.3.2 n edges are enough, see a directed circuit. $n-1$ edges cannot be enough since the underlying undirected graph should be a tree (due to the connectedness) and the points of degree one in that tree cannot have positive indegree and positive outdegree after the orientation.

1.3.3 If the graph has, say, no sinks then each time one reaches a point, the walk can be continued. Since the graph has a finite number of points only, sooner or later a point is repeated, hence a directed circuit is obtained.

1.3.4 One in each component G of the underlying undirected graph (see Fig. S.1.6). This trivially follows from $e(G) = v(G)$, see Exercise 1.3.1.

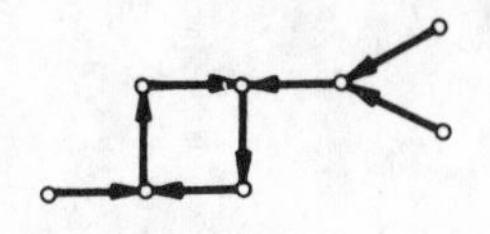

Fig. S.1.6

1.3.5 Consider a single directed circuit.

1.3.6 If $x \in A$ and $y \notin A$ then there exists a directed path from x to y by the strong connectivity. Consider the first edge of this path with tail in A and head in $V - A$.

On the other hand suppose that there were no directed path from u to v, i.e. that the set X of points which can be reached by directed paths from u would not contain the point v. Then $V - X$ is nonempty (it contains v). The directed edge with tail in X and head in $V - X$ would then contradict to the definition of X.

1.3.7 The directed graph has a directed cut set if and only if the condition of the previous exercise is violated.

1.3.8 If $e = (x, y)$ is a directed edge and there is a directed path from y to x then a directed circuit has been found which contains e. Otherwise let Y denote the set of those points which can be reached from y by a directed path. Clearly, $y \in Y$ and $x \in X = V(G) - Y$. The set K of edges, joining points of X to points of Y in the underlying undirected graph, is not empty ($e \in K$, for example). Hence K is the disjoint union of cut sets, by Problem 1.2.30. All these cut sets are directed, by the definition of Y, and one of them contains e.

1.3.9 If the original directed graph G contains directed circuits then the points of these circuits will never become sources or sinks, hence they will remain in the final graph. On the other hand, if G had no directed circuits then no new one will arise during the process which, by the result of Exercise 1.3.3, can really be continued until all the points are deleted.

1.3.10 No. This can easily be proved by the previous problem. If points 1,4,11 (sinks) and 9 (source) are deleted, 3 and 6 become sinks, 5,8 and 15 become sources. Deleting these points 7 becomes a sink, 2,10 and 12 become sources. Once they are also deleted, 13 becomes a sink and 14 becomes a source; their deletion leads to the empty graph.

1.3.11 Suppose that $p_1, p_2, \ldots, p_k$ are the points (in this order) of a directed circuit in the new graph. Hence there are points a_i, b_i in the corresponding strongly connected components $P_i (i = 1, 2, \ldots, k)$ of the original graph so that there are directed edges (a_i, b_{i+1}) for $i = 1, 2, \ldots, k - 1$ and a further edge (a_k, b_1), see Fig. S.1.7. Since there are directed paths in P_i from b_i to a_i, they and the above edges form a large directed circuit in the original graph, contradicting to the maximality of the P_i's.

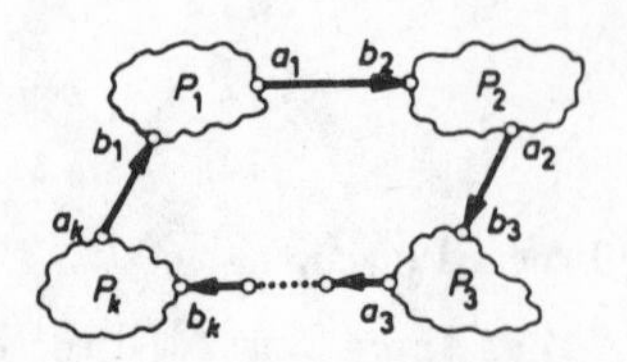

Fig. S.1.7

1.3.12 (a) Let V_0 be a point with maximum outdegree and let $v_1, v_2, \ldots, v_d$ be the heads of the edges, leaving v_0. If there were a further point v so that none of the edges (v_i, v) exist $(i = 0, 1, \ldots, d)$ then the outdegree of v would be at least $d+1$, a contradiction.

(b) Suppose that v is a point, not contained in a maximum length directed path $P = [v_1, v_2, \ldots, v_k]$. Colour the edge $\{v, v_i\}$ by red if its orientation is (v_i, v) and by blue otherwise. $\{v, v_1\}$ must be red and $\{v, v_k\}$ must be blue, by the maximality of P. Hence the sequence of colours must change at least once. Let $i < k$ so that $\{v, v_i\}$ is red and $\{v, v_{i+1}\}$ is blue. Then the directed path $[v_1, v_2, \ldots, v_i, v, v_{i+1}, \ldots, v_k]$ is longer than P, a contradiction.

(c) The strong connectedness implies the existence of directed circuits. Let C be one with maximum length and suppose that v is not included in it. Using the same colouring argument as in part (b), C can either be extended by v or all the edges $\{v, v_j\}$ are, say, red. v cannot be a sink (by the strong connectivity), hence there is a directed path from v to a point v_i of C (along further points outside C). Using (v_{i-1}, v) and this path instead of (v_{i-1}, v_i), C can be extended, a contradiction.

(d) We use induction over the number n of points. The statement is trivial for $n = 3$. If T has no directed circuit at all, it has a source x and a sink y, by Exercise 1.3.3. Then reversing the edge (x, y) one obtains a directed path $[a, y, x, b]$ from any point a to any point b. If T has directed circuits, consider the process of Problem 1.3.11. The result is a tournament again. Apply the inductive hypothesis.

1.3.13 Let $D \subseteq E(G)$ be a maximum edge set so that $G - D$ is not strongly connected. $V(G)$ can be decomposed into $A \cup B$ so that $G - A$ and $G - B$ are strongly connected and all the edges of D have tail in A and head in B. Reversing the orientation of the edges of D one cannot reach B from A any more.

1.3.14 Let a and b the tail and the head of e, respectively. Mark b. If x is a marked point and either there is a green directed edge (x, y) or a red edge $\{x, y\}$ with an arbitrary orientation, then mark y as well. Let Y denote the set of marked points after this process.

If $a \in X$ then a circuit with the requested property has also been found. Otherwise let D be the set of edges joining a point of X with a point not in X. All the edges in D are either blue or green, in the latter case with head in X. Hence D contains a cut set with the requested property.

A circuit and a cut set with the required property cannot simultaneously exist since their intersection would contain another green edge, different from e, and its orientation would contradict to one of the properties.

Our result reduces to Exercise 1.3.8 if all the edges are green.

1.3.15 Let $2k$ denote the number of points with odd degree. Add k edges to the graph so that the resulting graph G' has even degrees only. Since every component of G' is Eulerian, its edges can be oriented so that for every point, its outdegree equals its indegree. Then delete the k new edges.

1.4.1 A diagonal entry y_{ii} of $\mathbf{Y}$ is the degree of the ith point (counting the loops only once). An off-diagonal entry y_{ij} is the number of edges between the ith and the jth points.

An entry $x_{ij}^{(k)}$ of $\mathbf{X}_k$ is the number of edge sequences between the ith and the jth points. (Such an edge sequence can intersect itself; it may contain the ith or the jth point as internal point as well.)

1.4.2 Loops correspond to diagonal elements of $\mathbf{A}$, to columns with a single 1 entry in $\mathbf{B}$ and $\mathbf{C}$, and to zero columns in $\mathbf{Q}$.

A separating edge $e = \{i, j\}$ corresponds to a zero column in $\mathbf{C}$ and to a column with a single 1 entry in $\mathbf{Q}$. If G_1 and G_2 denote the two subgraphs of $G \backslash e$, with $i \in V(G_1)$ and $j \in V(G_2)$ then $\mathbf{A}$ and $\mathbf{B}$ of the original graph G can be decomposed as in Fig. S.1.8.

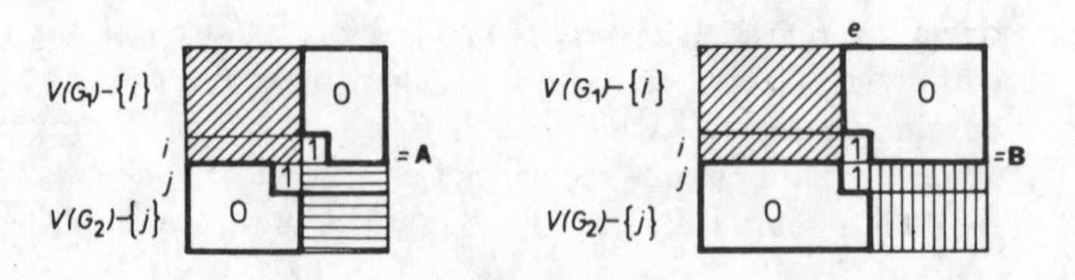

Fig. S.1.8

1.4.3 Cut the vector into two equal parts (to parts of length $\lfloor p/2 \rfloor$ and $\lceil p/2 \rceil$ if p is odd). If x is less than the first entry of the second part then continue the search in the first part, otherwise in the second part etc. The problem can be solved in $\lceil \log_2 p \rceil$ steps.

The base of the logarithm is of no interest. In order to change $\log_b p$ to $\log_2 p$, a multiplication by $\log_2 b$ (i.e. by a constant, not depending on p) is required only. Thus, if the total number of operations in an algorithm is proportional to $\log_2 p$, it is proportional to $\log_b p$ as well, for any b.

1.4.4 The three ones in the first column are obvious: the required number of steps for these three operations are constant (proportional to 1), independently of the size of the graph. Marking every edge requires a search all over $\mathbf{A}$ while the other two operations influence a row and a column each.

In the second column the first two entries are obvious. For the other four operations the required number of steps can be proportional to e (in the last three cases a part of the vector must be shifted to the right or to the left).

In the third column, $\log d$ is explained by the previous exercise. Adding an edge $\{i, j\}$ requires constant times $\log d$ steps to put i and j to the lists of the neighbours of j and i, respectively, but the above shifting requires more steps.

1.4.5 Marking every edge may require less steps in case of $\mathbf{B}$ (proportional to e rather than to v^2 in case of $\mathbf{A}$). The other five operations require more steps in case of $\mathbf{B}$ than in case of $\mathbf{A}$.

1.4.6 If the array where $\mathbf{A}$ will be stored, contains zeros only, the number of steps to convert $(\mathbf{e}, \mathbf{v})$ into $\mathbf{A}$ is proportional to $\max(e(G), v(G))$; otherwise it must also contain v^2 further initial steps. The required number of steps to convert $(\mathbf{e}, \mathbf{v})$ to $(\mathbf{e}, \mathbf{e}_c, \mathbf{v})$ is clearly proportional to $\max(e(G), v(G))$.

1.4.7 The first two columns $\mathbf{x} = (1,1,0)^T$ and $\mathbf{y} = (1,0,1)^T$ of the first matrix are clearly independent since each contains a nonzero entry in a row where the other's entry is zero. We try to obtain the third column $\mathbf{z} = (0,1,1)^T$ as a linear combination $\mathbf{z} = \lambda\mathbf{x} + \mu\mathbf{y}$ of the others. Clearly, $\lambda = 1$ and $\mu = 1$ is obtained, due to the second and to the third rows, respectively. Now, $\lambda + \mu = 1 + 1 = 0$ holds in $\mathbb{B}$ but not in $\mathbb{Q}$; hence $\mathbf{z}$ depends on $\{\mathbf{x}, \mathbf{y}\}$ in $\mathbb{B}$ but not in $\mathbb{Q}$.

Solve the system $\begin{pmatrix} 1 & 0 & 1 \\ 1 & 1 & 0 \\ 1 & 0 & 0 \end{pmatrix} \begin{pmatrix} a & b & c \\ d & e & f \\ g & h & i \end{pmatrix} = \begin{pmatrix} 1 & 0 & 0 \\ 0 & 1 & 0 \\ 0 & 0 & 1 \end{pmatrix}$ of equations to determine the inverse. $a = b = 0$, $c = 1$ are obvious (due to the last row of the leftmost matrix). By $a + g = b + e = 1$, $b + h = c + i = = a + d = c + f = 0$ one clearly obtains then $e = g = 1, d = h = 0$. In order to determine f and i, one must subtract 1 from 0. This is the only step where the result is different in $\mathbb{B}$ and in $\mathbb{Q}$.

1.4.8 $\mathbf{M}_3$ is clearly the circuit matrix of a graph with three parallel edges between two points. $\mathbf{M}_1$ and $\mathbf{M}_2$ cannot be circuit matrices. In the case of $\mathbf{M}_2$ the second and the third circuits were proper subsets of the first one, which is impossible. In the case of $\mathbf{M}_1$ there were two circuits $\{a, b, c\}$ and $\{a, b, d\}$ (see the first two rows). Then c, d were parallel edges, hence no circuit of form $\{a, c, d\}$ could be possible.

1.4.9 The circuits $\{1,2\}, \{2,3\}$ and $\{3,4\}$ in the first graph of Fig. S.1.9 form a maximal set of independent circuits but not a fundamental system of circuits since, for this latter, each circuit should contain an edge which is not contained in any other circuit, and this is violated by $\{2,3\}$.

The cut sets $\{1,2\}$, $\{2,3\}$ and $\{3,4\}$ in the second graph of Fig. S.1.9 have analogous properties.

1.4.10 (a) Every point in this subgraph has even degree, hence this subgraph is the edge-disjoint union of some circuits.

(b) Let F be the spanning forest and let C_e denote the fundamental circuit containing the edge $e \in E(G) - E(F)$. For an arbitrary

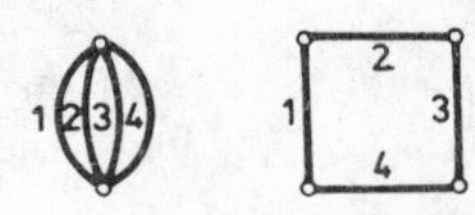

Fig. S.1.9

circuit C consider the sum of the rows, corresponding to $C_a, C_b, \ldots$ for $\{a, b, \ldots\} = E(C) - E(F)$. We claim that the subgraph H, determined by the ones of this sum, is just C. Clearly $a, b, \ldots \in$ $\in E(H)$ since each appeared once. An edge $f \in E(F) \cap E(C)$ appears each time when C intersects the fundamental cut set Q_f in an edge, different from f, hence in an odd number of times. We conclude that $E(C) \subseteq E(H)$. Should $E(C) \neq E(H)$, the remaining edges, all contained in F, would form circuits, which is impossible.

1.4.11 Let $\mathbf{M}$ be a matrix of rank t and $\mathbf{M}_r$ be another matrix, formed by t linearly independent rows of $\mathbf{M}$. We claim that a set of t columns of $\mathbf{M}$ is linearly independent if and only if the corresponding set of "shorter" columns in $\mathbf{M}_r$ is independent. If the t columns are dependent in $\mathbf{M}$, the "shorter" ones are clearly dependent in $\mathbf{M}_r$ as well. So suppose that the t columns are linearly independent and determine a submatrix $\mathbf{X}$ of size $n \times t$ and rank t in $\mathbf{M}$. The corresponding "shorter" columns determine a submatrix $\mathbf{X}_r$ with size $n_r \times t$ and rank $r(\mathbf{X}_r) \leq t$. We have to prove that this rank is exactly t. Since the rows of $\mathbf{M}_r$ were independent, there exists a matrix $\mathbf{D}$ so that $\mathbf{M}^T = \mathbf{M}_r^T \mathbf{D}$. Then $\mathbf{X}^T = \mathbf{X}_r^T \mathbf{D}$ also holds, hence $r(\mathbf{X}_r^T) \geq r(\mathbf{X}^T) = t$.

1.4.12 The entries in $\mathbf{B}(G)$, corresponding to the loops of G, could increase $r(\mathbf{B})$. For example, take a graph with n points, each incident to a single loop. Then $0 = r(G) \neq r(\mathbf{B}(G)) = n$.

1.4.13 An off-diagonal element x_{ij} of this matrix equals the number of edges oriented from j to i. The diagonal elements are zero.

1.4.14 Let $\mathbf{D}$ be a square submatrix of size $k \times k$ in $\mathbf{B}(G)$. We use induction on k to prove that det $\mathbf{D} = 0$ or ± 1. The assertion is obvious if $k = 1$.

The rows of $\mathbf{D}$ determine a subset V_1 of k points of G, the columns of $\mathbf{D}$ determine a subset E_1 of k edges of G. If there exists an edge of E_1 so that none of its end points is in V_1 then $\mathbf{D}$ has a column of zeros only, hence det $\mathbf{D} = 0$. If there exists an edge of E_1 so that only one of its end points is in V_1 then let us expand det $\mathbf{D}$ along the column of this edge. We obtain that det $\mathbf{D} = \pm$det (a smaller matrix) and can use the inductive hypothesis.

Finally, if every edge of E_1 is incident to points of V_1 only then these k edges form a subgraph with k points. This subgraph must contain a circuit C and then the sum of the columns corresponding to the edges of C equals zero, leading to det $\mathbf{D} = 0$.

1.4.15 Consider the circuits $\{1, 3, 2, 5\}$ and $\{1, 4, 2, 6\}$ of the graph in Fig. S.1.10. The corresponding rows of $\mathbf{C}$ are $(1\ 1\ 1\ 0\ 1\ 0)$ and $(1\ -1\ 0\ 1\ 0\ 1)$, respectively. Taking these two rows and the first two columns into consideration, a 2×2 matrix with determinant -2 is obtained. The same graph and the cut sets $\{1, 2, 4, 6\}$ and $\{3, 4, 5, 6\}$ show that $\mathbf{Q}$ is not totally unimodular either.

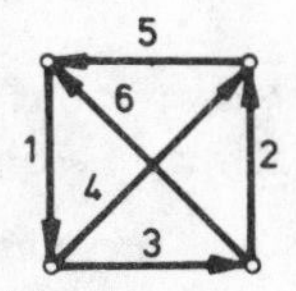

Fig. S.1.10

1.4.16 In order to show that the jth entry of the ith row of $\mathbf{CQ}^T$ is zero, let us consider the ith circuit C and the jth cut set Q of the graph. Let A, B denote the two point sets determined by Q and let us assign the orientation "from A towards B" to Q. Walking clockwise along C, there are k instances when we leave A for B and k ones *vice versa* ($k \geq 0$). In the former case the entries for $\mathbf{C}$ and $\mathbf{Q}$ are the same ($+1$ if the head of the edge is in B and -1 if it is in A), while in the latter case the entries are opposite.

1.4.17 By Exercise 1.4.1, the diagonal entry m_{ii} is the number of closed walks of length three, starting and ending at point i. If G is loop-free, all these walks are circuits. Hence tr $\mathbf{M}$ is the number of length 3 circuits, each counted exactly $3! = 6$ times.

1.4.18 If $d_1, d_2, \ldots$ denote the degrees of the points, the number of operations, required to the trivial conversion is clearly proportional to $\sum_{i=1}^{v(G)} d_i \log d_i < \log d \sum_{i=1}^{v(G)} d_i = 2e \log d$.

1.4.19 The adjacency lists of the points need not be ordered separately. Consider the neighbours $i_1, i_2, \ldots, i_k$ of the first point. (These numbers need not be in increasing order.) Then put point 1 to the first position of the ordered adjacency lists of $i_1, i_2, \ldots, i_k$. Now let $j_1, j_2, \ldots, j_\ell$ be the neighbours of the second point and put point 2 to the first free position of the ordered adjacency lists of $j_1, j_2, \ldots, j_\ell$ etc. Figure S.1.11 illustrates this process for the graph of **Box 1.1.**

1.4.20 The first five entries are obvious. The estimation d^2 is also clear since the point should be removed from the adjacency list of all of its neighbours.

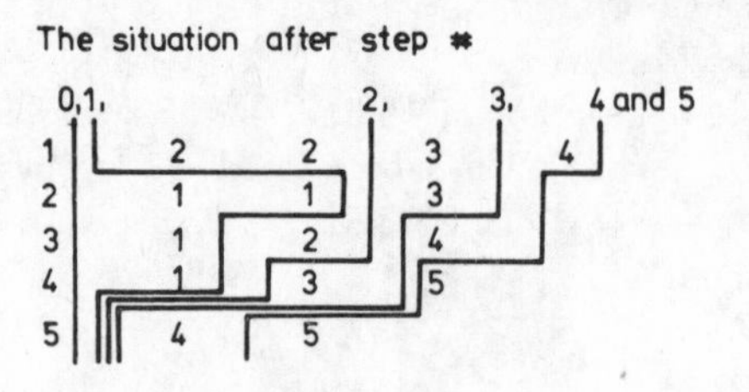

Fig. S.1.11

This may be much smaller than e but, say if the graph contains a large number of parallel edges, it may be quicker to check every edge.

1.4.21 Ordering the adjacency lists of the points in $(\mathbf{e}, \mathbf{e}_c, \mathbf{v})$ gives no advantage (in fact it increases the fourth entry of the last column of **Box 1.2** from 1 to d). The chain structure which makes deletion or addition of an edge much easier, does not allow one to jump "into the middle of a list".

1.4.22 $\mathbf{M}_3$ is the circuit matrix of a graph of three parallel edges. $\mathbf{M}_1$ and $\mathbf{M}_2$ are submatrices of the circuit matrices of the graphs of Fig. S.1.12 and S.1.13, respectively.

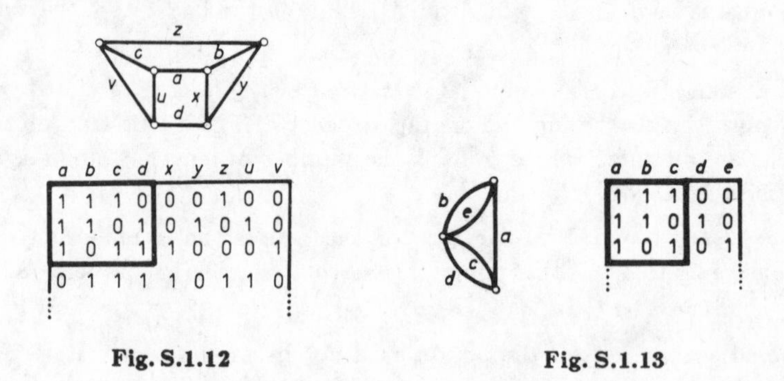

a	b	c	d	x	y	z	u	v
1	1	1	0	0	0	1	0	0
1	1	0	1	0	1	0	1	0
1	0	1	1	1	0	0	0	1
0	1	1	1	1	0	1	1	0

Fig. S.1.12

a	b	c	d	e
1	1	1	0	0
1	1	0	1	0
1	0	1	0	1

Fig. S.1.13

1.4.23 Let $e \in E(G)$. If the circuit matrix $\mathbf{C}$ of G is given, simply delete those rows whose entry was 1 in the column, corresponding to e, and then delete this column, as well. The resulting matrix is just $\mathbf{C}(G-e)$.

If the cut set matrix $\mathbf{Q}$ of G is given, the process is more complicated. If X is a cut set of G and $e \in X$ then $X - \{e\}$ is either empty or is a cut set of $G - e$. If $e \notin X$ then the deletion of X disconnects $G - e$ but need not be a cut set (i.e. need not be minimal). Hence the procedure is as follows:

Step 1 Delete from $\mathbf{Q}$ the column, corresponding to e.

Step 2 If a row of zeroes is obtained, delete it from the matrix.

Step 3 If there are two rows $\mathbf{r}_1$ and $\mathbf{r}_2$ so that, at the position of every one-entry of $\mathbf{r}_1$ the entry of $\mathbf{r}_2$ is also a one, delete the row $\mathbf{r}_2$. Repeat this step as long as there are such pairs of rows. Finally we obtain $\mathbf{Q}(G-e)$.

1.4.24 Let $\mathbf{MN} = \mathbf{P}$. After a suitable permutation of the rows of $\mathbf{M}$ and $\mathbf{P}$, the first $r(\mathbf{M})$ rows of $\mathbf{M}$ are linearly independent. Then after a suitable permutation of the columns of $\mathbf{M}$ and the corresponding rows of $\mathbf{N}$, the first $r(\mathbf{M})$ columns of $\mathbf{M}$ will also be linearly independent. The situation of Fig. S.1.14 is obtained where $\mathbf{M}_1$ is nonsingular. Clearly, $\mathbf{P} = \mathbf{M}_1\mathbf{N}_1 + \mathbf{M}_2\mathbf{N}_2$. Define $\mathbf{R} = \begin{pmatrix} \mathbf{E} & \mathbf{M}_1^{-1}\mathbf{M}_2 \\ \mathbf{O} & \mathbf{E} \end{pmatrix}$ and observe that $\mathbf{RN} = \begin{pmatrix} \mathbf{M}_1^{-1}\mathbf{P}_1 \\ \mathbf{N}_2 \end{pmatrix}$. Its rank can be estimated from above by the sum of the ranks of $\mathbf{M}_1^{-1}\mathbf{P}_1$ and $\mathbf{N}_2$. On the other hand, $r(\mathbf{RN}) = r(\mathbf{N})$ since $\mathbf{R}$ is nonsingular (it is triangular with nonzero diagonal entries). Hence $r(\mathbf{N}) = r(\mathbf{RN}) \le r(\mathbf{M}_1^{-1}\mathbf{P}_1) + r(\mathbf{N}_2) = r(\mathbf{P}_1) + r(\mathbf{N}_2) \le r(\mathbf{P}) + b - r(\mathbf{M})$, that is, $r(\mathbf{M}) + r(\mathbf{N}) \le b + r(\mathbf{P})$.

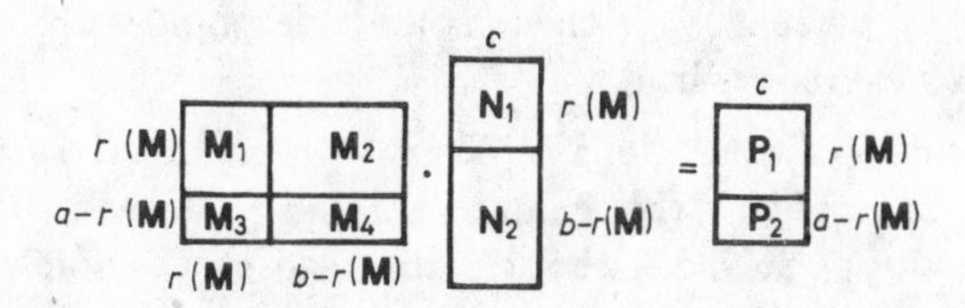

Fig. S.1.14

1.4.25 (a) Let $\mathbf{x}_0 = (1,1,0)^T$.

(b) Let M be the 1-dimensional subspace, generated by $\mathbf{x}_0$ of part (a). Then $(a,b,c) \in M^\perp$ if and only if $a+b=0, c$ is arbitrary. Hence $M^\perp$ is generated by $\mathbf{x}_0$ and $(0,0,1)^T$.

(c) M of part (b) can be applied here as well.

1.4.26 (a) Let $\mathbf{v}_1, \mathbf{v}_2, \dots, \mathbf{v}_k$ be a base of M and extend it to a base of the whole space by $\mathbf{v}_{k+1}, \mathbf{v}_{k+2} \dots, \mathbf{v}_n$. We prove dim $M^\perp = n-k$. Let the unit vectors of the space be $\mathbf{e}_1, \mathbf{e}_2, \dots, \mathbf{e}_n$ and let $\mathbf{A}$ denote the linear transformation satisfying $\mathbf{Ae}_i = \mathbf{v}_i$ for every i. Now $\mathbf{x} \in M^\perp$ holds if and only if, for $i = 1,2,\dots,k$, we have $0 = \mathbf{x}^T\mathbf{v}_i = (\mathbf{e}_i^T\mathbf{A}^T\mathbf{x})^T$.

This means that $\mathbf{A}^T\mathbf{x}$ is a linear combination of $\mathbf{e}_{k+1}, \ldots, \mathbf{e}_n$, that is, $M^\perp$ is generated by $(\mathbf{A}^T)^{-1}\mathbf{e}_{k+1} \ldots, (\mathbf{A}^T)^{-1}\mathbf{e}_n$ [Lovász, 1979].

(b) $M \cup M^\perp$ generates just the subspace $(M \cap M^\perp)^\perp$, hence all we have to show is that $\mathbf{j} = (1, 1, \ldots, 1)^T$ is perpendicular to every element $\mathbf{u}$ of $M \cap M^\perp$. But $\mathbf{u} \in M \cap M^\perp$ contains an even number of one-entries, hence $\mathbf{u}^T\mathbf{j} = 0$ [Gallai; Chen, 1970b].

1.4.27 Consider an arbitrary square matrix with integer entries. If each odd entry is changed to 1 and each even entry to 0, we say that the field $\mathbb{Q}$ is changed to $\mathbb{B}$. If the operations addition, subtraction, and multiplication (but not division) are performed on integers, changing $\mathbb{Q}$ to $\mathbb{B}$ before or after these operations is the same. In the case of X, if we change $\mathbb{Q}$ to $\mathbb{B}$ at first then the sum of the entries in every row becomes zero, hence $\det \mathbf{X} = 0$. This implies that $\det \mathbf{X}$ is even if we do not change $\mathbb{Q}$ to $\mathbb{B}$.

1.4.28 Each row of $\mathbf{B}$ corresponds to a star (see Exercise 1.2.11) which is either a cut set or the union of disjoint cut sets. Hence each row of $\mathbf{B}$ is either a row or the sum (over $\mathbb{B}$) of some rows of $\mathbf{Q}$. This immediately implies $r(\mathbf{B}) \leq r(\mathbf{Q})$ and $\mathbf{C}\mathbf{B}^T = \mathbf{O}$ and the "only if" part of the last statement. For $r(\mathbf{Q}) \geq r(\mathbf{B})$ observe that every cut set arises as the sum (over $\mathbb{B}$) of some stars. Finally, if X is the edge set of a spanning forest then the corresponding columns in $\mathbf{Q}$ form a base. Since they are linear combinations (over $\mathbb{B}$) of the corresponding columns of $\mathbf{B}$, these latter columns must also form a base.

1.4.29 Let the directed graph G be given by two adjacency lists (edges listed both by tails and by heads). $\mathbf{v}_{in}$ immediately tells us whether there exists any point of indegree zero. If there is none or more than one, G has no supersource. If there is a single point with indegree zero, we check its neighbours in $\mathbf{e}_{out}$.

The number of operations is clearly proportional to max (v, d) only. Unless d is close to e this is much smaller than max (v, e). That is, the problem can be solved without really reading the whole input about G.

1.4.30 We use induction on $v(G)$. The statement is trivial if $v(G) \leq 2$ and we may suppose that G is loopless. Let $C = [v_1, v_2, \ldots, v_k]$ be a directed circuit of minimum length ($k \geq 2$). This minimality assures that C has no "shortcuts".

These k points are united into a single new point v; the k edges of C are deleted and if an old edge connected v_i to or from another point u then join v to or from u, respectively, in the new graph G'. (In particular, if there were two parallel edges (v_i, v_{i+1}) then a loop is obtained.)

G' is also strongly connected, $v(G') = v(G) - k + 1$; $e(G') = e(G) - k$, hence $n(G') = n(G) - 1$. By the induction hypothesis, there are $n(G')$ linearly independent directed circuits in G'; these and C are linearly independent in G.

1.4.31 Apply the theorem of Binet and Cauchy for $\det \mathbf{M}\mathbf{M}^T$. Exactly those submatrices of $\mathbf{M}$ will be nonsingular which correspond to a spanning forest (Theorem 1.4.9). Their determinants are ± 1 (Exercise 1.4.14), hence their product with themselves are $+1$.

1.4.32 Let $\mathbf{B}_n$ be the incidence matrix of the complete directed graph $\mathbf{K}_n$ with an arbitrary orientation. Let $\mathbf{M}_n$ be obtained from $\mathbf{B}_n$ by deleting the last row. By the "directed analogue" of Exercise 1.4.1, all the off-diagonal entries of $\mathbf{M}_n\mathbf{M}_n^T$ are -1, all the diagonal entries are $n-1$. Addition of every row of $\mathbf{M}_n\mathbf{M}_n^T$ to the first one and then addition of the new first row to all the others give

$$\det \mathbf{M}_n\mathbf{M}_n^T = \begin{vmatrix} n-1 & -1 & \dots & -1 \\ -1 & n-1 & \dots & -1 \\ \vdots & \vdots & & \vdots \\ -1 & -1 & \dots & n-1 \end{vmatrix} = \begin{vmatrix} 1 & 1 & \dots & 1 \\ -1 & n-1 & \dots & -1 \\ \vdots & \vdots & & \vdots \\ -1 & -1 & \dots & n-1 \end{vmatrix} =$$

$$= \begin{vmatrix} 1 & 1 & \dots & 1 \\ 0 & n & \dots & 0 \\ \vdots & \vdots & & \vdots \\ 0 & 0 & \dots & n \end{vmatrix} = n^{n-2}.$$

1.4.33 Let F be a spanning forest and $\mathbf{C}_F$ be the submatrix of $\mathbf{C}$, determined by the rows of the fundamental circuits with respect to F. We shall prove at first that every submatrix $\mathbf{M}$ of $\mathbf{C}_F$ with size $n(G)$ has determinant 0 or ± 1. Clearly $\det \mathbf{M} \neq 0$ exactly if the columns of $\mathbf{M}$ correspond to the complement of a spanning forest F'. If $F = F'$ then $\det \mathbf{M} = \pm 1$. Otherwise consider $\mathbf{C}_{F'}$ as well and recall (Exercise 1.4.11) that there exists a nonsingular matrix $\mathbf{D}$ of size $n(G)$ satisfying $\mathbf{D}\mathbf{C}_{F'} = \mathbf{C}_F$. Observe $\det \mathbf{M} = \det \mathbf{D}$ (since the columns of M correspond to the unity submatrix of $\mathbf{C}_{F'}$). Finally the columns, corresponding to the complement of F, determine a submatrix $\mathbf{Y}$ of $\mathbf{C}_{F'}$. This is just the inverse of $\mathbf{D}$, hence $\det \mathbf{D} \det \mathbf{Y} = 1$. Since every entry of every matrix was integer, it is possible only if $\det \mathbf{D} = \det \mathbf{Y} = \det \mathbf{M} = \pm 1$.

Let finally $\mathbf{N}$ be a submatrix of $\mathbf{C}_F$ with size $n(G) - k$, $k \geq 1$. Deleting the other k rows from $\mathbf{C}_F$ we obtain k columns of zero entries only. If one of these columns belongs to $\mathbf{N}$ then $\det \mathbf{N} = 0$. Otherwise these k columns can be deleted from $\mathbf{C}_F$ and the fundamental circuit matrix of a graph H with $e(G) - k$ edges is obtained. $\mathbf{N}$ is now a submatrix of it with size $n(H)$, hence the above reasoning can be applied.

1.4.34 Let F be an arbitrary spanning forest of G and let $\mathbf{C}_F$ be the same as in the solution of the previous problem. We saw that there exists a nonsingular matrix $\mathbf{D}$ of size $n(G)$ satisfying $\mathbf{D}\mathbf{C}_F = \mathbf{X}$. Hence if $\mathbf{M}$ is a maximal nonsingular submatrix of $\mathbf{X}$ and $\mathbf{N}$ is a submatrix of $\mathbf{C}_F$, determined by the same columns, then $\det \mathbf{M} = \det \mathbf{D} \det \mathbf{N}$. Since $\mathbf{C}_F$ is totally unimodular, we obtain $\det \mathbf{M} = \pm k$ where $k = \det \mathbf{D}$ does depend on $\mathbf{X}$ but not on the actual choice of $\mathbf{M}$.

1.5.1 Yes, if C has other edges with the same weight. On the other hand, if all the other edges of C have larger weights then $e \in B$ is impossible (it could be exchanged to an edge of $C - B$, contradicting to the maximality of B).

1.5.2 If the weight of the edges are all different then every choice in the greedy algorithm is unique. Let B be the obtained spanning forest, let $b_0 \notin B$ and consider the unique circuit C in $B \cup \{b_0\}$. Clearly, $w(b_0)$ is smaller than the weight of any other edge in C. Hence, if another base B' were also of maximum weight then we should choose an edge $b_0 \in B' - B$ and exchange it with an edge of $C \cap B$. This leads to a base with weight greater than that of B', a contradiction.

1.5.3 F is clearly a spanning forest. Suppose that its weight is not minimum. Then there were two edges $e \notin F$ and $f \in F$ with $w(e) < w(f)$ so that f belongs to the unique circuit C contained in $F \cup \{e\}$. Since f was not deleted in Step 3, it was a separating edge. However, no edges of $C \cap F$ were deleted and e could be deleted after f only (since $w(e) < w(f)$), a contradiction.

1.5.4 G is shown in Fig. S.1.15 and let $X = \{a, c\}$. Then the "minimum spanning tree" F of H will have weight 3 while the other two edges form a spanning tree T of G with $w(T) = 2$.

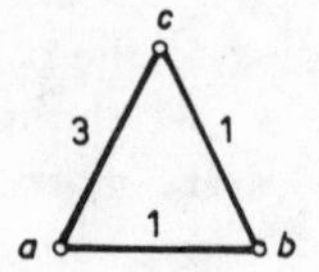

Fig. S.1.15

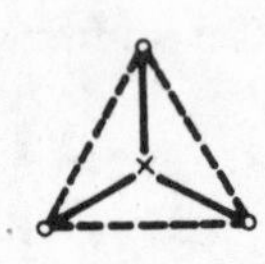

Fig. S.1.16

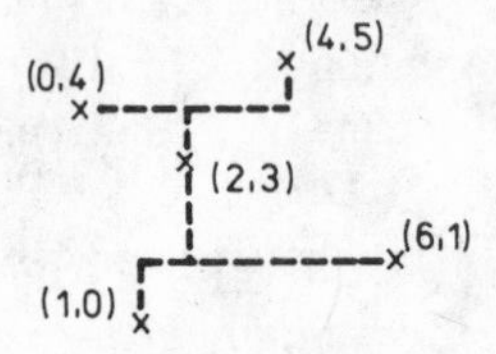

Fig. S.1.17

1.5.5 Any spanning tree of the length 3 circuit of Fig. S.1.16 has weight 2. However, if a junction is allowed in the middle then a Steiner tree of length $3(\sqrt{3}/3)$ is possible, a saving of 13.4%. Gilbert and Pollak [1968] conjecture that this ratio $\sqrt{3}/2$ is the best possible saving for any set of points in the plane.

Steiner trees play an important role in the wiring of printed circuit boards and integrated circuits (although their edge weights are usually not the "Euclidian" distances). For example, the minimum weight Steiner tree among the five points in Fig. S.1.17 has weight 14.

1.5.6 It need not be a minimum weight spanning tree of G, see Fig. S.1.18 with $w(e_1) = w(e_2) = 1, w(e_3) = w(e_4) = 2$ and $Q = \{e_1, e_2\}$.

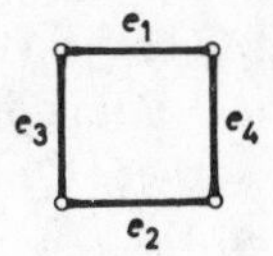

Fig. S.1.18

1.5.7 Consider the 2×2 chessboard

100	99
99	1

The greedy algorithm would place the first castle to the upper left corner, leading to a solution 101 instead of the maximum 198.

1.5.8 We need to find the minimum weight element at each instance of Step 3. Should we start finding it every time, the number of operations would be proportional to ve. Hence an ordering of the weights of all the edges should be done at first, requesting constant times $e \log e$ operations only. One can realize the whole algorithm so that the number of operations is proportional to $e \log e$.

1.5.9 F is clearly a spanning tree. If it were not of minimum weight, choose F_0 to be of minimum weight, satisfying the additional condition of $|E(F) \cap \cap E(F_0)|$ be maximum. Let $f \in E(F_0) - E(F)$, C be the unique circuit of $F \cup \{f\}$ and e be an edge of C so that $(F_0 - \{f\}) \cup \{e\}$ is also a spanning tree. $w(f) > w(e)$ is impossible since the weight of F_0 was minimum and $w(f) = w(e)$ is also impossible by the additional condition. But $w(f) < w(e)$ contradicts to the choice of e in Step 2.

1.5.10 Each instance of Step 2 could easily be performed in v^2 steps, leading to a total which were proportional to v^3. This could be reduced to the sometimes better bound of ve if G is stored in $(\mathbf{e}, \mathbf{v})$ rather than in $\mathbf{A}$.

However, the number of operations can further be reduced to proportional to v^2 in the following way. During the execution of Step 2 one determines the cheapest neighbour v_u of the "new" point u (i.e. of the point which was recently added to File 2). When we choose e, we have to compare only the weight of $\{u, v_u\}$ with the minimum of the previous "cheapest edges" of this type.

1.5.11 Suppose that the weights of the edges $e_1, e_2, e_3, \ldots$ are equal. Then increase $w(e_2)$ by a small positive number ϵ, increase $w(e_3)$ by 2ϵ etc. If ϵ is chosen so that the sum of all those increases is still less than the smallest difference among the original weights then the new minimum will serve as an optimum for the original weighting as well.

1.5.12 The resulting subgraph F is clearly a spanning forest. (Should it contain a circuit then its most expensive edge would never be chosen.) Suppose that F_0 were the cheapest spanning forest and let $e \in E(F_0) - E(F)$. Consider the unique circuit C of $F \cup \{e\}$ and let f be the most expensive edge of $E(C) - \{e\}$. Before building f, the points of C were in different components, in particular, so were the two end points of e as well. Hence $w(e) > w(f)$ since nobody has chosen e. But then $(F_0 - \{e\}) \cup \{f\}$ would be an even cheaper spanning forest, a contradiction.

1.5.13 If, for example, the cheapest edges form a circuit, one could build all of them in the first step. It is enough to require the following weaker condition: Edges of the same weight must form circuit free subgraphs.

1.5.14 Instead of showing that there exists a path between any two points a, b let us prove that the shortest such path has length $d_{ab} = |E(A) - E(B)| = |E(B) - E(A)|$. We use induction on d_{ab}. The case $d_{ab} = 1$ is trivial.

Let e be an arbitrary edge of $E(A) - E(B)$. Consider the unique circuit C of $B \cup \{e\}$. Let $f \in E(C) - E(A)$. Then the point of H_G, corresponding to $(B \cup \{e\}) - \{f\}$, is adjacent to b and its distance to a is $d_{ab} - 1$ only. Hence we can use the induction hypothesis.

1.5.15 Yes. Suppose the contrary, i.e. let X be the "greedy" subgraph and X_0 be a maximum weight subgraph (containing at most one circuit) with the additional property that $|E(X) \cap E(X_0)|$ is maximum. Let $e_0 \in E(X_0) - E(X)$ and let P be a path in X between the end points of e_0. Then $w(e) \geq w(e_0)$ for every $e \in E(P)$ since the greedy algorithm would choose e_0 instead of e otherwise. (We need that $(X \cup \{e_0\}) - \{e\}$ contains again at most one circuit. But this is clear since a subgraph H contains at most one circuit if and only if $n(H) \leq 1$.) Consider now $X_0 - \{e_0\}$. This can be extended by an edge f of P to a subgraph X_1 with $n(X_1) \leq 1$. If $w(f) > w(e_0)$ then $w(X_1) > w(X_0)$, contradicting to the maximality of $w(X_0)$; if $w(f) = w(e_0)$ then $|E(X) \cap E(X_0)|$ were not maximum, again a contradiction.

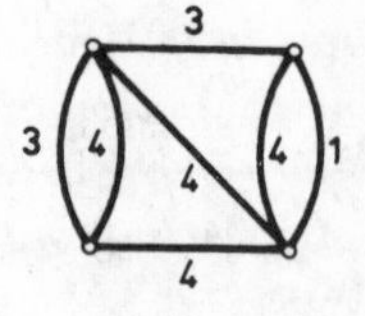

Fig. S.1.19

1.5.16 If we try to repeat the argument of the solution for the previous problem, we cannot modify the parenthesized sentences since a graph G with

$n(G) = 2$ may contain 3 circuits as well. This suggests the counterexample of Fig. S.1.19. The greedy algorithm chooses the four edges of weight 4 and then none of the weight 3 edges can be chosen and a total weight of 17 is obtained. Disregarding the "horizontal" edge of weight 4 a total of 18 can also be obtained.

Chapter 2

2.1.1 Let R_1 and R_2 be in series. Then their currents are equal (by KCL), and the voltage of the equivalent single resistor must be the sum of the voltages of R_1 and R_2 (by KVL, since every circuit either contains both of R_1 and R_2 or none of them). Hence $u_1 + u_2 = R_1 i_1 + R_2 i_2 = = (R_1 + R_2)i$. If R_1 and R_2 are in parallel then $i_1 + i_2 = R_1^{-1}u_1 + +R_2^{-1}u_2 = (R_1^{-1} + R_2^{-1})u$ similarly follows, leading to $R_p^{-1} = R_1^{-1} + +R_2^{-1}$.

2.1.2 Let u and i denote the voltage and the current, respectively, of the whole system. Then $u = u_2 = u_3 + u_1$ and $i = i_2 + i_3$, $i_3 = i_1$. Taking $u_3 = R_3 i_3$ also into consideration, $R_3(i - i_2) = u - u_1$, that is, $u = R_3 i + (u_1 - R_3 i_2)$. This shows that the current source i_2 can be deleted (i.e. substituted by an open circuit) if the voltage source u_1 is changed to $u_1' = u_1 - R_3 i_2$; or the voltage source u_1 can be contracted (i.e., substituted by a short circuit) if the current source i_2 is changed to $i_2' = i_2 - R_3^{-1}u_1$.

2.1.3 The resulting system, as a single network element, does not change if the voltage source u_1 is left only.

2.1.4 Kirchhoff's Current Law is expressed by $\mathbf{Qi} = \mathbf{0}$ (see Eq. (2.3)), the alternative form is just $\mathbf{Bi} = \mathbf{0}$ (where $\mathbf{B}$ is the incidence matrix of the directed graph). The rows of $\mathbf{Q}$ and $\mathbf{B}$ correspond to cut sets and to stars, respectively. Since every star is either a cut set or the union of disjoint cut sets, and since every cut set arises as the sum of some stars, each equation of both systems is implied by the other system. Hence the two systems are equivalent.

2.1.5 See Problem 1.2.29.

2.1.6 There is no way to distinguish two systems from outside, if one of them consists of a voltage source only while the other consists of a voltage source and a resistor in parallel. Similarly, a resistor cannot be observed if it is series to a current source.

2.1.7 (a) Consider Fig. S.2.1. The two systems can be described by the equations $u = u_1 + R_1 i$ (case a) and $u = R_2(i - i_2)$ (case b), see the solution of Exercise 2.1.2 as well. The two systems are equivalent if $R_1 = R_2$ and $u_1 + R_2 i_2 = 0$ hold. Hence, for example, the second system can be replaced by the first one putting $R_1 = R_2$, $u_1 = -R_2 i_2$.

(b) No circuit can be formed by voltage sources only, if each has a series resistor. Similarly, no cut set can be formed by current sources only, if each has a parallel resistor.

2.1.8 The left hand side expression is the resistance if two resistors of value $a + b$ and $c + d$ are in parallel (the switch is open); the right hand side gives the resistance if two resistors of value $\dfrac{ac}{a+c}$ and $\dfrac{bd}{b+d}$ are in series (the switch is closed).

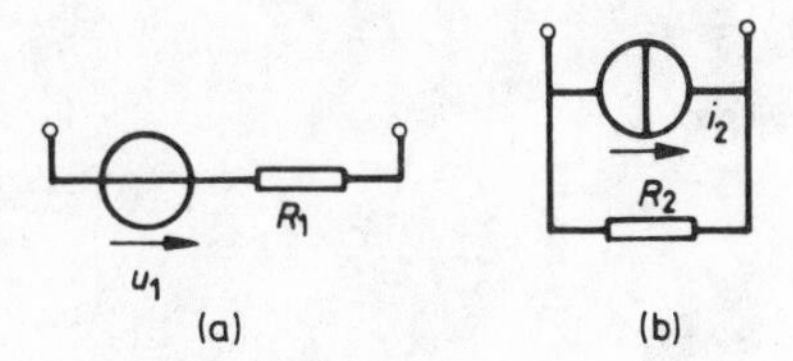

Fig. S.2.1

2.1.9 The resistances are $R_1 = 2$, $R_2 = 5/3$, $R_3 = 13/8$. In general, $R_k = 1+x$ where $x^{-1} = 1 + R_{k-1}^{-1}$, that is, $R_k = (2R_{k-1}+1)/(R_{k-1}+1)$ if $k \geq 2$. One can easily recognize that $R_k = F_{2k+1}/F_{2k}$ where F_t are the Fibonacci numbers, defined by $F_0 = 0$, $F_1 = 1$, $F_t = F_{t-1} + F_{t-2}$ for $t \geq 2$.

2.1.10 The resistance x does not change if the first resistor r in series and the next resistor R in parallel are removed. Hence $x = r + \dfrac{Rx}{R+x}$, leading to $x = \dfrac{1}{2}(r + \sqrt{r^2 + 4Rr})$, since x is clearly positive. In particular, $R = r = 1$ gives the limit of the systems of Problem 2.1.9. This implies the well known result $\lim_{k\to\infty} F_{2k+1}/F_{2k} = (\sqrt{5}+1)/2$ in number theory; this limit is the so called golden section.

2.1.11 The system is redrawn in Fig. S.2.2. The failure can be located if the ratio of $R - X$ and X is known. Hence, X is the root of a quadratic equation, obtained from $R_2 = 2(R-X) + R_1(R_0 + 2X)/(R_1 + R_0 + 2X)$.

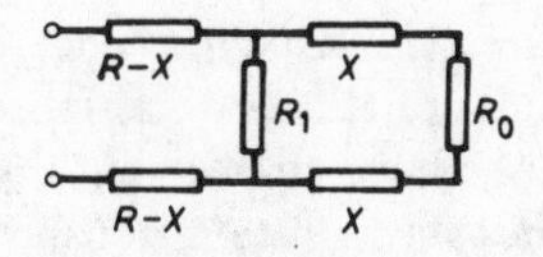

Fig. S.2.2

2.1.12 (a) If the system is measured between points A and B then the roles of points F and H, and those of points C and G, are symmetric (see Fig. S.2.3(a)). Hence the final resistance does not change if F and H are identified and C and G are identified. Thus the system of Fig. S.2.3(b) should be analyzed, leading to $r_{AB} = 7/12$.

(b) If the system is measured between points A and C then we can transform it to that of Fig. S.2.3(c) in a similar way. Then observe that the shaded resistor can be removed (no current will ever flow on it, due to its symmetric position), leading to $r_{AC} = 3/4$.

(c) $r_{AD} = 5/6$, see Fig. 5.2.3(d).

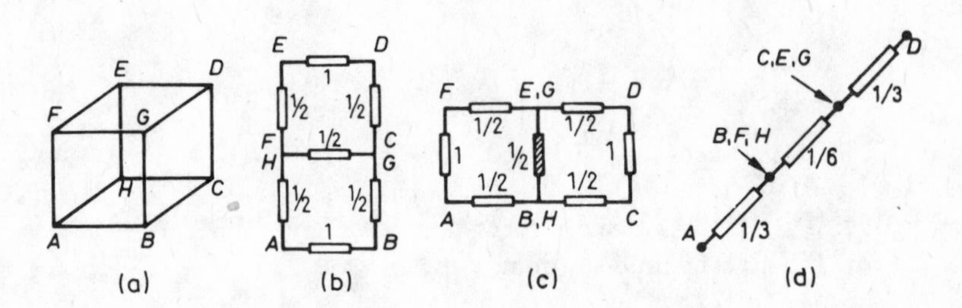

Fig. S.2.3

2.1.13 Let U denote the voltage of the resulting system and U'_j denote the voltage of the admittance G_j. Then clearly $U = U_j + U'_j$ for every j and the sum of the currents $\sum G_j U'_j$ must be zero. Hence $0 = \sum(G_j(U-U_j))$ leads to $U = (\sum G_j U_j)/(\sum G_j)$.

2.1.14 No, if we may suppose that G has neither loops nor separating edges (which is clearly the case if G is the graph of an electric network). For example, let e be an arbitrary nonseparating edge and $C_1, C_2, \ldots, C_{n(G)}$ a maximal system of linearly independent circuits. We have to show that $e \in E(C_1) \cup E(C_2) \cup \ldots \cup E(C_{n(G)})$. (The analogous statement for a nonloop edge and for cut sets can be proved similarly.)

e is contained in a circuit C which is contained in a fundamental system of circuits. The circuits of this system determine a submatrix $\mathbf{C}_1$ of the circuit matrix $\mathbf{C}$. The above circuits $C_1, C_2, \ldots, C_{n(G)}$ also determine a submatrix $\mathbf{C}_2$. Then there exists a nonsingular square matrix $\mathbf{D}$ satisfying $\mathbf{C}_1 = \mathbf{D}\mathbf{C}_2$. If $e \notin E(C_i)$ holds for every i then the corresponding column of $\mathbf{C}_2$ were $\mathbf{0}$. But then the column, corresponding to e, were $\mathbf{0}$ in $\mathbf{C}_1$ as well, while it must contain a one (in the row of C), a contradiction.

2.2.1 $E_X = E_Y = \emptyset$ means $E(G) = E_U \cup E_I$ and $E(F) = E_U$. Then $\mathbf{Cu} = \mathbf{0}$ and $\mathbf{Qi} = \mathbf{0}$ simply decompose into $\mathbf{C}_U\mathbf{u}_U + \mathbf{C}_I\mathbf{u}_I = \mathbf{0}$ and $\mathbf{Q}_U\mathbf{i}_U + \mathbf{Q}_I\mathbf{i}_I = \mathbf{0}$, respectively, where $\mathbf{C}_I$ and $\mathbf{Q}_U$ are unity matrices. Hence $\mathbf{u}_I = -\mathbf{C}_U\mathbf{u}_U$ and $\mathbf{i}_U = -\mathbf{Q}_I\mathbf{i}_I$ directly follow.

2.2.2 Let $E_Y = \emptyset$. Then $\mathbf{Cu} = \mathbf{0}$ and $\mathbf{Qi} = \mathbf{0}$ decompose into $\mathbf{C}_{21}\mathbf{u}_U +$ $+\mathbf{C}_{22}\mathbf{u}_X + \mathbf{u}_I = \mathbf{0}$ and $\begin{pmatrix} \mathbf{E} & \mathbf{O} & \mathbf{Q}_{12} \\ \mathbf{O} & \mathbf{E} & \mathbf{Q}_{22} \end{pmatrix} \begin{pmatrix} \mathbf{i}_U \\ \mathbf{i}_X \\ \mathbf{i}_I \end{pmatrix} = \begin{pmatrix} 0 \\ 0 \end{pmatrix}$, respectively. The solution of the network analysis is $\mathbf{i}_U = -\mathbf{Q}_{12}\mathbf{i}_I$; $\mathbf{i}_X = -\mathbf{Q}_{22}\mathbf{i}_I$ and $\mathbf{u}_I = -\mathbf{C}_{21}\mathbf{u}_U - \mathbf{C}_{22}\mathbf{D}_X\mathbf{i}_X = -\mathbf{C}_{21}\mathbf{u}_U + \mathbf{C}_{22}\mathbf{D}_X\mathbf{Q}_{22}\mathbf{i}_I$; $\mathbf{u}_X = - \mathbf{D}_X\mathbf{Q}_{22}\mathbf{i}_I$. Hence we did not invert any matrix. If $E_X = \emptyset$, we proceed similarly and only the diagonal matrix $\mathbf{D}_Y$ should be inverted.

2.2.3 $\left[R_3 - (-1\ 0)\begin{pmatrix} R_2 & 0 \\ 0 & R_4 \end{pmatrix}\begin{pmatrix} 1 \\ 0 \end{pmatrix}\right] i_3 = u_1 + (-1\ 0)\begin{pmatrix} R_2 & 0 \\ 0 & R_4 \end{pmatrix}\begin{pmatrix} 1 \\ -1 \end{pmatrix} i_5$ that is, $(R_2 + R_3)i_3 = u_1 - R_2 i_5$. If $R_2 + R_3 = 0$, the voltage source is short circuited hence the network is either not solvable (if $u_1 \neq 0$) or the solution is not unique (if $u_1 = 0$ then i_1 may be arbitrary).

2.2.4 Since matrix inversion is clearly more time-consuming than any other step in **Box 2.1**, the "right hand side method" is more advantageous if $n(G) - |E_I|$ is larger than $r(G) - |E_U|$. This is the case, for example, if the network graph is "almost complete", i.e. if the nullity of G is constant times $(v(G))^2$.

2.2.5 Consider the network of Fig. S.2.4. Here $n(G) - |E_I| = r(G) - |E_U| = 3$. However, one can check that choosing u_3 and i_8 as unknown quantities, everything else can be expressed as their linear combination (by Kirchhoff's Laws and Ohm's Law), hence a 2×2 matrix should be inverted only. The general problem of finding the minimum number of unknowns with this property will be considered as Problem 12.2.14.

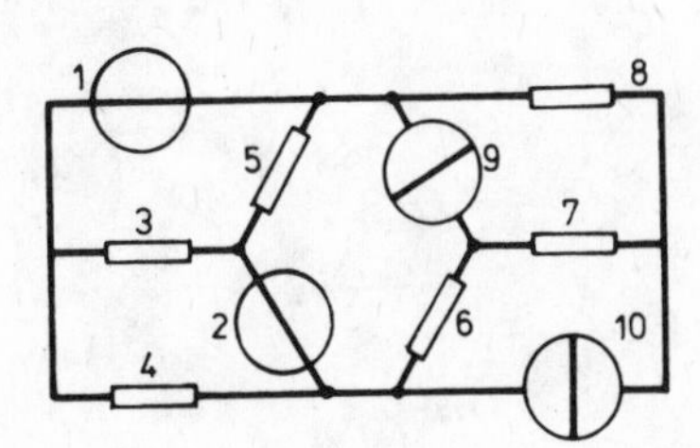

Fig. S.2.4

2.2.6 Let N and N' denote the original and the modified networks, respectively. Ohm's Law gives clearly the same equations for N and for N'. First we show that the voltages of the unmodified network elements do not change. Consider a circuit C of the graph of N. If $Q \cap E(C) = \emptyset$ then C is a circuit of N' as well and gives the same voltage equation. If $|Q \cap E(C)| = 2k$ (with $k > 0$) and $e \notin E(C)$ then we add the voltage of

e k times and subtract it k times as well, from the original equation, to obtain the new one. If $e \in E(C)$ then the new equation will contain the voltage of e k times with plus sign, $(k-1)$ times with minus sign, while the original equation contained it just once.

Finally, we have to consider the equations given by Kirchhoff's Current Law. Consider the fundamental system of cut sets, with respect to the spanning forest F, as in Theorem 2.2.1. Since $e \in E(F)$, exactly one cut set Q_e of the fundamental system contains e. The others give the same equations for N and N'. The "lost" equation served only for determining the current of e in N at the end of the analysis.

2.2.7 Let e be an edge, corresponding to a current source, and C be a cut set in the network graph, containing e. If the current source is replaced by an open circuit and a current source with the same value is added parallel to each of the edges of $C - \{e\}$ then the voltages and currents of the other network elements do not change.

2.2.8 If $f \in Q - \{e\}$ is a resistor then the process of Problem 2.2.6 really added a voltage source in series. But if there were a circuit C' of voltage sources in the network then short circuiting $e \in C'$ by this process leads to another circuit of voltages sources.

2.3.1 If two capacitors C_1 and C_2 are in parallel (or in series) then they can be substituted by a single capacitor $C_p = C_1 + C_2$ (or $C_s = (C_1+$ $+C_2)^{-1}C_1C_2$, respectively, provided that $C_1+C_2 \neq 0$). If two inductors L_1 and L_2 are in series (or in parallel) then we obtain $L_s = L_1 + L_2$ (or $L_p = (L_1 + L_2)^{-1}L_1L_2$, respectively, if $L_1 + L_2 \neq 0$).

2.3.2 The equations $u_1 = R_2 i_2 + u_3$; $-i_1 = i_2 = i_3 = C_3 \frac{d}{dt} u_3$ in the case of the first network give

$$\frac{d}{dt}u_3 = -\frac{1}{R_2C_3}u_3 + \frac{1}{R_2C_3}u_1$$

$$i_1 = \frac{1}{R_2}u_3 - \frac{1}{R_2}u_1$$

In the case of the second network the canonical state equations are

$$\frac{d}{dt}\begin{pmatrix} u_3 \\ u_4 \end{pmatrix} = \frac{-1}{R_2C_3}\begin{pmatrix} 1 & 1 \\ 1 & 1 \end{pmatrix}\begin{pmatrix} u_3 \\ u_4 \end{pmatrix} + \frac{1}{R_2C_3}\begin{pmatrix} 1 \\ 1 \end{pmatrix} u_1$$

$$i_1 = \frac{1}{R_2}(1 \quad 1)\begin{pmatrix} u_3 \\ u_4 \end{pmatrix} - \frac{1}{R_2}u_1$$

2.3.3 In the case of the first network no differential equation has to be solved at all (we have seen this in a similar example preceeding Theorem 2.3.1). This alone would not imply that the network has no canonical state equation ($\mathbf{A} = \mathbf{O}$ in Eq. (2.10) is not excluded) but in order to determine the output we need the derivative of the input as well ($i_1 = \frac{-1}{R_2}u_1 - C_3\frac{d}{dt}u_1$), unlike in Eq. (2.10).

The second network has a canonical state equation. It can be obtained from that of the first network of Fig. 2.18 by putting $C_3 + C_4$ instead of C_3.

2.3.4 The highly artificial networks of Fig. S.2.5 satisfy the condition of Theorem 2.3.2 (there exist spanning forests F with the requested property) yet they are not uniquely solvable if $C_2 + C_3 = 0$ or $L_2 + L_3 = 0$, respectively.

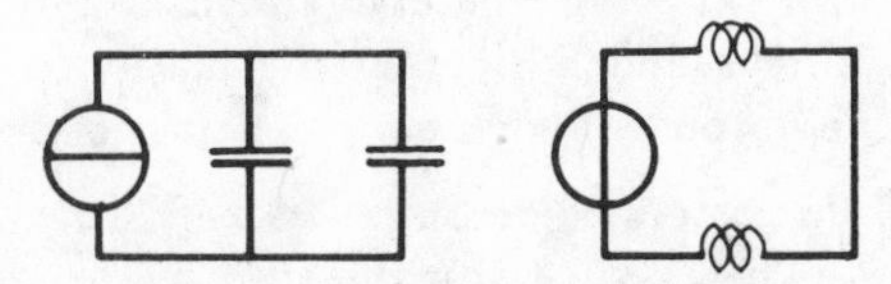

Fig. S.2.5

2.3.5 The network graph is shown in Fig. S.2.6. The role of F_0 in Theorem 2.3.2 can be played by the spanning tree with heavy edges. The differential equation will have order 4: the initial values of u_2, u_3, i_9 and i_{10} can independently be prescribed.

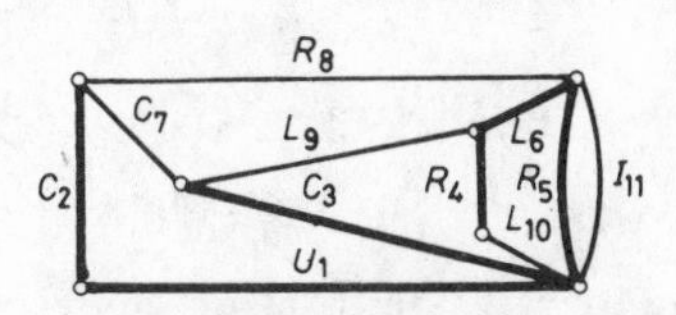

Fig. S.2.6

2.3.6 Let i_1 and i_2 denote the currents of the resistors, parallel to the capacitor and to the inductor, respectively. Let i denote the current flowing through the whole system and let $I = i_1 + i_2 - i$. If all the network elements are of unit value, $i = i_1' + i_1$ and $i' = i_2' + i_2$, due to the definitions of the capacitor and the inductor, respectively. (f' stands for $\frac{df}{dt}$.) The sum of these two equations leads to $I' + I = 0$ whose general solution is $I = c_0 e^{-t}$ with a constant c_0. If i is a direct current, $i_1 = i$ and $i_2 = 0$ clearly follows, hence $c_0 = 0$, leading to $I = 0$. The total voltage is $R_1 i_1 + R_2 i_2 = i_1 + i_2 = i$, hence the whole system is equivalent to a resistor with unit resistance.

2.3.7 The difference between the two networks of Fig. 2.18 is that, in the latter case, the voltages u_3 and u_4 of the two capacitors can be measured separately. If C_3 and C_4 are replaced by a single capacitor with $C_3' = \frac{C_3 C_4}{C_3 + C_4}$ then u_3' will be the sum of u_3 and u_4. Hence, in the first network, the initial value of a single voltage can be prescribed only, while in the second case the initial value of both capacitor voltages can independently be prescribed.

2.3.8 The order of the state equation decreases if capacitors form circuits or inductors form cut sets (see Theorem 2.3.2). This order does not decrease if capacitors form cut sets or inductors form circuits. But in this latter case one can always find a differential equation of smaller order, still determining the output (though not the state) of the system.

2.3.9 Let the set R of resistors be decomposed as $X \cup Y$ where the edges from X belong to F. Then Eq. (2.4) is replaced by

$$\begin{pmatrix} \mathbf{u}_Y \\ \mathbf{u}_L \\ \mathbf{u}_I \end{pmatrix} = - \begin{pmatrix} \mathbf{C}_{11} & \mathbf{C}_{12} & \mathbf{C}_{13} \\ \mathbf{C}_{21} & \mathbf{C}_{22} & \mathbf{C}_{23} \\ \mathbf{C}_{31} & \mathbf{C}_{32} & \mathbf{C}_{33} \end{pmatrix} \begin{pmatrix} \mathbf{u}_U \\ \mathbf{u}_C \\ \mathbf{u}_X \end{pmatrix} \quad \text{and}$$

$$\begin{pmatrix} \mathbf{i}_U \\ \mathbf{i}_C \\ \mathbf{i}_X \end{pmatrix} = - \begin{pmatrix} \mathbf{Q}_{11} & \mathbf{Q}_{12} & \mathbf{Q}_{13} \\ \mathbf{Q}_{21} & \mathbf{Q}_{22} & \mathbf{Q}_{23} \\ \mathbf{Q}_{31} & \mathbf{Q}_{32} & \mathbf{Q}_{33} \end{pmatrix} \begin{pmatrix} \mathbf{i}_Y \\ \mathbf{i}_L \\ \mathbf{i}_I \end{pmatrix}, \qquad (2.4')$$

while Eq. (2.5) is extended by $\mathbf{u}_L = \mathbf{D}_L \frac{d}{dt}\mathbf{i}_L$ and $\mathbf{i}_C = \mathbf{D}_C \frac{d}{dt}\mathbf{u}_C$. Then

$$\begin{aligned} \mathbf{u}_X = \mathbf{D}_X \mathbf{i}_X &= -\mathbf{D}_X \mathbf{Q}_{31}\mathbf{i}_Y - \mathbf{Q}_{32}\mathbf{i}_L - \mathbf{Q}_{33}\mathbf{i}_I = \\ &= -\mathbf{D}_X \mathbf{Q}_{31}\mathbf{D}_Y^{-1}\mathbf{u}_Y - \mathbf{Q}_{32}\mathbf{i}_L - \mathbf{Q}_{33}\mathbf{i}_I = \\ &= \mathbf{D}_X \mathbf{Q}_{31}\mathbf{D}_Y^{-1}(\mathbf{C}_{11}\mathbf{u}_U + \mathbf{C}_{12}\mathbf{u}_C + \mathbf{C}_{13}\mathbf{u}_X) - \mathbf{Q}_{32}\mathbf{i}_L - \mathbf{Q}_{33}\mathbf{i}_I . \end{aligned}$$

The coefficient matrix (to be inverted to determine $\mathbf{u}_X$) is

$$\mathbf{E} - \mathbf{D}_X \mathbf{Q}_{31}\mathbf{D}_Y^{-1}\mathbf{C}_{13}$$

and this is positive definite, due to $\mathbf{Q}_{31} = -\mathbf{C}_{13}^T$. Hence $\mathbf{u}_X$ (and $\mathbf{i}_Y$ in a similar way) are expressed as linear functions of the input ($\mathbf{u}_U$ and $\mathbf{i}_I$) and the state ($\mathbf{u}_C$ and $\mathbf{i}_L$). Then the output ($\mathbf{u}_I$ and $\mathbf{i}_U$) is directly obtained from Eq. (2.4′) and a similar calculation leads to the state equation as well.

2.3.10 Let F be a spanning forest containing the voltage sources and the elements of a subset X of the capacitors. Let Y denote the set of the capacitor edges not in F. Equation (2.4) is now replaced by

$$\begin{pmatrix} \mathbf{u}_Y \\ \mathbf{u}_I \end{pmatrix} = -\begin{pmatrix} \mathbf{C}_{11} & \mathbf{C}_{12} \\ \mathbf{C}_{21} & \mathbf{C}_{22} \end{pmatrix} \begin{pmatrix} \mathbf{u}_U \\ \mathbf{u}_X \end{pmatrix} \text{ and } \begin{pmatrix} \mathbf{i}_U \\ \mathbf{i}_X \end{pmatrix} = -\begin{pmatrix} \mathbf{Q}_{11} & \mathbf{Q}_{12} \\ \mathbf{Q}_{21} & \mathbf{Q}_{22} \end{pmatrix} \begin{pmatrix} \mathbf{i}_Y \\ \mathbf{i}_I \end{pmatrix}$$

and Eq. (2.5) becomes $\mathbf{i}_X = \mathbf{D}_X \frac{d}{dt}\mathbf{u}_X; \mathbf{i}_Y = \mathbf{D}_Y \frac{d}{dt}\mathbf{u}_Y$. Then $\frac{d}{dt}\mathbf{u}_X =$ $= \mathbf{D}_X^{-1}(-\mathbf{Q}_{21}\mathbf{i}_Y - \mathbf{Q}_{22}\mathbf{i}_I) = -\mathbf{D}_X^{-1}\mathbf{Q}_{21}\mathbf{D}_Y \frac{d}{dt}\mathbf{u}_Y - \mathbf{D}_X^{-1}\mathbf{Q}_{22}\mathbf{i}_I =$ $= \mathbf{D}_X^{-1}\mathbf{Q}_{21}\mathbf{D}_Y \frac{d}{dt}(\mathbf{C}_{11}\mathbf{u}_U + \mathbf{C}_{12}\mathbf{u}_X) - \mathbf{D}_X^{-1}\mathbf{Q}_{22}\mathbf{i}_I$. The coefficient matrix of $\frac{d}{dt}\mathbf{u}_X$ is $\mathbf{E} - \mathbf{D}_X^{-1}\mathbf{Q}_{21}\mathbf{D}_Y\mathbf{C}_{12}$ and this can be inverted since $\mathbf{Q}_{21} = -\mathbf{C}_{12}^T$. The output equation can be obtained in a straightforward way.

2.3.11 Let $E_C = E_{CX} \cup E_{CY}; E_R = E_{RX} \cup E_{RY}; E_L = E_{LX} \cup E_{LY}$ where subscript X refers to elements in F_0 and Y to those not in F_0. Then the decomposition of $\mathbf{C}$ is

$$\begin{pmatrix} \mathbf{u}_{CY} \\ \mathbf{u}_{RY} \\ \mathbf{u}_{LY} \\ \mathbf{u}_I \end{pmatrix} = -\begin{pmatrix} \mathbf{C}_{11} & \mathbf{C}_{12} & \mathbf{C}_{13} & \mathbf{C}_{14} \\ \mathbf{C}_{21} & \mathbf{C}_{22} & \mathbf{C}_{23} & \mathbf{C}_{24} \\ \mathbf{C}_{31} & \mathbf{C}_{32} & \mathbf{C}_{33} & \mathbf{C}_{34} \\ \mathbf{C}_{41} & \mathbf{C}_{42} & \mathbf{C}_{43} & \mathbf{C}_{44} \end{pmatrix} \begin{pmatrix} \mathbf{u}_U \\ \mathbf{u}_{CX} \\ \mathbf{u}_{RX} \\ \mathbf{u}_{LX} \end{pmatrix},$$

but the blocks $\mathbf{C}_{13}, \mathbf{C}_{14}$ and $\mathbf{C}_{24}$ are zero blocks each. (For example, if $\mathbf{C}_{13}$ contains a nonzero entry then a further capacitor could be added to F_0 by removing a resistor.)

2.4.1 Let, for example, the weight of F be 17 634. Then F contains 4 voltage source, 3 capacitors, 6 resistors, 7 inductors and 1 current source. If the weight is a five digit number, the set of current sources contains a cut set.

2.4.2 Yes. This is a trivial consequence of Problem 1.2.29.

2.4.3 **Step 1** Assign weight 0 to the voltage source edges, weight 1 to the capacitor edges and weight 2 to the rest. Using the subroutine we either find a circuit of voltage sources (and stop) or a maximal set C_X of capacitors so that C_X and the set U of voltage sources together is circuit-free.

Step 2 Assign weight 0 to the elements of $U \cup C_X$, weight 1 to the resistor edges and weight 2 to the rest. Thus obtain a maximal set R_X of resistors so that $U \cup C_X \cup R_X$ is circuit-free.

Step 3 Assign weight 0 to the elements of $U \cup C_X \cup R_X$, weight 1 to the inductor edges and 2 to the rest. If the resulting spanning forest of minimum weight contains an edge of weight 2 then the set of current sources contains a cut set. Otherwise this forest can play the role of F in Statement 2.4.2.

2.4.4 The idea of Exercise 2.4.1 can be applied if a bound s for the x_i's (like 9 in that exercise) is known:

Question 1 $(1, 1, \ldots, 1)$

Answer 1 $s = x_1 + x_2 + \ldots + x_k$

Question 2 $(1,\ (s+1),\ (s+1)^2, \ldots, (s+1)^{k-1})$

Then the digits of *Answer 2* in a number system of base $s+1$ give the solution.

2.5.1 Let G be the graph of the network and R_1 and R_2 be the two resistors in series. Replace them by a single resistor $R = R_1 + R_2$ and denote the new graph by G_0. The new determinant can be decomposed as $\sum_1 + \sum_2$ where summation in $\sum_1$ is over the trees containing R; and then each member of $\sum_2$ contains R.

On the other hand, a spanning forest F of G_0 can be transformed to a spanning forest of G as follows: If $R \in E(F)$, remove R and add both of R_1 and R_2; if $R \notin E(F)$, add either R_1 or R_2. Hence the original determinant is $\sum_1 + \frac{R_1}{R}\sum_2 + \frac{R_2}{R}\sum_2$ which is the same as above.

2.5.2 A spanning forest F of N becomes a spanning forest of N_o if $R_0 \notin E(F)$. If $R_0 \in E(F)$ then $F - \{R_0\}$ becomes a spanning forest of N_s. Hence, using the notation of the previous exercise, $D = \sum_1 + \sum_2$ with $\sum_1 = D_s$ and $\sum_2 / R_0 = D_o$.

2.5.3 Figure S.2.7 shows N_o and N_s if R_6 is replaced by an open circuit and by a short circuit, respectively. Then $D_s = R_1R_2R_3 + R_1R_2R_4 + R_1R_3R_4 + R_2R_3R_4 + R_5(R_1R_3 + R_1R_4 + R_2R_3 + R_2R_4)$ is obtained by short and open circuiting R_5, respectively, and similarly, $D_o = R_1R_2 + R_1R_4 + R_2R_3 + R_3R_4 + R_5(R_1 + R_2 + R_3 + R_4)$. Finally, $D = D_s + R_6D_o$.

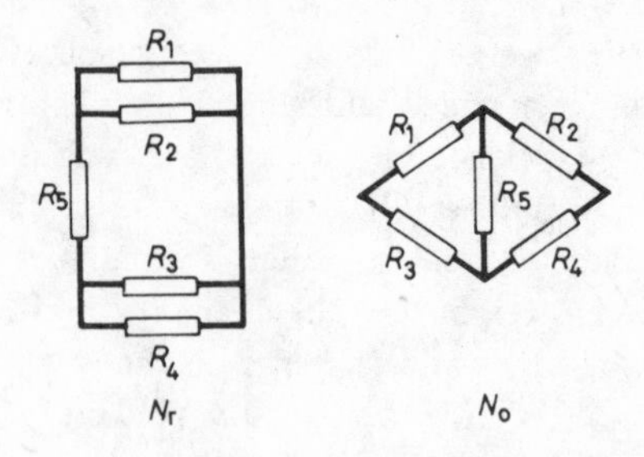

Fig. S.2.7

2.5.4 Let k and ℓ denote the number of rows of $\mathbf{A}$ and $\mathbf{B}$, respectively. In order to avoid zero columns, only such k-tuples of columns should be chosen for the upper k rows which belong entirely to $\mathbf{A}$. Then entries of $\mathbf{E}$ must be chosen for the remaining $\ell - k$ columns of $\mathbf{A}$ (see Fig. S.2.8). Hence the ith row of $\mathbf{B}$ is chosen if and only if so is the ith column of $\mathbf{A}$. Thus the expansion members are identical to those in the theorem of Binet and Cauchy, while the sign is $+$ or $-$ whether an even or an odd

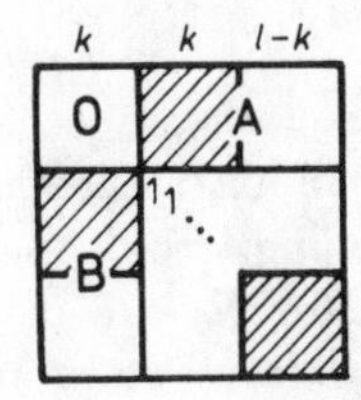

Fig. S.2.8

number of column inversions is needed to arrange the shaded blocks in a block-diagonal form.

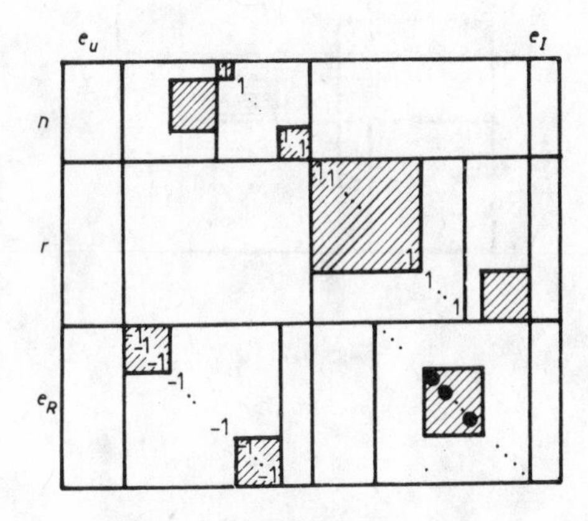

Fig. S.2.9

2.5.5 Consider a member of det **W**. We may re-arrange the rows and columns of the matrix of Fig. 2.22 so that the resistances appearing in this member follow consecutively as indicated by heavy dots in Fig. S.2.9. Then the location of the other expansion members can be identified in a straightforward way, see the shaded areas in Fig. S.2.9. The key observation is that the nonsingular submatrices of $\mathbf{C}_1$ and $\mathbf{Q}_1$ are just the negative transpose of each other, hence the product of their determinant is always -1, independently of the sign of these determinants. The number of row- and column-inversions in order to bring the shaded areas in a block-diagonal position can easily be computed and is also independent of the choice of the expansion member of det **W**.

A simpler proof can be found, combining the theorem of Binet and Cauchy (see Problem 1.4.31) and the method of the "node-potentials" (see **Box 4.2**). The interested reader is referred to pp. 156–157 of Seshu and Reed (1961).

2.5.6 Consider the system of equations for N, as in Fig. 2.22, with respect to a spanning tree F. Add an extra current source I_0 between the points x and y. F will be a spanning tree of the new network N_0 as well, hence the system of equations for N_0 can be written in the compact form of Fig. S.2.10 or in the alternative form of Fig. S.2.11 where only the unknown quantities are at the left hand side. For such an equation $\mathbf{Ax} = \mathbf{b}$ with det $\mathbf{A} \neq 0$ a particular unknown quantity x_i can be obtained as det $\mathbf{A}_i$/ det $\mathbf{A}$, where $\mathbf{A}_i$ is formed by replacing the ith column of $\mathbf{A}$ by $\mathbf{b}$.

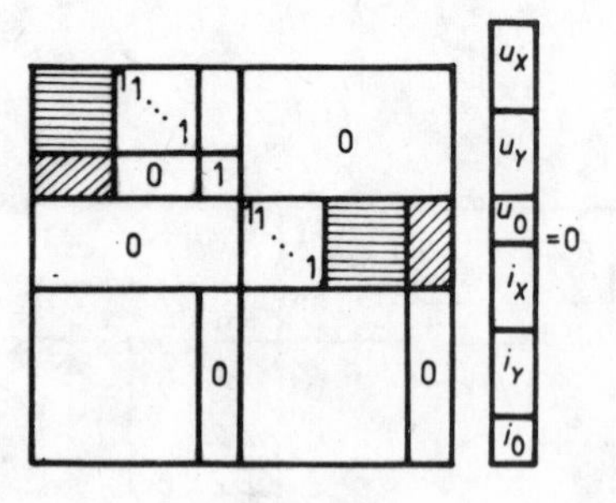

Fig. S.2.10

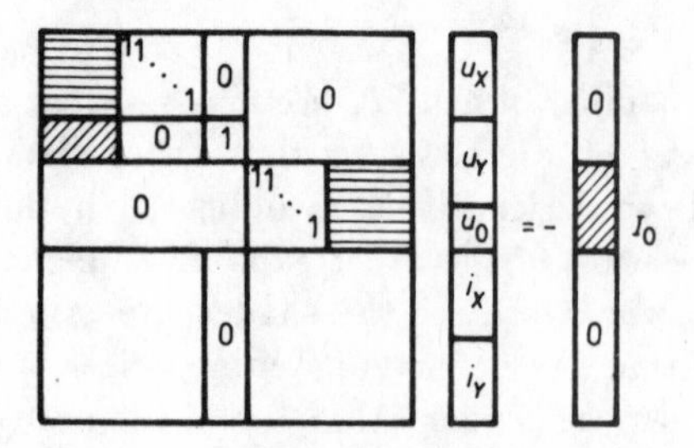

Fig. S.2.11

Clearly, det $\mathbf{A} = \pm D$ (expand det $\mathbf{A}$ along the column, corresponding to u_0). Using the Laplace expansion for $\mathbf{A}_i$ one immediately sees that the edge, corresponding to I_0, must belong to the spanning trees of N (and never to the complement of such a tree). Such trees just correspond to all the spanning trees of N_{xy}. Hence det $\mathbf{A}_i = D_{xy} \cdot I_0$, that is, the resistance in question is $u_0/I_0 = \det \mathbf{A}_i/(I_0 \cdot \det \mathbf{A}) = D_{xy}/D$.

2.5.7 The determinant D of this network was obtained in Exercise 2.5.3. The other determinant D_{xy} is R_6 times the determinant of N_s in Fig. S.2.7 (since R_6 becomes a loop and is, therefore, contained in the complement of every tree). Hence the total resistance is $R_6 D_s/(D_s + R_6 D_o)$, where D_o and D_s were obtained in Exercise 2.5.3.

2.5.8 Let D and D_j denote the determinant of N and N_j, respectively, where N_j is obtained from N by contracting the jth resistor. We know that $r_j = D_j/D$, by Problem 2.5.6. D_j is obtained from a summation along those trees which contained the jth edge (each multiplied by R_j since this edge becomes a loop), hence D_j/R_j equals the sum of those products in the expansion of D which contained the jth edge. Thus $\sum_{j=1}^{n}(D_j/R_j) = D \cdot r(G)$ since it counts each member as many times as the number of edges in each tree. Therefore

$$\sum_{j=1}^{n}(r_j/R_j) = \frac{1}{D}\sum_{j=1}^{n}(D_j/R_j) = r(G).$$

2.5.9 Consider an $n \times n$ finite portion G_n of the grid. Let ρ_n be the average of the total resistances between pairs of adjacent points of the grid. Then $v(G_n) = (n+1)^2$, $e(G_n) = 2n(n+1)$ and $r(G_n) = n^2 + 2n$. By the previous problem $\sum(r_j/R_j) = n^2 + 2n$ approximately equals to $e_n(\rho_n/1) = 2n(n+1)\rho_n$, hence $\lim_{n\to\infty} \rho_n = 1/2$.

2.5.10 For the triangular grids we use a finite triangular portion G_n. Then $v_n \approx n^2/2$, $e_n \approx 3n^2/2$ and $\lim_{n\to\infty} \rho_n = 1/3$. In the case of a hexagonal grid let G_n denote the graph formed by those hexagons whose "distance" from a given one is at most n (see Fig. S.2.12). The number of such hexagons is $1 + 6 + 12 + 18 + \ldots + 6n \approx 3n^2$, hence $v(G_n) \approx n^2$ and $e(G_n) \approx 3n^2/2$ (since each point and edge belongs to 3 and 2 hexagons, respectively), leading to $\lim_{n\to\infty} \rho_n = 2/3$.

2.5.11 Let t denote the number of spanning trees in G and let every edge be contained in k such trees. Then clearly $k \cdot e = t(v-1)$. On the other hand, the number of spanning trees in G/e_0 for any edge e_0 of G is the same as the number of those spanning trees of G which contain e_0. Hence the statement follows from Problem 2.5.6.

2.5.12 Let t denote the number of spanning trees in a graph G with n points. Then the total number of operations for only listing these trees is proportional to tn (and actually finding them at first may be an even longer job). We have seen several examples when t is growing at least exponentially as a function of n, namely in Problem 2.1.9 for ladder networks

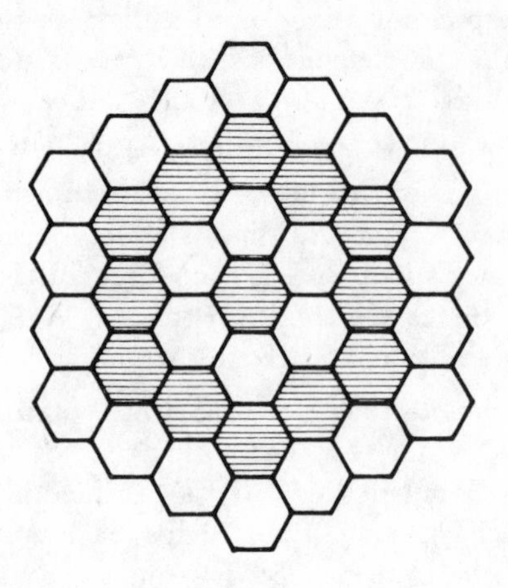

Fig. S.2.12

(recall that the Fibonacci numbers are exponential functions of n) or in Problem 1.2.23 for complete graphs. Hence the topological formulae cannot be suggested for practical network analysis.

2.6.1 The graphs of Fig. S.2.13 show that the first system is rigid and the second one is not. A deformation of the second one is shown in Fig. S.2.14.

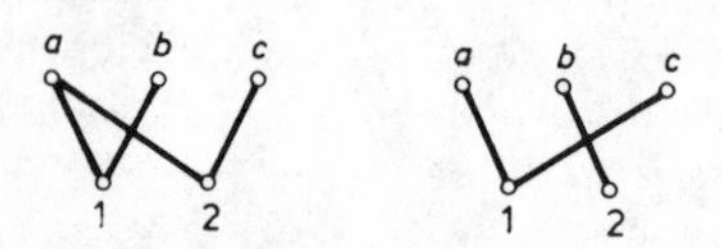

Fig. S.2.13

2.6.2 If the whole system is nonrigid, its graph contains several strongly connected components. If a room belongs to row i and column j then it can be deformed if and only if v_i and w_j are in different strongly connected components.

2.6.3 If a room can be deformed then v_i and w_j are in different connected components. If a deformation D did not effect this room then adding a diagonal rod to this room would make D still possible, although the new graph were connected, a contradiction.

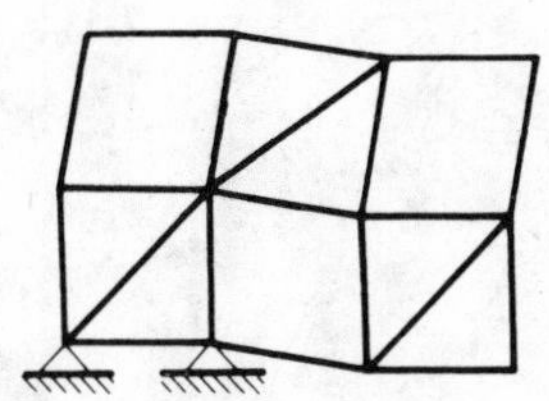

Fig. S.2.14

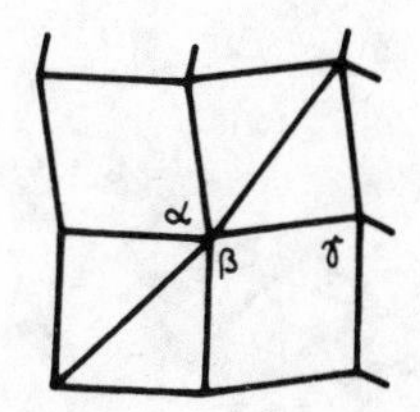

Fig. S.2.15

2.6.4 The adjacent angles β and γ of a rhombus give $\beta + \gamma = 180^o$, see Fig. S.2.15. On the other hand, $\alpha + \beta = 180^o$ since their sum plus the two right angles must give 360°. Hence $\alpha = \gamma$, so the two rhombi are congruent.

2.6.5 The graph of the system is shown in Fig. S.2.16. It is not strongly connected but can be made so by adding an edge from b to 2. Hence the present system is nonrigid (see a deformation in Fig. S.2.17) but the system in Fig. S.2.18 is rigid.

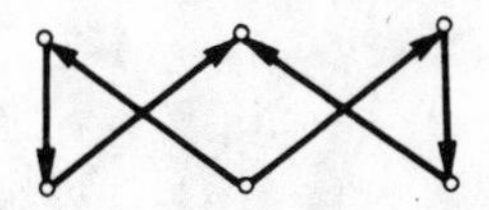

Fig. S.2.16

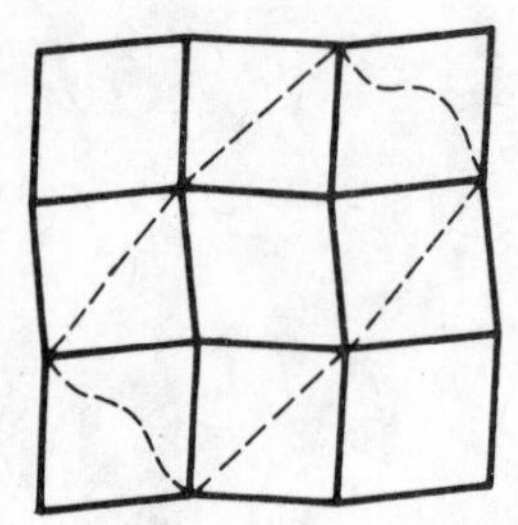

Fig. S.2.17

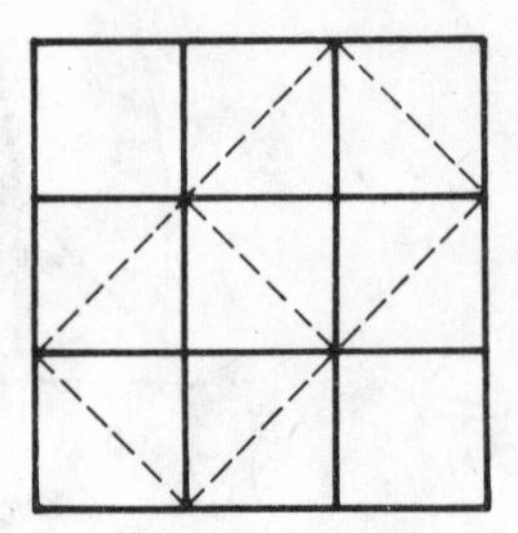

Fig. S.2.18

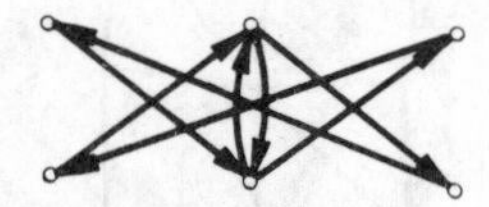

Fig. S.2.19

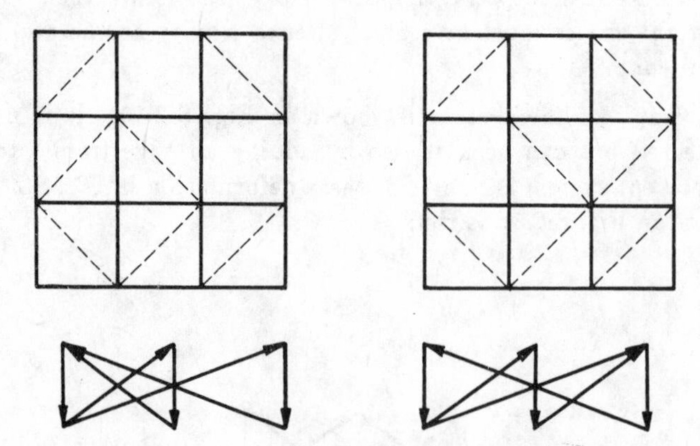

Fig. S.2.20

2.6.6 The graph of the system is shown in Fig. S.2.19. One can immediately see that even the deletion of the rod keeps the graph strongly connected.

2.6.7 Two of the possible solutions is shown in Fig. S.2.20, each requires 3 more cables. This number is clearly minimum since the graph of the original system has two isolated points, each should be incident to at least two edges, and these two pairs of edges cannot have more than one edge in common.

2.6.8 Let G be an undirected graph and let G' be obtained from G by replacing each edge by a pair of oppositely oriented edges. Then G' is strongly connected if and only if G is connected.

2.6.9 If H is a directed graph and H' is obtained from H by reversing the orientation of every edge then H' is strongly connected if and only if H is strongly connected.

2.6.10 The pairs a, b; c, d; e, f and g, h of rods are parallel, see Fig. S.2.21. b and f are perpendicular (since a is perpendicular to e) and so are c and g (due to d and h). Hence the angle formed by b and g, and the angle formed by c and f, are either equal or their sum is 180°. The congruence of the rhombi follows in both cases.

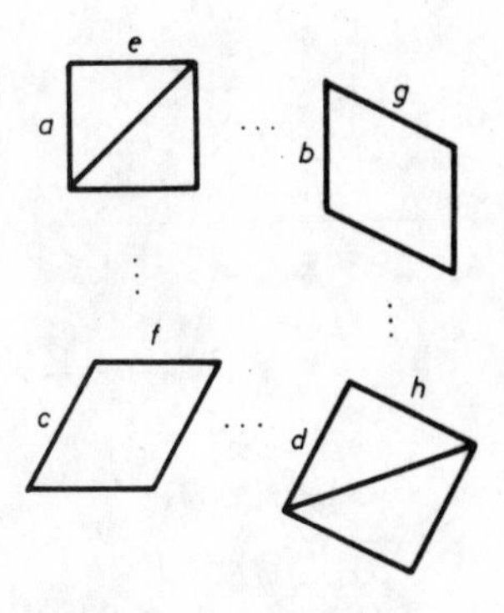

Fig. S.2.21

2.6.11 The graph G of the system is shown in Fig. S.2.22 where $V_1 \cup V_3$ and $V_2 \cup V_4$ must span connected subgraphs. The graph H of the new system contains every possible edge between V_1 and V_4 and between V_2 and V_3. Hence $c(H) = 2$ clearly holds if none of these four sets of points is empty. If, say, $V_4 = \emptyset$ then $|V_2| = 1$, since $c(G) = 2$. Then $c(H) = |V_1| + 1$.

2.6.12 Let, say, $k \geq \ell$. Then each of the k points should appear both as head and as tail in the strongly connected graph, and all these $2k$ edges must be different. Hence at least $2 \cdot \max(k, \ell)$ cables are really needed. Figure S.2.23 shows a strongly connected graph with exactly this number of edges.

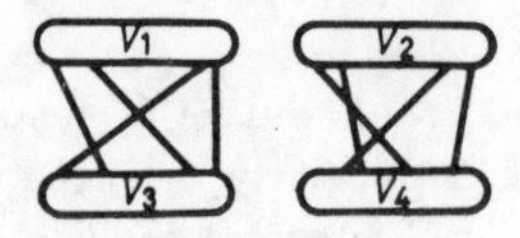

Fig. S.2.22

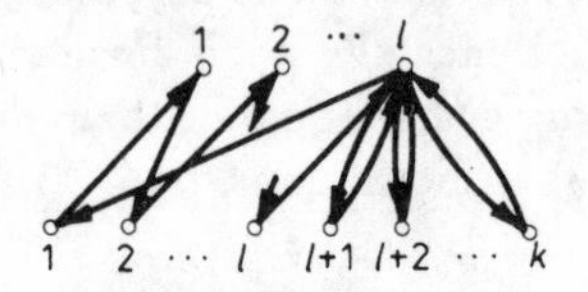

Fig. S.2.23

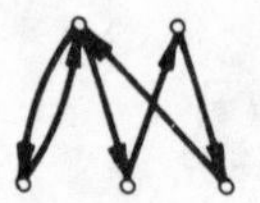

Fig. S.2.24

2.6.13 Figure S.2.24 shows that the answer is negative.

2.6.14 The graph of the system cannot have circuits of length three. Hence $k+\ell$ edges imply the existence of a circuit of length at least four. However, if we use $2 \cdot \max(k, \ell) + 1$ diagonal cables, the system may or may not have critical cables. For example, the two systems in Fig. S.2.20 contain 7 cables each and only the second contains a single noncritical cable (namely $c1$).

Chapter 3

3.1.1 The first and the fourth graphs are isomorphic (but their duals are 2-isomorphic only). The others are 2-isomorphic to one another (and their duals are isomorphic), moreover, the second and the fifth are isomorphic. The former two graphs and the latter three graphs are not 2-isomorphic (look for 3-element cut sets, for example).

3.1.2 Let C be a circuit of G. If it does not contain the points x and y then it remains unchanged in H. Otherwise C is the union of two xy paths of G and so is in H as well (the order of edges along one of the paths may change only). One can see in a similar way that if a set of edges does not form a circuit in G then it will not form a circuit in H either.

3.1.3 Yes. The requested one–one correspondence between their edge sets is shown in Fig. S.3.1.

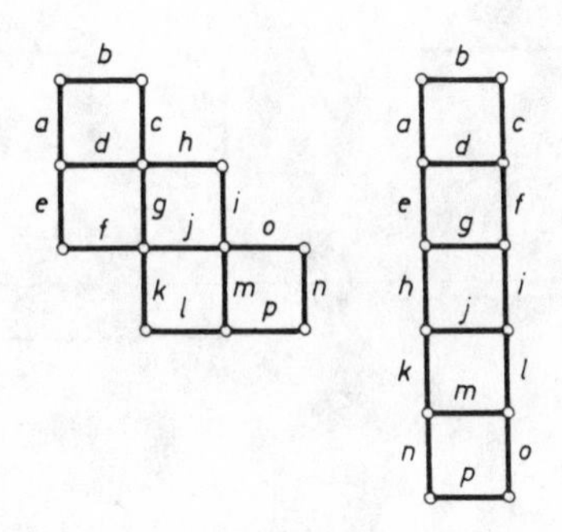

Fig. S.3.1

3.1.4 By our definition any two edges of, say, a circuit of length four are in series, not only the adjacent ones.

3.1.5 Two edges are in series if and only if they are not bridges and their entries in every row of $\mathbf{C}$ are either simultaneously 0 or simultaneously 1. Similarly, two edges are parallel if and only if they are not loops and have identical columns in $\mathbf{B}$ (and in $\mathbf{Q}$). Alternatively, two edges are parallel if and only if $\mathbf{C}$ has a row consisting of two ones only (in positions corresponding to these two edges) and series edges can similarly be recognized in $\mathbf{B}$ and $\mathbf{Q}$.

3.1.6 If F is an arbitrary spanning forest then either $e \in F$ (and then $F - \{e\}$ is a spanning forest in G/e) or $e \notin F$ and then F is a spanning forest in $G - e$.

3.1.7 An edge is a *loop* if it forms a single element circuit. Edges are *parallel* if any two form a circuit. Edges are *series* if they are not separating and, for every circuit C of the graph, either all the edges are contained in C or none of them. *Cut set* is a minimal nonempty set of edges, intersecting each circuit by an even number of edges. *A spanning forest* is a maximal circuit-free subgraph. *Rank* of a set is the number of edges in a spanning forest contained in this set. *Nullity* of a set is the number of edges in this set minus the rank of this set.

3.1.8 Let $e = \{v_i, v_j\}$. Parallel extension simply increase the entries a_{ij} and a_{ji} of $\mathbf{A}$ by one. The effect of a series extension (which introduces a new point v_k of degree 2) is shown in Fig. S.3.2. The column of e should be doubled in $\mathbf{B}$ and in $\mathbf{Q}$ in the case of parallel extension, and in $\mathbf{C}$ in the case of series extension. The effect of a series extension to $\mathbf{B}$ is shown in Fig. S.3.3 and that of a parallel extension to $\mathbf{C}$ in Fig. S.3.4. Finally, the series extension changes $\mathbf{Q}$ in the same way as shown in Fig. S.3.4.

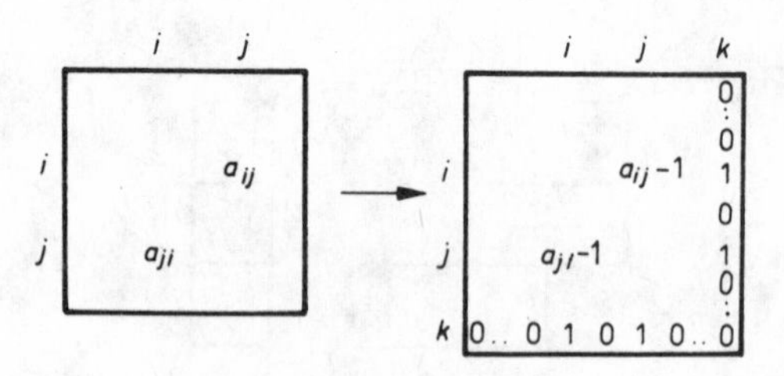

Fig. S.3.2

Series extension increases the rank of the graph and does not change its nullity, while it is *vice versa* for parallel extension.

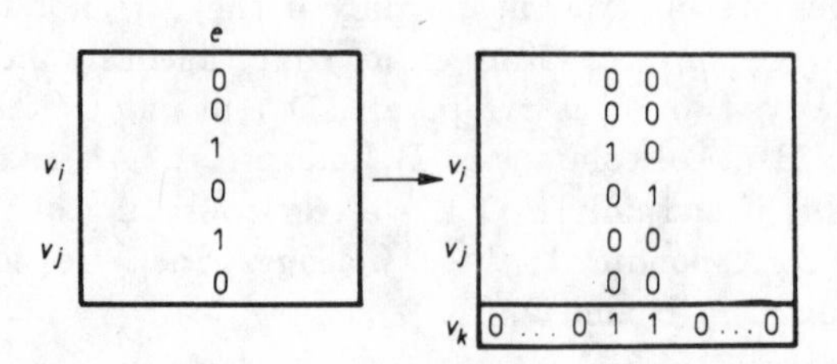

Fig. S.3.3

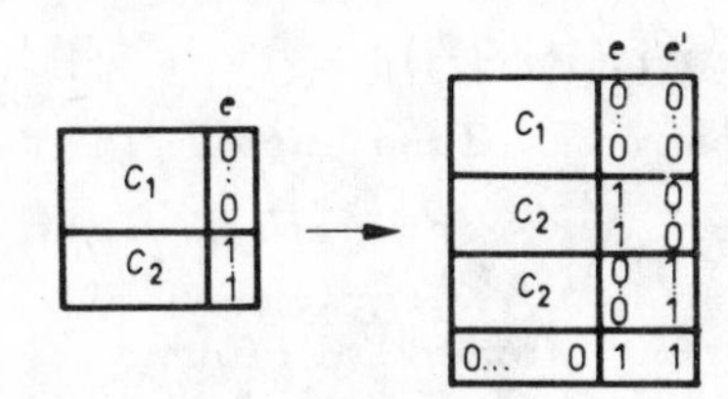

Fig. S.3.4

3.1.9 Let Q be a cut set of G and let Q' denote the corresponding set of edges in G'. The intersection of Q' with any circuit of G' has an even number of edges (apply Theorem 1.2.4 for G and the circuit preserving property of the one–one correspondence). Hence Q' is either a cut set of G' or properly contains a cut set Q'_1, by Problem 1.2.28. In the former case the proof is complete, all we have to show that the latter case is impossible. To see this let Q_1 denote the corresponding set of edges in G. Applying the same reasoning again, Q_1 is either a cut set of G or properly contains one. Both possibilities give a cut set, properly contained in Q, a contradiction.

3.1.10 We may suppose that G is connected since e, f and g are necessarily in the same connected component. Furthermore we may suppose that e, f and g are all different since the statement is trivial otherwise. Both e and g are separating edges in $G - f$ (see Fig. S.3.5, where any of the subgraphs A, B and C may be a single point as well). Then f must join a point of A to a point of C since G would contain e or g as a separating edge otherwise. Hence e and g are also related.

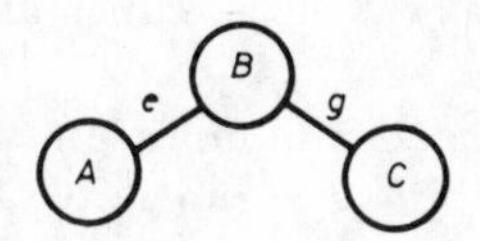

Fig. S.3.5

The edges $e, f, g \ldots$ are pairwise related if and only if, for every circuit C of G, either all of them or none of them is contained in C. The edges of such a class can be transformed to a path, using the "twisting operation" of Exercise 3.1.2 several times.

3.1.11 $\mathbf{C}(G/e)$ can be obtained from $\mathbf{C}(G)$ in the same way as $\mathbf{Q}(G-e)$ was obtained from $\mathbf{Q}(G)$ in Problem 1.4.23. Similarly, $\mathbf{Q}(G/e)$ can also be obtained from $\mathbf{Q}(G)$ as $\mathbf{C}(G-e)$ was obtained from $\mathbf{C}(e)$.

In order to determine $\mathbf{A}(G/e)$ let $e = \{v_i, v_j\}$.

Algorithm 3.3 *Edge contraction if the graph is given by its adjacency matrix.*

Input The matrix $\mathbf{A}$ of a graph G and an edge $e = \{v_i, v_j\}$.
Output The matrix $\mathbf{A}$ of G/e.
Step 1 Subtract 1 from a_{ij} and from a_{ji}.
Step 2 Let $\mathbf{x}$ be the sum of the ith and the jth row of $\mathbf{A}$, let $\mathbf{y}$ be the sum of the ith and the jth column of $\mathbf{A}$; let $\alpha = a_{ii}+a_{jj}+a_{ij}$.
Step 3 Delete the ith and the jth rows from $\mathbf{A}$ and from $\mathbf{y}$, delete the ith and the jth columns from $\mathbf{A}$ and from $\mathbf{x}$.
Step 4 Extend the resulting matrix with a last column $\mathbf{y}$ and with a last row $\mathbf{x}$. Finally put α to the intersection of the last row and the last column.

3.1.12 Choose a nonloop edge and contract it by the algorithm of the solution of the previous problem. Repeat this procedure as long as only loops are left. If a single loop remains, the graph was connected; if a diagonal matrix of size k is obtained, the graph had k connected components.

If the graph is simple, we can neglect the arising multiple edges during this procedure. Hence this algorithm is very fast because the summations in Step 2 can be performed by the Boolean OR operation $(0+0=0;\ 0+1=1+0=1;\ \ 1+1=1)$. The total number of steps can easily be seen to be proportional to v^2 only.

3.1.13 Let H denote the graph $G-B$. We shall prove that F is a spanning forest in H and contains A if and only if $F-A$ is a spanning forest of H/A. For simplicity we suppose that H is connected. If F was a tree of H, $F-A$ is clearly circuit-free in H/A and its connectedness also follows. Let now F' be a tree in H/A. Since $r(H)-r(H/A) = |A|$, if $F' \cup A$ were not a tree, it would contain a circuit C in H. But A was circuit-free, hence $C-A$ would then contain a circuit in H/A, a contradiction.

3.1.14 **Input** and **Output** as in **Algorithm** 1.4
Step 1 Initially let File 1 be empty and File 2 contain all edges of G.
Step 2 Let e be an edge of minimum weight among the elements of File 2. (If File 2 were empty, go to **End**.) If e is not a loop in G then put e to File 1.
Step 3 Delete e from File 2 and contract e in the graph G. Go to **Step 2**.
End The algorithm terminates, F is the subgraph, formed by the edges in File 1.

3.2.1 No. If G is disconnected or contains a separating point then a region may be bounded by several circuits of G.

3.2.2 If we glue the horizontal boundaries of the rectangle of Fig. S.3.6, we obtain a cylinder where points 1 and 3 are already connected. In order to obtain a torus, the two opposite circuits (resulted from the vertical boundaries) are also glued together, which makes point 5 adjacent to points 2 and 3. The other Kuratowski graph can also be drawn to the torus, see Fig. S.3.7. Finally Fig. S.3.8 shows the drawing of our graphs to the Möbius surface (the vertical boundaries are to be glued after a twist, hence the continuous and the dotted lines do not cross).

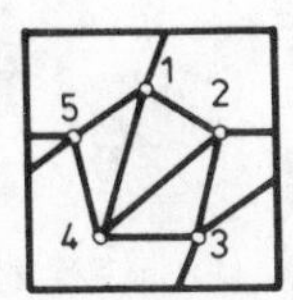

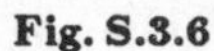
Fig. S.3.6

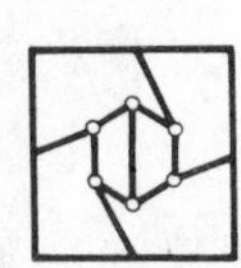
Fig. S.3.7

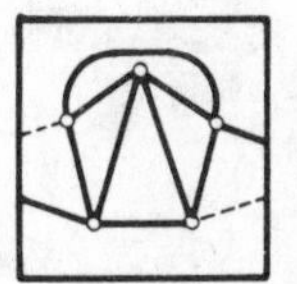
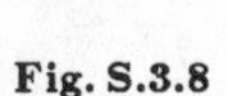
Fig. S.3.8

3.2.3 (a) Consider an arbitrary drawing of G at first. The point v lies on the boundary of some regions. Redraw G so that one of these regions be outside, as described in Statement 3.2.2.

(b) If the extended graph G' is planar (which is clearly a necessary condition), we can draw it so that $\{x, y\}$ belongs to the boundary of the outside region, hence both x and y are outside.

(c) The proof is a straightforward modification of that in part (b).

Remark. A graph is called *outerplanar* if it satisfies the requirement of part (c). The two graphs of Fig. S.3.9 are the smallest non-outerplanar graphs. They actually characterize outerplanarity in the same sense as the Kuratowski graphs characterize planarity, see [Chartrand and Harary, 1967].

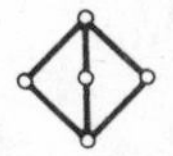

Fig. S.3.9

3.2.4 Suppose that G were the dual of K_5. Then $e(G) = 10$, $v(G) = 7$ should hold (since $10 - 5 + 1 = 6$ is the nullity of K_5 and hence the rank of G as well). Then G should have a point of degree at most 2, while K_5 has no cut set of cardinality 1 or 2, a contradiction.

Similarly, let H be the dual of the other Kuratowski graph. Since $e(H) = 9$, $v(H) = 5$ and the degree of every point of H should be at least 3, the degree sequence of H should necessarily be $3, 3, 4, 4, 4$. (Any other choice would give a degree 5 or 6 but multiple edges are clearly excluded.) But the only graph with this degree sequence is obtained from K_5 by deleting an edge, and its dual is clearly not our Kuratowski graph.

3.2.5 Figure S.3.10 shows some examples.

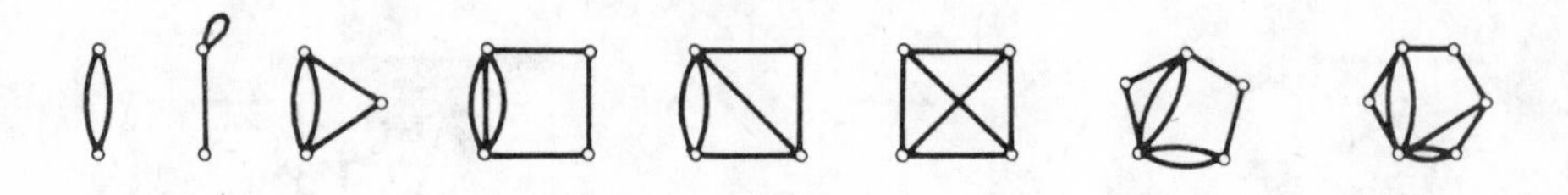

Fig. S.3.10

3.2.6 Clearly we need a statement which is true for nonplanar graphs only. For example: "There exists a graph G with $e(G) = 9$, $r(G) = 5$, $n(G) = 4$ so that the length of every circuit of G is at least 4."

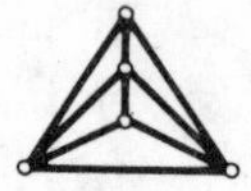

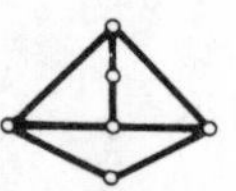

Fig. S.3.11

3.2.7 See Fig. S.3.11.

3.2.8 The graph of Fig. 1.8 can clearly be contracted to K_5 but cannot contain any series extension of it since it has no point of degree at least 4.

3.2.9 See Fig. S.3.12 for K_7 (see also Ringel, 1974) and Fig. S.3.13 for the Petersen graph.

3.2.10 There is no contradiction; the resulting graph does not meet the requirements for a dual. If C is a circuit (a closed line) on the surface of the

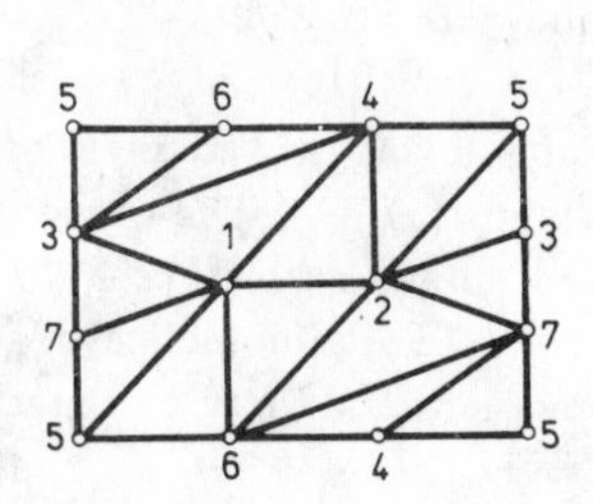

Fig. S.3.12

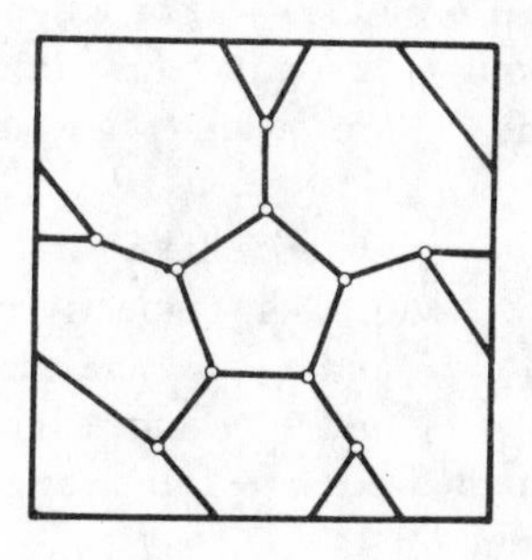

Fig. S.3.13

torus, and two points are on different "sides" of C, they still may be connected with a line which does not intersect C. Hence circuits will not necessarily correspond to cut sets after this process.

3.2.11 If the degree of every point is at least k then $kv \leq 2e$. We have seen that $e \leq 3v - 6$ holds for planar graphs (see the proof of Corollary 3.2.5). Thus $kv \leq 6v - 12$ leads to a contradiction if $k > 5$. If $k = 5$ then $v \geq 12$ clearly follows and this is sharp, see the graph of the icosahedron (Fig. 3.1).

3.2.12 Suppose that the graph is not planar. Then it contains either a copy or a series extension of one of the Kuratowski graphs. We have seen in Exercise 3.2.4 that the Kuratowski graphs have no dual. Hence the necessity follows from the trivial observation that if a graph has a dual then so does every subgraph of it.

3.2.13 The equivalence classes are $\{1\}, \{2\}, \{3,4,5\}, \{6,7,8\}$ and $\{9,10,11\}$. In order to see that we always have an equivalence relation, observe at first that every edge is C-equivalent to itself and if e is C-equivalent to f then so is f to e. Now let e be C-equivalent to f and f to g; we have to prove that e is C-equivalent to g. This is trivial if $e = f$ or $f = g$. Otherwise let P denote the path starting with e and ending with f, and let Q denote the path starting with f and ending with g. Walking along P from e towards f let v be the first internal point of P which belongs to Q as well. Then the path along P until v and then along Q shows that e is C-equivalent to g.

3.2.14 If $|V_1| \geq 4$ then H_1 and H_2 are either disjoint or crossing. (Observe that the definition of disjointness allowed $V_1 \cap V_2 = \{a, b\}$ if a and b divides C into two arcs.

If $|V_1| = 2$ then the statement is trivial again; H_1 and H_2 are either disjoint or crossing.

Finally let $V_1 = \{a,b,c\}, V_2 = \{u,v,w\}$ and $k = |V_1 \cup V_2|$. If $k = 3$ then H_1 and H_2 are equivalent. If $k = 4$ then let, say, $a = u$ and $b = v$. Then

H_1 and H_2 are either disjoint or crossing, depending on whether c and w are on the same arc or on different arcs. If $k = 5$ or $k = 6$ then let a and u be identical or adjacent along C. If the order of the other points along C is b and c first (and v and w thereafter) or *vice versa* then H_1 and H_2 are disjoint, in every other case they are crossing.

3.3.1 The equivalence of (a) and (c) to the definition was proved as Theorems 3.3.2 and 3.3.4, respectively. The implication $(c) \to (b)$ is obvious (let $e_1 = e$ and $e_2 \neq e$ be any edge, incident to v), so is $(b) \to (a)$ (let $v = v_1$ and e be any edge, incident to v_2). The equivalence of (e) to the definition is also clear since (e) only means that G has no separating points.

In order to see that (d) is satisfied by 2-connected graphs, add an edge e between v_1 and v_2 (if they were nonadjacent) and apply (b) to e and v_3. On the other hand, (d) implies 2-connectivity: should x be a separating point of G and y, z be two points in different components of $G - x$ then no path from x to y could contain z.

Finally the implication $(d) \to (f)$ can be proved in the same way as the necessity was in Theorem 3.3.4 and $(f) \to (d)$ is trivial.

3.3.2 If two edges are in the same cut set, they are clearly in the same block. If they are in the same block, they are contained in a common circuit (Theorem 3.3.4) and then in a common cut set as well (Exercise 1.2.14).

3.3.3 Not necessarily, see Fig. S.3.14 where P is indicated by heavy edges.

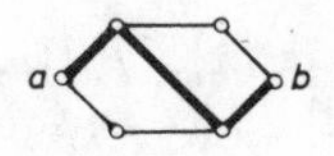

Fig. S.3.14

3.3.4 $H(G)$ is connected since every point (including the separating points) of G can be reached from any other in G. Should $(b_1, p_1, b_2, p_2, \ldots, b_n, p_n)$ be a circuit of $H(G)$, we could choose a path W_i in B_i between P_{i-1} and P_i (for every $i = 2, 3, \ldots, n$) and a path W_1 in B_1 between P_n and P_1. These W_1's together form a circuit of G which contains edges of different blocks of G, a contradiction.

3.3.5 If $G - e$ is not 2-connected then we show that G/e is 2-connected. Let $f, g \in E(G/e)$. We must find a circuit C_0 of G/e containing f and g. Consider a circuit C of G, containing f and g. If $e \in E(C)$ then let $E(C_0) = E(C) - \{e\}$. Otherwise the choice $C_0 = C$ is possible (should the contraction of e change C to two edge-disjoint circuits, the graph $G - e$ could not contain separating points).

3.3.6 Consider K_4 and three edges, all incident to the same point.

3.3.7 Let $e = \{x, y\}$. If $G - e$ were not $(k-1)$-connected, one could find a subset X of $k - 2$ points in $V(G - e) = V(G)$ so that $G' = (G - e) - X$ were disconnected. If u, v were in different components of G' then any uv path of $G - X$ would contain e. (Observe that $G - X$ is connected, due to the k-connectivity of G.) Then u, v were in different components of $G - X - \{x\}$ as well, contradicting the k-connectivity of G.

3.3.8 The graphs G_3 and G_5 of Fig. 3.20 are 2- and 3-edge-connected but not 2- or 3-connected, respectively.

3.3.9 Both the necessity and the sufficiency of this condition are obvious.

3.3.10 If G is 2-edge-connected and $G - e$ is not then, by the previous exercise, a separating edge must arise in $G - e$ which was clearly series to e.

3.3.11 The necessity of the condition is trivial (if there is a separating edge in G then no strongly connected orientation is possible).The sufficiency is proved by induction on the number of edges of G. The assertion holds for $e(G) \leq 3$ and for the case if G is a single circuit. Otherwise let $e \in E(G)$ and delete e and all the other edges, series to e. Every component of the resulting graph can be oriented in the required way. Finally consider a circuit C containing e. This C also contains all the other deleted edges. Assign the same orientation with respect to C to all the deleted edges. The reader can easily verify that this orientation makes G strongly connected.

3.3.12 This is obvious since every circuit of a graph is fully contained in one of its blocks.

3.3.13 Let G be nonplanar, we must find a subgraph of G which is either a copy or a series extension of one of the Kuratowski graphs. (The other direction of the theorem has already been proved.) We may suppose that G is connected since one of its components would be nonplanar otherwise. We prove by induction on the number of blocks that at least one of its blocks is nonplanar.

Suppose that every block of H were planar and consider the tree $H(G)$ as defined in Exercise 3.3.4. There must exist a block B of G so that the corresponding point of $H(G)$ is of degree one. B is planar by the indirect assumption and $G - B$ is planar by the induction assumption. Hence they have such planar representations, by Statement 3.2.2, where their common point lies on the outer face in both drawings. Then the two drawings can be "glued" together, hence G is planar, a contradiction.

3.3.14 Let G be a minimal nonplanar graph (that is, $G - e$ is planar for every edge e of G). We prove that G is 3-connected. G is clearly 2-connected by Exercise 3.3.13 hence it is either 3-connected or contains two points x, y so that $G - \{x, y\}$ has at least two components G_1 and G_2.

In this latter case let P_i be a path in G_i between x and y (for $i = 1, 2$) and let G_i' be obtained from G by deleting all the edges of $E(G_{3-i}) - E(P_{3-i})$.

If both G_1' and G_2' were planar, so were G (apply Exercise 3.2.3(b)). If a G' were nonplanar, G would not be minimal. Hence we have a contradiction in both cases.

3.3.15 (a) G is clearly connected. Its 2-connectivity was proved in Exercise 3.3.13 and its 3-connectivity in the previous problem.

(b) Let $[x_0, x_1, \ldots, x_k]$ be the sequence of points in a maximal path of G. Since the degree of x_0 is at least three, x_0 has at least two further neighbours which are of form $x_i, x_j (1 < i, j < k)$ due to the maximality of the path. If $i < j$ then $\{x_0, x_i\}$ is a chord of the circuit $[x_0, x_1, \ldots, x_j, x_0]$.

(c) Let $e = \{x, y\}$ be a chord of a circuit C. $G - e$ is planar, it may have several planar representations. Choose C so that embedding $G - e$ in the plane the number of faces inside C is as large as possible. Consider the C-components in G (cf., Problem 3.2.13) and call $V(H) \cap V(C) = S(H)$ the support of H.

First we show that every point of G is inside C. Otherwise there were a C-component H with $|S(H)| \geq 3$ since every point of G is of degree at least 3. Then choose two points of the support of H so that they do not cross e and replace one arc of C by a path of H between these points. The new circuit C' would contain a larger number of faces. Similarly, if a C-component has a 2-element support, it can be outside C only if it crosses e.

Consider the C-components inside C. If some of them do not cross any of the "outside" components, they may be redrawn outside, but some others must stay inside (and cross e) since G is nonplanar. Let B be such a C-component, crossing both $e = \{x, y\}$ and the two points a, v from the support of an outside component. Clearly, a, v and x, y must also be crossing.

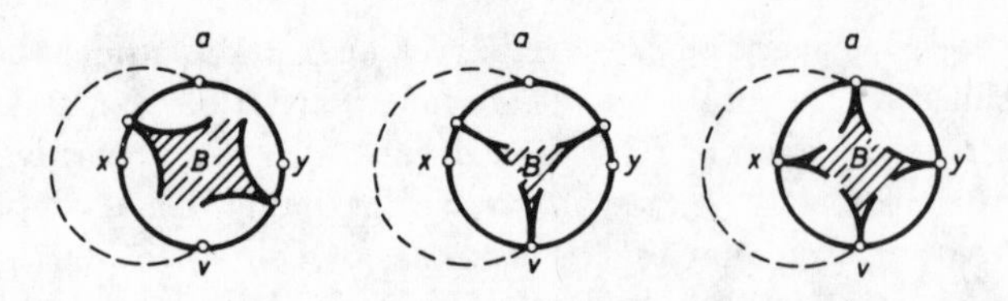

Fig. S.3.15

This is possible in three different ways only (see Fig. S.3.15):

— $S(B)$ either has internal points in both of the arcs (a, x) and (v, y);

— or $v \in S(B)$, and $S(B)$ has an internal point in the arc (a, x), and $S(B)$ has a third point either in the interior of the arc (a, y) or at y;

— or a, x, y, v all belong to $S(B)$.

(Every other possibility can be obtained from one of these by a cyclic permutation of the letters.)

In the first case one can immediately find a copy of the second Kuratowski graph (not a series extension of it since G was minimal). The same graph appears in the second case as well (observe that there exists a point inside B which is connected to all the three points of $S(B)$). In the third case $|S(B)| \geq 4$ and the paths inside B among the points of $S(B)$ can form a letter H or a letter X (see Fig. S.3.16). In the latter case a K_5 can be found, in the former case the second Kuratowski graph appears.

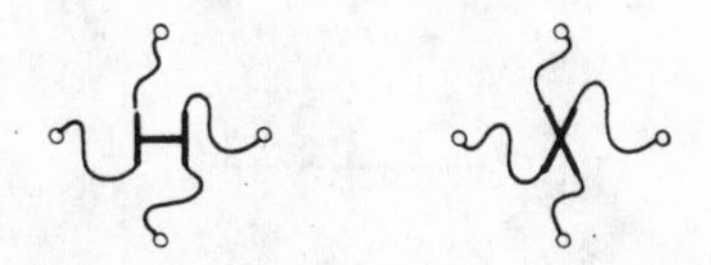

Fig. S.3.16

3.3.16 Let G be a minimal nonplanar graph. It must be 2-connected by Exercise 3.3.13 hence it has no point of degree one. If $v \in V(G)$ is of degree 2, contract one of its incident edges to obtain another minimal nonplanar graph. Repeating this process we reach a graph G' where the degree of every point is at least 3. Then apply part (c) of the previous problem.

3.3.17 Suppose the contrary, i.e. let C be a circuit of $G - x$ containing y and z, and consider the corresponding circuit C' of G'. Observe that neither e' nor f' can be a chord of C' (should, say, $C' \cup \{e'\}$ contain another circuit of G' then $C \cup \{e\}$ would also contain another circuit of G). However, $C \cup \{e, f\}$ does contain a new circuit in G hence so does $C' \cup \{e', f'\}$ in G'. This means that e' and f' are adjacent, a contradiction.

3.3.18 $G - x$ is 2-connected for every $x \in V(G)$. If $e_1, e_2, \ldots, e_k$ were incident to x then $e'_1, e'_2, \ldots e'_k$ are incident to a common point x', by the previous problem. (If $k = 3$, the three edges e'_1, e'_2 and e'_3 could form a circuit of length three as well, since its edges are also pairwise adjacent, but this would violate 2-isomorphism.) This map $x \longrightarrow x'$ is well defined (should x' be not unique, the edges e'_i were parallel) and can easily be seen to be one-to-one and adjacency-preserving.

3.3.19 Suppose that G' is nonplanar, hence it contains a forbidden subgraph. If the other edges of G' and the corresponding edges of G are deleted, the resulting graphs remain 2-isomorphic. Hence we may suppose that G' itself is a copy or a series extension of a Kuratowski graph.

If e', f' are series in G' then e, f can easily be shown to be series in G. Hence we can perform series contraction in G and in G' without destroying their 2-isomorphism. Thus G' may be supposed to be one of the two Kuratowski graphs.

But these graphs are 3-connected, hence G is isomorphic to G' by the previous problem. Thus G cannot be planar, a contradiction.

3.3.20 We may suppose that G and G' are 2-connected, just like in Exercise 3.3.13. If G and G' are not isomorphic, they can be made isomorphic by the twisting operation which clearly preserves planarity.

3.4.1 The first two trees of Fig. S.3.17 contain the required numberings. The statement is not true for K_5, see the third tree of the figure.

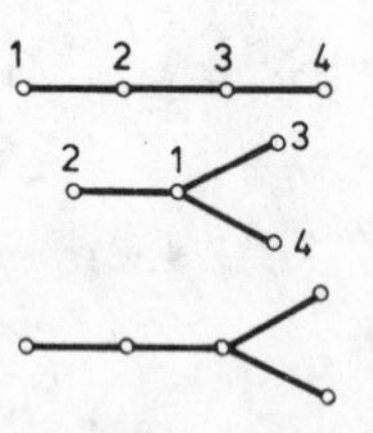

Fig. S.3.17

3.4.2 The heavy edges of Fig. S.3.18 illustrate a DFS tree of a directed graph. The values of the function f are shown next to the points. An unused edge (x, y) may be
(a) a *forward edge* if $f(x) < f(y)$ and y is a descendant of x;
(b) a *backward edge* if $f(x) > f(y)$ and x is a descendant of y;
(c) or a *cross edge* if $f(x) > f(y)$ and neither x nor y is a descendant of the other;
see the examples in Fig. S.3.18. If neither x, nor y is a descendant of the other, the case $f(x) < f(y)$ is impossible.

3.4.3 If G has a directed circuit then such a function t clearly cannot exist. Otherwise let $t(x) = 1$ for a source x (there must exist at least one, see Exercise 1.3.3), $t(y) = 2$ for a source y of $G - x$ etc.

3.4.4 Should we compare the weights $w(p, q)$ for every adjacent pair $p \in V_1$ and $q \in V_2$ (which could be a natural idea at first), the total number of steps could be as high as $\sum_{k=1}^{n} k(n-k)$ for a graph with n points. This is proportional to n^3, see the solution of Problem 1.2.17.

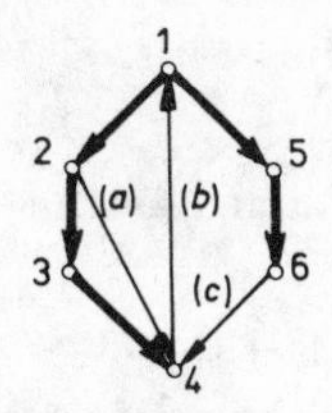

Fig. S.3.18

However, in Step 2 of Dijkstra's algorithm, we compare the distances of those points $q \in V_2$ only which are adjacent to z. Hence the total number of steps is estimated as $\sum_{k=1}^{n}(n-k)$ which is proportional to n^2 only.

Using a suitable data structure (see Aho *et al.*, 1974) the total number of operations can be reduced to constant times $e \log n$. This is more advantageous if n is large and e is much smaller than n^2.

3.4.5

Algorithm 3.4 *Shortest directed paths from a given point in an edge-weighted directed graph.*

Input A directed graph G with a given point v_0 and a nonnegative weight function w on the set of edges.

Output and **Step 1** as in Algorithm 3.2.

Step 2 Consider those edges whose tail is z and whose head is in V_2. If no such edge exists, go to **End**. If v_k is the head of such an edge, calculate $D + w(z, v_k)$ and if the result is smaller than d_k then put this sum instead of d_k.

Step 3 as in Algorithm 3.2.

End The algorithm terminates. If $V_2 \neq \emptyset$, only the points of V_1 can be reached from v_0 on directed paths. The values d_k are the distances of v_k from v_0 (and $d_k = \infty$ if $v_k \notin V_1$).

3.4.6

Algorithm 3.5 *Existence of directed circuits in a directed graph.*

Input A directed graph G.

Output A directed circuit, or the function $t(p)$ of Exercise 3.4.3 if there is no directed circuit in G.

Description Let $n = v(G)$. Perform the DFS algorithm. If v is the first point from which we wish to backtrack, check whether there is a backward edge e with tail v and head v'. If yes, output the obtained directed circuit (formed by e and by the directed path from v' to v in the DFS tree) and stop. If not, let $t(v) = n$. Then consider the second point v'' from which we wish to backtrack and either find a directed circuit using a backward edge with tail v'' (and stop) or let $t(v'') = n - 1$ etc.

3.4.7 The directed circuit, obtained as the output of Algorithm 3.5 clearly satisfies the requirement.

3.4.8 Let v_0 be the initial point of the DFS algorithm. If $\{u, v\}$ is an edge of the DFS tree then assign the orientation (u, v) to it if $f(u) = f(v) - 1$. If $\{u, v\}$ is a nontree edge then let us orient it as (u, v) with $f(u) > f(v)$.

For any point v consider the set $A(v)$ of those points which can be reached from v along a directed path, and let $F(v) = \min\{f(u) | u \in A(v)\}$. Since every point can be reached from v_0 along a directed path, strong connectivity only requires $F(v) = 1$ for every point v. Should $F(u) > 1$ for a point u, consider the point u' with $f(u') = F(u)$. Then the tree edge with head in u' is a separating edge.

(Observe that we gave another solution to Exercise 3.3.11.)

Algorithm 3.6 *Finding strongly connected orientation.*

Input A connected undirected graph G.

Output A strongly connected orientation of G if it is 2-edge-connected, or a separating edge otherwise.

Description Perform the DFS algorithm. The above function $F(v)$ need not be computed, we only need to check when backtracking from a point v whether $f(v) > g(v)$ where $g(v) = \min\{f(u) | u$ is the head of an edge whose tail is either v or a descendant of $v\}$. If $f(v) \leq g(v)$ then the tree edge with head in v is a separating edge.

The function $g(v)$ can be computed as follows. We can easily compute the function $g'(v) = \min\{f(u) | (v, u) \text{ is an edge}\}$. If v has no descendants then $g(v) = g'(v)$. Otherwise, if we backtrack from u to v then replace $g'(v)$ by $\min(g'(v), g(u))$.

Chapter 4

4.1.1 See Fig. S.4.1.

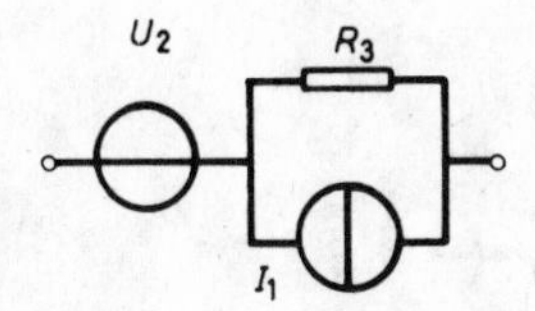

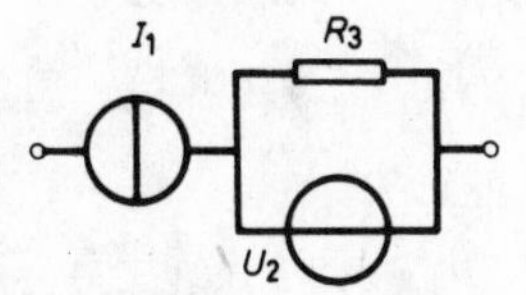

Fig. S.4.1

4.1.2 Consider the sequence of networks $N_1, N_2, N_3, \dots$ of Fig. S.4.2. The total resistance of N_k is F_{2n}/F_{2n+1}.

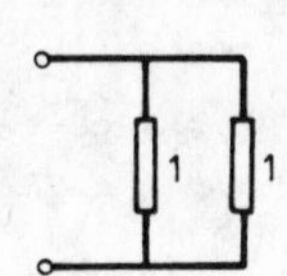

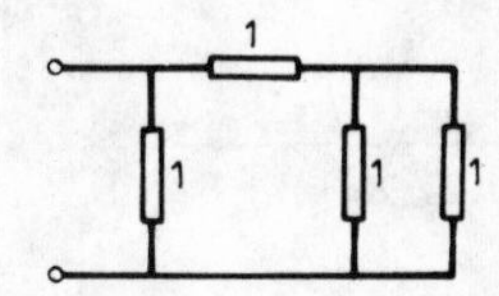

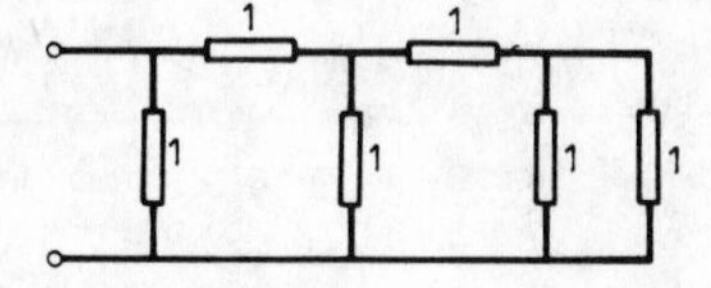

Fig. S.4.2

4.1.3 The current of the device of Fig. S.4.3 is $(\sum R_j I_j)/(\sum R_j)$.

4.1.4 Let $R_1, R_2, \dots, R_n$ be positive resistors of a network. For each resistor R_j do the following process: Insert a resistance meter series to R_j and denote the measured resistance by r_j. Clearly $r_j > R_j$ (since the measured resistance r_j was R_j *plus* something series to it). Then $\sum_{j=1}^{n}(R_j/r_j)$ equals the nullity of the graph of the network.

4.1.5 Suppose that the set of voltage sources contains a circuit. Then the method of the "loop currents" leads to a singular system of equations in Step 4. In the method of the "node potentials" the reduction in Step 2 will not work (the system of equations for the potentials of the points of the circuit is contradictory, or one of the equations is a consequence of the others).

If the set of current sources contains a cut set, similar problems arise in Step 2 of the method of the "loop currents" or in Step 4 of the method of the "node potentials".

4.1.6 By Euler's formula, k just equals the maximum number $n(G)$ of linearly independent circuits of G. Hence we only have to prove that any circuit C arises as the linear combination (over $\mathbb{B}$) of some circuits among $C_1, C_2, \ldots, C_k$. This circuit C determines a part of the plane in the given planar representation of G. This part is the union of certain bounded regions and the sum of the corresponding circuits (over $\mathbb{B}$) gives just C.

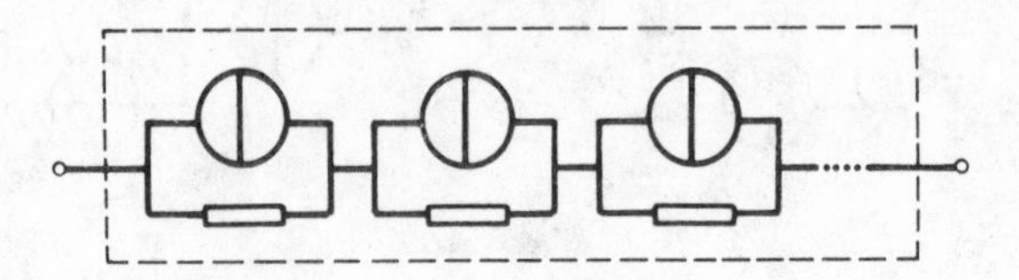

Fig. S.4.3

4.1.7 Let G^* denote the dual of G, obtained from the given planar representation. The submatrix in question is just the incidence matrix $\mathbf{B}$ of G^* (with one row deleted), which is totally unimodular by Exercise 1.4.14.

4.1.8 Add a special (say, red) edge between the two specified points. If the new graph is still planar, consider its dual. Then delete the "red" edge from the dual but consider its two end points as specified ones. (For example, this process gives the network of Fig. S.4.1 from those of Figs 2.6 and 2.7.)

4.1.9 The network is considered as the parallel connection of two simpler subnetworks whose inverses are determined separately. (The graph K_4 is self-dual and the one–one correspondence between the edge sets of K_4 and K_4^* interchanges two point-disjoint edges only.) Hence the solution is the second network of Fig. S.4.4.

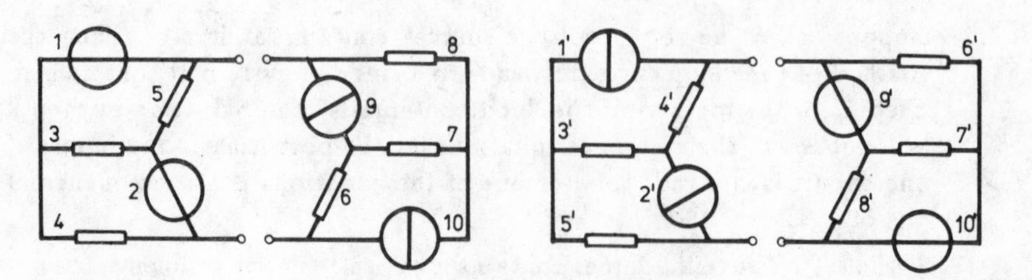

Fig. S.4.4

4.1.10 See Fig. S.4.5. Observe that the inverse of, say, a lowpass filter is another lowpass filter.

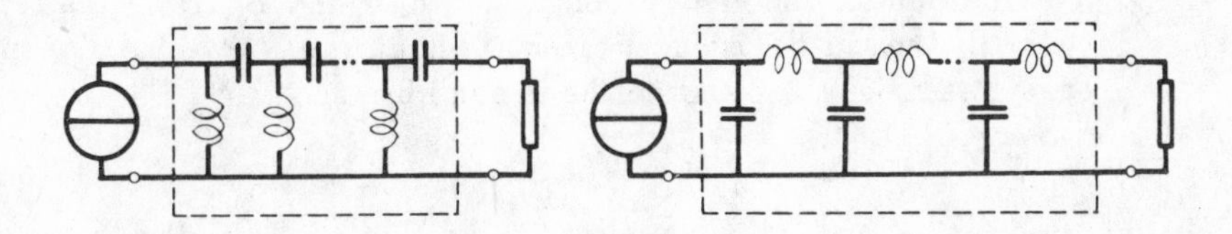

Fig. S.4.5

4.1.11 More generally, consider a subnetwork which keeps contact with the rest of the world *via* k pairs of specified points. Add k new edges (and call them, say, red) which connect such pairs. If the resulting graph is still planar, consider its dual. Then delete the "red" edges from the dual but consider their end points as specified pairs. Hence the inverse of such a "k-port" (see Section 8.1) is a "k-port" again.

4.1.12 Let G^* be a dual of G and $e' \in E(G^*)$ correspond to $e \in E(G)$. Assign such an orientation to e' that the resulting "arrow" be obtained from the "arrow" of e by a counterclockwise rotation with an angle less than 180°. This process is illustrated in Fig. S.4.6. Observe that a cut set of G^* is directed if and only if the corresponding circuit of G is directed.

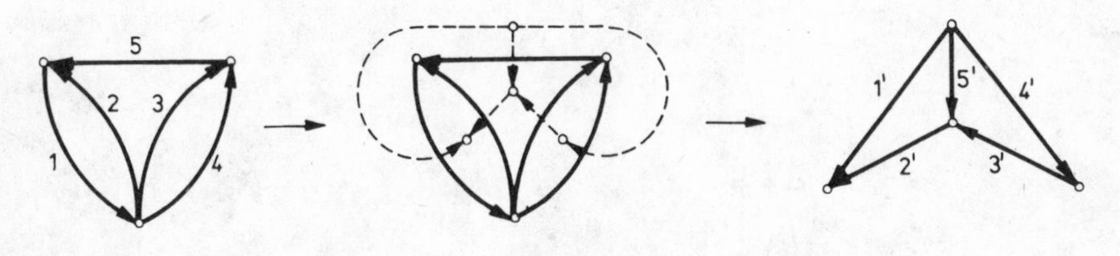

Fig. S.4.6

4.1.13 **Step 1** Assign weight 0 to the voltage source edges, weight 2 to the current source edges and weight 1 to the rest. Using the subroutine we can decide whether the set A of voltage sources is circuit-free and the set B of current sources is cut set free. If not, stop.

Step 2 Assign weight 0 to the capacitors, weight 1 to the resistors and weight 2 to the inductors, and use the subroutine to the graph $(G/A) - B$. The voltages of the capacitors in the tree and the currents of the inductors not in the tree form the state vector.

4.2.1 The solution for the first system is shown in Fig. S.4.7. In the second case only the rod BC is under tension. Should, say, the rod AB be under some stress, joint A could not be in equilibrium.

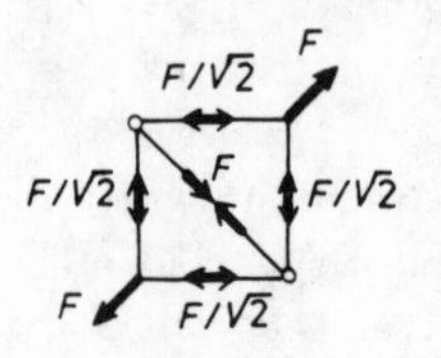

Fig. S.4.7

4.2.2 The equilibrium of joint B means that there is no stress in rod BC. Hence we could remove BC and repeat the argument for joint C etc. Thus, as long as the external forces attack at the joints A, H and I only, we can replace the framework of Fig. 4.22 with that of Fig. S.4.8. The rods AB, BD etc. are under tension of magnitude $\sqrt{17}/2$ times that of the horizontal force attacking at A; the rods AC, CE etc. are under compression of the same magnitude.

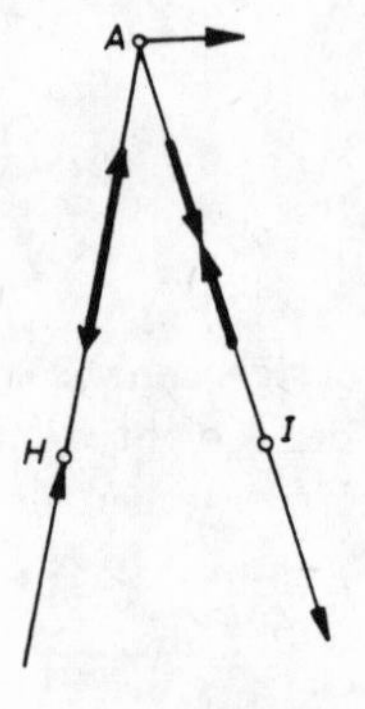

Fig. S.4.8

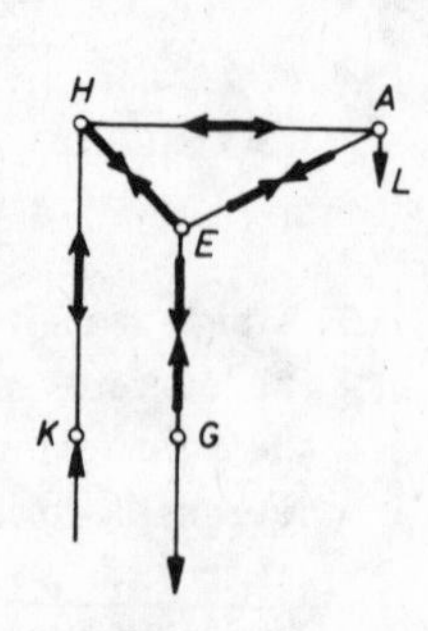

Fig. 4.9

4.2.3 Some rods can be removed here, too, see Fig. S.4.9. The rods AB, BD, DH, HI, IJ and JK are under tension T_1, AC and CE under compression C_2, HE under compression C_3 and EF and FG under compression C_4; the magnitude of these stresses are $2, \sqrt{5}, \sqrt{8}$ and 3 times, respectively, the magnitude of the vertical force, attacking at A.

4.2.4 The first system is redrawn as Fig. S.4.10 (BC and DE are removed, they have no stress under this loading). The equilibrium of the "subsystem" on the right hand side of the dotted line means that the stresses on b, c, d and the external force at A must also be in equilibrium. The sum of the forces in rods b and c attacks at joint X while the sum of the other two forces in point Y (see Fig. S.4.11). Hence these two sums must be of equal magnitude and opposite direction, necessarily along the line XY. Thus one can easily see that the magnitude of the stress in rod d is twice the magnitude of the force attacking at joint A.

The magnitude of the stress in the heavy rods of the second system is $\frac{3}{2m}$ times the magnitude of the loading. In order to calculate this, remove the rods without stress at first. (There are 15 such rods!)

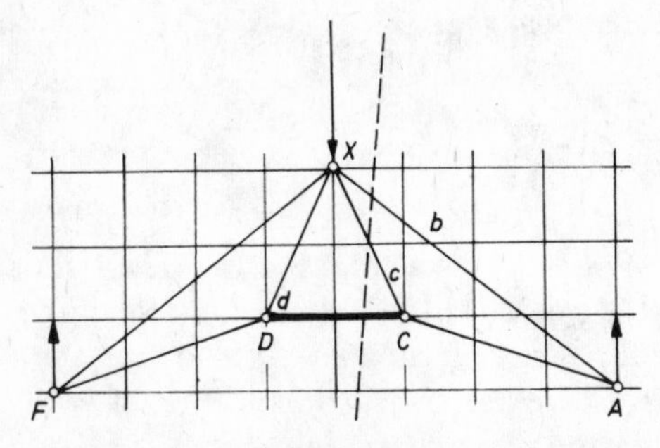

Fig. S.4.10

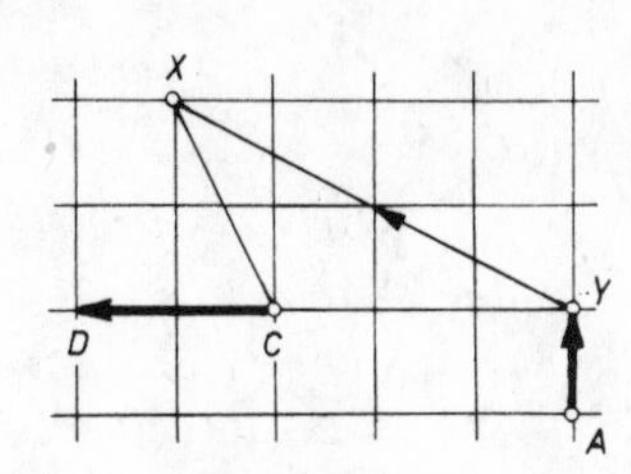

Fig. S.4.11

4.2.5 See Fig. S.4.12 for the first framework and Fig. S.4.13 for its Cremona diagram. Since the force F_B equals the stress in rod 2, they coincide on the diagram. The stress in rod 5 must also be drawn to the same line.

In the case of the second framework, there is no stress whatsoever in the two additional rods. Hence its Cremona diagram is the same, just two further vectors of length zero must be added at the point indicated by a heavy dot.

4.2.6 Figure S.4.14 shows the framework, the letters a, b, c, d indicate the regions for Bow's notation. The Cremona diagram (enlarged) is given in Fig. S.4.15. The external forces correspond to the dotted line.

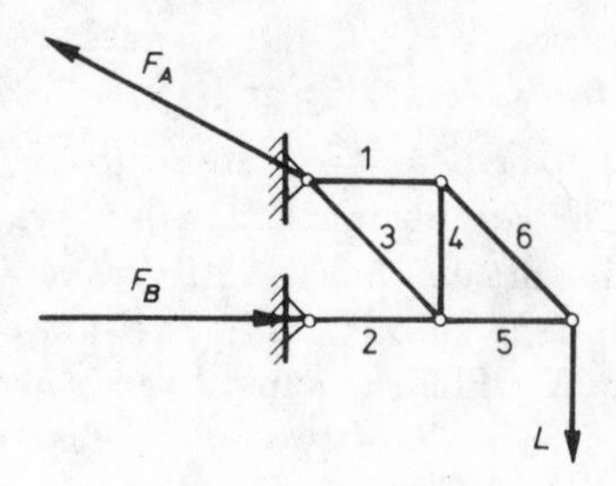

Fig. S.4.12

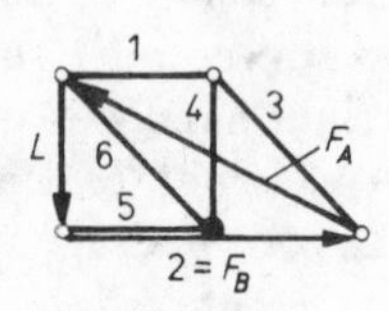

Fig. S.4.13

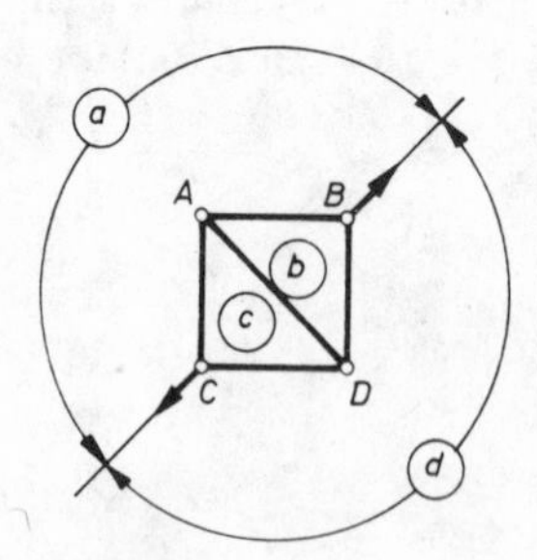

Fig. S.4.14

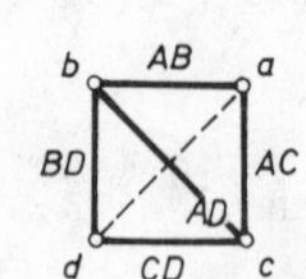

Fig. S.4.15

4.2.7 Some possible positions of joint A are shown in Fig. S.4.16. The Cremona diagram corresponding to the positions (b), (d) and (f) are given in Fig. S.4.17. Since the loading L, attacking at joint A, is always the same, these diagrams can be drawn together, fixing the position of vector L. Then the possible stresses in rod 3 are shown in Fig. S.4.18. The intersection of the vectors of the stresses in rods 3 and 4 is indicated by small circles for the various positions, that for rods 3 and 1 is indicated by small squares. These points determine a straight line and a parabola, respectively.

Figure S.4.18 immediately shows that there is no stress in rod 3 in positions (a) and (e), there is a compression between these two positions and a tension otherwise.

4.2.8 Figure S.4.19 and S.4.20 show the framework and its Cremona diagram, respectively. In order to construct the latter, P_1 and P_2 must be determined at first, from momentum equations like $3\ell F_3 + 6\ell F_2 + 9\ell F_1 = 12\ell P_1$. The parallel lines in Fig. S.4.20 actually coincide on the Cremona diagram.

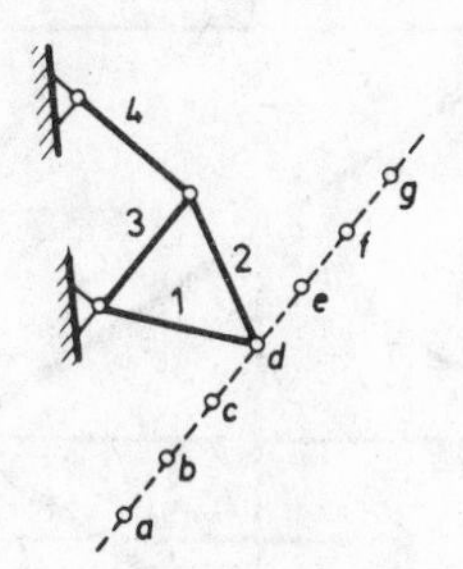

Fig. S.4.16

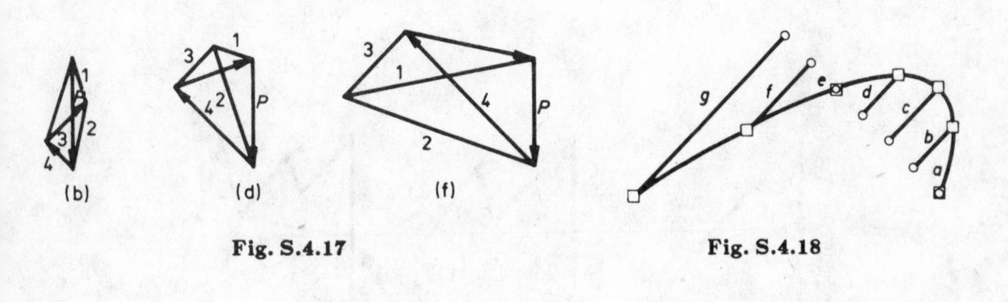

Fig. S.4.17

Fig. S.4.18

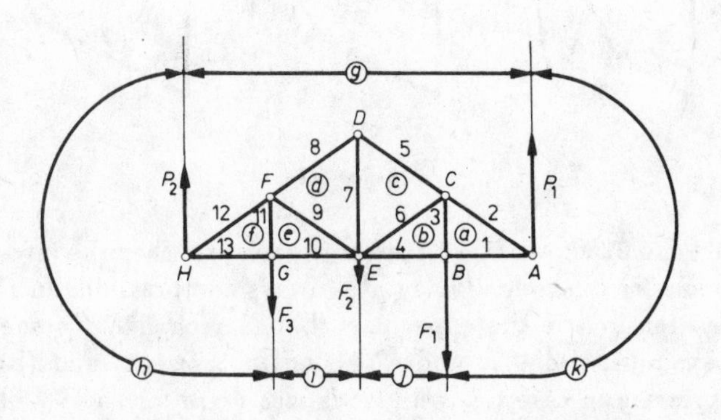

Fig. S.4.19

4.2.9 The Cremona diagrams are shown in Fig. S.4.21 (in case (a) there is no stress in bar 3).

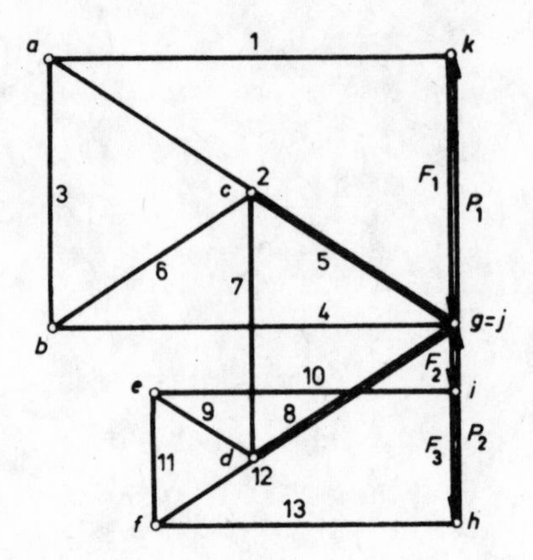

Fig. S.4.20

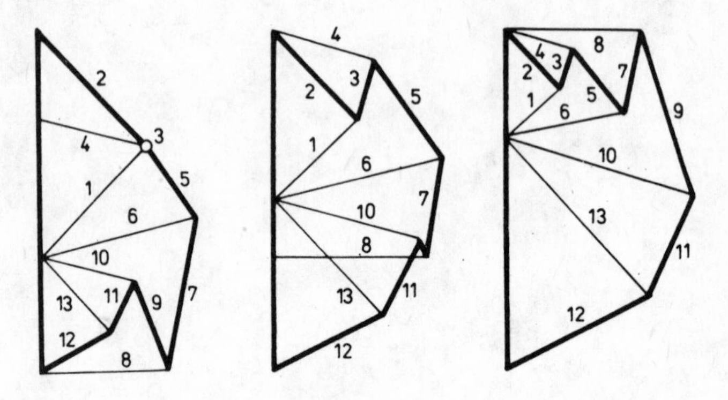

Fig. S.4.21

4.2.10 Figure S.4.22 shows in all the three cases whether the stresses are compressions or tensions. (They are always compressions in the lower and always tensions in the upper part, but may change in some other rods.) For example, rod 9 is under tension in cases (a) and (b), and under compression in case (c), while it is *vice versa* for rod 7. This difference is also reflected by the Cremona diagrams where the vectors for rods 9 and 7 form a $\vee$ shape in cases (a) and (b) and a $\wedge$ shape in case (c).

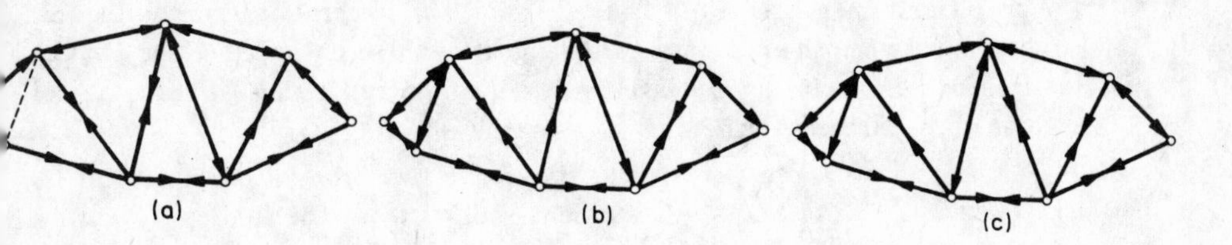

Fig. S.4.22

Chapter 5

5.1.1 Such a graph is, up to 2-isomorphism, the disjoint union of edges and odd circuits. Otherwise there are two circuits with a common edge in the graph and then the graph has two points with three internally disjoint paths between them. Among the lengths of these paths there are at least two odd numbers or at least two even numbers. Hence at least one of the three circuits formed by these paths is even.

5.1.2 Let a and b denote the cardinality of the two bipartition classes of G. Then $e(G) \leq ab$ and the statement follows from the inequality $\sqrt{ab} \leq$ $\leq (a+b)/2$ between the geometric and arithmetic mean, since $a+b =$ $=v(G)$.

5.1.3 Apply breadth-first-search to determine the distance $d(v)$ of every point v from a given point p. Let $V_1 = \{v|d(v)$ is odd $\}$ and $V_2 = \{p\} \cup \{v|d(v)$ is even $\}$. Check every edge whether it is incident to points of different classes.

5.1.4 If X_1, X_2 is the bipartition of G and, say, $X_1 = \{v_1, v_2, \ldots, v_k\}$ and $X_2 = \{v_{k+1}, \ldots, v_n\}$ then $a_{ij} = 0$ if either $i \leq k$ and $j \leq k$ or $i > k$ and $j > k$ simultaneously hold. The remaining two blocks are transpose of each other.

5.1.5 Each rectangle covers a black and a white square. Since the removed squares were of the same colour, the statement is trivial. Formally we may define a bipartite graph G with bipartition X_1, X_2 for the black and white squares, where edges correspond to the adjacencies of the squares. Then G has no perfect matching since $|X_1| \neq |X_2|$.

5.1.6 Let the common degree of every point of G be k. Then $|X_1| = |X_2|$ is obvious and if $X \subseteq X_1$ is of cardinality t then there are kt edges incident to points of X hence $|N(X)| \geq t$.

Thus there is a perfect matching M in G. Since $G - M$ is also regular of degree $k-1$, the argument can be repeated. We obtain that $E(G)$ is the union of k disjoint perfect matchings (cf., Problem 5.1.17).

5.1.7 Let G be a bipartite graph with bipartition X_1, X_2 so that $|X_1| = |X_2| =$ $=n$, the points of X_1 correspond to the subsets of the first decomposition, those of X_2 to the subsets of the second one, and let a point of X_1 and a point of X_2 be connected by t parallel edges if the corresponding subsets have t elements in common. G is regular of degree k hence the previous exercise can be applied.

The analogous statement for three decompositions may be false, even for $k = n = 2$, see the decompositions $\{a, b\} \cup \{c, d\}$; $\{a, c\} \cup \{b, d\}$ and $\{a, d\} \cup \{b, c\}$, for example.

5.1.8 Let F_1 and F_2 be two different 1-factors of a tree T. If an edge $e = \{x, y\}$ is incident to a point x of degree one, it must belong to both F_1 and F_2 and y must not be adjacent to any other point of degree one. Then $F_1 - \{e\}$ and $F_2 - \{e\}$ are different 1-factors of $G - \{x, y\}$. Finally we reach a tree with a single edge, a contradiction.

5.1.9 Consider the "spokes" of the Petersen graph, i.e. the edges $\{1,6\}, \{2,7\}$ etc. in Fig. 1.8. One can easily see that every 1-factor must contain 1 or 5 spokes. Hence one cannot find even two disjoint 1-factors.

5.1.10 No, such a 1-factor should contain the edge $\{4,5\}$ but deleting these two points results in two circuits of length 3 each. Hence one cannot find more than three disjoint edges.

5.1.11 Yes, see Fig. S.5.1, for example.

Fig. S.5.1

5.1.12 This is trivial since a set of points in $L(G)$ is independent if and only if the corresponding edges in G are independent.

5.1.13 Let G be strongly connected and suppose that the length of every directed circuit is even. At first we show that if P_1 is a directed path from x to y and P_2 is a directed path from y to x then their lengths are of the same parity (either both of them are odd or both of them are even). We use induction on the length ℓ of P_1. The statement is trivial for $\ell = 1$ and also if P_1 and P_2 are internally disjoint. Otherwise they have a common point z and we can apply the inductive hypothesis for the pairs x, z and z, y. Next, if Q_1 and Q_2 are directed paths both from x to y, their lengths are of the same parity (consider a path P from y to x).

Hence the sets $X_1 = \{u|$ every directed path from a given u_0 to u is of odd length$\}$ and $X_2 = \{u_0\} \cup \{u|$ every directed path from u_0 to u is of even length$\}$ give a bipartition of $V(G)$. If there were an edge (x,y) between two points x, y of the same set X_i then the path $u \to x$ and this edge would contradict to the definition of X_i. Hence G is bipartite, contradicting our assumption.

5.1.14 If G is disconnected, consider only the component H which contains Q. Let X_1, X_2 be a bipartition of H and let A, B be the point sets of the components of $H - Q$. Then every edge of H joined a point of $X_1 \cap A$ and a point of $X_2 \cap B$; or a point of $X_1 \cap B$ and a point of $X_2 \cap A$ (see Fig. S.5.2). Hence the dotted areas and the shaded areas determine a bipartition of H/Q as well .

5.1.15 (a) See Fig. S.5.3.
(b) $v(G) = 2^k$ is trivial. $e(G) = k \cdot 2^{k-1}$ follows from the observation that G is regular of degree k.

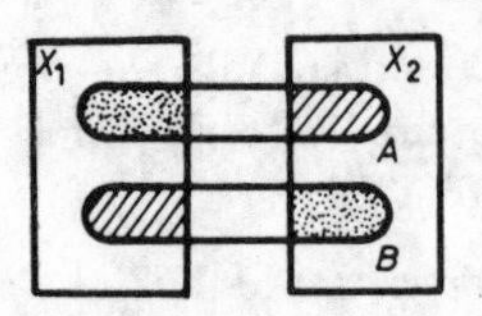

Fig. S.5.2

(c) The distance of two points is the number of coordinates where the corresponding sequences are different. (This is called *Hamming distance* in information theory.) Every "step" on every path changes exactly one coordinate, hence the length of every path between two fixed points is of the same parity. Hence G is bipartite, and since it is regular, it has a 1-factor as well (Exercise 5.1.6).
The direct construction of a perfect matching is even easier: match two points if the corresponding sequences differ in their last digit only.

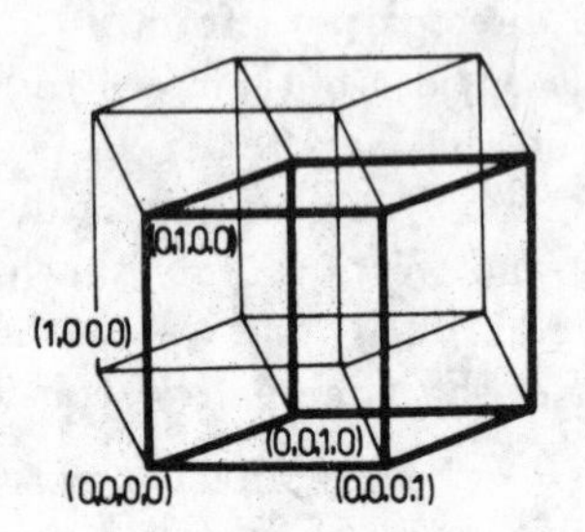

Fig. S.5.3

5.1.16 (1) → (3) $|X_1| = |X_2|$ is trivial since G has 1-factors. For the same reason, if $|N(X)| > (X)$ were violated for some X, we had $|N(X)| = |X|$. But then the connectedness of G implies that there exists an edge from $X \cup N(X)$ to the rest of $V(G)$ which is not contained in any 1-factor, a contradiction.

(3) → (2) Suppose that $G' = G - \{x_1, x_2\}$ has no 1-factor. Then there exists a subset X of $X_1 - \{x_1\}$ with less than $|X|$ neighbours in G', hence with at most $|X|$ neighbours in G, a contradiction.

(2) → (1) If G were disconnected, one could delete two points so that the resulting graph had two components with an odd number of points

in each, contradicting to (2). In order to show that G has a 1-factor let $e = \{x, y\}$ be an edge of G. Consider a 1-factor of $G - \{x, y\}$ and add e.

5.1.17 G will be extended to a bipartite graph which is regular of degree d. Then Exercise 5.1.6 can be applied.

If the cardinality of the bipartition classes are different, add some isolated points at first. Then join a point of degree less than d in the first class with an arbitrary point of degree less than d in the second class. This process is continued until we run out of such points (clearly at the same time in both classes).

5.1.18 (a) Different elements of an independent set X of points cannot be covered by the same edge hence $|X| \leq \rho(G)$ holds for every such X.
(b) X is a set of independent points if and only if $V(G) - X$ is a set of covering points. Hence $\tau(G) \leq |V(G) - X|$ for every set X of independent points, leading to $\tau(G) + \alpha(G) \leq v(G)$; and $\alpha(G) \geq$ $\geq |V(G) - Y|$ for every set Y of covering points, leading to $\tau(G)+$ $\alpha(G) \geq v(G)$.
(c) A set X of $\nu(G)$ disjoint edges cover $2\nu(G)$ points. The remaining points can be covered by $v(G) - 2\nu(G)$ further edges, hence $v(G) -$ $\nu(G) \geq \rho(G)$. On the other hand, let Y be a set of $\rho(G)$ edges which together cover every point. Y is the point-disjoint union of, say, k stars (should Y contain a path of length 3, the second edge of the path could be deleted). Hence $\rho(G) = v(G) - k$. If we choose one edge from each star, they are independent, hence $\nu(G) \geq v(G) - \rho(G)$.
(d) A trivial combination of parts (b), (c) and Theorem 5.1.5.

5.1.19 Let $o(X)$ denote the number of odd components in $G - X$. If M is a 1-factor of G and H is an odd component of $G - X$ then edges from $M \cap E(H)$ cannot cover every point of H. Hence at least one edge e_H of M connects a point of H to a point of X. Since these edges must be disjoint, $o(X) > |X|$ is impossible.

The graph of Fig. 5.8 cannot have a 1-factor since $X = \{4\}$ gives $o(X) = 3$.

5.1.20 Suppose that $o(X) = |X| - 1$ for some $X \subset V(G)$. If the union T of these $o(X)$ sets (of odd cardinality each) has t elements then t and $o(X)$ are of the same parity. Hence $|T \cup X|$ is odd. Since the remaining components of $G - X$ are even, $V(G)$ is odd, hence G cannot have a 1-factor, a contradiction.

5.1.21 (a) Any singleton satisfies the requirements. Hence $|Y| \geq 1$.
(b) Suppose that an even component H of $G - Y$ has no perfect matching. Clearly $v(H) < v(G)$, since $|Y| > 0$ and G was a minimum counterexample. Hence we can apply Tutte's theorem for H. Thus there exists a subset $Z \subset V(H)$ with $o(Z) > |Z|$ in H. But then $o(Y \cup Z) > |Y| + |Z|$ in G, a contradiction.
(c) Let $x \in V(H)$ be an arbitrary point in an odd component H of $G - Y$. If $H - x$ has no perfect matching, there exists a subset

$Z \subset V(H) - \{x\}$ with $o(Z) > |Z|$ in $H - x$. By Problem 5.1.20 we even have $o(Z) > |Z| + 1$.

If we delete $Y \cup Z \cup \{x\}$ from G, we obtain $o(Y) - 1$ odd components, different from H, plus more than $|Z| + 1$ odd components within $H - x$. Hence $o(Y \cup Z \cup \{x\}) > |Y| + |Z|$, that is, $o(Y \cup Z \cup \{x\}) \geq |Y \cup Z \cup \{x\}|$. Tutte's condition implies equality but then the maximality of Y is violated, a contradiction.

5.1.22 Let G be a counterexample (i.e., it satisfies Tutte's condition but has no 1-factor) with a minimum number of points. Let Y be defined in the same way as in the previous problem.

We define a bipartite graph G' with bipartition X_1, X_2 as follows. The points in X_1 correspond to those of Y, the points in X_2 correspond to the odd components of $G - Y$, and if $y \in Y$ and H is an odd component then the corresponding points be adjacent in G' if and only if there is an edge in G between y and a point of H.

$|X_1| = |X_2|$ follows from the definition of Y and $|N(X)| \geq |X|$ for every $X \subseteq X_2$ follows from Tutte's condition. Hence G' has a perfect matching which corresponds to a set M_1 of disjoint edges in G. The even components of $G - Y$ have perfect matchings (see part (b) of the previous problem), their edges together determine a set M_2 of disjoint edges. Finally, the edges in M_1 contain one point from each odd component of $G - Y$. Deleting these points the remaining subgraphs have perfect matchings by part (c) of the previous problem. Their edges together determine a set M_3 of disjoint edges. But then $M_1 \cup M_2 \cup M_3$ is a 1-factor, contradicting the assumption.

5.1.23 Let P be an alternating path with respect to a set M of independent edges. Delete the elements of $E(P) \cap M$ from M and then add the elements of $E(P) - M$ to the result. The final set of edges is independent and has cardinality $|M| + 1$.

On the other hand, if M, N are sets of independent edges and $|E(M)| < |E(N)|$ then there exists an alternating path with respect to M. To see this, consider the set $E(M) \triangle E(N)$ of those edges which are contained in exactly one of M and N. It determines a subgraph H of G where the degree of every point is at most 2. Hence H consists of disjoint paths and circuits only. Furthermore the length of every circuit must be even, consisting of the same number of edges from M and from N. Hence $e(M) < e(N)$ implies the existence of a path of odd length with more edges from N than from M. This is just an alternating path with respect to M.

5.1.24 The edge set of this graph can be covered by a closed edge sequence (since the graph is connected and every point has an even degree). Every second edge in this sequence gives a d-factor (since the number of edges is even).

5.2.1 A producer need not be a source but if it is the head of an edge, this edge can be deleted since it will never belong to any augmenting path. Similarly, a consumer may be thought as a sink since we can delete those edges whose tails are in it.

5.2.2 The value of the flow can be increased by one only if we use the augmenting paths containing the vertical edge. Hence the number K of operations is proportional to n. The size of the input is *not* $\max(v, e) = 5$ since the value of n also belongs to the input. Using, say, the binary representation of n, the size of the input is essentially $\log n$. Hence K is an exponential function of the size of the input.

5.2.3 If the edges of capacity x and y are not used, a flow of value 2 can be realized. Then we can augment by $d = \min(1, x) + \min(1, y)$. In the next step we can augment only if both x and y are greater than 1. The maximum flow is $4 + \min(x-1, y-1, 1)$ if $x, y > 1$ and $2 + d$ otherwise. This value z is shown as a function of x and y in Fig. S.5.4.

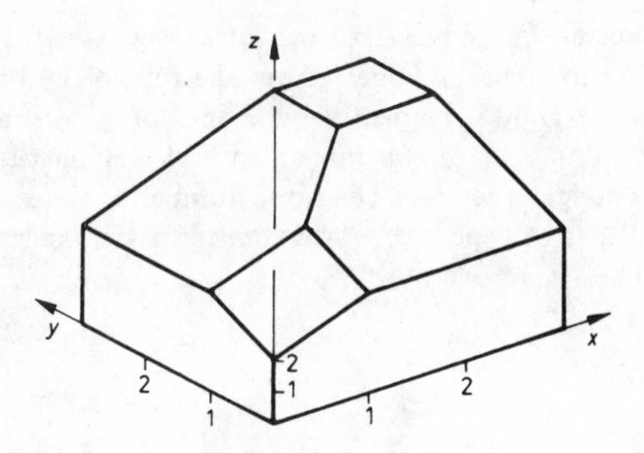

Fig. S.5.4

5.2.4 Extend the network with two more points S, T and with edges from S to $s_1, s_2, \ldots, s_k$ and from $t_1, t_2, \ldots, t_\ell$ to T (see Fig. S.5.5). If the new edges have very large capacities then the maximum flow from S to T determines the solution of the original problem.

5.2.5 Let G be a bipartite graph with bipartition X_1, X_2. Orient all the edges from a point of X_1 to a point of X_2. Extend G by two more points s, t and by edges (s, x) for every $x \in X_1$ and by edges (x, t) for every $x \in X_2$. Let G' denote the resulting directed graph.
If $k = \tau(G)$ then $G' - X$ has a directed path from s to t for every $X \subseteq V(G)$ with $|X| < k$. Hence G' contains k internally disjoint directed paths from s to t. Since the maximum number of such paths is clearly $\nu(G)$, we obtain the nontrivial direction $\nu(G) \geq \tau(G)$ of König's theorem.

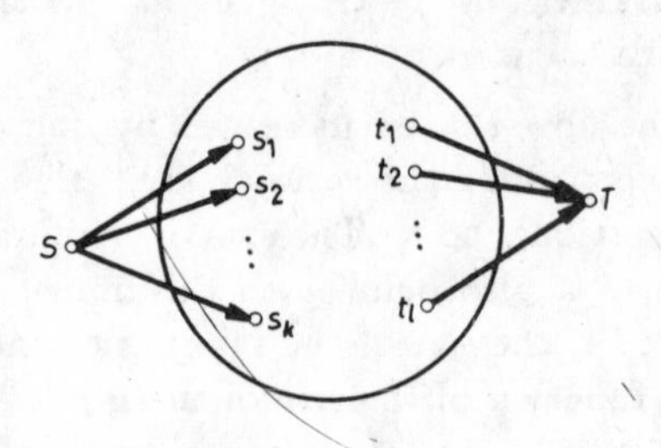

Fig. S.5.5

Suppose now that $|X_1| = |X_2| = k$. If there is no perfect matching in G then there exists a subset $X \subseteq V(G)$ with $|X| \leq k-1$ so that $G' - X$ has no directed path from s to t. Let $Y = X_1 - X$. Clearly Y cannot have any neighbour in $X_2 - X$, hence $N(Y) \subseteq X_2 \cap X$. Then $|X_2 \cap X| = |X| - |X_1 \cap X| < k - |X_1 \cap X| = |Y|$ leads to the nontrivial part of Hall's theorem.

5.2.6 Figure S.5.6 shows the relative position of the sets X, Y. Then $\delta(X)$ and $\delta(Y)$ are the numbers of edges "from the top to the bottom" and "from the left to the right", respectively. Edges of type a and b are counted at the left hand side of the inequality but not on the right hand side. Every other edge is counted the same number of times on both sides. For example, edges of type c or d are counted twice and once, respectively, on both sides.

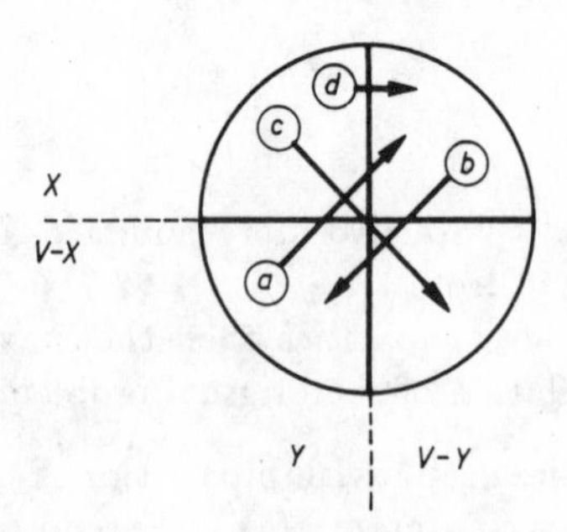

Fig. S.5.6

5.2.7 (a) If this statement were false for an $x_0 \in X$ then every path between s and t contained a point of X, different from x_0. Then $X - \{x_0\}$ would also separate s from t, contradicting to the minimality of X.

(b) Consider the path P with the property described in part (a). Then the neighbours of x on P are in G_s and in G_t, respectively.

5.2.8 Extend G with two more points s, t and join s to the points of A and join the points of B to t by edges (directed edges, respectively). The condition implies by Menger's theorem that there are k internally disjoint paths (directed paths, respectively) from s to t. Their subpaths just give the required paths.

5.2.9 We can augment by $x, x^2, x^3, \ldots$; the value of the flows on the various edges will be as follows:

after the kth step, where	edge $\{z,w\}$	edge $\{w,v\}$	edge $\{v,u\}$	edge $\{u,z\}$
$k=4m$	$1-x^{4m}$	0	$1-x^{4m+1}$	x^{4m+2}
$k=4m+1$	$1-x^{4m+3}$	x^{4m+1}	1	x^{4m+2}
$k=4m+2$	$1-x^{4m+3}$	x^{4m+4}	$1-x^{4m+2}$	0
$k=4m+3$	1	x^{4m+4}	$1-x^{4m+5}$	x^{4m+3}
	edge $\{s,z\}$	edge $\{s,v\}$	edge $\{w,t\}$	edge $\{u,t\}$
$k=4m$	σ_m	τ_m	π_m	ρ_m+x^{4m}
$k=4m+1$	σ_m+x^{4m+1}	τ_m	π_m	ρ_{m+1}
$k=4m+2$	σ_{m+1}	τ_m	π_m+x^{4m+2}	ρ_{m+1}
$k=4m+3$	σ_{m+1}	τ_m+x^{4m+3}	π_{m+1}	ρ_{m+1}

In order to check these data observe that the value of the flows on the edges $\{s,w\}, \{v,t\}, \{s,u\}$ and $\{z,t\}$ are always $0, 0, x^5$ and x^2, respectively. The augmentations sum up to $\sum_{i=1}^{\infty} x^i = \frac{1}{1-x} - 1 \approx 2.145$. The limit of the value of the maximum flow is $\lim(x^2+0+\pi_m+\rho_m) = 1 + x^2 + \frac{x(2-x)}{1-x+x^2} \approx 2.61$. However, the value of the maximum flow is 8 (namely 2 on each of the paths $[s,u,t], [s,v,t], [s,w,t]$ and $[s,z,t]$).

5.2.10 Extend the graph of the system with a single input s (like in Exercise 5.2.4) and determine the maximum flow from s to the single output t. Then assign weight 1 to those edges which are saturated by this maximum flow and assign weight 0 to the other edges. Finally construct a minimum weight directed path from s to t. The saturated edges of this path should be extended.

5.2.11 Figure S.5.7 shows a network with a single input s, with a single output t, with matchmakers $x_1, x_2, \ldots$, boys $y_1, y_2, \ldots$ and girls $z_1, z_2, \ldots$. Every edge of form (s, x_i) and (z_j, t) exists, an edge (x_i, y_j) means that the jth boy contacted the ith matchmaker and an edge (y_j, z_k) means that the jth boy and the kth girl are friends. Let the capacity of the (s, x_i) edges be m, that of the (z_j, t) edges be one and that of the other edges be very large. Consider the maximum flow from s to t and apply Corollary 5.2.3 to realize it by integer flows on every edge. This gives the solution of the South-Lilliputian version of the problem.

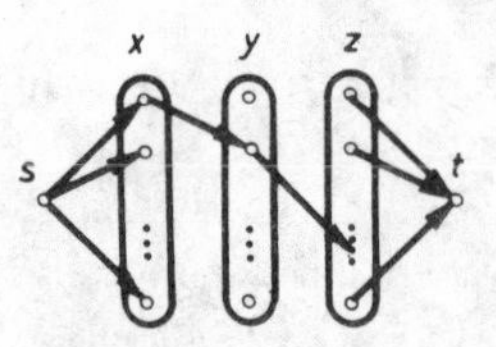

Fig. S.5.7

In order to solve the original problem modify the network as in Fig. S.5.8 by the splitting procedure (as in Fig. 5.14) and by prescribing capacity one to the new edges.

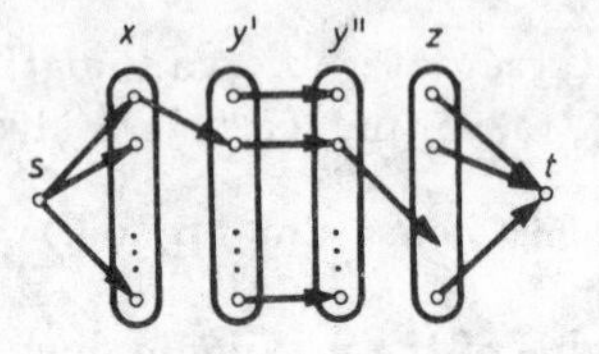

Fig. S.5.8

5.2.12 (a) The solution is the same as for Exercise 5.2.6 except that the capacities of the edges must be added rather than counting the edges.

(b) If C_1 and C_2 are minimum cuts then $c(C_1 \cup C_2) \geq c(C_1) = c(C_2)$, and a similar inequality holds for $c(C_1 \cap C_2)$. Hence the relation of part (a) is obtained with reversed inequality-sign, thus $C_1 \cup C_2$ and $C_1 \cap C_2$ are also minimum cuts. A point is in $\overline{C_0}$ if it is accessible from s at the end of Algorithm 5.1 (see the proof of Theorem 5.2.1).

5.2.13 Extend the graph G of the network by a new edge $e_f = (t, s)$ and let the value $m(f)$ of the flow be "sent back" along this edge. Then the quantity $m(v)$ as defined at the beginning of this section will be zero for every point including s and t. Hence the flow in this new graph G_0 satisfies Kirchoff's Current Law.

If G_0 is planar and G_0^* is a dual of it (see Problem 4.1.12) then let the edge $e_f^* = (t^*, s^*)$ of G_0^* correspond to e_f. Now $(G_0^* - e_f^*;\ s^*,\ t^*; c)$ can be defined as the dual of $(G; s, t; c)$, see also Problem 4.1.8. A flow in the dual also satisfies $f \leq c$ and Kirchhoff's Voltage Law (with respect to the circuits of the original graph). The dual of an (s,t)-cut is an $s-t$ path, see Fig. S.5.9.

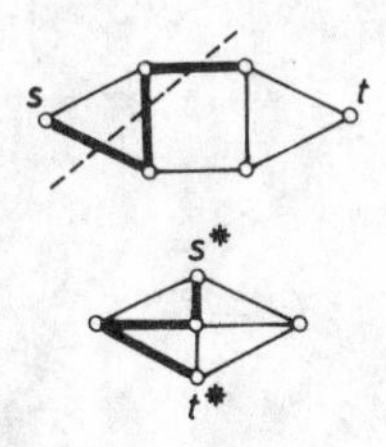

Fig. S.5.9

5.2.14 (a) See Fig. S.5.10.

(b) Suppose we have a lower and an upper bound on every edge, that is, $b(e) \leq f(e) \leq c(e)$ must hold for every edge e. Let us extend the network with two points s_a and t_a and if $e = (x, y)$ is an edge with $b(e) > 0$ then add two edges (x, t_a) and (s_a, y) with capacities $b(e)$ and reduce the capacity of edge e from $c(e)$ to $c(e) - b(e)$. Let us repeat this process for every edge with positive lower bound (using always the same pair s_a and t_a of points). Finally, connect t to s by a new edge e_0 of very large capacity and consider s_a and t_a to be the producer and the consumer, respectively, of the new network.

Every legitimate flow of the original network will correspond to such a legitimate flow of the new network which saturates every edge of form (x, t_a) and (s_a, y). Hence the original problem has a solution if and only if the maximum flow in the new network is $b(e_1)+b(e_2)+\ldots$.

s $0 \leq f \leq 1$ $2 \leq f \leq 5$ t

Fig. S.5.10

Our process is illustrated in Fig. S.5.11 for the example of part (a). The same network is redrawn in Fig. S.5.12. One can readily see that the value of the maximum flow is 1 (rather than 2) hence the original network of Fig. S.5.10 has no legitimate solution.

5.2.15 Replace each edge $e = \{x, y\}$ by two edges $e' = (x, y)$ and $e'' = (y, x)$ and define the new capacities $c(e') = c(e'')$ to be equal to $c(e)$. Apply Algorithm 5.1 to the new network to find a flow f with value $m(f)$. This f determines a flow f' with the same value in the original network, and we may additionally suppose that each undirected edge is used by f' in one direction only (giving $f'(e')+f'(e'') \leq c(e)$ as well) since if $f'(e')$ and $f'(e'')$ were both positive, it would be enough to transport the difference of these values only, in one of the directions.

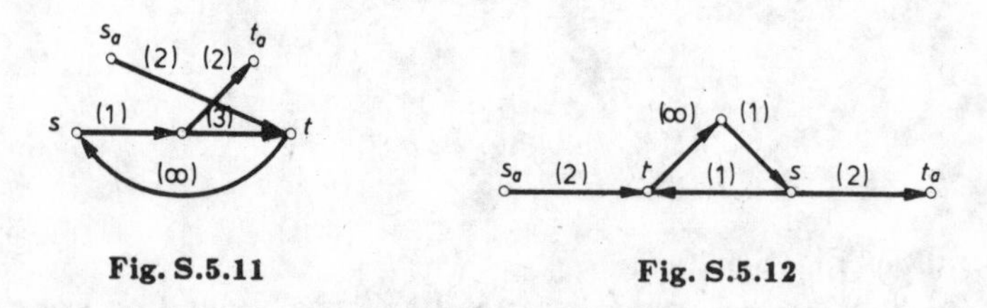

Fig. S.5.11 Fig. S.5.12

5.2.16 Let us transport 1 unit of the first produce from s_1 to t_1 and 1 unit of the second produce from s_2 to t_2 in the network of Fig. S.5.13. These are simultaneously possible if half of each produce is sent in different ways but impossible if the value of each flow on each edge must be integer.

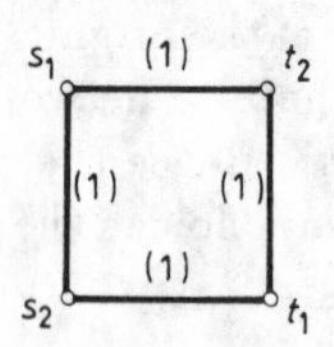

Fig. S.5.13

5.2.17 Replace each edge of the undirected graph G by two oppositely oriented parallel edges. There was a path between two points in G if and only if there are directed paths in both directions in the new directed graph H. If we delete at most $k-1$ edges from H, there will remain a directed path from s to t (since these $k-1$ directed edges correspond to at most $k-1$ edges of G and G is k-edge-connected). Hence Theorem 5.2.4 can be applied.

The only problem is that the obtained k edge-disjoint directed paths of H need not correspond to edge-disjoint paths of G. But if a directed path P_1 uses the edge (u,v) and another directed path P_2 uses the edge (v,u) in H then we can replace these paths by P_1' (using the subpath of P_1 from s to u and the subpath of P_2 from u to t) and P_2' (using P_2 from s to v and then P_1 from v to t). This replacement can be repeated, if necessary, until all the corresponding paths in G become edge-disjoint.

The statement for the internally disjoint paths can be deduced from this just like Theorem 5.2.5 was deduced from Theorem 5.2.4.

5.2.18 The statement is true for $k = 2$ (see Theorem 3.3.3) and we suppose it is true for every positive integer less than k. Consider the given points $x_1, x_2, \ldots, x_k$ and a circuit C containing the first $k-1$ of these points. If x_k happens to belong to $V(C)$, we are done. Otherwise by Exercise 5.2.8 we can either find k internally disjoint paths between x_k and points of C or a subset X of $k-1$ points so that $x_k \notin X$ and every path from x_k to any point of C contains a point of X.

The second alternative contradicts to k-connectivity unless $X = V(C)$. If $X = V(C)$ then simply replace an edge of form $\{x_i, x_j\}$ in C by two paths, one from x_i to x_k and one from x_k to x_j, to obtain the required circuit.

In the case of the first alternative consider the internally disjoint paths $P_1, P_2, \ldots, P_k$ from x_k to different points of C. The points $x_1, x_2, \ldots,$ x_{k-1} determine $k-1$ arcs of C. At least one of these arcs must contain the end points of at least two such paths, hence C can be extended to a circuit containing x_k as well.

5.3.1 This number is two for $\mathbf{A}_1$ (take the first column and the last row) but three for $\mathbf{A}_2$ (since x, z and v are all in different rows and in different columns). More generally, if $r_t(\mathbf{A}) = k$ then there exist k nonzero entries, all in different rows and in different columns, hence at least k lines are needed for covering them.

5.3.2 Let us define a bipartite graph $G(\mathbf{A})$ as follows. The two point sets correspond to the rows and to the columns of $\mathbf{A}$, respectively, and two edges are adjacent if and only if the corresponding entry of $\mathbf{A}$ is nonzero. The nonzero expansion members of det $\mathbf{A}$ correspond to the perfect matchings of $G(\mathbf{A})$. The condition of the exercise states that $\tau(G(\mathbf{A})) < n$. Hence $\nu(G(\mathbf{A})) < n$, that is, det $\mathbf{A}$ has no nonzero expansion member.

5.3.3 The sum, the difference and the product of such numbers are clearly of the form $a + b\sqrt{2}$ again. If at least one of a, b are nonzero then $(a+b\sqrt{2})^{-1} = (a-b\sqrt{2})/(a^2+2b^2)$, hence division can also be performed. Should $\sqrt{3} = a + b\sqrt{2}$ with $a, b \in \mathbb{Q}$ then a, b are clearly nonzero. Hence their square $3 = a^2 + 2b^2 + 2ab\sqrt{2}$ would give $\sqrt{2} = (3 - a^2 - 2b^2)/2ab$, contradicting to the irrationality of $\sqrt{2}$.

5.3.4 We have seen in the solution of Exercise 5.3.2 that the minimum number of lines covering all nonzero entries of the matrix $\mathbf{A}$ is $\tau(G(\mathbf{A}))$ and that $r_t(\mathbf{A}) = \nu(G(\mathbf{A}))$. Hence the statement follows from König's theorem.

5.3.5 The resulting matrix has term rank $e(G)$ and its expansion members correspond to the closed Eulerian walks of G (see Problem 1.1.28). Namely if an expansion member determines the kth entry from the ith row then we can orient the ith edge of G towards the point corresponding to the kth row. In this orientation the outdegree and the indegree of the jth point is $d_j/2$, hence it determines a closed Eulerian walk. In fact, the correspondence is one-to-one provided that two walks are considered as different if the same closed subwalks are visited in different order [Schrijver, 1982].

5.3.6 (a) The sum of every entry equals the number of rows and the number of columns, hence the matrix is square.

(b) Let $\mathbf{A}$ be such a matrix. Applying the construction of Exercise 5.3.2 and Problem 5.3.4 we have to prove that $G(\mathbf{A})$ has a perfect matching. Suppose the contrary, i.e. that there exists a set X of rows (i.e. "lower" points of $G(\mathbf{A})$), with $|X| > |N(X)|$. We may suppose that the rows of X are just the first k rows of $\mathbf{A}$ and that the columns not belonging to $N(X)$ are the first ℓ columns of $\mathbf{A}$. Then $a_{ij} = 0$ if $i \le k$ and $j \le \ell$, hence $\ell = \sum_{j=1}^{\ell}\sum_{i=1}^{n} a_{ij} = \sum_{j=1}^{\ell}\sum_{i=k+1}^{n} a_{ij} = \sum_{i=k+1}^{n}\sum_{j=1}^{\ell} a_{ij} \le \sum_{i=k+1}^{n}\sum_{j=1}^{\ell} a_{ij} = n-k$, but $|N(X)| = n - \ell < k = |X|$, a contradiction.

5.3.7 Suppose that there exists a nonidentically zero multivariable polynomial $P(\xi_1, \xi_2, \ldots)$ so that $\xi_1 = x_1, \xi_2 = x_2, \ldots$ gives zero. Let ξ_i be a variable which really arises (since P is nonidentically zero) and arrange the members of P according to its powers: $P(\xi_1, \xi_2, \ldots) = \sum_j P_j(\xi_1, \xi_2, \ldots, \xi_{i-1}, \xi_{i+1}, \ldots)\xi_i^j$. Putting $\xi_k = x_k$ for every k except i, we obtain a nonidentically zero polynomial of ξ_i whose coefficients are in $K(x_1, x_2, \ldots, x_{i-1}, x_{i+1}, \ldots)$ and which vanishes at $\xi_i = x_i$, hence x_i cannot be algebraically independent from the others over K.

On the other hand, if $x_1, x_2, \ldots$ are not algebraically independent over K then one of them, say x_1, would be a root of a polynomial $P(\xi) = \sum_i a_i \xi^i$, where the coefficients a_i were in $K(x_2, x_3, \ldots)$. Express every a_i as the ratio of two polynomials of $x_2, x_3, \ldots$ and multiply $P(\xi)$ by the least common multiplier of the denominators to obtain the requested polynomial.

5.3.8 (a) The term rank of all the three matrices is 2. This concept does not depend on the underlying field.

(b) Their rank over $\mathbb{R}$ is 1 since the second column is obtained by multiplying the first one by a constant c ($c = 2$, $\sqrt{2}$ and $\sqrt{27}$, respectively). Since $\sqrt{2}$ and $\sqrt{27}$ do not belong to $\mathbb{Q}$ and $\sqrt{27}$ does not belong to $\mathbb{Q}(\sqrt{2})$, one may be tempted to say that the respective ranks are 1, 2, 2 over $\mathbb{Q}$ and 1, 1, 2 over $\mathbb{Q}(\sqrt{2})$. However, rank of a matrix over a subfield cannot be defined in the usual way. For example, the columns of the matrix $\begin{pmatrix} 1 & \sqrt{2} \\ 2 & 2\sqrt{2} \end{pmatrix}$ are linearly independent and its rows are linearly dependent over $\mathbb{Q}$.

5.3.9 Let the vectors form the columns of a $k \times k$ matrix $\mathbf{A}$. By the condition we have $r(\mathbf{A}) = k$ over $\mathbb{Q}$, hence det $\mathbf{A} \neq 0$. Then $r(\mathbf{A}) = k$ must hold over $\mathbb{R}$ as well, since the determinant cannot vanish.

5.3.10 Let $\mathbf{M}_j$ denote the matrix, obtained from $\mathbf{M}$ by deleting $\mathbf{x}_j$. Suppose that, say, $\mathbf{m}_{k+1} = \sum_{i=1}^{k} \lambda_i \mathbf{m}_i$. Then det $\mathbf{M}_{k+1} \neq 0$ by the rank condition and $\lambda_i =$ det $\mathbf{M}_i/$det $\mathbf{M}_{k+1}$ by Cramer's rule. The right hand side clearly belongs to the mentioned field.

5.3.11 The necessity of the conditions is trivial. Let $n = |X_1| = |X_2|$, let $\mathbf{A}$ be a matrix so that $G = G(\mathbf{A})$, see Exercise 5.3.2, and all the nonzero entries of $\mathbf{A}$ be algebraically independent over $\mathbb{Q}$. If G has no perfect matching then $r_t(\mathbf{A}) < n$, hence $r(\mathbf{A}) < n$, by Corollary 5.3.3. Thus there exists a subset $X \subseteq X_1$ of columns of $\mathbf{A}$ which is minimally dependent over $\mathbb{R}$. Let $k = |X| - 1$. Let us delete the other columns of $\mathbf{A}$. For the remaining submatrix $\mathbf{B}$ clearly $r(\mathbf{B}) = k$ hence after some permutation of its rows $\mathbf{B}$ can be decomposed into two blocks $\begin{pmatrix} \mathbf{B}_1 \\ \mathbf{B}_2 \end{pmatrix}$ so that $\mathbf{B}_1$ has k linearly independent rows (over $\mathbb{R}$).

We claim that every entry of $\mathbf{B}_2$ is zero. For if, say, the jth column of $\mathbf{B}_2$ contained a nonzero entry b_{ij} then it must arise as a linear combination $\sum_t \lambda_t b_{it}$ of the other entries of the ith row (since the whole j^{th} column of $\mathbf{B}$ depends on the rest). But the λ_t's are in the smallest field, containing $\mathbb{Q}$ and the entries of $\mathbf{B}_1$, see the previous problem, contradicting to the algebraic independence of the entries of $\mathbf{B}$.

Hence the rows of $\mathbf{B}_1$ correspond to a subset $Y \subset X_2$ with $N(X) \subseteq Y$. By $|X| = k + 1$ and $|Y| = k$ this finishes the proof.

5.4.1 If C' is a minimum length circuit and $e \in E(C')$ then P_e of the algorithm clearly gives either $C' - \{e\}$ or another path of the same length. Hence the output C is of minimum length. We have seen that the number of steps required for Algorithm 3.2 is proportional to v^2, hence ev^2 (or, say, v^4) is clearly an upper bound for Algorithm 5.3.

5.4.2 Delete $k - 1$ points in all the possible $\binom{v}{k-1}$ ways from G. If the result is always connected (and $v \geq k + 1$) then G is k-connected. Since $\binom{v}{k-1} \approx \frac{1}{(k-1)!} v^{k-1}$ if k is fixed and v becomes large, and since connectivity-check requires constant times max $(e, v) \leq v^2$ steps, the number of steps required for this algorithm is proportional to at most v^{k+1}.

5.4.3 There is no contradiction. The calculation of $\alpha(H)$ for a graph H can be reduced to that of $\nu(G)$ for some graph G only if we have a graph G satisfying $H = L(G)$. Finding this graph may be a hard problem, and, more importantly, is not possible for every H. For example, if H is a star containing 3 edges incident to a point x then no such G exists. (The edge of G, corresponding to x, should be adjacent to three other edges, and since two of these latter must be adjacent, H should have a circuit of length 3.)

5.4.4 If a problem is in **NP**, its answer can be exhibited in polynomial time, using the help of the wizard. In case of a nondeterministic machine we can use this help to decide which copy of the machine should be followed at a particular step, when the machine replicates itself. Hence the answer is found in a polynomial number of steps and then every copy stops.

On the other hand, if a problem is answered in the affirmative by a nondeterministic machine then consider the copy which found the solution. Follow this copy backward until the start. Its "choices" at the steps where the machines replicated themselves are just the information we need from the wizard.

5.4.5 Let $P_1 \in \mathbf{P}$ and P_2 be an arbitrary decision problem. P_1 is clearly reducible to P_2 since this only means that the number of calls of the "subroutine P_2" is bounded by a polynomial (hence this number may be zero as well). Should $\mathbf{P} = \mathbf{NP}$ be true then every problem of this class would be reducible to every other one.

5.4.6 (a) The usual Gaussian elimination can be applied, which requires constant times n^3 steps.

(b) The nonzero entries are irrational, hence only symbolic manipulation is allowed (if roundoff errors must be excluded). However, this problem is equivalent to decide whether the bipartite graph $G(\mathbf{A})$ (see the solution of Exercise 5.3.2) has a perfect matching. This can be solved in constant times n^3 steps as well.

(c) We cannot use the idea of (b) now, since the polynomials of the nonzero entries can cancel out one another. Hence each "step" of the Gaussian elimination must be performed symbolically (the coefficients of every member of the polynomials must be calculated at every multiplication or addition). Although we still do not need more than constant times n^3 such "steps", one such "step" may mean symbolic manipulation by as much as $n!$ members. Hence the real number of steps may be an exponential or superexponential function of n as well.

5.4.7 The solution of Exercise 5.4.2 does not work now for (a) since the connectivity number must not appear in the exponent. We present another algorithm, immediately for the more difficult problem (b).

Let u, v be two points of the graph G with n points and e edges. Double all points except u and v, as in Fig. 5.14, then assign unit capacities for every edge and find the maximum flow f_{uv} between u and v. There are f_{uv} internally disjoint paths from u to v. The largest k for part (b) will clearly be min f_{uv}, where the minimum is taken for every pair u, v of points of G.

f_{uv} can be determined in constant times n^3 steps hence the number of steps in the whole algorithm is proportional to at most n^5. This bound can be reduced to $e^2\sqrt{n}$, see p. 215 of [Papadimitriou and Steiglitz, 1982], since one need not consider every pair of points. See also [Even, 1975].

5.4.8 If G has no Hamiltonian circuit, a call of (P) clarifies this. Otherwise try (P) for the input $G-e$ where $e \in E(G)$ is arbitrary. If $G-e$ still has a Hamiltonian circuit, delete e from G, otherwise try another edge. The algorithm terminates if the actual graph H has a Hamiltonian circuit but $H-e$ has not, for every $e \in E(H)$. Then H is just a Hamiltonian circuit of G. The number of deletions is less than $e(G)$ and two deletions cannot be separated by more than $e(G)$ calls of (P), hence the total number of calls of (P) is less than e^2.

5.4.9 If the graph G has a Hamiltonian circuit, it has Hamiltonian paths as well. On the other hand, let H be a subroutine which decides whether the input graph G has a Hamiltonian path, and H' be another one with input $(G; x, y)$ (where $x, y \in V(G)$) which decides whether G has a Hamiltonian path between x and y. The problem of the Hamiltonian circuit can clearly be reduced to H' by inputting $(G-e; x, y)$ for every $e = \{x, y\} \in E(G)$. Hence only H' must be reduced to H.

Let $(G; x, y)$ be an input. We may suppose that x and y are nonadjacent (since $\{x, y\}$ can be deleted from G otherwise). If G has no Hamiltonian path at all, the answer for H' is negative. Otherwise let d_x and d_y be the degree of x and y, respectively. Delete, in all the $d_x d_y$ different ways, all but one of the edges incident to x, and all but one of the edges incident to y. G has a Hamiltonian path from x to y if and only if at least one of the resulting graphs have a Hamiltonian path at all.

5.4.10 The undirected problems can trivially be reduced to the directed ones (replace each edge by a pair of oppositely oriented edges). For the other direction, double each point of the directed graph G in the usual way (Fig. 5.14) and replace every new edge by a directed path of length 2. Let G' be the resulting graph. Then G has a directed Hamiltonian path or circuit if and only if the underlying undirected graph of G' has a Hamiltonian path or circuit, respectively.

5.4.11 Should $P \in$ co-**NP** turn out to be **NP**-complete, then every problem of **NP** would be reducible to P, leading to **NP** $\subseteq$ co-**NP**. If the inclusion were proper, that is, if there were a problem Q so that $Q \in$ co-**NP** and $Q \notin$ **NP** then the complement of Q were in **NP** but not in co-**NP**, a contradiction.

Chapter 6

6.1.1 Consider the network of Fig. S.6.1. The initial voltages of C_3 and C_5 can independently be prescribed if $R_2 \neq 0$ and $R_4 + R_6 \neq 0$. If, say, $R_4 + R_6 = 0$ then two voltages are equal and only one of them can be prescribed.

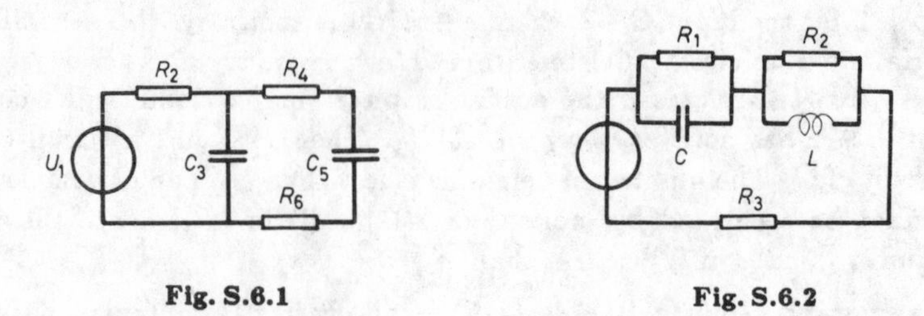

Fig. S.6.1 Fig. S.6.2

6.1.2 Consider the network of Fig. S.6.2. If the values of R_1, R_2, L and C are unity and $R_3 = -R_1$ then the network becomes unsolvable (see Problem 2.3.6).

6.2.1 Using the notation of Fig. S.6.3 we have $u_3 = u_1$; $u_4 = u_2$; $i_3 = -i_1$ and $i_4 = -i_2$ for the interconnection; $u_4 = Ri_4$ for the resistor; $u_2 = ku_1$ and $i_1 = -ki_2$ for the ideal transformer. Hence $u_3 = u_1 = k^{-1}u_2 = k^{-1}u_4 = k^{-1}Ri_4 = -k^{-1}Ri_2 = -k^{-2}Ri_1 = k^{-2}Ri_3$, that is, the whole system acts like a resistor of value R/k^2. Similarly, an inductor of value L and a capacitor of value C will be transformed to another inductor of value L/k^2 and to another capacitor of value Ck^2, respectively (see the first row of **Box 6.2** as well).

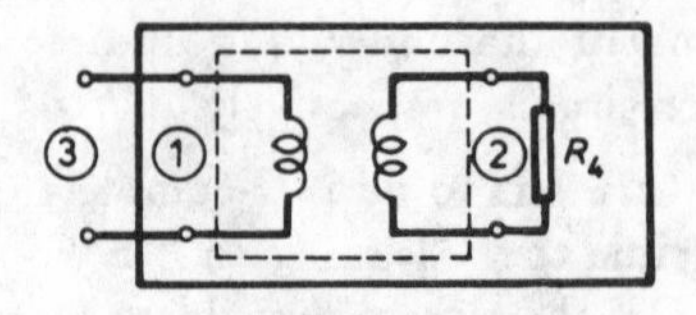

Fig. S.6.3

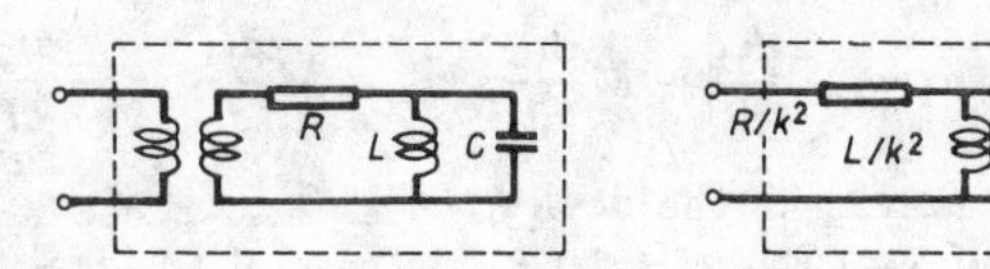

Fig. S.6.4

6.2.2 The other port acts like the original RLC subnetwork, with the only difference that each element of the original subnetwork must be replaced by the corresponding element from the first row of **Box 6.2.** See Fig. S.6.4 for an example.

6.2.3 See the second row of **Box 6.2.** For example, the second entry was calculated as follows: $i_1 = \frac{1}{\rho}u_2 = \frac{L}{\rho}\frac{d}{dt}i_L = -\frac{L}{\rho}\frac{d}{dt}i_2 = \frac{L}{\rho^2}\frac{d}{dt}u_1$.

6.2.4 The cascade connection of two ideal transformers gives a new ideal transformer whose transfer ratio is the product of the original transfer ratios. If the first 2-port is an ideal transformer (with transfer ratio k) and the second one is a gyrator with gyrator constant r then $u_1 = k^{-1}u_2 = k^{-1}u_3 = -rk^{-1}i_4$ and $i_1 = -ki_2 = ki_3 = kr^{-1}u_4$, hence the result is a gyrator with constant r/k. These and the other two possibilities are summarized in the upper left corner of **Box 6.3.**

6.2.5 The network graph is shown in Fig. S.6.5; the ideal transformers determine the pairs $\{1,2\}$ and $\{3,4\}$ of edges. If A is a voltage source then the graph has normal spanning trees ($\{A,1,4\}$ plus any two resistors). If A is a current source then there exists no normal spanning tree (if A is excluded from the tree, both 1 and 4 must be included and then neither 2 nor 3 can belong to the tree).

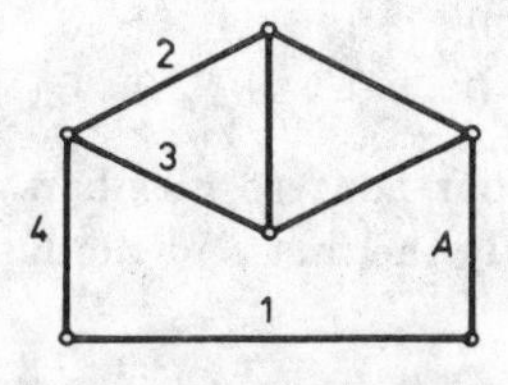

Fig. S.6.5

6.2.6 See the network of Fig. S.6.6 if $R_2 = -k^2 R_1$.

6.2.7 Replace the ideal transformer of Fig. S.6.6 by a gyrator and let $R_1R_2 = -R^2$.

6.2.8 The "real" transformer satisfies

$$u_1 = L_1\frac{di_1}{dt} + M\frac{di_2}{dt}; \quad u_2 = M\frac{di_1}{dt} + L_2\frac{di_2}{dt}.$$

Due to the passivity of the device, $L_1 \geq 0; L_2 \geq 0$ and $|M| \leq \sqrt{L_1L_2}$ will also follow (see Exercise 8.1.8). In case of an ideal transformer, M equals $\sqrt{L_1L_2}$ and then a linear combination of the above two equations gives $u_1 = ku_2$ with $k = \sqrt{L_1/L_2}$.

If both L_1 and L_2 tend to infinity but their ratio remains k^2 (which physically means that the magnetic permeability of the core is increased, keeping the ratio of the number of turns unaltered, see [Rohrer, 1970]) then $i_2 = -ki_1$ can also be obtained.

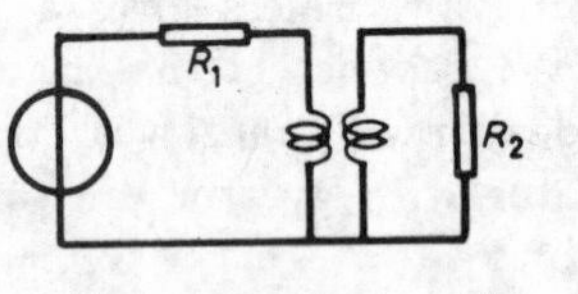

Fig. S.6.6

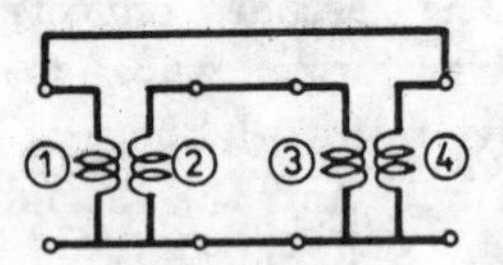

Fig. S.6.7

6.2.9 Figure S.6.7 shows the simplest, though highly artificial, example. The network graph has four spanning trees, namely $\{1,2\},\{1,3\},\{2,4\}$ and $\{3,4\}$. Only the second and the third are normal. The determinant of the network is

$$\begin{vmatrix} -1 & 0 & 0 & 1 & 0 & 0 & 0 & 0 \\ 0 & -1 & 1 & 0 & 0 & 0 & 0 & 0 \\ 0 & 0 & 0 & 0 & 1 & 0 & 0 & 1 \\ 0 & 0 & 0 & 0 & 0 & 1 & 1 & 0 \\ -1 & k & 0 & 0 & 0 & 0 & 0 & 0 \\ 0 & 0 & 0 & 0 & k & 1 & 0 & 0 \\ 0 & 0 & -1 & \ell & 0 & 0 & 0 & 0 \\ 0 & 0 & 0 & 0 & 0 & 0 & \ell & 1 \end{vmatrix}$$

Its expansion gives four nonzero members (of value $1, k\ell, k\ell$, and $k^2\ell^2$, respectively), and only the first and the last ones correspond to normal trees.

6.2.10 The difference between the resistive networks, and those with resistors and ideal transformers, is that the one-one correspondence between the nonzero expansion members and the normal trees holds in the former case only. However, the two columns are the same: if there are nonzero expansion members in a network with ideal transformer then, among these members, there are *some* which correspond to normal trees.

6.2.11 There are nonzero expansion members due to the unique solvability; such a member contains one element from each row of the coefficient matrix of Fig. 6.4. The first $n(G)$ rows determine a tree-complement (Theorem 1.4.6); let B_1 denote the corresponding tree. The next $r(G)$ rows determine a tree B_2. (Fig. 6.4 looks so simple because $B_1 = B_2$ but it need not be the case in general. For example, $B_1 = \{1,3\}$ and $B_2 = \{2,4\}$ can happen for the network of Fig. S.6.7.) The set U of voltage sources satisfies $U \subseteq B_1 \cap B_2$ and the set I of current sources satisfies $I \subseteq (E - B_1) \cap (E - B_2)$ where $E = E(G)$.

Consider now a resistor. The corresponding Ohm's law determines a row with two nonzero entries in the matrix. One of them appears in the expansion member under consideration. Whether this one corresponds to the voltage or to the current, the resistor is either in $B_1 \cap B_2$ or in $(E - B_1) \cap (E - B_2)$. Let R_1 and R_2 denote the sets of those resistors which are in both trees and in none of them, respectively (see Fig. S.6.8).

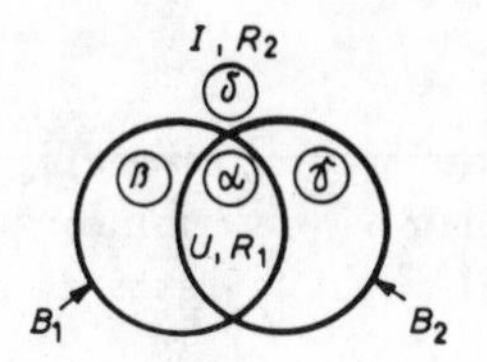

Fig. S.6.8

Finally, consider an ideal transformer and write its two equations symbolically as $\begin{pmatrix} x & y & 0 & 0 \\ 0 & 0 & s & t \end{pmatrix}$. If our expansion member contains x and t then the first port of the transformer is in $B_1 \cap B_2$, that is, in position α, since the column of x was not used in the first $n(G)$ rows but the column of s was used in the next $r(G)$ rows. Similarly, the second port is in position δ.

If y and s are in the expansion member then the first port is in position δ and the second one is in position α. Finally, if x and s (or y and t) appear in the expansion member then the first port is in position β and the second one in γ (or *vice versa*, respectively).

If $B_1 = B_2$ then we clearly have a normal tree. But the point is that $B_1 \neq B_2$ also implies that both of them are normal trees, since all we need is that they must intersect the pair of ports by one, for every ideal transformer, and this is always met, whether α and δ or β and γ describe the position of the ports in a pair.

6.2.12 Consider Fig. S.6.7 again and replace the ideal transformers by gyrators. Only $\{1,2\}$ and $\{3,4\}$ are normal trees, although the number of nonzero expansion members is four.

6.2.13 The absurd network of Fig. S.6.9 [Recski, 1978] has no unique solution, hence it has no normal tree either. However, the coefficient matrix of its system of equations is $\begin{pmatrix} -1 & 1 & 0 & 0 \\ 0 & 0 & 1 & 1 \\ 1 & 0 & 0 & R \\ 0 & -1 & R & 0 \end{pmatrix}$ and its determinant has two nonzero expansion members (which cancel out each other).

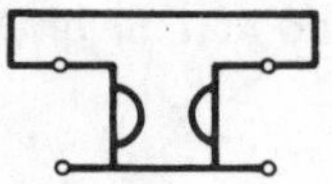

Fig. S.6.9

6.2.14 We may use the same way of reasoning as for Problem 6.2.11. Each nonzero expansion member determines two trees B_1, B_2, see Fig. S.6.8. The two equations of the gyrator can symbolically be written as $\begin{pmatrix} x & 0 & 0 & y \\ 0 & s & t & 0 \end{pmatrix}$. If x and t appear in our expansion member then β and γ are the positions of the first and the second edge of the gyrator, respectively. The same two positions appear in reversed order if the expansion member contains y and s. In the other two cases both edges are in position α or both in position δ.

The edge set of B_1 in G and that of B_2 in G' will be the same, hence Tow's result immediately follows. The positions β and γ interchange if G is changed to G'. If every gyrator edge is in position α or δ then $B_1 = B_2$. Hence every normal tree is "common" but not necessarily *vice versa*, see the network of Problem 6.2.13.

6.2.15 If the network is uniquely solvable and all the capacitor voltages can independently be prescribed as initial values then G has a tree F containing every capacitor edge and either 2 or 0 edges for each gyrator. Since F has $v(G) - 1$ edges, the number $v(G) - 1 - n_C$ must be even.

6.2.16 A spanning forest T satisfying $U \subseteq T$, $I \cap T = \emptyset$ is *normal* if it contains exactly one edge from every ideal transformer and either 2 or 0 edges from every gyrator.

Implication 1 is obviously true, implications 2, 3 and 4 are false (see the same counterexamples as for column 3). Implications 5 and 6 are true. The proof of the latter is the same as those of Theorems 6.2.1 and 6.2.4. Implication 5 can be reduced to Theorem 6.2.5 by replacing each ideal transformer by the cascade connection of two gyrators (see Exercise 6.2.4).

6.3.1 If the positions of two distinct points P, Q of the plane are fixed then every point of the plane is in fixed position. In order to bring P to the requested position, a horizontal and a vertical translations are enough. Then, if Q is not yet in the requested position, a further rotation around P will assure this. Hence every congruent transformation of the plane can be obtained by the combination of two translations and a rotation.

In the 3-dimensional space choose three not collinear points P, Q and R. Bring P to the final position (the combination of three translations) and then bring Q to the final position on the sphere around P. This latter is the combination of two rotations (recall the longitude and the latitude on the globe). Finally a rotation around the PQ axis brings R to the final position. Hence every congruent transformation of the space can be obtained as the combination of six elementary transformations.

6.3.2 Both systems of Fig. S.6.10 satisfy $e = 2v - 3$ but only the first one is rigid. In the 3-dimensional case fix points E and F to a tetrahedron $ABCD$ by 3 rods each (see the left hand side of Fig. S.6.11). If the rod AF is removed and another rod DE is added, $e = 3v - 6$ still holds but the framework (on the right hand side of Fig. S.6.11) is nonrigid.

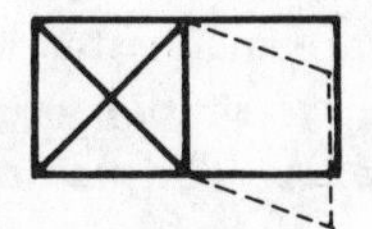

Fig. S.6.10

6.3.3 If the joints 1,2 and 3 are collinear (and 3 is connected to 1 and 2 only) then the other joints (if any) can be disregarded and the system can be replaced by that of Fig. S.6.12. If joint 3 is the origin and line 12 is the x-axis of the system of coordinates then the matrix $\mathbf{A}$ of the system is $\begin{pmatrix} x_1 - x_3 & 0 & x_3 - x_1 & y_1 - y_3 & 0 & y_3 - y_1 \\ 0 & x_2 - x_3 & x_3 - x_2 & 0 & y_2 - y_3 & y_3 - y_2 \end{pmatrix} =$ $= \begin{pmatrix} -\alpha & 0 & +\alpha & 0 & 0 & 0 \\ 0 & +\beta & -\beta & 0 & 0 & 0 \end{pmatrix}$ and $\mathbf{Au} = \mathbf{0}$; $\mathbf{u} \neq \mathbf{0}$ holds for $\mathbf{u} =$

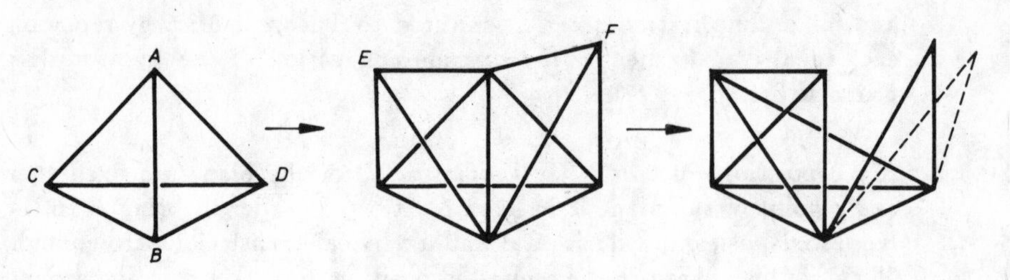

Fig. S.6.11

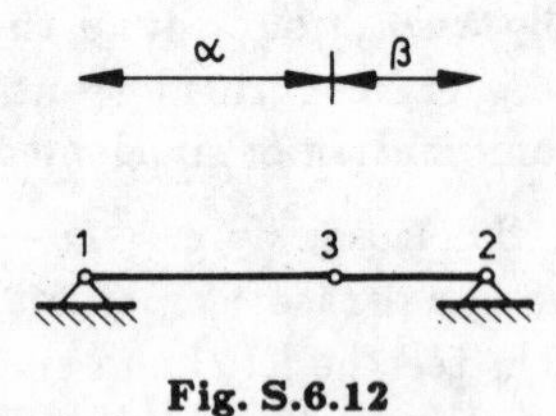

Fig. S.6.12

$=(0,0,0,0,0,v)$ with $v \neq 0$, which corresponds to a "motion" of joint 3 in the direction perpendicular to the line 12.

6.3.4 The first symbolic matrix of page 410 is the same for both systems. (The zero entries are not written for simplicity.) Let us choose joint 1 to be the origin of the system of coordinates and calculate the numerical values of the entries of $\mathbf{A}$. The results are

$$\begin{pmatrix}
-1 & 1 & & & & & -1 & 1 & & & & \\
-2 & & 2 & & & & 2 & & -2 & & & \\
 & -1 & 1 & & & & & 3 & -3 & & & \\
-3 & & & 3 & & & 1 & & & -1 & & \\
 & -3 & & & 3 & & & 0 & & & 0 & \\
 & & -3 & & & 3 & & & 0 & & & 0 \\
 & & & -1 & 1 & & & & & -1 & 1 & \\
 & & & -2 & & 2 & & & & 1 & & -1 \\
 & & & & -1 & 1 & & & & & 3 & -3
\end{pmatrix}$$

and

$$\begin{pmatrix}
-1 & 1 & & & & & -1 & 1 & & & & \\
-2 & & 2 & & & & 2 & & -2 & & & \\
 & -1 & 1 & & & & & 3 & -3 & & & \\
-3 & & & 3 & & & 0 & & & 0 & & \\
 & -3 & & & 3 & & & 0 & & & 0 & \\
 & & -3 & & & 3 & & & 0 & & & 0 \\
 & & & -1 & 1 & & & & & -1 & 1 & \\
 & & & -2 & & 2 & & & & 2 & & -2 \\
 & & & & -1 & 1 & & & & & 3 & -3
\end{pmatrix}$$

respectively. (Only those zeroes are explicitly written which were obtained by these substitutions.) The rank of these matrices, as can be calculated by usual numerical methods of linear algebra, are 9 and 8, respectively, hence the first system is rigid, the second one is not.

6.3.5 Suppose that the vertical move of the middle point is x and determine the change of the length of the rods. Since x is small relative to y (see Fig. S.6.13), put $x = cy$ (where c is a small positive number). In the case of the first system the lengths decrease by $\sqrt{2y^2} - \sqrt{y^2 + (1-c)^2 y^2} =$ $= y \cdot f(y)$, while in the case of the second one the lengths increase by $\sqrt{y^2 + c^2 y^2} - y = y \cdot g(y)$, using the law of Pythagoras.

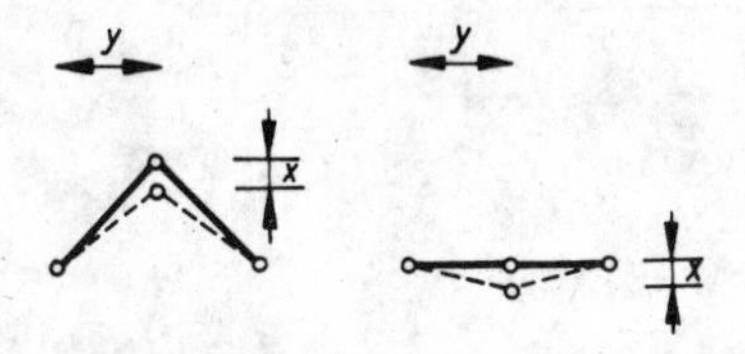

Fig. S.6.13

If a, b are two positive numbers, close to each other, then the identity $\sqrt{a} - \sqrt{b} = \dfrac{a-b}{\sqrt{a} + \sqrt{b}}$ helps to give an estimation. In our case $f(y) \approx$ $\approx c/\sqrt{2}$ and $g(y) \approx c^2/2$ is obtained, giving that the required force is much smaller (by a whole order of magnitude) in the second case.

6.3.6 Recall at first that the position of a point on the surface of an n-dimensional globe is determined by $n-1$ data. For example, we may use the n-dimensional polar coordinates $x_1 = r\cos\phi_1$; $x_2 = r\sin\phi_1\cos\phi_2$; $x_3 = r\sin\phi_1\sin\phi_2\cos\phi_3$; $\ldots$ $x_n = r\sin\phi_1\sin\phi_2\ldots\sin\phi_{n-1}$ for these points $(x_1, x_2, \ldots, x_n)$ since if they are on the surface of a globe of radius r and centrum at the origin then $x_1^2 + x_2^2 + \ldots + x_n^2 = r^2$.

$$
\begin{pmatrix}
x_1-x_2 & x_2-x_1 & & & & & y_1-y_2 & y_2-y_1 & & & & \\
x_1-x_3 & & x_3-x_1 & & & & y_1-y_3 & & y_3-y_1 & & & \\
 & x_2-x_3 & x_3-x_2 & & & & & y_2-y_3 & y_3-y_2 & & & \\
x_1-x_4 & & & x_4-x_1 & & & y_1-y_4 & & & y_4-y_1 & & \\
 & x_2-x_5 & & & x_5-x_2 & & & y_2-y_5 & & & y_5-y_2 & \\
 & & x_3-x_6 & & & x_6-x_3 & & & y_3-y_6 & & & y_6-y_3 \\
 & & & x_4-x_5 & x_5-x_4 & & & & & y_4-y_5 & y_5-y_4 & \\
 & & & x_4-x_6 & & x_6-x_4 & & & & y_4-y_6 & & y_6-y_4 \\
 & & & & x_5-x_6 & x_6-x_5 & & & & & y_5-y_6 & y_6-y_5
\end{pmatrix}
$$

The symbolic matrix for Exercise 6.3.4.

$$
\begin{pmatrix}
x_5-x_1 & y_5-y_1 & z_5-z_1 & & & & & & & & & \\
x_5-x_2 & y_5-y_2 & z_5-z_2 & & & & & & & & & \\
 & & & x_6-x_2 & y_6-y_2 & z_6-z_2 & & & & & & \\
 & & & x_6-x_3 & y_6-y_3 & z_6-z_3 & & & & & & \\
 & & & & & & x_7-x_3 & y_7-y_3 & z_7-z_3 & & & \\
 & & & & & & x_7-x_4 & y_7-y_4 & z_7-z_4 & & & \\
 & & & & & & & & & x_8-x_1 & y_8-y_1 & z_8-z_1 \\
 & & & & & & & & & x_8-x_4 & y_8-y_4 & z_8-z_4 \\
x_5-x_6 & y_5-y_6 & z_5-z_6 & x_6-x_5 & y_6-y_5 & z_6-z_5 & & & & & & \\
 & & & x_6-x_7 & y_6-y_7 & z_6-z_7 & x_7-x_6 & y_7-y_6 & z_7-z_6 & & & \\
 & & & & & & x_7-x_8 & y_7-y_8 & z_7-z_8 & x_8-x_7 & y_8-y_7 & z_8-z_7 \\
x_5-x_8 & y_5-y_8 & z_5-z_8 & & & & & & & x_8-x_5 & y_8-y_5 & z_8-z_5
\end{pmatrix}
$$

The symbolic matrix for Problem 6.3.8.

Now we choose n points $P_1, P_2, \ldots, P_n$ in the n-dimensional space so that they are not all in the same $(n-2)$-dimensional hyperplane. The combination of n translations is enough to place P_1 to its final position. Then the combination of $n-1$ rotations is required to place P_2 to its final position on the surface of the globe around P_1. Project now the space to the $(n-1)$-dimensional hyperplane perpendicular to the line P_1P_2. (Observe that none of $P_3, P_4, \ldots$ is on this line.) The combination of $n-2$ rotations is required to place P_3 to its final position on the surface of the $(n-1)$- dimensional globe around P_1 etc. Hence $n + (n-1) + $ $+ (n-2) + \ldots = n(n+1)/2$ elementary transformations are required to obtain every congruent transformations of the n-dimensional space.

6.3.7 One can directly obtain the matrix **A** for both systems. For example, the row corresponding to the rod between joints 1 and 7 is $(-10, 0, 0, 0, 0, 0, 10, -2, 0, 0, 0, 0, 0, 2)$ for the first system and $(-9, 0, 0, 0, 0, 0, 9, -3, 0, 0, 0, 0, 0, 3)$ for the second one. The rank of the first matrix is 10, that of the second one is 11, hence only the second system is rigid. (Cf. Exercise 14.3.3 for an alternative solution.)

6.3.8 Let the origin of our system of coordinates be at joint 5, let the directions of the x, y and z axes be towards the joints 8,6 and 1, respectively (see Fig. S.6.14). The first "one-third" of the matrix **A** for both systems is

$$\begin{pmatrix}
x_1-x_5 & & & & x_5-x_1 & & & \\
x_1-x_8 & & & & & & & x_8-x_1 \\
 & x_2-x_5 & & & x_5-x_2 & & & \\
 & x_2-x_6 & & & & x_6-x_2 & & \\
 & & x_3-x_6 & & & x_6-x_3 & & \\
 & & x_3-x_7 & & & & x_7-x_3 & \\
 & & & x_4-x_7 & & & x_7-x_4 & \\
 & & & x_4-x_8 & & & & x_8-x_4 \\
 & & & & x_5-x_6 & x_6-x_5 & & \\
 & & & & x_5-x_8 & & & x_8-x_5 \\
 & & & & & x_6-x_7 & x_7-x_6 & \\
 & & & & & & x_7-x_8 & x_8-x_7
\end{pmatrix}$$

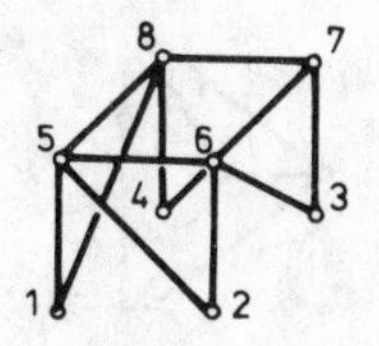

Fig. S.6.14

We delete the columns, corresponding to joints 1, 2, 3 and 4. The remaining 12×12 matrix is shown as the second matrix on page 410 (after a permutation of the rows).

Since the first three columns have nonzero entries in four rows only, the determinant of this matrix is the difference of two products; one of them is

$$\begin{vmatrix} x_5 - x_1 & y_5 - y_1 & z_5 - z_1 \\ x_5 - x_2 & y_5 - y_2 & z_5 - z_2 \\ x_5 - x_6 & y_5 - y_6 & z_5 - z_6 \end{vmatrix} \begin{vmatrix} x_6 - x_2 & y_6 - y_2 & z_6 - z_2 \\ x_6 - x_3 & y_6 - y_3 & z_6 - z_3 \\ x_6 - x_7 & y_6 - y_7 & z_6 - z_7 \end{vmatrix} \times$$

$$\times \begin{vmatrix} x_7 - x_3 & y_7 - y_3 & z_7 - z_3 \\ x_7 - x_4 & y_7 - y_4 & z_7 - z_4 \\ x_7 - x_8 & y_7 - y_8 & z_7 - z_8 \end{vmatrix} \begin{vmatrix} x_8 - x_1 & y_8 - y_1 & z_8 - z_1 \\ x_8 - x_4 & y_8 - y_4 & z_8 - z_4 \\ x_8 - x_5 & y_8 - y_5 & z_8 - z_5 \end{vmatrix}$$

and the other is obtained from this by replacing the last row of each 3×3 determinant by that of the (cyclically) previous one.

In the case of the first system, the first of these products vanishes while the other does not. Apply the well known theorem from geometry that $\begin{vmatrix} x_a - x_b & y_a - y_b & z_a - z_b \\ x_a - x_c & y_a - y_c & z_a - z_c \\ x_a - x_d & y_a - y_d & z_a - z_d \end{vmatrix}$ equals zero if and only if a, b, c and d are coplanar (see [Hajós, 1964], for example). Hence the first system is rigid.

Instead of the second system one may consider the more general question when the square 5, 6, 7, 8 is rotated by an angle ϕ from the position of the square 1, 2, 3, 4. (The second system is obtained by $\phi = 45°$.) The two products (of four determinants each) will equal to $(\frac{1}{2}\cos\phi)^4$ and $(\frac{1}{2}\sin\phi)^4$, respectively, and their difference vanishes if and only if $\phi = 45°$.

Hence the second system is nonrigid; it were rigid for a rotation of $\phi \neq 45°$ but then its aesthetic advantage disappears.

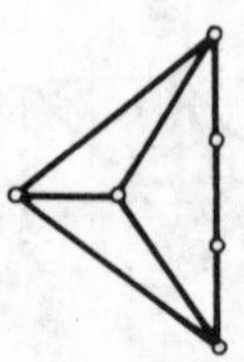

Fig. S.6.15

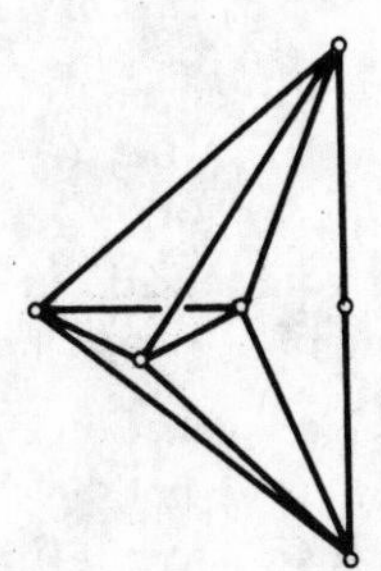

Fig. S.6.16

6.3.9 Consider the planar framework of Fig. S.6.15. It has infinitesimal motions only, and $e = 8$ is less than $2v - 3$. Similarly, the 3-dimensional framework of Fig. S.6.16 violates $e \geq 3v - 6$.

Chapter 7

7.1.1 Yes. If $x = y$ then $(S, \mathbf{F})$ is the free matroid. Otherwise it corresponds to a graph consisting of a pair x, y of parallel edges and $|S| - 2$ further disjoint edges.

7.1.2 Let $X, Y \in \mathbf{F}$ and $|X| > |Y|$. Let G_X and G_Y denote the subgraphs of G, determined by the edge sets X and Y, respectively. The number of points and connected components in G_X are denoted by v_X and c_X, respectively, the meaning of v_Y and c_Y is the same for G_Y.

G_X and G_Y are circuit-free, hence $v_X - c_X = |X|$ and $v_Y - c_Y = |Y|$. If there exists a point of G_X, not belonging to G_Y, then any edge of X which is incident to this point can be added to Y without forming a circuit. If not then $v_X \leq v_Y$ and $|X| > |Y|$ imply $c_X < c_Y$. Hence there exists an edge in X joining different components of G_Y and then this edge can be added to Y.

7.1.3 Let $G_{n,0}$ denote a graph with n loops, $G_{n,1}$ a graph with a single edge of multiplicity n. The graph $G_{n,n-1}$ is a circuit of length n and $G_{n,n}$ is a tree with n edges. Then $\mathbf{U}_{n,k}$ is isomorphic to the matroid, corresponding to $G_{n,k}$ for $k = 0$, 1, $n - 1$ or n.

7.1.4 Properties (F1) and (F2) are clearly satisfied. Let $X, Y \in \mathbf{F}'$ and $|X| > >|Y|$. Clearly, $X, Y \in \mathbf{F}$ as well, hence there exists an $x \in X - Y$ so that $Y \cup \{x\} \in \mathbf{F}$. But $Y \cup \{x\} \in \mathbf{F}'$ as well, since $|Y \cup \{x\}| = |Y| + 1 \leq \leq |X| \leq p - 1$.

The truncation of the matroid, corresponding to K_4, is $\mathbf{U}_{6,2}$ (since K_4 has no parallel edges).

7.1.5 $X \subseteq S$ is independent if and only if there exists a base B of $\mathbf{M}$ with $X \subseteq B$. Alternatively, $X \subseteq S$ is independent if and only if $r(X) = |X|$ holds for the rank function r of $\mathbf{M}$.

7.1.6 We can replace (F2) by (F2′). If $Y \subseteq X$ and $|X| - |Y| > 1$ for some $X \in \mathbf{F}$ then let us delete the elements of $X - Y$ one by one. We obtain elements of $\mathbf{F}$ in every step, by (F2′), hence $Y \in \mathbf{F}$.

On the other hand, (F2″) cannot replace (F2). Let $S = \{a, b\}$ and $\mathbf{F} = \{\emptyset, \{a\}, \{a, b\}\}$. Then (F1),(F2″) and (F3) are satisfied, yet $(S, \mathbf{F})$ is not a matroid.

7.1.7 We can replace (F3) by (F3′) provided that (F1) and (F2) are still prescribed. If $|X| - |Y| = p > 1$ for some $X, Y \in \mathbf{F}$ then delete $p - 1$ elements from $X - Y$. The resulting subset X' belongs to $\mathbf{F}$ by (F2). Now apply (F3′) for X' and for Y.

On the other hand, (F3″) cannot replace (F3), see the counterexample of the previous exercise.

7.1.8 All the matroids with underlying set of cardinality 1,2 or 3 are graphic and are shown in Fig. S.7.1. There is a single nongraphic one among

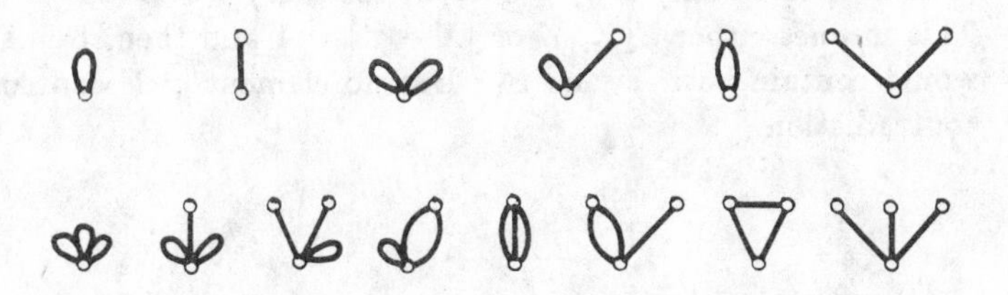

Fig. S.7.1

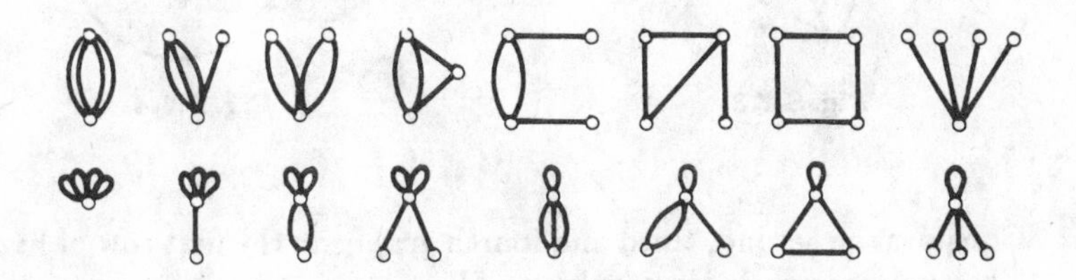

Fig. S.7.2

the matroids on a 4-element set, namely $\mathbf{U}_{4,2}$. Figure S.7.2 shows the others.

7.1.9 An n-element set S has 2^n subsets. Each of them is either independent or not, hence $f(n) \leq 2^{2^n}$ is clearly an upper bound for the number $f(n)$ of nonisomorphic matroids on S.

This trivial estimation turns out to be fairly good. One can prove [Piff, 1973; Piff and Welsh, 1971] that $n - \frac{5}{2}\log n \leq \log\log f(n) \leq n - (1+\epsilon)\log n$ for any $\epsilon > 0$, if n is sufficiently large.

7.1.10 (F1) and (F2) are clearly satisfied. In order to see (F3), let $|X| > |Y|$ for some X, Y which are independent in our matroid. They were independent in $\mathbf{U}_{n,k}$ as well, hence there exists an $x \in X - Y$ so that $Y \cup \{x\}$ is independent in $\mathbf{U}_{n,k}$. Suppose that $Y \cup \{x\}$ is just the k-element set T which became dependent. If $X - Y$ has some other element y as well then $Y \cup \{y\}$ is independent, otherwise $X - Y = \{x\}$ means $X = T$, contradicting to the independence of X in our matroid. See Fig. S.7.3 for the case $n = 4$, $k = 2$.

7.1.11 (F1) and (F2) are clearly satisfied. In order to see (F3), let $|X| > |Y|$ for some $X, Y \in \mathbf{F}$. Since $|X| \leq n - 3$, if we add an element $x \in X - Y$ to Y, the result has at most $n - 3$ elements so it is either independent or contains S_1 or S_2.

If $x_1 \in X - Y$ and $S_1 \subseteq Y \cup \{x_1\}$ then $S_2 \subseteq Y \cup \{x_2\}$ for some other $x_2 \in X - Y$ is impossible (since it would imply $|Y| \geq n - 2$). Hence (F3) is violated only if $S_1 \subseteq Y \cup \{x\}$ for every $x \in X - Y$, see Fig. S.7.4. This implies either $S_1 \subseteq Y$ or $|X - Y| = 1$ and then, by $|X| > |Y|$, X would contain both Y and S_1. But no element of $\mathbf{F}$ can contain S_1, a contradiction.

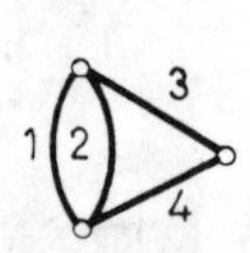

Fig. S.7.3

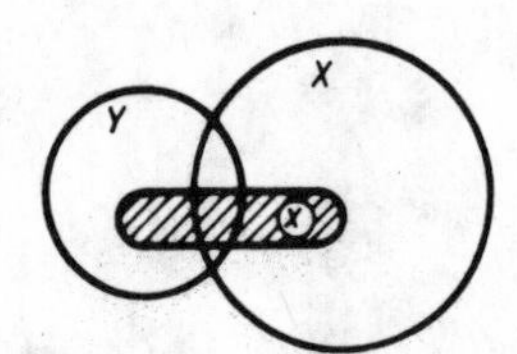

Fig. S.7.4

7.1.12 Consider the second, third and fourth graphs in the first row of Fig. S.7.2. Their truncation is isomorphic to $\mathbf{U}_{4,1}$.

7.1.13 The answer is positive in both cases. For example, the truncation of $\mathbf{U}_{n,k}$ is $\mathbf{U}_{n,k-1}$ and we have seen that $\mathbf{U}_{4,2}$ is nongraphic while $\mathbf{U}_{4,3}$ and $\mathbf{U}_{4,1}$ are graphic.

7.1.14 If such a matroid exists then clearly $\mathbf{F} = \{X | X$ is contained in an element of $\mathbf{B}\}$. This obviously satisfies (F1) and (F2) (observe that (B1) is needed to deduce (F1)). In order to see (F3) let $X, Y \in \mathbf{F}$ and $|X| > |Y|$. There are some members B_X, B_Y of $\mathbf{B}$, containing X and Y, respectively. In fact, this can happen in several different ways. We choose one so that $|B_X \cap B_Y|$ be as large as possible.

If $X \cap (B_Y - Y) \neq \emptyset$ then for any element x of it we have $Y \cup \{x\} \subseteq B_Y$ and hence $Y \cup \{x\} \in \mathbf{F}$ (see the dotted area in Fig. S.7.5). This is nonempty since $|B_X| = |B_Y|$ by (B2) and $|X| > |Y|$.

By (B3) there exists a $t \in B_X - B_Y$ so that $B'_Y = (B_Y - \{z\}) \cup \{t\}$ also belongs to $\mathbf{B}$. Then $|B'_Y \cap B_X| > |B_Y \cap B_X|$, contradicting to the choice of B_X and B_Y.

7.1.15 Properties (B1) and (B2) are obvious for $\mathbf{M}^p$ hence by the previous problem it is enough to show (B3). Let B_1, B_2 be bases of $\mathbf{M}^p$, i.e. let $T_1 \subset B_1$ and $T_2 \subset B_2$ be bases in $\mathbf{M}$. Let $x \in B_1$ and find an element $y \in B_2$ so that $(B_1 - \{x\}) \cup \{y\}$ is also a base of $\mathbf{M}^p$.

If x happens to belong to B_2, let $y = x$. Otherwise if $x \notin T_1$ then any element of $B_2 - B_1$ can be chosen for y. Finally, if $x \in T_1 - B_2$ then apply (B3) for $T_1 - \{x\}$ and T_2 in the matroid $\mathbf{M}$.

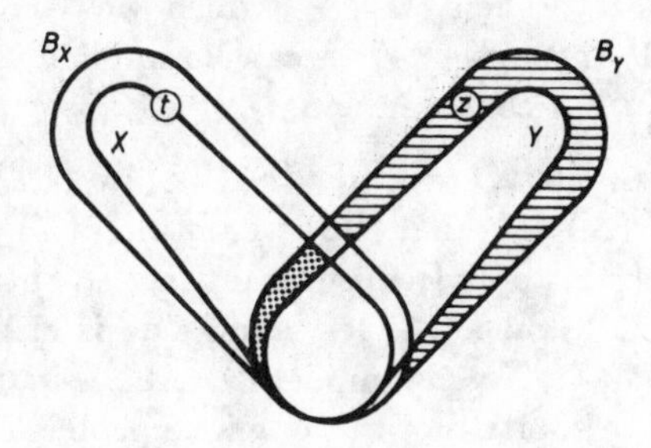

Fig. S.7.5

For example, $\mathbf{M} = \mathbf{U}_{4,1}$ is graphic and $\mathbf{M}^2$ is not; $\mathbf{M} = \mathbf{U}_{4,2}$ is non-graphic but $\mathbf{M}^3$ is graphic; if $\mathbf{M}$ is the graphic matroid corresponding to K_4 then $\mathbf{M}^5$ is also graphic.

7.1.16 Consider the bases $B_1 = \{1,2,3\}$ and $B_2 = \{4,5,6\}$ of the matroid, corresponding to K_4 (see Fig. S.7.6). 6 cannot be the pair of 2 (since 1,3,6 is not a base) and 2 cannot be the pair of 5 for similar reason. Hence the only candidate for the pair of 2 is 4 but so is it for the pair of 3, hence the bijection is impossible.

This fact does not contradict to (B3′) since a $y \in B_2$ can belong to several elements of B_1.

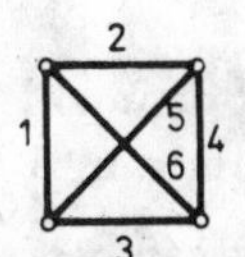

Fig. S.7.6

7.1.17 $r_1(X) - r_1(X \cap Y) \geq r_1(X \cup Y) - r_1(Y)$ holds by (R4) of Theorem 7.1.4. Hence we only have to show $r_1(X \cup Y) - r_1(Y) \geq r_2(X \cup Y) - r_2(Y)$. If Y contains a base of $\mathbf{M}_1$ then both sides of this inequality is zero. Otherwise it follows from $r_1(Y) = r_2(Y)$.

7.1.18 We show (F1), (F2) and (F3) for $\mathbf{F} = \{X | r(X) = |X|\}$. (F1) is trivial by (R2). If $X \in \mathbf{F}$ and $Y \subset X$ then apply (R4) for $X - Y$ and Y to obtain $r(Y) \geq |X| - r(X - Y)$ at first. (R2) gives $r(X - Y) \leq |X - Y|$, hence we have $r(Y) \geq |Y|$. Since $r(Y) \leq |Y|$ is given by (R2), we obtain $Y \in \mathbf{F}$.

In order to show (F3) let $X, Y \in \mathbf{F}$ and $|X| > |Y|$. Suppose indirectly that $Y \cup \{x\} \notin \mathbf{F}$ for every $x \in X - Y$, that is, that $r(Y \cup \{x\}) = r(Y)$ for every such x. Then (R7) — which was shown to be implied by properties (R1) through (R4) — can be applied several times to obtain $r(X \cup Y) = r(Y)$, a contradiction.

7.1.19 Let again $\mathbf{F} = \{X | r(X) = |X|\}$. (F1) is just (R5). Applying (R5) and $r(X \cup \{y\}) \leq r(X) + 1$ from (R6) we obtain $r(X) \leq |X|$ for every X. In order to show (F2) apply (R6) several times to obtain $r(Y \cup (X - Y)) \leq$ $\leq r(Y) + |X - Y|$. Since the left hand side is $r(X)$ and the right hand side is at most $|X|, X \in \mathbf{F}$ implies $Y \in \mathbf{F}$. Finally (F3) was obtained from (R7) in the solution of the previous problem.

7.1.20 Apply (R4) for $X \cup Y$ and $X \cup Z$. Then $r(X \cup Y) + r(X \cup Z) - r(X \cup Y \cup Z) \geq r((X \cup Y) \cap (X \cup Z)) = r(X \cup (Y \cap Z))$. The right hand side is greater than or equal to $r(X)$ by (R3), hence (R′) follows.

On the other hand, (R4) follows from (R′) by putting $A \cap B = X$, $A - B = Y$ and $B - A = Z$, and (R3) also follows by letting $Y = Z$ [Ingleton, 1971b].

7.1.21 (R″) implies (R′) by putting $W = Y$ [Ingleton, 1971b].

7.1.22 We should verify, by Problem 7.1.19, that $f(X)$ satisfies (R5) through (R7). The first two properties are trivially met. In order to see (R7) observe that $f(X) = f(X \cup \{y\})$ is possible only if $r(X) = r(X \cup \{y\})$ holds.

7.2.1 If $x \neq z$, we can apply (C2) to the circuits $X = \{x, y\}, Y = \{y, z\}$ and to the element $u = y$. Hence $\{x, z\}$ must contain a circuit and none of its elements can be a single element circuit by the definition of parallel elements.

7.2.2 X must be the complement of an independent set since otherwise it were disjoint from circuits contained in $S - X$.

7.2.3 $C - \{x\}$ is independent and can be extended to a base. That base B has the required property.

7.2.4 If x is contained in a circuit C then $C - \{x\}$ can be extended to a base not containing x. On the other hand, if x is contained in no circuit then it is contained in every base B. For otherwise $B \cup \{x\}$ would contain a circuit C and $x \notin C$ would then imply $C \subseteq B$ which is impossible.

7.2.5 $X = \{a, b, c\}$ and $Y = \{a, b, d\}$ are circuits in $\mathbf{U}_{4,2}$ and $X \triangle Y$ is independent. On the other hand, the statement is true, for example, for the cut set matroid of a nonplanar graph.

7.2.6 The statement is trivial since loops and parallel elements in a graphic matroid correspond to loops and parallel edges of the graph, respectively.

If G is a tree, $\mathbf{M}(G)$ is simple but $\mathbf{M}^*(G)$ is not.

7.2.7 The dual of the circuit matroid is not necessarily graphic hence the graph need not have a dual among the graphs (only among the matroids) ; just like every complex number (including reals) has square roots among the complex numbers but not necessarily among the reals.

7.2.8 Yes. Any singleton in $\mathbf{U}_{2,1}$ or any triple in $\mathbf{U}_{4,2}$ have this property.

7.2.9 If a cographic matroid $\mathbf{N}$ is the cut set matroid of a graph G then the rank in $\mathbf{N}$ is the nullity in G.

7.2.10 Extend A to a base X of $\mathbf{M}$ and B to a base Y of $\mathbf{M}^*$, see Fig. S.7.7. Then $S-Y$ is a base of $\mathbf{M}$ and is disjoint of B. If $X \cap B = \emptyset$ then choose $K = X$. Otherwise delete an element of $X \cap B$ from X and extend the result to a base X' from an element of $S - Y$. If $X' \cap B = \emptyset$ then let $K = X'$. Otherwise continue the process which will terminate since each time we deleted an element of B and added an element of $S - Y$ which is disjoint of B.

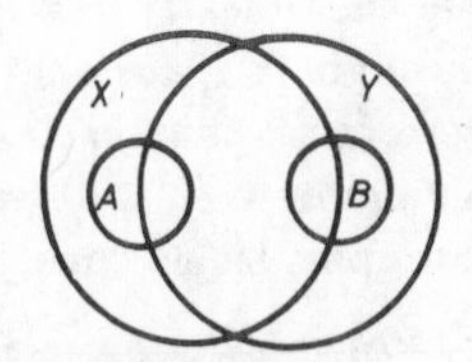

Fig. S.7.7

7.2.11 Suppose that $C \cap D = \{x\}$ and apply the result of the previous exercise for $A = C - \{x\}$ and $B = D - \{x\}$ to obtain a base K which separates them. $x \in K$ is impossible since D is a circuit of $\mathbf{M}^*$ and therefore it cannot be contained in a base (namely $S - K$) of $\mathbf{M}^*$. $x \notin K$ is also impossible since then C would be a subset of K, hence we have a contradiction.

$|C \cap D| = 3$ is possible, see any triple of $\mathbf{U}_{4,2}$.

7.2.12 Let B be a base of the matroid $\mathbf{M} = (S, \mathbf{F})$. For any $x \in S - B$ the set $B \cup \{x\}$ contains a unique circuit C_x. The collection $\{C_x | x \in S - B\}$ is called the *fundamental system of circuits* of $\mathbf{M}$ with respect to the base B. The *fundamental system of cut sets* of $\mathbf{M}$ with respect to the base B is just the fundamental system of circuits of $\mathbf{M}^*$ with respect to $S - B$.

7.2.13 Should $x_1, x_2, \ldots$ contain a circuit C then any $x_i \in C$ would imply $|C \cap D| = 1$, contradicting to the result of Exercise 7.2.11.

7.2.14 If x is a loop, it must not be contained in any independent set. Otherwise $\{x\}$ is independent and can be extended to a base. (Observe that this is the dual of Exercise 7.2.4.)

7.2.15 A subset $X \subseteq S$ is independent if and only if it does not contain any circuit.

If the cut sets of a matroid are known, consider those subsets of S which are minimal with respect to the property that they intersect every cut set. These are just the bases of the matroid.

7.2.16 Let $x \in S$ be a nonloop element of the matroid $\mathbf{M} = (S, \mathbf{F})$. The *parallel extension* of x in $\mathbf{M}$ is a matroid over the set $S \cup \{y\}$ where $X \subseteq S \cup \{y\}$ is independent if and only if *either* $y \notin X$ and $X \in \mathbf{F}$ *or* $y \in X$ and then $x \notin X$ and $(X - \{y\}) \cup \{x\} \in \mathbf{F}$.

Let $x \in S$ be a nonbridge element of $\mathbf{M}$. Then the *series extension* of x in $\mathbf{M}$ is a matroid over $S \cup \{y\}$ where $X \subseteq S \cup \{y\}$ is independent if and only if *either* $|\{x, y\} \cap X| < 2$ and $X - \{x, y\} \in \mathbf{F}$ *or* $x, y \in X$ and $X - \{y\} \in \mathbf{F}$. Alternatively, the series extension of x in $\mathbf{M}$ could be defined as the dual of the parallel extension of x in $\mathbf{M}^*$.

7.2.17 Let $\mathbf{M}_1 = \mathbf{U}_{4,2}$ and $\mathbf{M}_2$ be the circuit matroid of the graph of Fig. S.7.3. $X = \{3, 4\}$ is a common base of them with an identical fundamental system of circuits.

7.2.18 If the statement is true then clearly $\mathbf{F} = \{X | T \subseteq X \text{ for no } T \in \mathbf{C}\}$. We have to prove (F1) through (F3) for this $\mathbf{F}$. (F1) and (F2) are trivial. Suppose that (F3) were false, that is, let $X, Y \in \mathbf{F}, |X| > |Y|$ so that $Y \cup \{x\}$ would contain a circuit $T_x \in \mathbf{C}$ for every $x \in X - Y$. Among all the possible counterexamples let us consider one with $|X \cap Y|$ maximal.

Observe that each T_x contains some elements of $Y - X$ (since X is circuit-free). Let $z \in T_x \cap (Y - X)$ and consider $Y' = (Y \cup \{x\}) - \{z\}$, see Fig.S.7.8. Since $|Y'| = |Y| < |X|$ and $|Y' \cap X| > |Y \cap X|$, the pair X, Y' is not a counterexample, so let $y \in X$ so that $Y' \cup \{y\} \in \mathbf{F}$. Then the set $(Y' \cup \{y\}) \cup \{z\}$ cannot contain more than one circuit and this circuit (namely T_x) does not contain y. Hence $Y \cup \{y\}$ is circuit-free, a contradiction.

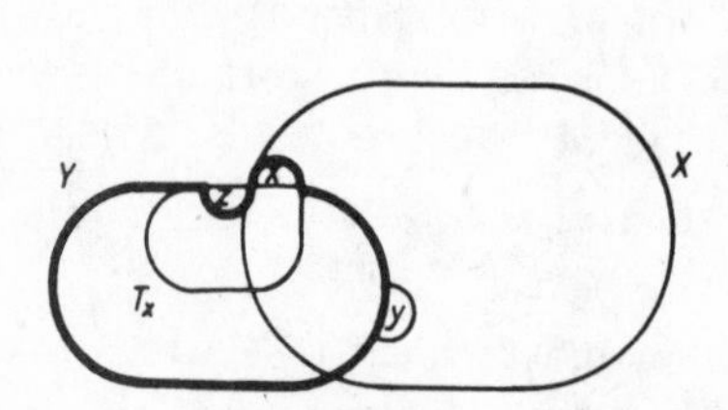

Fig. S.7.8

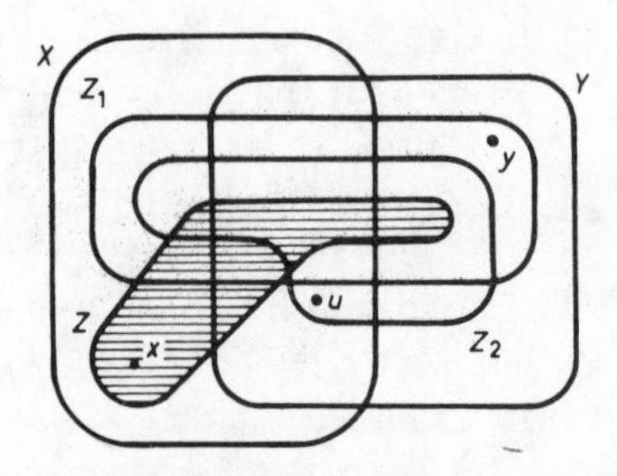

Fig. S.7.9

7.2.19 Suppose the contrary, that is, let X, Y be two intersecting circuits, $u \in$ $\in X \cap Y$, $x \in X - Y$ and suppose that there is no circuit Z satisfying $Z \subseteq (X \cup Y) - \{u\}$ and $x \in Z$. We also suppose that $|X \cup Y|$ is minimal among all counterexamples.

At least there exists a circuit $Z_1 \subseteq (X \cup Y) - \{u\}$ with $x \notin Z_1$, by (C2). Since $Z_1 \subseteq X$ is impossible, there exists a $y \in (Y \cap Z_1) - X$, see Fig. S.7.9. Since $|Z_1 \cup Y| < |X \cup Y|$, the pair Z_1, Y is not a counterexample. Hence there exists a circuit $Z_2 \subseteq (Z_1 \cup Y) - \{y\}$ with $u \in Z_2$. Finally, by $|Z_2 \cup X| < |X \cup Y|$ there exists a circuit $Z \subseteq (Z_2 \cup X) - \{u\}$ with $x \in Z$. But then the pair X, Y was not a counterexample, which is a contradiction.

7.2.20 Suppose the contrary, that is, let X, Y be two intersecting circuits and $x \in X - Y$, $y \in Y - X$ so that there exists no circuit in $X \cup Y$ containing both x and y. We suppose that $|X \cup Y|$ is minimal among all counterexamples.

Let $z \in X \cap Y$ and, by (C3), let Z_1 be a circuit in $(X \cup Y) - \{z\}$, containing y. We may suppose $Y \subset X \cup Z_1$ (see Fig. S.7.10) since otherwise $|Y \cup Z_1| < |X \cup Y|$ implies the existence of a circuit containing x and y.

Consider now a circuit $Z_2 \subset (X \cup Z_1) - \{z\}$ with $x \in Z_2$. Observe that $z \notin Z_1 \cup Z_2$ and $Z_1 \cap Z_2 \neq \emptyset$. (For this latter statement observe that $Z_2 \subset X$ is impossible and if $t \in (Y - X) \cap Z_2$ then $t \in Z_1 \cap Z_2$ since $Y \subset X \cup Z_1$.) Hence by $|Z_1 \cup Z_2| < |X \cup Y|$ we have a circuit Z in $Z_1 \cup Z_2$, containing x and y, a contradiction.

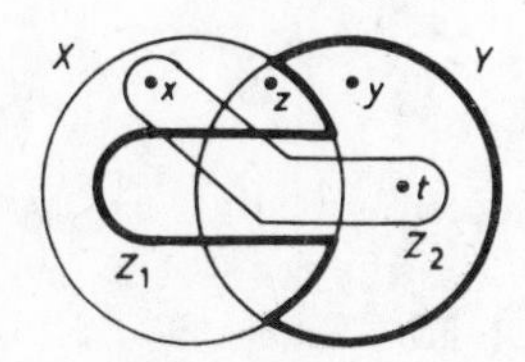

Fig. S.7.10

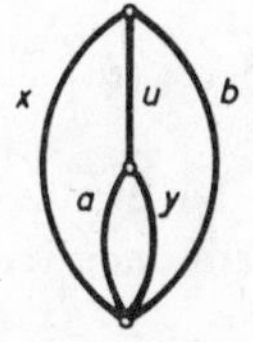

Fig. S.7.11

7.2.21 Consider the circuit matroid of the graph in Fig. S.7.11. If $X = \{a, x, u\}$ and $Y = \{b, y, u\}$ then the only circuit containing both x and y contains u as well.

7.2.22 If $\mathbf{N}$ has no circuit of cardinality $r(S)+1$ then define $\mathbf{M}$ as the truncation of $\mathbf{N}$. All the circuits of $\mathbf{N}$ will clearly remain circuits in $\mathbf{M}$ as well.

On the other hand, let $\mathbf{N}$ have a circuit of length $r(S) + 1$. It must remain a circuit of $\mathbf{M}$ as well. Hence the rank of $\mathbf{M}$ cannot be smaller than that of **N**. Neither can it be greater since then a base of **M** should contain a circuit of **N**. Then every independent set of **M** is independent in **N** as well (since $X \in \mathbf{M} - \mathbf{N}$ would imply that X contains a circuit of **N** which were a circuit of **M** as well). Now let Y be a circuit in **M** and independent in **N**, and let X be a base of **M** with $Y - X = \{y\}$. X must be a base of **N** as well, hence the only circuit of **N** in $X \cup \{y\}$ can be neither Y (by the choice of Y) nor something else (since it could not be a circuit of **M**), a contradiction.

7.2.23 Extend the independent set $X - \{y\}$ into a base B and consider the (unique) cut set Y contained in $(S - B) \cup \{x\}$. Clearly, $x \in Y$ and $|X \cap Y| \neq 1$ then implies that Y must contain other element(s) of X as well. But $(S - B) \cap X = \{y\}$ then gives $X \cap Y = \{x, y\}$.

7.2.24 Let $X \subseteq S$ and let X_0 be a maximal cut set free subset of X. Hence $|X_0| = r^*(X)$. Let, furthermore, A be a maximal circuit free subset of $S - X$, hence $|A| = r(S - X)$. Since $A \cap X_0 = \emptyset$, apply Exercise 7.2.10, so let Y be a base with $A \subseteq Y \subseteq S - X_0$, see Fig. S.7.12. The shaded areas in the figure are empty (by the maximalities of X_0 and A, respectively), hence $r^*(X) = |X| - |X - X_0| = |X| - |Y - A| = |X| - [r(S) - r(S - X)]$.

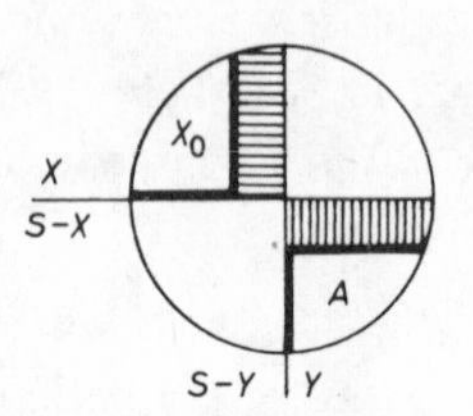

Fig. S.7.12

7.2.25 Let A_0 be a maximum size independent subset of A. If $A_0 \cup \{z\}$ is dependent then its only circuit X has the requested property. Otherwise extend $A_0 \cup \{z\}$ to a base K of **M**. Since $S - K$ is a base of $\mathbf{M}^*$, there is a unique cut set Y of **M** contained in $(S - K) \cup \{z\}$. This Y is disjoint to A since a common element could be exchanged by z otherwise, contradicting to the maximality of A_0.

A circuit X and a cut set Y with the required properties cannot simultaneously exist, see Exercise 7.2.11.

7.2.26 Let B_1, B_2 be two bases and $x \in B_1$. We may suppose $x \in B_1 - B_2$ since the statement is trivial otherwise, by $y = x$. Let $y_1, y_2, \ldots, y_k$ be those elements of the unique circuit C in $B_2 \cup \{x\}$, which are in $B_2 - B_1$. Each set $B_1 \cup \{y_i\}$ contains a unique circuit C_i. Clearly $k > 1$ (since $C \subseteq B_1$ is impossible) and if $x \in C_i$ then the choice $y = y_i$ will do.

Hence our indirect assumption is $x \notin C_i$ for every i. Since $y_k \in C \cap C_k$ and $x \in C - C_k$, there exists, by (C3), a circuit $C' \subset (C \cup C_k) - \{y_k\}$ with $x \in C'$. $C' \subseteq B_1$ is impossible, hence some of the elements $y_1, y_2, \ldots, y_{k-1}$ must belong to C'. For such a y_j let C'' be a circuit in $(C' \cup C_j) - \{y_j\}$ with $x \in C''$ etc. After some steps no y_i remains, hence a subset (which contains x and is, therefore, nonempty) is a circuit and a subset of B_1, a contradiction.

7.2.27 Let r^* denote the rank function of $\mathbf{M}^*$. Then the rank of a subset X in $\mathbf{N}^*$ is $\min [r^*(X), r^*(S) - 1]$; in particular, the rank of S is $|S| - r(S) - 1$. Applying Theorem 7.2.5 we obtain that the rank of a subset X in $\mathbf{N}$ is $\min (|X|, r(X) + 1)$.

Observe that this is the function $f(X) = \min (|X|, r(X) + c)$ of Problem 7.1.22 with $c = 1$. More generally, $f(X)$ is the solution of our problem if $\mathbf{M}^*$ is truncated c times before forming its dual.

7.2.28 Let $X \subseteq S$ be a cut set of $\mathbf{M}$. Then (a) is obvious and (b) was shown as Exercise 7.2.11. In order to prove (c) let $Y \subset X$ and $x \in X - Y$, $y \in Y$. By Problem 7.2.23 there exists a circuit C with $C \cap X = \{x, y\}$. Hence $|C \cap Y| = 1$ so Y cannot be a cut set.

On the other hand suppose that (a), (b) and (c) are met by a subset $X \subseteq S$. If X were cut set free, it could be extended to a base B of $\mathbf{M}^*$. Let $x \in X$ (such an x exists by (a)). Then the unique circuit of $(S - B) \cup \{x\}$ violates (b), a contradiction. Hence X contains a cut set Q and then, by (c), X must equal just Q itself.

7.3.1 Yes. Deletion or contraction of any element of $\mathbf{U}_{n,k}$ gives $\mathbf{U}_{n-1,k}$ and $\mathbf{U}_{n-1,k-1}$, respectively.

7.3.2 Let Y be a circuit of $\mathbf{M}$, intersecting both A and $S - A$. Then $Y \cap (S - A)$ is clearly dependent in $\mathbf{M}/A$; the minimal such subsets are just circuits.

7.3.3 Not necessarily. Let $\mathbf{M} = \mathbf{U}_{4,2}$ and $C = S - \{x\}$.

7.3.4 Let $x_1, x_2, \ldots, x_k \in S$ be series elements in the matroid $\mathbf{M} = (S, \mathbf{F})$. The contraction $\mathbf{M}/A$ is a *series contraction* of $\mathbf{M}$ if A is a proper subset of $\{x_1, x_2, \ldots, x_k\}$. For brevity, the result of several such series contractions (with different series classes) will also be called a series contraction of $\mathbf{M}$.

A subset $X \subseteq S - A$ is a base of $\mathbf{M}/A$ if and only if $A \cup X$ is a base of $\mathbf{M}$; a subset $Y \subseteq S - A$ is a circuit of $\mathbf{M}/A$ if and only if $A \cup Y$ is a circuit of $\mathbf{M}$.

7.3.5 Let $\mathbf{M}_1 = (S_1, \mathbf{F}_1)$ and $\mathbf{M}_2 = (S_2, \mathbf{F}_2)$ be two matroids and let $\mathbf{F}$ denote the collection of the independent subsets of their direct sum $\mathbf{M}_1 \oplus \mathbf{M}_2$. (F1) and (F2) are clearly satisfied by $\mathbf{F}$. In order to show (F3) let $X, Y \in \mathbf{F}$ and $|X| > |Y|$. Then $|X \cap S_i| > |Y \cap S_i|$ holds for at least one of the values $i = 1$ and $i = 2$. If, say, this holds for $i = 1$ then there is an element $x \in (X - Y) \cap S_1$ so that $(S_1 \cap Y) \cup \{x\} \in \mathbf{F}_1$. Then $Y \cup \{x\} \in \mathbf{F}$.

7.3.6 (a) Let $s_i = |S_i|$. Then $(S, \mathbf{F})$ is the direct sum of uniform matroids of the form $\mathbf{U}_{s_i, a_i}$.

(b) The circuit matroid $\mathbf{M}(G)$ of the graph G of Fig. S.7.3 is the smallest such example. G is 2-connected hence $\mathbf{M}(G)$ is connected but is not uniform.

(c) $\mathbf{V}_p$ is just the truncation of the partitional matroid $\mathbf{U}_{p,p-1} \oplus \mathbf{U}_{p,p-1}$.

7.3.7 A subset $X \subseteq S_1 \cup S_2 \cup \ldots \cup S_k$ is a base of $\mathbf{M}$ if $X \subseteq S_i$ is a base in $\mathbf{M}_i$ for every $i = 1, 2, \ldots, k$. The rank function $R(X)$ of $\mathbf{M}$ is obtained from the respective rank functions r_i of $\mathbf{M}_i$ by $R(X) = \sum_{i=1}^{k} r_i(X \cap S_i)$. Finally, Y is a circuit of $\mathbf{M}$ if and only if there exists a subscript $1 \leq i \leq k$ so that $Y \subseteq S_i$ and Y is a circuit of $\mathbf{M}_i$.

7.3.8 $\mathbf{M}^* = \mathbf{M}_1^* \oplus \mathbf{M}_2^* \oplus \ldots \oplus \mathbf{M}_k^*$ follows from the above characterization of the bases. Then a subset X is a cut set of $\mathbf{M}$ if and only if there exists a subscript $1 \leq i \leq k$ so that $X \subseteq S_i$ and X is a cut set of $\mathbf{M}_i$.

7.3.9 (1) $\rightarrow$ (4) follows from Exercise 7.3.7. (4) $\rightarrow$ (3) can be proved by considering the rank functions of $\mathbf{M} \backslash T$ and $\mathbf{M}/T$. (3) $\rightarrow$ (2) is obvious since if a circuit C intersects both T and $S - T$ then $C \cap (S - T)$ is independent in $\mathbf{M} \backslash T$ and dependent in $\mathbf{M}/T$. Finally (2) $\rightarrow$ (1) was shown in Statement 7.3.5.

7.3.10 The condition is necessary since X were a separator otherwise, see part (4) of the previous exercise. It is also sufficient for the same reason: If $\mathbf{M}$ were disconnected then choosing X to be one of its separators gives a contradiction.

7.3.11 This is trivial since, by Exercise 7.3.8, direct sum can be dualized for each summand separately.

7.3.12 If $X_0 \cup Y$ is independent in $\mathbf{M}$ then Y is independent in $\mathbf{M}/X$ since $R(Y) = r(X \cup Y) - r(X) \geq r(X_0 \cup Y) - r(X) = |X_0 \cup Y| - |X_0| =$ $= |Y| \geq R(Y)$.

On the other hand, if Y is independent in $\mathbf{M}/X$ then $|Y| = r(X \cup Y) -$ $- r(X)$. Hence if $X \cup Y$ has a circuit, it is fully contained in X. Thus, if $X_0 \subseteq X$ is independent then so is $X_0 \cup Y$ as well.

7.3.13 Let T_1, T_2 be two separators. Then $T = T_1 \cap T_2$ is also a separator (for if there were a circuit $C \subseteq T_1$ containing elements from both T and $T_2 - T$ then T_2 could not be a separator). Hence if we have different decompositions, consider their common refinement. The subsets of the finest decomposition (with respect to set intersection) are thus separators. Therefore they give connected direct summands in a unique way.

7.3.14 Suppose, indirectly, that both $\mathbf{M} \backslash \{x\}$ and $\mathbf{M}/\{x\}$ are disconnected, with separators T and U, respectively (which are nonempty and different from $S - \{x\}$). At first we show that no pair a, b of elements can arise in "opposite quarters", as in Fig. S.7.13; that is, a pair cannot be separated both in $\mathbf{M} \backslash \{x\}$ and in $\mathbf{M}/\{x\}$.

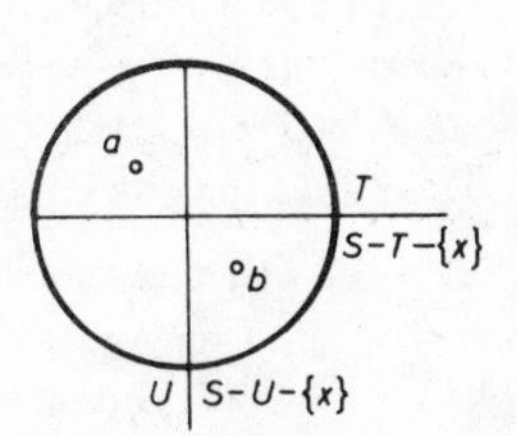

Fig. S.7.13

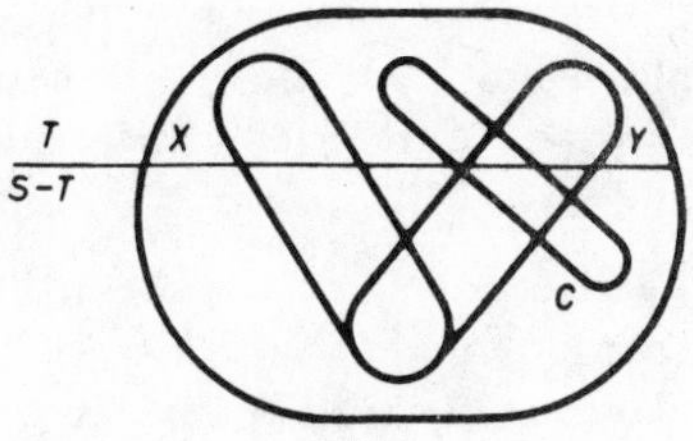

Fig. S.7.14

Namely, if $a, b \in S - \{x\}$ are in different components of $\mathbf{M}\backslash\{x\}$ then consider a circuit X of $\mathbf{M}$, containing both a and b. (Such an X exists since $\mathbf{M}$ is connected.) This X must contain x as well, hence $X - \{x\}$ is a circuit of $\mathbf{M}/\{x\}$.

Hence if b belongs to, say, $(S - \{x\} - U) \cap (S - \{x\} - T)$ then all the other elements of $S - \{x\}$ must be either in $(S - \{x\} - T) \cap U$ or in $(S - \{x\} - U) \cap T$. One of these subsets must be empty by the above reasoning, but then either T or U is empty, a contradiction.

7.3.15 Let $\mathbf{M}$ be the circuit matroid of the graph of Fig. S.7.3 and let $A = \{1, 3\}$. Then 2 is a loop in $\mathbf{M}/A$ and a bridge in $\mathbf{M}\backslash A$, hence both matroids are disconnected.

7.3.16 Let x, y be two elements of a connected matroid $\mathbf{M}$ and suppose that no circuit contains both x and y. Let X be the union of all the circuits containing x. Then $y \notin X$ by the indirect assumption.

X cannot be a separator of $\mathbf{M}$ hence there exists a circuit Y containing an element a of X and another element b of $S - X$. By the definition of X there exists a circuit Z containing both a and x. Hence, by Problem 7.2.20, there exists a circuit containing x and b, a contradiction.

7.3.17 If $S = S_1 \cup S_2$ is a partition of S and $x_1 \in S_1$, $x_2 \in S_2$ are not bridges then each is contained in a circuit. In the sequence of the circuits connecting them, there will be one intersecting both S_1 and S_2. Hence this condition is sufficient for connectivity.

7.3.18 Suppose the contrary, that is, let T be a separator of $\mathbf{M}\backslash(X \cap Y)$. Then $T \cap (X \cap Y) = \emptyset$ hence at least one of $T - X$ and $T - Y$ is nonempty. Let, say, $T - X \neq \emptyset$. $T - X$ could not be a separator of $\mathbf{M}\backslash X$ (since $\mathbf{M}\backslash X$ is connected) thus there exists a circuit C of $\mathbf{M}\backslash X$, intersecting both $T - X$ and $S - T - X$ (see Fig. S.7.14). But C is a circuit of $\mathbf{M}\backslash(X \cap Y)$ as well, hence T cannot be a separator of this latter matroid, a contradiction.

Alternatively, we may say that if $\mathbf{M}$ restricted to two sets are connected and these sets intersect each other then the restriction of $\mathbf{M}$ to their union is also connected.

7.4.1 (a) Let X be the fundamental cut set (with respect to B_m) containing x. If X had an element y with $w(y) > w(x)$ then the weight of $(B_m - \{x\}) \cup \{y\}$ would be greater than that of B_m, a contradiction.

(b) If x were the minimum weight element of a circuit C then the greedy algorithm would not ever choose x. Namely if every element of $C - \{x\}$ is in File 1 then adding x leads to a circuit; otherwise adding any element of $C - \{x\}$ is more advantageous than adding x.

7.4.2 (a) There exists a cut set where the weight of every other element is less than or equal to $w(x)$.

(b) There exists no circuit where the weight of every other element is greater than $w(x)$.

7.4.3 Let $X = \{x_1, x_2, \ldots, x_k\}$ be the output of our algorithm, its elements chosen in this order. X is clearly independent, even a base since its contraction leaves loops only. $w(x_1) \geq w(x_2) \geq \ldots \geq w(x_k)$ is also obvious.

Suppose there were a base Y with $w(X) < w(Y)$ and let the elements of Y be arranged so that $w(y_1) \geq w(y_2) \geq \ldots \geq w(y_k)$. Let i be the first subscript with $w(y_i) > w(x_i)$ and consider the element $y_j (j \leq i)$ which can be added to $\{x_1, \ldots, x_{i-1}\}$. This y_j cannot be a loop after contracting $\{x_1, \ldots, x_{i-1}\}$ and $w(y_j) > w(x_i)$ also holds, hence the choice of x_i was wrong.

7.4.4 Let S be the set of squares of the chessboard and define $\mathbf{F} = \{X | X \subseteq S$ and no two elements of S share the same row or the same column $\}$. By Theorem 7.4.2 we only have to prove that $(S, \mathbf{F})$ is not a matroid. Consider the subsets, corresponding to the two "diagonals". They are clearly maximum and violate (B3).

7.4.5 The greedy algorithm can be applied since the edge sets of the considered subgraphs are just the independent sets of a partitional matroid (defined in Exercise 7.3.6).

7.4.6 Let Q be a cut set of $\mathbf{M}$ and suppose that none of the maximum weight elements of Q is contained in B. Add such an element x to B. If C denotes the unique circuit of $B \cup \{x\}$ then $x \in C \cap Q$ implies the existence of at least one further element y in this intersection. Since $w(y) < w(x)$ must hold, the weight of $(B \cup \{x\}) - \{y\}$ is greater than that of B, a contradiction. The other statement can be obtained in a similar way.

7.4.7 (a) This is just a reformulation of the statements of the previous problem.

(b) Let X, X' and Y denote the subsets of elements in Files 1,2 and 3, respectively. If $X \cup Y$ is circuit-free then a subset of X' is a base of $\mathbf{M}^*$ and even this subset plus Y contains a cut set. Hence one of the alternatives is always met in Step 3, unless $Y = \emptyset$.

(c) At first we show that X is a base. If it contained a circuit C then consider that element of C which was joined to File 1 at last. This were the only common element of a cut set of $X' \cup Y$ and a circuit (namely C) which is clearly impossible, see Theorem 7.2.7(b). Thus X is circuit-free; should it be not a base, we could find a cut set in $S - X$ in the same way.

Suppose finally that X were not maximum. Let B be a maximum weight base; if $\mathbf{M}$ has several such bases then let $|X \cap B|$ be maximum among them. Consider an arbitrary $x \in X - B$. When it was placed to File 1, there was a cut set Q so that x was one of the maximum elements of Q. The only circuit C of $B \cup \{x\}$ contains at least one element y, different from x, satisfying $y \in C \cap Q$. If $w(x) > w(y)$ then the weight of $B' = (B \cup \{x\}) - \{y\}$ is greater than $w(B)$, contradicting to the maximality of B. Hence $w(x) = w(y)$, a contradiction again, since $|B' \cap X| > |B \cap X|$.

7.4.8 Consider the matroid $\mathbf{M}^p$ of Problem 7.1.15 for the circuit matroid $\mathbf{M}$ of the given graph with $p = k + 1$.

7.4.9 Let $\mathbf{F}$ and $\mathbf{B}$ denote the collections of the independent sets and the bases, respectively, of a matroid $\mathbf{M}$. If all the weights are nonnegative then $\max_{X \in \mathbf{B}} w(X)$ and $\max_{X \in \mathbf{F}} w(X)$ are the same. Otherwise the first quantity (which can still be obtained by Algorithm 7.1) may be smaller than the second one. In order to find a maximum weight independent set in a matroid (if the weights may be negative) modify Algorithm 7.1 by replacing Step 3 with the following one:

If $Y = \emptyset$, go to **End**. Otherwise choose an element $x \in Y$ with maximum weight (if there are several such elements, choose randomly among them). If $w(x) \leq 0$, go to **End**. If $w(x) > 0$, add x to File 1 and go to **Step 2.**

7.4.10 Let $S = \{a, b, c, d\}$ and $\mathbf{G} = \{\emptyset, \{a\}, \{c\}, \{a, b\}, \{a, c\}, \{c, d\}\}$. If the weights of a, b, c and d are 2,1,1 and 3, respectively, then the greedy algorithm finds $\{a, b\}$ or $\{a, c\}$ instead of the optimal $\{c, d\}$.

7.4.11 The condition is violated. Put $A = \emptyset$, $B = \{c\}$, $x = a$ and $y = d$.

Chapter 8

8.1.1 The equation of such a 1-port is $au+bi=c$ where at least one of a and b is nonzero. If $a \neq 0$ then there exists a Thévenin-equivalent with $U_0 = c/a$ and $R_0 = -b/a$. Otherwise $b \neq 0$ and there exists a Norton-equivalent with $I_0 = c/b$ and $G_0 = 0$.

8.1.2 We obtain a 2-port with two open circuits if $g = 0$, and that with two short circuits if $r = 0$. If c or d becomes zero, a 2-port with an open and a short circuit is obtained.

8.1.3 $\begin{pmatrix} -1 & 0 & 0 \\ 0 & -1 & 0 \\ 0 & 0 & -1 \end{pmatrix} \mathbf{u} + \begin{pmatrix} R_1 & 0 & 0 \\ 0 & R_2 & 0 \\ 0 & 0 & R_3 \end{pmatrix} \mathbf{i} = \mathbf{0}$ for the first 3-port. The coefficient of $\mathbf{i}$ changes to $\begin{pmatrix} R_1+R_2+R_3 & R_2+R_3 & R_3 \\ R_2+R_3 & R_2+R_3 & R_3 \\ R_3 & R_3 & R_3 \end{pmatrix}$ in the case of the second 3-port.

8.1.4 (a) If the 1-port is given by $au+bi=0$ and, say, $a \neq 0$ then substituting $u_1 = -(b/a)i_1$ and $u_2 = -(b/a)i_2$ gives $u_1 i_2 - i_1 u_2 = -(b/a)i_1 i_2 + $ $+(b/a)i_1 i_2 = 0$.

(b) Let $\mathbf{u}_j = \begin{pmatrix} u_{j1} \\ u_{j2} \end{pmatrix}$ and $\mathbf{i}_j = \begin{pmatrix} i_{j1} \\ i_{j2} \end{pmatrix}$ for $j = 1, 2$. Then $u_{j2} = k u_{j1}$ and $i_{j1} = -k i_{j2}$ gives $(u_{11}i_{21} + u_{12}i_{22}) - (u_{21}i_{11} + u_{22}i_{12}) = (-u_{11}k i_{22} +$ $+k u_{11} i_{22}) - (u_{21}(-k)i_{12} + k u_{21} i_{12}) = 0$.

(c) Since $\mathbf{A} = -\mathbf{E}$, we have $\mathbf{u}_j = \mathbf{B}\mathbf{i}_j$. Thus $\mathbf{u}_1^T \mathbf{i}_2 - \mathbf{u}_2^T \mathbf{i}_1 = \mathbf{i}_1^T \mathbf{B}^T \mathbf{i}_2 -$ $-\mathbf{i}_2^T \mathbf{B}^T \mathbf{i}_1 = \mathbf{i}_1^T \mathbf{B}^T \mathbf{i}_2 - \mathbf{i}_1^T \mathbf{B} \mathbf{i}_2 = \mathbf{i}_1^T (\mathbf{B}^T - \mathbf{B}) \mathbf{i}_2$, and this latter vanishes if $\mathbf{B}$ is symmetric.

8.1.5 (a) Using the same notation as in part (b) of the previous exercise,

$$(u_{11}i_{21} + u_{12}i_{22}) - (u_{21}i_{11} + u_{22}i_{12}) =$$

$$= (-Ri_{12}i_{21} + Ri_{11}i_{22}) - (-Ri_{22}i_{11} + Ri_{21}i_{12}) = 2R(i_{11}i_{22} - i_{21}i_{12})$$

and this need not be zero if $\mathbf{i}_1, \mathbf{i}_2$ are arbitrary.

(b) In the case of the voltage controlled current source $i_1 = 0$, $i_2 = g u_1$ we have $\mathbf{u}_1^T \mathbf{i}_2 - \mathbf{u}_2^T \mathbf{i}_1 = (u_{11} \cdot 0 + u_{12} \cdot g u_{21}) - (u_{21} \cdot 0 + u_{22} \cdot g u_{11}) =$ $= g(u_{12}u_{21} - u_{11}u_{22})$, which need not vanish if u_{j2} is arbitrary. The solutions for the other three cases are similar.

(c) Substituting $u_j = (a_j + b_j)/2$ and $i_j = (a_j - b_j)/2$, a longer calculation gives $\mathbf{u}_1^T \mathbf{i}_2 - \mathbf{u}_2^T \mathbf{i}_1 = -\begin{vmatrix} b_{11} & b_{12} & b_{13} \\ b_{21} & b_{22} & b_{23} \\ 1 & 1 & 1 \end{vmatrix}$, which need not be zero.

(A shorter solution can be obtained by the combination of Exercise 10.1.11 and Problem 10.1.16.)

8.1.6 The short circuit $u = 0$ and the open circuit $i = 0$ are lossless. If the 1-port is given by $au + bi = 0$, $a \neq 0$, $b \neq 0$ then it is passive if and only if $b/a < 0$. Hence the "real" (positive) resistor is passive, the negative one is active.

8.1.7 $u_1 i_1 + u_2 i_2 = u_1(-k i_2) + k u_1 i_2 = 0$ and $-R i_2 i_1 + R i_1 i_2 = 0$ prove the first two assertions.

The controlled sources are active. For example, the voltage controlled current source has $i_1 = 0$, $i_2 = g u_1$, leading to $u_1 i_1 + u_2 i_2 = g u_1 u_2$, and the sign of this latter cannot be determined since u_2 is arbitrary.

The circulator is lossless since, by the same substitution as in Exercise 8.1.5(c), $\mathbf{u}^T \mathbf{i} = \frac{1}{4}\sum(a_j^2 - b_j^2) = \frac{1}{4}(a_1^2 + a_2^2 + a_3^2 - b_1^2 - b_2^2 - b_3^2) = 0$.

8.1.8 $L_1 \geq 0$, $L_2 \geq 0$ and $M \leq \sqrt{L_1 L_2}$ are required. The necessity of these relations can be obtained by putting $i_2 = 0$, $i_1 = 0$ and $i_1 = -M i_2 / L_1$, respectively. In order to see their sufficiency consider the equations in the solution of Problem 6.2.1. They give $\mathbf{u}^T \mathbf{i} = \frac{1}{2}\frac{d}{dt}(L_1 i_1^2 + 2M i_1 i_2 + L_2 i_2^2)$ and its integral is positive if the matrix $\begin{pmatrix} L_1 & M \\ M & L_2 \end{pmatrix}$ is positive definite.

8.1.9 A 1-port with equation $au + bi = 0$ has an impedance or admittance description if and only if $a \neq 0$ and $b \neq 0$, respectively. The ideal transformer has neither an impedance nor an admittance description, the gyrator has both descriptions. Among the controlled sources, only the current controlled voltage source has an impedance description and only the voltage controlled current source has an admittance description. Finally, the 3-port circulator has an admittance description with $\mathbf{Y} = \begin{pmatrix} 0 & 1 & -1 \\ -1 & 0 & 1 \\ 1 & -1 & 0 \end{pmatrix}$ but no impedance description.

8.1.10 This 2-port has no impedance description, it has an admittance description with $\mathbf{Y} = \begin{pmatrix} 1/R & 0 \\ a/R & 0 \end{pmatrix}$. The voltage of the second port cannot be determined unless u_2 is part of the input. Hence the admissible inputs are $\{u_1, u_2\}$ and $\{i_1, i_2\}$.

8.1.11 Using the matrix in the solution of Exercise 8.1.9 one can readily see that the inputs $\{u_2, u_3, i_1\}, \{u_1, u_3, i_2\}, \{u_1, u_2, i_3\}$ and $\{i_1, i_2, i_3\}$ are the nonadmissible ones. Their complements are circuits and those of the other triples are bases in the matroid $\mathbf{M}_C$ of the circulator C. Hence $\mathbf{M}_C$ is the circuit matroid of the graph of Fig. S.8.1.

8.1.12 Let $u_1 = a i_1 + b i_2$, $u_2 = c i_1 + d i_2$ with a, b, c, d all different from zero and $\begin{vmatrix} a & b \\ c & d \end{vmatrix} \neq 0$. The matroid of this 2-port is clearly isomorphic to $\mathbf{U}_{4,2}$.

8.1.13 $\{u_1, i_1\}$ is a base in the matroid of the ideal transformer, yet $\{u_2, i_2\}$ is not an "input", according to our definition.

8.1.14 The subnetworks are redrawn as 2-ports in Fig. S.8.2; instead of the resistors A, B and C, their reciprocal values α, β and γ, respectively, are shown as admittances. (These 2-ports are often called T- and Π-structures, respectively.) The impedance description of the

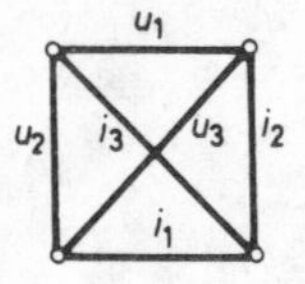

Fig. S.8.1

first one and the admittance description of the second one can easily be obtained: $\begin{pmatrix} u_{12} \\ u_{13} \end{pmatrix} = \begin{pmatrix} a+b & a \\ a & a+c \end{pmatrix} \begin{pmatrix} i_{12} \\ i_{13} \end{pmatrix}$ and $\begin{pmatrix} i_{12} \\ i_{13} \end{pmatrix} = = \begin{pmatrix} \alpha+\gamma & -\alpha \\ -\alpha & \alpha+\beta \end{pmatrix} \begin{pmatrix} u_{12} \\ u_{13} \end{pmatrix}$. Hence all what one has to verify is that these two matrices are inverses of each other if and only if the given relations among the numbers a, b, c and A, B, C are met.

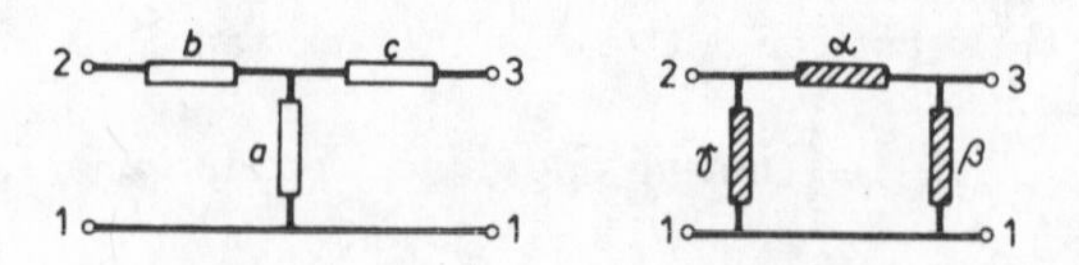

Fig. S.8.2

8.1.15 Using a Δ-connection one can obtain an open circuit between points 2 and 3 (at a given frequency): Let, for example, A be a capacitor of value C_0; B and C be two inductors of value $L_0/2$ each; and put a voltage source with voltage $U(t) = U_0 \sin \omega t$ between points 2 and 3, as an input, with $\omega = 1/\sqrt{L_0 C_0}$. Then the current of the voltage source is zero. However, if there is no input at point 1 then an open circuit between 2 and 3 can be realized with Y-connection at every frequency only.

8.1.16 Suppose that the multiport has a **Z**-description. Then $\mathbf{u}_j = \mathbf{Z}\mathbf{i}_j$ holds for every admissible pair $\mathbf{u}_j, \mathbf{i}_j$. Hence reciprocity means that $0 = \mathbf{u}_1^T \mathbf{i}_2 - -\mathbf{u}_2^T \mathbf{i}_1 = \mathbf{i}_1^T \mathbf{Z}^T \mathbf{i}_2 - \mathbf{i}_2^T \mathbf{Z}^T \mathbf{i}_1 = (\mathbf{Z}\mathbf{i}_1)^T \mathbf{i}_2 - \mathbf{i}_2 \mathbf{Z}^T \mathbf{i}_1 = \mathbf{i}_2^T (\mathbf{Z} - \mathbf{Z}^T)\mathbf{i}_1$ holds for every pair $\mathbf{i}_1, \mathbf{i}_2$, that is, **Z** must be symmetric.

Similarly, passivity means that **Z** is positive definite and if the multiport is lossless then **Z** is antisymmetric (that is, $\mathbf{Z}^T = -\mathbf{Z}$).

If the admittance description exists, the same properties are required for the **Y** matrix.

8.1.17 If the 2-port has an impedance description then $\mathbf{Z}$ must be of form $\begin{pmatrix} 0 & a \\ -a & 0 \end{pmatrix}$ by the previous problem. This is a gyrator if $a \neq 0$ and a pair of short circuits if $a = 0$. If the admittance description exists, we obtain the gyrator again, and a pair of open circuits.

If neither the impedance nor the admittance matrix exists then $\mathbf{Au} + \mathbf{Bi} = \mathbf{0}$ satisfies not only $r(\mathbf{A}|\mathbf{B}) = 2$ but $r(\mathbf{A}) = r(\mathbf{B}) = 1$ as well. Hence this system of equations reduces to $au_1 + bu_2 = 0$, $ci_1 + di_2 = 0$; where at most one of a and b and at most one of c and d vanishes, and where $ac + bd = 0$ since the 2-port is lossless. If none of a, b, c and d is zero then we obtain the ideal transformer, otherwise a 2-port with one open and one short circuit.

8.1.18 The system $u_1 = Ri_1$, $i_2 = ai_1$ of equations can be imagined like a current source controlled by the current of a resistor R. This can be realized by controlled sources in two different ways, see Fig. S.8.3.

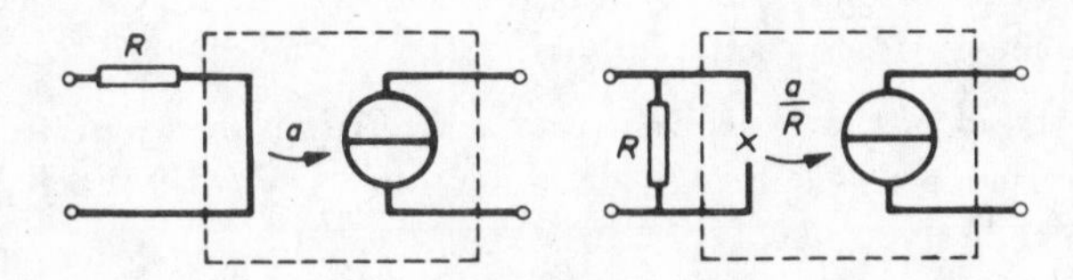

Fig. S.8.3

8.1.19 Put a voltage source U_3 parallel to a resistor r. This can also be drawn in the unnecessarily complicated way of Fig. S.8.4 as well, with the control equation $u_2 = ri_1$. This has unique solution, in spite of the circuit formed by voltage sources.

This network does not violate Statement 2.1.1. The circuit contains a controlled voltage source as well, its voltage is somehow implied by the rest so that the sum of the voltages in the circuit is always zero. Furthermore, the current of one of the voltage sources controls something, hence its value is not arbitrary.

Observe that what we call controlled sources are just models, there is no causality type connection between the left and right hand sides of a controlled source.

8.1.20 This is now obvious, by the solution of the previous problem. Should, for example, independent and controlled voltage sources form a circuit so that their currents do not control anything then the current flowing in this circuit could not uniquely be determined.

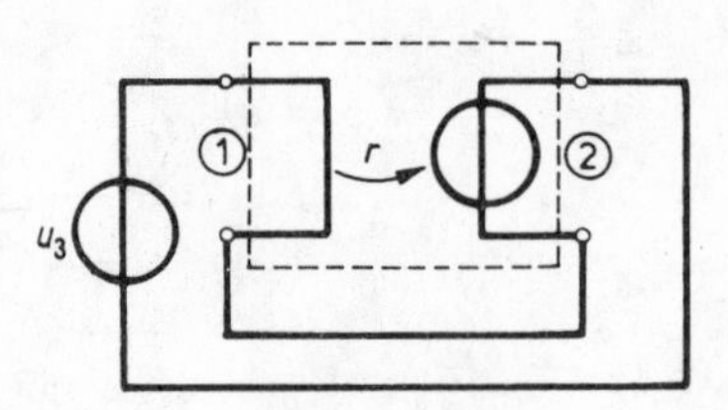

Fig. S.8.4

8.1.21 If $\mathbf{u}_1^T\mathbf{i}_2 + \mathbf{u}_2^T\mathbf{i}_1 = 0$ holds for every pair $(\mathbf{u}_1, \mathbf{i}_1)$ and $(\mathbf{u}_2, \mathbf{i}_2)$ then put $\mathbf{u}_1 = \mathbf{u}_2$ and $\mathbf{i}_1 = \mathbf{i}_2$. This special case directly gives that the multiport is lossless.

On the other hand, let $\mathbf{A}, \mathbf{B}$ describe a multiport, that is, let $\mathbf{A}\mathbf{u}_k + \mathbf{B}\mathbf{i}_k = =\mathbf{0}$, $k = 1, 2$ hold for the admissible pairs $(\mathbf{u}_1, \mathbf{i}_1)$ and $(\mathbf{u}_2, \mathbf{i}_2)$. Then their sum $(\mathbf{u}_1 + \mathbf{u}_2, \mathbf{i}_1 + \mathbf{i}_2)$ is also admissible. If the multiport is lossless, we have $\mathbf{u}_k^T\mathbf{i}_k = 0$ for $k = 1, 2$ and similarly, $(\mathbf{u}_1 + \mathbf{u}_2)^T(\mathbf{i}_1 + \mathbf{i}_2) = =\mathbf{u}_1^T\mathbf{i}_1 + \mathbf{u}_1^T\mathbf{i}_2 + \mathbf{u}_2^T\mathbf{i}_1 + \mathbf{u}_2^T\mathbf{i}_2 = 0$ holds. Since the first and last members vanish, the antireciprocity follows.

8.1.22 The matroid of the resistive 1-port $u = Ri$ is isomorphic to $\mathbf{U}_{2,1}$ unless $R = 0$. Hence the sign of R does not play any role; the 1-port may be passive or active alike.

Similarly, let a 2-port N have an impedance description $\mathbf{Z} = \begin{pmatrix} a & b \\ b & c \end{pmatrix}$. If $a, b, c \neq 0$ and $ac \neq b^2$ then $\mathbf{M}_N \cong \mathbf{U}_{4,2}$, see Exercise 8.1.12 as well. The matroid remains the same if one of the b's slightly changes, yet reciprocity is destroyed.

Finally, one cannot decide from $\mathbf{M}_N$ whether N is lossless. For example, replace an ideal transformer by a 2-port with $u_1 = ku_2$, $i_1 = mi_2$ and $km \neq -1$.

8.2.1 See Fig. S.8.5.

8.2.2 $\dot{x}_i = 0$ means that the ith joint can move only in a plane S parallel to that of the coordinate axes y and z. We may say that the joint is now not on a track but on plane S, hence it still can move in 2 dimensions.

8.2.3 A track now means that two of the coordinates of the joint are fixed. Using the terminology of the previous exercise we can also say that we place the joint to two perpendicular planes, their line of intersection is the track.

8.2.4 $\mathbf{M}_F$ is a rank 3 matroid, its circuits are those of $\mathbf{V}_3$ (that is, $\{\dot{x}_1, \dot{x}_2, \dot{x}_3\}$ and $\{\dot{y}_1, \dot{y}_2, \dot{y}_3\}$) plus $\{\dot{x}_3, \dot{y}_i, \dot{y}_j\}$ for every $i, j \in \{1, 2, 3\}$, $i \neq j$. On the

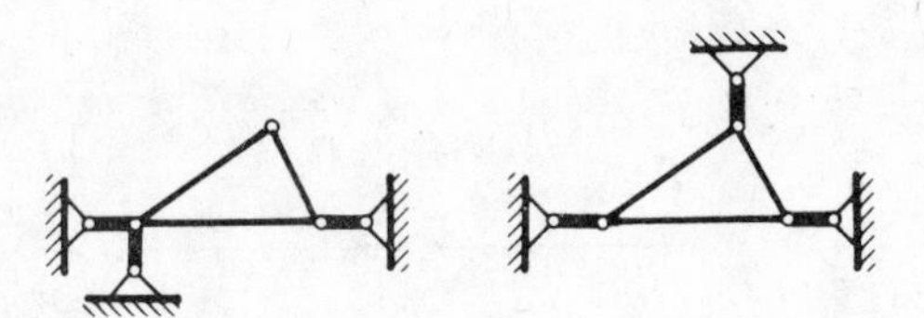

Fig. S.8.5

other hand, the elements $\dot{x}_3$ and $\dot{y}_2$ are parallel in $\mathbf{M}_G$, and any of them forms a circuit with either $\{\dot{x}_1, \dot{x}_2\}$ or $\{\dot{y}_1, \dot{y}_3\}$.

The affine representation of both matroids is shown in Fig. S.8.6. The latter happens to be graphic, see the graph of Fig. S.8.7.

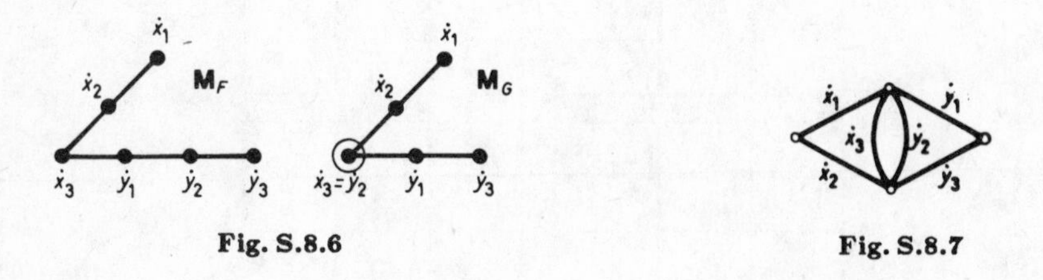

Fig. S.8.6 **Fig. S.8.7**

8.2.5 Fix the first joint and place the second one to a horizontal track (Fig.S.8.8). Then F is fixed but G still has an infinitesimal motion.

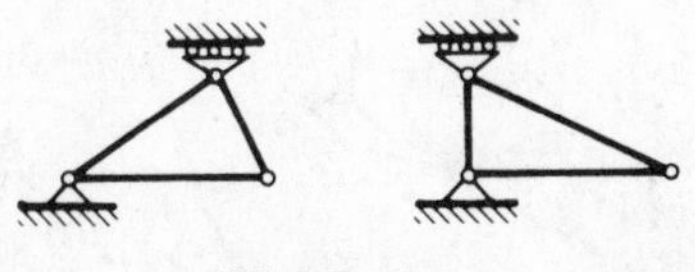

Fig. S.8.8

8.2.6 We saw the difference between placing a joint to a track or to a plane (see Exercises 8.2.2–3). If we introduce a new rod, fixing one end of it to a joint and the other to a point of the space then this corresponds to placing the joint to a plane. If the joint should be placed to a track then two additional rods must be introduced.

8.2.7 Let $(\dot{x}_i, \dot{y}_i, \dot{z}_i)$ be the velocity vector of joint i ($i = 1, 2, 3$). In the arrangement

$$\begin{array}{ccc} \dot{x}_1 & \dot{x}_2 & \dot{x}_3 \\ \dot{y}_1 & \dot{y}_2 & \dot{y}_3 \\ \dot{z}_1 & \dot{z}_2 & \dot{z}_3 \end{array}$$

of these nine quantities each column corresponds to a joint, each row to a direction. Since rigidity means $r(\mathbf{A}) = 3v - 6 = 3$, we should fix 6 of these 9 elements. The 3 "remaining" elements must not form a column (if we fix two joints, the third one can still rotate around the line of the former two), neither a row (the whole framework could still be translated into the "free" direction). Hence there are only 4 essentially different arrangements of these 3 elements: two of them may be in a common row and/or in a common column, as shown in the diagram below.

X	X	
X		

X	X	
		X

X		
X		
	X	

X		
	X	
		X

The respective realizations are shown in Fig. S.8.9 (with tracks and planes) and in Fig. S.8.10 (with introducing new rods).

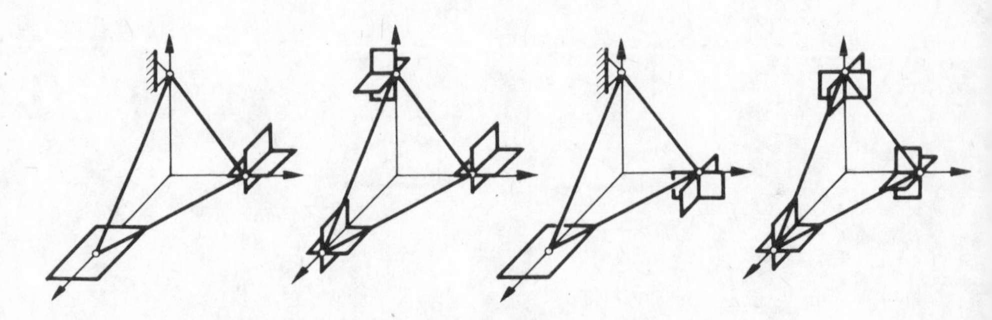

Fig. S.8.9

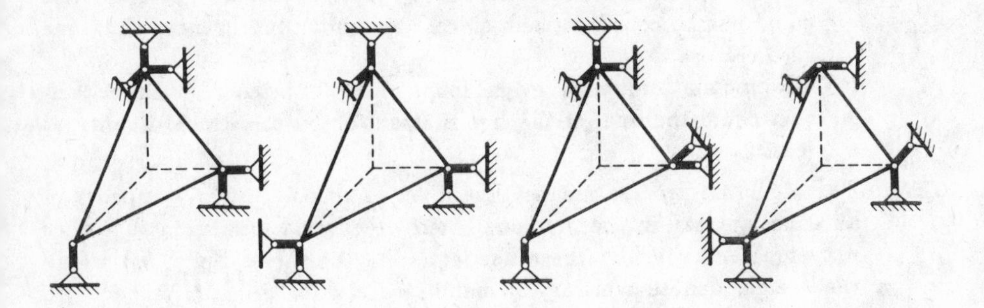

Fig. S.8.10

8.2.8 The matroids $\mathbf{V}_3, \mathbf{M}_F$ and $\mathbf{M}_G$ (in this order) are less and less "free", that is, if a subset of a matroid is independent, it is also independent in the previous matroid(-s) as well, but not necessarily *vice versa.* $\mathbf{M}_G$ is graphic (Fig. S.8.7), the other two are not (contract $\dot{x}_1$ and delete $\dot{x}_2$ to obtain $\mathbf{U}_{4,2}$).

Chapter 9

9.1.1 (a) One-one correspondence would imply that matroids with several loops could not be coordinatized, since V_0 cannot contain more than one zero vectors.

(b) Graphic matroids with several loops could not be coordinatized. Neither could the dual of the free matroid on a set with cardinality at least 2.

9.1.2 The algebraic process means that if an unknown (say x_1) appears in an equation (say in the first one) then we eliminate x_1 from the other equations and consider these new equations only (see Fig. S.9.1 where the + signs denote arbitrary elements).

Since x was not a loop, it can be extended to a base. Hence $\mathbf{M}$ can easily be coordinatized by a matrix of form $\mathbf{A}_1$. We prove that $\mathbf{A}_2$ represents $\mathbf{M}/\{x\}$.

If a subset X of column vectors of $\mathbf{A}_2$ is linearly independent then adding the unit vector, corresponding to x, leads to a set of independent vectors again (the dimension of the subspace increased by one). If X was dependent then the same linear combination of the "longer" vectors gives either zero or the constant multiple of the unit vector corresponding to x. Hence either X itself or $X \cup \{x\}$ determine a set of dependent vectors of $\mathbf{A}_1$.

9.1.3 $\mathbf{U}_{4,3}$ is graphic, hence it is representable over $\mathbb{B}$ as well. Its truncation $\mathbf{U}_{4,2}$ is nonbinary.

9.1.4 Let G be a bipartite graph and assign such a vector to an edge $\{v_i, v_j\}$ whose i^{th} and j^{th} entries are 1 and all the other entries are 0. This coordinatization will also do since every circuit of G contains an even number of edges hence a linear combination of their vectors with alternating $+1$ and -1 coefficients gives the zero vector.

9.1.5 No. $\begin{pmatrix} 1 & 0 & 1 \\ 0 & 1 & 1 \end{pmatrix}$ is a proper representation of the graph consisting of a single circuit of length three.

9.1.6 A nonternary matroid $\mathbf{M} = (S, \mathbf{F})$ must satisfy $|S| \geq 5$ since we saw in the solution of Exercise 7.1.8 that $|S| \leq 4$ implies binarity (in fact graphicness for every matroid except $\mathbf{U}_{4,2}$). It is a natural idea to try $\mathbf{U}_{5,2}$ with a representation $\begin{pmatrix} 1 & 0 & x & x & x \\ 0 & 1 & x & x & x \end{pmatrix}$. For the last three columns we may try $\begin{pmatrix} 1 \\ 1 \end{pmatrix}, \begin{pmatrix} 2 \\ 2 \end{pmatrix}, \begin{pmatrix} 1 \\ 2 \end{pmatrix}$ and $\begin{pmatrix} 2 \\ 1 \end{pmatrix}$. The first two vectors are clearly parallel and so are the last two vectors as well (since $2 \cdot 2 = 1$ in GF(3)). Hence $\mathbf{U}_{5,2}$ (and $\mathbf{U}_{5,3}$) are not ternary.

9.1.7 If $n \geq 4$ and $2 \leq k \leq n-2$ then $\mathbf{U}_{4,2}$ appears as a minor of $\mathbf{U}_{n,k}$ by deleting $n-k-2$ elements and contracting $k-2$ elements, cf. Exercise 7.3.1. Hence $\mathbf{U}_{n,k}$ cannot be binary.

9.1.8 Let the $0-1$ matrix $\mathbf{A}_1$ coordinatize $\mathbf{M}$ over $\mathbb{B}$ and let $\mathbf{a}_y$ denote the column vector of $\mathbf{A}_1$ corresponding to $y \in S$. If C is a circuit then $\sum_{y\in C} \mathbf{a}_y = \mathbf{0}$ since every nonzero coefficient in a linear combination over $\mathbb{B}$ is 1. Consider now the representation of $\mathbf{M}/\{x\}$ over $\mathbb{B}$ by $\mathbf{A}_2$ (see Fig. S.9.1 for Exercise 9.1.2). The sum of the C-columns of $\mathbf{A}_2$ will be $\mathbf{0}$ as well. Hence if a subset C_1 of C is a circuit of $\mathbf{M}/\{x\}$ then the sum of the remaining columns of $\mathbf{A}_2$ must be still $\mathbf{0}$. Thus $C - C_1$ is either a circuit or contains another circuit C_2 of $\mathbf{M}/\{x\}$ etc. However, the total number of disjoint circuits in C cannot be greater than 2 since the rank of C cannot decrease by more than one after contracting a single element.

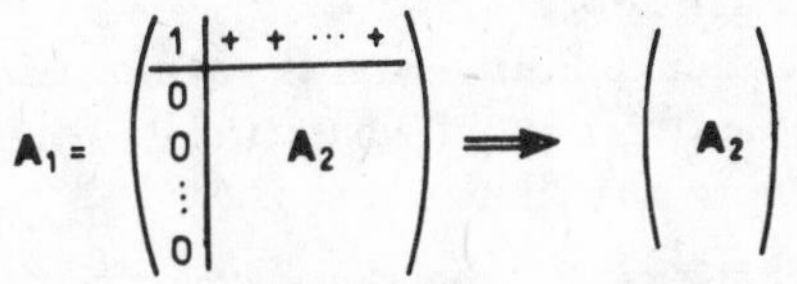

Fig. S.9.1

9.1.9 $X_i = \{a, b, c_i\}$ are circuits for $i = 1, 2$. Hence if $\mathbf{M}$ is a matroid then $(X_1 \cup X_2) - \{a\} = \{b, c_1, c_2\}$ must contain a circuit X. But X is a proper subset of the circuit $S - \{a\}$ if $k > 2$, a contradiction. If $k = 2$ then $\mathbf{M}$ is just $\mathbf{U}_{4,2}$.

9.1.10 Let C be a circuit and Q be a cut set in the binary matroid $\mathbf{M}$. Choose an element $x \in C$ and extend $C - \{x\}$ to a base B. Represent $\mathbf{M}$ by a $0-1$ matrix $\mathbf{A}$ where the elements of B correspond to the unity matrix. The first matrix of Fig. S.9.2 shows this situation: the columns of $B - C$ are listed first, those of $C - \{x\}$ thereafter, and the elements of $S - B$ are listed after the heavy vertical line, starting with x. Then the other matrix of the figure represents $\mathbf{M}^*$. The sum of the Q-columns in this latter matrix is $\mathbf{0}$ (since Q is a cut set of $\mathbf{M}^*$), hence the number of those Q-columns which have a one entry in a given row is even. In particular, the ones of the first row correspond to C hence $|Q \cap C|$ is even.

9.1.11 Let $\mathbf{A}$ represent $\mathbf{M}$ over $\mathbb{B}$ and denote the column corresponding to $x \in S$ by $\mathbf{a}_x$. If $x \in C_1 \cap C_2$ then $\mathbf{a}_x = \sum\{\mathbf{a}_y | y \in C_1 - \{x\}\} = \sum\{\mathbf{a}_y | y \in C_2 - \{x\}\}$. The difference of these two sums, that is, $\mathbf{0} = \sum\{\mathbf{a}_y | y \in C_1 \triangle C_2\}$ is a nontrivial linear combination, hence $C_1 \triangle C_2$ is dependent.

9.1.12 Let C be a circuit, B be a base and let $C - B = \{x_1, x_2, \ldots, x_k\}$. Consider the symmetric difference $X = C_1 \triangle C_2 \triangle \ldots \triangle C_k$ of the fundamental circuits with respect to B. Clearly $x_i \in X$ for every i since $x_i \in C_j$ for

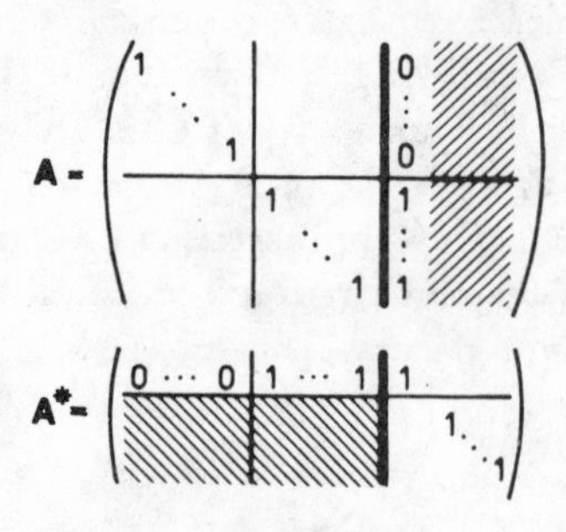

Fig. S.9.2

$i = j$ only. Consider now $Y = X \triangle C$. $x_i \in X \cap C$ implies $Y \subseteq B$. B is circuit-free, hence our condition that the symmetric difference of circuits is the disjoint union of a set of *some* circuits can be met for Y only if this set is the empty one. Hence $X = C$.

9.1.13 Possible loops of the rank 2 matroid $\mathbf{M}$ can be disregarded (and some zero vectors added to the representation at the end). Similarly, if a, b are parallel elements of $\mathbf{M}$ then it is enough to represent $\mathbf{M}\backslash\{a\}$ and assign thereafter the same vector to a as to b. Hence $\mathbf{M}$ may be supposed to be simple and then isomorphic to the uniform matroid $\mathbf{U}_{n,2}$ for some n. If our field is of sufficiently large cardinality then the vectors $\begin{pmatrix}1\\0\end{pmatrix}, \begin{pmatrix}1\\1\end{pmatrix}, \dots \begin{pmatrix}1\\n-1\end{pmatrix}$ are all different and coordinatize $\mathbf{U}_{n,2}$.

9.1.14 Such a field is obtained on the set $\{0, 1, a, b\}$ if operations are defined as

$+$	0	1	a	b
0	0	1	a	b
1	1	0	b	a
a	a	b	0	1
b	b	a	1	0

$\cdot$	0	1	a	b
0	0	0	0	0
1	0	1	a	b
a	0	a	b	1
b	0	b	1	a

$\mathbf{U}_{4,2}$ is represented over this field by $\begin{pmatrix}1 & 0 & 1 & 1\\0 & 1 & 1 & a\end{pmatrix}$.

9.1.15 Suppose indirectly that C is an odd circuit in $\mathbf{N} = \mathbf{M}/Q$. Then $C' = C \cup X$ were a circuit of $\mathbf{M}$ for a suitable subset X of Q. Then $|X|$ is odd since $\mathbf{M}$ is bipartite, but $|C' \cap Q| = |C|$ were then odd, contradicting to the binarity of $\mathbf{M}$.

9.1.16 If $\mathbf{M}$ is binary and S is the union of disjoint circuits then every circuit X of $\mathbf{M}^*$ (that is, every cut set of $\mathbf{M}$) intersects all these circuits by an even number of elements. Hence $|X|$ is the sum of even numbers, so it is even itself.

On the other hand, let $\mathbf{M}^*$ be bipartite. If C_1 is one of its cut sets then $\mathbf{M}^*/C_1$ is also bipartite by the previous problem; if C_2 is a cut set of this latter then $(\mathbf{M}^*/C_1)/C_2$ is bipartite again etc. Since C_2 is a circuit not only in $\mathbf{M}\backslash C_1$ but in $\mathbf{M}$ as well, S is decomposed into $C_1 \cup C_2 \cup \ldots$ as requested. (We applied that one can find a cut set in $\mathbf{M}^*$, in $\mathbf{M}^*/C_1$ etc. The only matroids without cut sets are those consisting of loops only, but they are not bipartite.)

9.1.17 (a) Suppose that $\mathbf{M}$ (and its minors) have no minors isomorphic to $\mathbf{U}_{4,2}$. Let C_1, C_2 be two intersecting circuits and suppose that $C_1 \Delta C_2$ were independent. Among all the counterexamples choose one where $k = |C_1 \cap C_2|$ is minimal. Restrict $\mathbf{M}$ to $C_1 \cup C_2$ and if $C_1 - C_2$ or $C_2 - C_1$ is of cardinality greater than 1 then contract all but one elements from these series classes. The result is a k-element set $X = C_1 \cap C_2$ and two further elements y_1, y_2 so that $C_1' = X \cup \{y_1\}$ and $C_2' = X \cup \{y_2\}$ are circuits and $\{y_1, y_2\}$ is independent.
Clearly, $k > 1$ (see Exercise 7.2.1). If $\{x, y_1, y_2\}$ were independent for some $x \in X$ then contracting x would lead to a smaller counterexample. Otherwise all these subsets $\{x, y_1, y_2\}$ are circuits. Applying Exercise 9.1.9 this implies $k = 2$ and the existence of a $\mathbf{U}_{4,2}$ minor, a contradiction.

(b) Let T be the symmetric difference of the circuits $C_1, C_2, \ldots$. At first we show that if T is nonempty then it is dependent. We have seen this for two circuits. Let k be the smallest integer so that $T = C_1 \Delta C_2 \Delta \ldots \Delta C_k$ were independent. Consider $T' = C_1 \Delta C_2 \Delta \ldots \Delta C_{k-1}$. This is dependent hence it contains a circuit C. Denoting $T' - C$ by D we have $T = C_k \Delta (C \cup D) = (C_k \Delta C) \cup (C_k \Delta D)$. Hence T is the union of two disjoint sets and the first one alone is dependent already, a contradiction.
So T is dependent, hence it contains a circuit C'. Applying the same argument to $T \Delta C' = T - C'$ we either obtain an empty set or a dependent one, containing a circuit C'' etc.

(c) Let $B = \{b_1, b_2, \ldots, b_r\}$ be a base of $\mathbf{M}$ and let $S - B = \{e_1, e_2, \ldots, e_q\}$. Define the matrix $\mathbf{A} = (a_{ij} | i = 1, 2, \ldots, r; j = 1, 2, \ldots, q)$ by

$$a_{ij} = \begin{cases} 1 & \text{if } b_i \text{ belongs to the fundamental circuit of } e_j \\ 0 & \text{otherwise.} \end{cases}$$

We claim that the matroid $\mathbf{N}$ represented by the matrix $(\mathbf{I}|\mathbf{A})$ over $\mathbf{B}$ is identical to $\mathbf{M}$.
If C is a circuit of $\mathbf{M}$ then it is the symmetric difference of the fundamental circuits of the elements of $C - B$, by our assumption. Hence the C-columns are dependent in $\mathbf{N}$ as well. On the other hand, $\mathbf{N}$ is trivially binary, hence its circuits also arise as symmetric differences of fundamental circuits. Thus a circuit of $\mathbf{N}$ is also dependent in $\mathbf{M}$. Therefore $\mathbf{M} = \mathbf{N}$ (should a circuit C_1 of $\mathbf{N}$ properly contain a circuit C_2 of $\mathbf{M}$ then we had a circuit $C_3 \subseteq C_2$ of $\mathbf{N}$ but $C_3 \subset C_1$ is impossible).

9.1.18 If $|C \cap Q|$ is even for every circuit C and for every cut set Q of $\mathbf{M} = (S, \mathbf{F})$ then in particular $|C \cap Q| = 3$ is impossible. We show that $\mathbf{M}$ cannot have a $\mathbf{U}_{4,2}$ minor in this case. Suppose that A, B were two disjoint subsets of S so that $\mathbf{M}/A \backslash B = \mathbf{U}_{4,2}$ and denote by X an arbitrary 3-element subset of $S - A - B$. X is clearly a circuit and a cut set of $\mathbf{U}_{4,2}$.

Since X is a circuit of $\mathbf{M}/A$ as well, there exists a subset $A_0 \subseteq A$ so that $C = X \cup A_0$ is a circuit of $\mathbf{M}$. Similarly, since X is a cut set of $\mathbf{M}\backslash B$ as well, there exists a subset $B_0 \subseteq B$ so that $Q = X \cup B_0$ is a cut set of $\mathbf{M}$. Finally $|Q \cap C| = |X| = 3$, a contradiction.

9.1.19 The necessity of (C5) for binarity is clear from Statement 5 of **Box 9.2**: $X \triangle Y$ is dependent, hence it contains a circuit Z and neither a nor b belongs to Z.

On the other hand, let $\mathbf{M} = (S, \mathbf{F})$ be nonbinary, let A and B be two disjoint subsets of S with $\mathbf{M}/A \backslash B = \mathbf{U}_{4,2}$. Denote the elements of $S_0 = S - A - B$ by s_1, s_2, s_3 and s_4 and the circuits of form $S_0 - \{s_i\}$ in $\mathbf{U}_{4,2}$ by C_i. Choose a maximal independent subset A_0 of A, that is, $r(A) = r(A_0) = |A_0|$. There exists a subset $A_i \subseteq A_0$ for every i so that $C_i \cup A_i$ is a circuit of $\mathbf{M}$. Finally put $X = C_1 \cup A_1, Y = C_2 \cup A_2$ and consider the elements $s_3, s_4 \in X \cap Y$. The subset $(X \cup Y) - \{s_3, s_4\}$ is independent in $\mathbf{M}$ (being a subset of $A_0 \cup \{s_1, s_2\}$), violating (C5).

9.1.20 (a) In this case $(\mathbf{H}, \mathbf{F})$ is a graphic matroid without loops.
(b) The resulting matroid is the direct sum of a circuit of length three and a bridge.
(c) Let $(S, \mathbf{F})$ be a binary matroid and let $\mathbf{P} = \{Q_1, Q_2, \ldots\}$ be the collection of its cut sets. Consider those subsets $\mathbf{P}_i$ of $\mathbf{P}$ where $Q_j \in \mathbf{P}_i$ if and only if Q_j contains the ith element of S.

9.1.21 We may suppose, possibly after renumbering the elements, that $\{x_1, x_2, \ldots, x_r\} \subseteq S_1$ is a base of $\mathbf{M}_1$ and $\{y_1, y_2, \ldots, y_r\} \subseteq S_2$ is a base of $\mathbf{M}_2$. Then let $(\mathbf{I}|\mathbf{A}_1)$ and $(\mathbf{I}|\mathbf{A}_2)$ represent $\mathbf{M}_1$ and $\mathbf{M}_2$, respectively, where $\mathbf{A}_t$ has $n_t - r$ columns ($n_t = |S_t|$) for $t = 1, 2$. Let us denote the entries of $\mathbf{A}_t$ by $a_{pq}^{(t)}$. Consider the matrix $\begin{pmatrix} \mathbf{I} & \mathbf{A}_1 \\ \mathbf{A}_2^T & \mathbf{B} \end{pmatrix}$ with $b_{ij} = \sum_{k=1}^{r} a_{ki}^{(2)} a_{kj}^{(1)}$. One can easily see that its column space matroid is still $\mathbf{M}_1$ and its rows represent $\mathbf{M}_2$.

9.1.22 Apply the previous solution with $\mathbf{M}_1 = \mathbf{M}_2$ and $\mathbf{A}_1 = \mathbf{A}_2$.

9.2.1 The first matroid is graphic, it is the circuit matroid of the graph of Fig. S.9.3. The other two are nongraphic, both contain a $\mathbf{U}_{5,2}$ minor (delete 6 and contract 7).

9.2.2 The rank of $\mathbf{F}_7$ is three, hence if it were the circuit matroid of a graph G then $v(G) = 4$ would hold, implying that either $e(G) \leq 6$ or G contains loops or parallel edges. Similarly, if $\mathbf{F}_7^* \cong \mathbf{M}(H)$ then $v(H) = 5$ implies either $e(H) \geq 8$ or the existence of a point of degree at most 2.

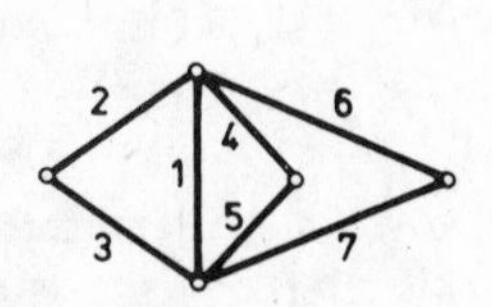

Fig. S.9.3

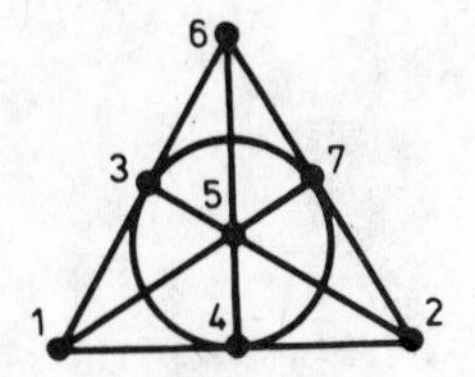

Fig. S.9.4

9.2.3 Assign the unit vectors for the elements of the base $\{1,2,3\}$, see Fig. S.9.4, and write formally

$$\begin{array}{ccccccc} 1 & 2 & 3 & 4 & 5 & 6 & 7 \end{array}$$
$$\begin{pmatrix} 1 & 0 & 0 & a & 0 & e & x \\ 0 & 1 & 0 & b & c & 0 & y \\ 0 & 0 & 1 & 0 & d & f & z \end{pmatrix}$$

The zeroes in columns 4, 5 and 6 are implied by the lines $\{1,2,4\},\{2,3,5\}$ and $\{1,3,6\}$, respectively, and all the letters are nonzero. The line $\{4,5,6\}$ gives an equation $acf + bde = 0$.

Point 7 is contained in the lines $\{1,5,7\},\{2,6,7\}$ and $\{3,4,7\}$, giving the equations $cz = dy, ez = fx$ and $ay = bx$, respectively. This system of homogeneous equations has a solution, different from the trivial one $x = y = z = 0$ if and only if the determinant of the coefficients, that is, $acf - bde$, equals zero.

Since neither acf nor bde is zero, their sum and difference can simultaneously vanish only if the characteristic of the field is 2. Every such field will really do, for example $\begin{pmatrix} 1 & 0 & 0 & 1 & 0 & 1 & 1 \\ 0 & 1 & 0 & 1 & 1 & 0 & 1 \\ 0 & 0 & 1 & 0 & 1 & 1 & 1 \end{pmatrix}$ represents $\mathbf{F}_7$ over $\mathbf{B}$. Similarly, the matrix $\begin{pmatrix} 1 & 1 & 0 & 1 & 0 & 0 & 0 \\ 0 & 1 & 1 & 0 & 1 & 0 & 0 \\ 1 & 0 & 1 & 0 & 0 & 1 & 0 \\ 1 & 1 & 1 & 0 & 0 & 0 & 1 \end{pmatrix}$ represents $\mathbf{F}_7^*$ over $\mathbf{B}$.

9.2.4 Let $\mathbf{M} = (S, \mathbf{F})$ be a matroid of rank $n+1$, given by an affine representation, and project the points from $x \in S$ to an $(n-1)$-dimensional subspace of general position. Let the result be the affine representation of a matroid $\mathbf{M}'$. We prove $\mathbf{M}' = \mathbf{M}/\{x\}$.

If $X \subset S$ with cardinality $|X| = k$ was independent in $\mathbf{M}$ and $x \in X$ then the points of $X - \{x\}$ span a $(k-1)$- dimensional subspace. Their projection will also have this property since $k \leq n$, hence $X - \{x\}$ will be independent in $\mathbf{M}'$. On the other hand, if Y is independent in $\mathbf{M}'$ then

$Y \cup \{x\} \in \mathbf{F}$ since x was outside the subspace, spanned by the elements of Y.

9.2.5 If we proceed in the same way as in the solution of Exercise 9.2.3 then $acf + bde = 0$ (or $cf/de = -b/a$) will follow again. However, this time we have $cz = dy$, $ez = fx$ and $ay \neq bx$ (since $\{3,4,7\}$ is independent). Hence $cf/de = y/x \neq b/a$ is obtained and $b/a \neq -b/a$ (with $b \neq 0$) holds if and only if the characteristic of our field is *not* 2.

9.2.6 Since $\mathbf{F}_7$ is representable over T if and only if the characteristic of T is 2 and $\mathbf{F}_7^-$ is representable if and only if this characteristic is not 2, their direct sum is not representable over any field.

9.2.7 Figure S.9.5 shows the matroids $\mathbf{M}, \mathbf{M}/\{x\}$ and $\mathbf{M}/\{x,z\}$. This latter shows that $\mathbf{M}/\{x,z\}\backslash\{v,w\} = \mathbf{U}_{4,2}$ hence $\mathbf{M}$ is not binary.

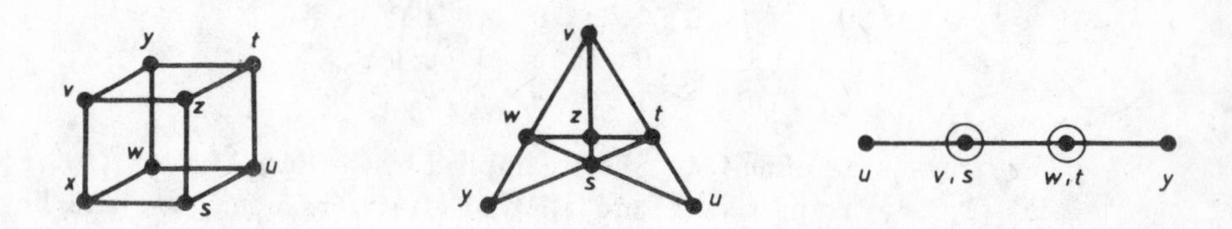

Fig. S.9.5

9.2.8 Let us denote the first $p+1$ elements of $\mathbf{M}_p$ by $a_1, a_2, \ldots, a_{p+1}$, the next $p+1$ elements by $b_1, b_2, \ldots, b_{p+1}$ and the last element by c. Try to represent $\mathbf{M}_p$ over a field T. We may suppose that the first $p+1$ elements correspond to unit vectors and let $(c_1, c_2, \ldots, c_{p+1})^T$ be the vector corresponding to the element c. Since $\{a_i, b_i, c\}$ is a circuit for every i, we may represent b_1 by $(0, \lambda_1 c_2, \lambda_1 c_3, \ldots, \lambda_1 c_{p+1})^T$; b_2 by $(\lambda_2 c_1, 0, \lambda_2 c_3, \ldots, \lambda_2 c_{p+1})$ etc. Now $\{b_1, b_2, \ldots, b_{p+1}\}$ is dependent if and only if the determinant D of these $p+1$ vectors vanishes. However, if we divide the first, second etc. rows of D by c_1, c_2, etc., respectively, and the first, second etc. columns of D by λ_1, λ_2, etc., respectively, then we obtain a $(p+1) \times (p+1)$ determinant where every diagonal entry is 0 and every other entry is 1. This determinant vanishes only if the characteristic of T is p.

9.2.9 $\mathbf{M}$ is isomorphic to $\mathbf{M}(K_5)$, see Fig. S.9.6.

9.2.10 The Petersen graph, see Fig. S.9.7.

9.2.11 The circuits of the non-Desarguesian matroid are of form $\{A_i, B_i, P\}$, $\{A_i, A_j, Q_{ij}\}$, $\{B_i, B_j, Q_{ij}\}$, $\{A_i, A_j, B_i, B_j\}$, $\{A_i, B_j, Q_{ij}, P\}$ and $\{A_i, B_j, Q_{ij}, A_k, B_k\}$ where the values of i, j and k are all different and either 1 or 2 or 3. A lengthy but routine verification of Properties (C1) and (C2) shows that this is a matroid.

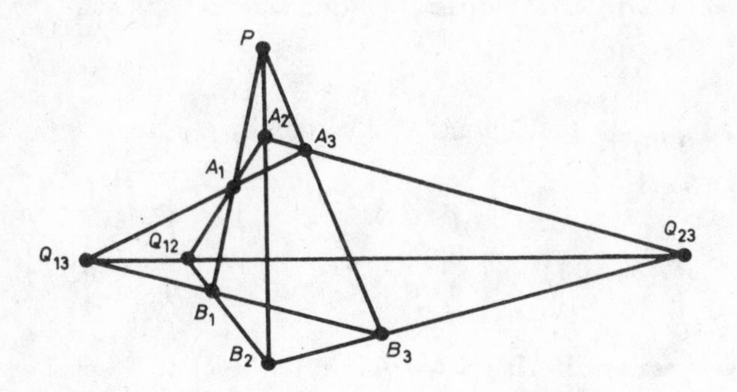

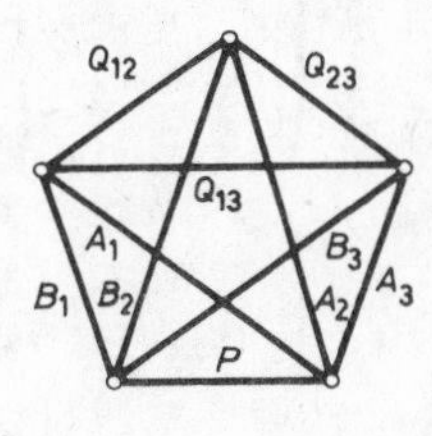

Fig. S.9.6

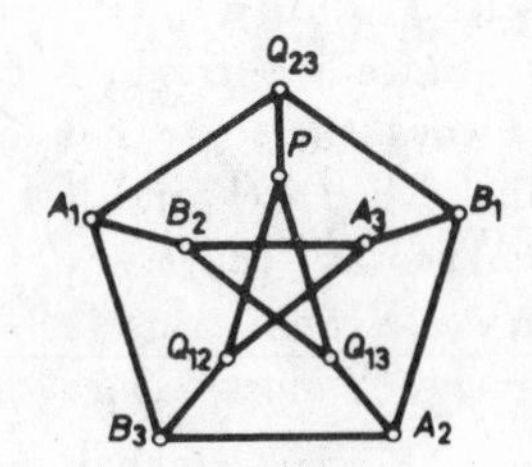

Fig. S.9.7

9.2.12 Every triple is a base in the non-Papposian matroid, except $\{A_1, B_1, C_1\}$, $\{A_2, B_2, C_2\}$ and the triples $\{A_i, B_j, C_k\}$ where i, j and k are all different. The properties (B1), (B2) and (B3) can easily be verified for this set of bases.

We only mention that this matroid can be represented over noncommutative fields (or *skew fields*), which is not the case for the non-Desarguesian matroid. Representability over skew fields raises other questions which are not covered by this book. For example, Theorem 9.1.3 does not remain true.

9.2.13 Should K_8 be representable over some field, the following geometric argument could be given: If $A \cup B$, $B \cup C$ and $C \cup A$ are dependent then the lines aa', bb' and cc' pairwise intersect one another (possibly in an ideal point, if they are parallel). Should these three points of intersection be all different, the set $\{a, a', b, c\}$ were dependent. Hence all three lines intersect at a common point P. Since the same argument works for the lines aa', bb' and dd', finally we obtain that cc' and dd' intersect at P, hence $C \cup D$ is dependent, contradicting to the definition of K_8.

9.2.14 Put $X = A$, $Y = B$, $Z = C$ and $W = D$. Then $2+2+4+4+4 > 5 \cdot 3$.

9.2.15 Applying Statement 9.2.1 one can directly obtain a representation

$$\begin{matrix} 1 & 2 & 3 & 4 & 5 & 6 & 7 \end{matrix}$$
$$\begin{pmatrix} 1 & 0 & 0 & 0 & 1 & 1 & 0 \\ 0 & 1 & 0 & 1 & 0 & 1 & 0 \\ 0 & 0 & 1 & 1 & 1 & 0 & 0 \\ 1 & 1 & 1 & 1 & 1 & 1 & 1 \end{pmatrix}.$$ Multiplying this matrix by

$$\begin{pmatrix} 1 & 0 & 0 & 0 \\ 0 & 1 & 0 & 0 \\ 0 & 0 & 0 & 1 \\ -1 & -1 & -1 & 1 \end{pmatrix} \text{ we obtain } \begin{pmatrix} 1 & 0 & 0 & 0 & 1 & 1 & 0 \\ 0 & 1 & 0 & 1 & 0 & 1 & 0 \\ 0 & 0 & 1 & 1 & 1 & 0 & 0 \\ 0 & 0 & 0 & -1 & -1 & -1 & 1 \end{pmatrix},$$ cf., the solution of Exercise 9.2.3. Hence we obtain the dual of $\mathbf{F}_7$ or $\mathbf{F}_7^-$, depending on whether the characteristic of our field is 2 or not.

9.2.16 We shall prove that this matroid is representable over a field if and only if the equation $x^2 + x + 1 = 0$ is solvable in this field. In particular, $\mathbb{C}$ and GF(3) have this property but $\mathbb{Q}$ or $\mathbb{R}$ have not.
We assign unit vectors to three independent points (see Fig. S.9.8). Then let $(a,1,0)^T$, $(0,b,1)^T$ and $(c,0,1)^T$ correspond to the points A, B and C, respectively. (Since we are in a field where division can be performed, we can always suppose that one of the nonzero entries equal 1.) The circuits $\{A,B,D\}$ and $\{A,C,E\}$ enable us to represent D and E by the vectors $(ad, b+d, 1)^T$ and $(ae+c, e, 1)^T$, respectively. So far, a,b,c,d and e are arbitrary nonzero elements of the field, their proper choice must assure that the unit vectors are on the lines BE, CD and ED, respectively.Hence we have three equations

$$\begin{vmatrix} 1 & 0 & 0 \\ 0 & b & 1 \\ ae+c & e & 1 \end{vmatrix} = 0, \quad \begin{vmatrix} 0 & 1 & 0 \\ c & 0 & 1 \\ ad & b+d & 1 \end{vmatrix} = 0 \text{ and } \begin{vmatrix} 0 & 0 & 1 \\ ae+c & e & 1 \\ ad & b+d & 1 \end{vmatrix} = 0,$$

leading to $e = b$, $c = ad$ and $(b+d)c + abe = 0$. Then $x = b/d$ must satisfy $x^2 + x + 1 = 0$.

9.2.17 Consider an arbitrary graphic matroid for case (a), and the matroids of Exercises 9.2.6, 9.2.3 and 9.2.5, respectively, for the cases (b), (c) and (d).

9.2.18 If the matroid $\mathbf{M}$ of rank k is affinely represented in the $(k-1)$-dimensional space then an affine representation of its truncation can be obtained by projecting every point to a $(k-2)$-dimensional subspace of general position. In order to see this all we have to observe that "generality" means that if the projection of a subset X of cardinality p (with $p \le k-1$) spans a subset of dimension less than $p-1$ then X was already dependent in $\mathbf{M}$.

9.2.19 For example, 1, 2, 3 and 7 are noncollinear. The first two properties are obviously met.

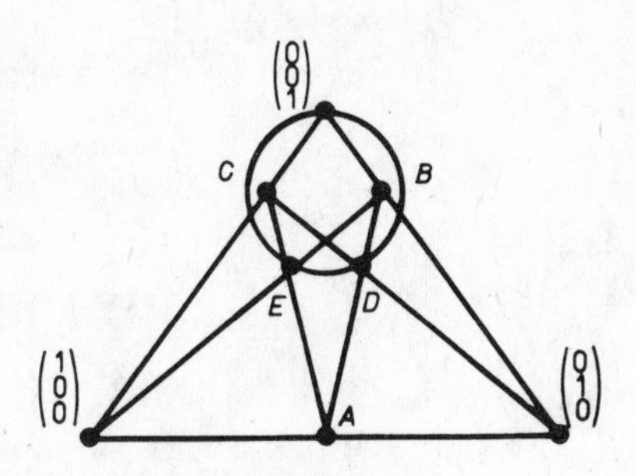

Fig. S.9.8

9.2.20 Let e, f be two lines and let p denote their common point. Choose a point q not contained in $e \cup f$. For every $r \in e$, $r \neq p$ determine the unique common point r' of the lines f and qr. This gives a one–one correspondence $r \to r'$ between the points of $e - \{p\}$ and $f - \{p\}$, hence $|e| = |f|$.

Let $k+1$ denote the common cardinality of every line. The number of lines containing any point p is also $k+1$ (choose a line e not passing through p and take all the lines of form pq with $q \in e$).

9.2.21 The $k+1$ lines containing p cover each point (namely they cover p exactly $(k+1)$ times and all the other $k(k+1)$ points once). Hence $|P| = k^2 + k + 1$.

9.2.22 Denote the original ("shorter") column vectors by $\mathbf{u}_1, \mathbf{u}_2, \ldots$ and the extended ones by $\mathbf{u}'_1, \mathbf{u}'_2, \ldots$. If the former ones are affinely independent then $\sum \lambda_i \mathbf{u}'_i = \mathbf{0}$ can hold only if $\sum \lambda_i = 0$ and $\sum \lambda_i \mathbf{u}_i = \mathbf{0}$, which implies $\lambda_i = 0$ for every i. Hence the $\mathbf{u}'_i$ vectors are linearly independent. On the other hand, if the original ones are not affinely independent then there exists a linear combination $\sum \lambda_i \mathbf{u}_i = \mathbf{0}$ satisfying $\sum \lambda_i = 0$ but with some λ_i's different from zero. Then $\sum \lambda_i \mathbf{u}'_i = \mathbf{0}$ holds in a nontrivial way, hence the $\mathbf{u}'_i$ vectors are linearly dependent.

9.2.23 Yes. Start from the representation of $\mathbf{F}_7$ over $\mathbb{B}$ (see the solution of Exercise 9.2.3) and consider it as a representation over a larger field T of characteristic 2 (for example, T can be the field shown in the solution of Problem 9.1.14). Then the last element can also be represented using a further element of T.

9.3.1 If x is a loop, $r(X) = r(X \cup \{x\})$ follows for every X, hence (1) implies the other three statements. (2) $\to$ (1) is trivial and (2) follows from either (3) or (4) by putting $X = \sigma(\emptyset)$.

9.3.2 Suppose $r(X \cup \{y\}) > r(X)$. This is clearly impossible if $x \in X$. Otherwise $r(X \cup \{y\}) = r(X \cup \{x\})$ would imply $x \notin \sigma(X)$, a contradiction.

9.3.3 Let $C - X = \{x\}$ and X be closed. Then $r(X \cup \{x\}) > r(X)$. Choose an independent subset Y of X which has cardinality $r(X)$ and contains every element of $C - \{x\}$. Then $r(X) = r(Y) = r(Y \cup \{x\})$ and this latter equals to $r(X \cup \{x\})$, by the repeated use of property (R7) of Corollary 7.1.5, a contradiction.

On the other hand, if X is not closed then there exists an element $x \notin X$ with $r(X \cup \{x\}) = r(X)$. Choose an independent subset Y of X with cardinality $r(X)$ and let C be the unique circuit of $Y \cup \{x\}$. Now $|C - X| = 1$.

9.3.4 An element x is a loop of $\mathbf{M}/X$ if and only if there exists a circuit C of $\mathbf{M}$ satisfying $C - X = \{x\}$. Hence our assertion follows from that of the previous exercise.

9.3.5 Let $x \in \sigma(X_1) \cap \sigma(X_2)$. Then there exist two circuits C_1, C_2 with $C_1 - X_1 = C_2 - X_2 = \{x\}$. Let us extend $C_1 - \{x\}$ and $C_2 - \{x\}$ to maximal independent subsets I_1 and I_2 of X_1 and of X_2, respectively. Clearly, $r(X_1) = |I_1|$ and $r(X_2) = |I_2|$, and our condition implies that $I_1 \cup I_2$ is independent. However, if x is not a loop, $(C_1 \cup C_2) - \{x\} \subseteq I_1 \cup I_2$ must contain a circuit, a contradiction.

9.3.6 Let B be a base of the matroid. If X_1, X_2 are distinct subsets of B then $\sigma(X_1) \neq \sigma(X_2)$ since if $x \in X_1 - X_2$ then $x \in \sigma(X_1) - \sigma(X_2)$ holds as well. Hence the closure of all the 2^r different subsets of B are different.

There are matroids with exactly 2^r closed sets (e.g. the free matroid); some others like $\mathbf{U}_{2r,r}$ have much more.

9.3.7 $X \subseteq S$ is independent if and only if $x \notin \sigma(X - \{x\})$ holds for every $x \in X$.

9.3.8 Suppose $X \cap Y$ is not closed and choose an $x \notin X \cap Y$ and a circuit C satisfying $C - (X \cap Y) = \{x\}$. At least one of X and Y, say the latter, does not contain x, hence $C - Y = \{x\}$ also holds, but Y is closed, a contradiction.

On the other hand, let X and Y be two distinct single element subsets of $\mathbf{U}_{3,2}$. They are closed but their union is not.

9.3.9 Let $\mathbf{M}$ be the circuit matroid of the graph of Fig. S.9.9. $\emptyset$ is closed (since $\mathbf{M}$ has no loops), yet the whole underlying set (the complement of $\emptyset$) cannot be covered by disjoint cut sets.

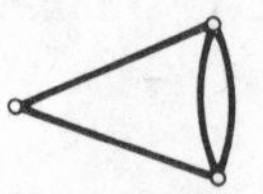

Fig. S.9.9

9.3.10 Property (C1) is trivially met. Let now X and Y be two distinct minimal supports and let $s_0 \in X \cap Y$. If $\mathbf{x} = (x_1, x_2, \ldots, x_n)$ and $\mathbf{y} = (y_1, y_2, \ldots y_n)$ are the row vectors corresponding to X and to Y, respectively, and x_0, y_0 are the coordinates corresponding to s_0 then consider $\mathbf{u} = y_0 \cdot \mathbf{x} - x_0 \cdot \mathbf{y}$. The support of $\mathbf{u}$ is nonempty and is included in $(X \cup Y) - \{s_0\}$.

9.3.11 (a) $\mathbf{L}(\mathbf{U}_{n,n})$ is the lattice of all the subsets of the n-element set, ordered by inclusion, since every subset of $\mathbf{U}_{n,n}$ is closed. This is sometimes also called an n-dimensional cube. Now $\mathbf{L}(\mathbf{U}_{n,k})$ can be obtained from this by deleting the $k^{th}, (k+1)^{th}, \ldots, (n-1)^{st}$ "levels" of this lattice and joining every element of the $(k-1)^{st}$ level to the top.

(b) See Fig. S.9.10.

(c) The simplest example is a chain of three points. For if there were a matroid $\mathbf{M} = (S, \mathbf{F})$ with a nonloop element $x \in S$ then the three points must correspond to $\sigma(\emptyset)$, to $\sigma(\{x\})$ and to S, respectively. But then there must exist a $y \in S - \sigma(\{x\})$ as well, thus giving a fourth point $\sigma(\{y\})$ for the lattice, a contradiction.

For a characterization of the "matroid lattices" see [Birkhoff, 1967] or Chapter 3 of [Welsh, 1976].

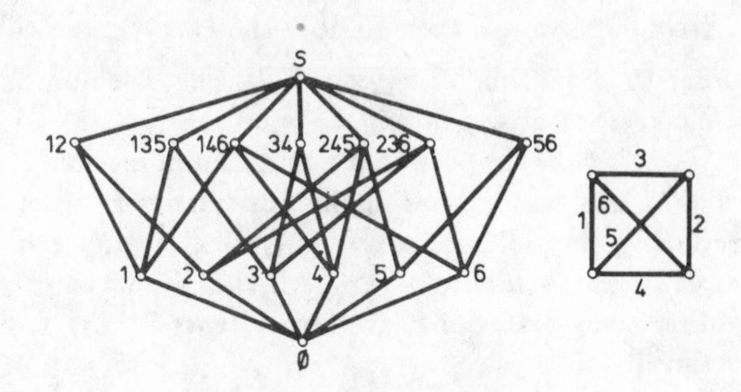

Fig. S.9.10

9.3.12 If X, Y are two such closed sets then $\sigma(X \cup Y)$ also has this property since the operation $X \vee Y$ remains the same. $X \wedge Y$ will be defined as the union of the circuits contained in $X \cap Y$ (or as the empty set if there is no circuit contained in $X \cap Y$). The lattice corresponding to $\mathbf{M}(K_4)$ is shown in Fig. S.9.11.

This construction [Brylawski, 1975b], [Ingleton, 1977] also describes the matroid if not only the elements of the lattice but their ranks are also given. This gives a useful presentation of matroids, see [Acketa, 1979] as well.

9.3.13 $\mathbf{M}'$ is just the truncation of $\mathbf{M}$.

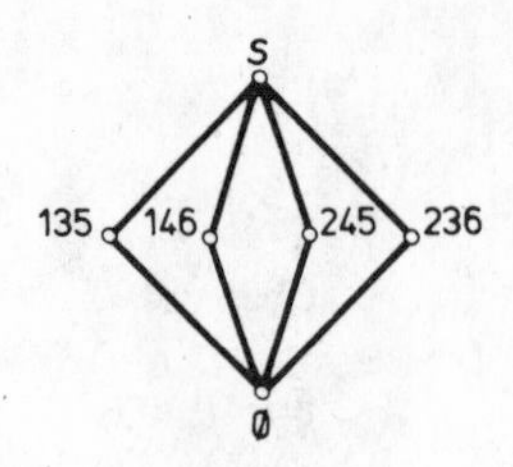

Fig. S.9.11

9.3.14 The circuits of $\mathbf{M}$ (the elements of Q) must be proved to be just the cut sets of $\mathbf{N}$, that is, that they satisfy the conditions of Theorem 7.2.7. Conditions (a) and (c) are clearly met. In order to prove (b) suppose indirectly that $X \in Q$ and there exists a circuit Y of $\mathbf{N}$ with $X \cap Y = \{s\}$. Then the rows of $\mathbf{A}$ have a linear combination whose entries in the Y-columns all vanish, except in the column of s. But Y is a circuit, hence any column of Y can be expressed as a linear combination of the others, a contradiction.

9.3.15 Let $X \subset S$. We claim that an element $x \notin X$ is contained in $\sigma(X)$ if and only if, for every linear combination of the rows of $\mathbf{A}$, if all the entries of the X-columns vanish then so does the entry in the column of x as well.

In order to show this let $\mathbf{x}_1, \mathbf{x}_2, \ldots$ be the X-columns and let $\mathbf{x}$ be the column vector corresponding to x. Clearly, $x \in \sigma(X)$ if and only if $\mathbf{x} = \sum \nu_i \mathbf{x}_i$. Now suppose that a linear combination $\mathbf{y}^T = \sum \mu_j \mathbf{a}_j^T$ of the rows of $\mathbf{A}$ has zero entries in the positions corresponding to $\mathbf{x}_1, \mathbf{x}_2, \ldots$. Introducing the notation $\mathbf{m} = (\mu_1, \mu_2 \ldots)^T$ and $\mathbf{n} = (\nu_1, \nu_2, \ldots)^T$ we have $\mathbf{y}^T = \mathbf{m}^T \mathbf{A}$ and $\mathbf{x} = \mathbf{A}\mathbf{n}$. All the X-entries of $\mathbf{y}$ are zero and all the other coordinates of $\mathbf{n}$ are zero. Hence $0 = \mathbf{y}^T \mathbf{n} = \mathbf{m}^T \mathbf{A}\mathbf{n} = \mathbf{m}^T \mathbf{x}$, as claimed.

9.3.16 Let C be a circuit and Q be a cut set in the binary matroid $\mathbf{M}$. Using Exercise 9.3.10 and Problem 9.3.14 we can coordinatize $\mathbf{M}$ over $\mathbf{B}$ by such a matrix where the support of one of the rows is Q. Recall that every nonzero entry of the matrix is one and that the mod 2 sum of the C-columns is zero. Hence $|C \cap Q|$ is even.

9.3.17 Let r and r^* denote the rank functions of $\mathbf{M}$ and $\mathbf{M}^*$, respectively, and let $k = |X|$. Clearly, $r(S) > k-1$ and $r^*(S) > n-k-1$, where $n = |S|$, since X is a circuit in $\mathbf{M}$ and $S-X$ is a circuit in $\mathbf{M}^*$. Taking $r(S)+r^*(S) = n$ into consideration, this gives $r(S) = k$ and $r^*(S) = n-k$.

Hence X is a base (not just an independent set) in $\mathbf{M}'$ (if $\mathbf{M}'$ is a matroid at all). Let Y be another base. We have to prove that for every $x \in X$ there exists a $y \in Y$ so that $(X-\{x\}) \cup \{y\}$ is a base and for every $y \in Y$ there exists an $x \in X$ so that $(Y-\{y\}) \cup \{x\}$ is a base. (The exchange property among the "old" bases is clearly known.) The former assertion is obvious since Y and $X - \{x\}$ are in $\mathbf{F}$.

Let us denote $Y - \{y\}$ by Z. We may suppose that $Z \subset X$ does not hold (since the assertion is trivial otherwise). If $Z \cup \{x\}$ were dependent for every $x \in X$ then $X \subseteq \sigma(Z)$ would hold. Now $|Z| = r(S) - 1$ implies that $S - \sigma(Z)$ is a cut set. But it cannot be a proper subset of $S - X$, hence $\sigma(Z) = X$ and then $Z \subset X$, a contradiction.

Chapter 10

10.1.1 The graph of the cascade connection of Fig. S.10.1 is shown in Fig. S.10.2. The results are shown in **Box 6.3.**

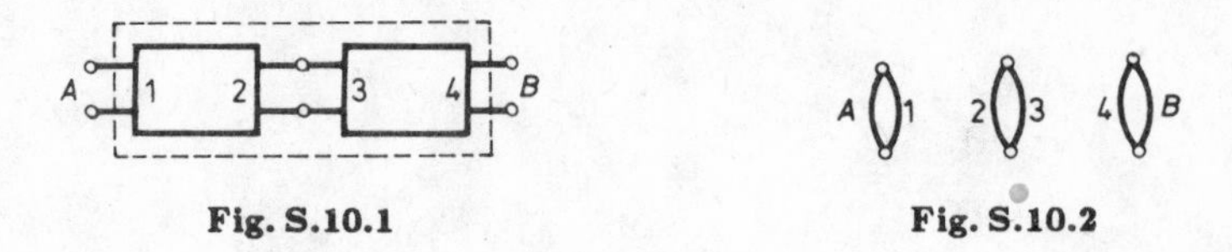

Fig. S.10.1 **Fig. S.10.2**

10.1.2 (a) The resulting 2-ports are shown in Fig. S.10.3, where $\rho_a = (\rho_1^{-1} + \rho_2^{-1} + r_2^{-1})^{-1}$ and $\rho_c = (\rho_1^{-1} + \rho_2^{-1} + (r_1 + r_2)^{-1})^{-1}$.

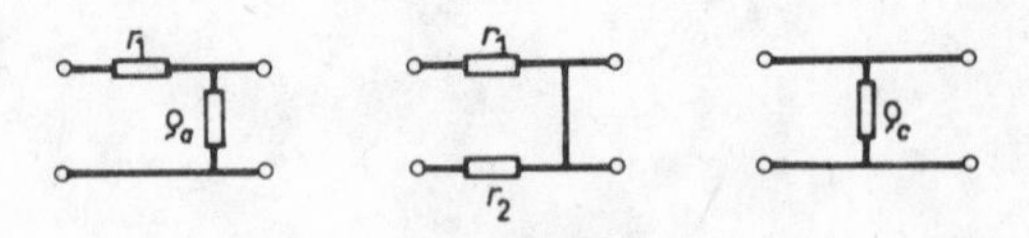

Fig. S.10.3

(b) The matrix descriptions of N_k' and N_k'' are the same: for example their $\mathbf{Z}$ matrix is $\begin{pmatrix} r_k + \rho_k & \rho_k \\ \rho_k & \rho_k \end{pmatrix}$. However, if they are interconnected with other multiports, they will short-circuit points in different ways. That is why the three answers in part (a) were all different.

(c) If $1a$ and $1b$ denote the first and the second port of the first 2-port, respectively, and $2a$ and $2b$ denote those of the second 2-port, then the graph of the series-parallel interconnection is shown in Fig. S.10.4.

If the 2-ports contain no internal short-circuits then P_1, P_2, P_3, P_4 and P_5 are all different. In case (1), P_3 and P_5 become equipotential, due to the short circuit in N_1' and P_1 and P_5 become equipotential due to N_2'. Hence G_1 of Fig. S.10.5 is obtained. Cases (2) and (3) lead to G_2 and G_3, respectively, in a similar way.

10.1.3 We present solutions with 1-ports. Let N and N' be resistors with value R and R', respectively. If $R \neq 0$ and $R' \neq 0$ then $\mathbf{M}_N$, $\mathbf{M}_{N'}$, $\varphi(\mathbf{M}_{N'})$ and $\mathbf{M}^*_{N'}$ are all equal (and isomorphic to $\mathbf{U}_{2,1}$). Yet, R and R' can be

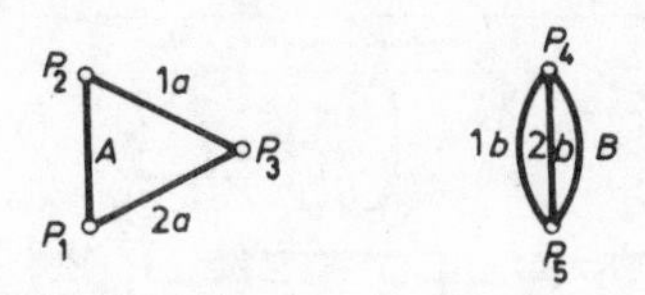

Fig. S.10.4

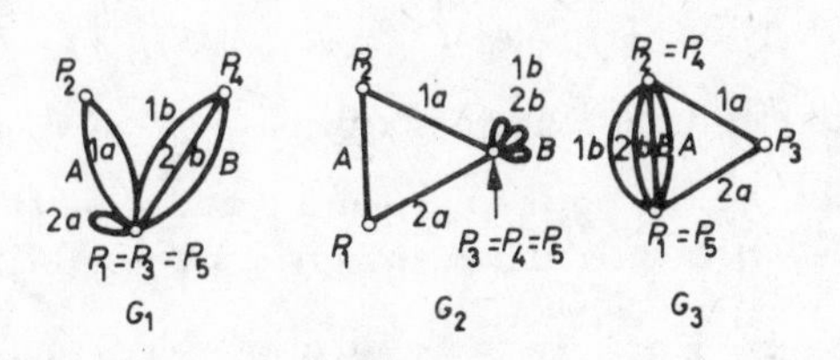

Fig. S.10.5

chosen so that none of the relations $R = R'$; $R = 1/R'$; $R = -1/R'$ be valid.

10.1.4 Let N_k be given by $\mathbf{A}_k\mathbf{u} + \mathbf{B}_k\mathbf{i} = \mathbf{0}$ for $k = 1, 2$. Let $\mathbf{u}_k, \mathbf{i}_k$ be arbitrary pairs of vectors, satisfying $\mathbf{A}_k\mathbf{u}_k + \mathbf{B}_k\mathbf{i}_k = \mathbf{0}$. Consider now the following property (P_k)

(P_k) $\quad \mathbf{u}^T\mathbf{i}_k = \mathbf{u}_k^T\mathbf{i}$ for every pair $\mathbf{u}$, $\mathbf{i}$ satisfying $\mathbf{A}_1\mathbf{u} + \mathbf{B}_1\mathbf{i} = \mathbf{0}$

What does this property mean? For $k = 1$ it means that N_1 is reciprocal; for $k = 2$ it means that $N_2 \cong \text{adj } N_1$. Putting $N_1 \cong N_2$ we obtain that reciprocity of N is equivalent to $N \cong \text{adj } N$.

Using Corollary 10.1.5 we obtain that reciprocity is closed with respect to interconnection.

10.1.5 (a) This is trivial (let the constants of the controls be reciprocal to each other).

(b) See Fig. S.10.6 (the sum of the constants of the controls must be zero).

(c) There is no contradiction: The independent and controlled voltage sources (including the short circuit) do form a circuit but the current of one of them controls something else.

10.1.6 Consider the 2-ports of Fig. 10.9 with $\rho_1 = \rho_2 = 2$, $r_2 = -1$ and prepare their series-parallel interconnection. The resulting 2-port is described by

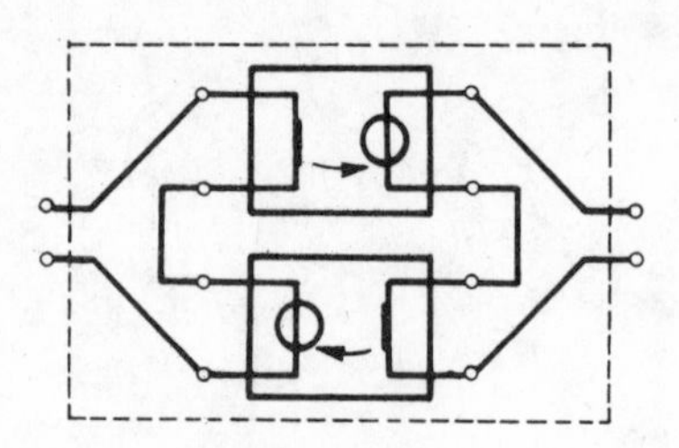

Fig. S.10.6

$$\begin{pmatrix} -1 & 1 \\ 0 & 0 \end{pmatrix} \begin{pmatrix} u_A \\ u_B \end{pmatrix} + \begin{pmatrix} r_1 & -r_1 \\ 1 & 1 \end{pmatrix} \begin{pmatrix} i_A \\ i_B \end{pmatrix} = \begin{pmatrix} 0 \\ 0 \end{pmatrix}$$

hence it has no impedance description.

10.1.7 Let 1 and 2 be the ports of N_1 and 3 and 4 be the ports of N_2. Then their series-series interconnection gives a 2-port with

$$u_A = u_1 + u_3; \quad u_B = u_2 + u_4; \quad i_A = i_1 = i_3; \quad i_B = i_2 = i_4.$$

Hence $\begin{pmatrix} u_A \\ u_B \end{pmatrix} = \begin{pmatrix} u_1 \\ u_2 \end{pmatrix} + \begin{pmatrix} u_3 \\ u_4 \end{pmatrix} = \mathbf{Z}_1 \begin{pmatrix} i_1 \\ i_2 \end{pmatrix} + \mathbf{Z}_2 \begin{pmatrix} i_3 \\ i_4 \end{pmatrix} =$

$= (\mathbf{Z}_1 + \mathbf{Z}_2) \begin{pmatrix} i_A \\ i_B \end{pmatrix}$. Thus the resulting 2-port has an impedance matrix (which is, in fact, the sum of those of the two 2-ports).

10.1.8 Consider the cascade connection (see Fig. S.10.1) of the 2-ports given by $u_1 = ri_2$, $i_1 = 0$ and $i_3 = 0$, $u_4 = 0$. The result can be described by 3 independent equations only, namely $u_A = 0$, $i_A = 0$ and $u_B = 0$.

Should we interconnect the 2-ports $u_1 = ri_2$, $u_2 = 0$ and $u_3 = 0$, $u_4 = 0$ in cascade, we had only one equation $u_B = 0$.

10.1.9 The relation $\mathbf{M}_N = \varphi(\mathbf{M}_N^*)$ is clearly necessary for the reciprocity of the multiport N. However, this relation holds for the matroid of the gyrator as well.

10.1.10 (a) This is trivial (cf. the previous problem).

(b) Yes. If N is the 3-port circulator then $\mathbf{M}_N$ satisfies $\mathbf{M}_N = \varphi(\mathbf{M}_N^*)$. However, we prove that no reciprocal 3-port N' has the same matroid. We saw (Exercise 8.1.9) that N has an admittance description $\mathbf{Y} = \begin{pmatrix} 0 & 1 & -1 \\ -1 & 0 & 1 \\ 1 & -1 & 0 \end{pmatrix}$. Then N' should have $\mathbf{Y}' = \begin{pmatrix} 0 & a & b \\ a & 0 & c \\ b & c & 0 \end{pmatrix}$ with $a, b, c \neq 0$ (recall from Problem 8.1.16 that reciprocity of N' means the symmetry of $\mathbf{Y}'$). However, det $\mathbf{Y}' = 2abc \neq 0$ and det $\mathbf{Y} = 0$ contradicts to $\mathbf{M}_N = \mathbf{M}_{N'}$.

10.1.11 The graph of the port edges must be connected since a possible internal connection could not be detected otherwise. (Exercise 10.1.2 illustrates that the way of the interconnection and the matrices of the multiports together do not determine yet the properties of the final network.) Then the $k-1$ port edges necessarily form a tree. If it is not a star, the multiport description will be different (see Exercise 8.1.3, for example).

10.2.1 The bipartite graphs are shown in Fig. S.10.7. The second and the fourth systems determine asymmetric 2-component forests, hence they are rigid. The other two are nonrigid, see their deformations in Fig. S.10.8.

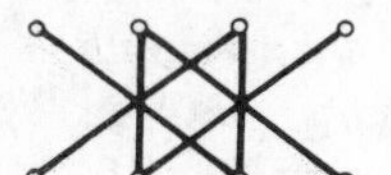
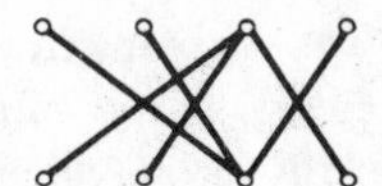
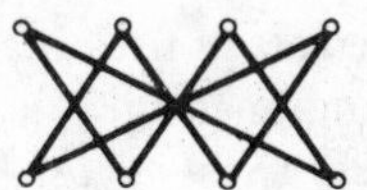
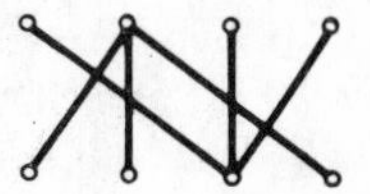

Fig. S.10.7

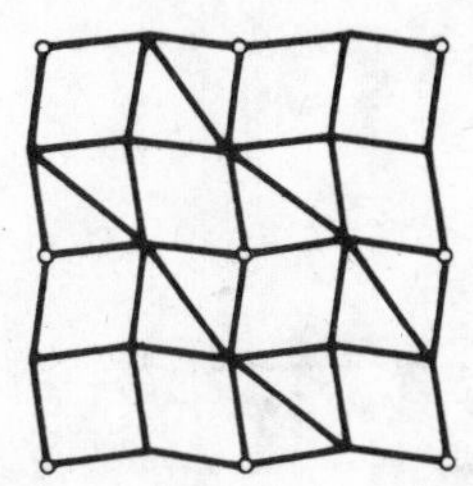
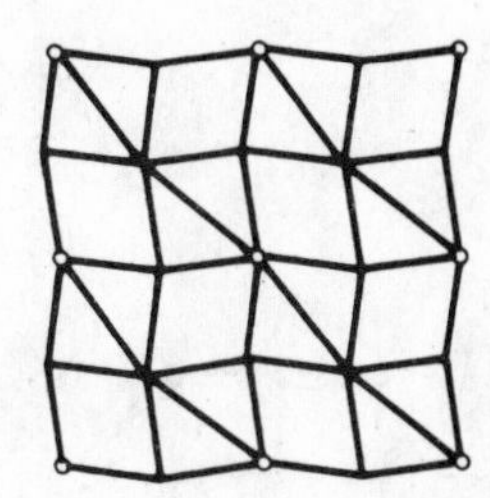

Fig. S.10.8

10.2.2 If each vertical wall contains one diagonal then every joint is fixed. The number of necessary diagonals is $(k+1)+(\ell+1)=k+\ell+2$; just like in the case when 4 diagonals are used in vertical walls and $k+\ell-2$ ones in the horizontal plane.

10.2.3 In case of $k=\ell=2$ every pair except $\{u,y\}$ and $\{v,x\}$ is a base (see the notation of Fig. S.10.9), hence $\mathbf{M}_{2,2}$ is graphic. On the other hand, $\mathbf{M}_{2,3}$ is isomorphic to $\mathbf{U}_{6,3}$ since every triple is a base (every 2-component forest is asymmetric).

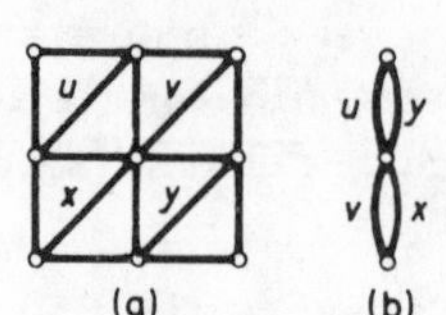

Fig. S.10.9

10.2.4 If we delete or contract certain elements then the corresponding diagonals must not be applied or must be applied, respectively, for a minimal system in order to rigidify the building.

10.2.5 The $k+\ell-1$ edges of the bipartite graph correspond either to a spanning tree or a 2-component graph with a circuit in one of the components. In the former case those edges (if any) are critical whose deletion leads to a symmetric 2-component forest. In the latter case the two components are asymmetric (since the system was rigid by the assumption) and every edge except those in the circuit is critical.

One can also show that the four diagonal rods in the vertical walls are not critical in the former case but are critical in the latter one.

10.2.6 The affine representation of these matroids is shown in Fig. S.10.10.

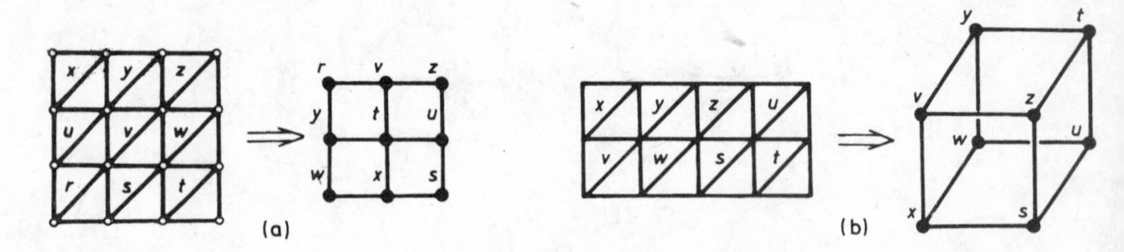

Fig. S.10.10

10.2.7 One can easily see from Fig. S.9.5 that $\mathbf{M}_{2,4}/\{x\}\backslash\{z\} \cong \mathbf{M}(K_4)$.

10.2.8 Apply 4 vertical and $4+4-2=6$ horizontal diagonals to make the "building" rigid. Then the roof of the terrace could still have a motion, see Fig. S.10.11, hence one of its four squares should also be braced.

10.2.9 Three such diagonals can be enough if the vertical walls containing them are not all parallel. Then the diagonals in the roof must determine a

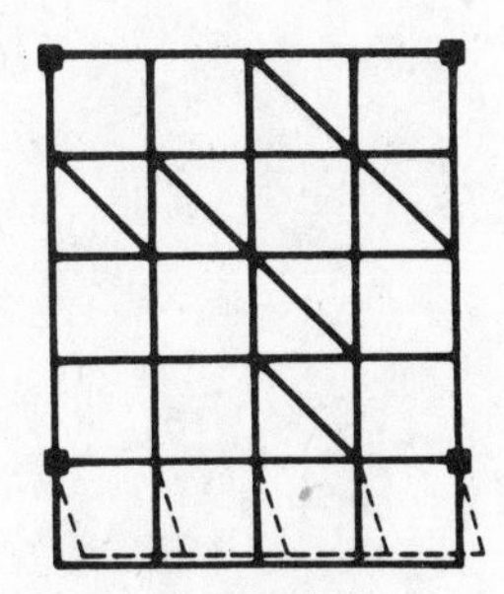

Fig. S.10.11

connected graph in our model. Diagonals in two vertical walls were not enough, since a rotation of the roof (like in Fig. 10.15) could not be prevented. See also [Whiteley, 1983].

10.2.10 All the rods cannot be critical. In the solution of Exercise 10.2.5 we distinguished two possibilities. In the second one the edges of the circuit are noncritical. In the first case there must be some edges (like those, incident to points of degree one) whose deletion cannot lead to symmetric 2-component forests.

It is possible that none of the rods is critical, for example if k or ℓ is a prime.

10.2.11 Let G be the bipartite graph and $e = \{x, y\} \in E(G)$ correspond to a critical rod. Both components of $G - e$ contain even number of points (by the symmetricity of the forest). If the statement were false, the degree of every point of G, except x and y, would be odd. But then the components of $G - e$ had one point of odd degree each, contradicting to Problem 1.1.15.

Chapter 11

11.1.1 See Fig. S.11.1.

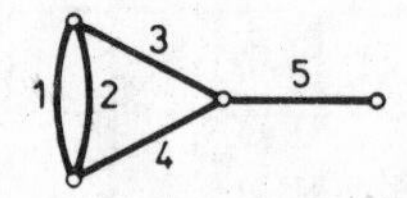

Fig. S.11.1

11.1.2 Suppose, indirectly, that C is a circuit of $\mathbf{M}_1 \vee \mathbf{M}_2$, containing x. Then $C - \{x\} = X_1 \cup X_2$, where X_i is independent in $\mathbf{M}_i$ (for $i = 1, 2$). However, $X_1 \cup \{x\}$ is also independent in $\mathbf{M}_1$, hence C is independent in $\mathbf{M}_1 \vee \mathbf{M}_2$, a contradiction.

11.1.3 A subset $Y \subseteq S - X$ is independent in $(\vee_{i=1}^k \mathbf{M}_i)\backslash X$ if and only if $Y = \cup_{i=1}^k Y_i$ where $Y_i \subseteq S - X$ and Y_i is independent in $\mathbf{M}_i$ (for $i = 1, 2, \ldots, k$). These latter conditions are met if and only if Y_i is independent in $\mathbf{M}_i \backslash X$ for every i.

The analogous statement for contractions may be false. If $S = \{a, b\}$ and $\mathbf{M}_1 = \mathbf{M}_2 = \mathbf{U}_{2,1}$ then put $X = \{b\}$. The element a is a loop in $(\mathbf{M}_1/X) \vee (\mathbf{M}_2/X)$ and a bridge in $(\mathbf{M}_1 \vee \mathbf{M}_2)/X$.

11.1.4 No. The example in the solution of the previous exercise shows that series or parallel elements may become bridges, for example.

However, if x, y were series in $\mathbf{M}_1$ and are not series in $\mathbf{M}_1 \vee \mathbf{M}_2$ then they are bridges in $\mathbf{M}_1 \vee \mathbf{M}_2$. If x, y were parallel in $\mathbf{M}_1$, they never remain parallel except if both of them were loops in $\mathbf{M}_2$.

11.1.5 Let $S_1 \cap S_2 = \emptyset$ and $\mathbf{M}_i = (S_i, \mathbf{F}_i)$ for $i = 1, 2$. If $\mathbf{M}_i$ is extended to $\mathbf{M}'_i$ by loops on S_{3-i} — formally $\mathbf{M}'_1 = \mathbf{M}_1 \oplus (S_2, \{\emptyset\})$ and $\mathbf{M}'_2 = \mathbf{M}_2 \oplus (S_1, \{\emptyset\})$ — then $\mathbf{M}_1 \oplus \mathbf{M}_2 = \mathbf{M}'_1 \vee \mathbf{M}'_2$.

11.1.6 Let $R(S) = k$ and consider a k-element subset $X \subseteq S$ which is independent in $\vee_{i=1}^k \mathbf{M}_i$. Thus $X = \bigcup_{i=1}^k X_i$, where X_i is independent in $\mathbf{M}_i$, that is, $|X_i| = r_i(X) \le r_i(S)$ holds for every $i = 1, 2, \ldots, k$. Hence $R(S) = k = |X| \le \sum_{i=1}^k |X_i| \le \sum_{i=1}^k r_i(S)$. For example, $r_1(S) = r_2(S) = 2$ but $R(S) = 3$ for the two matroids of Exercise 11.1.1.

11.1.7 X can be decomposed as $X_1 \cup X_2$ where X_1 is independent in $\mathbf{M}_1$ and X_2 is independent in $\mathbf{M}_2$. Should $X_1 \cap X_2 \neq \emptyset$, replace X_2 by $X_2 - X_1$. This is also independent in $\mathbf{M}_2$ and is disjoint of X_1.

11.1.8 X can be decomposed as $X_1 \cup X_2$ where X_i is independent in $\mathbf{M}_i$ for $i = 1, 2$. If X_2 is not a base of $\mathbf{M}_2$, extend it to a base $X_2 \cup Y$ of $\mathbf{M}_2$. Now Y is fully contained in X_1 since otherwise $X_1 \cup X_2 \cup Y$ were a larger independent set of $\mathbf{M}_1 \vee \mathbf{M}_2$. Hence $(X_1 - Y) \cup (X_2 \cup Y)$ is a decomposition with the required property.

11.1.9 Let $\mathbf{M}_1$ and $\mathbf{M}_2$ be the circuit matroids of the two graphs of Fig. S.11.2 and consider the base $\{a\}$ of the second matroid.

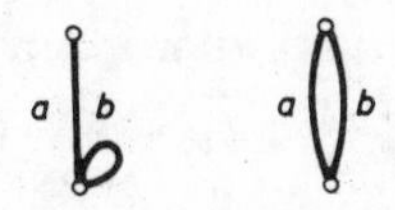

Fig. S.11.2

11.1.10 $\mathbf{M} \vee \mathbf{M}^*$ is clearly the free matroid since the complement of every base of $\mathbf{M}$ is a base of $\mathbf{M}^*$.

11.1.11 Let $\mathbf{M}_1 = (S, \mathbf{F}_1)$ and let $S = S_1 \cup S_2$ where S_1 is the (possibly empty) set of bridges of $\mathbf{M}_1$. We claim that $\mathbf{M}_1 \vee \mathbf{M}_2 = \mathbf{M}_1$ if and only if $\mathbf{M}_2 \backslash S_1$ contains loops only (while $\mathbf{M}_2 \backslash S_2$ may be arbitrary).

In order to see this let $x \in S_2$ be a nonloop element of $\mathbf{M}_2$. Then any circuit of $\mathbf{M}_1$ containing x becomes independent in $\mathbf{M}_1 \vee \mathbf{M}_2$. The sufficiency of the condition is straightforward (cf. Exercise 11.1.2 as well).

11.1.12 Let us decompose X as $Y \cup (X - Y)$. The relation $R(X) \leq R(Y) + R(X - Y)$ is obvious. Since $R(Y) \leq \sum_{i=1}^{k} r_i(Y)$ follows (like in the solution of Exercise 11.1.6) and $R(X-Y) \leq |X-Y|$, we obtain $R(X) \leq \sum_{i=1}^{k} r_i(Y) + |X - Y|$ for every $Y \subseteq X$. Then it holds for the minimum of these sums as well.

11.1.13 Let Y be a common independent subset with cardinality t. Then $S - Y$ contains a base X of $\mathbf{M}_2^*$; hence $X \cup Y$ is independent in $\mathbf{M}_1 \vee \mathbf{M}_2^*$ and its cardinality is $t + r_2^*(S)$.

On the other hand if X is a base of $\mathbf{M}_1 \vee \mathbf{M}_2^*$ then it can be decomposed into the disjoint union $X = X_1 \cup X_2$ so that X_1 is independent in $\mathbf{M}_1$ and X_2 is a base of $\mathbf{M}_2^*$ (see Exercise 11.1.8). X_1 is clearly of cardinality at least t and is independent in $\mathbf{M}_2$ as well.

11.1.14 $\mathbf{M}_1$ and $\mathbf{M}_2$ are defined on the set $S = \{1, 2, 3\}$ by listing their bases $\mathbf{B}_1 = \{\{1, 2, \}, \{1, 3\}\}$ and $\mathbf{B}_2 = \{\{1, 2\}, \{2, 3\}\}$. Then $\mathbf{F}_1 \cap \mathbf{F}_2$ contains $\{1, 2\}$ but $\{3\}$ cannot be extended to an "independent" set of cardinality 2.

11.1.15 Consider matroids $\mathbf{M}_i = (S, \mathbf{F}_i)$ with $i = 1, 2, \ldots, k$ on the common underlying set $S = \{s_1, s_2, \ldots, s_n\}$. We prepare k disjoint copies $S^{(1)}, S^{(2)}, \ldots, S^{(k)}$ of S, using the notation $S^{(i)} = \{s_1^{(i)}, s_2^{(i)}, \ldots, s_n^{(i)}\}$ for $i = 1, 2, \ldots, k$.

We define two matroids on the set $T = S^{(1)} \cup S^{(2)} \cup \ldots \cup S^{(k)}$. One of them is the direct sum of the matroids $(S^{(i)}, \mathbf{F}_i)$. The other is a partitional matroid where a subset $X \subseteq T$ is independent if and only if $|X \cap \{s_j^{(1)}, s_j^{(2)}, \ldots, s_j^{(k)}\}| \leq 1$ holds for $j = 1, 2, \ldots, n$. Now these two matroids have a common base if and only if $\vee_{i=1}^{k} \mathbf{M}_i$ is the free matroid.

11.2.1 If $\mathbf{M}_1, \mathbf{M}_2$ are represented over a field T by the matrices $\mathbf{A}_1$ and $\mathbf{A}_2$, respectively, then their direct sum is represented over T by the matrix $\begin{pmatrix} \mathbf{A}_1 & \mathbf{O} \\ \mathbf{O} & \mathbf{A}_2 \end{pmatrix}$. Hence all we have to prove is that every uniform matroid $\mathbf{U}_{n,k}$ is representable over $\mathbf{R}$. A $k \times n$ matrix whose entries are algebraically independent over $\mathbb{Q}$ will clearly do since none of its $k \times k$ submatrices can be singular.

11.2.2 The affine representation of the two matroids is shown on the two skew lines of Fig. S.11.3. Their sum is then represented in Fig. S.11.4. One immediately sees that no four of the points are coplanar, hence the sum equals $\mathbf{U}_{6,4}$.

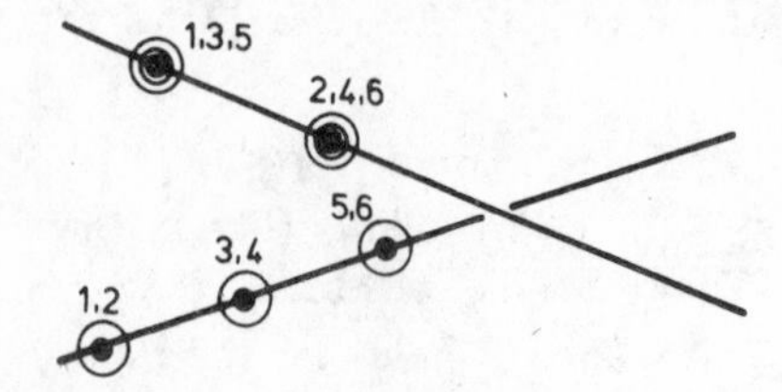

Fig. S.11.3

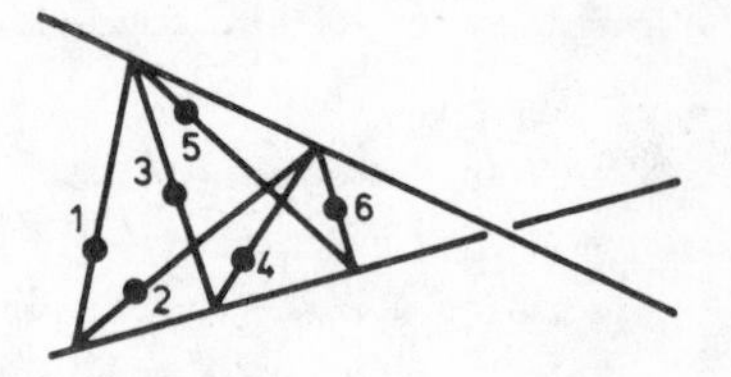

Fig. S.11.4

11.2.3 The sum is $\mathbf{U}_{6,4}$ again. Its affine representation is shown in Fig. S.11.5 (originally the two matroids were represented on the skew lines ab and cd).

11.2.4 Let $k > 2$, let $\mathbf{N}_0 = \mathbf{M}_1 \vee \mathbf{M}_2 \vee \ldots \vee \mathbf{M}_{k-1}$ and let $\mathbf{A}_0$ denote the matrix, obtained from the matrices $\mathbf{A}_1, \mathbf{A}_2, \ldots, \mathbf{A}_{k-1}$ by placing them one under another. If we place $\mathbf{A}_k$ under $\mathbf{A}_0$, we obtain the matrix $\mathbf{A}$. If X is independent in the column space matroid of $\mathbf{A}$ then it is independent in $\mathbf{N}_0 \vee \mathbf{M}_k$ as well (see Statement 11.2.1 for $k = 2$) hence $X = X_0 \cup X_k$ where X_0 and X_k are independent in $\mathbf{N}_0$ and in $\mathbf{M}_k$, respectively. If $k-1$

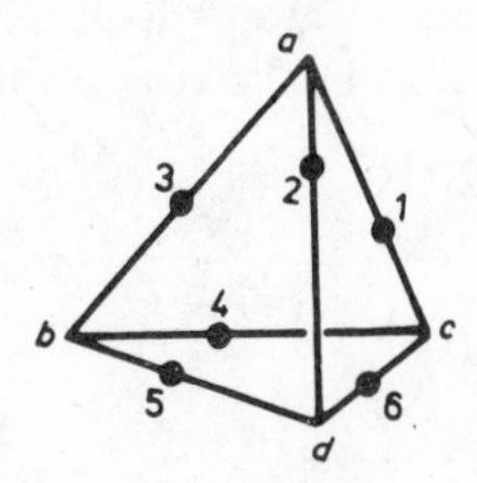

Fig. S.11.5

is still greater than 2, we repeat this reasoning. Finally X is decomposed into $X_1 \cup X_2 \cup \ldots \cup X_k$, where X_i is independent in $\mathbf{M}_i$ for every i.

11.2.5 The condition of the theorem required algebraic independence of all nonzero entries. However, the structure of a matroid $\mathbf{M}_i$ is reflected by linear dependencies among the columns of a single matrix $\mathbf{A}_i$. Such dependencies need not be excluded, so the weaker assumption may be: An algebraic relation among the nonzero entries of $\mathbf{A}$ is permitted only if it is reflected by the structure of one of the matroids $\mathbf{M}_i$. (We shall see the importance of this weakening in Section 14.1.)

11.2.6 $\mathbf{M}$ is shown in Fig. S.11.6. One can easily decompose it into the sum of two matroids with rank 2 each, starting from the skew lines AB and CD, for example. The results are the circuit matroids of the graphs of Fig. S.11.7.

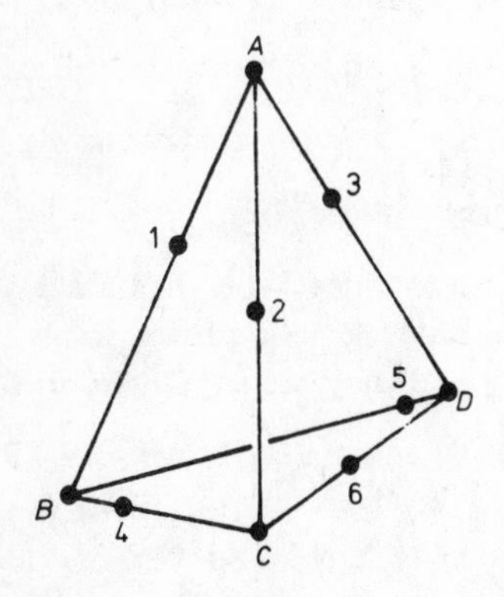

Fig. S.11.6

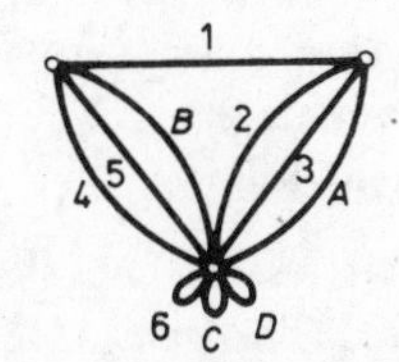

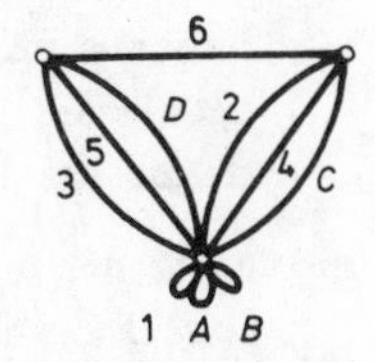

Fig. S.11.7

11.2.7 Figure S.11.6 of the previous problem is "redrawn" in Fig. S.11.8 but now points 1,3,4 and 6 "happen to be" coplanar. Of course, this is not the sum of the circuit matroids of the graphs of Fig. S.11.7 since $\{1,3\}$ and $\{4,6\}$ are independent in the two graphic matroids, respectively.

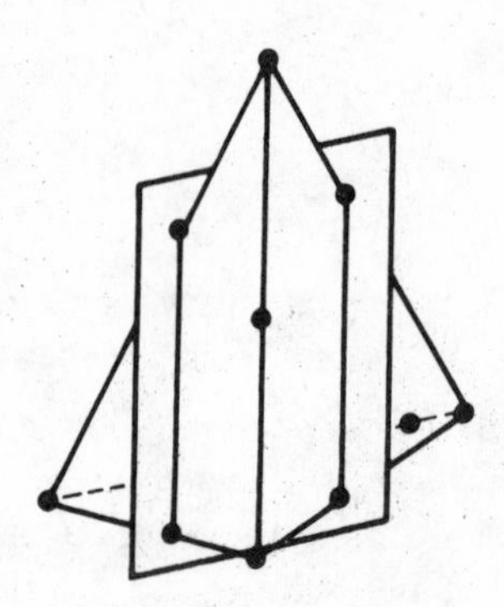

Fig. S.11.8

11.3.1 Suppose that the maximum for the induced subgraphs of G only were smaller than that for every subgraph of G. Hence the "real" maximum M were attained at such a subgraph H of G which is not induced. If H is disconnected, consider only that component H_i of it for which $M_i = \left\lceil \frac{e(H_i)}{v(H_i) - c(H_i)} \right\rceil$ is maximum. Clearly $M_i = M$, since if $\frac{a}{b} \geq \frac{c}{d}$ then $\frac{a+c}{b+d} < \frac{a}{b}$ holds for any positive numbers a, b, c and d. The subgraph H_i' induced by $V(H_i)$ has at least as many edges as H_i (and the same number of points and components), hence $\left\lceil \frac{e(H_i')}{v(H_i') - c(H_i')} \right\rceil \geq M$, contradicting either to the maximality of M or to the indirect assumption.

11.3.2 A spanning tree T_1 is connected, thus it interconnect the m classes. Hence it must contain at least $m-1$ edges connecting points from different classes. The same is true for $T_2, T_3, \dots, T_k$ and they are edge-disjoint. Hence the statement follows.

11.3.3 If $\mathbf{M}$ has k disjoint bases then the rank of S in the sum $\vee_{i=1}^{k}\mathbf{M}$ is at least $k \cdot r(S)$. Then Theorem 11.1.3 gives $k \cdot r(S) \leq k \cdot r(A) + |S - A|$ for every $A \subseteq S$.

11.3.4 The condition means that edges of each colour class should be circuit-free. Hence the answer is the maximum in Theorem 11.3.1.

11.3.5 (a) Let G be a bipartite graph with bipartition $X_1 \cup X_2$. Let a subset T of $E(G)$ be independent in $\mathbf{M}_i$ if the edges of T are all incident to different points in X_i. Both $\mathbf{M}_1$ and $\mathbf{M}_2$ are partitional matroids and the common independent sets of them are just the sets of point-disjoint edges.

(b) Let G be a directed graph. Let a subset T of $E(G)$ be independent in $\mathbf{N}$ if the outdegree of every point is at most one in the subgraph determined by T. $\mathbf{N}$ is a partitional matroid. The common independent sets of $\mathbf{N}$ and the circuit matroid $\mathbf{M}(G)$ of the graph are the appropriate subgraphs.

(c) This problem is **NP**-hard (for $k = 2$ it contains the problem of finding a Hamiltonian path, which was shown (Problem 5.4.9) to be **NP**-complete). Should we again call $T \subseteq E(G)$ "independent" if the degree of every point is at most k in the subgraph determined by T, these "independent" sets would not form a matroid (try with $k = 1$ and a circuit of length 4).

(d) This time the construction of part (c) gives a matroid (edges not covered by any point of X become bridges, the rest is partitioned by the stars of the points of X). The common independent sets of this matroid and the circuit matroid of the graph are the appropriate subgraphs.

11.3.6 Define $\mathbf{M}$ on the set S by linear independence in the usual way. Then we should decide whether $\vee_{i=1}^{k} \mathbf{M}_i$ is the free matroid and the condition is just identical to that of Theorem 11.1.3.

11.3.7 Suppose that G does not contain at least k edge-disjoint spanning trees. Let $S = E(G)$. By Exercise 11.3.3 there exists then a subset $A \subseteq S$ with $|S - A| < k(r(S) - r(A))$. Let us delete the edges of $S - A$ from G and denote the obtained graph by G_0. Clearly $r(S) = v(G) - 1$ and $r(A) = v(G) - p$ where p denotes the number of connected components of G_0 (including the possible isolated points as well). Decompose $V(G)$ into these p classes. Then the number n of edges connecting points from different classes is at most $|S - A|$. Hence $n \leq |S - A| <$ $< k[v(G) - 1 - (v(G) - p)] = k(p - 1)$, that is, the condition of Theorem 11.3.2 is violated.

11.3.8 Let $S = E(K_n)$ and $A \subseteq S$ be the edge set of an arbitrary subgraph G. We may suppose that G is induced (see Exercise 11.3.1), hence it is a complete graph with $v(G)$ points. Then the maximum in Theorem 11.3.1 is simply $\max_{2 \leq k \leq n} \left\lceil \frac{\binom{k}{2}}{k-1} \right\rceil = \left\lceil \frac{n}{2} \right\rceil$.

Chapter 12

12.1.1 After a possible renumbering of the ports we may suppose that $\mathbf{x} = \mathbf{H}\mathbf{y}$ with $\mathbf{x} = (u_1, u_2, \ldots, u_k, i_{k+1}, i_{k+2}, \ldots, i_n)^T$ and $\mathbf{y} = (i_1, i_2, \ldots, i_k, u_{k+1}, u_{k+2}, \ldots, u_n)^T$. Put shortly $\mathbf{x} = \begin{pmatrix} \mathbf{u}' \\ \mathbf{i}'' \end{pmatrix}$ and $\mathbf{y} = \begin{pmatrix} \mathbf{i}' \\ \mathbf{u}'' \end{pmatrix}$ and decompose $\mathbf{H}$ into four blocks as follows: $\begin{pmatrix} \mathbf{u}' \\ \mathbf{i}'' \end{pmatrix} = \begin{pmatrix} \mathbf{Z} & \mathbf{U} \\ \mathbf{V} & \mathbf{Y} \end{pmatrix} \begin{pmatrix} \mathbf{i}' \\ \mathbf{u}'' \end{pmatrix}$. Then the hybrid description of the other four multiports are given in **Box 12.1.**

12.1.2 The equations of the pair of nullator and norator is $\begin{pmatrix} 1 & 0 \\ 0 & 0 \end{pmatrix} \begin{pmatrix} u_1 \\ u_2 \end{pmatrix} + \begin{pmatrix} 0 & 0 \\ 1 & 0 \end{pmatrix} \begin{pmatrix} i_1 \\ i_2 \end{pmatrix} = \begin{pmatrix} 0 \\ 0 \end{pmatrix}$. The only way to describe it by a 2×2 matrix is $\begin{pmatrix} u_1 \\ i_1 \end{pmatrix} = \begin{pmatrix} 0 & 0 \\ 0 & 0 \end{pmatrix} \begin{pmatrix} u_2 \\ i_2 \end{pmatrix}$, and this is not a hybrid description.

If $\mathbf{A} \begin{pmatrix} u_1 \\ u_2 \end{pmatrix} + \mathbf{B} \begin{pmatrix} i_1 \\ i_2 \end{pmatrix} = \mathbf{0}$ describes a 2-port N, with $r(\mathbf{A}|\mathbf{B}) = 2$, then N either has a hybrid description or is a pair of nullator and norator. In order to show this denote by $\mathbf{a}_1, \mathbf{a}_2, \mathbf{b}_1, \mathbf{b}_2$ the columns of $(\mathbf{A}|\mathbf{B})$ in this order and observe that each pair $\{\mathbf{a}_1, \mathbf{a}_2\}, \{\mathbf{a}_1, \mathbf{b}_2\}, \{\mathbf{b}_1, \mathbf{a}_2\}$ and $\{\mathbf{b}_1, \mathbf{b}_2\}$ must be linearly dependent if N has no hybrid description. Thus one of $\{\mathbf{a}_1, \mathbf{b}_1\}$ and $\{\mathbf{a}_2, \mathbf{b}_2\}$ must be independent. (Both of them cannot, by the exchange property of bases.) Hence one of the two ports is a nullator and the other is a norator.

12.1.3 A pair of nullator and norator has no hybrid description. The controlled sources (Fig. 8.11) have one hybrid description each (for example, the current controlled voltage source has $\mathbf{Z}$ description only). Two hybrid descriptions are possessed by the gyrator ($\mathbf{Z}$ and $\mathbf{Y}$) and by the ideal transformer (just the other two descriptions). Finally the two 2-ports of Fig. S.12.1 have 3 and 4 hybrid descriptions, respectively (the former has no $\mathbf{Z}$ description).

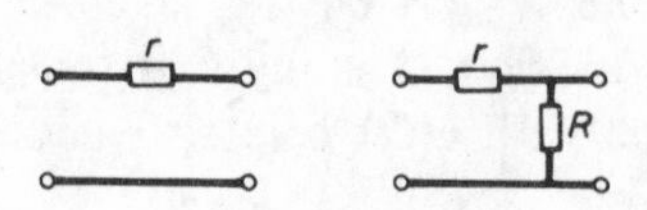

Fig. S.12.1

12.1.4 Yes, they are described by the respective equations $\begin{pmatrix} i_1 \\ i_2 \end{pmatrix} = \begin{pmatrix} 0 & 0 \\ 0 & 0 \end{pmatrix} \begin{pmatrix} u_1 \\ u_2 \end{pmatrix}$ and $\begin{pmatrix} i_1 \\ u_2 \end{pmatrix} = \begin{pmatrix} 0 & 0 \\ 1 & 0 \end{pmatrix} \begin{pmatrix} u_1 \\ i_2 \end{pmatrix}$.

12.1.5 Since the resistors R and r have the same current i, we have $u_2 = (R+r)i$ and $u_1 = Ri$. Thus the 2-port can be described by the equations $i_1 = 0,\ u_2 = (1+\frac{r}{R})u_1$.

12.1.6 A possible solution is shown in Fig. S.12.2.

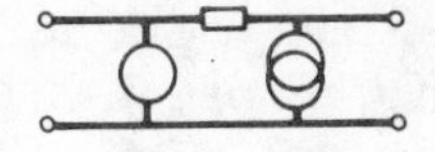

Fig. S.12.2

12.1.7 The "incoming" and "outgoing" currents of port 2 (and hence those of port 1 as well) are equal, due to the ideal transformer. Let i' and i'' denote the currents of the positive and the negative resistors, respectively, measured from the left to the right. Then $i'' = -i_1$ and $i' = -i_2$ (since the current of the nullator is zero). Furthermore $u_1 = Ri'$ and $u_2 = -Ri''$ (since the voltage of the nullator is also zero). Hence $u_1 = -Ri_2$ and $u_2 = Ri_1$, that is, the 2-port is a gyrator.

12.1.8 Points A and C are equipotential and so are points B and D. Hence the voltage of the resistor is the same as that between C and D. However, the current will flow along the path $CBAD$, that is, in "reversed order". Hence the 1-port is a resistor of value $-R$.

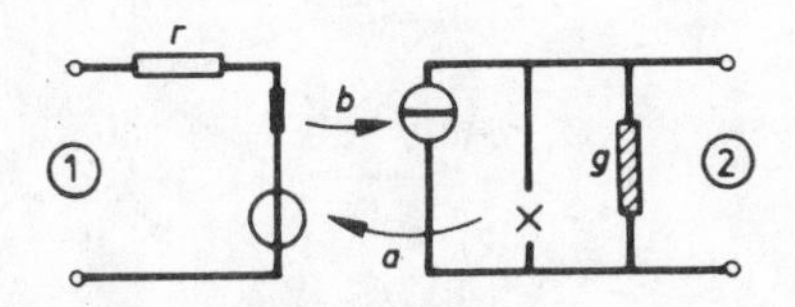

Fig. S.12.3

12.1.9 Consider at first the 2-port given by the hybrid immitance description $\begin{pmatrix} u_1 \\ i_2 \end{pmatrix} = \begin{pmatrix} r & a \\ b & g \end{pmatrix} \begin{pmatrix} i_1 \\ u_2 \end{pmatrix}$. Its realization is shown in Fig. S.12.3. In general, if a multiport is described by $(\mathbf{u}', \mathbf{i}'')^T = \mathbf{H}(\mathbf{i}', \mathbf{u}'')^T$ then the ports with single bar are realized by series connections of an impedance and some controlled voltage sources, and the ports with double bar by parallel connections of an admittance and some controlled current sources. Every nonzero diagonal entry of $\mathbf{H}$ corresponds to a resistor and every nonzero off-diagonal entry corresponds to a controlled source.

12.1.10 Consider, for example, the 3-port given by the equations $u_1 = Ri_2$, $u_2 = ri_2$, $i_1 = \alpha i_2$.

12.1.11 If N has no hybrid description then $\mathbf{M}_N \vee \mathbf{B}_n$ is not the free matroid. Then their common underlying set S has, by Theorem 11.1.3, a subset X with $r_M(X) + r_B(X) + |S - X| < |S|$, that is, $r_M(X) + r_B(X) < |X|$. We show that among these subsets X there is at least one containing both the voltage and the current of some ports and neither the voltage nor the current of the other ports.

Let, for example, $u_k \in X$ and $i_k \notin X$ hold for such an X which satisfies $r_M(X) + r_B(X) < |X|$. Changing X to $X \cup \{i_k\}$ does not increase the rank in $\mathbf{B}_n$ and increases that in $\mathbf{M}_N$ by at most one. Hence $X \cup \{i_k\}$ is also appropriate. Repeating this extension, if necessary, we obtain a set X_0 with $r_M(X_0) < |X_0| - r_B(X_0) = \frac{1}{2}|X_0|$, as requested.

12.1.12 Such a 4-port is given by the equations $u_1 + u_2 = R(i_1 + i_2 + i_4)$, $u_3 = ri_4$, $u_4 = \rho i_4$, $i_3 = i_4$. No 3-port with this property exists. By the previous problem an n-port with no hybrid immitance description has k ports whose voltages and currents appear in less than k equations. Clearly $k < n$. If $k = 1$ then this port is a norator and if $k = 2$ then the third port is a nullator.

In the case of the above 4-port we had $k = 2$ but the remaining 3 equations referred to two ports and none of them was a nullator. The matroid of this 4-port is the circuit matroid of the graph of Fig. S.12.4.

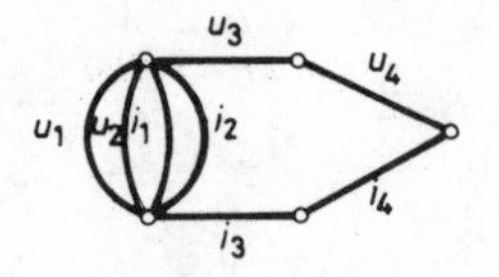

Fig. S.12.4

12.1.13 Suppose that N has no hybrid immitance description. Then by Problem 12.1.11 there is a subset $X \subset S$ with $r(X) < \frac{1}{2}|X|$ where X is the set of voltages and currents of some ports. Let B be such a base of $\mathbf{M}_N$ where $|B \cap X|$ is maximum (and hence equals to $r(X)$). Then $r^*(X) \geq r^*(X - B) = |X - B| > \frac{1}{2}|X|$ implies $r(X) \neq r^*(X)$. On the other hand, using the function φ introduced in Section 10.1, we obtain $\varphi(X) = X$, hence $\mathbf{M}_N = \varphi(\mathbf{M}_N^*)$ is impossible.

12.1.14 Proceed like in the solution of Exercise 6.2.4.

12.1.15 The realization of the 2-port given by $\mathbf{u} = \begin{pmatrix} z_{11} & z_{12} \\ z_{21} & z_{22} \end{pmatrix} \mathbf{i}$ is shown in Fig. S.12.5. A simpler realization is shown in Fig. 12.4 (cf. Exercise 12.1.7).

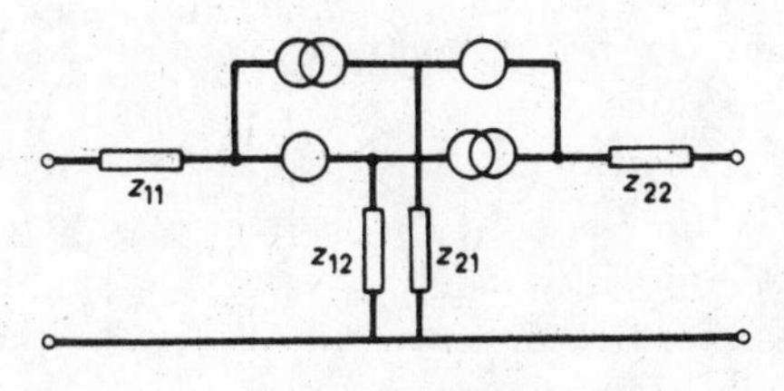

Fig. S.12.5

12.1.16 The result is a nullator if $g_x = 1$, $g_y = -1$; and a norator if $g_x = -1$ and $g_y = 1$. Otherwise a "usual" 1-port is obtained.

12.1.17 We obtain a pair of nullator and norator (port 3 is the norator and port 1 is the nullator).

12.1.18 See Fig. S.12.6 with the controlled sources $u_1 = 0$, $u_2 = Ri_1$ and $u_3 = 0$, $u_4 = ri_3$.

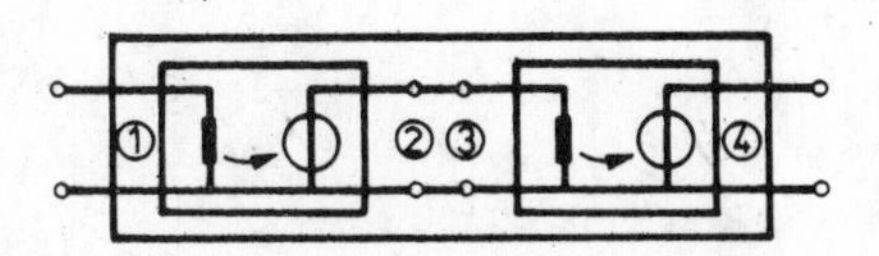

Fig. S.12.6

12.2.1 (a) The coefficient matrix of the 2-port equations is $\begin{pmatrix} 1 & -1 & 0 & r \\ 0 & 0 & k & -1 \end{pmatrix}$, hence matroid $\mathbf{A}$ is the circuit matroid of the graph of Fig. S.12.7. If current sources are placed both to A and B then i_1, i_2 are bridges, u_1, u_2 are loops in $\mathbf{G}$. Thus $\mathbf{G} \vee \mathbf{A}$ is not the free matroid ($\{u_1, u_2\}$ being a circuit). Then the network has no unique solution. However, $\mathbf{G} \vee \mathbf{A}$ is the free matroid in the other three cases.

(b) Now $\mathbf{A} \cong \mathbf{U}_{4,2}$, hence $\mathbf{G} \vee \mathbf{A}$ is the free matroid in any case.

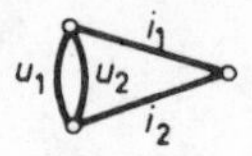

Fig. S.12.7

12.2.2 (a) i_1 and u_2 are bridges, u_1 and i_2 are loops in **A**. Hence if A and B are both voltage sources or both current sources then either i_1 and i_2 or u_1 and u_2 are both loops in **G**. Thus $\mathbf{G} \vee \mathbf{A}$ is not the free matroid. On the other hand, if the two sources at A and B are different then $\mathbf{G} \vee \mathbf{A}$ is the free matroid.

(b) Now u_2 is a loop in **A**, hence $\mathbf{G} \vee \mathbf{A}$ will be the free matroid if and only if B is a voltage source.

12.2.3 These matroids happen to be graphic, **G** and **A** are isomorphic to the circuit matroids of the graphs of Figures S.12.8 and S.12.9, respectively.

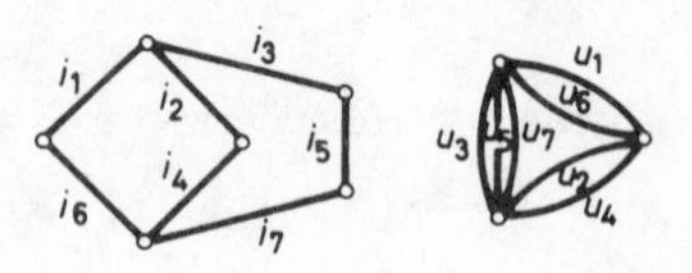

Fig. S.12.8

12.2.4 After contracting edge 5 and deleting edge 6, **G** will become the circuit matroid of the graph of Fig. S.12.10. If the 2-port has at least one hybrid immitance description then at least one of $\{u_1, u_2\}$, $\{u_1, i_2\}$, $\{i_1, u_2\}$ and $\{i_1, i_2\}$ is independent in **A** and can be extended to a base of **A** by any of $\{u_3, u_4\}$, $\{u_3, i_4\}$, $\{i_3, u_4\}$ and $\{i_3, i_4\}$. Then $\mathbf{G} \vee \mathbf{A}$ becomes the free matroid.

Fig. S.12.9

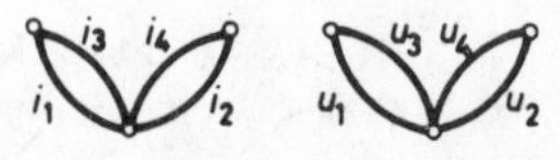

Fig. S.12.10

12.2.5 If the 2-port equations are $u_1 = R_1 i_1$, $u_2 = R_2 i_2$ then the network becomes singular if $R_1 = -R_3$ or if $R_2 = -R_4$.

12.2.6 Let T be a normal forest in the network consisting of resistors and ideal transformers only. (The sources can be disregarded, by Statement 12.2.1.) Let X denote the set, formed by the current copies of the T-edges and by the voltage copies of the edges not in T. Clearly X is independent in $\mathbf{G}$.

If $x \in S - X$ then one of the following cases will hold:
(a) $x = i_k$ for some resistor R_k. Then R_k is not in T, hence $u_k \in X$.
(b) $x = u_k$ for some resistor R_k. Then R_k is in T, hence $i_k \in X$.
(c) $x = i_p$ where $\{e_p, e_q\}$ are the edges for an ideal transformer. Then e_q is in T, hence $i_q \in X$.
(d) $x = u_p$ for a transformer like above. Then e_q is not in T hence $u_q \in X$.

Now observe that $\mathbf{A}$ is the direct sum of circuits of length 2 and each $x \in S - X$ shares such a circuit with an element in X. Hence $S - X$ is independent in $\mathbf{A}$. Thus X and $S - X$ determine nonsingular submatrices in the upper and in the lower parts of $\mathbf{W}$, respectively.

The proof of Theorem 6.2.4 is very similar. For example, if $\{e_p, e_q\}$ are the edges for a gyrator then $i_p, i_q \in X$ and $u_p, u_q \in S - X$.

12.2.7 A tree was called normal in Section 6.2 if it contained exactly one of the two edges of an ideal transformer. This definition must be modified so that in case of such a generalized ideal transformer the tree must contain exactly two of the three edges.

12.2.8 Let $\mathbf{Au} + \mathbf{Bi} = \mathbf{0}$ be the description of the 2-port. One can calculate that the network has unique solution if and only if

$$R_3 R_4 \begin{vmatrix} a_{11} & a_{12} \\ a_{21} & a_{22} \end{vmatrix} - R_3 \begin{vmatrix} a_{11} & b_{12} \\ a_{21} & b_{22} \end{vmatrix} + R_4 \begin{vmatrix} a_{12} & b_{11} \\ a_{22} & b_{21} \end{vmatrix} + \begin{vmatrix} b_{11} & b_{12} \\ b_{21} & b_{22} \end{vmatrix} \neq 0.$$

The necessary condition in Exercise 12.2.4 means that at least one of the four members is nonzero.

In case of the 2-port given in the solution of Exercise 12.2.5 all the entries of $\mathbf{A}$ and $\mathbf{B}$ are zero except $a_{11} = a_{22} = -1$, $b_{11} = R_1$, $b_{22} = R_2$. Then the above condition means $(R_1 + R_3)(R_2 + R_4) \neq 0$.

12.2.9 Let $\{e_p, e_q\}$ correspond to the edges of the pair of nullator and norator, with equations $i_p = 0$, $u_p = 0$. Then the requirement $S = S_1 \cup S_2$ (where S_1 is independent in $\mathbf{A}$ and S_2 is in $\mathbf{G}$) implies $\{u_q, i_q\} \subseteq S_1$. Hence the tree $S_1 \cap E_i$ and the tree complement $S_1 \cap E_u$ contain the norator edges but do not contain the nullator edges. (Saying it in another way: contract the norator edges and delete the nullator edges both in the graph of the current copy and in the dual graph of the voltage copy.)

12.2.10 Consider, say, a voltage controlled voltage source $i_p = 0$, $u_q = \alpha u_p$. Then $\{u_q, i_p\}$ is independent in $\mathbf{A}$ and — by the assumption — $\{u_p, i_q\}$

is independent in **G**. The other three types of controlled sources can be treated in a similar way and the unions of the independent subsets are also independent in **A** and in **G**, respectively. Hence the statement follows.

12.2.11 Let G be the graph of a network, consisting of multiports, capacitors and inductors. Accordingly, $E(G)$ can be decomposed into $E^M \cup E^C \cup E^L$. The generalization of Theorem 12.2.2 states that a necessary condition of the unique solvability is the existence of two subsets $X^C \subseteq E^C, X^L \subseteq E^L$ with the property that the complement of $X^C_u \cup (E^L - X^L)_u \cup X^L_i \cup$ $\cup (E^C - X^C)_i$ be a base of $\mathbf{G} \vee \mathbf{A}$.

12.2.12 Define a matroid $\mathbf{B}$ on $S = E_u \cup E_i$ so that the elements of $E^M_u \cup E^M_i$ be loops and the rest be the direct sum of circuits of form $\{u_k, i_k\}$. The condition is met if and only if $\mathbf{G} \vee \mathbf{A} \vee \mathbf{B} = (S, 2^S)$.

12.2.13 Instead of the examples of the quoted articles, we give a simpler — although artificial — one. Consider the first network of Fig. S.12.11. The voltages of both capacitors can independently be prescribed as initial values. The second network is the same, just resistor 4 is replaced by a controlled source. Hence σ still equals two while σ' is only one (after contracting the short circuit we obtain a circuit containing a voltage source and two capacitors).

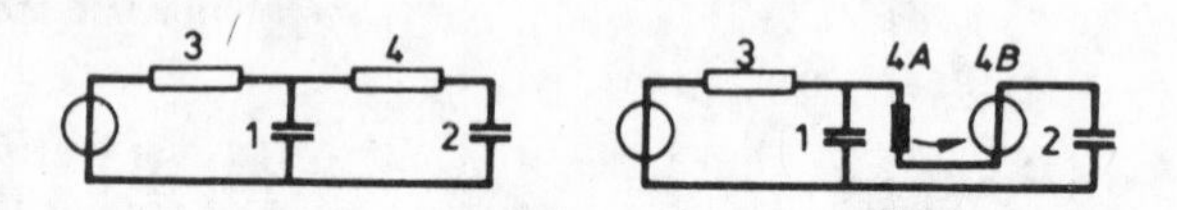

Fig. S.12.11

12.2.14 We have seen (**Boxes 2.1** and **4.2**) that the size of the matrices to be inverted during the two types of network analysis methods was essentially $r(G)$ and $n(G) = r^*(G)$. Hence if $E(G)$ is decomposed into two subsets (as in Problem 2.2.5) then $r(X) + r^*(E - X)$ can be smaller than min $[r(G), n(G)]$. The topological degree of freedom is $\min_{X \subseteq E}[r(X) + r^*(E - X)] = \min_{X \subseteq E}[2r(X) + |E - X|] - r(E)$. This latter shows that a base must be found in $\mathbf{M}(G) \vee \mathbf{M}(G)$, using the matroid partition algorithm.

The original problem was posed as early as in 1939 [Kron], and was solved at first without matroids [Iri, 1968c; Kishi and Kajitani, 1968; Ohtsuki et al., 1968]. The present solution is due to Iri and Tomizawa, [1975], see also [Iri, 1982].

12.3.1 We may suppose that the graph has no point of degree zero or one, and let us denote by v_2, v_3 and v_4 the number of points with degree 2, 3 and *at least* 4, respectively. Then $2e \geq 2v_2 + 3v_3 + 4v_4$. On the other hand, $e = 2v - 3 = 2(v_2 + v_3 + v_4) - 3$ also holds, leading to $2v_2 + v_3 \geq 6$. Hence there are at least three points with degree less than 4.

12.3.2 The condition of Theorem 12.3.3 must be checked. Since all the edges play equal roles, there is only one case to be considered, see Fig. S.12.12.

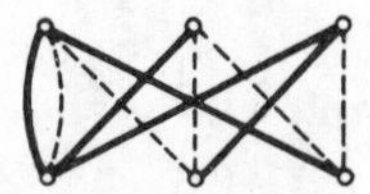

Fig. S.12.12

12.3.3 Suppose P were a point of degree 2 in G. Deleting P and its incident edges e, f we obtain a graph H, see Fig. S.12.13. We claim that H satisfies the conditions of Laman's theorem. Let namely $x \in E(H)$. G_x was the union of two spanning trees and clearly one of them had to contain e and the other had to contain f. Hence deleting them from the respective spanning trees we obtain the decomposition of H_x.

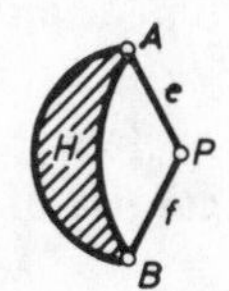

Fig. S.12.13

The generic system corresponding to H is rigid since G was a minimal counterexample. But then G is rigid unless P is on the line AB, a contradiction to genericity.

A similar but longer argument implies that G cannot have points of degree 3 either (which, by Exercise 12.3.1, gives a complete proof of Laman's theorem). A key step is that if the system of Fig. S.12.14 is rigid then so is at least one of the systems of Fig. S.12.15 as well [Henneberg, 1911]. Another difference is that the "forbidden" positions for point P are determined not by a linear algebraic equation (which gave line AB above) but by one of higher degree.

Fig. S.12.14

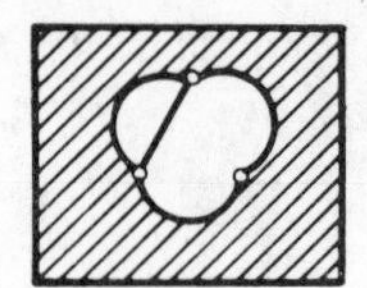

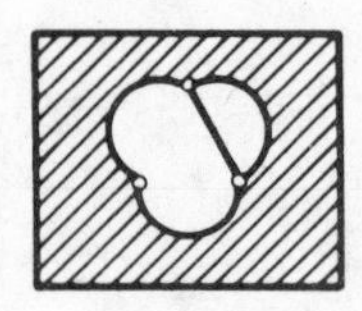

Fig. S.12.15

12.3.4 Indirectly, let G be a graph with v points and $e = 2v - 3$ edges and let j, k be two nonadjacent points so that the edge set of $G_0 = G + \{j, k\}$ cannot be decomposed into the union of two trees, although G satisfies the condition of Theorem 12.3.3. Then $V(G_0)$ can be decomposed into $V_1 \cup V_2 \cup \ldots \cup V_m$ so that the number t_0 of edges between points of distinct V_i's is at most $2m - 3$ (see Theorem 11.3.2).

On the other hand, doubling any edge x of G the edge set of the resulting graph G_x is the union of two trees, hence the number t of edges between points of distinct V_i's of the above decomposition is at least $2m - 2$. Since all but one of these edges are contained in G_0 as well, we have $t_0 = 2m - 3$ and $t = 2m - 2$, which also shows that j, k are in the same subset V_i (hence, in particular, $m < v$). Moreover, this edge x must connect points of distinct V_i's for every $x \in E(G)$, hence $e = t$. But $m \leq v - 1$ means $t = 2m - 2 \leq 2v - 4 = e - 1$, a contradiction [Recski, 1984b].

12.3.5 Let G be the graph of a 3-dimensional framework F, with v points and $e = 3v - 6$ edges. A necessary condition of the rigidity of F is that adding 3 new edges to G in any way (except if all three of them are parallel) the resulting graph should consist of three edge-disjoint spanning trees [Recski, 1984b].

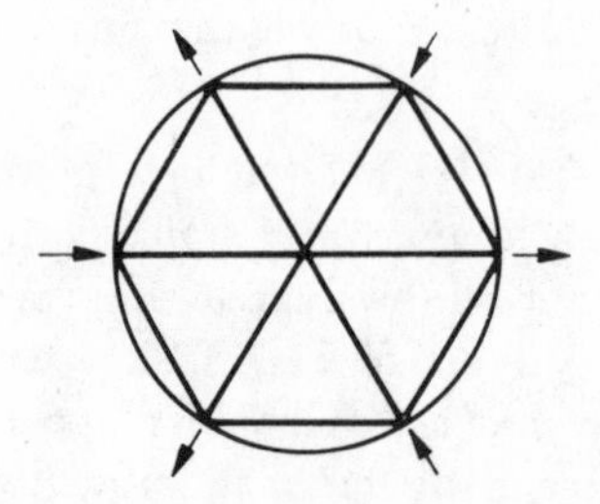

Fig. S.12.16

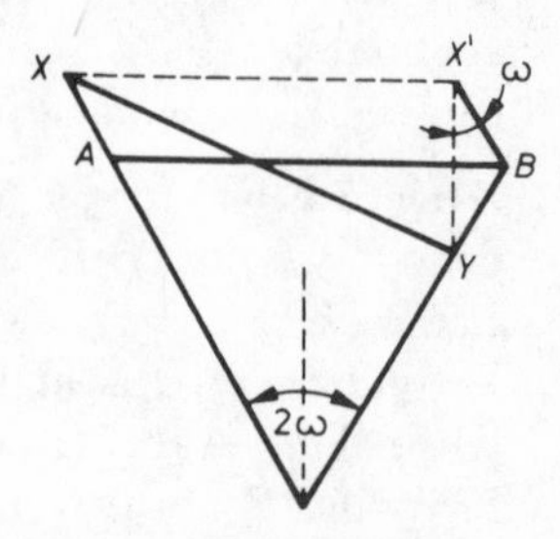

Fig. S.12.17

12.3.6 The effect of the forces shown in Fig. S.12.16 is an infinitesimal motion. Figure S.12.17 shows that the infinitesimal motion of $A \longrightarrow X$, $B \longrightarrow Y$ is the same as an infinitesimal translation $A \longrightarrow X$, $B \longrightarrow X'$ followed by an infinitesimal rotation $X' \longrightarrow Y$ around X (since the lines $X'Y$ and AB are perpendicular). The actual value of the angle ω did not play any role. More generally, if the points of a bipartite graph are on a conic section, the graph as a 2-dimensional framework is nonrigid [Bolker and Roth, 1980].

12.3.7 Only if the extra rods are "diagonals" in one of the horizontal grids; since if the system was nonrigid, further rods between the parallel planes do not help.

Chapter 13

13.1.1 (a) **M** contains at most $n-r$ loops. Hence in the worst case we need $n-r+1$ steps even for finding an independent singleton. If an element could not be added to an independent set, it will be useless in future steps as well, hence the total number of steps for finding a base is at most n. (This bound is best possible. If, say, $r=0$, we must ask whether $\{x\} \in \mathbf{F}$ for every $x \in S$.)

(b) Let $|X| = k$. We wish to decide whether there exists a base of $\mathbf{M}^*$ containing X, hence we must find a base of $\mathbf{M}$ in $S-X$. Hence at first we construct a maximal independent subset T of $S-X$ (this requires $n-k$ steps, see part (a) above) and then we check whether T is a base of $\mathbf{M}$ by asking if $T \cup \{x\}$ is independent (for every $x \in X$.) X is independent in $\mathbf{M}^*$ if and only if the answer to all these questions is negative. Hence we need at most k further steps, leading to n steps altogether.

13.1.2 We prove that $|I \cap T| = r_1(T)$, while $|I \cap (S-T)| = r_2(S-T)$ can be shown in a similar way. $|I \cap T| \le r_1(T)$ is obvious since $I \in \mathbf{F}_1$. Let $a \in T - I$ and suppose $(I \cap T) \cup \{a\} \in \mathbf{F}_1$ would hold. Then either $I \cup \{a\} \in \mathbf{F}_1$, implying the existence of an edge (a,t), a contradiction, or the circuit of $\mathbf{M}_1$, contained in $I \cup \{a\}$, would contain an element of $S-T$, contradicting to the definition of T.

13.1.3 Let $|S| = n$. The number of steps to construct the graph is proportional to n^2 (decide for any fixed x and for every y whether (x,y) is an edge). Dijkstra's algorithm requires cn^2 steps as well, hence the total number of steps in the algorithm is proportional to n^3.

13.1.4 Set A of the neighbours of s at the end of the algorithm is just the complement of B. If $x \in A$ then the set of those points which can be reached from x along a directed path (including x itself) is just the unique circuit of $B \cup \{x\}$.

13.1.5 $W' \ge W$ is clear since the subset X where the maximum W is attained also belongs to $\mathbf{F}_1 \cap \mathbf{F}_2$. They need not be equal. For example, consider the circuit matroids of the graphs of Fig. 13.5. If $w_1 = 5$, $w_2 = 1$ and $w_3 = 2$ then $W' = 5$ and $W = 3$.

13.1.6 Let $A(\mathbf{M}_1, \mathbf{M}_2)$ denote the polynomial algorithm which finds the maximum cardinality of the common independent sets of $\mathbf{M}_1$ and $\mathbf{M}_2$ and the maximum weight $W(\mathbf{M}_1, \mathbf{M}_2)$ among these sets. Apply this algorithm not only for the pair $\mathbf{M}_1, \mathbf{M}_2$ but for $\mathbf{M}_1^{(i)}, \mathbf{M}_2^{(i)}$ as well, where $\mathbf{M}_j^{(i)}$ is obtained from $\mathbf{M}_j$ by truncating it as long as its rank becomes i. Then $W' = \max_{i \le k} W(\mathbf{M}_1^{(i)}, \mathbf{M}_2^{(i)})$.

13.1.7 At the first instance of Step 2 we obtain the graph of Fig. S.13.1 (dotted edges are not shown), hence $I = \{1\}$ for $k=1$. At the second instance the graph is that of Fig. S.13.2 (including a dotted edge of weight 3).

Fig. S.13.1 **Fig. S.13.2**

This edge also becomes continuous after the modification in Step 4, then $I = \{2,3\}$ for $k = 2$ and the algorithm terminates.

13.1.8 The number of operations for Step 2 is proportional to n^2, just like in Exercise 13.1.3. Step 4 increases the cardinality of T, hence we can go from Step 2 to Step 4 (without increasing I) at most n times. Thus $k = |I|$ increases after at most cn^3 steps, and the total number of steps of the algorithm is proportional to n^4.

For a practical realization let us remark that when we go from Step 4 to Step 2, we need not start Dijkstra's algorithm from the beginning. The old values of d_i (see Algorithm 3.2) can be applied again.

13.1.9 If the length of the minimal directed path is less than 4, or if this path is (s, x_1, x_2, x_3, t) so that $(x_1, x_2), (x_2, x_3)$ and (x_3, t) denote augmentations in three distinct matroids then the statement is obvious. Suppose therefore that the first and third of the above edges refer to $\mathbf{M}_1$ and the second one to $\mathbf{M}_2$. Then $x_2 \in I_1$, $x_3 \in I_2$, $I_1 \cup \{x_3\} \in \mathbf{F}_1$, but $I_1 \cup \{x_1\}$ is not independent in $\mathbf{M}_1$ (since (x_1, t) is not an edge), it rather contains a circuit C_1 of $\mathbf{M}_1$ with $x_1, x_2 \in C_1$. We must prove that $(I_1 \cup \{x_1, x_3\}) - \{x_2\} \in \mathbf{F}_1$. Otherwise it contained a circuit C_2 of $\mathbf{M}_1$ with $x_1, x_3 \in C_2$, $x_2 \notin C_2$. But then there exists a circuit $C_3 \subseteq (C_1 \cup C_2) - \{x_1\}$ with $x_3 \in C_3$, implying that C_3 is contained in $I_1 \cup \{x_3\}$, a contradiction.

13.1.10 We use induction on the length of the path. Paths of length at most four were already treated. Let $(s, x_1, x_2, \ldots, t)$ be a minimal path and suppose that the edge (x_1, x_2) refers to an augmention in matroid $\mathbf{M}_1$. Let $I_1' = (I_1 \cup \{x_1\}) - \{x_2\}$ and $I_k' = I_k$ for $k > 1$. The new directed graph (with respect to this partition) contains a directed path whose length is smaller by one, hence the inductive hypothesis can be applied. All we need to check is that if an edge (x_k, x_{k+1}) or (x_k, t) refers to an augmentation or to an extension in $\mathbf{M}_1$ then this change is possible with I_1' as well; and that the path remains minimal. Their proof follows the same line as in the solution of the previous problem.

13.1.11 Let $I_1 = \{2,4\}$, $I_2 = \{3\}$ for the circuit matroids of the graphs of Fig. S.13.3. The directed graph is shown in Fig. S.13.4. The augmentation along (s, x_1, x_4, t) is possible, but it is not the case along the longer path, since $(I_1 \cup \{1,3\}) - \{2,4\} \notin \mathbf{F}_1$.

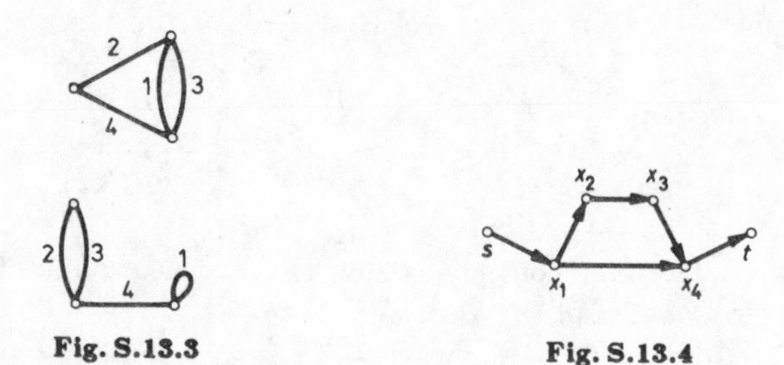

Fig. S.13.3 **Fig. S.13.4**

13.1.12 The number of operations is proportional to n^3, just like in case of Algorithm 13.2. Its proof is similar to that in the solution of Exercise 13.1.3. However, the estimation of the number of operations for constructing the graph is more complicated (only kn^2 follows in a straightforward way, although n^2 is required).

Let $|I_i| = x_i$. The graph can be prepared in $\sum_{i=1}^{k} x_i(n - x_i)$ steps as follows: For every i, check whether there exists any element $a \in S - I_i$ which can be added to I_i in the matroid $\mathbf{M}_i$. If yes, we obtain an edge (a, t); otherwise we obtain edges of form (a, b) after at most x_i steps. Now $\sum_{i=1}^{k} x_i \le n$ leads to $\sum_{i=1}^{k} x_i(n - x_i) \le \sum_{i=1}^{k} x_i n \le n^2$.

13.1.13 Suppose that we consider $\mathbf{M}_1 \cap \mathbf{M}_2^*$ rather than $\mathbf{M}_1 \vee \mathbf{M}_2$. We saw in Exercise 13.1.1(b) that an independence test in the dual matroid requires n steps. Hence the graph can be constructed in cn^3 (instead of cn^2) steps and the total number of operations is proportional to n^4 rather than to n^3.

13.1.14 The number of operations is proportional to n_0^3 where n_0 is the cardinality of the underlying set T (see the solution of Problem 11.1.15) of the new matroid. Hence $n_0 = kn$ and the total number is $k^3 n^3$ rather than n^3.

13.1.15 The difference $s_k - s_{k-1}$ is just the actual value of $m_1 + m_2$ in Algorithm 13.3. Since neither m_1 nor m_2 can increase during the algorithm, the statement is obvious.

13.2.1 (a) Let $x \in S - T_0$. If x were not a bridge in $\mathbf{N}$, the relation $R(S - \{x\}) = R(S)$ would hold, that is, $R(S) = \sum_{i=1}^{k} r_i(T_0) + |S - T_0| = \min_{T \subseteq S - \{x\}} [\sum_{i=1}^{k} r_i(T) + |S - T| - 1] \le \sum_{i=1}^{k} r_i(T_0) + |S - T_0| - 1$, a contradiction.

(b) If x is a bridge then $R(S - \{x\}) = R(S) - 1$, hence the sign $\le$ above changes to equality.

13.2.2 Let $T \subseteq S$ be a subset satisfying $R(S) = \sum_{i=1}^{k} r_i(T) + |S - T|$. The previous exercise implies $S - T = \emptyset$, hence the statement follows.

13.2.3 Let $\mathbf{M}_1$ be the direct sum of a circuit $\{a, b\}$ and a bridge c. Let the construction of $\mathbf{M}_2$ and $\mathbf{M}_3$ be similar, except that the bridge is a and b, respectively. Then the left hand side equals 1 and the right hand side equals 2.

13.2.4 A subset X is independent in $\mathbf{M}_\phi \vee \mathbf{N}_\phi$ if and only if there exist two subsets $Y_1, Y_2 \subseteq S$ so that $\phi(Y_1) = X_1$, $\phi(Y_2) = X_2$, $X_1 \cup X_2 = X$, Y_1 is independent in $\mathbf{M}$ and Y_2 is independent in $\mathbf{N}$. This latter condition is equivalent to requiring that $\phi(Y) = X$ and Y is independent in $\mathbf{M} \vee \mathbf{N}$, where $Y = Y_1 \cup Y_2$. This exactly means that X is independent in $(\mathbf{M} \vee \mathbf{N})_\phi$.

13.2.5 Since the rank of the matroids $\mathbf{N}_i = \mathbf{M}_i \vee \mathbf{M}_i \vee \ldots \vee \mathbf{M}_i$ (k times) is $k \cdot \ell$ (both for $i = 1$ and $i = 2$), Theorem 13.2.1 implies $k\ell \leq kr_i(X) + |S - X|$ for every $X \subseteq S$. Apply this for the choice $X = X_0$; $i = 1$; and also for $X = S - X_0$; $i = 2$. Then $r_1(X_0) + r_2(S - X_0) \geq \ell$ follows. Since X_0 was arbitrary, the assertion follows from Theorem 13.2.2.

13.2.6 Following Woodall [1974] define two matroids on X_2 as follows. Contract A_1 in $\mathbf{M}$ and restrict the result to X_2. Denote the obtained matroid by $\mathbf{M}_1$. (If $X_2 \cap A_1$ is nonempty, its elements become loops.) Similarly, $\mathbf{M}_2$ is the restriction of $\mathbf{M}/(X_1 - A_1)$ to X_2. Now (B3″) means that X_2 can be decomposed into $(X_2 - A_2) \cup A_2$ where $X_2 - A_2$ is independent in $\mathbf{M}_1$ and A_2 is independent in $\mathbf{M}_2$. That is, if R denotes the rank function of $\mathbf{M}_1 \vee \mathbf{M}_2$ then $R(X_2) = |X_2|$ must hold.

Let r, r_1 and r_2 denote the rank functions of $\mathbf{M}$, $\mathbf{M}_1$ and $\mathbf{M}_2$, respectively. Then $R(X_2) = \min_{Y \subseteq X_2}[r_1(Y) + r_2(Y) + |X_2 - Y|] =$ $= \min_{Y \subseteq X_2}[r(Y \cup A_1) - r(A_1) + r(Y \cup (X_1 - A_1)) - r(X_1 - A_1) +$ $+ |X_2 - Y|]$. Now $|X_2| = r(A_1) + r(X_1 - A_1)$ — since X_1 and X_2 are bases of $\mathbf{M}$ —, hence all we have to prove is $r(Y \cup A_1) + r(Y \cup (X_1 - A_1)) - |Y| \geq$ $\geq |X_2|$ for every $Y \subseteq X_2$ (since $Y = \emptyset$ assures then equality). But this is just the submodularity since $|Y| = r(Y)$.

13.2.7 We prove that if $\mathbf{M}$ is coordinatizable over $\mathbf{R}$ then so is its homomorphic image $\mathbf{N}$. (We saw during the solution of Exercise 11.2.1 that direct sum of representable matroids is representable again and that the sum is the homomorphic image of direct sum.)
Moreover, the homomorphic image should be considered for the function shown in Fig. S.13.5 only, since every function is a composition of such ones. Let $\mathbf{x}_0, \mathbf{x}_1, \ldots, \mathbf{x}_k$ represent the elements of $\mathbf{M}$. We prove that the elements of $\mathbf{N}$ can be represented by $\mathbf{y}_1, \mathbf{y}_2, \ldots, \mathbf{y}_k$ so that $\mathbf{y}_1 = \lambda \mathbf{x}_0 + \mu \mathbf{x}_1$ (for suitable λ and μ) and $\mathbf{y}_i = \mathbf{x}_i$ for $i > 1$.

In order to check this we must consider such sets T which contain y_1. Let V denote the subset which corresponds to $T - \{y_1\}$ in $\mathbf{M}$. If $V \cup \{x_0\}$ and $V \cup \{x_1\}$ are both dependent in $\mathbf{M}$ then the subspace spanned by the vectors of V contains $\mathbf{x}_0$ and $\mathbf{x}_1$, hence their linear combination as well, that is, T is dependent in $\mathbf{N}$. On the other hand, T is independent

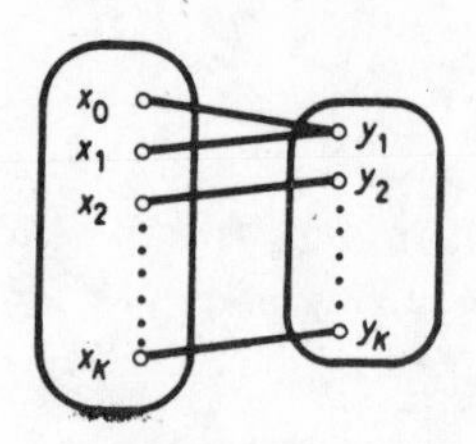

Fig. S.13.5

in $\mathbf{N}$ if exactly one of $V \cup \{x_0\}$ and $V \cup \{x_1\}$ is independent in $\mathbf{M}$. (We need $\lambda \neq 0$, $\mu \neq 0$ for this latter statement.)

Finally, if $V \cup \{x_0\}$ and $V \cup \{x_1\}$ are both independent in $\mathbf{M}$ then neither $\mathbf{x}_0$, nor $\mathbf{x}_1$ arise as a linear combination of the vectors of V. However, $\lambda\mathbf{x}_0 + \mu\mathbf{x}_1$ could still be a linear combination of these vectors. In order to avoid this (in the case of a given set V) a particular ratio $\lambda : \mu$ of the coefficients must be excluded. The number of such forbidden ratios is at most 2^{k-1} (the number of different subsets of the set $\{x_2, x_3, \ldots, x_k\}$), hence we can always find such coefficients from the infinite set of real numbers.

13.2.8 Let the matroids $\mathbf{M}_1, \mathbf{M}_2$ be coordinatizable over the fields F_1 and F_2, respectively. If $F_1 = F_2$ then by the above proof $\mathbf{M}_1 \vee \mathbf{M}_2$ is coordinatizable over a sufficiently large extension of F_1. If $F_1 \neq F_2$ but they have a common extension F then $\mathbf{M}_1 \vee \mathbf{M}_2$ can be coordinatized over a suitable extension of F. Finally, if F_1 and F_2 have no common extension then $\mathbf{M}_1 \vee \mathbf{M}_2$ may or may not be coordinatized at all. For example, $\mathbf{F}_7 \vee \mathbf{F}_7^-$ is graphic but the non-coordinatizable example of Exercise 9.2.6 can also be considered as a sum.

13.2.9 The inequality $\leq$ is obtained in the usual way, while $\geq$ is proved by induction on $|X|$. If there exists a subspace F_i with $r(X - \{x_i\}) < r(X)$ then apply the formula for $X - \{x_i\}$ to find a subset $Y \subseteq X - \{x_i\}$ with $r(X - \{x_i\}) = r(\cup_{x_j \in Y} F_j) + |X - \{x_i\} - Y|$. This Y will be appropriate for $R(X)$ as well since $r(X - \{x_i\}) = r(X) - 1$. On the other hand, if no such F_i exists, that is, if $x_i \in <X - \{x_i\}>$ for every i then, by the assumption on the general position, the whole subspace F_i is contained in $<X - \{x_i\}>$. Thus $<X>$ contains every subspace F_i, leading to $r(X) = r(\cup_{x_i \in X} F_i)$, that is, the choice $Y = X$ is appropriate.

13.3.1 The same matroid $\mathbf{M}$. In order to see this let $X \subseteq Y$ be two subsets of A. Then $f(Y) - f(X)$ is at most $|Y - X|$, while the difference $|A - X| - |A - Y|$ is exactly $|Y - X|$. Hence the minimum in $\min_{A_0 \subseteq A}[f(A_0) + |A - A_0|]$ is attained if $A_0 = A$.

13.3.2 Yes. The proof requires the submodularity of r and that $f(X \cup Y) = r((S-X) \cap (S-Y))$, $f(X \cap Y) = r((S-X) \cup (S-Y))$.

13.3.3 If $r(X)$ and $r(Y)$ are less than k then

$$f(X)+f(Y) = r(X)+r(Y) \geq r(X \cup Y)+r(X \cap Y) \geq f(X \cup Y)+f(X \cap Y).$$

If, say, $r(X) \geq k$ then $r(X \cup Y) \geq k$, hence $f(X) = f(X \cup Y) = k$. Now

$$f(X) + f(Y) = k + \min[k, r(Y)] = f(X \cup Y) + \min[k, r(Y)],$$

and $\min[k, r(Y)] \geq \min[k, r(X \cap Y)] = f(X \cap Y)$.

13.3.4 Let $P = \cup_{i \in X} A_i$ and $Q = \cup_{i \in Y} A_i$. Consider the sets $R = \cup_{i \in X \cup Y} A_i$ and $S = \cup_{i \in X \cap Y} A_i$. Clearly, $P \cup Q = R$ but only $S \subseteq P \cap Q$ follows in general. Hence, using the submodularity of r, $f(X) + f(Y) = r(P) + r(Q) \geq r(P \cup Q) + r(P \cap Q) \geq r(R) + r(S) = f(X \cup Y) + f(X \cap Y)$.

13.3.5 f is clearly modular. On the other hand, if m is an arbitrary modular function then put $m(\emptyset) = c$ and define the function ρ by $\rho(a) = m(\{a\}) - c$. Then $m(X) = c + \sum_{a \in X} \rho(a)$ follows. This can be shown by induction on $|X|$, since $m(X) = m(X - \{x\}) + m(\{x\}) - m(\emptyset)$ must hold for every $x \in X$.

13.3.6 Check at first that

$$|A - A_1| + |B - B_1| = |(A \cup B) - (A_1 \cup B_1)| + |(A \cap B) - (A_1 \cap B_1)|$$

holds for any $A_1 \subseteq A$ and $B_1 \subseteq B$. (Observe that $A \cap B - (A_1 \cup B_1)$ is counted twice on both sides of the equality.) Then, by the submodularity of f,

$$\tilde{f}(A) + \tilde{f}(B) \geq \min_{\substack{A_1 \subseteq A \\ B_1 \subseteq B}} [f(A_1 \cup B_1) + f(A_1 \cap B_1) +$$

$$+ |(A \cup B) - (A_1 \cup B_1)| + |(A \cap B) - (A_1 \cap B_1)||,$$

and this is greater than or equal to

$$\min_{\substack{C \subseteq A \cup B \\ D \subseteq A \cap B}} [f(C) + |(A \cup B) - C| + f(D) + |(A \cap B) - D|] = \tilde{f}(A \cup B) + \tilde{f}(A \cap B).$$

13.3.7 (a) Let $S = \{a, b, c\}$, $f(\emptyset) = 0$, $f(S) = f(\{a, b\}) = 2$ and $f(X) = 1$ for every other subset X. This is not submodular (let $X = \{a, c\}$, $Y = \{b, c\}$) and $f = \tilde{f}$, hence $\tilde{f}$ does not arise as the rank function of any matroid.

(b) Let $S = \{a, b, c\}$, $f(\emptyset) = 0$, $f(S) = 6$ and $f(X) = 3$ for every other subset X. This is not submodular either (let $X = \{a, c\}$, $Y = \{b, c\}$) but $\tilde{f}(A) = |A|$, hence it gives the rank function of the free matroid on S.

13.3.8 (a) $f(\emptyset) = 0$, $f \geq 0$ implies that $\overline{f}(\emptyset) = 0$. Monotonicity is also obvious: If $A \subseteq B$ then $\overline{f}(B) = f(X)$ for some $X \supseteq B$ and then $\overline{f}(A) \leq f(X)$. In order to prove submodularity let $\overline{f}(A) = f(X)$ and

$\overline{f}(B) = f(Y)$ for some $X \supseteq A$ and $Y \supseteq B$. Then $\overline{f}(A) + \overline{f}(B) = = f(X) + f(Y) \geq f(X \cup Y) + f(X \cap Y) \geq \overline{f}(A \cup B) + \overline{f}(A \cap B)$, since $X \cup Y \supseteq A \cup B$ and $X \cap Y \supseteq A \cap B$.

(b) Apply the operation $g(X) \to \tilde{g}(X)$ of Theorem 13.3.1 to the function $g(X) = \overline{f}(X)$ of part (a) of this problem and write f'' instead of $\tilde{\overline{f}}$. We claim that $f'(A) = f''(A)$ for every $A \subseteq S$.

At first observe that if $f''(A) = \min_{Y \subseteq A}[\min_{Y \subseteq X} f(X) + |A - Y|]$ is attained at some $Y_0 \subseteq A$ and at some $X_0 \supseteq Y_0$ then $A - X_0 \subseteq A - Y_0$. Hence

$$f''(A) = f(X_0) + |A - Y_0| \geq f(X_0) + |A - X_0| \geq$$

$$\geq \min_{X \subseteq S}[f(X) + |A - X|] = f'(A).$$

For the other direction $f' \geq f''$ let T be a subset of S so that $f'(A) = = f(T) + |A - T|$. With the choice $Y = T \cap A$ and $X = T$ we have $f'(A) = f(T) + |A - T| \geq \min_{Y \subseteq A}[\min_{Y \subseteq X} f(X) + |A - Y|] = f''(A)$.

13.3.9 Let $k = \min_{X \subseteq S}[r_1(X) + r_2(S - X)]$, we must find a common independent subset of $\mathbf{M}_1$ and $\mathbf{M}_2$ with cardinality k. Let $f(x) = \min[k, r_1(X)]$ and $g(X) = \max[0, k - r_2(S - X)]$. $f(X)$ is submodular and $g(X)$ is supermodular (see Exercises 13.3.2–3); furthermore $g(X) \leq f(X)$ for every X, by the definition of k. Applying the sandwich theorem, there exists a modular, integer valued function $h(X)$ with $g(X) \leq h(X) \leq f(X)$ for every X.

Let A denote the set of those elements x of S which satisfy $h(\{x\}) = 1$. Then $h(X) = |X \cap A|$, by Exercise 13.3.5. We claim that A is independent in $\mathbf{M}_1$ and in $\mathbf{M}_2$ and has cardinality k. $h(A) = |A| \leq f(A)$ implies $|A| \leq k$ and that A is independent in $\mathbf{M}_1$. On the other hand, $h(S - A) = 0 \geq g(S - A)$ implies $r_2(A) \geq k$, so A is independent in $\mathbf{M}_2$ as well.

13.3.10 The necessity of the theorem is trivial (apply the submodularity for the sets X and $(X \cup Y) - \{a\}$). The sufficiency will be proved indirectly, so suppose that $f(X) + f(Y) < f(X \cup Y) + f(X \cap Y)$ holds for a pair X, Y, and that among all counterexamples let $|X|$ be minimum. We may suppose that $X - Y \neq \emptyset$ and then, by the minimality, $f(X - \{a\}) + f(Y) \geq \geq f((X - \{a\}) \cup Y) + f((X - \{a\}) \cap Y)$ holds for any $a \in X - Y$. Furthermore, $f(X) - f(X - \{a\}) \geq f(X \cup Y) - f((X \cup Y) - \{a\})$ holds by our condition. The sum of these two inequalities contradicts to the indirect assumption since $(X - \{a\}) \cup Y = (X \cup Y) - \{a\}$ and $(X - \{a\}) \cap Y = X \cap Y$ hold, due to $a \notin Y$.

13.3.11 This is a trivial consequence of the previous problem.

13.3.12 Let X, Y be two subsets. If $f(X) = f_i(X)$ and $f(Y) = f_i(Y)$ for the same i then simply

$$f(X) + f(Y) \geq f_i(X \cup Y) + f_i(X \cap Y) \geq f(X \cup Y) + f(X \cap Y).$$

Otherwise $f(X) = f_1(X) < f_2(X)$ and $f(Y) = f_2(Y) < f_1(Y)$. Then $\phi(X) > 0$ and $\phi(Y) < 0$ for $\phi = f_1 - f_2$ which contradicts to monotonicity unless X or Y is the empty set. Finally, if X or Y is the empty set, the submodularity obviously holds.

Chapter 14

14.1.1 The equations of the 2-port are $u_1 = 2u_2$, $i_2 = (u_2/1) + (-2)i_1$. These give the rows of $\mathbf{A}'$ and the second row of $\mathbf{A}''$, while the first row of $\mathbf{A}''$ is the sum of the rows of $\mathbf{A}'$. Hence both matrices describe the same 2-port . $\mathbf{M}'$ is the circuit matroid of the graph of Fig. S.14.1 while $\mathbf{M}''$ is the uniform matroid $\mathbf{U}_{4,2}$. The strong generality assumption excludes this 2-port in case of description $\mathbf{A}''$.

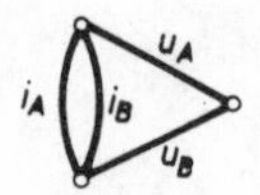

Fig. S.14.1

14.1.2 The "strong" assumption excludes every relation (including those which are excluded by the "weak" one), hence it cannot be weaker than the "weak" one. The previous exercise shows that it is actually stronger than the "weak" assumption (since the 2-port is excluded by the "weak" assumption neither in case of $\mathbf{A}'$ nor in case of $\mathbf{A}''$).

14.1.3 $\sigma = \max\{|X^C| + |X^L|\}$, where the maximum is taken for every $X^C \subseteq$ $\subseteq E^C$, $X^L \subseteq E^L$ satisfying the condition that the complement of $X_u^C \cup$ $\cup (E^L - X^L)_u \cup X_i^L \cup (E^C - X^C)_i$ is a base of $\mathbf{G} \vee \mathbf{A}$.

14.1.4 Prescribe small weights for the elements of $E_u^C \cup E_i^L$ and large weights for the others. Then find a maximum weight common base of $\mathbf{G} \vee \mathbf{A}$ and $\mathbf{B}^*$, where $\mathbf{B}$ is the matroid introduced in the solution of Problem 12.2.12.

14.1.5 Replacing the voltage source by a short circuit, the resulting matroids $\mathbf{G}$ and $\mathbf{A}$ are shown as the circuit matroids of the graphs of Figs S.14.2 and S.14.3, respectively. If we try $X^C = E^C$ at first, we shall immediately see that the complement of $\{u_1, u_2\}$ is a base of $\mathbf{G} \vee \mathbf{A}$ (for example, $\{i_1, i_2, i_B, u_3, u_B\}$ is a base of $\mathbf{G}$ and $\{i_3, u_A, i_A\}$ is a base of $\mathbf{A}$). Hence the initial values of the voltages of both capacitors can independently be prescribed.

14.1.6 Two matrices $\mathbf{A}'$ and $\mathbf{A}''$ describe the same n-port if there exists a nonsingular $n \times n$ matrix $\mathbf{T}$ satisfying $\mathbf{A}' = \mathbf{T}\mathbf{A}''$. If there are some algebraic relations among the nonzero entries of $\mathbf{A}''$, they must refer to linear dependencies among the columns of $\mathbf{A}''$, by the assumption. These dependencies are valid among the columns of $\mathbf{A}'$ as well.

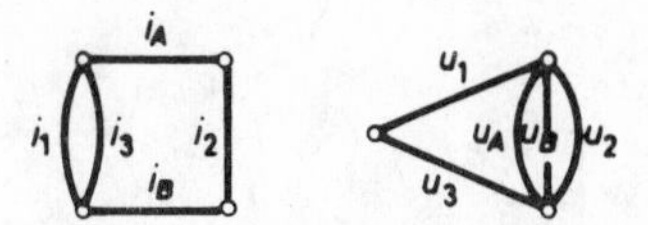

Fig. S.14.2

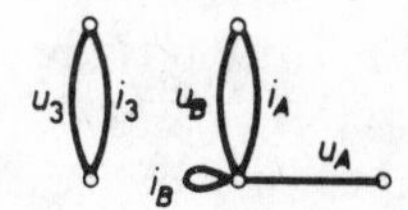

Fig. S.14.3

14.1.7 The three quantities, corresponding to the vector $\mathbf{x}$ in case of a matrix description $\mathbf{x} = \mathbf{Hy}$, must correspond to a tree of the graph K_4 of Fig. S.14.4. Every tree is either a star, like $\{i_1, i_2, i_3\}$ or a path like $\{u_3, i_1, i_3\}$. Hence there are only two essentially different descriptions, namely $\begin{pmatrix} i_1 \\ i_2 \\ i_3 \end{pmatrix} = \begin{pmatrix} 0 & * & * \\ * & 0 & * \\ * & * & 0 \end{pmatrix} \begin{pmatrix} u_1 \\ u_2 \\ u_3 \end{pmatrix}$ and $\begin{pmatrix} i_1 \\ i_3 \\ u_3 \end{pmatrix} = \begin{pmatrix} * & * & * \\ 0 & * & * \\ * & * & 0 \end{pmatrix} \begin{pmatrix} i_2 \\ u_1 \\ u_2 \end{pmatrix}$.
If the $*$'s are algebraically independent transcendentals then these descriptions correspond to matroids $\mathbf{M}_1$ and $\mathbf{M}_2$, respectively, given by their affine representations in Fig. S.14.5. Since u_1, u_2 and u_3 are not collinear on these drawings, neither $\mathbf{M}_1$ nor $\mathbf{M}_2$ is isomorphic to $\mathbf{M}(K_4)$.

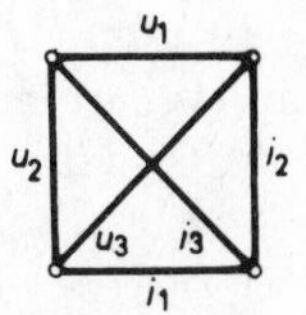

Fig. S.14.4

14.1.8 Kirchhoff's equations for this "network" are $u_1 = u_2$, $i_1 = -i_2$; together with the equations $u_1 = R_1 i_2$, $u_2 = R_2 i_1$ it is uniquely solvable if and

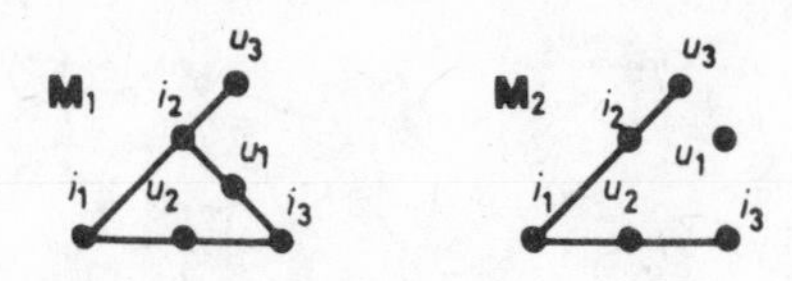

Fig. S.14.5

only if $R_1+R_2 \neq 0$. Since R_1, R_2 are nonzero, this means that the 2-port must not be a gyrator.

14.1.9 We saw (Exercise 6.2.3) that the interconnection of a gyrator and an inductor is equivalent to a capacitor. Hence Fig. S.14.6 is essentially two capacitors or two inductors in parallel. Thus u_1 or i_2 can initially be prescribed but both of them independently cannot.

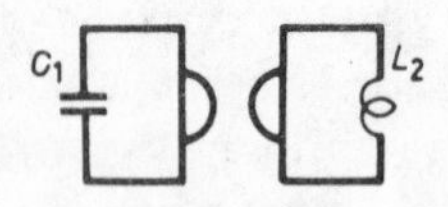

Fig. S.14.6

14.1.10 First of all the voltage source U_7 is replaced by a short circuit, and further short circuits (denoted by $1A$ and $2A$) are placed series to the inductors C_1 and C_2, respectively. Their currents will control the voltage sources U_8 and U_9, respectively. Now the graph of the network is shown in Fig. S.14.7 while **A** is visualized as the circuit matroid of the graph of Fig. S.14.8. Trying $X^C = E^C$ and $X^L = E^L$ at first, we find that $\{u_4, u_8, u_9, i_1, i_2, i_3, i_5, i_6, i_8, i_9\}$ is a base of **G** and $\{u_{1A}, u_{2A}, u_5, u_6, i_{1A}, i_{2A}\}$ is a base of **A**. Hence the initial voltages of every capacitor and the initial currents of every inductor can independently be prescribed.

14.1.11 Just like in the solution of the previous problem, introduce two new edges $1A, 2A$ series to 1 and to 2, respectively, and a new edge $5A$ parallel to 5. The graph of the network is shown in Fig. S.14.9 and **A** is visualized as the circuit matroid of the graph of Fig. S.14.10. Since i_7 is a loop in **A**, it must appear in the independent subset of **G** (where i_3 cannot appear therefore). Similarly, i_{5A} must appear in **A**, hence one of i_4 and i_5 in **G**. Thus the initial value of u_3 cannot be prescribed,

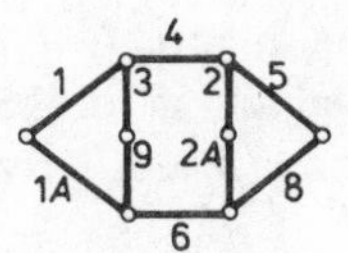

Fig. S.14.7

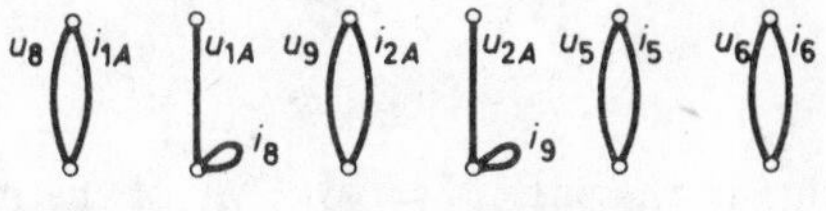

Fig. S.14.8

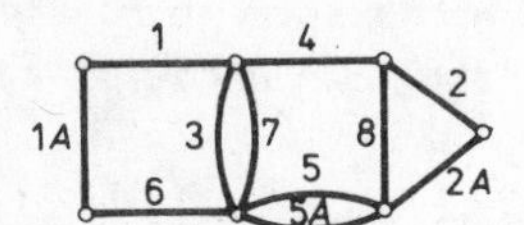

Fig. S.14.9.

Fig. S.14.10

neither can both i_4 and i_5 independently. Now one can easily check that $X_u^C = \{u_1, u_2\}$, $X_i^L = \{i_5\}$ is a right choice.

14.2.1 Yes. Let, for example, N be an operational amplifier given by $i_2 = 0$, $u_1 = au_2$.

14.2.2 (a) Let $R_1 \neq 0$ and $R_2 \neq 0$ be the resistances of two resistors in series. The matroid $\mathbf{M}_N$ obtained by Statement 14.2.1 is a circuit of length two. However, in case of $R_1 = -R_2$ it should consist of a bridge and a loop.

(b) Every independent subset of the matroid $\mathbf{M}_N$ will be independent in $(\mathbf{G} \vee \mathbf{A})/B$ as well (cf., Statement 11.2.1).

14.2.3 u_1 and u_2 are proportional in the result, and $i_1 = 0$ also holds. In the first case we have a hybrid description ($i_1 = 0$; $u_2 = k\ell u_1$). In the second case we have not, since $i_2 = 0$ must also hold.

14.2.4 Proceed like in the solution of Exercises 8.1.5, 8.1.7. and 12.1.2.

14.2.5 See **Box 14.1.**

14.2.6 The cascade connection of the pair of nullator and norator (given by $u_1 = 0$, $i_1 = 0$) and the 2-port N (see Fig. S.14.11) leads to a pair of nullator and norator again if and only if u_B and i_B are loops in $(\mathbf{G} \vee \mathbf{A})/B$. This is the case if and only if $\{u_3, i_3\}$ is a base of $\mathbf{M}_N$.

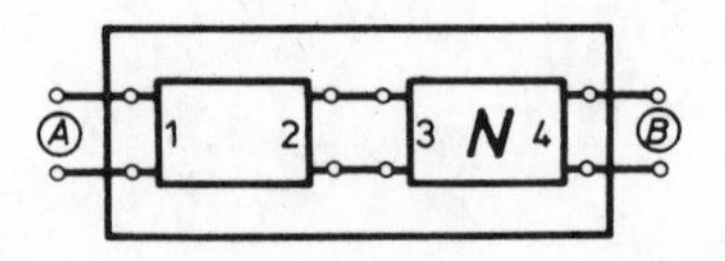

Fig. S.14.11

14.2.7 $x \in S$ is a bridge in the sum $\mathbf{N} = \vee_{i=1}^{k}\mathbf{M}_i$ of the matroids $\mathbf{M}_i = (S, \mathbf{F}_i)$ if and only if the rank of $S - \{x\}$ in $\mathbf{N}\backslash\{x\}$, that is, in $\vee_{i=1}^{k}(\mathbf{M}_i\backslash\{x\})$, is less than the rank of S in $\mathbf{N}$. Hence one can imagine a "brute force" approach of applying the matroid partition algorithm for $\mathbf{N}$ (to find a base B) and then for $\mathbf{N}\backslash\{x\}$ with every $x \in B$ as well.

A quicker solution is if we modify the matroid partition algorithm to obtain the fundamental system of circuits in $\mathbf{N}$ with respect to B as well (see Exercise 13.1.4). The union of these circuits is just the complement of the set of the bridges of $\mathbf{N}$ [Recski, 1981].

14.2.8 For the first condition see the solution of the previous exercise. In order to check the second condition apply the matroid partition algorithm for $(\mathbf{G}\backslash K) \vee (\mathbf{A}\backslash K)$ and for $\mathbf{G} \vee \mathbf{A}$. (For a practical realization apply the output of the first algorithm (i.e., the partition of $S - K$) as the input of the second algorithm.)

14.2.9 Consider the cascade connection of the 2-ports $u_1 = 0$, $i_1 = i_2$ and $i_3 = 0$, $i_4 = 0$. The result can be described by three independent equations only. The second network of Exercise 14.2.3 has similar properties.

14.2.10 The definitions of the negative and inverse of a multiport do not change. However, in addition to the equations $\mathbf{u}^T\mathbf{u}' + \mathbf{i}^T\mathbf{i}' = 0$ and $\mathbf{u}^T\mathbf{i}' = \mathbf{u}'^T\mathbf{i}$ for the definitions of dual and adjoint multiports, respectively, we require that the rank $r(\mathbf{A}|\mathbf{B})$ of the multiport plus that of the dual or adjoint must equal to $2n$. Otherwise some statements (like that of Exercise 10.1.4) would become false.

14.2.11 Yes. Even the proof is the same in case of the first three statements. However, the proof of Theorem 10.1.4 (and hence, implicitly, that of Corollary 10.1.5) is more involved. When we obtain $\mathbf{u}_k^T\mathbf{u}_k' + \mathbf{i}_k^T\mathbf{i}_k' = 0$ as the difference of Equations (10.3) and (10.4), this trivially implies

$r(\mathbf{A}|\mathbf{B}) + r(\mathbf{A}'|\mathbf{B}') \leq 2n$ only [Belevitch, 1968]. For a possible proof of the equality the reader is referred to [Iri and Recski, 1979].

14.2.12 Consider the cascade of the pairs of nullators and norators with the equations $u_1 = i_1 = 0$ and $u_4 = i_4 = 0$. The sets $\{i_2, i_3\}$ and $\{u_2, u_3\}$ are circuits in $\mathbf{G} \vee \mathbf{A}$, hence they are circuits in $\mathbf{G} \vee \mathbf{A} \vee \mathbf{B}_n$ as well, although $[(\mathbf{G} \vee \mathbf{A})/B] \vee \mathbf{B}_n$ is the free matroid on the set K.

14.2.13 Applying Eq. (7.2) for the rank function in a matroid after contraction, we immediately obtain that the rank of $(\mathbf{G} \vee \mathbf{A})/B$ equals to n if and only if $r(S) - r(B) = n$ holds for the rank function r of the matroid $\mathbf{G} \vee \mathbf{A}$. Hence the first condition of Statement 14.2.2 is equivalent to the second condition of Theorem 14.2.4.

If the rank of $(\mathbf{G} \vee \mathbf{A})/B$ is n then the sum of this matroid and $\mathbf{B}_n$ can be $(K, 2^K)$ only if K (with cardinality $2n$) is the disjoint union of a base B_1 of $(\mathbf{G} \vee \mathbf{A})/B$ and a base B_2 of $\mathbf{B}_n$. Let us extend B_1 to a base $B_1 \cup X$ of $\mathbf{G} \vee \mathbf{A}$. Then $B_1 \cup X \cup B_2$ is a base of $\mathbf{G} \vee \mathbf{A} \vee \mathbf{B}_n$, and every base arises in this way. Hence every element of $B_1 \cup B_2 = K$ arises as a bridge. The reverse statement can be proved in a similar way.

14.2.14 The second interconnection in Exercise 14.2.3 leads to a network where all the four elements are bridges, yet $r(S) - r(B) = 10 - 7 \neq 2$ for the rank function r of $\mathbf{G} \vee \mathbf{A}$. On the other hand, $r(S) - r(B) = 10 - 8 = 2$ holds for the network of Fig. S.12.6, yet u_B and i_B are not bridges in $\mathbf{G} \vee \mathbf{A} \vee \mathbf{B}_n$.

14.3.1 $\mathbf{A}_p$ contains ℓ triples of columns. Due to the block-diagonal structure we only have to prove that elements of such triples are linearly independent. If a triple belongs to a face F and V_i, V_j, V_k are non-collinear vertices of F then the triple contains the submatrix $\begin{pmatrix} x_i & y_i & 1 \\ x_j & y_j & 1 \\ x_k & y_k & 1 \end{pmatrix}$ which is nonsingular.

14.3.2 If four vertices are coplanar then the z-coordinates of at most three of them can be prescribed independently. If the z-coordinates of four noncoplanar vertices are given, we can determine the z-coordinate of the common intersection of the lines 14, 25, 36; and then all the remaining data as well. Hence the matroid really has all the 4-tuples as bases except $\{1,2,4,5\}, \{1,3,4,6\}, \{2,3,5,6\}$.

14.3.3 Using Theorem 14.3.5 we have to decide whether the drawings of Fig. 6.19 can arise as projections of a polyhedron P. If yes, P should be a truncated pyramid, see Fig. S.14.12. Now the points H (the intersection of the lines BC and EF) and I (the intersection of CD and FG) both belong to the planes $ABCD$ and $AEFG$, just like point A. Then these points A, H and I must be collinear since the intersection of the two planes is a line. This holds for the first drawing of Fig. 6.19 only

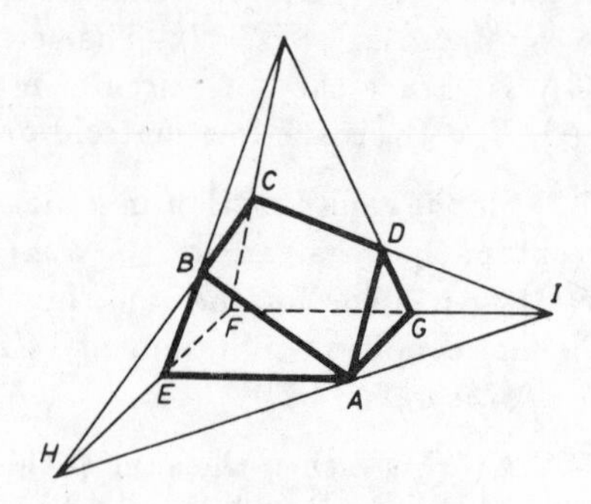

Fig. S.14.12

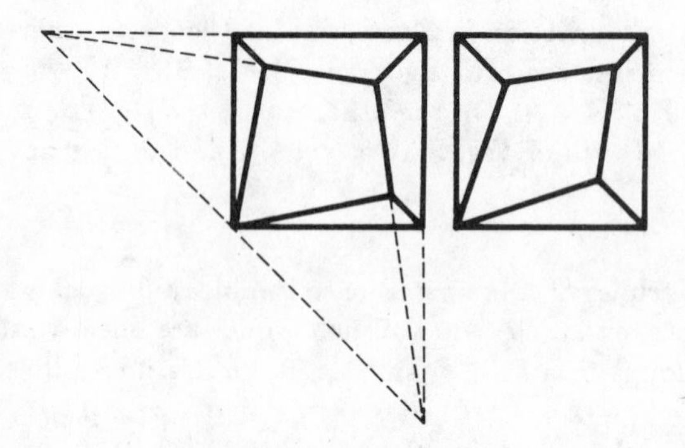

Fig. S.14.13

(see Fig. S.14.13). Accordingly, the second system is rigid, the first one is not.

14.3.4 Consider the transformation $\phi : \begin{pmatrix} x \\ y \\ z \end{pmatrix} \rightarrow \begin{pmatrix} 1 & 0 & 0 \\ 0 & 1 & 0 \\ a & b & c \end{pmatrix} \begin{pmatrix} x \\ y \\ z \end{pmatrix} + \begin{pmatrix} 0 \\ 0 \\ d \end{pmatrix}$.

This transformation acts like identity on the xy plane. Hence if P is a polyhedron with the given projection then so is $\phi(P)$. Thus at least four z-coordinates must be prescribed to determine a, b, c and d in a unique way.

14.3.5 The tetrahedron of Fig. 14.12 is an example for the first case, while the polyhedron of Fig. S.14.14 can be given for the second one (where $v_X = 5$, $r_X = 18$ and $|X| = 6$ for the choice $X = L$).

If Eq. (14.2) is met by equality for every subset then every solution is obtained by the transformation ϕ (see the solution of the previous problem). This gives the assertion: four points are coplanar if and only if so are their images under ϕ. On the other hand, if an inequality $<$ can also arise in Eq. (14.2) then coplanarity cannot uniquely be detected from the projection. For example, if we place a tetrahedron to each face of a pyramid (Fig. S.14.15) then the question whether the new vertices are coplanar cannot be decided from the projection.

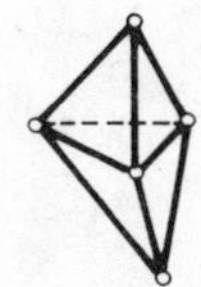

Fig. S.14.14

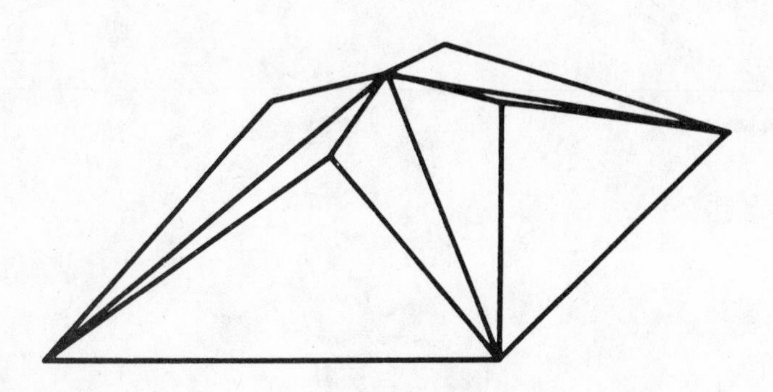

Fig. S.14.15

14.3.6 Let $\mathbf{M}$ denote the resulting matrix after deleting the four columns. We must show that the union of any c rows of $\mathbf{M}$ contains nonzero entries in at least c columns. If these c rows correspond to equations, referring to a single face, then these rows contain a unity matrix in the $\mathbf{B}_P$-part of $(\mathbf{A}_P|\mathbf{B}_P)$. In this case the unity matrix can be found in $\mathbf{M}$ as well since all the four columns were deleted from $\mathbf{A}_P$.

Suppose therefore that the c rows correspond to equations referring to several faces F_1, F_2, ... Then the set of these faces satisfies $X \subseteq F$, $|X| \geq 2$, hence $3|X| + v_X \geq r_X + 4$ holds. Thus, after the deletion of four columns, still at least r_X columns of $\mathbf{M}$ contain nonzero entries. Let us now delete one by one the $r_X - c$ "unnecessary" rows. Each time at most one column (belonging to $\mathbf{B}_P$) can become zero. Hence the final c rows contain nonzero entries of at least c columns of $\mathbf{M}$.

Chapter 15

15.1.1 If 1 and 2 were parallel elements in a transversal matroid then the corresponding points in the bipartite graph representation had a single common neighbour a_{12} only. Similarly, we obtain two more points a_{34} and a_{56} but they cannot be all distinct since $\{1,3,5\}$ is dependent. However, if two of them, say a_{12} and a_{34}, are the same then $\{1,3\}$ cannot be independent, a contradiction.

15.1.2 Let G be the bipartite graph representing the transversal matroid $\mathbf{T}$, and let $v \in V(G)$ correspond to an element x of $\mathbf{T}$. Then the bipartite graph $G - v$ represents $\mathbf{T}\backslash\{x\}$, hence this matroid is also transversal.

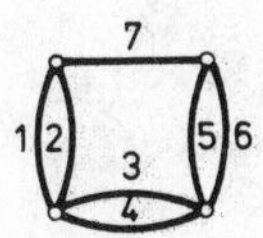

Fig. S.15.1

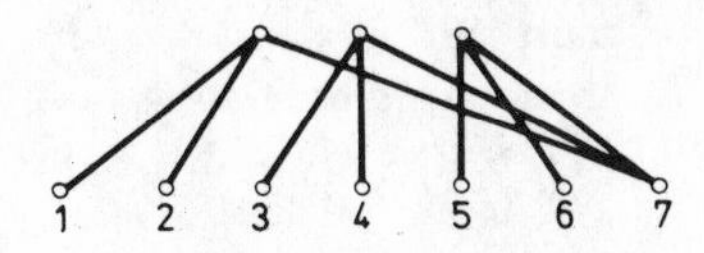

Fig. S.15.2

The circuit matroid of the graph of Fig. S.15.1 is transversal (see Fig. S.15.2) but if we contract edge 7, the result is not transversal, see the previous exercise.

15.1.3 No. Deleting any element of $\mathbf{F}_7$ we obtain $\mathbf{M}(K_4)$ which is known to be nontransversal, see the remark before Theorem 15.1.4. Alternatively, if $\mathbf{F}_7$ were transversal, it were representable over $\mathbb{R}$ by Theorem 15.1.2, contradicting to the result of Exercise 9.2.3.

15.1.4 (a) Consider the transversal matroid which is defined by the bipartite graph of Fig. S.15.3. One can easily see that its dual is just the matroid discussed in Exercise 15.1.1.

(b) Let $\mathbf{M}$ be the direct sum of three circuits of length two. $\mathbf{M}$ is clearly transversal while its truncation is not.

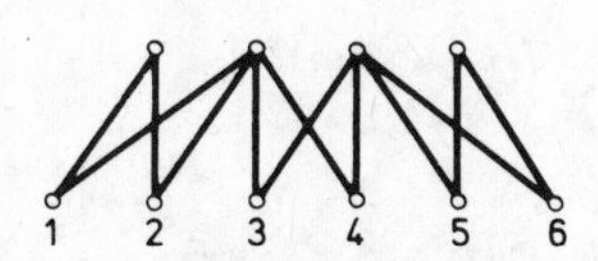

Fig. S.15.3

15.1.5 The partitional matroids arise as sums of rank one matroids and every rank one matroid is partitional, hence the statement follows from Corollary 15.1.3.

15.1.6 A trivial consequence of Corollary 15.1.3, by the associativity and commutativity of the sum.

15.1.7 Let $\mathbf{T}$ be a transversal matroid, represented by the bipartite graph G, and let $v \in V(G)$ be the point, corresponding to the element to be extended in series. Add another point v' to the graph (in the same bipartition class) and join v' to the neighbours of v. Finally add one more point and join it to v and v' only. The new bipartite graph will represent the requested matroid.

On the other hand, the parallel extension of a transversal matroid need not be transversal, see Fig. 15.4.

15.1.8 The transversality is obvious by the previous exercise. If it were a fundamental transversal matroid, the starting base would contain one of the pairs 12, 34 and 56, and one element from each of the other two pairs. If, say, this base were $\{1,2,3,5\}$, we would obtain the bipartite graph of Fig. S.15.4, but $\{3,4,5,6\}$ would then also be independent.

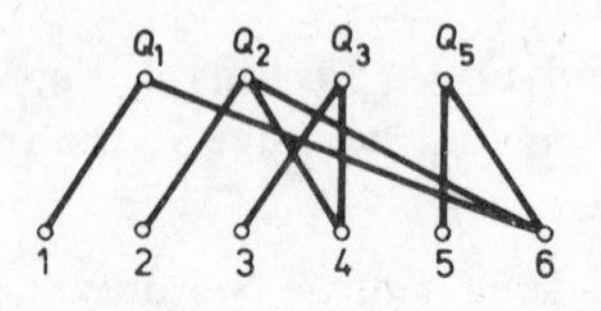

Fig. S.15.4

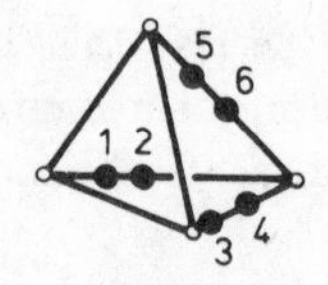

Fig. S.15.5

15.1.9 The matrix $\begin{pmatrix} * & * & 0 & 0 & 0 & 0 \\ 0 & 0 & * & * & 0 & 0 \\ 0 & 0 & 0 & 0 & * & * \\ * & * & * & * & * & * \end{pmatrix}$ coordinatizes this matroid if asterisks ($*$) stand for algebraically independent transversals. The geometric representation is shown in Fig. S.15.5.

15.1.10 Consider the affine representation of an arbitrary transversal matroid $\mathbf{T}$. If some vertices of the simplex do no belong to $\mathbf{T}$, add them to $\mathbf{T}$. Hence the affine representation of a fundamental transversal matroid $\mathbf{F}$ is obtained and $\mathbf{T}$ clearly arises from $\mathbf{F}$ by deleting the "unnecessary" elements.

15.1.11 Let $V_A(X)$ denote the set of end points of the X-edges in A. We must prove that the function $r(X) = |V_A(X)|$ is the rank function of a matroid. The submodularity easily follows from $V(X) \cap V(Y) \supseteq V(X \cap Y)$ and from $V(X) \cup V(Y) = V(X \cup Y)$ while the other properties are trivial.

15.1.12 Contracting element 7 leads to the circuit matroid of the second graph of Fig. 15.4 which was shown to be nontransversal. However, this observation does not prove the assertion since the contraction of a transversal matroid may also be nontransversal.

Hence observe first that if this matroid were transversal, the upper point set of its bipartite graph representation would contain 3 points only. The points 1,2 cannot be adjacent to all three of these points since 7 is not a loop but $\{1, 2, 7\}$ is dependent. Hence one of the upper points is not adjacent to either 1 or 2, we denote this point by a_{12}. Two further points a_{34} and a_{56} can be obtained in a similar way and they must be all distinct (since, say, $a_{12} = a_{34}$ would imply that the rank of $\{1, 2, 3, 4\}$ were two only). But now 7 cannot be adjacent to any of them, a contradiction.

15.1.13 One can easily check that f is monotone, normalized and submodular. Should f arise as the sum of rank functions, this were the sum of rank functions of two matroids $\mathbf{M}_1, \mathbf{M}_2$ of rank one each. Then $\mathbf{M}_1 \vee \mathbf{M}_2$ were transversal and its rank function were $\tilde{f}$. However, $(\mathbf{M}_1 \vee \mathbf{M}_2) \setminus \{7\}$ were nontransversal (by Exercise 15.1.1), contradicting to the statement of Exercise 15.1.2.

15.1.14 The first two axioms for the independent sets of matroids are trivially met. The third one can be proved in the same way as in the first proof of Theorem 15.1.1. (The length of the circuits of H are still even, although G need not be a bipartite graph, since H is the symmetric difference of two sets of independent edges.)

15.1.15 A trivial consequence of Exercise 13.2.4 and Corollary 15.1.3 since the homomorphic image of a rank one matroid is of rank 0 or 1.

15.1.16 A transversal matroid $\mathbf{T}$ of rank k is a fundamental transversal matroid if and only if its matrix $\mathbf{M}$ (see the proof of Theorem 15.1.2) contains a diagonal submatrix of order k.

Alternatively, $\mathbf{T}$ is a fundamental transversal matroid if and only if its affine representation (as described after the proof of Corollary 15.1.3) can be performed by such a simplex whose vertices are all elements of $\mathbf{T}$.

15.1.17 Let $(\mathbf{D}|\mathbf{X})$ coordinatize over $\mathbb{R}$ a fundamental transversal matroid $\mathbf{M}$, as described in the previous problem. Hence $\mathbf{D}$ is diagonal and we may suppose that it is the unity matrix (since multiplying the columns of the representing matrix by nonzero reals does not change the matroid). $\mathbf{M}^*$ is then coordinatized by $(\mathbf{X}^T|\mathbf{D})$, see Theorem 9.1.3, thus $\mathbf{M}^*$ is also a fundamental transversal matroid by the previous problem.

15.1.18 Let t be the number of square submatrices of $\mathbf{A}$ with full term rank. $\mathbf{M} \neq \mathbf{M}'$ means that at least one of these submatrices has zero determinant. This can happen in N^{m^2-1} ways if m is the order of the submatrix. Hence the probability in question is less than $t \cdot N^{m^2-1}/N^{nk}$. This leads to the assertion since $t < 2^n$ and $m \leq \min(n, k)$.

15.2.1 The proof is the same as it was for Exercise 15.1.11, except that $r(X)$ is defined as $r_A(V_A(X))$, where r_A is the rank function of $\mathbf{A}$. Obviously, the submodularity of r_A must also be applied.

15.2.2 Let $A = \{1,2,3\}$, $B = \{a,b,c\}$, $E(G) = \{\{1,a\},\{2,b\},\{3,c\}\}$, furthermore let $\mathbf{A}$ and $\mathbf{B}$ be defined by their sets of bases $\{\{1,2\},\{1,3\}\}$ and $\{\{a,b\},\{b,c\}\}$, respectively.

15.2.3 Let $\mathbf{M}$ be base-orderable. If we consider two bases of $\mathbf{M}\backslash T$ then they can be extended by the same elements to bases of $\mathbf{M}$. The bijection of these latter bases should be restricted to the original ones.

If X_1, X_2 are bases of $\mathbf{M}^*$ then consider the bijection between their complements. This bijection assigns "pairs" to the elements of $X_1 - X_2$. Then, for the elements of $X_1 \cap X_2$, we assign themselves.

Now the statement for the contraction follows from the above two statements and from Theorem 7.3.4.

15.2.4 Since the class of graphic transversal matroids is closed with respect to deletion and series contraction, we only have to prove that $\mathbf{M}(K_4)$ and $\mathbf{M}(C_n^2)$ are nontransversal. The former was shown in Section 15.1 already, while the proof of the latter is a straightforward modification of that of Exercise 15.1.1.

15.2.5 This follows from the solution of Problem 7.1.16 and from Exercise 15.2.3.

15.2.6 Since r_A is submodular and monotone, we only need $N(X) \cup N(Y) = N(X \cup Y)$ and $N(X) \cap N(Y) \supseteq N(X \cap Y)$.

15.2.7 Let $\mathbf{M}_1 \vee \mathbf{M}_2 = \mathbf{N}$. We may suppose that $\mathbf{N}$ has no bridges (since they are contained in every base and can clearly be assigned to themselves). Then the bases of $\mathbf{N}$ are disjoint unions of bases of $\mathbf{M}_1$ and $\mathbf{M}_2$ (see Exercise 13.2.2). Hence the bijection for the base of $\mathbf{N}$ is the "union" of the bijections for the two bases.

15.2.8 The rank one matroids are base-orderable. Hence the statement trivially follows from Corollary 15.1.3 and from the previous problem.

15.3.1 All pairs except $\{1,2\},\{3,4\}$ and $\{5,6\}$. Hence we obtain the bases of the circuit matroid of the second graph of Fig. 15.4.

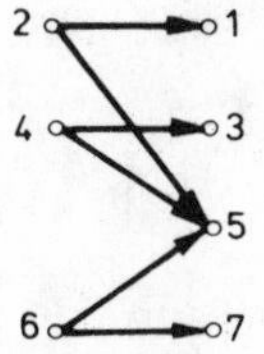

Fig. S.15.6

15.3.2 This matroid $\mathbf{M}$ is a strict gammoid (consider the subsets which can be linked onto $\{1,3,5,7\}$ in the directed graph of Fig. S.15.6) but $\mathbf{M}/\{5\}$ is not.

15.3.3 Consider the duals of these statements (by Theorem 15.3.2). They are just Exercise 15.1.2, 15.1.7 and 15.1.4, respectively.

15.3.4 A consequence of Theorems 15.1.2 and 9.1.3 since gammoids can be obtained by deletion from the duals of transversal matroids.

15.3.5 Let us represent a transversal matroid $\mathbf{T}$ by a bipartite graph with bipartition $A \cup B$ and orient every edge from B towards A. If $\mathbf{G}$ denotes the strict gammoid whose bases are just the subsets which can be linked in this graph onto A then $\mathbf{T} = \mathbf{G} \backslash A$, hence $\mathbf{T}$ is a gammoid.

The circuit matroid of the graph of Fig. 15.11 is the simplest transversal matroid which is not a strict gammoid.

15.3.6 Let $\mathbf{A}$ be a gammoid, that is, let $\mathbf{B}$ be a strict gammoid and Y be a subset with $\mathbf{A} = \mathbf{B} \backslash Y$. If X is a subset of the underlying set of $\mathbf{A}$ then $\mathbf{A} \backslash X$ is trivially a gammoid. For the contraction apply $\mathbf{A}/X = (\mathbf{B}/X) \backslash Y$ and recall (Exercise 15.3.3) that $\mathbf{B}/X$ is also a strict gammoid.

Finally, $\mathbf{A}^* = \mathbf{B}^*/Y$, $\mathbf{B}^*$ is transversal, hence it is a gammoid (by the previous problem) and then so are its contractions as well (by the first part of the previous problem).

15.3.7 Let $\mathbf{M}_i = (S, \mathbf{F}_i)$ for $i = 1, 2, \ldots, k$ and consider the common underlying set S of them in k copies $S_1, S_2, \ldots, S_k$. Let $|S| = n$. Since $\mathbf{M}_i$ is a gammoid, it arises by the deletion of a subset T_i from a strict gammoid $\mathbf{N}_i$ with the underlying set $S_i \cup T_i$. These $\mathbf{N}_i$'s can be represented by directed graphs G_i with $V(G_i) = S_i \cup T_i$, where a subset $X_i \subseteq V(G_i)$ is specified for each i (namely the subset onto which the bases of $\mathbf{N}_i$ can be linked).

Draw these graphs with disjoint point sets and add a set S_0 of n further points $v_1, v_2, \ldots, v_n$. Join v_1 by directed edges to the first points of $S_1, S_2, \ldots, S_k$, join v_2 to their second points etc. Now a subset of S_0 can be linked into $X_1 \cup X_2 \cup \ldots \cup X_k$ in the new graph if and only if it is independent in $\mathbf{M}_1 \vee \mathbf{M}_2 \vee \ldots \vee \mathbf{M}_k$. Hence the sum of these matroids is the restriction (to S_0) of a strict gammoid.

15.3.8 The contraction of a transversal matroid is a gammoid (Problems 15.3.5–6). On the other hand, if $\mathbf{A}$ is a gammoid then so is $\mathbf{A}^*$ (Problem 15.3.6). Hence there exists a strict gammoid $\mathbf{B}$ and a subset Z with $\mathbf{A}^* = \mathbf{B} \backslash Z$. However, $\mathbf{B}^*$ is transversal by Theorem 15.3.2 and then $\mathbf{A} = \mathbf{B}^*/Z$ gives the assertion.

15.3.9 Combine Theorem 15.3.2 with Exercise 15.2.3 and Problem 15.2.8.

15.3.10 We must prove that the class of gammoids is closed with respect to series and parallel extensions. We prove the latter only; the former will then follow by Problem 15.3.6. If a point v of a directed graph corresponds to that element of the strict gammoid which should be parallelly extended then apply the operation shown in Fig. 5.14.

15.3.11 K_4 is not series-parallel (it has neither parallel, nor series elements hence none of its edges could be obtained during the last step). $\mathbf{M}(K_4)$ is not base orderable thus it cannot be a gammoid by Problem 15.3.9. The previous problem and Problem 15.3.6 then imply that $\mathbf{M}(K_4)$ cannot arise as a minor of circuit matroids of series-parallel graphs.

15.3.12 Trivial, since both Kuratowski graphs contain subdivisions of K_4.

Chapter 16

16.1.1 Let the control equations be $i_a = gu_b$, $u_c = ri_d$ and $u_f = si_e$. Then the usual description of the 4-port is

$$\begin{pmatrix} R_1^{-1} & 0 & 0 & g \\ 0 & -1 & 0 & 0 \\ 0 & 0 & -1 & 0 \\ 0 & 0 & 0 & -1 \end{pmatrix} \begin{pmatrix} u_1 \\ u_2 \\ u_3 \\ u_4 \end{pmatrix} + \begin{pmatrix} -1 & 0 & 0 & 0 \\ 0 & R_2 & 0 & r \\ 0 & 0 & R_3 & 0 \\ 0 & 0 & s & R_4 \end{pmatrix} \begin{pmatrix} i_1 \\ i_2 \\ i_3 \\ i_4 \end{pmatrix} = \begin{pmatrix} 0 \\ 0 \\ 0 \\ 0 \end{pmatrix},$$

and there is no algebraic relation among the entries, different from 0 or -1.

16.1.2 The matroid of this 4-port is just the circuit matroid of the graph of Fig. S.16.1 which is transversal but not fundamental transversal (since it contains a subdivision of C_3^2, cf. Exercise 15.2.4).

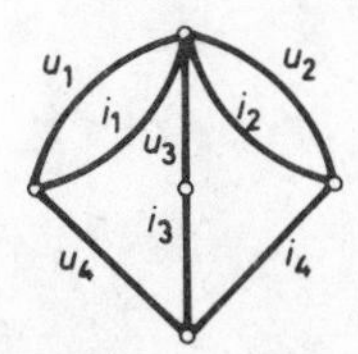

Fig. S.16.1

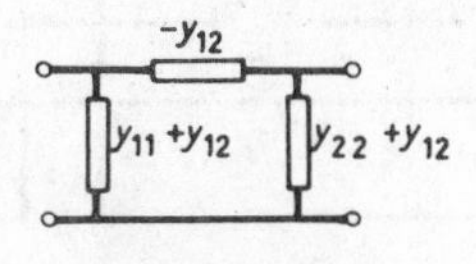

Fig. S.16.2

16.2.1 See Fig. S.16.2.

16.2.2 The impedance matrix of the 2-port is

$$\frac{1}{R_a + R_b + R_c + R_d} \begin{pmatrix} (R_a + R_d)(R_b + R_c) & R_c R_d - R_a R_b \\ R_c R_d - R_a R_b & (R_a + R_c)(R_b + R_d) \end{pmatrix}.$$

For example, if $R_a = R_b = 2$ and $R_c = R_d = 1$ then $\mathbf{Z} = \begin{pmatrix} 1.5 & -0.5 \\ -0.5 & 1.5 \end{pmatrix}$.

16.2.3 Let $\mathbf{M}_k$ denote the matroid of the k-port circulator. One can easily show that $\mathbf{M}_{k-1} = \mathbf{M}_k/\{u_k\}\backslash\{i_k\}$. Hence the assertion follows from Problem 15.3.6.

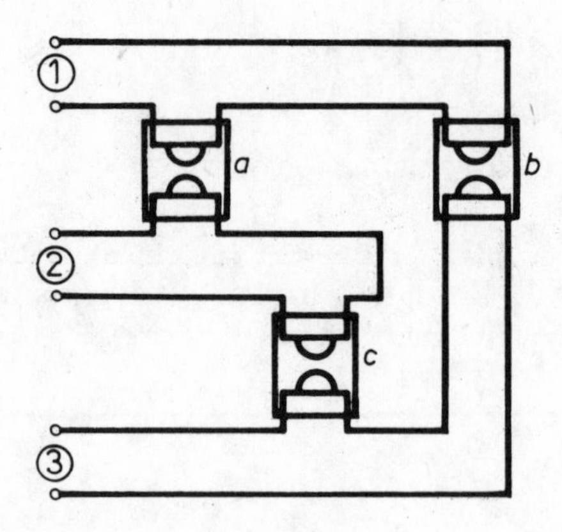

Fig. S.16.3

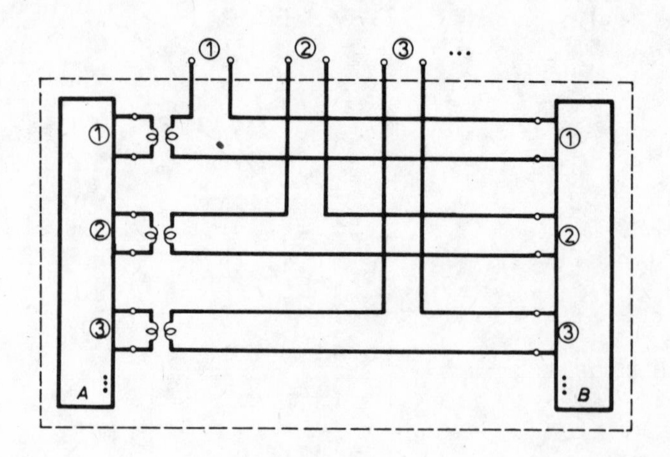

Fig. S.16.4

16.2.4 Since the condition $\mathbf{M}^* = \phi(\mathbf{M})$ holds both for $\mathbf{G}$ and for $\mathbf{A}$, we only have to prove $(\mathbf{G} \vee \mathbf{A}^*)\backslash B = (\mathbf{G}^* \vee \mathbf{A}^*)/B$. A subset $X \subseteq K$ is a base in the right hand side matroid if and only if $X \cup B$ is a base in $\mathbf{G}^* \vee \mathbf{A}^*$

which means that B can be decomposed as $B_1 \cup B_2$ so that $X \cup B_1$ is a base of $\mathbf{G}^*$ and B_2 is a base of $\mathbf{A}^*$. If B_1 and B_2 are disjoint then this is true if and only if $B_2 \cup (K - X)$ is a base of $\mathbf{G}$ and B_1 is a base of $\mathbf{A}$, that is, if the complement of $B \cup (K - X)$ is a base of $(\mathbf{G} \vee \mathbf{A})^*$. If B_1 and B_2 are nondisjoint, the proof is somewhat longer, see [Cunningham, 1973].

16.2.5 See the solution in Fig. S.16.3 for the case $\begin{pmatrix} 0 & a & b \\ -a & 0 & c \\ -b & -c & 0 \end{pmatrix}$. One can then easily imagine the general solution.

16.2.6 Let $\mathbf{Z}_1 = \frac{1}{2}(\mathbf{Z} + \mathbf{Z}^T)$ and $\mathbf{Z}_2 = \frac{1}{2}(\mathbf{Z} - \mathbf{Z}^T)$. Then $\mathbf{Z}$ arises as the sum of a symmetric matrix $\mathbf{Z}_1$ and an antisymmetric matrix $\mathbf{Z}_2$. Let A be a reciprocal multiport with impedance matrix $\mathbf{Z}_1$ and let B be formed of gyrators to realize $\mathbf{Z}_2$, as in the previous problem. Then the solution is shown in Fig. S.16.4.

Chapter 17

17.1.1 The closure $\sigma(X)$ of a subset $X \subseteq S$ contains, by definition, the elements of X and those elements y for which $r(X \cup \{y\}) = r(X)$ holds. Hence $\sigma(X)$ can be obtained by at most n calls of the rank-oracle.

On the other hand, at most $n+1$ calls of the closure oracle is enough to determine whether a subset $X \subseteq S$ is independent. Check for every $x \in X$ whether $\sigma(X - \{x\}) = \sigma(X)$. X is independent if and only if the answer is "no" for every x.

Since the independence and the rank oracles were polynomially equivalent, the statement follows.

17.1.2 Let $\mathbf{M}_1 = \mathbf{U}_{n,k}$ and $\mathbf{M}_2$ be the matroid of Problem 7.1.10 (which is almost $\mathbf{U}_{n,k}$ but contains a single circuit of cardinality k). The girth of these matroids are $k+1$ and k, respectively, but deciding this we may require $\binom{n}{k}$ calls of the independence oracle. This number is nearly 2^n if $k = \left\lfloor \frac{n}{2} \right\rfloor$. On the other hand, the independence oracle can be realized by a single call of the girth oracle.

17.1.3 If we only wish to know whether an element $x \in S$ is a loop, we must ask every subset of S, containing x, whether it is a base. Now let $S = A \cup B \cup \{x\}$ with $|S| = n$, $|A| = \left\lfloor \frac{n}{2} \right\rfloor$ and $|B| = \left\lfloor \frac{n-1}{2} \right\rfloor$. Let every element of A be a bridge both in $\mathbf{M}_1$ and in $\mathbf{M}_2$. Let the elements of B be parallel in both matroids, let x be a loop in $\mathbf{M}_1$ and parallel to the B-elements in $\mathbf{M}_2$. In order to distinguish these two matroids we may need nearly 2^n questions.

17.1.4 Consider the matroids $\mathbf{M}_1, \mathbf{M}_2$ of the solution of Exercise 17.1.2. They cannot be distinguished with less than an exponential number of questions. This immediately proves (1) since only $\mathbf{M}_1$ is uniform, and also (7) if k is odd since then $\mathbf{M}_1$ is bipartite but $\mathbf{M}_2$ is not.

17.1.5 The first question is equivalent to "Does $\min_{Y \subseteq X}[\sum_{i=1}^{k} r_i(Y) - |Y|] = 0$ hold?", see Theorem 13.2.1, and this function is submodular, being the sum of submodular functions minus a modular one. By Theorem 13.2.2 the second question is equivalent to "Does $\min_{X \subseteq S}[r_1(X) + r_2(S - X)] = r$ hold?" and we saw (Exercise 13.3.2) that $r_2(S - X)$ is also submodular.

17.1.6 We wish to decide whether $X \subseteq S$ is independent and we know that $B \subseteq S$ is a base. $X_0 = X \cap B$ is clearly independent, let $x_1, x_2, \ldots$ be the other elements of X. In order to check whether $X_0 \cup \{x_1\}$ is independent, we ask whether $(B \cup \{x_1\}) - \{y\}$ is a base (for every $y \in B - X_0$). If all these answers are negative, $X_0 \cup \{x_1\}$ is dependent (hence so is X as well), otherwise we have a new base with one more element in common with B and we can repeat the procedure.

17.1.7 We can decide whether $Y \subseteq S$ is closed by a single call of the closure oracle (just check whether $Y = \sigma(Y)$ holds). However, if we can only distinguish closed sets from the rest, we cannot always determine the closure of the empty set by a polynomial number of calls. Let, for example, $\mathbf{M}_1$ be the trivial matroid (only the whole set S is closed) and let $\mathbf{M}_2$ contain $\left\lfloor \frac{n}{2} \right\rfloor$ loops (where $n = |S|$) and let the other elements of $\mathbf{M}_2$ be all parallel. $\mathbf{M}_2$ has only three closed sets, hence $\mathbf{M}_1$ and $\mathbf{M}_2$ can be distinguished by nearly 2^n questions only.

17.1.8 (a) Let A, B, C, D be four disjoint k-element subsets of the underlying set of $\mathbf{M}_1$ and $\mathbf{M}_2$ and let every $2k$-element subset be a base in $\mathbf{M}_2$ except $A \cup B$, $A \cup C$ and $A \cup D$. $\mathbf{M}_1$ is self-dual, $\mathbf{M}_2$ is not, but nearly $\binom{4k}{2k}$ questions are required to distinguish them [Jensen and Korte, 1982].

(b) Let X, Y, T be three disjoint subsets of the underlying set of $\mathbf{M}_1$ and $\mathbf{M}_2$, with cardinality $r-4$, $r-4$ and 8, respectively. Let every r-element subset of $\mathbf{M}_2$ be a base except those which arise in form $X \cup K$ where K is a circuit of $\mathbf{K}_8$. Then $\mathbf{M}_1$ is representable, $\mathbf{M}_2$ is not (since $\mathbf{M}_2 \backslash Y/X = \mathbf{K}_8$) and they cannot be distinguished by a polynomial number of questions [Jensen and Korte, 1982].

17.1.9 Prove property (6) of **Box 9.2** (the symmetric difference of circuits is the disjoint union of circuits).

17.1.10 Let $\mathbf{M}_1$ be the matroid of the previous problem and define $\mathbf{M}_2$ by changing one of the k-element subsets T to be independent. Then $\mathbf{M}_1$ is binary but $\mathbf{M}_2$ is not and we need 2^{k-1} questions to distinguish them.

17.2.1 Let $a, b, c, d, e, f, g, h, j$ and k denote the elements of the underlying set of $\mathbf{R}_{10}$ in the same order as the columns of the 5×10 matrix (at the beginning of this section) follow. One can verify that $\mathbf{R}_{10} \backslash \{k\}$ is just $\mathbf{M}(K_{3,3})$ with the bijection shown at Fig. S.17.1.

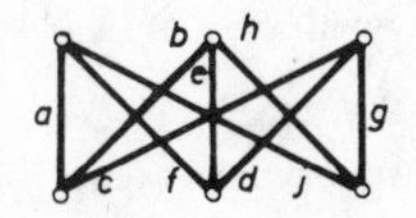

Fig. S.17.1

17.2.2 Consider the base $\{a, b, c, d, g\}$ from the previous exercise. Then $\mathbf{R}_{10}$ can alternatively be coordinatized over $\mathbf{B}$ as follows:

$$\begin{array}{ccccccccccc} a & b & c & d & g & & e & f & h & j & k \end{array}$$
$$\begin{pmatrix} 1 & 0 & 0 & 0 & 0 & & 0 & 1 & 0 & 1 & 1 \\ 0 & 1 & 0 & 0 & 0 & & 1 & 0 & 1 & 0 & 1 \\ 0 & 0 & 1 & 0 & 0 & & 1 & 1 & 1 & 1 & 1 \\ 0 & 0 & 0 & 1 & 0 & & 1 & 1 & 0 & 0 & 1 \\ 0 & 0 & 0 & 0 & 1 & & 0 & 0 & 1 & 1 & 1 \end{pmatrix}$$

After a suitable permutation of the rows and columns of this matrix we obtain

$$\begin{pmatrix} 1 & & & & & 1 & 1 & 1 & 1 & 1 \\ & 1 & & \mathbf{0} & & 1 & 0 & 1 & 1 & 0 \\ & & 1 & & & 1 & 0 & 0 & 1 & 1 \\ & \mathbf{0} & & 1 & & 1 & 1 & 0 & 0 & 1 \\ & & & & 1 & 1 & 1 & 1 & 0 & 0 \end{pmatrix}$$

which shows that $\mathbf{R}_{10}$ is self-dual (since the submatrix, obtained after deleting the unity block, is the same as its transpose, after permuting its columns).

However, it is not identically self-dual. For example, $\{a, c, d, f\}$ is a circuit, but its complement is of rank 5, hence it cannot be a cut set.

17.2.3 $\mathbf{R}_{10}$ cannot be cographic since we saw that even $\mathbf{R}_{10} \backslash \{k\}$ was not. It cannot be graphic either, due to the previous exercise.

17.2.4 $\mathbf{R}_{10}$ is clearly binary (by its definition), hence, by Theorem 17.2.1, we only have to show that it does not contain $\mathbf{F}_7$ or $\mathbf{F}_7^*$ as a minor. If it were the case, we should delete or contract three elements but even after the first one the result were graphic or cographic.

17.2.5 Let $\mathbf{N}_k$ denote the k-sum of the matroids $\mathbf{M}_1$ and $\mathbf{M}_2$ for $k = 1, 2, 3$. X is a circuit of $\mathbf{N}_1$ if and only if it is a circuit in $\mathbf{M}_1$ or in $\mathbf{M}_2$. In case of $k = 2$ let e_1, e_2 be the elements of $\mathbf{M}_1$ and of $\mathbf{M}_2$, respectively, along which 2-sum was formed. X is a circuit of $\mathbf{N}_2$ if and only if either it is a circuit of $\mathbf{M}_i$ (for some i) with $e_i \notin X$ or it arises in form $X_1 \cup X_2$ where $X_i \cup \{e_i\}$ is a circuit of $\mathbf{M}_i$ for both $i = 1$ and $i = 2$.

Finally, for $k = 3$ let the 3-sum be formed along the circuits $C_i = \{e_i^1, e_i^2, e_i^3\}$ of the matroids $\mathbf{M}_i$ $(i = 1, 2)$. X is a circuit of $\mathbf{N}_3$ if and only if it is either a circuit of $\mathbf{M}_i$ (for some i) with $X \cap C_i = \emptyset$ or X arises in form $X_1 \cup X_2$ and there exists a $j \in \{1, 2, 3\}$ so that, both for $i = 1$ and $i = 2$, at least one of $X_i \cup \{e_i^j\}$ and $X_i \cup (C_i - \{e_i^j\})$ is a circuit of $\mathbf{M}_i$.

(The circuits of the k-sum can be described in a simpler way if $\mathbf{M}_1, \mathbf{M}_2$ are binary.)

17.2.6 Let $\mathbf{M}_i$ be coordinatized over an arbitrary field F by a matrix $\mathbf{A}_i$ (for $i = 1$ and $i = 2$). We prove that their k-sum $(k = 1, 2, 3)$ is also coordinatizable over F. If $k = 1$, consider $\left(\begin{array}{c|c} \mathbf{A}_1 & \mathbf{0} \\ \hline \mathbf{0} & \mathbf{A}_2 \end{array}\right)$. If $k = 2$, suppose that e_1 corresponds to the last column of $\mathbf{A}_1$ and is of form $\begin{pmatrix} 0 \\ 0 \\ \vdots \\ 1 \end{pmatrix}$ and that e_2 is the first column of $\mathbf{A}_2$ and is of form $\begin{pmatrix} 1 \\ 0 \\ \vdots \\ 0 \end{pmatrix}$. Then, applying the notation $\mathbf{A}_1 = \left(\begin{array}{c|c} \mathbf{X} & \mathbf{0} \\ \hline \mathbf{x} & 1 \end{array}\right)$ and $\mathbf{A}_2 = \left(\begin{array}{c|c} 1 & \mathbf{y} \\ \hline \mathbf{0} & \mathbf{Y} \end{array}\right)$ one obtains that $\left(\begin{array}{c|c} \mathbf{X} & \mathbf{0} \\ \hline \mathbf{x} & \mathbf{y} \\ \hline \mathbf{0} & \mathbf{Y} \end{array}\right)$ coordinatizes the 2-sum of $\mathbf{M}_1$ and $\mathbf{M}_2$. The argument is similar for $k = 3$ (instead of a common entry 1, form a common block $\begin{pmatrix} 1 & 1 & 0 \\ 1 & 0 & 1 \end{pmatrix}$ of $\mathbf{A}_1$ and $\mathbf{A}_2$.

17.2.7 If a matroid contains $\mathbf{R}_{10}$ as a series minor then deleting a further element we find (see Exercise 17.2.1) that $\mathbf{M}(K_{3,3})$ is another series minor. This appears on the list of excluded series minors when cographic matroids are characterized. (We may delete any element but cannot contract but series ones. Hence the same argument cannot be used for graphic matroids.)

17.2.8 Let $\mathbf{A}$ be a totally unimodular matrix and let $\mathbf{M}$ denote the matroid which is coordinatized by $\mathbf{A}$ over $\mathbb{Q}$. We show that $\mathbf{M}$ can be coordinatized over any field F by a matrix $\mathbf{A}_0$ which is defined as follows: If the characteristic of F is not 2 then $\mathbf{A}_0 = \mathbf{A}$; if it equals 2, change the -1 entries (if any) to $+1$ to obtain $\mathbf{A}_0$.

Let $X \subseteq S$ be an arbitrary subset of cardinality k. If X is dependent in $\mathbf{M}$ then every $k \times k$ submatrix of its columns has zero determinant in $\mathbf{A}$ and then in $\mathbf{A}_0$ as well. If X is independent in $\mathbf{M}$ then its columns determine a nonsingular $k \times k$ submatrix of $\mathbf{A}$ with determinant ± 1. Expanding the determinant requires additions and multiplications only, and these can be performed by the residue classes (modulo the characteristic of F) as well. Hence the determinant in $\mathbf{A}_0$ will also be ± 1.

17.2.9 If $\mathbf{M}$ is not regular, it must contain $\mathbf{U}_{4,2}$, $\mathbf{F}_7$ or $\mathbf{F}_7^*$ as a minor. The former is impossible if $\mathbf{M}$ is binary, the latter two cases are impossible if $\mathbf{M}$ is ternary (cf. Exercise 9.2.3).

17.3.1 Since the last row of the matrix consists of ones only, the affine representation of $\mathbf{M}$ can trivially be drawn by Statement 9.2.1 (see Fig. S.17.2). The corresponding 2-polymatroid function equals 4 on the sets $\{1,3\}$ and $\{1,2,3\}$, 3 on $\{1,2\}$ and $\{2,3\}$ and 0 on the empty set. The only maximum matching is $\{1,3\}$.

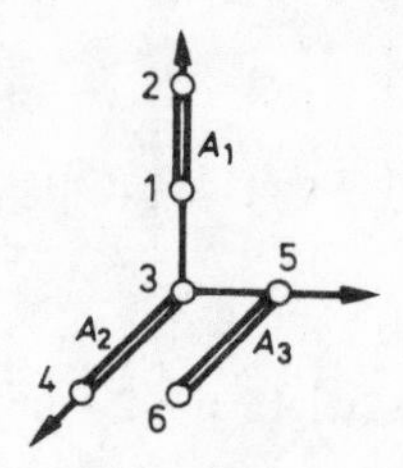

Fig. S.17.2

17.3.2
$$\mathbf{W}_1 = \begin{pmatrix} 0 & 0 & 0 & 0 \\ 0 & 0 & 0 & 0 \\ 0 & 0 & 0 & -1 \\ 0 & 0 & 1 & 0 \end{pmatrix}; \quad \mathbf{W}_2 = \begin{pmatrix} 0 & 0 & 0 & -1 \\ 0 & 0 & 0 & 0 \\ 0 & 0 & 0 & 0 \\ 1 & 0 & 0 & 0 \end{pmatrix};$$
$$\mathbf{W}_3 = \begin{pmatrix} 0 & -1 & 0 & -1 \\ 1 & 0 & 0 & 0 \\ 0 & 0 & 0 & 0 \\ 1 & 0 & 0 & 0 \end{pmatrix},$$
and then $c_1\mathbf{W}_1 + c_2\mathbf{W}_2 + c_3\mathbf{W}_3 = \begin{pmatrix} 0 & -c_3 & 0 & -c_2 - c_3 \\ c_3 & 0 & 0 & 0 \\ 0 & 0 & 0 & -c_1 \\ c_2 + c_3 & 0 & c_1 & 0 \end{pmatrix}$.
Its determinant is $c_1^2 c_3^2$, hence its maximum rank is four.

17.3.3 Let f_1 and f_2 be two functions as constructed in Problem 13.3.11 so that $f_1(X) = 2n+1$ for every X of cardinality $n+1$ but $f_2(X_0) = 2n+2$ for a fixed subset X_0 of cardinality $n+1$ and $f_2(X) = 2n+1$ for the others. Then the size of a maximum matching is n for f_1 and $n+1$ for f_2. Hence the algorithm should ask the value of f for every $(n+1)$-element subset (in the worst case), and there are exponentially many of them if, say, $|S| = 2n+2$.

17.3.4 Let r denote the rank function of the matroid $\mathbf{M} = (S, \mathbf{F})$. Let $\mathbf{M}$ be coordinatized over $\mathbb{R}$ and let $A_1, A_2, \ldots$ be disjoint pairs in S. The (≤ 2)-polymatroid f arises on the set $T = \{1, 2, \ldots\}$ as $f(X) = r(\cup_{i \in X} A_i)$. If $r(A_i) < 2$ then no matching will contain i, hence these i's can simply be deleted from T and Theorem 17.3.5 can be applied to the new 2-polymatroid.

17.3.5 Since $f(X) \leq 2|X|$ holds for every X if f is a 2-polymatroid function, the maximum size of a matching is equal to the maximum of the submodular set function $f(X) - |X|$.

17.3.6 If we use large random numbers (instead of algebraically independent ones) for $c_1, c_2, \ldots$ of Theorem 17.3.6, the rank of the matrix $c_1\mathbf{W}_1+$

$+c_2\mathbf{W}_2+\ldots$ will be equal to the cardinality of the maximum matching with large probability. This time the wedge-products are not known either, since the representation of $\mathbf{M}_1 \vee \mathbf{M}_2$ is not available. However, let us prepare the matrix

$$\left(\begin{array}{c|c|c|c} \multicolumn{4}{c}{\mathbf{A}_1} \\ \hline \lambda_1\mathbf{a}_1 & \lambda_2\mathbf{a}_2 & \ldots & \lambda_n\mathbf{a}_n \end{array}\right)$$

where $\mathbf{A}_1$ and $\mathbf{A}_2$ represent $\mathbf{M}_1$ and $\mathbf{M}_2$, respectively, and the columns of $\mathbf{A}_2$ are $\mathbf{a}_1, \mathbf{a}_2, \ldots, \mathbf{a}_n$. If $\lambda_1, \lambda_2 \ldots$ were algebraically independent transcendentals then this would be a representation of $\mathbf{M}_1 \vee \mathbf{M}_2$. Thus, using large random numbers instead of the λ_i's as well, we obtain a representation which is correct with large probability.

17.4.1 Consider a region in a planar representation of the oriented graph. A point P of the dual graph will correspond to this region. Orient the incident edges towards P if and only if the corresponding edge in the original graph has an orientation clockwise around the region (see Fig. S.17.3).

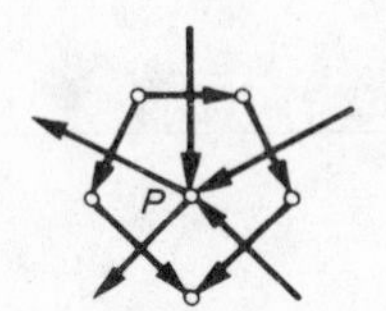

Fig. S.17.3

17.4.2 Indirectly suppose that Y were a circuit of $\mathbf{M}^*$ with $Y^- = \emptyset$. Let $y \in Y$ and X be a circuit of $\mathbf{M}$ containing y, with $X^- = \emptyset$. Then $X^+ \cap Y^+$ is nonempty but $(X^+ \cap Y^-) \cup (X^- \cap Y^+) = \emptyset$, contradicting to Theorem 17.4.4.

For graphic matroids we obtain that strongly connected graphs have no directed cuts (see Exercise 1.3.7).

17.4.3 We needed the signs of the entries only. Hence the proof works for any matroid, representable over an ordered field.

17.4.4 We must find the bipartition $Y^+ \cup Y^-$ of every cut set Y of $\mathbf{M}$. By (OC1) we only have to say whether a pair x, y of elements of Y are in the same class or not. By Problem 7.2.23 there exists a circuit X of $\mathbf{M}$ with $X \cap Y = \{x, y\}$. Hence if both x and y are in X^+ (or both in X^-) then one of them must be in Y^+ and the other in Y^-; if one of them is in X^+ and the other in X^-, they must be in the same subset of Y.

Clearly, no other bipartition satisfies the property described in Theorem 17.4.4. Proving that this one is really good is more difficult.

Chapter 18

18.1.1 If $(\mathbf{A}|\mathbf{B})$ is the description of an n-port then every other description of this n-port arises in form $\mathbf{T}(\mathbf{A}|\mathbf{B})$ where $\mathbf{T}$ is a nonsingular matrix of size $n \times n$. Hence the relations of Eq. (18.1) do not depend on the actual description since the matroidal structure is always the same (see Statement 9.1.1). The relation of (18.2) means that det $(\mathbf{x}+\mathbf{y}|\mathbf{u}-\mathbf{v}) = 0$ where $\mathbf{x}, \mathbf{y}, \mathbf{u}, \mathbf{v}$ stand for $\begin{pmatrix} a \\ c \end{pmatrix}, \begin{pmatrix} b \\ d \end{pmatrix}, \begin{pmatrix} \alpha \\ \gamma \end{pmatrix}$ and $\begin{pmatrix} \beta \\ \delta \end{pmatrix}$, respectively. Since det $(\mathbf{Tx}+\mathbf{Ty}|\mathbf{Tu}-\mathbf{Tv}) = \det \mathbf{T} \cdot \det (\mathbf{x}+\mathbf{y}|\mathbf{u}-\mathbf{v})$, the statement follows from det $\mathbf{T} \neq 0$.

18.1.2 Consider the 2-port of Fig. S.18.1 with arbitrary R and r.

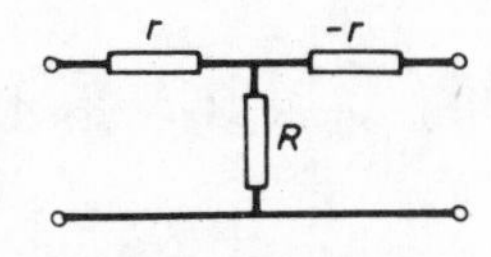

Fig. S.18.1

18.1.3 The 2-port, corresponding to ports 3 and 4 satisfies two equations from (18.1) while the 2-port, corresponding to ports 5 and 6 satisfies three of them.

18.1.4 Let the underlying set S of the matroid $\mathbf{M}$ be of cardinality $2k+\ell$, that is, suppose that the network consists of k gyrators and ℓ resistors. We need an r-element subset which is independent in $\mathbf{M}$ and intersects each pair of gyrator-edges in 0 or 2 elements. Let λ denote the number of further elements of this subset $(0 \leq \lambda \leq \min(\ell, r))$. We introduce a subset T of ℓ elements which will be in pairs with the resistor elements of S. Then define a uniform matroid $\mathbf{N}_\lambda = \mathbf{U}_{\ell,\lambda}$ on T and then consider the matroid $\mathbf{M} \oplus \mathbf{N}_\lambda$. This has an $(r+\lambda)$-element independent subset, intersecting every pair with 0 or 2 elements, if and only if the original problem has a solution. Hence the 2-polymatroid matching problem must be applied at most $\min(\ell, r)$ times.

18.1.5 An ideal transformer with transfer ratio k can be replaced by two gyrators in cascade (see **Box 6.3**) if their gyrator constants are R and kR. Hence such a spanning forest is required which intersects every gyrator-pair by 0 or 2 and every transformer-pair by exactly 1 edges. This condition becomes sufficient as well if the above constants R are chosen in a suitable way (see the next problem).

18.1.6 Let $x_1, x_2, \ldots, y_1, y_2, \ldots$ be algebraically independent transcendentals over the field $\mathbf{Q}$ of the rationals. Let $\mathbf{Q}'$ denote the field, obtained from $\mathbf{Q}$ by extending it with these elements. Finally let $z_1, z_2, \ldots$ be algebraically independent over $\mathbf{Q}'$. Then

$$x_1, x_2, \ldots,\ y_1, y_2, \ldots,\ z_1 x_1,\ z_2 x_2, \ldots$$

are algebraically independent over $\mathbf{Q}$. (Its proof is left to the reader: suppose the contrary and arrange the members of the polynomial according to the powers of the z_i's.)

18.2.1 Joint 2 must be fixed by all means (it would have an infinitesimal motion otherwise, even if all the other joints were fixed). Then one more joint (4 or 5) must still be fixed.

18.2.2 $\{\dot{x}_1, \dot{x}_2, \dot{x}_3, \dot{x}_4\}$ and $\{\dot{y}_1, \dot{y}_2, \dot{y}_3, \dot{y}_4\}$ are dependent (place the joints to, say, a horizontal track). Among the six sets of form

$$\{\{\dot{x}_i, \dot{y}_i, \dot{x}_j,\ \dot{y}_j\} | 1 \le i < j \le 4\}$$

there are only two independent ones ($i = 1,\ j = 3$ or $i = 2,\ j = 4$).

18.2.3 See the affine representation in Fig. S.18.2.

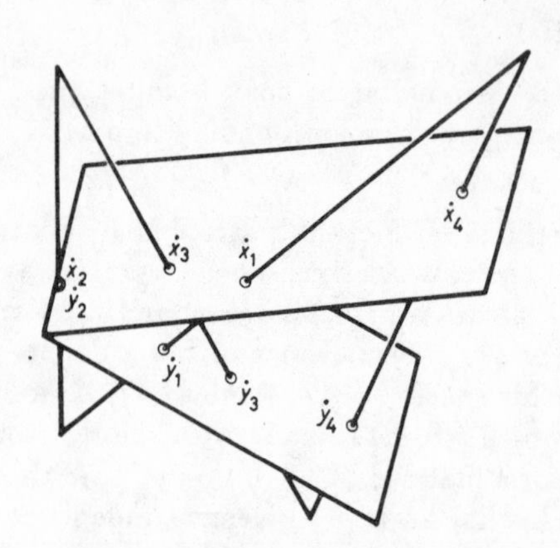

Fig. S.18.2

18.2.4 Let $j_1, j_2, \ldots$ be the joints, adjacent to the ith joint. If $P_i = \{\dot{x}_i, \dot{y}_i\}$ is dependent then the vectors $(x_i - x_{j_1}, x_i - x_{j_2}, \ldots)^T$ and $(y_i - y_{j_1}, y_i - y_{j_2}, \ldots)^T$ are parallel (or one of them is $\mathbf{0}$). This holds if and only if the joints $i, j_1, j_2, \ldots$ are all collinear. Then we obtain a (≤ 2)-polymatroid (cf. Problem 17.3.4) and, accordingly, P_i is not contained in any matching, that is, joint i must be fixed by all means.

18.3.1 The square formed by four rods becomes rigid due to the two diagonal cables. If a rod is removed, the distance of its two end points can decrease only (see Fig. S.18.3), hence the rods can be replaced by struts as well. Alternatively, the four edges can be cables and the two diagonals be struts.

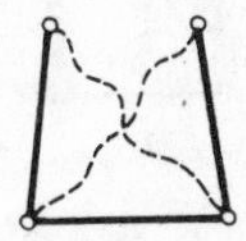

Fig. S.18.3

18.3.2 The matrix $\mathbf{W}$ of this system is $\begin{pmatrix} x_4 - x_1 & y_4 - y_1 \\ x_4 - x_2 & y_4 - y_2 \\ x_4 - x_3 & y_4 - y_3 \end{pmatrix}$. We try to find the zero vector as the linear combination of the rows with coefficients α, β and γ. In case (a) the sign of α, β and γ are the same (1,1 and 2 is a possible solution), hence all the rods can be replaced by cables, or all by struts. In case (b) the sign of α is different from that of the other two (3, -3 and -2 is a possible solution), hence rod $\{1,4\}$ should be replaced by a cable and the other two rods by struts (or *vice versa*). Finally, in case (c), $\alpha = \beta$ and $\gamma = 0$. Hence rod $\{3,4\}$ can be replaced neither by a cable nor by a strut, while the other two can both be cables or both be struts.

18.3.3 See the directed graphs of Fig. S.18.4.

18.3.4 (a) Rods $\{1,5\}$ and $\{4,5\}$ can be replaced by cables (or both by struts). The other two rods can also be replaced (either both by cables or both by struts), independently of the former pair.

(b) Replace $\{3,5\}$ by cable (or strut) and rods $\{1,5\}$ and $\{4,5\}$ by strut (by cable, respectively). There will be no stress in rod $\{2,5\}$, it can be replaced by anything (it can even be deleted).

18.3.5 See the directed graphs of Fig. S.18.5.

18.3.6 It is not necessarily true, see Fig. S.18.6.

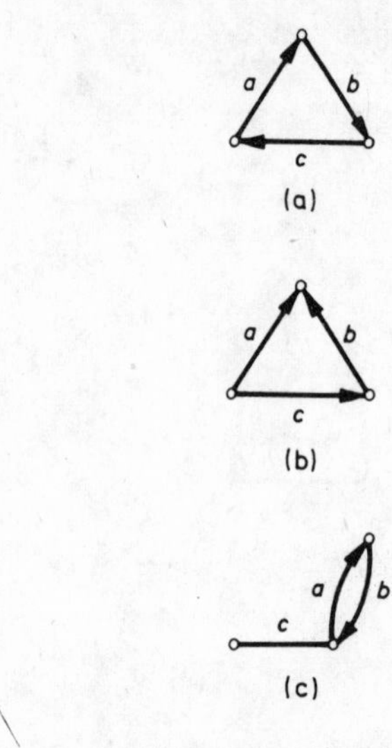

Fig. S.18.4

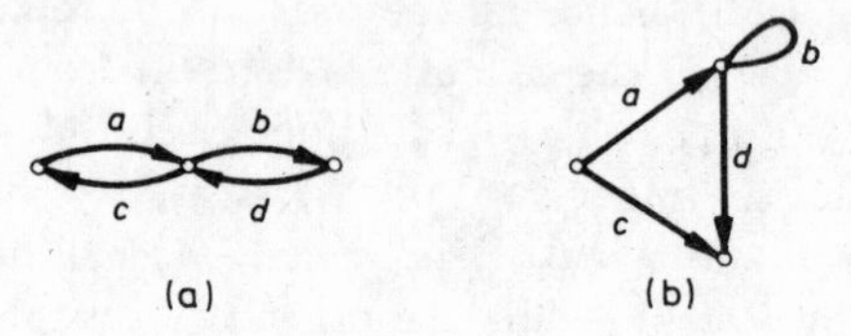

Fig. S.18.5

Fig. S.18.6

18.3.7 Let $\mathbf{a}, \mathbf{b}, \mathbf{c}, \mathbf{d}, \mathbf{e}$ and $\mathbf{f}$ denote the rows of the matrix

$$\mathbf{W} = \begin{pmatrix} x_2 - x_1 & 0 & 0 & 0 & 0 \\ 0 & x_3 - x_1 & 0 & y_3 - y_1 & 0 \\ 0 & 0 & x_4 - x_1 & 0 & y_4 - y_1 \\ x_2 - x_3 & x_3 - x_2 & 0 & y_3 - y_2 & 0 \\ x_2 - x_4 & 0 & x_4 - x_2 & 0 & y_4 - y_2 \\ 0 & x_3 - x_4 & x_4 - x_3 & y_3 - y_4 & y_4 - y_3 \end{pmatrix} =$$

$$= \begin{pmatrix} 1 & 0 & 0 & 0 & 0 \\ 0 & 0 & 0 & 1 & 0 \\ 0 & 0 & 0.5 & 0 & 0.5 \\ 1 & -1 & 0 & 1 & 0 \\ 0.5 & 0 & -0.5 & 0 & 0.5 \\ 0 & -0.5 & 0.5 & 0.5 & -0.5 \end{pmatrix}.$$

A linear combination $\alpha\mathbf{a} + \beta\mathbf{b} + \gamma\mathbf{c} + \delta\mathbf{d} + \epsilon\mathbf{e} + \varsigma\mathbf{f} = \mathbf{0}$ implies $\alpha = \beta = \gamma = 0$, $\epsilon = \varsigma = -2\delta$. The oriented matroid of the row space of $\mathbf{W}$ is shown by the directed graph of Fig. S.18.7. Rods $\{2,4\}$ and $\{3,4\}$ can be replaced by cables (or struts) and rod $\{2,3\}$ by strut (by cable, respectively), while the others cannot be replaced by cables or struts.

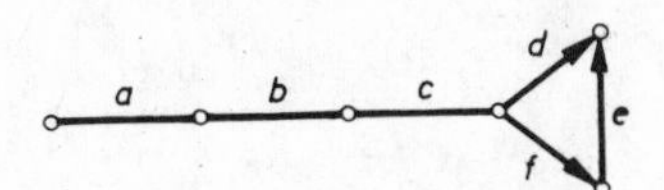

Fig. S.18.7

18.3.8 If α, β, γ and δ denote the coefficients in a linear combination of the row vectors which gives the zero vector then $\gamma = (\alpha + 3\beta)/2$, $\delta = (3\alpha + \beta)/2$ are obtained. Hence the oriented matroid of the row space of $\mathbf{W}$ is $\mathbf{U}_{4,2}$. If $\{1,5\}$ and $\{2,5\}$ are replaced by cables (or by struts) then the other two cannot be replaced by struts (by cables, respectively). A similar statement holds for the pairs $\{1,5\}$ and $\{3,5\}$. The other 12 possibilities preserve rigidity.

18.3.9 Consider the two tensegrity frameworks of Fig. S.18.8 [Roth and Whiteley, 1981]. Both are rigid as 2-dimensional frameworks and nonrigid as 3-dimensional ones. However, in this latter case the first framework has an infinitesimal deformation only (the central joint moves perpendicularly to the plane) while the second framework has a real motion.

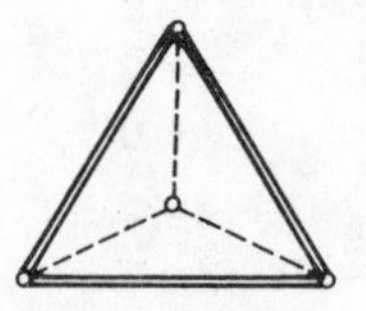

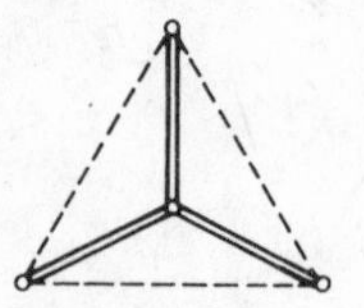

Fig. S.18.8

18.3.10 We may clearly suppose that each joint is incident to at least one cable. Then there are 6 different cases, see Fig. S.18.9. The first three are rigid, the others are not (a deformation of each is indicated by arrows). The reader is referred to Roth and Whiteley [1981] for more general results and remarks on the history of this problem (Cauchy, Grünbaum).

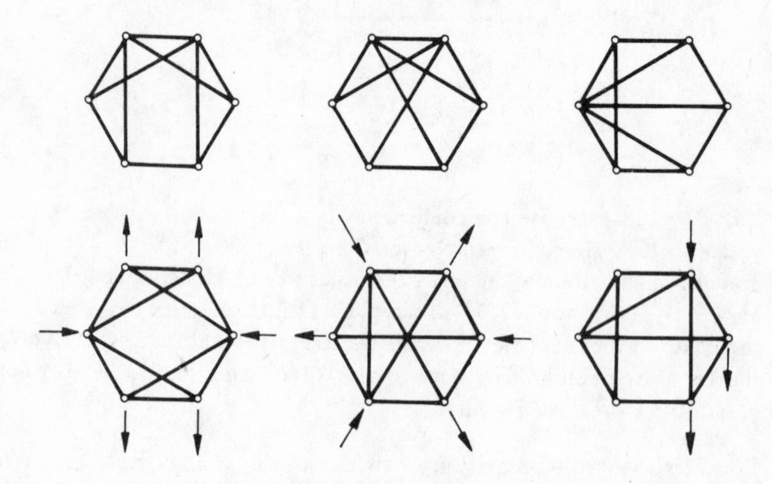

Fig. S.18.9

Bibliography

Abdullah, K., 1972. A necessary condition for complete solvability of RLCT networks, *IEEE Trans. Circuit Theory*, CT-19, No. 5, 492–493.

Abdullah, K. and Y. Tokad, 1972. On the existence of mathematical models for multiterminal RCΓ networks, *IEEE Trans. Circuit Theory*, CT-19, No. 5, 419–424.

Acketa, D.M., 1979. On the essential flats of geometric lattices, *Publications de l'Institut Mathématique, Nouvelle série*, 26, No. 40, 11–17.

Aho, A.V., J.E. Hopcroft and J.D. Ullman, 1974. *The Design and Analysis of Computer Algorithms*, Addison-Wesley, Reading, Mass.

Aigner, M., 1979. *Combinatorial Theory*, Springer, Berlin.

Aigner, M. and T.A. Dowling, 1970. Matching theorems for combinatorial geometries, *Bull. Am. Math. Soc.*, 76, 57–60.

Aigner, M. and T.A.Dowling, 1971. Matching theory for combinatorial geometries, *Trans. Am. Math. Soc.*, 158, 231–245.

Andrásfai, B., 1977. *Introductory Graph Theory*, Akadémiai Kiadó, Budapest & Adam Hilger, Bristol.

Andrásfai, B., 1983. *Graph Theory: Flows, Matrices* (in Hungarian), Akadémiai Kiadó, Budapest.

Asano, T., T. Nishizeki and P. Seymour, 1984. A note on nongraphic matroids, *J. Comb. Theory Ser. B*, 37, 290–293.

Asimow, L. and B. Roth, 1978. The rigidity of graphs I, *Trans. Am. Math. Soc.*, 245, 279–289.

Asimow, L. and B. Roth, 1979. The rigidity of graphs II, *J. Math. Anal. & Appl.*, 68, 171–190.

Atkin, A.O.L., 1972. Remark on a paper of Piff and Welsh, *J. Comb. Theory Ser.B*, 13, No. 2, 179–182.

Auslander, L. and S.V.Parter, 1961. On imbedding graphs in the plane, *J. Math. & Mech.*, 10, 517–523.

Baglivo, J.A. and J.E. Graver, 1983. *Incidence and Symmetry in Design and Architecture*, Cambridge University Press, Cambridge.

Balabanian, N. and T.A. Bickart, 1969. *Electrical Network Theory*, Wiley, New York.

Bean, D.T., 1972. Refinements of a matroid, *Proc. Third Southeastern Conf. on Combin. Graph Theory and Comput.*, 69–72.

Belevitch, V., 1968. *Classical Network Theory*, Holden-Day, San Francisco.

Bendik, J., 1967. Equivalent gyrator networks with nullators and norators, *IEEE Trans. Circuit Theory*, CT-14, 98.

Berge, C., 1957. Two theorems in graph theory, *Proc. Natl. Acad. Sci. USA*, 43, 842–844.

Berge, C., 1958. *Théorie des Graphes et Ses Applications*, Dunod, Paris.

Berge, C., 1970. *Graphes et Hypergraphes*, Dunod, Paris.

Birkhoff, G., 1935. Abstract linear dependence and lattices, *Am. J. Math.*, 57, 800–804.

Birkhoff, G., 1967. *Lattice theory*, Am. Math. Soc., Providence.

Bixby, R.E., 1975. A composition for matroids, *J. Comb. Theory Ser.B*, 18, 1, 59–72.

Bixby, R.E., 1976. A strengthened form of Tutte's characterization of regular matroids, *J. Comb. Theory Ser.B*, 20, 216–221.

Bixby R.E., 1977a. Kuratowski's and Wagner's theorems for matroids, *J. Comb. Theory Ser.B*, 22, 31–53.

Bixby, R.E., 1977b. A simple proof that every matroid is an intersection of fundamental transversal matroids, *Discrete Math.*, 18, 311–312 .

Bixby, R.E., 1979. On Reid's characterization of the ternary matroids, *J. Comb. Theory Ser.B*, 26, 174–204.

Bixby, R.E. and W.H. Cunningham, 1980. Converting linear programs to network problems, *Math. Oper. Res*, 5, No. 3, 321–357.

Bland, R.G., 1977. A combinatorial abstraction of linear programming, *J. Comb. Theory Ser.B*, 23, 33–57.

Bland, R.G. and M. Las Vergnas, 1978. Orientability of matroids, *J. Comb. Theory Ser.B*, 24, 94–123.

Bland R.G. and M. Las Vergnas, 1979. Minty colorings and orientations of matroids, *Ann. New York Acad. Sci.*, 319, 86–92.

Bolker, E.D. and H. Crapo, 1977. How to brace a one-story building, *Environ. Plan.B*, 4, 125–152.

Bolker, E.D. and B. Roth, 1980. When is a bipartite graph a rigid framework? *Pacific J. Math.*, 90, No. 1, 27–44.

Bondy, J.A, 1972. Transversal matroids, base-orderable matroids and graphs, *Quart. J. Math.*, 23, No. 89, 81–89.

Bondy, J.A. and U.S.R. Murty, 1976. *Graph Theory with Applications*, Macmillan Press, London.

Bondy, J.A. and D.J.A. Welsh, 1971. Some results on transversal matroids and constructions for identically self-dual matroids, *Quart. J. Math.*, 22, No. 87, 435–451.

Booth, K.S. and G.S. Leuker, 1976. Testing for the consecutive ones property, interval graphs and graph planarity using PQ-tree algorithms, *J. Comput. & Syst. Sci.*, 13, 335–379.

Bordewijk, J.L., 1956. Inter-reciprocity applied to electrical networks, *Appl. Sci. Res. (Netherlands)*, 6B, No. 1–2, 1–74.

Borůvka, O., 1926. O jistém problému minimálním, *Acta Soc. Sci. Natl. Moraviae III*, 3, F23, 37–58.

Bouchet, A., 1987. Greedy algorithm and symmetric matroids, *Math. Programming*, 38, 147–159.

Bow, R.H., 1873. *On the Economics of Construction in Relation to Framed Structures*, E. & F. N. Spon, London.

Brezovec, C., G. Cornuéjols and F. Glover, 1986. Two algorithms for weighted matroid intersection, *Math. Programming*, 36, 39–53.

Brualdi, R.A., 1969. Comments on bases in dependence structures, *Bull. Austr. Math. Soc.*, 50, No. 1, 161–167.

Brualdi, R.A., 1970a. Admissible mappings between dependence spaces, *Proc. London Math. Soc.*, 21, 296–312.

Brualdi, R.A., 1970b. Common transversals and strong exchange systems, *J. Comb. Theory Ser.B.*, 8, 307–329.

Brualdi, R.A., 1971a. Menger's theorem and matroids, *J. London Math. Soc.*, 2, No. 4, 46–50.

Brualdi, R.A., 1971b. Induced matroids, *Proc. Am. Math. Soc.*, 29, 213–221.

Brualdi, R.A., 1974. On fundamental transversal matroids, *Proc. Am. Math. Soc.*, 45, No. 1, 151–156.

Brualdi, R.A., 1975. Matroids induced by directed graphs, a survey, *Recent Advances in Graph Theory, Proc. Symp. Prague*, June 1974, Academia Praha, 115–134.

Brualdi, R.A. and G.W. Dinolt, 1972. Characterizations of transversal matroids and their presentations, *J. Comb. Theory Ser.B.*, 12, No. 3, 268–286.

Brualdi, R.A. and J. Mason, 1972. Transversal matroids and Hall's theorem, *Pacific J. Math.*, 41, 601–613.

Brualdi, R.A. and E.B. Scrimger, 1968. Exchange systems, matchings and transversals, *J. Comb. Theory*, 5, 244–257.

Bryant, P.R., 1959. The order of complexity of electrical networks, *Proc. IEE(GB)*, 106, 174–188.

Bryant, P.R., 1962. The explicit form of Bashkow's A matrix, *Trans. IRE*, CT-8, 303–306.

Bryant, P.R. and J. Tow, 1972. The A-matrix of linear passive reciprocal networks, *J. Franklin Inst.*, 293, No. 6, 401–419.

Brylawski, T.H., 1971. A combinatorial model for series-parallel networks, *Trans. Am. Math. Soc.*, 154, 1–22.

Brylawski, T.H., 1975a. A note on Tutte's unimodular representation theorem, *Proc. Am. Math. Soc.*, 52, 499–502.

Brylawski, T.H., 1975b. An affine representation for transversal geometries, *Stud. Appl. Math.*, 54, 143–160.

Brylawski, T.H., 1986a. Constructions, in N. White (ed), *Theory of Matroids*, Cambridge University Press, Cambridge, 127–223.

Brylawski, T.H., 1986b. Appendix of Matroid Cryptomorphisms, in N. White (ed), *Theory of Matroids*, Cambridge University Press, Cambridge, 298–312.

Brylawski, T.H. and D. Kelly, 1980. *Matroids and Combinatorial Geometries*, University of North Carolina at Chapel Hill.

Busacker, R.B. and T.L. Saaty, 1965. *Finite Graphs and Networks: An Introduction with Applications*, McGraw–Hill, New York.

Calladine, C.R., 1978. Buckminster Fuller's "tensegrity" structures and Clerk Maxwell's rules for the construction of stiff frames, *Int. J. Solids Structures*, 14, 161–172.

Camion, P., 1959. Chemins et circuits hamiltoniens des graphes complets, *Compt. rend.*, 249, 2151–2152.

Carlin, H.J., 1960. General n-port synthesis with negative resistors, *Proc. IRE*, 48, 1174–75.

Carlin, H.J., 1964. Singular network elements, *IEEE Trans. Circuit Theory*, CT-11, 67–72.

Carlin, H.J. and D.C. Youla, 1961. Network synthesis with negative resistors, *Proc. IRE*, 49, 907–920.

Cauer, W., 1954. *Theorie der linearen Wechselstromschaltungen*, Akademie Verlag, Berlin.

Cayley, A., 1889. A theorem on trees, *Quart. J. Pure Appl. Math.*, 23, 376–378.

Cederbaum, I., 1956. Invariance and mutual relations of electrical network determinants, *J. Math. Phys.*, 34, No. 4, 236–244.

Cederbaum, I., 1959. Applications of matrix algebra to network theory, *IRE Trans. Circuit Theory*, CT-6, 127–137.

Cederbaum, I., 1961. Topological considerations in the realization of resistive n-port networks, *IRE Trans. Circuit Theory*, CT-8, 324–329.

Cederbaum, I., 1963. Paramount matrices and realization of resistive 3-port networks, *Proc. IEE(GB)*, 110, No. 11, 1960–1964.

Chaiken, S., 1982. A combinatorial proof of the all minors matrix tree theorem, *SIAM J. Alg. Discrete Methods*, 3, 319–329.

Chakravarty, N., G. Holman, S. McGuinness and A. Recski, 1986. One-story buildings as tensegrity frameworks, *Structural Topology*, 12, 11–18.

Chartrand, G. and F. Harary, 1967. Planar permutation graphs, *Ann. Inst. Henri Poincaré, Sect.B*, 3, 433–438.

Chen, Wai-Kai, 1970a. Nonsingular submatrices of the incidence matrix of a graph over the real field, *J. Franklin Inst.*, 289, 155–166.

Chen, Wai-Kai, 1970b. On vector spaces associated with a graph, *SIAM J. Appl. Math.*, 20, 526–529.

Chvátal, V., 1973. Edmonds polytopes and a hierarchy of combinatorial problems, *Discrete Math.*, 4, 305–337.

Coatés, C.L., 1958. General topological formulas for linear network functions, *IRE Trans.*, CT-5, 42–54.

Conforti, M. and G. Cornuéjols, 1984. Submodular set functions, matroids and the greedy algorithm: Tight worst-case bounds and some generalizations of the Rado-Edmonds theorem, *Discrete Appl. Math.*, 7, 251–274.

Connelly, R., 1975. An attack on rigidity, I, II, *Bull. Am. Math. Soc.*, 81, No. 3, 566–569.

Cook, S.A., 1971. The complexity of theorem-proving procedures, *Proc. Third ACM Symp. Th. Comput.*, 151–158.

Cordovil, R., 1982. Sur les matroïdes orientés de rang 3 et les arrangements de pseudodroites dans le plan projectif réel, *Eur. J. Comb.*, 3, 307–318.

Crapo, H.H., 1970. Erecting geometries, *Proc. 2nd Chapel Hill Conf. Comb. Math.*, University of North Carolina, 74–99.

Crapo, H.H., 1979. Structural rigidity, *Structural Topology*, 1, 26–45.

Crapo, H.H., 1977. More on the bracing of one story buildings, *Environ. Plan.B*, 4, 153–156.

Crapo, H.H., 1982. The tetrahedral-octahedral truss, *Structural Topology*, 7, 51–60.

Crapo, H.H. and G.C. Rota, 1968. *Combinatorial Geometries* (On the Foundations of Combinatorial Theory II), M.I.T., preprint.

Cremona, L., 1874. *Elementi di calcolo grafico*, Torino.

Cremona, L., 1879. *Le figure reciproche nella statica grafica*, Milano. (English translation): *Graphical Statics*, Oxford University Press, London 1890.

Cunningham, W.H., 1973. *A Combinatorial Decomposition Theory*, PhD Thesis, Waterloo.

Cunningham, W.H., 1981. On matroid connectivity, *J. Comb. Theory Ser.B*, 30, 94–99.

Cunningham, W.H., 1986. Improved bounds for matroid partition and intersection algorithms, *SIAM J. Comput.*, 15, 948–957.

Cunningham, W.H. and J. Edmonds, 1980. A combinatorial decomposition theory, *Can. J. Math.*, 32, No. 3, 734–765.

Csurgay, A., 1971. Conditions on the existence of frequency-domain description of linear distributed active networks, *Proc. Summer School on Circuit Th.*, Talé, Czechoslovakia, 2, 189–193.

Csurgay, A., Z. Kovács and A. Recski, 1974. On the transient analysis of lumped-distributed nonlinear networks, *Proc. Fifth Int. Coll. Microwave Commun.*, Budapest.

Dervişoğlu, A., 1964. Bashkow's A matrix for active RLC networks, *Trans. IRE*, CT-10, 404–406.

Dervişoğlu, A., 1969. Comments "On the existence of the A state matrix", *Trans. IEEE*, CT-16, 242.

De Sousa, J. and D.J.A. Welsh, 1972. A characterization of binary transversal structures, *J. Math. Anal. & Appl.*, 40, 55–59.

Dijkstra, E.W., 1959. A note on two problems in connection with graphs, *Numer. Math.*, 1, 269–271.

Dinic, E.A., 1970. An algorithm for the solution of the problem of maximal flow in a network with power estimation, *Dokl. Akad. Nauk SSSR*, 194, 754–757, (English translation *Soviet Math. Dokl.*, 11, 1277–1280).

Dirac, G.A., 1952. A property of 4-chromatic graphs and some remarks on critical graphs, *J. London Math. Soc.*, 27, 85–92.

Dirac, G.A., 1960. 4-chrome Graphen und vollständige 4-Graphen, *Math. Nachr.*, 22, 51–60.

Doignon, J.P., 1981. On characterizations of binary and graphic matroids, *Discrete Math.*, 37, 299–301.

Dowling, T.A. and D.G. Kelly, 1974. Elementary strong maps and transversal geometries, *Discrete Math.*, 7, No. 3–4, 209–225.

Dress, A., 1986. Chirotops and oriented matroids, *Bayreuth Math. Schr.*, 21, 14–68.

Duffin, R.J., 1965. Topology of series-parallel networks, *J. Math. Anal. & Appl.*, 10, 303–318.

Duffin, R.J., 1975. Electrical Network Models, *Studies in Graph Theory, Part I, MAA, Stud. Math.*, 11, 94–138.

Duke, R., 1981. *Freedom in Matroids*, PhD Thesis, The Open University, Milton Keynes.

Duncan Luce, R., 1952. Two decomposition theorems for a class of finite oriented graphs, *Am. J. Math.*, 74, 701–722.

Edmonds, J.R., 1965a. Minimum partition of a matroid into independent subsets, *J. Res. Natl. Bur. Stand.*, 69B, 67–72.

Edmonds, J.R., 1965b. Paths, trees and flowers, *Can. J. Math.*, 17, 449–467.

Edmonds, J.R., 1967. Systems of distinct representatives and linear algebra, *J. Res. Natl. Bur. Stand.*, 71B, 241–245.

Edmonds, J.R.,1968. Matroid partition, *Mathematics of the Decision Sciences, Part I.*, Lectures in Appl. Math., 11, 335–345.

Edmonds, J.R., 1970. Submodular functions, matroids and certain polyhedra, *Combinatorial Structures and Their Applications*, Gordon and Breach, New York, 69–87.

Edmonds, J.R., 1971. Matroids and the greedy algorithm, *Math. Program.*, 1, No. 2, 127–136.

Edmonds, J.R., 1979. Matroid intersection, *Ann. Discrete Math.* 4, 39–49.

Edmonds, J.R. and D.R. Fulkerson, 1965. Transversals and matroid partition, *J. Res. Natl. Bur. Stand.*, 69B, 147–153.

Edmonds, J.R. and E.L. Johnson, 1970. Matching: a well-solved class of integer linear programs, *Combinatorial Structures and Their Applications*, Gordon and Breach, New York, 89–92.

Edmonds, J.R. and R.M. Karp, 1972. Theoretical improvements in algorithmic efficiency for network flow problems, *J. Assoc. Comput. Mach.* (USA) 19, 248–264.

Egervári, J., 1931. Combinatorial properties of matrices (in Hungarian), *Mat. Fiz. Lapok*, 38, 16–28.

Erol Emre and Ozay Hüseyin, 1973. On the order of complexity of active RLC networks, *IEEE Trans.*, CT-20, No. 5, 615.

Euler, L., 1736. Solutio problematis ad geometriam situs pertinentis, *Acad. Petropolitanae*, 8, 128–140.

Even, S., 1975. An algorithm for detemining whether the connectivity of a graph is at least *k*, *SIAM J. Comput.*, 4, 393–396.

Even, S., 1979. *Graph Algorithms*, Computer Science Press, Potomac, MD.

Even, S. and R.E. Tarjan, 1976. Computing an *st*-numbering, *Theor. Comput. Sci.*, 2, 339–344.

Fáry, I., 1948. On the straight line representation of planar graphs, *Acta Sci. Math.*, 11/4, 229–233.

Feldtkeller, R., 1943. *Einführung in die Vierpoltheorie der elektrischen Nachrichtentechnik*, S. Hirzel, Leipzig.

Fettweis, A., 1969. On the algebraic derivation of the state equations, *IEEE Trans.*, CT-16, 171–175.

Feussner, W., 1902. Über Stromverzweigung in netzformigen Leitern, *Ann. Phys.*, 9, 1304–1329.

Feussner, W., 1904. Zur Berechnung der Stromstärke in netzformigen Leitern, *Ann. Phys.*, 15, 385–394.

Fischer, H.D., 1975. On the unique solvability of RLCT networks, *Int. J. Circuit Theory and Appl.*, 3, No. 4, 391–394.

Foeppl, A., 1888. Über das räumliche Fachwerk, *Schweizerische Bauzeitung*, 11, 115–117.

Folkman, J. and J. Lawrence, 1978. Oriented matroids, *J. Comb. Theory Ser.B*, 25, 199–236.

Fonlupt, J. and A. Zemirline, 1983. On the number of common bases of two matroids, *Discrete Math.*, 45, 217–228.

Ford, L.R. and D.R. Fulkerson, 1956. Maximal flow through a network, *Can. J. Math.*, 8, 399–404.

Ford, L.R. and D.R. Fulkerson, 1962. *Flows in Networks*, Princeton University Press, Princeton.

Foster, R.M., 1949. The average impedance of an electrical network, *Contributions to Applied Mechanics*, J. Edwards, Ann Arbor, 330–340.

Fournier, J.C., 1981. A characterization of binary geometries by a double elimination axiom, *J. Comb. Theory Ser.B*, 31, No. 2, 249–250.

Frank, A., 1981. A weighted matroid intersection algorithm, *J. Algorithms*, 2, 328–336.

Frank, A., 1982. An algorithm for submodular functions on graphs, *Ann. Discrete Math.*, 16, 189–212.

Frank, A. and É. Tardos, 1984. Matroids from crossing families, Coll. Math. Soc. János Bolyai 37. *Finite and Infinite Sets*, Eger, Hungary, North-Holland, Amsterdam, 295–304.

Frobenius, F.G., 1875. Über das Pfaffsche Problem, *J. für die reine und angewandte Math.*, 82, 230–315, see also in F.G. Frobenius, 1968. *Gesammelte Abhandlungen*, Vol. I, Springer, Berlin.

Fukuda, K., 1982. *Oriented Matroid Programming*, PhD Thesis, University of Waterloo.

Gabow, H.N. and M. Stallmann, 1986. An augmenting path algorithm for linear matroid parity, *Combinatorica*, 6, 123–150.

Gabow, H.N. and R.E. Tarjan, 1984. Efficient algorithms for a family of matroid intersection problems, *J. Algorithms*, 5, 80–131.

Gallai, T., 1959. Über extreme Punkt- und Kantenmengen, *Ann. Univ. Sci. Budapest Eötvös*, Sect. Math. 2, 133–138.

Gallai, T., 1963. Neuer Beweis eines Tutte'schen Satzes, *MTA Mat. Kutató Int. Közleményei*, 8, 135–139.

Garey, M.R. and D.S. Johnson, 1979. *Computers and Intractability: A Guide to the Theory of NP-Completeness*, W.H. Freeman, San Francisco.

Gel'fand, I.M. and V.V. Serganova, 1987. On the general definition of a matroid and a greedoid, *Soviet Math. Dokl.*, 35, 6–10.

Gibbs, N.E., 1975. Basic cycle generation, *Commun. ACM*, 18, 275–276.

Gilbert, E.N. and H.O. Pollak, 1968. Steiner minimal trees, *SIAM J. Appl. Math.*, 16, 1–29.

Goldstein, A.J., 1963. An efficient and constructive algorithm for testing whether a graph can be embedded in a plane, report, Princeton University.

Golumbic, M.C., 1980. *Algorithmic Graph Theory and Perfect Graphs*, Academic Press, New York.

Grimbleby, J.B., 1981. Symbolic analysis of circuits containing active elements, *Electron. Lett.*, 17, 754–756.

Gröflin, H. and A.J. Hoffman, 1981. On matroid intersections, *Combinatorica*, 1, 43–47.

Grötschel, M., L. Lovász and A. Schrijver, 1981. The ellipsoid method and its consequences in combinatorial optimization, *Combinatorica*, 1, 169–197.

Grötschel, M., L. Lovász and A. Schrijver, 1984. Geometric methods in combinatorial optimatization, in W.R. Pulleyblank (ed), *Progress in Combinatorial Optimatization*, Academic Press, London, 167–183.

Grötschel, M., L. Lovász and A. Schrijver, 1988. *Geometric Algorithms and Combinatorial Optimization*, Springer, Berlin.

Guan Mei Gu and Chen Quing Hua, 1985. A simple algorithm for finding the second-order constrained base of matroid, *Kezue Tongbao* (English ed.), 30, 862–865.

Guillemin, E., 1936. *Communication Networks I-II*, Wiley, New York.

Hajós, Gy., 1961. Über eine Konstruktion nicht n-farbbarer Graphen, *Wiss. Z. Martin-Luther-Univ. Halle-Wittenberg*, A10, 116–117.

Hajós, Gy., 1964. *Introduction to Geometry* (in Hungarian), Tankönyvkiadó, Budapest.

Hall, P., 1935. On representatives of subsets, *J. London Math. Soc.*, 10, 26–30.

Hausmann, D. and B. Korte, 1978. Lower bounds on the worst case complexity of some oracle algorithms, *Discrete Math.*, 24, 261–276.

Hausmann, D. and B. Korte, 1981. Algorithmic versus axiomatic definitions of matroids, *Math. Program. Stud.*, 14, 98–111.

Hausmann, D., B. Korte and T.A. Jenkyns, 1980. Worst case analysis of greedy type algorithms for independence systems, *Math. Program. Stud.*, 12, 120–131.

Helgason, T., 1974. Aspects of the theory of hypermatroids, *Hypergraph Seminar*, Springer Lecture Notes 411, Springer, Berlin, 191–214.

Henneberg, L., 1911. *Die graphische Statik der starren Systeme*, Leipzig.

Hetyei, G., 1964. On Covering by 2×1 Rectangles (in Hungarian), *Pécsi Tanárképző Főisk. Közl.* 351–368.

Hirano, K., F. Nishi and S. Tomiyama, 1974. Some considerations on state equations of linear active networks, *Int. J. Circuit Theory and Appl.*, 2, No. 1, 39–50.

Holzmann, C.A., 1979. Binary netoids, *Proc. IEEE ISCAS, Tokyo*, 1000–1003.

Hopcroft, J. and R. Tarjan, 1973. Efficient algorithms for graph manipulation, *Commun. ACM*, 16, No. 6, 372–378.

Hopcroft, J. and R. Tarjan, 1974. Efficient planarity testing, *J. Assoc. Comput. Mach.*, 21, No. 4, 549–568.

Hu, T.C., 1969. *Integer Programming and Network Flows*, Addison-Wesley, Reading, Mass.

Imai, H., 1985. On combinatorial structures of line drawings of polyhedra, *Discrete Applied Math.*, 10, 79–92.

Ingleton, A.W., 1959. A note on independence functions and rank, *J. London Math. Soc.*, 34, 49–56.

Ingleton, A.W., 1971a. A geometrical characterization of transversal independence structures, *Bull. London Math. Soc.*, 3, 47–51.

Ingleton, A.W., 1971b. Conditions for representability and transversality of matroids, in C.P. Bruter (ed), *Théorie des Matroïdes*, Lec. Notes in Math. 211, Springer, Berlin, 62–66.

Ingleton, A.W., 1976. Non base-orderable matroids, Proceedings of the Fifth British Combinatorial Conference, *Congressus Numerantium XV*, Utilitas Math., Winnipeg, 355–359.

Ingleton, A.W., 1977. Transversal matroids and related structures, in M. Aigner (ed), *Higher Combinatorics*, D. Reidel, Dordrecht, 117–131.

Ingleton, A.W. and R.A. Main, 1975. Non-algebraic matroids exist, *Bull. London Math. Soc.*, 7, 144–146.

Ingleton, A.W. and M.J. Piff, 1973. Gammoids and transversal matroids, *J. Comb. Theory Ser.B*, 15, 51–68.

Iri, M., 1968a. On the synthesis of loop and cut set matrices and the related problems, *RAAG Memoirs IV*, A-XIII, 4–38.

Iri, M., 1968b. On topological conditions for the realization of multiport networks, *RAAG Memoirs IV*, A-XV, 47–62.

Iri, M., 1968c. A min–max theorem for the ranks and term–ranks of a class of matrices: An algebraic approach to the problem of the topological degrees of freedom of a network (in Japanese), *Trans. IECEJ*, 51A, 180–187.

Iri, M., 1969. *Network Flow, Transportation and Scheduling; Theory and Algorithms*, Academic Press, New York.

Iri, M., 1979a. Survey of recent trends in applications of matroids, *Proc. IEEE ISCAS, Tokyo*, 987.

Iri, M., 1979b. A review of recent work in Japan on principal partitions of matroids and their applications, *Ann. New York Acad. Sci.*, 319, 306–319.

Iri, M., 1980. "Dualities" in Graph Theory and the Related Fields viewed from the Methatheoretical Standpoint, *Tohoku Univ. Tsuken Symp. "Graph Theory and Algorithms"*, Sendai, 139–151.

Iri, M., 1983. Applications of matroid theory, *Mathematical Programming: The State of the Art*, Springer, Berlin, 158–201.

Iri, M., 1984. Structural theory for the combinatorial systems characterized by submodular functions, in W.R. Pulleyblank(ed), *Progress in Combinatorial Optimization*, Academic Press, New York, 197–219.

Iri, M. and S. Fujishige, 1981. Use of matroid theory in operations research, circuits and systems theory, *Int. J. Syst. Sci.*, 12, 27–54.

Iri, M. and A. Recski, 1979. Reflections on the concepts of dual, inverse and adjoint networks (in Japanese), *Proc. IECEJ Coll. Circuits and Systems, Tokyo*, CAS-79-78, 19–25.

Iri, M. and A. Recski, 1980. What does duality really mean? *Int. J. Circuit Theory and Appl.*, 8, 317–324.

Iri, M. and A. Recski, 1982. Duality and reciprocity – a qualitative approach, *Proc. IEEE ISCAS, Rome, II*, 415–418.

Iri, M. and N. Tomizawa, 1974. A practical criterion for the existence of the unique solution in a linear electrical network with mutual couplings, *Trans. Inst. Electron. & Commun. Eng. Jpn.*, 57-A, 8, 35–41.

Iri, M. and N. Tomizawa, 1975. A unifying approach to fundamental problems in network theory by means of matroids, *Trans. Inst. Electron. & Commun. Eng. Jpn.*, 58-A, No. 1, 33–40.

Iri, M., J. Tsunekawa and K. Murota, 1982. Graph theoretical approach to large-scale systems – Structural solvability and block-triangularization, *Trans. Inform. Process. Soc. Japan*, 23, 88–95.

Isachenko, A.N., 1984. Polynomial convergence of matroid oracles, *Vestsī Akad. Navuk. BSSR Ser. Fīz.-Mat. Navuk.*, 6, 125.

Jaeger, F., 1983. Symmetric representation of binary matroids, *Ann. Discrete Math.* 17, 371–376.

Jensen,P.M. and B. Korte, 1982. Complexity of matroid property algorithms, *SIAM J. Comput.*, 11. No. 1, 184–190.

Kahn, J., 1982. Characteristic sets of matroids, *J. London Math. Soc.*, 2, 26, 207–217.

Kahn, J., 1984. A geometric approach to forbidden minors for GF(3), *J. Comb. Theory Ser.A*, 37, 1-12.

Kahn, J., 1985. A problem of P. Seymour on nonbinary matroids, *Combinatorica*, 5, 319–323.

Kalai, G., 1985. Hyperconnectivity of graphs, *Graphs and Combinatorics*, 1, 65–79.

Kalman, R.E. and N. DeClaris, 1971. *Aspects of Network and System Theory*, Holt, Rinehart and Winston, New York.

Karp, R.M., 1972. Reducibility among combinatorial problems, in R.E. Miller, J.W. Thatcher (eds), *Complexity of Computer Computations*, Plenum Press, New York, 85–104.

Karp, R.M., 1975. On the computational complexity of combinatorial problems, *Networks*, 5, No. 1, 45–68.

Kelmans, A.K., M.V. Lomonosov and V.P. Polesskij, 1976. Minimal coverings in a matroid, *Probl. Peredachi Inf.*, 12, 94–107; English translation *Probl. Inf. Transm.*, 12 (1977), 231–241.

Khachiyan, L.G., 1979. A polynomial algorithm in linear programming, *Dokl. Akad. Nauk SSSR*, 244, 1093–6, English translation *Sov. Math. Dokl.*, 191–194.

Kirchhoff, G., 1847. Ueber die Auflösung der Gleichungen, auf welche man bei der Untersuchungen der linearen Vertheilung galvanischer Ströme geführt wird, *Poggendorff Annalen der Physik und Chemie* LXXII. No. 12, Leipzig, pp. 497–508.

Kishi, G. and Kajitani, 1968. Maximally distinct trees in a linear graph (in Japanese), *Trans. Inst. Electron. & Commun. Eng. Jpn.*, 51A, 196–203.

Knuth, D.E., 1973. Matroid partitioning. Report, Stan-CS-73-342, Stanford University.

Korte, B. and D. Hausmann, 1978. An analysis of the greedy heuristic for independence systems. *Ann. Discrete Math.*, 2, 65–74.

Korte, B. and L. Lovász, 1981. Mathematical structures underlying greedy algorithms, in F. Gécseg (ed), *Fundamentals of Computation Theory*, Springer, Berlin, 205–209.

Korte, B. and L. Lovász, 1984. Greedoids – A structural framework for the greedy algorithm, in W.R. Pulleyblank (ed), *Progress in Combinatorial Optimization*, Academic Press, New York, 221–243.

König, D., 1931. On graphs and matrices (in Hungarian), *Mat. Fiz. Lapok*, 38, 116–119.

König, D., 1936. *Theorie der endlichen und unendlichen Graphen*, Leipzig, reprinted: Chelsea, New York, 1950.

Krogdahl, S., 1974. A combinatorial base for some optimal matroid intersection algorithms. Report, Stan-CS-74-468, Stanford University.

Krogdahl, S., 1975. A combinatorial proof of Lawler's matroid intersection algorithm, unpublished.

Kron, G., 1939. *Tensor Analysis of Networks*, J. Wiley, New York.

Kruskal, F.B. Jr., 1956. On the shortest spanning subtree of a graph and the travelling salesman problem, *Proc. Amer. Math. Soc.*, 7, 48–50.

Kuh, E.S., D.M. Layton and J. Tow, 1968. Network analysis and synthesis *via* state variables, in G. Biorci (ed), *Network and Switching Theory*, a NATO Advanced Study Institute, Academic Press, New York, 135–155.

Kuh, E.S. and R.A. Rohrer, 1965. The state-variable approach to network analysis. *Proc. IEEE*, 53, 672–686.

Kung, J.P.S., 1986a. *A Source Book in Matroid Theory*, Birkhäuser, Boston.

Kung, J.P.S., 1986b. Basis-exchange properties, in N. White (ed), *Theory of Matroids*, Cambridge University Press, Cambridge, 62–75.

Kuratowski, C., 1930. Sur le Probléme des Courbes Gauches en Topologie, *Fundamenta Mathematicae*, 15–16, 271–283.

Laman, G., 1970. On graphs and rigidity of plane skeletal structures, *Eng. Math.*, 4, 331–340.

Las Vergnas, M., 1970. Sur les systèmes de représentants distincts d'une famille d'ensembles, *Compt. rend.* 270, Ser.A, B, No. 8/23, 501–503.

Las Vergnas, M., 1975. Matroïdes orientables, *Compt. rend.*, Ser.A, 280, 61–64.

Las Vergnas, M., 1977. Acyclic and totally cyclic orientations of combinatorial geometries, *Discrete Math.*, 20, 51–61.

Las Vergnas, M.,1978. Bases in oriented matroids, *J. Comb. Theory Ser.B*, 25, 283–289.

Las Vergnas, M., 1980. Fundamental circuits and a characterization of binary matroids, *Discrete Math.*, 31, No. 3, 327.

Lawler, E.L., 1975. Matroid intersection algorithms, *Math. Program.*, 9, 31–56.

Lawler, E.L., 1976. *Combinatorial Optimization: Networks and Matroids*, Holt, Rinehart and Winston, New York.

Lazarson, T., 1958. The representation problem for independence functions, *J. London Math. Soc.*, 33, 21–25.

Lehman, A., 1979. On the width-length inequality, *Math. Program.*, 16, 245–259.

Lempel, A., S. Even and I. Cederbaum, 1967. An algorithm for planarity testing of graphs, *Theory of Graphs*, Symp., Rome, 1966, Gordon and Breach, New York, 215–232.

Lhuilier, S.A.J., 1812–13. Mémoire sur la polyédometrie, *Annales de Mathématiques*, 3, 169–189.

Lin, Pen-Min, 1980. Complementary trees in circuit theory, *IEEE Trans. Circuits & Syst.*, CAS-27, No. 10, 921–928.

Lindström, B., 1984. On algebraic representations of matroids, Report, Dept. Math., Univ. Stockholm.

Liu Gui Zhen, 1985. Welsh's conjecture is true for simple binary matroids, *Kezue Tongbao (English ed.)*, 30, 1147–1150.

Lovász, L., 1971. Brief survey of matroid theory (in Hungarian), *Matematikai Lapok*, 22, 249–267.

Lovász, L., 1977. Flats in matroids and geometric graphs, in P.J. Cameron (ed), *Combinatorial Surveys*, Proceedings of the Sixth British Combinatorial Conference, Academic Press, London, 45–86.

Lovász, L., 1979a. *Combinatorial Problems and Exercises*, North-Holland, Amsterdam, Akadémiai Kiadó, Budapest.

Lovász, L., 1979b. On determinants, matching and random algorithms, *Proceedings of Fundamentals of Computation Theory*, Akademie Verlag, Berlin, 565–574.

Lovász, L., 1980a. Selecting independent lines from a family of lines in a space, *Acta Sci. Math. Univ. Szeged*, 42, 121–131.

Lovász, L., 1980b. Matroid matching and some applications, *J. Comb. Theory Ser.B*, 28, 208–236.

Lovász, L., 1981. The matroid matching problem. Coll. Math. Soc. János Bolyai 25, *Algebraic Methods in Graph Theory*, Szeged, Hungary, North-Holland, Amsterdam, 495–517.

Lovász, L.,1983. Submodular functions and convexity, in A. Bachem, M. Grötschel and B. Korte (eds), *Mathematical Programming, The State of the Art*, Springer, Berlin, 235–257.

Lovász, L. and M. Plummer, 1986. *Matching Theory*, North-Holland, Amsterdam, Akadémiai Kiadó, Budapest.

Lovász, L. and A. Recski, 1973. On the sum of matroids, *Acta Math. Acad. Sci. Hung.*, 24, No. 3–4, 329–333.

Lovász, L. and Y. Yemini, 1982. On generic rigidity in the plane, *SIAM J. Algebraic and Discrete Methods*, 3, No. 1, 91–98.

Liu Gui Zhen, 1985. Welsh's conjecture is true for simple binary matroids, *Kezue Tongbao* (English ed.), 30, 1147–1150.

McDiarmid, C., 1972. Strict gammoids and rank functions, *Bull. London Math. Soc.*, 4, 196–198.

McDiarmid, C., 1975. Rado's theorem for polymatroids, *Math. Proc. Cambridge Philos. Soc.*, 78, 263–281.

MacLane, S., 1936. Some interpretations of abstract linear dependence in terms of projective geometry, *Am. J. Math.*, 58, 236–240.

Macmillan, R.H., 1950. The freedom of linkages, *The Mathematical Gazette*, 34, No. 307, 26–37.

Mandel, A., 1982. *Topology of Oriented Matroids*, PhD Thesis, University of Waterloo.

Mansfield, A., 1981. On the computational complexity of a rigidity problem, *IMA J. Appl. Math.*, 27, 423–429.

Mark, S.K. and M.N.S. Swamy, 1976. The generalized tree for state variables in linear active networks, *Int. J. Circuit Theory and Appl.*, 4, 87–92.

Mason, J.H., 1972. On a class of matroids arising from paths in graphs, *Proc. London Math. Soc.*, Ser 3, 25, 55–74.

Mason, J.H., 1977. Matroids as the study of geometrical configurations, in M. Aigner (ed), *Higher Combinatorics*, D. Reidel, Dordrecht, 133–176.

Mason, J.H. and J.G. Oxley, 1980. A circuit covering result for matroids, *Math. Proc. Cambridge Philos. Soc.*, 87, 25–27.

Maurer, S.B., 1975. A maximum-rank minimum-term-rank theorem for matroids, *Linear Algebra & Appl.*, 10, No. 2, 129–137.

Maxwell, J.C.,1864a. On reciprocal figures and diagrams of forces, *Philos. Mag.* Ser.(4), 27, 250–261.

Maxwell, J.C., 1864b. On the calculation of the equilibrium and stiffness of frames, *Philos. Mag.* Ser.(4), 27, 294–299.

Maxwell, J.C., 1869–72. On reciprocal figures, frames and diagrams of forces, *Trans. R. Soc. Edinburgh*, 26, 1–40.

Maxwell, J.C., 1876. On Bow's method of drawing diagrams in graphical statics, with illustrations from Peauceiller's linkage, *Proc. Cambridge Philos. Soc.*, 2, 407–414.

Maxwell, J.C., 1892. *Electricity and Magnetism*, Clarendon Press, Oxford.

Mayeda, W., 1958. Topological formulas for active networks, Interim Technical Report No. 8, US Army Contract No. DA-11-022-ORD-1983, University of Illinois.

Mayeda, W., 1972. *Graph Theory*, Wiley, Chichester.

Menger, K., 1927. Zur allgemeinen Kurventheorie, *Fund. Math.*, 10, 96–115.

Milić, M.M., 1974. General passive networks – solvability, degeneracies and order of complexity, *IEEE Trans. Circuits & Syst.*, CAS-21, 177-183.

Minty, G., 1966. On the axiomatic foundations for the theories of directed linear graphs, electrical networks and network programming, *J. Math. & Mech.*, 15, 485–520.

Mirsky, L., 1971. *Transversal Theory*, Academic Press, London.

Morgan-Voyce, A.M., 1959. Ladder-network analysis using Fibonacci numbers, *IRE Trans. Circuit Theory*, CT-6, 321–322.

Murota, K., 1987. *Systems Analysis by Graphs and Matroids – Structural Solvability and Controllability*, Springer, Berlin.

Murota, K. and M. Iri, 1985. Structural solvability of systems of equations – a mathematical formulation for distinguishing accurate and inaccurate numbers in structural analysis of systems, *Japan J. Appl. Math.*, 2, 247–271.

Murty, U.S.R. and I. Simon, 1976. A β-function that is not a sum of rank functions of matroids, *Colloques internationaux C.N.R.S.*, No. 260, Problèmes combinatoires et théorie des graphes, Orsay, 305–306.

Nakamura, M., 1978. A note on fundamental transversal matroids, preprint.

Nakamura, M., 1979. Private communication.

Nakamura, M., 1986. On the representation of the rigid subsystems of a plane link system, *J. Oper. Res. Soc. Japan*, 29, 305–318.

Narayanan, H., 1974. *Theory of Matroids and Network Analysis*, PhD Thesis, Indian Institute of Technology, Bombay.

Narayanan, H. and M.N. Vartak, 1973. Gammoids, base-orderable matroids and series–parallel graphs, preprint.

Nash-Williams, C.St. J.A., 1961. Edge-disjoint spanning trees of finite graphs, *J. London Math. Soc.*, 36, 445–450.

Nash-Williams, C.St. J.A., 1964. Decomposition of finite graphs into forests, *J. London Math. Soc.*, 39, 12.

Nash-Williams, C.St. J.A., 1967. On applications of matroids to graph theory, *Theory of Graphs*, Int. Symp. Rome, 1966, Gordon and Breach, New York, 263–265.

Nemhauser, G.L. and L.A. Wolsey, 1988. *Integer and Combinatorial Optimization*, Wiley, New York.

Nešetřil, J., S. Poljak and D. Turzík, 1985. Special amalgams and Ramsey matroids, Coll. Math. Soc. János Bolyai 40. *Matroid Theory*, Szeged, Hungary, North-Holland, Amsterdam, 267–298.

Nitta, T., 1977. *State Variable Approach to Linear Electrical Network Analysis*, PhD Thesis, Faculty of Engineering, Kyoto University.

Noltemeier, H., 1976. *Graphentheorie mit Algorithmen und Anwendungen*, Walter de Gruyter, Berlin.

O'hEigeartaigh, M., J.K. Lenstra and A.H.G. Rinnooy Kan (eds), 1985. *Combinatorial Optimization, Annotated Bibliographies*, Centre for Mathematics and Computer Science, Amsterdam, Wiley, Chichester.

Ohtsuki, T., Y. Ishizaki and H. Watanabe, 1968. Network analysis and topological degrees of freedom (in Japanese), *Trans. Inst. Electron. & Commun. Eng. Jpn*, 51A, 238–245.

Oono, Y., 1960. Formal realizability of linear networks, *Proceedings of the Symposium on Active Networks and Feedback Systems*, Polytechnik Inst. Brooklyn, 475–486.

Oono, Y., 1972. Minimum-gyrator synthesis of n-ports, *IEEE Trans. Circuit Theory*, CT-19, 313–316.

Orlin, J.B. and J.H. Vande Vate (with special contributions from E. Gugenheim and J. Hammond), 1986. An algorithm for the linear matroid parity problem, to appear.

Oxley, J., 1981. On a matroid generalization of graph connectivity, *Math. Proc. Cambridge Philos. Soc.*, 90, 207–214.

Oxley, J., 1986. Graphs and series-parallel networks, in N. White (ed), *Theory of Matroids*, Cambridge University Press, Cambridge, 97–126.

Ozawa, T., 1975. Solvability of linear electric networks, *Memoirs of the Faculty of Engineering*, Kyoto University, 37, 299–315.

Ozawa, T., 1976. Topological conditions for the solvability of active linear networks, *Int. J. Circuit Theory and Appl.*, 4, 125–136.

Papadimitriou, C.H. and K. Steiglitz, 1982. *Combinatorial Optimization: Algorithms and Complexity*, Prentice-Hall, Englewood Cliffs.

Paton, K., 1969. An algorithm for finding a fundamental set of cycles of a graph, *Commun. ACM*, 12, 514–518.

Perfect, H., 1969. Independence spaces and combinatorial problems, *Proc. London Math. Soc.*, 3, 19, 17–30.

Perfect, H., 1981. Independence theory and matroids, *The Mathematical Gazette*, 65, No. 432, 103–111.

Petersen, B., 1979. Investigating solvability and complexity of linear active networks by means of matroids, *IEEE Trans. Circuits & Syst.*, CAS-26, No. 5, 330–342.

Petersen, J., 1891. Die Theorie der regulären Graphs, *Acta Math.*, 15, 193–220.

Piff, M.J., 1973. An upper bound for the number of matroids, *J. Comb. Theory Ser.B*, 14, 241–245.

Piff, M.J. and D.J.A. Welsh, 1970. On the vector representation of matroids, *J. London Math. Soc.*, 2, 284–288.

Piff, M.J. and D.J.A. Welsh, 1971. On the number of combinatorial geometries, *Bull. London Math. Soc.*, 3, 55–56.

Prüfer, H., 1918. Neuer Beweis eines Satzes über Permutationen, *Arch. Math. Phys.*, 27, 742–744.

Purslow, E.J., 1970. Solvability and analysis of linear active networks by use of the state equations, *IEEE Trans. Circuit Theory*, CT-17, 469–475.

Purslow, E.J. and R. Spence, 1967. Order of complexity of active networks, *Proc. IEE (GB)*, 114, 195–198.

Pym, J.S., 1969a. The linking of sets in graphs, *J. London Math. Soc.*, 44, 542–550.

Pym, J.S., 1969b. A proof of the linkage theorem, *J. Math. Anal. & Appl.*, 27, 636-638.

Pym, J.S. and H. Perfect, 1979. Submodular functions and independence structures, *J. Math. Anal. & Appl.*, 30, 1-31.

Rabin, M.O. and D. Scott, 1959. Finite automata and their decision problems, *IBM J. Res. & Dev.*, 3, 114-125.

Rado, R., 1942. A theorem on independence relations, *Quart. J. Math. Oxford Ser.*, 13, 83–89.

Rado, R., 1966. Abstract linear dependence, *Colloq. Math.*, Vol. 14, 258–264.

von Randow, R., 1976. *Introduction to the Theory of Matroids*, Springer, Berlin.

Rankine, W.F.M., 1864. Principle of equilibrium of polyhedral frames, *Philos. Mag.*, Ser. 4, 27, 92.

Read, R.C. and R.E. Tarjan, 1975. Bounds on backtrack algorithms for listing cycles, paths and spanning trees, *Networks*, 5, No. 3, 237–252.

Recski, A., 1973, 1975. Application of matroids in electric network analysis (in Hungarian), *Matematikai Lapok*, Part I, Vol. 24, No. 3–4, 311–328. Part II. Vol. 26, No. 3–4, 287–303.

Recski, A., 1973. On partitional matroids with applications. Coll. Math. Soc. János Bolyai 10, *Infinite and Finite Sets*, Keszthely, Hungary, North-Holland, Amsterdam, Vol. III, 1169–1179.

Recski, A., 1975. On the generalization of the Fibonacci numbers, *The Fibonacci Quart.*, 13, 315–317.

Recski, A., 1978. Contributions to the n-port interconnection problem by means of matroids. Coll. Math. Soc. János Bolyai 18, *Combinatorics*, Keszthely, Hungary, North-Holland, Amsterdam, 877–892.

Recski, A., 1979. Terminal solvability and the n-port interconnection problem, *Proceedings of the 1979 IEEE International Symposium on Circuits and Systems*, Tokyo, 988–991.

Recski, A., 1980a. Sufficient conditions for the unique solvability of linear networks containing memoryless 2-ports, *Int. J. Circuit Theory and Appl.*, 8, 95–103.

Recski, A., 1980b. Matroids and network synthesis, *Proceedings of the 1980 European Conference on Circuit Theory and Design*, Warsaw, Poland, 192–197.

Recski, A., 1981. An algorithm to determine whether the sum of some graphic matroids is graphic. Coll. Math. Soc. János Bolyai 25, *Algebraic Methods in Graph Theory*, Szeged, Hungary, North-Holland, Amsterdam, Vol. II. 647–656.

Recski, A., 1982. A practical remark on the minimal synthesis of resistive n-ports, *IEEE Trans. Circuits & Syst.*, CAS-29, 267–269.

Recski, A., 1983a. On the generalization of the matroid parity and the matroid partition problems, with applications. *Ann. Discrete Math.*, 17, 567–574.

Recski, A., 1983b. Local and global inconsistencies in the n-port interconnection problem, *Int. J. Circuit Theory and Appl.*, 11, 371–375.

Recski, A., 1984a. A network theory approach to the rigidity of skeletal structures. Part I. Modelling and interconnection, *Discrete Appl. Math.*, 7, 313–324.

Recski, A., 1984b. A network theory approach to the rigidity of skeletal structures, Part II. Laman's theorem and topological formulae, *Discrete Appl. Math.*, 8, 63–68.

Recski, A., 1984c. A network theory approach to the rigidity of skeletal structures. Part III. An electric model of planar frameworks. *Structural Topology*, 9, 59–71.

Recski, A. and M. Iri, 1980. Network theory and transversal matroids, *Discrete Appl. Math.*, 2, 311–326.

Recski, A. and J. Takács, 1981. On the combinatorial sufficient conditions for linear network solvability, *Int. J. Circuit Theory and Appl.*, 9, 351–354.

Rédei, L., 1934. Ein kombinatorischer Satz, *Acta Litt. Sci.*, Szeged 7, 39–43.

Reingold, E.M., J. Nievergeld and N. Deo, 1977. *Combinatorial Algorithms: Theory and Practice*, Prentice-Hall, Englewood Cliffs, New Jersey.

Ringel, G., 1974. *Map Color Theorem*, Springer, Berlin.

Robinson, G.C. and D.J.A. Welsh, 1980. The computational complexity of matroid properties, *Math. Proc. Cambridge Philos. Soc.*, 87, 29–45.

Rockafellar, R.T., 1970. *Convex Analysis*, Princeton University Press, Princeton, N.J.

Rohrer, R.A., 1970. *Circuit Theory: An Introduction to the State Variable Approach*, McGraw–Hill, New York.

Rosenberg, I.G., 1975. Structural rigidity in plane, preprint CRM-510, Université de Montréal.

Rosenberg, I.G., 1980. Structural rigidity I, Foundations and rigidity criteria, *Ann. Discrete Math.*, 8, 143–161.

Roth, B., 1980. Questions on the rigidity of structures, *Structural Topology*, 4, 67–71.

Roth, B., 1981. Rigid and flexible frameworks, *Am. Math. Monthly*, 88, No. 1, 6–21.

Roth, B. and W. Whiteley, 1981. Tensegrity frameworks, *Trans. Am. Math. Soc.*, 265, No. 2, 419–446.

Schrijver, A., 1979. Matroids and linking systems, *J. Comb. Theory Ser.B*, 26, 349–369.

Schrijver, A., 1983. Bounds on permanents, and the number of 1-factors and 1-factorizations of bipartite graphs, in E. Keith Lloyd (ed), *Surveys in Combinatorics*, Cambridge University Press, Cambridge.

Schrijver, A., 1986. *Theory of Linear and Integer Programming*, Wiley, Chichester.

Seshu, S. and M.B. Reed, 1961. *Linear Graphs and Electric Networks*, Addison-Wesley, Reading, Mass., London.

Seymour, P.D., 1977. The matroids with the max-flow min-cut property, *J. Comb. Theory Ser. B*, 23, 189–222.

Seymour, P.D., 1979. Matroid representation over GF(3), *J. Comb. Theory Ser.B*, 26, 159–173.

Seymour, P.D., 1980. Decomposition of regular matroids, *J. Comb. Theory Ser.B*, 28, 305–359.

Seymour, P.D., 1981. Recognizing graphic matroids, *Combinatorica*, 1, 75–78.

Seymour, P.D. and P.N. Walton, 1981. Detecting matroid minors, *J. London Math. Soc.*, 2, 23, No. 2, 193–203.

Shameeva, O.V., 1985. Algebraic representability of matroids, *Vestnik Mosk. Univ. Matem.*, 40, 29–32.

Shor, N.Z., 1970. Convergence rate of the gradient descent method with dilation of the space, *Kibernetika*, 2, 80–85; English translation *Cybernetics*, 6, 102–108.

Snay, R.A., 1978. Solvability analysis of horizontal networks by the identification of rigidity components, *Proceedings of the Second International Symposium on Problems Related to the Redefinition of North American Geodetic Networks*, Arlington, VA, 347–355.

Spriggs, J. and R.A. Snay, 1982. An algorithm for testing the solvability of horizontal networks, *Proceedings of the 1982 Spring Meeting American Geophysical Union*, Philadelphia, PA.

Stallman, M. and H.N. Gabow, 1984. An augmenting path algorithm for the parity problem on linear matroids, *Proc. 25th Ann. IEEE Symp. on Foundations of Computer Science*, 217–228.

Su, K.L., 1965. *Active Network Synthesis*, McGraw–Hill, New York.

Sugihara, K., 1979. Studies on mathematical structures of line drawings of polyhedra and their applications to scene analysis, *Res. Electrotech. Lab.*, No. 800.

Sugihara, K., 1980. On redundant bracing in plane skeletal structures, *Bull. Electrotech. Lab.*, 44, No. 5–6, 78–88.

Sugihara, K., 1982. Mathematical structures of line drawings of polyhedra–toward man-machine communications by means of line-drawings, *IEEE Trans. Pattern Analysis and Machine Intelligence*, PAMI-4, No. 5, 458–469.

Sugihara, K., 1983. A unifying approach to descriptive geometry and mechanisms, *Discrete Appl. Math.*, 5, 313–328.

Sugihara, K., 1984. An algebraic and combinatorial approach to the analysis of line drawings of polyhedra, *Discrete Appl. Math.*, 9, 77–104.

Sugihara, K., 1986. *Machine Interpretation of Line Drawings*, MIT Press, Cambridge, Mass.

Sussman-Fort, S.E. and J. Orchard, 1980. Canonic structures for lossless one-ports, *IEEE Trans. Circuits & Syst.*, CAS-27, No. 9, 772–778.

Szabó, J. and B. Roller, 1978. *Anwendung der Matrizenrechnung auf Stabwerke*, Akadémiai Kiadó, Budapest.

Tarjan, R.E., 1972. Depth-first search and linear graph algorithms, *SIAM J. Comput.*, 1, 146–160.

Tarnai, T., 1980. Simultaneous static and kinematic indeterminacy of space trusses with cyclic symmetry, *Int. J. Solids and Structures*, 16, 347–359.

Tarnai, T., 1980. Problems concerning spherical polyhedra and structural rigidity, *Structural Topology*, 4, 61–66.

Tarry, G., 1895. Le problème des labyrinthes, *Nouv. Ann. Math.*, 14, 187–189.

Tiong-Seng Tay, 1984. Rigidity of multigraphs I, Linking rigid bodies in n-space, *J. Comb. Theory Ser.B*, 36, 95–112.

Tokad, Y., 1981. On the existence of mathematical models of multiterminal linear lumped time-invariant networks, *Proceedings of the 1981 European Conference on Circuit Theory and Design*, The Hague, 886–891.

Tomizawa, N., 1973. On a specialization sequence from general matroids to ladder graphs with special emphasis on the characterization of ladder matroids, *RAAG Research Notes*, 191.

Tow, J., 1968. Order of complexity of linear active networks, *Proc. IEE (GB)*, 115, 1259–1262.

Tow, J., 1970. The explicit form of Bashkow's A matrix for a class of linear passive networks, *IEEE Trans. Circuit Theory*, CT-17, 113—115.

Truemper, K., 1982a. Alpha-balanced graphs and matrices and GF(3)-representability of matroids, *J. Comb. Theory Ser.B*, 32, 112–139.

Truemper, K., 1982b. On the efficiency of representability tests for matroids, *Eur. J. Comb.*, 3, 275–291.

Truemper, K., 1984. Elements for a decomposition theory for matroids, in A. Bondy (ed), *Progress in Graph Theory*, Academic Press, Toronto, 439–475.

Tutte, W.T., 1947. The factorization of linear graphs, *J. London Math. Soc.*, 22, 107–111.

Tutte, W.T., 1960. An algorithm to determine whether a given binary matroid is graphic, *Proc. Am. Math. Soc.*, 11, 905–917.

Tutte, W.T., 1961. On the problem of decomposing a graph into n-connected factors, *J. London Math. Soc.*, 36, 221–230.

Tutte, W.T., 1965. Lectures on matroids, *J. Res. Natl. Bur. Stand.*, 69B, 1–47.

Tutte, W.T., 1967. On even matroids, *J. Res. Natl. Bur. Stand.*, 71B, 213–214.

Tutte, W.T.,1971. *Introduction to the Theory of Matroids*, American Elsevier, New York.

Ünver, Z. and Y. Ceyhun, 1983. On a graphical representation for matroids, *Proc. IEEE ISCAS*, Newport Beach, California, 335–337.

Vágó, I., 1985. *Graph Theory – Application to the Calculation of Electric Networks*, Akadémiai Kiadó, Budapest, Elsevier, Amsterdam.

Vizing, V.G., 1964. On an estimate of the chromatic class of a p-graph (in Russian), *Diskret. Analiz.*, 3, 25–30.

Wagner, K., 1936. Bemerkungen zum Vierfarbenproblem, *Jber. Deutsch. Math.-Verein*, 46, 26–32.

Wagner,K., 1937. Über eine Eigenschaft der ebenen Komplexe, *Mathematische Annalen*, 114, 570–590.

Weinberg, L., 1962. *Network Analysis and Synthesis*, McGraw–Hill, New York.

Welsh, D.J.A., 1969a. On the hyperplanes of a matroid, *Proc. Cambridge Philos. Soc. (Mathematical and Physical Sciences)*, 65, Part I, 11–18.

Welsh, D.J.A., 1969b. Euler and bipartite matroids, *J. Comb. Theory*, 6, 375–377.

Welsh, D.J.A., 1970. On matroid theorems of Edmonds and Rado, *J. London Math. Soc. Ser. 2*, 2, 251–256.

Welsh, D.J.A., 1971. Generalized versions of Hall's theorem,*J. Comb. Theory Ser.B*, 10, No. 2, 95–101.

Welsh, D.J.A., 1976. *Matroid Theory*, Academic Press, London.

White, N.L.,(ed), 1986. *Theory of Matroids*, Cambridge University Press, Cambridge.

White, N.L. and W. Whiteley, 1983. The algebraic geometry of stresses in frameworks, *SIAM J. Algebraic and Discrete Methods*, 4, 481–511.

Whiteley, W., 1982a. Motions, stresses and projected polyhedra, *Structural Topology*, 7, 13–38.

Whiteley, W., 1982b. Motions of trusses and bipartite frameworks, *Structural Topology*, 7, 61–68.

Whiteley, W., 1983. Cones, infinity and 1-storey buildings, *Structural Topology*, 8, 53–70.

Whitney, H., 1932. Non-separable and planar graphs, *Trans. Amer. Math. Soc.*, 34, 339–362.

Whitney, H., 1933a. On the classification of graphs, *Am. J. Math*, 55, 236–244.

Whitney, H., 1933b. 2-isomorphic graphs, *Am. J. Math.*, 55, 245–254.

Whitney, H., 1935. On the abstract properties of linear dependence, *Am. J. Math.*, 57, 509–533.

Williams, J.W.J., 1964. Algorithm 232: Heapsort, *Commun. ACM*, 7, 6, 347–348.

Wilson, R.J., 1973. An introduction to matroid theory, *Am. Math. Monthly*, 80, No. 5, 500–525.

Woodall, D.R., 1974. An exchange theorem for bases of matroids, *J. Comb. Theory Ser.B*, 16, 227–228.

Woodall, D.R., 1978. Minimax theorems in graph theory, in L.W. Beineke, R.J. Wilson (eds), *Selected Topics in Graph Theory*, Academic Press, London, 237–269.

Zemanian, A.H., 1981. The status of research on infinite electrical networks, *Proceedings of the 1981 European Conference on Circuit Theory and Design*, The Hague, 186–189.

Subject index*

* Alternative names for the same concepts, used by other authors, are indicated in parentheses.

Algorithms and Combinatorics

Editors: R. L. Graham, B. Korte, L. Lovász

Combinatorial mathematics has substantially influenced recent trends and developments in the theory of algorithms and its applications. Conversely, research on algorithms and their complexity has established new perspectives in discrete mathematics. This new series is devoted to the mathematics of these rapidly growing fields with special emphasis on their mutual interactions.

The series will cover areas in pure and applied mathematics as well as computer science, including: combinatorial and discrete optimization, polyhedral combinatorics, graph theory and its algorithmic aspects, network flows, matroids and their applications, algorithms in number theory, group theory etc., coding theory, algorithmic complexity of combinatorial problems, and combinatorial methods in computer science and related areas.

The main body of this series will be monographs ranging in level from first-year graduate up to advanced state-of-the-art research. The books will be conventionally type-set and bound in hard covers. In new and rapidly growing areas, collections of carefully edited monographic articles are also appropriate for this series. Occasionally there will also be "lecture-notes-type" volumes within the series, published as *Study and Research Texts* in soft cover and camera-ready form. This will be primarily an outlet for seminar notes, drafts of textbooks with essential novelty in their presentation, and preliminary drafts of monographs.

Prospective readers of the series ALGORITHMS AND COMBINATORICS include scientists and graduate students working in discrete mathematics, operations research and computer science and their applications.

Volume 1

K. H. Borgwardt

The Simplex Method

A Probabilistic Analysis

1987. 42 figures in 115 separate illustrations. XI, 268 pages. ISBN 3-540-17096-0

Contents: Introduction. - The Shadow-Vertex Algorithm. - The Average Number of Pivot Steps. - The Polynomiality of the Expected Number of Steps. - Asymptotic Results. - Problems with Nonnegativity Constraints. - Appendix. - References. - Subject Index.

Springer-Verlag Berlin Heidelberg New York London Paris Tokyo Hong Kong

Springer